AF804061

Applied Nonsingular Astrodynamics

Optimal Low-Thrust Orbit Transfer

This essential book describes the mathematical formulations and subsequent computer simulations required to accurately project the trajectory of spacecraft and rockets in space, using the formalism of optimal control for minimum-time transfer in general elliptic orbit. The material will aid research students in aerospace engineering, as well as practitioners in the field of spaceflight dynamics, in developing simulation software to carry out trade studies useful in vehicle and mission design. It will also teach them to develop flight software for operational applications in autonomous mode, to actually transfer space vehicles from one orbit to another. The hands-on real-life applications discussed will give all readers a clear understanding of the mathematics of orbit transfer, allow them to develop their own operational software to fly actual missions, and use the contents as a research tool to carry out even more complex analyses.

Jean Albert Kéchichian is a retired Engineering Specialist from *The Aerospace Corporation*. His career has included senior level engineering positions at *NASA's Jet Propulsion Laboratory* and at *Ford Aerospace*. His main areas of contribution are in spaceflight guidance and navigation. He is a Fellow of *The American Astronautical Society*, and his work has regularly appeared in *Acta Astronautica, Journal of Guidance Control and Dynamics, Journal of the Astronautical Sciences*, and *Journal of Spacecraft and Rockets*. He holds Degrees in Aeronautical and Mechanical Engineering from *l'Université de Liège, UC Berkeley*, and a PhD in Aeronautics and Astronautics from *Stanford University*.

Cambridge Aerospace Series

Editors: Wei Shyy and Vigor Yang

1. J. M. Rolfe and K. J. Staples (eds.): *Flight Simulation*
2. P. Berlin: *The Geostationary Applications Satellite*
3. M. J. T. Smith: *Aircraft Noise*
4. N. X. Vinh: *Flight Mechanics of High-Performance Aircraft*
5. W. A. Mair and D. L. Birdsall: *Aircraft Performance*
6. M. J. Abzug and E. E. Larrabee: *Airplane Stability and Control*
7. M. J. Sidi: *Spacecraft Dynamics and Control*
8. J. D. Anderson: *A History of Aerodynamics*
9. A. M. Cruise, J. A. Bowles, C. V. Goodall, and T. J. Patrick: *Principles of Space Instrument Design*
10. G. A. Khoury (ed.): *Airship Technology*, Second Edition
11. J. P. Fielding: *Introduction to Aircraft Design*
12. J. G. Leishman: *Principles of Helicopter Aerodynamics*, Second Edition
13. J. Katz and A. Plotkin: *Low-Speed Aerodynamics*, Second Edition
14. M. J. Abzug and E. E. Larrabee: *Airplane Stability and Control: A History of the Technologies that Made Aviation Possible*, Second Edition
15. D. H. Hodges and G. A. Pierce: *Introduction to Structural Dynamics and Aeroelasticity*, Second Edition
16. W. Fehse: *Automatic Rendezvous and Docking of Spacecraft*
17. R. D. Flack: *Fundamentals of Jet Propulsion with Applications*
18. E. A. Baskharone: *Principles of Turbomachinery in Air-Breathing Engines*
19. D. D. Knight: *Numerical Methods for High-Speed Flows*
20. C. A. Wagner, T. Hüttl, and P. Sagaut (eds.): *Large-Eddy Simulation for Acoustics*
21. D. D. Joseph, T. Funada, and J. Wang: *Potential Flows of Viscous and Viscoelastic Fluids*
22. W. Shyy, Y. Lian, H. Liu, J. Tang, and D. Viieru: *Aerodynamics of Low Reynolds Number Flyers*
23. J. H. Saleh: *Analyses for Durability and System Design Lifetime*
24. B. K. Donaldson: *Analysis of Aircraft Structures*, Second Edition
25. C. Segal: *The Scramjet Engine: Processes and Characteristics*
26. J. F. Doyle: *Guided Explorations of the Mechanics of Solids and Structures*
27. A. K. Kundu: *Aircraft Design*
28. M. I. Friswell, J. E. T. Penny, S. D. Garvey, and A. W. Lees: *Dynamics of Rotating Machines*
29. B. A. Conway (ed): *Spacecraft Trajectory Optimization*
30. R. J. Adrian and J. Westerweel: *Particle Image Velocimetry*
31. G. A. Flandro, H. M. McMahon, and R. L. Roach: *Basic Aerodynamics*
32. H. Babinsky and J. K. Harvey: *Shock Wave–Boundary-Layer Interactions*
33. C. K. W. Tam: *Computational Aeroacoustics: A Wave Number Approach*
34. A. Filippone: *Advanced Aircraft Flight Performance*
35. I. Chopra and J. Sirohi: *Smart Structures Theory*
36. W. Johnson: *Rotorcraft Aeromechanics vol. 3*
37. W. Shyy, H. Aono, C. K. Kang, and H. Liu: *An Introduction to Flapping Wing Aerodynamics*
38. T. C. Lieuwen and V. Yang: *Gas Turbine Emissions*
39. P. Kabamba and A. Girard: *Fundamentals of Aerospace Navigation and Guidance*
40. R. M. Cummings, W. H. Mason, S. A. Morton, and D. R. McDaniel: *Applied Computational Aerodynamics*
41. P. G. Tucker: *Advanced Computational Fluid and Aerodynamics*
42. Iain D. Boyd and Thomas E. Schwartzentruber: *Nonequilibrium Gas Dynamics and Molecular Simulation*
43. Joseph J. S. Shang and Sergey T. Surzhikov: *Plasma Dynamics for Aerospace Engineering*
44. Bijay K. Sultanian: *Gas Turbines: Internal Flow Systems Modeling*
45. J. A. Kéchichian: *Applied Nonsingular Astrodynamics: Optimal Low-Thrust Orbit Transfer*

Applied Nonsingular Astrodynamics

Optimal Low-Thrust Orbit Transfer

JEAN ALBERT KÉCHICHIAN
The Aerospace Corporation (Retired)

CAMBRIDGE
UNIVERSITY PRESS

CAMBRIDGE
UNIVERSITY PRESS

Shaftesbury Road, Cambridge CB2 8EA, United Kingdom

One Liberty Plaza, 20th Floor, New York, NY 10006, USA

477 Williamstown Road, Port Melbourne, VIC 3207, Australia

314–321, 3rd Floor, Plot 3, Splendor Forum, Jasola District Centre, New Delhi – 110025, India

103 Penang Road, #05–06/07, Visioncrest Commercial, Singapore 238467

Cambridge University Press is part of Cambridge University Press & Assessment, a department of the University of Cambridge.

We share the University's mission to contribute to society through the pursuit of education, learning and research at the highest international levels of excellence.

www.cambridge.org
Information on this title: www.cambridge.org/9781108472364

DOI: 10.1017/9781108560061

© Jean Albert Kéchichian 2018

This publication is in copyright. Subject to statutory exception and to the provisions of relevant collective licensing agreements, no reproduction of any part may take place without the written permission of Cambridge University Press & Assessment.

First published 2018

A catalogue record for this publication is available from the British Library

Library of Congress Cataloging-in-Publication data
Names: Kechichian, Jean Albert, 1945– author.
Title: Applied nonsingular astrodynamics optimal low-thrust orbit transfer / Jean Albert Kechichian (The Aerospace Corporation (retired)).
Description: Cambridge : Cambridge University Press, [2018] | Series: Cambridge aerospace series | Includes bibliographical references and index.
Identifiers: LCCN 2018015860 | ISBN 9781108472364 (alk. paper)
Subjects: LCSH: Orbital transfer (Space flight) | Trajectory optimization. | Space trajectories.
Classification: LCC TL1075 .K43 2018 | DDC 629.4/113–dc23
LC record available at https://lccn.loc.gov/2018015860

ISBN 978-1-108-47236-4 Hardback

Cambridge University Press & Assessment has no responsibility for the persistence or accuracy of URLs for external or third-party internet websites referred to in this publication and does not guarantee that any content on such websites is, or will remain, accurate or appropriate.

To the memory of my beloved parents
Albert Kéchichian and Joséphine Séraydarian
blessed with extraordinary wisdom, and who placed selflessly my siblings and me at the gravitational center of their lives.

Contents

Preface

The objective of this comprehensive book is to lay out the mathematics of optimal transfer and rendezvous between any two general elliptic orbits, based on the use of nonsingular orbital elements in the form of the equinoctial elements, that lead to the casting of the system differential equations and their adjoints in a form that is free of singularities at the crucial zero eccentricity and zero inclination conditions, allowing for the numerical integration of these equations without computer overflow and accuracy loss.

As such, the contents of this book will assist research students in aerospace engineering, as well as practitioners in the field of spaceflight dynamics, in developing simulation software to carry out trade studies useful in vehicle and mission design, and develop flight software for operational applications in autonomous mode, to actually transfer space vehicles from one orbit to another – the most important applications being for transfers between low Earth orbit (LEO) and geostationary Earth orbit (GEO), using electric ion or arcjet thrusters, being clear that the use of the popular classical elements will inevitably lead to errors and even computer run terminations because these orbits exhibit circular and equatorial conditions leading to singularities in the description of the system equations.

The main contributions to the solution of optimal low-thrust orbit transfers were historically provided by a team of Scientists at the Massachusets Institute of Technology's Draper Laboratory in Cambridge, Massachusets, led by Theodore Edelbaum, who made use of the nonsingular dynamic equations of orbital flight developed by Dr. Paul Cefola also of the Draper Laboratory, and Professor Roger Broucke of the University of Texas, at Austin.

The long duration of these powered flights necessitated the use of the averaging technique in the development of fast computer codes in which large integration steps spanning several orbital revolutions were made possible, in order to solve the two-point boundary value problem defining the transfer between two given orbits.

This rapid integration and consequent iterations using Newton's scheme, coupled to an indirect optimization method, needed to converge onto the initial values of the Lagrange multipliers or adjoints, and therefore onto the optimal solutions, were made possible by the use of the equinoctial elements, and because these elements were orbital elements, the technique of averaging could be applied. It was then sufficient to disregard the actual position of the space vehicle along its current orbit due to the averaging of the equations of motion which then described only the size, shape and orientation of the running orbit.

This five-element system of dynamic and adjoint differential equations is described in the first chapter with all its details and derivations along with many extensions, especially concerning the certain additional real-world spacecraft attitude constraints that affect the thrust vector orientation at each instant of time during an actual transfer.

The full six-state system of dynamic and adjoint equations for precision integration is then developed in the following chapters, while still using the same rotating orbital frame, also called the equinoctial frame, for the component resolution of the thrust acceleration vector, as well as the particular set of the equinoctial elements used by Edelbaum. It is thus possible to simulate "exact" transfer trajectories by taking advantage of the increased computational speed of modern computers, instead of generating only the approximate solutions of the averaged system of equations.

Chapters 2, 3 and 4 revisit earlier contributions by providing more accurate and error-free minimum-time and minimum-fuel simulation examples with additional numerical results and graphical representations, as well as performance comparisons with both the simplified analytic Edelbaum theory and also the averaged solutions. Both constant thrust and variable bounded thrust examples are thus depicted.

Chapters 5, 6 and 7 introduce the use of the epoch mean longitude and epoch eccentric longitude variables as the sixth orbital element, showing how the system and adjoint equations are related with the original current-time counterparts, for both mutual verification of the mathematics, the numerical simulations, and for possibly additional benefits in computation times for minimum-fuel transfers involving several coasting arcs. The savings in computation time would occur because the running orbit in the fundamental epoch-based systems as well as the epoch longitudes stay constant during coast in the absence of any thrust. The epoch-based systems are also convenient in orbit determination applications.

Chapters 8 and 9 introduce the use of the rotating Euler–Hill frame for component resolution of the thrust acceleration vector, as well as the adoption of the true longitude as the sixth orbital element for further simplification of the mathematics and corresponding software, while Chapters 10, 11 and 12 show how the second zonal harmonic J_2 of the Earth gravity model is accounted for, both within the frameworks of precision integration and averaging, using a variety of different sets of equinoctial elements for mutual verification and validation.

Chapter 13 derives the augmented system by also considering the perturbation effects of the higher zonal harmonics, namely J_3 and J_4, while the remaining chapters develop the mathematics of the further augmented system encompassing the gravity perturbations due to the Sun and the Moon, by adopting the analytic lunar theory of de Pontécoulant. The accelerations due to these third bodies are given in mathematical form by series expansions that converge rapidly at the third or at most fourth terms of the expansions, as is shown in the last chapter. The precession of the current orbit during the transfer of the space vehicle, due to the gravity perturbations imparted by the third bodies, can help reduce the effort spent to carry out the transfer, as shown by the examples in the last chapters.

A special effort was made to render each chapter as self-contained as possible for ease of reading, even though this necessitated the repetition of some definitions, crucial equations, and other relevant material, with the aim of preventing the reader from needing to turn pages back and forth too often between various chapters.

That is also the reason why references are kept at the end of each chapter instead of displaying them at the end of the book. Thus the relevant references for each chapter are shown at the end of the particular chapter.

In order to validate each formulation mathematically, it is necessary to compare results for mutual verification and validation of the mathematics. For example, when the J_2 perturbation is first discussed in the book, the mean longitude formulation is used in Chapters 10 and 11, and the examples are compared later with formulations using the eccentric longitude in Chapter 12, and also with both the mean longitude and the true longitude formulations by also including the higher zonal harmonics J_3 and J_4 in Chapter 13, again for mutual verification of the mathematics. The true longitude formulation is what most current researchers are using in their studies. For Chapters 15 and 16, being rather large, it was decided to leave them separate. Readers would readily know that they are indeed related.

The analyses and simulations presented in this book were carried out in the Flight Mechanics and Astrodynamics Departments at the Aerospace Corporation in El Segundo, California between 1989 and 2011, under contract with the United States Air Force Space and Missile Systems Center.

Finally, it is appropriate to recognize the extraordinary teachers I was fortunate to be associated with, starting with the late Professors Baudouin Fraeijs de Veubeke, Jean Smolderen and Charles Massonnet at the Faculté des Sciences Appliquées de L'Université de Liège in Belgium, and later Professor John V. Breakwell at the Department of Aeronautics and Astronautics at Stanford University, who directed my thesis. The two classic theoretical books on Astronautics, by Dr. Richard Battin of the MIT Draper Laboratory and Dr. Jean-Pierre Marec of the French ONERA, namely *Astronautical Guidance* and *Optimal Space Trajectories*, respectively, with the latter initially published in 1973 as Lecture Notes at the Ecole Nationale Supérieure de L'Aéronautique et de L'Espace under the title *Trajectoires Spatiales Optimales*, were extremely influential and inspirational.

The challenging and difficult work of formatting the contents of this book were carried out by Mary Villanueva of the Aerospace Corporation in El Segundo, California, and the transcription of many of the figures into electronic Adobe Illustrator files was expertly carried out by Yvonne Crane of the Rand Corporation in Arlington, Virginia. Their contribution is gratefully acknowledged.

Jean Albert Kéchichian
Analytic Astronautics
La Canada, California

Nomenclature

θ^*	true anomaly
a	semimajor axis
β	$1/(1+G)$
e	eccentricity
E	eccentric anomaly
$\mathbf{f}$	acceleration vector due to J_2 (context dependent)
$\mathbf{f}$	thrust vector, N
F	eccentric longitude $= E + \omega + \Omega$, rad
f	thrust magnitude, N
F'	$1 + G = 1/\beta$
$\hat{\mathbf{f}}, \hat{\mathbf{g}}, \hat{\mathbf{w}}$	unit vectors along axes of direct equinoctial frame
f_t	thrust acceleration, km/s^2
G	$(1 - h^2 - k^2)^{1/2}$
$\mathbf{g}$	gravity acceleration vector, km/s^2
h	$e \sin(\omega + \Omega)$
i	orbit inclination
J_2	Earth second zonal harmonic $= 1.08263 \times 10^{-3}$
K	$1 + p^2 + q^2$
k	$e \cos(\omega + \Omega)$
λ	mean longitude = $M + \omega + \Omega$
λ_0	mean longitude at epoch, $M_0 + \omega + \Omega$
L	true longitude, $\theta^* + \omega + \Omega$
M	mean anomaly
m	spacecraft mass, kg
μ	Earth gravity constant $= 398,601.3$ (km^3/s^2)
M_0	mean anomaly at epoch
n	orbit mean motion $= \mu^{1/2} a^{-3/2}$ (rad/s)
p	$\tan(i/2) \sin\Omega$
p'	semi-latus rectum $= a(1 - e^2)$
q	$\tan(i/2) \cos\Omega$
R	equatorial radius of the Earth $= 6378.14$ km
r	radial distance, km
$\mathbf{r}$, $\dot{\mathbf{r}}$	position and velocity vectors, km and km/s

$\ddot{\mathbf{r}}$	acceleration vector, km/s^2
$\hat{\mathbf{r}}, \hat{\boldsymbol{\theta}}, \hat{\mathbf{h}}$	unit vectors along axes of the Euler–Hill frame
s_F, c_F	sin F, cos F, etc.
$\hat{\mathbf{u}}$	unit vector in the direction of thrust
v	$\lvert\dot{\mathbf{r}}\rvert$, velocity vector magnitude, km/s
ω	argument of perigee
Ω	right ascension of ascending node

1 The Fundamental Classic Analysis of Edelbaum, Sackett and Malchow, with Additional Detailed Derivations and Extensions

1.1 Introduction

Important advances in the development of realistic simulation tools for the analysis of optimal transfers between general elliptic orbits using low-thrust electric ion engines were made in the early 1970s by a team of inspired scientists at the Massachusetts Institute of Technology (MIT) Draper Laboratory. This analysis and the resulting computer software are exposed in this chapter, and built on in the succeeding chapters, in particular for the creation of more accurate tools taking full advantage of the increasing speed of modern computers. Edelbaum [1] and his collaborators have used the method of averaging in order to eliminate the quickly varying orbit element for the spacecraft position along its orbit, leaving only the slowly varying elements that describe the instantaneous shape and orientation of the orbit during the transfer of the spacecraft. This is made possible due to the low acceleration imparted by the electric engines. The averaged rates are calculated by numerical quadrature, representing thereby the dynamical system quite accurately for numerical integration. A set of nonsingular orbit elements called the equinoctial elements are adopted in order to avoid any singularities for zero eccentricity and zero inclination during the numerical integrations. These singularities are unavoidable if the classical orbit elements are used instead.

Optimal control methodology is applied to develop the adjoint differential equations associated with each element for simultaneous integration with the dynamic equations, and the transfer trajectory is optimized by orienting the thrust acceleration vector in the optimal direction at each moment in time, leading to the creation of minimum-time trajectories. The indirect shooting method is used to converge on the guessed initial values of the adjoint or Lagrange multipliers in order to generate those initial values that meet the desired final conditions, solving thereby the two-point boundary value problem.

The important perturbations due to the earth's oblateness which precess the orbit, and the power degradation resulting from the damage incurred by the solar panels during the transit through the Van Allen belts, which in effect reduces the thrust magnitude of the thrusters, are also carried out by the three pioneers mentioned above. The averaging technique is thus employed for these perturbations because they too change or affect the trajectory, slowly, as was the case for the five slowly varying orbit elements.

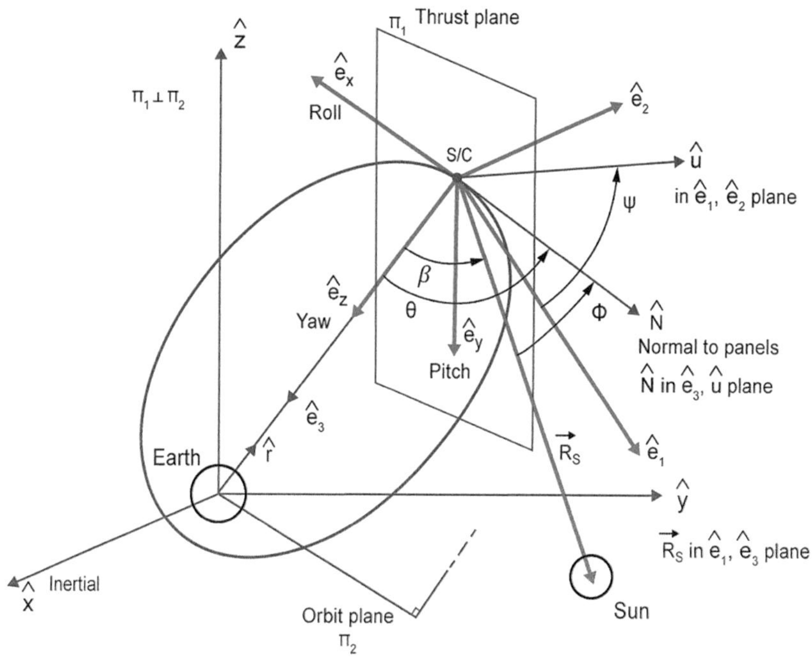

Figure 1.1 General configuration of spacecraft axes.

Spacecraft attitude constraints are also included to accommodate certain spacecraft configurations, such as for the case of the ion thrusters mounted on the negative roll face of the spacecraft and directed parallel to the roll axis. The solar panels are assumed to be flat and capable of rotation about their longitudinal axis. A proper combination of spacecraft roll and panels axis rotation can align the panels to fully face the Sun vector for maximum power output. The nominal attitude of the principle body-centered axes of the spacecraft shown in Figure 1.1 is such that these axes are aligned with a coordinate system made of the roll axis, the yaw axis and the pitch axis, with the roll axis in the orbit plane and orthogonal to the spacecraft–Earth radius vector, the yaw axis along the radius vector and the pitch axis normal to the orbit plane, with each positive direction specified in more detail in what follows.

Three attitude-constrained cases are investigated, namely, the zero roll and pitch case, free roll and zero pitch case, and the free roll, yaw and pitch case, also called the unconstrained case. The constrained case of zero roll is such that the panels cannot be oriented normal to the Sun vector, but must instead adopt the best orientation possible for maximum power extraction under the circumstances. Finally, in their analysis, Edelbaum et al. [1] assume that thrust is proportional to power with constant specific impulse I_{sp}, and efficiency.

The following sections describe many of these elements in some detail, with some new additional analysis for the zero roll case, which is important in its own right for trade studies.

The next chapters introduce the use of different orbit elements by also accounting for the fast variable for more accurate precision integration simulations, the use of the

Euler–Hill rotating frame for the resolution of the thrust acceleration vector components, and the use of gravitational perturbations due to the Sun and the Moon, among other additions and refinements.

Figures 1.1 through 1.6 are enhanced versions of the ones found in [1–3].

1.2 The Technique of Averaging

The variational equations for the classical elements in the Gaussian form with the components of the disturbing acceleration in the radial, transverse and out-of-plane directions are given by $\dot{a} = \frac{2a^2}{j}\left(es_{\theta^*}f_r + \frac{b}{r}f_\theta\right)$, $\dot{e} = \frac{1}{j}[bs_{\theta^*}f_r + \{(b+r)c_{\theta^*} + re\}f_\theta]$, $\dot{i} = \frac{rc_\theta}{j}f_h$, $\dot{\Omega} = \frac{rs_\theta}{js_i}f_h$, $\dot{\omega} = \frac{1}{je}[-bc_{\theta^*}f_r + (b+r)s_{\theta^*}f_\theta] - \frac{rs_\theta c_i}{js_i}f_h$, $\dot{M} = n + \frac{(1-e^2)^{1/2}}{je} \times [(bc_{\theta^*} - 2re)f_r - (b+r)s_{\theta^*}f_\theta]$ where a, e, i, Ω, ω and M are the classical orbit elements; $b = a(1-e^2)$ is the orbit parameter, $j = [\mu a(1-e^2)]^{1/2}$ is the orbit angular momentum, r is the radial distance defined by $r = b/(1 + ec_{\theta^*})$, θ is the angular position defined by $\theta = \omega + \theta^*$, with θ^* the true anomaly, and f_r, f_θ and f_h are the components of the disturbing acceleration vector resolved along the rotating Euler–Hill coordinate axes.

These differential equations are singular for $e = 0$, and $i = 0$, and are therefore not suitable for integration when the running orbit is at, or very near the circular and/or the equatorial conditions. For example, the important application of transferring a spacecraft from a circular low orbit to a geostationary orbit requires the adoption of a set of equations that are nonsingular for these two conditions for proper integration. This book develops several such nonsingular sets for use by mission analysts and designers, allowing the development of flight guidance software to optimally carry out such important transfers. Chapter 8 shows how the singular equations depicted above are converted into the nonsingular form using a particular set of equinoctial elements. Analysts and software developers can be better equipped by using several different sets of the dynamic and multiplier theories from the simple thrust-only formulations, to more involved models such as with oblateness and luni-solar gravity perturbations, both in precision integration mode for exact simulations, and in averaged mode for approximate applications. This chapter begins by showing the averaging method used by Edelbaum et al. in the 1970s, with subsequent chapters developing more refined models for both academic and practical interests, and further research endeavors by specialists in space flight guidance.

The averaging technique is used to consider only the first approximation to the state and costate or adjoint differential equations for rapid numerical integration of these equations as opposed to the much slower precision integration of the unaveraged equations. The averaging technique of [4] effectively eliminates the short period variations in the state and costate variables and assumes that these variables vary slowly. This is the case in low-thrust transfer problems where the sixth orbital element, which varies rapidly as it describes the location of the spacecraft in its orbit, is effectively eliminated from the equations of interest, which concern then only the size, shape and orientation of the orbit, but not the actual location of the spacecraft along its orbit, as the first five state variables vary slowly over time.

The averaged Hamiltonian $\tilde{H}$ is then obtained from the unaveraged Hamiltonian H by way of

$$\tilde{H} = \frac{1}{T}\int_{t-\frac{T}{2}}^{t+\frac{T}{2}} H dt \tag{1.1}$$

where t stands for the current time and T for the current orbital period. The state and adjoint variables are held constant during the integration, which is carried out by numerical quadrature, such that the spacecraft is considered to ride a Keplerian orbit during the averaging integration procedure. Thus, the first approximations to the state and adjoint variables are obtained from the integration of the coupled Euler–Hill differential equations

$$\dot{\tilde{\mathbf{x}}}^T = \frac{\partial \tilde{H}}{\partial \tilde{\boldsymbol{\lambda}}} \tag{1.2}$$

$$\dot{\tilde{\boldsymbol{\lambda}}}^T = -\frac{\partial \tilde{H}}{\partial \tilde{\mathbf{x}}} \tag{1.3}$$

with $\tilde{\mathbf{x}}$ and $\tilde{\boldsymbol{\lambda}}$ standing for the averaged generic state and corresponding adjoint variables. These equations can then be integrated numerically with large time steps spanning several orbital periods.

1.3 Summary of the Mechanics of the Equinoctial Orbit Elements

A summary of the various calculations of the position and velocity vector components along the slowly rotating $\hat{\mathbf{f}}$, $\hat{\mathbf{g}}$, $\hat{\mathbf{w}}$ coordinate system shown in Figure 1.2 is shown next and can be found in [1, 2, 5]. The details of their derivations are shown in [5] and also in Chapter 2.

Considering only the first five elements a, h, k, p and q, the relations between these elements and the classical elements are given by

$$\begin{aligned} a &= a \\ h &= e\sin(\omega+\Omega) \\ k &= e\cos(\omega+\Omega) \\ p &= \tan\left(\frac{i}{2}\right)\sin\Omega \\ q &= \tan\left(\frac{i}{2}\right)\cos\Omega \end{aligned} \tag{1.4}$$

and

$$\begin{aligned} a &= a \\ e &= \left(h^2+k^2\right)^{1/2} \\ i &= 2\tan^{-1}\left(p^2+q^2\right)^{1/2} \\ \Omega &= \tan^{-1}(p/q) \\ \omega &= \tan^{-1}(h/k) - \tan^{-1}(p/q) \end{aligned} \tag{1.5}$$

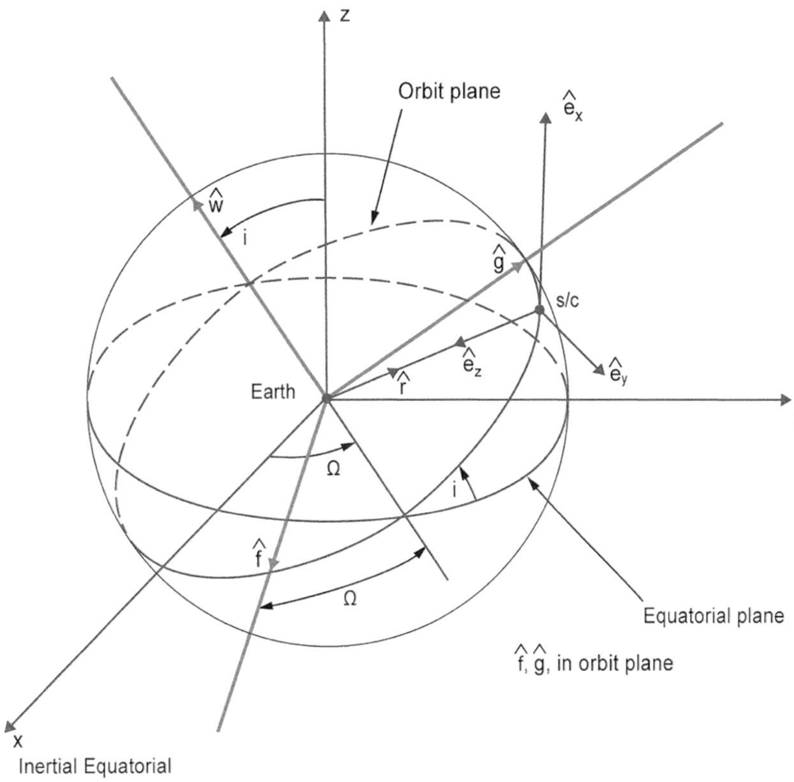

Figure 1.2 The Earth-centered equinoctial coordinate frame.

The fast variable or sixth element is selected as the eccentric longitude F which is related to the eccentric anomaly E by

$$F = E + \omega + \Omega \tag{1.6}$$

This element is eliminated from the dynamic equations by the averaging process. Figure 1.2 shows how the Earth-centered $\hat{\mathbf{f}}$, $\hat{\mathbf{g}}$, $\hat{\mathbf{w}}$ rotating frame is constructed by moving backwards by an angle Ω from the orbit ascending node for $\hat{\mathbf{f}}$, with $\hat{\mathbf{g}}$ in the orbital plane and 90 deg ahead, and with $\hat{\mathbf{w}}$ completing the triad. These unit vectors are given with respect to an inertial equatorial frame $\hat{\mathbf{x}}$, $\hat{\mathbf{y}}$, $\hat{\mathbf{z}}$ by

$$\hat{\mathbf{f}} = \frac{1}{(1+p^2+q^2)}\begin{bmatrix}(1-p^2+q^2)\\ 2pq\\ -2p\end{bmatrix} = \begin{pmatrix}f_1\\ f_2\\ f_3\end{pmatrix} = f_1\hat{\mathbf{x}} + f_2\hat{\mathbf{y}} + f_3\hat{\mathbf{z}}$$

$$\hat{\mathbf{g}} = \frac{1}{(1+p^2+q^2)}\begin{bmatrix}2pq\\ (1+p^2-q^2)\\ 2q\end{bmatrix} = \begin{pmatrix}g_1\\ g_2\\ g_3\end{pmatrix} = g_1\hat{\mathbf{x}} + g_2\hat{\mathbf{y}} + g_3\hat{\mathbf{z}}$$

$$\hat{\mathbf{w}} = \frac{1}{(1+p^2+q^2)}\begin{bmatrix}2p\\ -2q\\ (1-p^2+q^2)\end{bmatrix} = \begin{pmatrix}w_1\\ w_2\\ w_3\end{pmatrix} = w_1\hat{\mathbf{x}} + w_2\hat{\mathbf{y}} + w_3\hat{\mathbf{z}} \tag{1.7}$$

Given position and velocity vectors $\mathbf{r}$ and $\dot{\mathbf{r}}$, the equinoctial elements are calculated from

$$a = \left(\frac{2}{|\mathbf{r}|} - \frac{|\dot{\mathbf{r}}|^2}{\mu}\right)^{-1} \tag{1.8}$$

$$\mathbf{e} = -\frac{\mathbf{r}}{|\mathbf{r}|} - \frac{(\mathbf{r} \times \dot{\mathbf{r}}) \times \dot{\mathbf{r}}}{\mu} \tag{1.9}$$

$$\hat{\mathbf{w}} = \frac{\mathbf{r} \times \dot{\mathbf{r}}}{|\mathbf{r} \times \dot{\mathbf{r}}|} \tag{1.10}$$

$$p = \frac{w_1}{1 + w_3} \tag{1.11}$$

$$q = \frac{w_2}{1 + w_3} \tag{1.12}$$

μ being the Earth gravity constant, such that

$$h = \mathbf{e} \cdot \hat{\mathbf{g}} \tag{1.13}$$

$$k = \mathbf{e} \cdot \hat{\mathbf{f}} \tag{1.14}$$

The position coordinates X_1 and Y_1 are given in the $\hat{\mathbf{f}}$, $\hat{\mathbf{g}}$ frame by

$$X_1 = \mathbf{r} \cdot \hat{\mathbf{f}} \tag{1.15}$$

$$Y_1 = \mathbf{r} \cdot \hat{\mathbf{g}} \tag{1.16}$$

and the eccentric longitude F by

$$\cos F = k + \frac{\left(1 - k^2\beta\right) X_1 - hk\beta Y_1}{a\left(1 - h^2 - k^2\right)^{1/2}} \tag{1.17}$$

$$\sin F = h + \frac{\left(1 - h^2\beta\right) Y_1 - hk\beta X_1}{a\left(1 - h^2 - k^2\right)^{1/2}} \tag{1.18}$$

where

$$\beta = \frac{1}{1 + \left(1 - h^2 - k^2\right)^{1/2}} \tag{1.19}$$

From the definition of the mean longitude λ, in terms of the mean anomaly M namely,

$$\lambda = M + \omega + \Omega \tag{1.20}$$

the eccentric longitude can be extracted from Kepler's transcendental equation

$$\lambda = F - k \sin F + h \cos F \tag{1.21}$$

Thus,

$$\mathbf{r} = X_1\hat{\mathbf{f}} + Y_1\hat{\mathbf{g}} \tag{1.22}$$

$$\dot{\mathbf{r}} = \dot{X}_1\hat{\mathbf{f}} + \dot{Y}_1\hat{\mathbf{g}} \tag{1.23}$$

with

$$X_1 = a\left[\left(1 - h^2\beta\right)\cos F + hk\beta\sin F - k\right] \tag{1.24}$$

$$Y_1 = a\left[\left(1 - k^2\beta\right)\sin F + hk\beta\cos F - h\right] \tag{1.25}$$

$$\dot{X}_1 = \frac{na^2}{r}\left[hk\beta\cos F - \left(1 - h^2\beta\right)\sin F\right] \tag{1.26}$$

$$\dot{Y}_1 = \frac{na^2}{r}\left[\left(1 - k^2\beta\right)\cos F - hk\beta\sin F\right] \tag{1.27}$$

where the current mean motion n is given by

$$n = \left(\frac{\mu}{a^3}\right)^{1/2} \tag{1.28}$$

and the orbit equation

$$r = a\left(1 - kc_F - hs_F\right) \tag{1.29}$$

Letting $\boldsymbol{\lambda}$ and $\boldsymbol{\psi}$ represent the adjoint vectors in the equinoctial and classical representations, they are related by way of

$$\begin{pmatrix}\lambda_a\\ \lambda_h\\ \lambda_k\\ \lambda_p\\ \lambda_q\end{pmatrix} = \begin{pmatrix} 1 & 0 & 0 & 0 & 0\\ 0 & \sin(\omega+\Omega) & 0 & 0 & \frac{\cos(\omega+\Omega)}{e}\\ 0 & \cos(\omega+\Omega) & 0 & 0 & \frac{-\sin(\omega+\Omega)}{e}\\ 0 & 0 & 2\sin\Omega\cos^2\frac{i}{2} & \frac{c_\Omega}{\tan\left(\frac{i}{2}\right)} & \frac{-c_\Omega}{\tan\left(\frac{i}{2}\right)}\\ 0 & 0 & 2\cos\Omega\cos^2\frac{i}{2} & \frac{-s_\Omega}{\tan\left(\frac{i}{2}\right)} & \frac{s_\Omega}{\tan\left(\frac{i}{2}\right)} \end{pmatrix} \begin{pmatrix}\psi_a\\ \psi_e\\ \psi_i\\ \psi_\Omega\\ \psi_\omega\end{pmatrix} \tag{1.30}$$

which can also be written in terms of the equinoctial elements

$$\begin{pmatrix}\lambda_a\\ \lambda_h\\ \lambda_k\\ \lambda_p\\ \lambda_q\end{pmatrix} = \begin{pmatrix} 1 & 0 & 0 & 0 & 0\\ 0 & \frac{h}{(h^2+k^2)^{1/2}} & 0 & 0 & \frac{k}{(h^2+k^2)}\\ 0 & \frac{k}{(h^2+k^2)^{1/2}} & 0 & 0 & \frac{-h}{(h^2+k^2)}\\ 0 & 0 & \frac{2p}{(p^2+q^2)^{1/2}(1+p^2+q^2)} & \frac{q}{(p^2+q^2)} & \frac{-q}{(p^2+q^2)}\\ 0 & 0 & \frac{2q}{(p^2+q^2)^{1/2}(1+p^2+q^2)} & \frac{-p}{(p^2+q^2)} & \frac{p}{(p^2+q^2)} \end{pmatrix} \begin{pmatrix}\psi_a\\ \psi_e\\ \psi_i\\ \psi_\Omega\\ \psi_\omega\end{pmatrix} \tag{1.31}$$

The inverse relationships both in terms of the classical and equinoctial elements are obtained as

$$\begin{pmatrix}\psi_a\\ \psi_e\\ \psi_i\\ \psi_\Omega\\ \psi_\omega\end{pmatrix} = \begin{pmatrix}1 & 0 & 0 & 0 & 0\\ 0 & \sin(\omega+\Omega) & \cos(\omega+\Omega) & 0 & 0\\ 0 & 0 & 0 & \frac{1}{2}\sec^2\left(\frac{i}{2}\right)\sin\Omega & \frac{1}{2}\sec^2\left(\frac{i}{2}\right)\cos\Omega\\ 0 & e\cos(\omega+\Omega) & -e\sin(\omega+\Omega) & \tan\left(\frac{i}{2}\right)\cos\Omega & -\tan\left(\frac{i}{2}\right)\sin\Omega\\ 0 & e\cos(\omega+\Omega) & -e\sin(\omega+\Omega) & 0 & 0\end{pmatrix}\begin{pmatrix}\lambda_a\\ \lambda_h\\ \lambda_k\\ \lambda_p\\ \lambda_q\end{pmatrix} \tag{1.32}$$

$$\begin{pmatrix}\psi_a\\ \psi_e\\ \psi_i\\ \psi_\Omega\\ \psi_\omega\end{pmatrix} = \begin{pmatrix}1 & 0 & 0 & 0 & 0\\ 0 & \frac{h}{(h^2+k^2)^{1/2}} & \frac{k}{(h^2+k^2)^{1/2}} & 0 & 0\\ 0 & 0 & 0 & \frac{\frac{1}{2}p(1+p^2+q^2)}{(p^2+q^2)^{1/2}} & \frac{\frac{1}{2}q(1+p^2+q^2)}{(p^2+q^2)^{1/2}}\\ 0 & k & -h & q & -p\\ 0 & k & -h & 0 & 0\end{pmatrix}\begin{pmatrix}\lambda_a\\ \lambda_h\\ \lambda_k\\ \lambda_p\\ \lambda_q\end{pmatrix} \tag{1.33}$$

1.4 The Equations of Motion

The unaveraged differential equations for the equinoctial orbit elements and the variable spacecraft mass are derived in more detail in the next chapter. They have the following form:

$$\dot{\mathbf{z}} = \frac{2P}{mc} M(\mathbf{z}, F)\,\hat{\mathbf{u}} \tag{1.34}$$

$$\dot{m} = \frac{-2P}{c^2} \tag{1.35}$$

with $\mathbf{z}$ standing for the five elements a, h, k, p and q, m for spacecraft mass and P for thruster beam power given by

$$P = \eta P_0 D(N) \frac{1}{R_s^2} \cos\phi \tag{1.36}$$

where P_0 is the maximum array power at 1 au, η the power efficiency factor, $D(N)$ the damage factor as a function of fluence N, with maximum value of 1, R_s the spacecraft–Sun distance in au, and ϕ the angle between the normal to the solar panels and the spacecraft–Sun vector. In the following and without any loss of generality, $\eta = 1, R_s = 1$ is assumed in [2] for the sake of the analysis such that

$$P = P_0 \cos\phi \tag{1.37}$$

It is also assumed that the exhaust velocity c is constant,

$$c = I_{\text{sp}} g_0 \tag{1.38}$$

with constant specific impulse I_{sp}, and with g_0, the acceleration of gravity at the equator. The unit vector along the thrust direction is depicted by $\hat{\mathbf{u}}$, and the (5×3) matrix M is derived in Chapter 6 and given below as

$$M(\mathbf{z}, F) = \frac{\partial \mathbf{z}}{\partial \dot{\mathbf{r}}} \tag{1.39}$$

where F was selected as the sixth independent orbital element in [2]. The seventh variable is the high-energy particle fluence $N(\mathbf{z})$ such that

$$\dot{N} = f(\mathbf{z}, F) \tag{1.40}$$

Letting H represent the unaveraged Hamiltonian and $\mathbf{x}$ the full seven-state vector, a first-order approximation to the state and costate is obtained by forming an averaged Hamiltonian $\tilde{H}$ from which the averaged canonical equations for the state and costate are derived:

$$\begin{aligned}\tilde{H}\left(\tilde{\mathbf{x}}, \tilde{\boldsymbol{\lambda}}, t\right) &= \frac{1}{T_0}\int_{t-\frac{T_0}{2}}^{t+\frac{T_0}{2}} H\left[\tilde{\mathbf{x}}, \tilde{\boldsymbol{\lambda}}, t, F(t), \hat{\mathbf{u}}(F(t))\right] dt \\ &= \frac{1}{T_0}\int_{-\pi}^{\pi} H\left[\tilde{\mathbf{x}}, \tilde{\boldsymbol{\lambda}}, t, F, \hat{\mathbf{u}}(F)\right] \frac{dt}{dF} dF = \int_{-\pi}^{\pi} HsdF \end{aligned} \tag{1.41}$$

From Kepler's equation, $\lambda = F - k\sin F + h\cos F = ndt$, we have

$$\begin{aligned} d\lambda &= dF - k\cos F dF - h\sin F dF \\ &= (1 - k\cos F - h\sin F)\, dF = ndt = \frac{2\pi}{T_0} dt \end{aligned} \tag{1.42}$$

where T_0 is the current Keplerian orbital period, $T_0 = \frac{2\pi}{n}$, such that

$$\frac{dt}{dF} = \frac{T_0}{2\pi}\left(1 - \tilde{k}\cos F - \tilde{h}\sin F\right) \tag{1.43}$$

The integrations are carried out with respect to F such that letting

$$s(\mathbf{z}, F) = \frac{1}{T_0}\frac{dt}{dF} = \frac{1}{2\pi}\frac{r}{a} \tag{1.44}$$

the canonical equations take the form

$$\dot{\tilde{\mathbf{x}}} = \left(\frac{\partial \tilde{H}}{\partial \tilde{\boldsymbol{\lambda}}}\right)^T = \int_{-\pi}^{\pi}\left(\frac{\partial H}{\partial \tilde{\boldsymbol{\lambda}}}\right)^T sdF \tag{1.45}$$

$$\dot{\tilde{\boldsymbol{\lambda}}} = -\left(\frac{\partial \tilde{H}}{\partial \tilde{\mathbf{x}}}\right)^T = -\int_{-\pi}^{\pi}\left[\left(\frac{\partial H}{\partial \tilde{\mathbf{x}}}\right)^T s + H\left(\frac{\partial s}{\partial \tilde{\mathbf{x}}}\right)^T\right] dF \tag{1.46}$$

the state and costate being held constant during the integrations. As noted in [2], the $\frac{\partial H}{\partial \tilde{\boldsymbol{\lambda}}}$ and $-\frac{\partial H}{\partial \tilde{x}}$ are the unaveraged state and costate equations, and the $\frac{\partial s}{\partial \tilde{\mathbf{x}}}$ partials are just

$$\frac{\partial s}{\partial \tilde{h}} = -\frac{1}{2\pi}\sin F \tag{1.47}$$

$$\frac{\partial s}{\partial \tilde{k}} = -\frac{1}{2\pi}\cos F \tag{1.48}$$

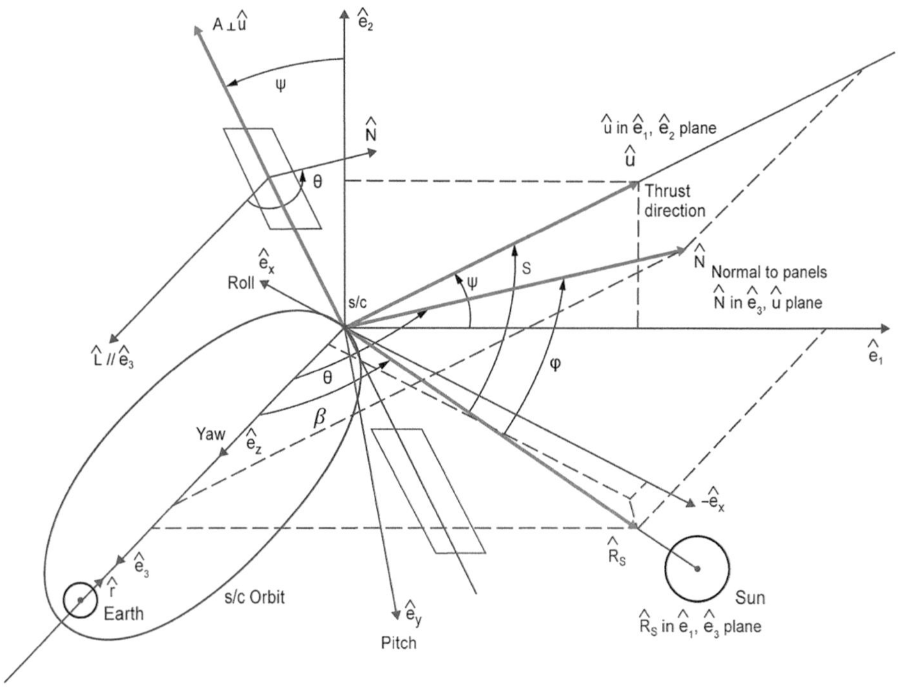

Figure 1.3 Thrust and solar panels' orientation angles.

1.5 The Averaged State and Costate Differential Equations for the Thrust-Constrained Case $\hat{\mathbf{u}} \perp \hat{\mathbf{r}}$

In [2], an important operational constraint is considered whereby the thrust vector is orthogonal to the spacecraft–Earth vector, such that $\hat{\mathbf{u}} \perp \hat{\mathbf{r}}$ at all times. It is then convenient to express $\hat{\mathbf{u}}$ as a function of a single control angle ψ which lies in a plane orthogonal to the spacecraft–Earth vector. This plane $(\hat{\mathbf{e}}_1, \hat{\mathbf{e}}_2)$ is shown in Figure 1.3, where $\hat{\mathbf{e}}_2$ is obtained from the $\hat{\mathbf{e}}_3, \hat{\mathbf{R}}_s$ vectors, with $\hat{\mathbf{e}}_1$ completing the triad. The angle ψ is shown as the angle between $\hat{\mathbf{e}}_1$ and $\hat{\mathbf{u}}$ with $\hat{\mathbf{u}}$ in the $(\hat{\mathbf{e}}_1, \hat{\mathbf{e}}_2)$ plane. The figure also shows the roll, pitch and yaw axes $\hat{\mathbf{e}}_x, \hat{\mathbf{e}}_y, \hat{\mathbf{e}}_z$ at the spacecraft, as well as the solar panels' orientation given by the normal vector $\hat{\mathbf{N}}$ to the panels.

Letting $T_1 = [\hat{\mathbf{e}}_1, \hat{\mathbf{e}}_2]$, the unaveraged state and costate differential equations can be written as

$$\dot{\mathbf{z}} = \frac{2P}{mc} (\mathbf{z}, F, \psi) M (\mathbf{z}, F) T_1 (\mathbf{z}, F) \begin{bmatrix} \cos\psi \\ \sin\psi \end{bmatrix} \tag{1.49}$$

$$\dot{m} = -\frac{2P}{c^2} (\mathbf{z}, F, \psi) \tag{1.50}$$

with $\hat{\mathbf{u}}$ replaced by $T_1 (\mathbf{z}, F) = \begin{bmatrix} \cos\psi \\ \sin\psi \end{bmatrix}$ and from

$$H = \boldsymbol{\lambda}_z^T \frac{2P}{mc} M T_1 \begin{bmatrix} \cos\psi \\ \sin\psi \end{bmatrix} - \lambda_m \frac{2P}{c^2} = \boldsymbol{\lambda}_z^T \dot{\mathbf{z}} + \lambda_m \dot{m} = H_z + H_m$$

such that

$$\dot{\boldsymbol{\lambda}}_z = -\frac{\partial H}{\partial \mathbf{z}} = -\frac{\partial P}{\partial \mathbf{z}} \frac{H}{P} - \boldsymbol{\lambda}_z^T \frac{2P}{mc} \left[\frac{\partial M}{\partial \mathbf{z}} T_1 + M \frac{\partial T_1}{\partial \mathbf{z}} \right] \begin{pmatrix} \cos\psi \\ \sin\psi \end{pmatrix} = -\boldsymbol{\lambda}_z^T \frac{\partial \dot{\mathbf{z}}}{\partial \mathbf{z}} \tag{1.51}$$

$$\dot{\lambda}_m = -\frac{\partial H}{\partial m} = \boldsymbol{\lambda}_z^T \frac{2P}{m^2 c} M T_1 \begin{pmatrix} \cos\psi \\ \sin\psi \end{pmatrix} = \frac{H_z}{m} \tag{1.52}$$

The shadowing and oblateness effects are ignored here to simplify the exposition of the governing equations.

In [2], a constraint equation, namely

$$C = \mathbf{r}^T \hat{\mathbf{u}} = 0 \tag{1.53}$$

is used as an alternate method to continue the analysis, and with the use of a Lagrange multiplier μ, the constraint is adjoined to the Hamiltonian H with the use of the original equations

$$\dot{\mathbf{z}} = \frac{2P}{mc} M \hat{\mathbf{u}} \tag{1.54}$$

$$\dot{m} = -\frac{2P}{c^2} \tag{1.55}$$

such that,

$$H = \boldsymbol{\lambda}_z^T \frac{2P}{mc} M \hat{\mathbf{u}} + \mu \mathbf{r}^T \hat{\mathbf{u}} - \frac{2P}{c^2} \lambda_m \tag{1.56}$$

The costate equations reduce to the following form:

$$\dot{\lambda}_z = -\frac{\partial H}{\partial \mathbf{z}} = -\frac{\partial P}{\partial \mathbf{z}} \left[\boldsymbol{\lambda}_z^T \frac{2}{mc} M \hat{\mathbf{u}} - \frac{2\lambda_m}{c^2} \right] + \boldsymbol{\lambda}_z^T \frac{2P}{mc} \frac{\partial M}{\partial \mathbf{z}} \hat{\mathbf{u}} + \mu \frac{\partial \mathbf{r}^T}{\partial \mathbf{z}} \hat{\mathbf{u}} \tag{1.57}$$

$$\dot{\lambda}_m = -\frac{\partial H}{\partial m} = \boldsymbol{\lambda}_z^T \frac{2P}{m^2 c} M \hat{\mathbf{u}} \tag{1.58}$$

Letting $T = [\hat{\mathbf{e}}_1, \hat{\mathbf{e}}_2, \hat{\mathbf{e}}_3]$, $\hat{\mathbf{e}}_3 = -\frac{\mathbf{r}}{|\mathbf{r}|}$, $T^T \hat{\mathbf{u}}$ represents the control $\hat{\mathbf{u}}$ in $\hat{\mathbf{e}}_1, \hat{\mathbf{e}}_2, \hat{\mathbf{e}}_3$ system such that with

$$\hat{\mathbf{u}}_T = T^T \hat{\mathbf{u}} = \begin{pmatrix} \cos\psi \cos\xi \\ \sin\psi \cos\xi \\ \sin\xi \end{pmatrix} \tag{1.59}$$

with ξ the angle between $\hat{\mathbf{u}}_T$ and the $(\hat{\mathbf{e}}_1, \hat{\mathbf{e}}_2)$ plane, and ψ the angle between the projection of $\hat{\mathbf{u}}_T$ onto the $(\hat{\mathbf{e}}_1, \hat{\mathbf{e}}_2)$ plane and $\hat{\mathbf{e}}_1$, the Hamiltonian of the system can be written alternately as

$$H = \left[\boldsymbol{\lambda}_z^T \frac{2P}{mc} M + \mu \mathbf{r}^T \right] \hat{\mathbf{u}} - \frac{2P}{c^2} \lambda_m = \left[\boldsymbol{\lambda}_z^T \frac{2P}{mc} M + \mu \mathbf{r}^T \right] T T^T \hat{\mathbf{u}} - \frac{2P}{c^2} \lambda_m \tag{1.60}$$

$$H = \left[\frac{2P\boldsymbol{\lambda}_z^T}{mc} M T + \mu \mathbf{r}^T T \right] \hat{\mathbf{u}}_T - \frac{2P}{c^2} \lambda_m \tag{1.61}$$

$$H = \frac{2P}{mc}\boldsymbol{\lambda}_z^T M\left[\hat{\mathbf{e}}_1 \cos\psi + \mathbf{e}_2 \sin\psi\right]\cos\xi + \left[\frac{2P}{mc}\boldsymbol{\lambda}_z^T M\hat{\mathbf{e}}_3 - \mu\,|\mathbf{r}|\right]\sin\xi - \frac{2P}{c^2}\lambda_m \tag{1.62}$$

The long axis of the panels being along the spacecraft pitch axis, their orientation is therefore independent of the pitch angle such that $P \neq f(pitch)$. The optimal ξ angle is seen to be equal to zero if we set $\frac{\partial H}{\partial \xi} = 0$, which results in

$$\tan\xi = \frac{\frac{2P}{mc}\boldsymbol{\lambda}_z^T M\hat{\mathbf{e}}_3 - \mu\,|\mathbf{r}|}{\frac{2P}{mc}\boldsymbol{\lambda}_z^T M\left[\hat{\mathbf{e}}_1 \cos\psi + \mathbf{e}_2 \sin\psi\right]} \tag{1.63}$$

and because C, the constraint, can be written as $C = \mathbf{r}^T T\, T^T\hat{\mathbf{u}} = 0$ due to $T\,T^T = I$, the identity matrix, it is seen from $T = [\hat{\mathbf{e}}_1, \hat{\mathbf{e}}_2, \hat{\mathbf{e}}_3]$ and Equation (1.59) for $\hat{\mathbf{u}}_T$ that $\sin\xi = 0$. Therefore $\xi = 0$ is true, such that setting $\tan\xi$ in Equation (1.63) also to 0 provides the expression for μ as

$$\mu = -\frac{2P}{mc}\boldsymbol{\lambda}_z^T M\frac{\mathbf{r}}{|\mathbf{r}|^2} \tag{1.64}$$

where the definition of $\hat{\mathbf{e}}_3 = \frac{-\mathbf{r}}{|\mathbf{r}|}$ is put to use. For free roll, P is independent of $\hat{\mathbf{u}}$, i.e., ψ, and for pitch = 0, the constraint $C = 0$ still holds such that P is still independent of $\hat{\mathbf{u}}$ because of the free roll freedom, and therefore the optimal $\hat{\mathbf{u}}$ must be aligned with $\boldsymbol{\lambda}_z^T\frac{2P}{mc}M + \mu\mathbf{r}^T$, yielding

$$\hat{\mathbf{u}} = \frac{\left(\boldsymbol{\lambda}_z^T\frac{2P}{mc}M + \mu\mathbf{r}^T\right)^T}{\left|\boldsymbol{\lambda}_z^T\frac{2P}{mc}M + \mu\mathbf{r}^T\right|} \tag{1.65}$$

Applying the constraint $C = \hat{\mathbf{u}}^T\mathbf{r} = 0$ to the above optimal $\hat{\mathbf{u}}$ yields $\boldsymbol{\lambda}_z^T\frac{2P}{mc}M\mathbf{r} + \mu\mathbf{r}^T\mathbf{r} = 0$, which yields the same μ as in Equation (1.64). In the absence of any constraints, $\mu = 0$, and $\hat{\mathbf{u}}$ reduces to

$$\hat{\mathbf{u}} = \frac{M^T\boldsymbol{\lambda}_z}{\left|M^T\lambda_z\right|} \tag{1.66}$$

In summary, the canonical equations are given by direct application of Equations (1.45) and (1.46) using H in Equation (1.56):

$$\dot{\tilde{\mathbf{z}}} = \int_{-\pi}^{\pi}\left[\frac{2P\left(\tilde{\mathbf{z}}, \hat{\mathbf{u}}\right)}{mc}M\left(\tilde{\mathbf{z}}, F\right) + \mu\mathbf{r}^T\right]\hat{\mathbf{u}}s\left(\tilde{\mathbf{z}}, F\right)dF \tag{1.67}$$

$$\dot{\tilde{m}} = -\int_{-\pi}^{\pi}\frac{2P\left(\tilde{\mathbf{z}}, \hat{\mathbf{u}}\right)}{c^2}s\left(\tilde{\mathbf{z}}, F\right)dF \tag{1.68}$$

$$\begin{aligned}\dot{\tilde{\boldsymbol{\lambda}}}_z = -\int_{-\pi}^{\pi}\Bigg[\Bigg\{&\frac{\partial P\left(\tilde{\mathbf{z}}, \hat{\mathbf{u}}\right)}{\partial\tilde{\mathbf{z}}}\left(\boldsymbol{\lambda}_z^T\frac{2}{\tilde{m}c}M\left(\tilde{\mathbf{z}}, F\right)\hat{\mathbf{u}} - \frac{2\lambda_m}{c^2}\right)\\ &+ \tilde{\boldsymbol{\lambda}}_z^T\frac{2P\left(\tilde{\mathbf{z}}, \hat{\mathbf{u}}\right)}{\tilde{m}c}\left(\frac{\partial M\left(\tilde{\mathbf{z}}, F\right)}{\partial\tilde{\mathbf{z}}} + \mu\frac{\partial\mathbf{r}^T\left(\tilde{\mathbf{z}}, F\right)}{\partial\tilde{\mathbf{z}}}\right)\hat{\mathbf{u}}\Bigg\}s\left(\tilde{\mathbf{z}}, F\right)\\ &+ H\left(\tilde{\mathbf{z}}, \boldsymbol{\lambda}_z, \hat{\mathbf{u}}\right)\frac{\partial s\left(\tilde{\mathbf{z}}, F\right)}{\partial\tilde{\mathbf{z}}}\Bigg]dF\end{aligned} \tag{1.69}$$

$$\dot{\boldsymbol{\lambda}}_m = \int_{-\pi}^{\pi} \tilde{\boldsymbol{\lambda}}_z^T \frac{2P\left(\tilde{\mathbf{z}}, \hat{\mathbf{u}}\right)}{m^2 c} M\left(\tilde{\mathbf{z}}, F\right) \hat{\mathbf{u}}_S\left(\tilde{\mathbf{z}}, F\right) dF \tag{1.70}$$

and with μ given by Equation (1.64). These equations are valid for the free roll, zero pitch case, in which case $\frac{\partial P}{\partial \tilde{\mathbf{z}}} = 0$ is used because then P is not a function of $\tilde{\mathbf{z}}$. The yaw and pitch control angles are also readily available if both roll and pitch are free and unconstrained. In all these cases, yaw is free to rotate the orbit plane. The case of zero roll and zero pitch is discussed later in this chapter following the analysis shown in [2] and [3].

Note that $\frac{\partial \mathbf{r}^T}{\partial \tilde{\mathbf{z}}}$ partials are obtained from the expression for $\mathbf{r} = X_1\hat{\mathbf{f}} + Y_1\hat{\mathbf{g}}$, involving the $\frac{\partial X_1}{\partial h}$, $\frac{\partial Y_1}{\partial h}$, $\frac{\partial X_1}{\partial k}$ and $\frac{\partial Y_1}{\partial k}$ given in Chapter 6, which also shows the full six-state formulation for the case where F is considered as an independent variable and the sixth orbital equinoctial element adopted in [1] and [2].

1.6 A Discussion of the Zero Roll and Zero Pitch Case and the Maximization of the Solar Panels' Power Output

Sackett et al. [2] and [3] address the problem of the optimization of the thrust orientation as well as the solar panels' orientation for maximum power output, for the constrained case of both zero roll and zero pitch, and provide a suboptimal solution that is easier to compute than the fully optimized one for the minimum-time transfer between general elliptic or circular orbits. A short exposition of Sackett's et al.'s analysis [2] is shown in this section without all the detailed derivations.

The coordinate system used is shown in Figure 1.3 with $\hat{\mathbf{u}}$ in the $(\hat{\mathbf{e}}_1, \hat{\mathbf{e}}_2)$ plane representing the thrust plane orthogonal to the spacecraft–Earth vector $\hat{\mathbf{e}}_3$, as well as the spacecraft–Sun vector $\hat{\mathbf{R}}_s$ in the $\hat{\mathbf{e}}_1, \hat{\mathbf{e}}_3$ plane with the angle β measured from $\hat{\mathbf{e}}_3$. The thrust angle ψ is the angle between $\hat{\mathbf{e}}_1$ and $\hat{\mathbf{u}}$, which is also the angle between $\hat{\mathbf{e}}_2$ and the longitudinal axis of the solar panels. This axis, depicted by $\hat{\mathbf{A}}$, is orthogonal to the thrust vector $\hat{\mathbf{u}}$. The normal to the panels $\hat{\mathbf{N}}$ makes an angle θ with vector $\hat{\mathbf{L}}$ parallel to $\hat{\mathbf{e}}_3$, and the angle between $\hat{\mathbf{N}}$ and $\hat{\mathbf{R}}_s$ is called ϕ. It is shown that θ can be cast in terms of ψ such that $\theta = f(\psi)$, and the power P cast in terms of ψ and β such that $P = f(\psi, \beta)$. As in all the analysis shown thus far, the thrust vector is contained in the plane $(\hat{\mathbf{e}}_1, \hat{\mathbf{e}}_2)$ normal to $\hat{\mathbf{e}}_3$.

From $\hat{\mathbf{e}}_3 = -\hat{\mathbf{r}}$, $\hat{\mathbf{e}}_2 = (\hat{\mathbf{e}}_3 \times \hat{\mathbf{R}}_s)/|\hat{\mathbf{e}}_3 \times \hat{\mathbf{R}}_s|$, $\hat{\mathbf{e}}_1 = \hat{\mathbf{e}}_2 \times \hat{\mathbf{e}}_3$, it is seen that $\hat{\mathbf{R}}_s$ and $\hat{\mathbf{u}}$ can be expressed as

$$\hat{\mathbf{R}}_s = \cos\beta\hat{\mathbf{e}}_3 + \sin\beta\hat{\mathbf{e}}_1 \tag{1.71}$$

$$\hat{\mathbf{u}} = \cos\psi\hat{\mathbf{e}}_1 + \sin\psi\hat{\mathbf{e}}_2 \tag{1.72}$$

Because $\hat{\mathbf{N}}$ is lying in the $\hat{\mathbf{e}}_3, \hat{\mathbf{u}}$ plane, it is clear that it can be expressed in terms of the angles θ and ψ,

$$\hat{\mathbf{N}} = \cos\theta\hat{\mathbf{e}}_3 + \sin\theta\hat{\mathbf{u}} = \cos\psi\,\sin\theta\hat{\mathbf{e}}_1 + \sin\psi\,\sin\theta\hat{\mathbf{e}}_2 + \cos\theta\hat{\mathbf{e}}_3 \tag{1.73}$$

Maximum power output is achieved by maximizing the expression for $\cos\phi$, namely

$$\cos\phi = \hat{\mathbf{N}}^T \cdot \hat{\mathbf{R}}_s = \sin\beta\,\sin\theta\,\cos\psi + \cos\theta\,\cos\beta \tag{1.74}$$

The maximizing value of θ for a given β is obtained from $\partial \cos\phi/\partial\theta = 0$, such that

$$\tan\theta = \frac{\sin\beta\cos\psi}{\cos\beta} \tag{1.75}$$

which yields

$$\sin\theta = \frac{\sin\beta\ \cos\psi}{\left(\cos^2\beta + \sin^2\beta\ \cos^2\psi\right)^{1/2}} \tag{1.76}$$

$$\cos\theta = \frac{\cos\beta}{\left(\cos^2\beta + \sin^2\beta\ \cos^2\psi\right)^{1/2}} \tag{1.77}$$

By replacing $\sin\theta$ and $\cos\theta$ in Equation (1.74), $\cos\phi$ takes the form

$$\cos\phi = \left(\cos^2\beta + \sin^2\beta\ \cos^2\psi\right)^{1/2} = \left(1 - \sin^2\beta\sin^2\psi\right)^{1/2} \tag{1.78}$$

The power output is therefore given by

$$P = P_0\cos\phi = P_0\left(1 - \sin^2\beta\sin^2\psi\right)^{1/2} \tag{1.79}$$

which is only a function of the thrust angle ψ. The partial $\partial P/\partial\mathbf{z}$ can then be expressed as

$$\frac{\partial P}{\partial\mathbf{z}} = -P_0\left(1 - \sin^2\beta\sin^2\psi\right)^{-1/2}\sin^2\psi\sin\beta\cos\beta\frac{\partial\cos\beta}{\partial\mathbf{z}} \tag{1.80}$$

such that from $\cos\beta = \hat{\mathbf{e}}_3\cdot\hat{\mathbf{R}}_s$,

$$\frac{\partial\cos\beta}{\partial\mathbf{z}} = \frac{\partial\hat{\mathbf{e}}_3}{\partial\mathbf{z}}\cdot\hat{\mathbf{R}}_s + \hat{\mathbf{e}}_3\cdot\frac{\partial\hat{\mathbf{R}}_s}{\partial\mathbf{z}} \tag{1.81}$$

From $\mathbf{r} = X_1\hat{\mathbf{f}} + Y_1\hat{\mathbf{g}} + 0\hat{\mathbf{w}}$, it follows that $|\mathbf{r}| = (X_1^2\hat{\mathbf{f}} + Y_1^2\hat{\mathbf{g}})^{1/2}$ and because $\hat{\mathbf{e}}_3 = \frac{-\mathbf{r}}{|\mathbf{r}|}$,

$$\hat{\mathbf{e}}_3 = \left(-X_1\hat{\mathbf{f}} - Y_1\hat{\mathbf{g}}\right)/\left(X_1^2 + Y_1^2\right)^{1/2} \tag{1.82}$$

$$\frac{\partial\hat{\mathbf{e}}_3}{\partial z_i} = -\frac{1}{|\mathbf{r}|^3}\begin{bmatrix}\left(Y_1^2\frac{\partial X_1}{\partial z_i} - X_1Y_1\frac{\partial Y_1}{\partial z_i}\right)\\ \left(-X_1Y_1\frac{\partial X_1}{\partial z_i} + X_1^2\frac{\partial Y_1}{\partial z_i}\right)\\ 0\end{bmatrix} \tag{1.83}$$

X_1 and Y_1 are functions of h and k, and they are given in Chapter 6, z_i standing here for the equinoctial orbit elements. Also,

$$\frac{\partial\hat{\mathbf{R}}_s}{\partial z_i} = \begin{pmatrix}\partial R_f/\partial z_i\\ \partial R_g/\partial z_i\\ \partial R_w/\partial z_i\end{pmatrix} \tag{1.84}$$

The apparent orbit of the Sun with respect to the center of the Earth can be assumed to be a conic, and if the $\hat{\mathbf{R}}_s$ vector is assumed to be the same as the Earth–Sun vector as a further simplification, which is rather a good assumption for Earth-orbiting applications, then the $\hat{\mathbf{R}}_s$ vector can be obtained from its inertial equatorial counterpart $\hat{\mathbf{R}}_s'$ by $\hat{\mathbf{R}}_s = (\hat{\mathbf{f}}\,\hat{\mathbf{g}}\hat{\mathbf{w}})^T \hat{\mathbf{R}}_s'$, such that in the equinoctial coordinates

$$\hat{\mathbf{R}}_s = X_s\hat{\mathbf{f}} + Y_s\hat{\mathbf{g}} + Z_s\hat{\mathbf{w}} = R_f\hat{\mathbf{f}} + R_g\hat{\mathbf{g}} + R_w\hat{\mathbf{w}} \tag{1.85}$$

$\hat{\mathbf{f}}$, $\hat{\mathbf{g}}$ and $\hat{\mathbf{w}}$ being only a function of p and q, it follows that in the general case the p partials are given by

$$\frac{\partial X_s}{\partial p} = \hat{\mathbf{R}}_s \cdot \frac{\partial \hat{\mathbf{f}}}{\partial p} = -\frac{2}{\left(1+p^2+q^2\right)}(qY_s + Z_s) \tag{1.86}$$

$$\frac{\partial Y_s}{\partial p} = \hat{\mathbf{R}}_s \cdot \frac{\partial \hat{\mathbf{g}}}{\partial p} = \frac{2qX_s}{\left(1+p^2+q^2\right)} \tag{1.87}$$

$$\frac{\partial Z_s}{\partial p} = \hat{\mathbf{R}}_s \cdot \frac{\partial \hat{\mathbf{w}}}{\partial p} = \frac{2X_s}{\left(1+p^2+q^2\right)} \tag{1.88}$$

and the q partials by

$$\frac{\partial X_s}{\partial q} = \hat{\mathbf{R}}_s \cdot \frac{\partial \hat{\mathbf{f}}}{\partial q} = \frac{2pY_s}{\left(1+p^2+q^2\right)} \tag{1.89}$$

$$\frac{\partial Y_s}{\partial q} = \hat{\mathbf{R}}_s \cdot \frac{\partial \hat{\mathbf{g}}}{\partial q} = \frac{2(-pX_s + Z_s)}{\left(1+p^2+q^2\right)} \tag{1.90}$$

$$\frac{\partial Z_s}{\partial q} = \hat{\mathbf{R}}_s \cdot \frac{\partial \hat{\mathbf{w}}}{\partial q} = -\frac{2Y_s}{\left(1+p^2+q^2\right)} \tag{1.91}$$

The partials of $\hat{\mathbf{f}}$, $\hat{\mathbf{g}}$ and $\hat{\mathbf{w}}$ with respect to p and q are given by

$$\frac{\partial \hat{\mathbf{f}}}{\partial p} = -\frac{2}{\left(1+p^2+q^2\right)}\left(q\hat{\mathbf{g}} + \hat{\mathbf{w}}\right) \tag{1.92}$$

$$\frac{\partial \hat{\mathbf{f}}}{\partial q} = \frac{2p}{\left(1+p^2+q^2\right)}\hat{\mathbf{g}} \tag{1.93}$$

$$\frac{\partial \hat{\mathbf{g}}}{\partial p} = \frac{2q}{\left(1+p^2+q^2\right)}\hat{\mathbf{f}} \tag{1.94}$$

$$\frac{\partial \hat{\mathbf{g}}}{\partial q} = \frac{2}{\left(1+p^2+q^2\right)}\left(-p\hat{\mathbf{f}} + \hat{\mathbf{w}}\right) \tag{1.95}$$

$$\frac{\partial \hat{\mathbf{w}}}{\partial p} = \frac{2}{\left(1+p^2+q^2\right)}\hat{\mathbf{f}} \tag{1.96}$$

$$\frac{\partial \hat{\mathbf{w}}}{\partial q} = -\frac{2}{\left(1+p^2+q^2\right)}\hat{\mathbf{g}} \tag{1.97}$$

Thus, the $\partial P/\partial \mathbf{z}$ partials needed in the canonical equations of the previous section can be evaluated. The optimization of the thrust angle ψ can be obtained from the Hamiltonian H in Equation (1.51), which is written as

$$H = \frac{2P_0}{c}\left(1 - \sin^2\beta \sin^2\psi\right)^{1/2}\left[\boldsymbol{\lambda}_z^T \frac{MT}{m}\begin{pmatrix}\cos\psi \\ \sin\psi \\ 0\end{pmatrix} - \frac{\lambda_m}{c}\right] \tag{1.98}$$

such that setting $\partial H/\partial \psi = 0$ to maximize H with respect to ψ for a minimum-time solution, and after much algebraic manipulations and expressing the resulting equations in terms of $\cos\psi$, [2] and [3] arrive at a sixth-order polynomial in $\cos\psi$ for the minimizing solution. This solution requires dumping of propellant without using it in the thruster to provide thrust until $\lambda_m = 0$ is reached, after which thrust is generated to effect the desired transfer. This waste of propellant is avoided by ignoring the λ_m term in the Hamiltonian, resulting in a suboptimal solution given by solving a third-order polynomial in $\cos^2\psi$ that can be generated analytically instead [2, 3]. A detailed discussion and derivation is available in [2] and [3], and is not shown here.

1.7 The Analysis of the Zero Roll Constraint with Otherwise Free Pitch and Free Yaw

The following analysis derives the mathematics of the problem of minimum-time low-thrust transfer in which the instantaneous thrust direction is constrained in roll. This means that only pitch and yaw are free while roll is held constant at zero.

This represents another attitude-constrained case not considered by Edelbaum and Sackett [1–3]. It is more important than the zero pitch cases of SECKSPOT because a LEO–GEO (low Earth orbit; geostationary Earth orbit) pitch constrained transfer is essentially identical in propellant expenditure to the unconstrained transfer. This is due to the fact that the intermediate transfer orbits remain essentially near-circular during the continuous thrust transfer requiring pitch angle variations from zero of only some 10 deg at most.

In order to find out the tradeoff between RCS (reaction control system) propellant and main propulsion system propellant expenditures, it is necessary to consider various attitude constrained transfers. For example, for the zero roll case, snap-roll maneuvers needed to orient the solar panels normal to the Sun could be eliminated, saving RCS propellant at the expense of additional transfer ΔV and time due to the inability to orient the panels normal to the Sun, which results in reduced power output and therefore thrust.

There are now two control angles instead of one, because pitch is free and the thrust vector is no longer constrained in a plane orthogonal to the instantaneous radius vector. The thrust vector specification requires the specification of two polar angles in three-dimensional space, and the optimal panel orientation angle is now a function of these two control angles, which are derived themselves from the optimality conditions.

The thrust vector $\hat{\mathbf{u}}$ is defined by the two angles ξ and ψ, such that from Figure 1.4

$$\hat{\mathbf{u}} = c_\psi c_\xi \hat{\mathbf{e}}_1 + s_\psi c_\xi \hat{\mathbf{e}}_2 + s_\xi \hat{\mathbf{e}}_3 \tag{1.99}$$

where $c_a = \cos a$ and $s_a = \sin a$ are now used for compact writing.

Nominally, the spacecraft attitude is such that the yaw axis is along $\hat{\mathbf{e}}_3$, the roll axis along $\hat{\mathbf{e}}_x$, and the positive pitch axis is along $\hat{\mathbf{e}}_y$. The orbit plane is contained in the $\hat{\mathbf{e}}_x$, $\hat{\mathbf{e}}_z$ plane. The thrusters are on the negative roll face of the vehicle such that the thrust vector is along the positive roll axis or $\hat{\mathbf{u}}$ direction, which can be achieved in two ways. One way would be to yaw the vehicle by $\psi_y = \alpha + \psi$ before pitching by an angle ξ, while the

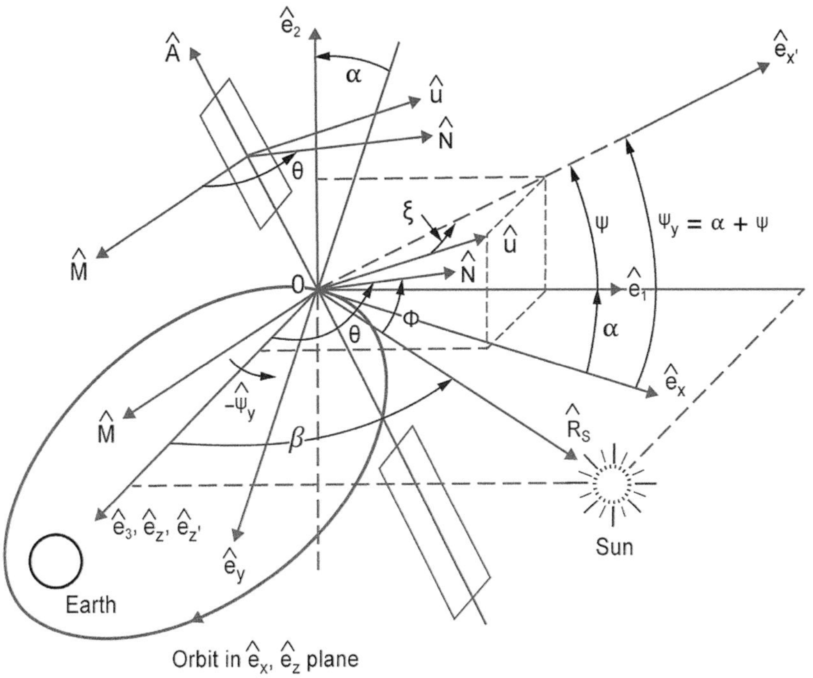

Figure 1.4 Spacecraft yaw and pitch rotation sequence.

reverse sequence of pitch and yaw will also result in the right $\hat{\mathbf{u}}$ orientation. However, the spacecraft ends up with a different roll orientation or rather panel axis orientation $\hat{\mathbf{A}}$, with respect to the $\hat{\mathbf{e}}_1$, $\hat{\mathbf{e}}_2$, $\hat{\mathbf{e}}_3$ frame according to which sequence is selected for the thrust orientation line-up. This makes a difference in how the panel orientation angle θ is computed to get maximum power from the array.

We shall explore both scenarios. First, let the sequence consist of pitch followed by yaw. In this case, $\hat{\mathbf{A}}$ is contained in the $(\hat{\mathbf{u}}, \hat{\mathbf{e}}_2)$ plane such that with $\alpha = 0$ without loss of generality,

$$\hat{\mathbf{M}} = \frac{\hat{\mathbf{u}} \times \hat{\mathbf{e}}_2}{|\hat{\mathbf{u}} \times \hat{\mathbf{e}}_2|} = \frac{-s_\xi \hat{\mathbf{e}}_1 + c_\psi c_\xi \hat{\mathbf{e}}_3}{\left(c_\psi^2 c_\xi^2 + s_\xi^2\right)^{1/2}} \tag{1.100}$$

The two vectors $\hat{\mathbf{u}}$ and $\hat{\mathbf{A}}$ are orthogonal and $\hat{\mathbf{M}} = \hat{\mathbf{u}} \times \hat{\mathbf{A}}$ defines the direction from which the panel normal $\hat{\mathbf{N}}$ is measured through the angle θ. This normal vector $\hat{\mathbf{N}}$ can therefore be expressed in the $\hat{\mathbf{M}}, \hat{\mathbf{u}}$ space by $\hat{\mathbf{N}} = c_\theta \hat{\mathbf{M}} + s_\theta \hat{\mathbf{u}}$, which reduces to

$$\begin{aligned}\hat{\mathbf{N}} = &\left[s_\theta c_\psi c_\xi - c_\theta s_\xi \left(c_\psi^2 c_\xi^2 + s_\xi^2\right)^{-1/2}\right] \hat{\mathbf{e}}_1 \\ &+ \left(s_\theta s_\psi c_\xi\right) \hat{\mathbf{e}}_2 \\ &+ \left[c_\theta c_\psi c_\xi \left(c_\psi^2 c_\xi^2 + s_\xi^2\right)^{-1/2} + s_\theta s_\xi\right] \hat{\mathbf{e}}_3\end{aligned} \tag{1.101}$$

Let ϕ define, as before, the angle between $\hat{\mathbf{N}}$ and $\hat{\mathbf{R}}_s$, then

$$\begin{aligned} c_\phi &= \hat{\mathbf{N}} \bullet \hat{\mathbf{R}}_s \\ &= s_\beta \left[s_\theta c_\psi c_\xi - c_\theta s_\xi \left(c_\psi^2 c_\xi^2 + s_\xi^2 \right)^{-1/2} \right] \\ &\quad + c_\beta \left[s_\theta s_\xi + c_\theta c_\psi c_\xi \left(c_\psi^2 c_\xi^2 + s_\xi^2 \right)^{-1/2} \right] \end{aligned} \tag{1.102}$$

Because the beam power P is proportional to c_ϕ and P_0, the maximum array power is given by $P = P_0 c_\phi$. The maximization of P is achieved by the condition $\frac{\partial c_\phi}{\partial \theta} = 0$, which yields the optimal panel orientation angle θ:

$$\tan\theta = \frac{\left(-s_\beta c_\psi c_\xi - c_\beta s_\xi\right)}{\left(s_\beta s_\xi - c_\beta c_\psi c_\xi\right)\left(c_\psi^2 c_\xi^2 + s_\xi^2\right)^{-1/2}} \tag{1.103}$$

Squaring this quantity and expressing it in terms of s_θ^2 and c_θ^2, respectively, yields expressions for s_θ and c_θ as follows:

$$s_\theta = \frac{\left(s_\beta c_\psi c_\xi + c_\beta s_\xi\right)\left(c_\psi^2 c_\xi^2 + s_\xi^2\right)^{1/2}}{\left[\left(c_\beta c_\psi c_\xi - s_\beta s_\xi\right)^2 + \left(s_\xi^2 + c_\psi^2 c_\xi^2\right)\left(s_\beta c_\psi c_\xi + c_\beta s_\xi\right)^2\right]^{1/2}} \tag{1.104}$$

$$c_\theta = \frac{\left(c_\beta c_\psi c_\xi - s_\beta s_\xi\right)}{\left[\left(s_\xi^2 + c_\psi^2 c_\xi^2\right)\left(s_\beta c_\psi c_\xi + c_\beta s_\xi\right)^2 + \left(c_\beta c_\psi c_\xi - s_\beta s_\xi\right)^2\right]^{1/2}} \tag{1.105}$$

If we replace these quantities in Equation (1.102), the c_ϕ will be expressed in terms of β and the control angles ψ and ξ only, such that

$$c_\phi = \frac{\left[\left(s_\xi^2 + c_\psi^2 c_\xi^2\right)\left(s_\beta c_\psi c_\xi + c_\beta s_\xi\right)^2 + \left(c_\beta c_\psi c_\xi - s_\beta s_\xi\right)^2\right]^{1/2}}{\left(s_\xi^2 + c_\psi^2 c_\xi^2\right)^{1/2}} \tag{1.106}$$

Incidentally, the orientation of $\hat{\mathbf{A}} = a_1\hat{\mathbf{e}}_1 + a_2\hat{\mathbf{e}}_2 + a_3\hat{\mathbf{e}}_3$ can be determined from $\hat{\mathbf{A}} = \hat{\mathbf{M}} \times \hat{\mathbf{u}}$, which results in

$$a_1 = -s_\psi c_\psi c_\xi^2 \left(s_\xi^2 + c_\psi^2 c_\xi^2\right)^{-1/2} \tag{1.107}$$

$$a_2 = \left(s_\xi^2 + c_\psi^2 c_\xi^2\right) \tag{1.108}$$

$$a_3 = -s_\xi c_\xi s_\psi \left(s_\xi^2 + c_\psi^2 c_\xi^2\right)^{-1/2} \tag{1.109}$$

For the case of zero pitch and zero roll, [3] derives the optimal value of θ and the corresponding c_ϕ as also shown in the previous section, as

$$s_\theta = s_\beta c_\psi \left(c_\beta^2 + s_\beta^2 c_\psi^2\right)^{-1/2} \tag{1.110}$$

$$c_\theta = c_\beta \left(c_\beta^2 + s_\beta^2 c_\psi^2\right)^{-1/2} \tag{1.111}$$

$$c_\phi = \left(1 - s_\beta^2 s_\psi^2\right)^{1/2} \tag{1.112}$$

The last three relationships are readily recovered from the more general formulae in Equations (1.104), (1.105) and (1.106) as soon as $\xi = 0$, which corresponds to the zero pitch condition.

If the thrust vector alignment is carried out by using the yaw, pitch sequence, then $\hat{\mathbf{A}}$ will be contained in the $(\hat{\mathbf{e}}_1, \hat{\mathbf{e}}_2)$ plane such that it is orthogonal to the projection $\hat{\mathbf{u}}_1$ of the vector $\hat{\mathbf{u}}$ onto the $\hat{\mathbf{e}}_1, \hat{\mathbf{e}}_2$ plane with

$$\hat{\mathbf{u}}_1 = c_\psi c_\xi \hat{\mathbf{e}}_1 + s_\psi c_\xi \hat{\mathbf{e}}_2 \tag{1.113}$$

$$\hat{\mathbf{A}} = -s_\psi c_\xi \hat{\mathbf{e}}_1 + c_\psi c_\xi \hat{\mathbf{e}}_2 = a_1 \hat{\mathbf{e}}_1 + a_2 \hat{\mathbf{e}}_2 \tag{1.114}$$

Now, $\hat{\mathbf{M}} = \hat{\mathbf{u}} \times \hat{\mathbf{A}}$ is reduced to

$$\hat{\mathbf{M}} = -c_\psi s_\xi c_\xi \hat{\mathbf{e}}_1 - s_\psi s_\xi c_\xi \hat{\mathbf{e}}_2 + c_\xi^2 \hat{\mathbf{e}}_3 \tag{1.115}$$

and the angle θ can now be measured from the $\hat{\mathbf{M}}$ direction such that the normal $\hat{\mathbf{N}}$ to the panels is given by $\hat{\mathbf{N}} = c_\theta \hat{\mathbf{M}} + s_\theta \hat{\mathbf{u}}$:

$$\hat{\mathbf{N}} = \left(s_\theta c_\psi c_\xi - c_\theta c_\psi s_\xi c_\xi\right) \hat{\mathbf{e}}_1 + \left(s_\theta s_\psi c_\xi - c_\theta s_\psi s_\xi c_\xi\right) \hat{\mathbf{e}}_2 + \left(s_\theta s_\xi + c_\theta c_\xi^2\right) \hat{\mathbf{e}}_3 \tag{1.116}$$

Therefore $c_\phi = \hat{\mathbf{N}} \bullet \hat{\mathbf{R}}_s = s_\beta \left(s_\theta c_\psi c_\xi - c_\theta c_\psi s_\xi c_\xi\right) + c_\beta \left(s_\theta s_\xi + c_\theta c_\xi^2\right)$, and the optimal angle θ is obtained as before from $\partial c_\phi / \partial \theta = 0$, which yields

$$\tan\theta = \frac{\left(s_\beta c_\psi c_\xi + c_\beta s_\xi\right)}{\left(c_\beta c_\xi^2 - s_\beta c_\psi s_\xi c_\xi\right)} \tag{1.117}$$

$$s_\theta = \frac{\left(s_\beta c_\psi c_\xi + c_\beta s_\xi\right)}{\left[\left(c_\beta c_\xi^2 - s_\beta c_\psi s_\xi c_\xi\right)^2 + \left(s_\beta c_\psi c_\xi + c_\beta s_\xi\right)^2\right]^{1/2}} \tag{1.118}$$

$$c_\theta = \frac{\left(c_\beta c_\xi^2 - s_\beta c_\psi s_\xi c_\xi\right)}{\left[\left(s_\beta c_\psi c_\xi + c_\beta s_\xi\right)^2 + \left(c_\beta c_\xi^2 - s_\beta c_\psi s_\xi c_\xi\right)^2\right]^{1/2}} \tag{1.119}$$

$$c_\phi = \left[\left(s_\beta c_\psi c_\xi + c_\beta s_\xi\right)^2 + \left(c_\beta c_\xi^2 - s_\beta c_\psi s_\xi c_\xi\right)^2\right]^{1/2} \tag{1.120}$$

The last three equations, in turn, reduce to the simpler expression in Equations (1.110), (1.111) and (1.112) as soon as $\xi = 0$ for zero pitch. However, Equations (1.118), (1.119) and (1.120) are different from Equations (1.104), (1.105) and (1.106) due to the difference in the orientation sequence mentioned earlier. SECKSPOT adopts the sequence yaw, pitch and roll exclusively.

The optimization of the thrust vector orientation can now be carried out starting from the system differential equations used by SECKSPOT [1–3] given by:

$$\dot{\mathbf{z}} = \frac{2P}{mc} \frac{\partial \mathbf{z}}{\partial \dot{\mathbf{r}}} \hat{\mathbf{u}} \tag{1.121}$$

$$\dot{m} = \frac{-2P}{c^2} \tag{1.122}$$

The thrust vector $\hat{\mathbf{u}}$ can be expressed in the $(\hat{\mathbf{e}}_1,\ \hat{\mathbf{e}}_2,\ \hat{\mathbf{e}}_3)$ frame by way of the matrix T:

$$T = \begin{bmatrix} \hat{\mathbf{e}}_1 & \hat{\mathbf{e}}_2 & \hat{\mathbf{e}}_3 \end{bmatrix} \tag{1.123}$$

$$\hat{\mathbf{u}}_T = T^T\hat{\mathbf{u}} = \begin{pmatrix} c_\psi c_\xi \\ s_\psi c_\xi \\ s_\xi \end{pmatrix} \tag{1.124}$$

The Hamiltonian of the system can be written as

$$H = \boldsymbol{\lambda}_z^T\dot{\mathbf{z}} + \lambda_m\dot{m} = \frac{2P_0}{c}c_\phi\left[\boldsymbol{\lambda}_z^T\frac{MT}{m}\begin{pmatrix} c_\psi c_\xi \\ s_\psi c_\xi \\ s_\xi \end{pmatrix} - \frac{\lambda_m}{c}\right] \tag{1.125}$$

where $M = \frac{\partial \mathbf{z}}{\partial \dot{\mathbf{r}}}$, and $\hat{\mathbf{u}} = (T^T)^{-1}(c_\psi c_\xi\ s_\psi c_\xi\ s_\xi)^T$, with $(T^T)^{-1} = T$ due to $T^{-1} = T^T$, because T represents a transformation between two orthonormal systems. The optimality condition is given by

$$\frac{\partial H}{\partial \psi} = \frac{\partial P}{\partial \psi}\frac{H}{P} + \frac{2P}{c}\left[\boldsymbol{\lambda}_z^T\frac{MT}{m}\begin{pmatrix} -s_\psi c_\xi \\ c_\psi c_\xi \\ 0 \end{pmatrix}\right] = 0 \tag{1.126}$$

$$\frac{\partial H}{\partial \xi} = \frac{\partial P}{\partial \xi}\frac{H}{P} + \frac{2P}{c}\left[\boldsymbol{\lambda}_z^T\frac{MT}{m}\begin{pmatrix} -c_\psi s_\xi \\ -s_\psi s_\xi \\ c_\xi \end{pmatrix}\right] = 0 \tag{1.127}$$

The Euler–Lagrange equations are given by

$$\dot{\boldsymbol{\lambda}}_z = \frac{-\partial H}{\partial \mathbf{z}} = -\frac{\partial P}{\partial \mathbf{z}}\frac{H}{P} - \boldsymbol{\lambda}_z^T\frac{2P}{mc}\left(\frac{\partial M}{\partial \mathbf{z}}T + M\frac{\partial T}{\partial \mathbf{z}}\right)\begin{pmatrix} c_\psi c_\xi \\ s_\psi s_\xi \\ s_\xi \end{pmatrix} \tag{1.128}$$

$$\dot{\lambda}_m = \frac{-\partial H}{\partial m} = \boldsymbol{\lambda}_z^T\frac{2P}{m^2c}MT\begin{pmatrix} c_\psi c_\xi \\ s_\psi c_\xi \\ s_\xi \end{pmatrix} \tag{1.129}$$

Because $P = P_0 c_\phi$, $\partial P/\partial\psi = P_0\ \partial c_\phi/\partial\psi$ and $\partial P/\partial\xi = P_0\ \partial c_\phi/\partial\xi$, and considering only the yaw, pitch sequence of SECKSPOT, c_ϕ is then as given in Equation (1.120), such that

$$\frac{\partial P}{\partial \psi} = \frac{P_0}{c_\phi}\left[\left(-s_\beta^2 s_\psi c_\psi c_\xi^2\left(1+s_\xi^2\right) - s_\beta c_\beta s_\psi s_\xi^3 c_\xi\right)\right] \tag{1.130}$$

$$\begin{aligned}\frac{\partial P}{\partial \xi} = \frac{P_0}{c_\phi}\Big[&\left(s_\beta c_\psi c_\xi + c_\beta s_\xi\right)\left(-s_\beta c_\psi s_\xi + c_\beta c_\xi\right) \\ &+ \left(c_\beta c_\xi^2 - s_\beta c_\psi s_\xi c_\xi\right)\left(-c_\beta s_{2\xi} - s_\beta c_\psi c_{2\xi}\right)\Big]\end{aligned} \tag{1.131}$$

Also, from the c_ϕ relation,

$$\frac{\partial P}{\partial \mathbf{z}} = \frac{P_0}{c_\phi}\Big[\left(s_\beta c_\psi c_\xi + c_\beta s_\xi\right)\left(c_\psi c_\xi c_\beta - s_\beta s_\xi\right) - \left(c_\beta c_\xi^2 - s_\beta c_\psi s_\xi c_\xi\right)\left(s_\beta c_\xi^2 + c_\beta c_\psi s_\xi c_\xi\right)\Big]\frac{\partial \beta}{\partial \mathbf{z}} \tag{1.132}$$

and because $\partial c_\beta / \partial \mathbf{z} = -s_\beta \partial\beta/\partial\mathbf{z}$, we can write $\frac{\partial \beta}{\partial \mathbf{z}} = \frac{-1}{s_\beta}\frac{\partial c_\beta}{\partial \mathbf{z}}$, and using the definition of $c_\beta = \hat{\mathbf{e}}_3 \bullet \hat{\mathbf{R}}_s$,

$$\frac{\partial c_\beta}{\partial \mathbf{z}} = \frac{\partial \hat{\mathbf{e}}_3}{\partial \mathbf{z}} \bullet \hat{\mathbf{R}}_s + \hat{\mathbf{e}}_3 \bullet \frac{\partial \hat{\mathbf{R}}_s}{\partial \mathbf{z}} \tag{1.133}$$

The expressions for $\hat{\mathbf{e}}_3$ and $\frac{\partial \hat{\mathbf{e}}_3}{\partial z_i}$ are as in Equations (1.82) and (1.83).

The expressions for X_1 and Y_1 and their partial derivatives are once again shown in Chapter 6 for F, the sixth orbit element.

Finally, in order to derive the partials $\frac{\partial \hat{\mathbf{R}}_s}{\partial z_i}$ in more detail than in the previous section,

$$\frac{\partial \hat{\mathbf{R}}_s}{\partial z_i} = \begin{pmatrix} \frac{\partial R_f}{\partial z_i} \\ \frac{\partial R_g}{\partial z_i} \\ \frac{\partial R_w}{\partial z_i} \end{pmatrix} \tag{1.134}$$

let us first observe that $\hat{\mathbf{R}}_s$ can be written in the inertial and equinoctial systems as

$$\hat{\mathbf{R}}_s = R_x \hat{\mathbf{x}} + R_y \hat{\mathbf{y}} + R_z \hat{\mathbf{z}} \tag{1.135}$$

$$\hat{\mathbf{R}}_s = R_f \hat{\mathbf{f}} + R_g \hat{\mathbf{g}} + R_w \hat{\mathbf{w}} \tag{1.136}$$

The unit vectors $\hat{\mathbf{f}}$, $\hat{\mathbf{g}}$, $\hat{\mathbf{w}}$ in the inertial frame are obtained in terms of the equinoctial elements in Equation (1.7), such that replacing in Equation (1.136) and assuming that R_x, R_y and R_z are known, expressions for R_f, R_g and R_w can be obtained by regrouping terms and comparing with Equation (1.135). The system of three algebraic equations is then solved to yield

$$R_f = \frac{\left(1 - p^2 + q^2\right)}{\left(1 + p^2 + q^2\right)} R_x + \frac{2pq}{\left(1 + p^2 + q^2\right)} R_y - \frac{2p}{\left(1 + p^2 + q^2\right)} R_z \tag{1.137}$$

$$R_g = \frac{2pq}{\left(1 + p^2 + q^2\right)} R_x + \frac{\left(1 + p^2 - q^2\right)}{\left(1 + p^2 + q^2\right)} R_y + \frac{2q}{\left(1 + p^2 + q^2\right)} R_z \tag{1.138}$$

$$R_w = \frac{2p}{\left(1 + p^2 + q^2\right)} R_x - \frac{2q}{\left(1 + p^2 + q^2\right)} R_y + \frac{\left(1 - p^2 - q^2\right)}{\left(1 + p^2 + q^2\right)} R_z \tag{1.139}$$

These expressions could also have been obtained directly from the transformation between the $(\hat{\mathbf{f}}, \hat{\mathbf{g}}, \hat{\mathbf{w}})$ and $(\hat{\mathbf{x}}, \hat{\mathbf{y}}, \hat{\mathbf{z}})$ systems shown in Chapter 2, and also noted here:

$$\begin{pmatrix} \hat{\mathbf{f}} \\ \hat{\mathbf{g}} \\ \hat{\mathbf{w}} \end{pmatrix} = \frac{1}{\left(1 + p^2 + q^2\right)} \begin{pmatrix} \left(1 - p^2 + q^2\right) & 2pq & -2p \\ 2pq & \left(1 + p^2 - q^2\right) & 2q \\ 2p & -2q & \left(1 - p^2 - q^2\right) \end{pmatrix} \begin{pmatrix} \hat{\mathbf{x}} \\ \hat{\mathbf{y}} \\ \hat{\mathbf{z}} \end{pmatrix} \tag{1.140}$$

$$\begin{pmatrix}\hat{\mathbf{x}}\\ \hat{\mathbf{y}}\\ \hat{\mathbf{z}}\end{pmatrix} = \frac{1}{(1+p^2+q^2)}\begin{pmatrix}(1-p^2+q^2) & 2pq & 2p\\ 2pq & (1+p^2-q^2) & -2q\\ -2p & 2q & (1-p^2-q^2)\end{pmatrix}\begin{pmatrix}\hat{\mathbf{f}}\\ \hat{\mathbf{g}}\\ \hat{\mathbf{w}}\end{pmatrix} \tag{1.141}$$

R_f, R_g and R_w are functions of p and q only, such that from

$$R_f = \hat{\mathbf{R}}_s \bullet \hat{\mathbf{f}} \tag{1.142}$$

$$R_g = \hat{\mathbf{R}}_s \bullet \hat{\mathbf{g}} \tag{1.143}$$

$$R_w = \hat{\mathbf{R}}_s \bullet \hat{\mathbf{w}} \tag{1.144}$$

it follows that

$$\frac{\partial R_f}{\partial p} = \hat{\mathbf{R}}_s \bullet \frac{\partial \hat{\mathbf{f}}}{\partial p} \tag{1.145}$$

$$\frac{\partial R_g}{\partial p} = \hat{\mathbf{R}}_s \bullet \frac{\partial \hat{\mathbf{g}}}{\partial p} \tag{1.146}$$

$$\frac{\partial R_w}{\partial p} = \hat{\mathbf{R}}_s \bullet \frac{\partial \hat{\mathbf{w}}}{\partial p} \tag{1.147}$$

and

$$\frac{\partial R_f}{\partial q} = \hat{\mathbf{R}}_s \bullet \frac{\partial \hat{\mathbf{f}}}{\partial q} \tag{1.148}$$

$$\frac{\partial R_g}{\partial q} = \hat{\mathbf{R}}_s \bullet \frac{\partial \hat{\mathbf{g}}}{\partial q} \tag{1.149}$$

$$\frac{\partial R_w}{\partial q} = \hat{\mathbf{R}}_s \bullet \frac{\partial \hat{\mathbf{w}}}{\partial q} \tag{1.150}$$

The various partials appearing on the right-hand side of Equations (1.145) to (1.150) are derived by using the transformation Equation (1.140). From

$$\hat{\mathbf{f}} = \frac{1}{(1+p^2+q^2)}\left[\left(1-p^2+q^2\right)\hat{\mathbf{x}} + 2pq\hat{\mathbf{y}} - 2p\hat{\mathbf{z}}\right] \tag{1.151}$$

$$\frac{\partial \hat{\mathbf{f}}}{\partial p} = \frac{1}{(1+p^2+q^2)^2}\left[-4p\left(1+q^2\right)\hat{\mathbf{x}} + 2\left(1-p^2+q^2\right)q\hat{\mathbf{y}} - 2\left(1-p^2+q^2\right)\hat{\mathbf{z}}\right] \tag{1.152}$$

Replacing $\hat{\mathbf{x}}$, $\hat{\mathbf{y}}$ and $\hat{\mathbf{z}}$ in terms of $\hat{\mathbf{f}}$, $\hat{\mathbf{g}}$ and $\hat{\mathbf{w}}$ by way of Equation (1.141), the above partial can be reduced after regrouping, to the following form:

$$\frac{\partial \hat{\mathbf{f}}}{\partial p} = -\frac{2q}{(1+p^2+q^2)}\hat{\mathbf{g}} - \frac{2}{(1+p^2+q^2)}\hat{\mathbf{w}} \tag{1.153}$$

This is also shown in Equation (1.92).

In a similar way, the partials $\frac{\partial \hat{\mathbf{f}}}{\partial q}$, $\frac{\partial \hat{\mathbf{g}}}{\partial p}$, $\frac{\partial \hat{\mathbf{g}}}{\partial q}$, $\frac{\partial \hat{\mathbf{w}}}{\partial p}$ and $\frac{\partial \hat{\mathbf{w}}}{\partial q}$ are also derived to yield the expressions in Equations (1.93)–(1.97).

Because $\hat{\mathbf{R}}_s = R_f\hat{\mathbf{f}} + R_g\hat{\mathbf{g}} + R_w\hat{\mathbf{w}}$, the partials in Equations (1.145) through (1.150) can now be reduced to the form shown in Equations (1.86)–(1.91).

These partials can also be obtained in a different manner. For example, for the case of $\partial R_f/\partial p$, from Equation (1.137),

$$R_f = \frac{1}{\left(1+p^2+q^2\right)}\left[\left(1-p^2+q^2\right)R_x + 2pqR_y - 2pR_z\right] \tag{1.154}$$

$$\frac{\partial R_f}{\partial p} = \frac{1}{\left(1+p^2+q^2\right)^2}\left[-4p\left(1+q^2\right)R_x + 2q\left(1-p^2+q^2\right)R_y - 2\left(1-p^2+q^2\right)R_z\right] \tag{1.155}$$

Using the orthonormal transformation in Equation (1.141), the inertial components are reduced to the following form:

$$R_x = \frac{1}{\left(1+p^2+q^2\right)}\left[\left(1-p^2+q^2\right)R_f + 2pqR_g + 2pR_w\right] \tag{1.156}$$

$$R_y = \frac{1}{\left(1+p^2+q^2\right)}\left[2pqR_f + \left(1+p^2-q^2\right)R_g - 2qR_w\right] \tag{1.157}$$

$$R_z = \frac{1}{\left(1+p^2+q^2\right)}\left[-2pR_f + 2qR_g + \left(1-p^2-q^2\right)R_w\right] \tag{1.158}$$

Substituting in $\partial R_f/\partial p$ results in Equation (1.159), which is identical to Equation (1.86):

$$\frac{\partial R_f}{\partial p} = \frac{-2}{\left(1+p^2+q^2\right)}\left[qR_g + R_w\right] \tag{1.159}$$

The partial in Equation (1.132), namely $\partial p/\partial \mathbf{z}$, is fully determined because $\partial\beta/\partial\mathbf{z}$ can now be obtained by simply evaluating Equation (1.133) in view of $\partial\hat{\mathbf{e}}_3/\partial\mathbf{z}$ and $\partial\hat{\mathbf{R}}_s/\partial\mathbf{z}$ having been derived. Therefore

$$\frac{\partial\hat{\mathbf{e}}_3}{\partial z_i}\bullet\hat{\mathbf{R}}_s = -\frac{1}{|\mathbf{r}|^3}\left[\left(Y_1^2\frac{\partial X_1}{\partial z_i} - X_1Y_1\frac{\partial Y_1}{\partial z_i}\right)R_f + \left(-X_1Y_1\frac{\partial X_1}{\partial z_i} + X_1^2\frac{\partial Y_1}{\partial z_i}\right)R_g\right] \tag{1.160}$$

$$\hat{\mathbf{e}}_3\bullet\frac{\partial\hat{\mathbf{R}}_s}{\partial z_i} = -\frac{1}{|\mathbf{r}|}\left(X_1\frac{\partial R_f}{\partial z_i} + Y_1\frac{\partial R_g}{\partial z_i}\right) \tag{1.161}$$

and

$$\begin{aligned}\frac{\partial\beta}{\partial z_i} = -\frac{1}{s_\beta}\frac{\partial c_\beta}{\partial z_i} = \frac{1}{s_\beta}\Bigg\{\frac{1}{|\mathbf{r}|^3}\Bigg[\left(Y_1^2\frac{\partial X_1}{\partial z_i} - X_1Y_1\frac{\partial Y_1}{\partial z_i}\right)R_f \\ + \left(-X_1Y_1\frac{\partial X_1}{\partial z_i} + X_1^2\frac{\partial Y_1}{\partial z_i}\right)R_g\Bigg] + \frac{1}{|\mathbf{r}|}\left(X_1\frac{\partial R_f}{\partial z_i} + Y_1\frac{\partial R_g}{\partial z_i}\right)\Bigg\}\end{aligned} \tag{1.162}$$

This is then used in Equation (1.132) to produce the $\partial P/\partial z_i$ partial.

Going back to Equation (1.128) for $\dot{\boldsymbol{\lambda}}_z$, the partials $\partial M/\partial\mathbf{z}$ and $\partial T/\partial\mathbf{z}$ have to be derived. The partials $\partial M/\partial\mathbf{z}$ are derived in Chapter 6; however, the $\partial T/\partial\mathbf{z}$ partials can be obtained as follows.

From the definition of T,

$$\frac{\partial T}{\partial \mathbf{z}} = \begin{pmatrix} \dfrac{\partial \hat{\mathbf{e}}_1}{\partial \mathbf{z}} & \dfrac{\partial \hat{\mathbf{e}}_2}{\partial \mathbf{z}} & \dfrac{\partial \hat{\mathbf{e}}_3}{\partial \mathbf{z}} \end{pmatrix} \tag{1.163}$$

the expressions of $\hat{\mathbf{e}}_1$, $\hat{\mathbf{e}}_2$ and $\hat{\mathbf{e}}_3$ in terms of the equinoctial frame ($\hat{\mathbf{f}}$, $\hat{\mathbf{g}}$, $\hat{\mathbf{w}}$) are needed.

$$\hat{\mathbf{e}}_3 = -\hat{\mathbf{r}} = -\frac{X_1}{|\mathbf{r}|}\hat{\mathbf{f}} - \frac{Y_1}{|\mathbf{r}|}\hat{\mathbf{g}} \tag{1.164}$$

$$\hat{\mathbf{e}}_2 = \frac{\hat{\mathbf{e}}_3 \times \hat{\mathbf{R}}_s}{\left|\hat{\mathbf{e}}_3 \times \hat{\mathbf{R}}_s\right|} = \frac{-Y_1 R_w \hat{\mathbf{f}} + X_1 R_w \hat{\mathbf{g}} + \left(Y_1 R_f - X_1 R_g\right)\hat{\mathbf{w}}}{\left[Y_1^2 R_w^2 + X_1^2 R_w^2 + \left(Y_1 R_f - X_1 R_g\right)^2\right]^{1/2}} \tag{1.165}$$

$$\hat{\mathbf{e}}_1 = \hat{\mathbf{e}}_2 \times \hat{\mathbf{e}}_3 = \frac{1}{|\mathbf{r}|}\left[\left(X_1^2 + Y_1^2\right)R_w^2 + \left(Y_1 R_f - X_1 R_g\right)^2\right]^{-1/2}\left[Y_1\left(Y_1 R_f - X_1 R_g\right)\hat{\mathbf{f}}\right.$$
$$\left. - X_1\left(Y_1 R_f - X_1 R_g\right)\hat{\mathbf{g}} + \left(X_1^2 + Y_1^2\right)R_w \hat{\mathbf{w}}\right] \tag{1.166}$$

These unit vectors can also be written as

$$\hat{\mathbf{e}}_3 = -\left(X_1^2 + Y_1^2\right)^{-1/2}\left(X_1\hat{\mathbf{f}} + Y_1\hat{\mathbf{g}}\right) \tag{1.167}$$

$$\hat{\mathbf{e}}_2 = \left[\left(X_1^2 + Y_1^2\right)R_w^2 + \left(Y_1 R_f - X_1 R_g\right)^2\right]^{-1/2}\left[-Y_1 R_w \hat{\mathbf{f}} + X_1 R_w \hat{\mathbf{g}} + \left(Y_1 R_f - X_1 R_g\right)\hat{\mathbf{w}}\right] \tag{1.168}$$

$$\hat{\mathbf{e}}_1 = \left[\left(X_1^2 + Y_1^2\right)^2 R_w^2 + \left(X_1^2 + Y_1^2\right)\left(Y_1 R_f - X_1 R_g\right)^2\right]^{-1/2} \times$$
$$\left[\left(Y_1^2 R_f - X_1 Y_1 R_g\right)\hat{\mathbf{f}} + \left(X_1^2 R_g - X_1 Y_1 R_f\right)\hat{\mathbf{g}} + \left(X_1^2 + Y_1^2\right)R_w \hat{\mathbf{w}}\right] \tag{1.169}$$

The angle α is easily obtained from $c_\alpha = \hat{\mathbf{e}}_x \bullet \hat{\mathbf{e}}_1$, with $\hat{\mathbf{e}}_1$ given in Equation (1.169) and $\hat{\mathbf{e}}_x$ derived later in Equation (1.224), also resolved along the $\hat{\mathbf{f}}$, $\hat{\mathbf{g}}$, $\hat{\mathbf{w}}$ frame.

The partial derivatives are then obtained as

$$\frac{\partial \hat{\mathbf{e}}_3}{\partial z_i} = \left(X_1^2 + Y_1^2\right)^{-3/2}\left(X_1\frac{\partial X_1}{\partial z_i} + Y_1\frac{\partial Y_1}{\partial z_i}\right)\left(X_1\hat{\mathbf{f}} + Y_1\hat{\mathbf{g}}\right)$$
$$- \left(X_1^2 + Y_1^2\right)^{-1/2}\left(\frac{\partial X_1}{\partial z_i}\hat{\mathbf{f}} + \frac{\partial Y_1}{\partial z_i}\hat{\mathbf{g}}\right) \tag{1.170}$$

$$\frac{\partial \hat{\mathbf{e}}_2}{\partial z_i} = -\left[\left(X_1^2 + Y_1^2\right)R_w^2 + \left(Y_1 R_f - X_1 R_g\right)^2\right]^{-3/2}$$
$$\times\left[\left(X_1\frac{\partial X_1}{\partial z_i} + Y_1\frac{\partial Y_1}{\partial z_i}\right)R_w^2 + \left(X_1^2 + Y_1^2\right)R_w\frac{\partial R_w}{\partial z_i}\right.$$
$$\left. + \left(Y_1 R_f - X_1 R_g\right)\left(\frac{\partial Y_1}{\partial z_i}R_f - \frac{\partial X_1}{\partial z_i}R_g + Y_1\frac{\partial R_f}{\partial z_i} - X_1\frac{\partial R_g}{\partial z_i}\right)\right]$$
$$\times\left[-Y_1 R_w \hat{\mathbf{f}} + X_1 R_w \hat{\mathbf{g}} + \left(Y_1 R_f - X_1 R_g\right)\hat{\mathbf{w}}\right]$$

$$+\left[\left(X_1^2+Y_1^2\right)R_w^2+\left(Y_1R_f-X_1R_g\right)^2\right]^{-1/2}\left[\left(\frac{-\partial Y_1}{\partial z_i}R_w-Y_1\frac{\partial R_w}{\partial z_i}\right)\hat{\mathbf{f}}\right.$$
$$\left.+\left(\frac{\partial X_1}{\partial z_i}R_w+X_1\frac{\partial R_w}{\partial z_i}\right)\hat{\mathbf{g}}+\left(\frac{\partial Y_1}{\partial z_i}R_f-\frac{\partial X_1}{\partial z_i}R_g+Y_1\frac{\partial R_f}{\partial z_i}-X_1\frac{\partial R_g}{\partial z_i}\right)\hat{\mathbf{w}}\right] \tag{1.171}$$

$$\frac{\partial\hat{\mathbf{e}}_1}{\partial z_i}=-\left[\left(X_1^2+Y_1^2\right)^2R_w^2+\left(X_1^2+Y_1^2\right)\left(Y_1R_f-X_1R_g\right)^2\right]^{-3/2}$$
$$\times\left\{\left(X_1\frac{\partial X_1}{\partial z_i}+Y_1\frac{\partial Y_1}{\partial z_i}\right)\left[2\left(X_1^2+Y_1^2\right)R_w^2+\left(Y_1R_f-X_1R_g\right)^2\right]\right.$$
$$+\left(X_1^2+Y_1^2\right)\left[\left(X_1^2+Y_1^2\right)R_w\frac{\partial R_w}{\partial z_i}+\left(Y_1R_f-X_1R_g\right)\left(\frac{\partial Y_1}{\partial z_i}R_f-\frac{\partial X_1}{\partial z_i}R_g+Y_1\frac{\partial R_f}{\partial z_i}\right.\right.$$
$$\left.\left.\left.-X_1\frac{\partial R_g}{\partial z_i}\right)\right]\right\}\left[\left(Y_1^2R_f-X_1Y_1R_g\right)\hat{\mathbf{f}}+\left(X_1^2R_g-X_1Y_1R_f\right)\hat{\mathbf{g}}+\left(X_1^2+Y_1^2\right)R_w\hat{\mathbf{w}}\right]$$
$$+\left[\left(X_1^2+Y_1^2\right)^2R_w^2+\left(X_1^2+Y_1^2\right)\left(Y_1R_f-X_1R_g\right)^2\right]^{-1/2}$$
$$\times\left\{\left(2Y_1\frac{\partial Y_1}{\partial z_i}R_f+Y_1^2\frac{\partial R_f}{\partial z_i}-\frac{\partial X_1}{\partial z_i}Y_1R_g-\frac{\partial Y_1}{\partial z_i}X_1R_g-X_1Y_1\frac{\partial R_g}{\partial z_i}\right)\hat{\mathbf{f}}\right.$$
$$+\left(2X_1\frac{\partial X_1}{\partial z_i}R_g+X_1^2\frac{\partial R_g}{\partial z_i}-\frac{\partial X_1}{\partial z_i}Y_1R_f-\frac{\partial Y_1}{\partial z_i}X_1R_f-X_1Y_1\frac{\partial R_f}{\partial z_i}\right)\hat{\mathbf{g}}$$
$$\left.+\left[\left(2X_1\frac{\partial X_1}{\partial z_i}+2Y_1\frac{\partial Y_1}{\partial z_i}\right)R_w+\left(X_1^2+Y_1^2\right)\frac{\partial R_w}{\partial z_i}\right]\hat{\mathbf{w}}\right\} \tag{1.172}$$

The various partials appearing in these expressions have already been derived such that $\partial T/\partial\mathbf{z}$ is now fully defined.

1.8 The Perturbations Due to the Oblateness of the Earth

Cefola [5] provided the equations for the variation of parameters for the equinoctial elements h, k, p and q due to the second zonal harmonic J_2 in order to account for the perturbing effects of that zonal on the motion of the spacecraft in Earth orbit, after evaluating the single-averaged perturbing potential due to J_2 in terms of the equinoctial elements. Detailed derivations are shown in Chapters 10–13 for various sets of equinoctial elements and rotating frames. A summary of the additional differential equations used in [1–3] to augment the state as well as the adjoint equations is provided next.

Denoting by $\dot{\mathbf{z}}_t$ the state equations due to thrust and by $\dot{\mathbf{z}}_{J_2}$ the contribution due to J_2, the averaged Hamiltonian $\tilde{H}$ can be written as

$$\tilde{H}=\boldsymbol{\lambda}_z^T\dot{\tilde{\mathbf{z}}}_t+\boldsymbol{\lambda}_z^T\dot{\tilde{\mathbf{z}}}_{J_2} \tag{1.173}$$

such that

$$\dot{\tilde{\mathbf{z}}}=\dot{\tilde{\mathbf{z}}}_t+\dot{\tilde{\mathbf{z}}}_{J_2} \tag{1.174}$$

$$\dot{\boldsymbol{\lambda}}_z^T = \frac{-\partial \tilde{H}}{\partial \tilde{\mathbf{z}}} = -\tilde{\boldsymbol{\lambda}}_z^T \frac{\partial \dot{\tilde{\mathbf{z}}}_{J_2}}{\partial \tilde{\mathbf{z}}} - \int_{-\pi}^{\pi} \left[\tilde{\boldsymbol{\lambda}}_z^T \frac{\partial \dot{\tilde{\mathbf{z}}}_t}{\partial \tilde{\mathbf{z}}} s + \tilde{\boldsymbol{\lambda}}_z^T \dot{\tilde{\mathbf{z}}}_t \frac{\partial s}{\partial \tilde{\mathbf{z}}} \right] dF \tag{1.175}$$

The contributions to $\dot{\mathbf{z}}_t$ emanate from $\dot{h}_{J_2}$, $\dot{k}_{J_2}$, $\dot{p}_{J_2}$ and $\dot{q}_{J_2}$, with $R_e = 6314.16$ km for the equatorial radius of the Earth, and $J_2 = 1.0827 \times 10^{-3}$, and given by

$$\dot{h}_{J_2} = \frac{3\mu R_e^2 J_2 k \left[1 - 6\left(p^2+q^2\right) + 3\left(p^2+q^2\right)^2\right]}{2na^5\left(1-h^2-k^2\right)^2\left(1+p^2+q^2\right)^2} \tag{1.176}$$

$$\dot{k}_{J_2} = -\frac{3\mu R_e^2 J_2 h \left[1 - 6\left(p^2+q^2\right) + 3\left(p^2+q^2\right)^2\right]}{2na^5\left(1-h^2-k^2\right)^2\left(1+p^2+q^2\right)^2} \tag{1.177}$$

$$\dot{p}_{J_2} = -\frac{3\mu R_e^2 J_2 q \left(1-p^2-q^2\right)}{2na^5\left(1-h^2-k^2\right)^2\left(1+p^2+q^2\right)} \tag{1.178}$$

$$\dot{q}_{J_2} = \frac{3\mu R_e^2 J_2 p \left(1-p^2-q^2\right)}{2na^5\left(1-h^2-k^2\right)^2\left(1+p^2+q^2\right)} \tag{1.179}$$

The partial derivatives of these differential equations needed in the adjoint equations in Equation (1.175) are given by the following. For the partials with respect to a,

$$\left.\begin{aligned} \frac{\partial \dot{h}_{J_2}}{\partial a} &= -\frac{7}{2a}\dot{h}_{J_2} & \frac{\partial \dot{k}_{J_2}}{\partial a} &= -\frac{7}{2a}\dot{k}_{J_2} \\ \frac{\partial \dot{p}_{J_2}}{\partial a} &= -\frac{7}{2a}\dot{p}_{J_2} & \frac{\partial \dot{q}_{J_2}}{\partial a} &= -\frac{7}{2a}\dot{q}_{J_2} \end{aligned}\right\} \tag{1.180}$$

The following partials are with respect to h, and k:

$$\left.\begin{aligned} \frac{\partial \dot{h}_{J_2}}{\partial h} &= \frac{4h\,\dot{h}_{J_2}}{\left(1-h^2-k^2\right)} & \frac{\partial \dot{k}_{J_2}}{\partial h} &= \frac{\dot{k}_{J_2}}{h} + \frac{4h\,\dot{k}_{J_2}}{\left(1-h^2-k^2\right)} \\ \frac{\partial \dot{p}_{J_2}}{\partial h} &= \frac{4h\,\dot{p}_{J_2}}{\left(1-h^2-k^2\right)} & \frac{\partial \dot{q}_{J_2}}{\partial h} &= \frac{4h\,\dot{q}_{J_2}}{\left(1-h^2-k^2\right)} \end{aligned}\right\} \tag{1.181}$$

$$\left.\begin{aligned} \frac{\partial \dot{h}_{J_2}}{\partial k} &= \frac{\dot{h}_{J_2}}{k} + \frac{4k\,\dot{h}_{J_2}}{\left(1-h^2-k^2\right)} & \frac{\partial \dot{k}_{J_2}}{\partial k} &= \frac{4k\,\dot{k}_{J_2}}{\left(1-h^2-k^2\right)} \\ \frac{\partial \dot{p}_{J_2}}{\partial k} &= \frac{4k\,\dot{p}_{J_2}}{\left(1-h^2-k^2\right)} & \frac{\partial \dot{q}_{J_2}}{\partial k} &= \frac{4k\,\dot{q}_{J_2}}{\left(1-h^2-k^2\right)} \end{aligned}\right\} \tag{1.182}$$

Finally, the partials with respect to p and q, in their corrected version, are shown to be given by

$$\frac{\partial \dot{h}_{J_2}}{\partial p} = \frac{12\mu R_e^2 J_2}{na^5} \frac{kp\left[3\left(p^2+q^2\right)-2\right]}{\left(1-h^2-k^2\right)^2\left(1+p^2+q^2\right)^3} \tag{1.183}$$

$$\frac{\partial \dot{k}_{J_2}}{\partial p} = -\frac{12\mu R_e^2 J_2}{na^5} \frac{hp\left[3\left(p^2+q^2\right)-2\right]}{\left(1-h^2-k^2\right)^2\left(1+p^2+q^2\right)^3} \tag{1.184}$$

$$\frac{\partial \dot{p}_{J_2}}{\partial p} = -\frac{4p\,\dot{p}_{J_2}}{1-\left(p^2+q^2\right)^2} \tag{1.185}$$

$$\frac{\partial \dot{q}_{J_2}}{\partial p} = \frac{\dot{q}_{J_2}}{p}\left[\frac{1-\left(p^2+q^2\right)^2-4p^2}{1-\left(p^2+q^2\right)^2}\right] \tag{1.186}$$

$$\frac{\partial \dot{h}_{J_2}}{\partial q} = \frac{12\mu R_e^2 J_2}{na^5}\frac{kq\left[3\left(p^2+q^2\right)-2\right]}{\left(1-h^2-k^2\right)^2\left(1+p^2+q^2\right)^3} \tag{1.187}$$

$$\frac{\partial \dot{k}_{J_2}}{\partial q} = -\frac{12\mu R_e^2 J_2}{na^5}\frac{hq\left[3\left(p^2+q^2\right)-2\right]}{\left(1-h^2-k^2\right)^2\left(1+p^2+q^2\right)^3} \tag{1.188}$$

$$\frac{\partial \dot{p}_{J_2}}{\partial q} = \frac{\dot{p}_{J_2}}{q}\left[\frac{1-\left(p^2+q^2\right)^2-4q^2}{1-\left(p^2+q^2\right)^2}\right] \tag{1.189}$$

$$\frac{\partial \dot{q}_{J_2}}{\partial q} = -\frac{4q\,\dot{q}_{J_2}}{1-\left(p^2+q^2\right)^2} \tag{1.190}$$

1.9 The Flux Model of SECKSPOT and the Augmented State and Adjoint Equations

The damage function $D(N)$ as a function of N, the accumulated equivalent 1 MEV electron fluence, is given in SECKSPOT as

$$D\left(N\right) = e^{C_1+C_2(log_{10}N)^{12}} \tag{1.191}$$

where C_1 and C_2 are certain constants dependent on the spacecraft, being functions of cell thickness and base resistivity of the solar panel in use. Because $\frac{d}{dN}\log_{10}N = \frac{1}{N}\log_{10}e$, the partial derivative of $D(N)$ with respect to N is then given by

$$\frac{\partial D}{\partial N} = \frac{12}{N}\log_{10}eC_1\left(\log_{10}N\right)^{11}D\left(N\right) \tag{1.192}$$

The power P being a function of $D(N)$, $\frac{\partial P}{\partial N}$ will be needed in $\dot{\lambda}_N$. The flux equation as a function of spatial coordinates is given in analytic form in [3] as

$$\dot{N} = e^{\sum\limits_{i=1}^{5}A_iS_i^{i-1}} \tag{1.193}$$

where $\dot{N}$ represents the flux in equivalent 1 MEV particles, and where S is given by

$$S = \ell n\left(R-1\right) \tag{1.194}$$

with R the radial distance from the Earth center in units of Earth radii $R_\oplus$, and

$$A_i = \sum_{j=1}^{5}\sum_{m=0}^{2}K_{i,j+5m}\ell^{j-1}t_s^m \tag{1.195}$$

Here, ℓ is the geographic latitude in degrees, t_s the shield thickness in mils and $K_{i,j+5m}$ are certain coefficients given in tabular form in [3]. The A_i can be cast in a more compact form in terms of the latitude w in radians as

$$A_i = \sum_{j=1}^{5} \tilde{K}_{ij} w^{j-1} \qquad i = 1, \ldots, 5 \tag{1.196}$$

with

$$\tilde{K}_{ij} = \frac{\sum_{m=0}^{2} K_{i,j+5m} t_s^m}{C_f^{j-1}} \qquad i = 1, \ldots, 5, j = 1, \ldots, 5 \tag{1.197}$$

with C_f the conversion factor from degrees to radians. Thus $\dot{N}$ can be calculated at any point along the orbit in terms of ℓ, R and $\tilde{K}_{i,j}$, with $R = (X_1^2 + Y_1^2)^{1/2}/R_\oplus$. From $\mathbf{r} \bullet \hat{\mathbf{z}} = rs_w$, we have $\left(X_1\hat{\mathbf{f}} + Y_1\hat{\mathbf{g}}\right) \cdot \hat{\mathbf{z}} = rs_w$ with $r = RR_\oplus$, $\hat{\mathbf{f}} = f_1\hat{\mathbf{x}} + f_2\hat{\mathbf{y}} + f_3\hat{\mathbf{z}}$, $\hat{\mathbf{g}} = g_1\hat{\mathbf{x}} + g_2\hat{\mathbf{y}} + g_3\hat{\mathbf{z}}$ shown in Figure 1.2, such that $\hat{\mathbf{f}} \cdot \hat{\mathbf{z}} = f_3$, $\hat{\mathbf{g}} \cdot \hat{\mathbf{z}} = g_3$ and hence $X_1 f_3 + Y_1 g_3 = rs_w$, leading to

$$w = \sin^{-1}\left(\frac{f_3 X_1 + g_3 Y_1}{r}\right) \tag{1.198}$$

f_3 and g_3 are $f(p, q)$ from Equation (1.7) and X_1, Y_1 are $f(a, h, k)$.

Letting u stand for S, the partial of $\dot{N}$ with respect to the orbit elements $\mathbf{z}$ to be used in the adjoint equations will have the form

$$\frac{\partial \dot{N}}{\partial \mathbf{z}} = \dot{N}\left[\sum_{i=1}^{5} \frac{\partial A_i}{\partial \mathbf{z}} u^{i-1} + \sum_{i=2}^{5} A_i u^{i-2} (i-1) \frac{\partial u}{\partial \mathbf{z}}\right] \tag{1.199}$$

with

$$\frac{\partial A_i}{\partial \mathbf{z}} = \sum_{j=2}^{5} \tilde{K}_{ij} w^{j-2} (j-1) \frac{\partial w}{\partial \mathbf{z}} \tag{1.200}$$

If X_1 and Y_1 are normalized with respect to $R_\oplus$, then from

$$\frac{\partial u}{\partial \mathbf{z}} = \frac{1}{(R-1)} \frac{\partial R}{\partial \mathbf{z}}$$

$$\frac{\partial u}{\partial \mathbf{z}} = \frac{1}{R(R-1)} \left(X_1 \frac{\partial X_1}{\partial \mathbf{z}} + Y_1 \frac{\partial Y_1}{\partial \mathbf{z}}\right) \tag{1.201}$$

$\frac{\partial u}{\partial h}$, $\frac{\partial u}{\partial k}$ are known from the expressions for X_1 and Y_1 in Equations (1.24) and (1.25), and $\frac{\partial u}{\partial a}$ is simply given by

$$\frac{\partial u}{\partial a} = \frac{R}{a(R-1)} \tag{1.202}$$

From Equation (1.198) for w, the latitude of the spacecraft, and using the normalized X_1 and Y_1 such that $R = \left(X_1^2 + Y_1^2\right)^{1/2}$, it follows that

$$\frac{\partial w}{\partial \mathbf{z}} = \frac{1}{Rc_w}\left[\frac{\partial X_1}{\partial \mathbf{z}} f_3 + \frac{\partial Y_1}{\partial \mathbf{z}} g_3 + X_1 \frac{\partial f_3}{\partial \mathbf{z}} + Y_1 \frac{\partial g_3}{\partial \mathbf{z}} \right.$$
$$\left. - \frac{(X_1 f_3 + Y_1 g_3)}{R^2}\left(X_1 \frac{\partial X_1}{\partial \mathbf{z}} + Y_1 \frac{\partial Y_1}{\partial \mathbf{z}}\right)\right] \tag{1.203}$$

Of course, $\frac{\partial w}{\partial a} = 0$ from Equation (1.198), and from

$$f_3 = \frac{-2p}{\left(1 + p^2 + q^2\right)}$$
$$g_3 = \frac{2q}{\left(1 + p^2 + q^2\right)}$$
$$w_3 = \frac{\left(1 - p^2 - q^2\right)}{\left(1 + p^2 + q^2\right)}$$

it follows that

$$\frac{\partial f_3}{\partial p} = \frac{-2\left(q g_3 + w_3\right)}{\left(1 + p^2 + q^2\right)} \tag{1.204}$$
$$\frac{\partial f_3}{\partial q} = \frac{2 p g_3}{\left(1 + p^2 + q^2\right)} \tag{1.205}$$
$$\frac{\partial g_3}{\partial p} = \frac{2 q f_3}{\left(1 + p^2 + q^2\right)} \tag{1.206}$$
$$\frac{\partial g_3}{\partial q} = \frac{2\left(-p f_3 + w_3\right)}{\left(1 + p^2 + q^2\right)} \tag{1.207}$$

to be used in Equation (1.203). The partials of X_1 and Y_1 with respect to h and k being known from Chapter 6 for the case in which F is considered as an independent orbital element, the $\frac{\partial w}{\partial \mathbf{z}}$ partials above are therefore fully derived, such that $\frac{\partial \dot{N}}{\partial \mathbf{z}}$ is now known.

The state and adjoint differential equations can now be extended to the seven-dimensional case, namely the five orbit elements, the spacecraft mass and the accumulated particle flux, such that with

$$\mathbf{x} = \begin{pmatrix} \mathbf{z} \\ m \\ N \end{pmatrix} \tag{1.208}$$

the unaveraged Hamiltonian is given by

$$H = \boldsymbol{\lambda}^T \dot{\mathbf{x}} = H_z + H_m + H_N \tag{1.209}$$

where

$$H_z = \boldsymbol{\lambda}_z^T \dot{\mathbf{z}} \tag{1.210}$$
$$H_m = \lambda_m \dot{m} \tag{1.211}$$
$$H_N = \lambda_N \dot{N} \tag{1.212}$$

with

$$\dot{\mathbf{z}} = \frac{2P}{mc}\left(\mathbf{x}, t, N, F, \hat{\mathbf{u}}\right) M\left(\mathbf{z}, F\right) \hat{\mathbf{u}} \tag{1.213}$$

$$\dot{m} = -\frac{2P}{c^2}\left(\mathbf{x}, t, N, F, \hat{\mathbf{u}}\right) \tag{1.214}$$

$$\dot{N} = f\left(\mathbf{z}, F\right) \tag{1.215}$$

and where the power P is given by

$$P = \frac{\eta P_0 D\left(N\right)}{R_s^2\left(t\right)} \cos\left[\phi\left(\mathbf{z}, \hat{\mathbf{u}}\right)\right] \tag{1.216}$$

The averaged Hamiltonian $\tilde{H}$ can now be written as

$$\begin{aligned}\tilde{H} &= \int_{-\pi}^{\pi} H\left(\tilde{\mathbf{x}}, \tilde{\boldsymbol{\lambda}}, \tilde{t}, F, \hat{\mathbf{u}}\right) s\left(\tilde{h}, \tilde{k}, F\right) dF \\ &= \int_{-\pi}^{\pi} \left[\tilde{\boldsymbol{\lambda}}_z^T \dot{\mathbf{z}}\left(\tilde{\mathbf{x}}, \tilde{t}, F, \hat{\mathbf{u}}\right) + \tilde{\lambda}_m \dot{m}\left(\tilde{\mathbf{x}}, \tilde{t}, F, \hat{\mathbf{u}}\right)\right] s\left(\tilde{h}, \tilde{k}, F\right) dF \\ &+ \int_{-\pi}^{\pi} \tilde{\lambda}_N \dot{N}\left(\tilde{\mathbf{z}}, F\right) s\left(\tilde{h}, \tilde{k}, F\right) dF\end{aligned} \tag{1.217}$$

The thrust being considered between the anomalies F_1 and F_2, the state equations are shown to take the following form:

$$\begin{aligned}\dot{\tilde{\mathbf{z}}} &= \frac{\partial \tilde{H}^T}{\partial \tilde{\boldsymbol{\lambda}}_z} = \int_{F_1}^{F_2} \dot{\mathbf{z}}\left(\tilde{\mathbf{x}}, \tilde{t}, F, \hat{\mathbf{u}}\right) s\left(\tilde{h}, \tilde{k}, F\right) dF \\ &= \frac{2}{\tilde{m}c} \int_{F_1}^{F_2} P\left(\tilde{\mathbf{x}}, \tilde{t}, F, \hat{\mathbf{u}}\right) M\left(\tilde{\mathbf{z}}, F\right) \hat{\mathbf{u}} s\left(\tilde{h}, \tilde{k}, F\right) dF\end{aligned} \tag{1.218}$$

$$\begin{aligned}\dot{\tilde{m}} &= \frac{\partial \tilde{H}}{\partial \tilde{\lambda}_m} = \int_{F_1}^{F_2} \dot{m}\left(\tilde{\mathbf{x}}, \tilde{t}, F, \hat{\mathbf{u}}\right) s\left(\tilde{h}, \tilde{k}, F\right) dF \\ &= -\frac{2}{c^2} \int_{F_1}^{F_2} P\left(\tilde{\mathbf{x}}, \tilde{t}, F, \hat{\mathbf{u}}\right) s\left(\tilde{h}, \tilde{k}, F\right) dF\end{aligned} \tag{1.219}$$

$$\dot{N} = \frac{\partial \tilde{H}}{\partial \tilde{\lambda}_N} \int_{-\pi}^{\pi} \dot{N}\left(\tilde{\mathbf{z}}, F\right) s\left(\tilde{h}, \tilde{k}, F\right) dF \tag{1.220}$$

Letting $\tilde{H} = \tilde{H}_z + \tilde{H}_m + \tilde{H}_N$ with $\tilde{H}_z = \tilde{\boldsymbol{\lambda}}_z^T \dot{\tilde{\mathbf{z}}}$, $\tilde{H}_m = \tilde{\lambda}_m \dot{\tilde{m}}$, and $\tilde{H}_N = \tilde{\lambda}_N \dot{\tilde{N}}$ represent the averaged values, then the averaged adjoint equations take the final form

$$\begin{aligned}\dot{\tilde{\boldsymbol{\lambda}}}_z^T &= -\frac{\partial \tilde{H}}{\partial \tilde{\mathbf{z}}} = -\int_{F_1}^{F_2} \left\{\tilde{\boldsymbol{\lambda}}_z^T \left(\frac{\partial M}{\partial \tilde{\mathbf{z}}} \hat{\mathbf{u}} + M \frac{\partial \hat{\mathbf{u}}}{\partial \tilde{\mathbf{z}}}\right)\left(\tilde{\mathbf{x}}, \tilde{t}, F, \hat{\mathbf{u}}\right) s\left(\tilde{h}, \tilde{k}, F\right)\right. \\ &+ \left.\left[\tilde{\boldsymbol{\lambda}}_z^T \dot{\mathbf{z}}\left(\tilde{\mathbf{x}}, \tilde{t}, F, \hat{\mathbf{u}}\right) + \lambda_m \dot{m}\left(\tilde{\mathbf{x}}, \tilde{t}, F, \hat{\mathbf{u}}\right)\right] \frac{\partial s}{\partial \tilde{\mathbf{z}}}\left(\tilde{h}, \tilde{k}, F\right)\right\} dF \\ &- \left.\frac{dF}{d\tilde{\mathbf{z}}}\right|_{F_2} \left[\tilde{\boldsymbol{\lambda}}_z^T \dot{\mathbf{z}}\left(\tilde{\mathbf{x}}, \tilde{t}, F_2, \hat{\mathbf{u}}\right) + \tilde{\lambda}_m \dot{m}\left(\tilde{\mathbf{x}}, \tilde{t}, F_2, \hat{\mathbf{u}}\right)\right] s\left(\tilde{h}, \tilde{k}, F_2\right) \\ &+ \left.\frac{dF}{d\tilde{\mathbf{z}}}\right|_{F_1} \left[\tilde{\boldsymbol{\lambda}}_z^T \dot{\mathbf{z}}\left(\tilde{\mathbf{x}}, \tilde{t}, F_1, \hat{\mathbf{u}}\right) + \tilde{\lambda}_m \dot{m}\left(\tilde{\mathbf{x}}, \tilde{t}, F_1, \hat{\mathbf{u}}\right)\right] s\left(\tilde{h}, \tilde{k}, F_1\right) \\ &- \int_{-\pi}^{\pi} \left\{\tilde{\lambda}_N \frac{\partial \dot{N}}{\partial \tilde{\mathbf{z}}}\left(\tilde{\mathbf{z}}, F\right) s\left(\tilde{h}, \tilde{k}, F\right) + \tilde{\lambda}_N \dot{N}\left(\tilde{\mathbf{z}}, F\right) \frac{\partial s}{\partial \tilde{\mathbf{z}}}\left(\tilde{h}, \tilde{k}, F\right)\right\} dF\end{aligned} \tag{1.221}$$

$$\dot{\tilde{\lambda}}_m = \frac{-\partial \tilde{H}}{\partial \tilde{m}} = \frac{2}{\tilde{m}^2 c} \int_{F_1}^{F_2} \tilde{\boldsymbol{\lambda}}_z^T P\left(\tilde{\mathbf{x}}, \tilde{t}, F, \hat{\mathbf{u}}\right) M\left(\tilde{\mathbf{z}}, F\right) \hat{\mathbf{u}} s\left(\tilde{h}, \tilde{k}, F\right) dF \tag{1.222}$$

$$\dot{\tilde{\lambda}}_N = \frac{-\partial \tilde{H}}{\partial \tilde{N}} = -\frac{2}{c} \int_{F_1}^{F_2} \frac{\partial P\left(\tilde{\mathbf{x}}, \tilde{t}, F, \hat{\mathbf{u}}\right)}{\partial \tilde{N}} \left[\frac{\tilde{\boldsymbol{\lambda}}_z^T M\left(\tilde{\mathbf{z}}, F\right) \hat{\mathbf{u}}}{m} - \frac{\tilde{\lambda}_m}{c}\right] s\left(\tilde{h}, \tilde{k}, F\right) dF \tag{1.223}$$

where $\frac{\partial P}{\partial \tilde{N}} = \frac{\partial D(\tilde{N})}{\partial \tilde{N}} \frac{P}{D(\tilde{N})}$

1.10 The Calculations of the Thrust Angle, the Panel Orientation Angle, the Sun Incidence Angle on the Panel, and the Three Sides of the Spacecraft Body, in SECKSPOT

The unit vector along the thrust direction is already available in the equinoctial frame, such that it is straightforward to calculate in the $\hat{\mathbf{e}}_x, \hat{\mathbf{e}}_y, \hat{\mathbf{e}}_z$ frame discussed in Figure 1.2, because as seen in Figure 1.5, $\hat{\mathbf{e}}_z$ is opposite to $\hat{\mathbf{e}}_r$ such that $\hat{\mathbf{e}}_z = -\hat{\mathbf{r}} = -\frac{\mathbf{r}}{r} = -\frac{X_1}{r}\hat{\mathbf{f}} - \frac{Y_1}{r}\hat{\mathbf{g}}$ from the definition of $\mathbf{r}$. Also $\hat{\mathbf{e}}_h = -\hat{\mathbf{e}}_y$ and $\hat{\mathbf{e}}_x$ is obtained by a rotation of $\pi/2$ from $\hat{\mathbf{e}}_y$. Thus,

$$\hat{\mathbf{e}}_x = \frac{-Y_1}{r}\hat{\mathbf{f}} + \frac{X_1}{r}\hat{\mathbf{g}} \tag{1.224}$$

$$\hat{\mathbf{e}}_y = -\hat{\mathbf{w}} \tag{1.225}$$

$$\hat{\mathbf{e}}_z = \frac{-X_1}{r}\hat{\mathbf{f}} - \frac{Y_1}{r}\hat{\mathbf{g}} \tag{1.226}$$

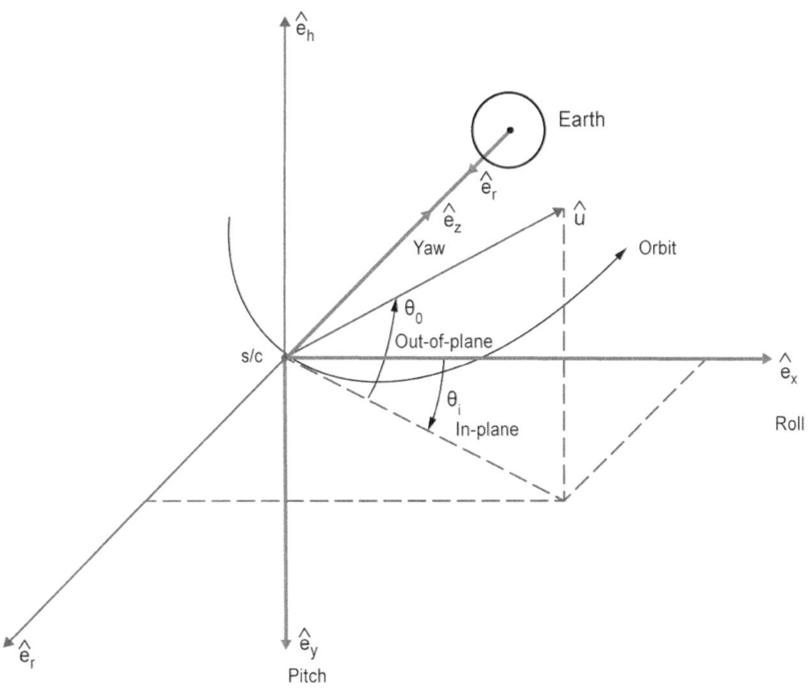

Figure 1.5 Thrust vector in-plane and out-of-plane angles.

The thrust vector $\hat{\mathbf{u}}$ can now be cast in the $\hat{\mathbf{e}}_x, \hat{\mathbf{e}}_y, \hat{\mathbf{e}}_z$ system, $\hat{\mathbf{u}} = u_f\hat{\mathbf{f}} + u_g\hat{\mathbf{g}} + u_w\hat{\mathbf{w}}$ being already known

$$\hat{\mathbf{u}} = c_{\theta_0}c_{\theta_i}\hat{\mathbf{e}}_x - s_{\theta_0}\hat{\mathbf{e}}_y - c_{\theta_0}s_{\theta_i}\hat{\mathbf{e}}_z = u_x\hat{\mathbf{e}}_x + u_y\hat{\mathbf{e}}_y + u_z\hat{\mathbf{e}}_z \tag{1.227}$$

as illustrated in Figure 1.5 which shows the in-plane and out-of-plane angles θ_i and θ_0 of interest.

Thus u_x, u_y and u_z are also known, leading to

$$s_{\theta_0} = -u_y \qquad |\theta_0| \leq \frac{\pi}{2} \tag{1.228}$$

$$c_{\theta_i} = \frac{u_x}{c_{\theta_0}} \tag{1.229}$$

$$s_{\theta_i} = -\frac{u_z}{c_{\theta_0}} \tag{1.230}$$

The angle $\theta_i = 0$ if pitch and roll are constrained and held at 0.

The roll, pitch and yaw axes $\hat{\mathbf{e}}_x, \hat{\mathbf{e}}_y, \hat{\mathbf{e}}_z$ can now be rotated by the Euler angle sequence involving the yaw ψ_y, pitch θ' and roll σ as shown in Figure 1.6, such that

$$\hat{\mathbf{e}}_x, \hat{\mathbf{e}}_y, \hat{\mathbf{e}}_z \xrightarrow{\psi_y} \hat{\mathbf{e}}_{x'}, \hat{\mathbf{e}}_{y'}, \hat{\mathbf{e}}_{z'} \xrightarrow{-\theta'} \hat{\mathbf{e}}_{x''}, \hat{\mathbf{e}}_{y''}, \hat{\mathbf{e}}_{z''} \xrightarrow{\sigma} \hat{\mathbf{e}}_{x'''}, \hat{\mathbf{e}}_{y'''}, \hat{\mathbf{e}}_{z'''}$$

The angles θ' and ψ_y can now be calculated in the roll, pitch, yaw frame $\hat{\mathbf{e}}_x, \hat{\mathbf{e}}_y, \hat{\mathbf{e}}_z$ because

$$\hat{\mathbf{u}} = c_{\theta'}c_{\psi_y}\hat{\mathbf{e}}_x + c_{\theta'}s_{\psi_y}\hat{\mathbf{e}}_y - s_{\theta'}\hat{\mathbf{e}}_z = u_x\hat{\mathbf{e}}_x + u_y\hat{\mathbf{e}}_y + u_z\hat{\mathbf{e}}_z \tag{1.231}$$

Thus u_x, u_y and u_z being known as before, from the knowledge of $\hat{\mathbf{u}}$ in the equinoctial system,

$$s_{\theta'} = -u_z \qquad |\theta'| \leq \frac{\pi}{2} \tag{1.232}$$

$$c_{\psi_y} = \frac{u_x}{c_{\theta'}} \tag{1.233}$$

$$s_{\psi_y} = \frac{u_y}{c_{\theta'}} \tag{1.234}$$

If pitch is constrained at 0, then $\theta' = 0$ and $\psi_y = \theta_0$.

When roll is free, i.e., not constrained at 0, a roll rotation σ followed by a panel rotation about its longitudinal axis will orient the panel normal to the Sun vector for maximum power output. The σ rotation will line up the panel axis normal to the Sun vector, after which the panel will rotate about its axis to orient itself normal to the Sun vector.

The calculation of the third Eulerian angle σ is carried out as follows. Carrying out the yaw rotation ψ_y about $\hat{\mathbf{e}}_z$, the pitch rotation $-\theta'$ about $\hat{\mathbf{e}}_{y'}$, and the roll rotation σ about $\hat{\mathbf{e}}_{x''}$ and denoting these three rotations by their rotation matrices, namely

$$[\psi_y] = \begin{pmatrix} c_{\psi_y} & s_{\psi_y} & 0 \\ -s_{\psi_y} & c_{\psi_y} & 0 \\ 0 & 0 & 1 \end{pmatrix} \tag{1.235}$$

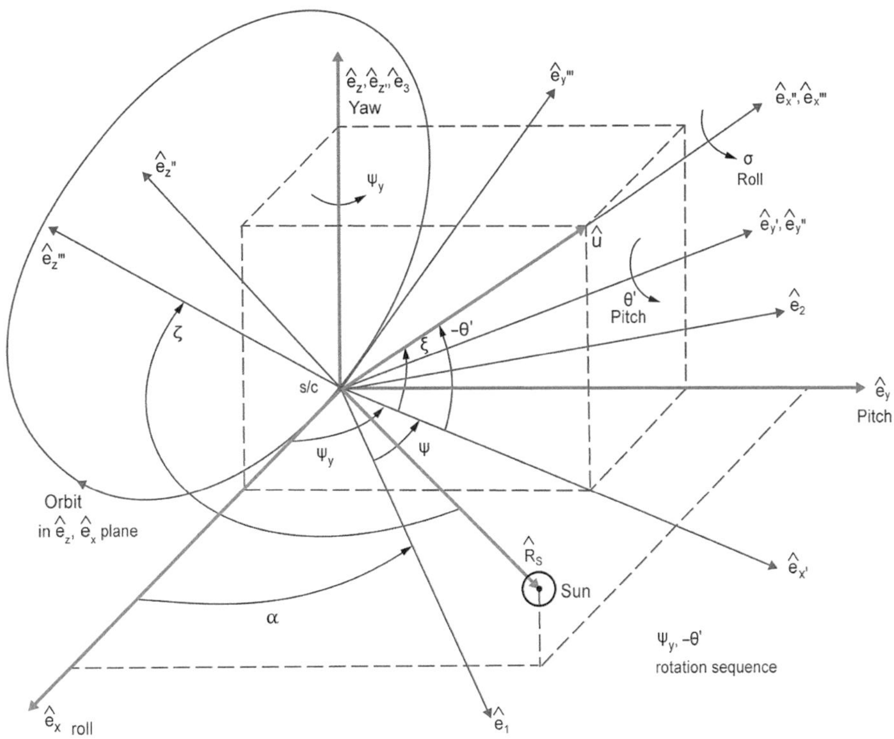

Figure 1.6 Spacecraft rotation sequence angles.

$$[\theta'] = \begin{pmatrix} c_{\theta'} & 0 & -s_{\theta'} \\ 0 & 1 & 0 \\ s_{\theta'} & 0 & c_{\theta'} \end{pmatrix} \tag{1.236}$$

$$[\sigma] = \begin{pmatrix} 1 & 0 & 0 \\ 0 & c_\sigma & s_\sigma \\ 0 & -s_\sigma & c_\sigma \end{pmatrix} \tag{1.237}$$

the unit vector along the panel longitudinal axis, i.e., the pitch axis $\hat{\mathbf{e}}_y$, must be orthogonal to the Sun vector $\hat{\mathbf{R}}_s$ after the three Eulerian rotations. Thus, in the $\hat{\mathbf{e}}_x, \hat{\mathbf{e}}_y, \hat{\mathbf{e}}_z$ system,

$$\left(\hat{\mathbf{R}}_s\right)^T [\psi_y]^T [\theta']^T [\sigma]^T \begin{pmatrix} 0 \\ 1 \\ 0 \end{pmatrix} = 0 \tag{1.238}$$

can be written in a more compact form without showing the various brackets for simplicity, as

$$\left(\sigma\theta'\psi_y\hat{\mathbf{R}}_s\right)^T = \sigma\hat{\mathbf{R}}_s'^T = \hat{\mathbf{R}}_s^T\psi_y^T\theta'^T\sigma^T \tag{1.239}$$

such that with ψ_y, θ' known from Equations (1.232)–(1.234),

$$\hat{\mathbf{R}}_s' = \theta' \psi_y \hat{\mathbf{R}}_s \tag{1.240}$$

is also known.

Now using the expanded form,

$$\begin{pmatrix} 1 & 0 & 0 \\ 0 & c_\sigma & s_\sigma \\ 0 & -s_\sigma & c_\sigma \end{pmatrix} \begin{pmatrix} R_x' \\ R_y' \\ R_z' \end{pmatrix} = \begin{pmatrix} R_x' \\ R_y' c_\sigma + R_z' s_\sigma \\ -R_y' s_\sigma + R_z' c_\sigma \end{pmatrix} \tag{1.241}$$

the inner product in Equation (1.238) will yield

$$R_y' c_\sigma + R_z' s_\sigma = 0 \tag{1.242}$$

such that σ will be calculated from

$$\tan\sigma = -\frac{R_y'}{R_z'} \tag{1.243}$$

in the $(0, 2\pi)$ range.

In SECKSPOT, it is assumed that the panel orientation angle is zero when the normal to the panel is along the z direction. When roll is constrained at zero, the panels cannot rotate to be normal to the Sun vector $\hat{\mathbf{R}}_s$, but the incidence angle can be chosen in such a way that the angle between the normal to the panel and the Sun vector is at its minimum for maximum power possible under these circumstances.

Letting $\hat{\mathbf{R}}'''$ represent the Sun vector in the $(\hat{\mathbf{e}}_{x'''}, \hat{\mathbf{e}}_{y'''}, \hat{\mathbf{e}}_{z'''})$ frame after the three Eulerian rotations, such that

$$\begin{pmatrix} R_x''' \\ R_y''' \\ R_z''' \end{pmatrix} = [\sigma]\,[\theta']\,[\psi_y]\,\hat{\mathbf{R}}_s \tag{1.244}$$

Because the normal $\hat{\mathbf{N}}$ to the panel is along the $\hat{\mathbf{e}}_{z'''}$ direction, i.e., the new $\hat{\mathbf{e}}_z$ yaw axis,

$$\hat{\mathbf{R}}''' \cdot \hat{\mathbf{e}}_{z'''} = c_\zeta = \left[R_x'''\ R_y'''\ R_z'''\right] \begin{pmatrix} 0 \\ 0 \\ 1 \end{pmatrix} = R_z''' \tag{1.245}$$

will yield the desired rotation of $\hat{\mathbf{N}}$, i.e., the angle ζ as seen in Figure 1.6,

$$c_\zeta = R_z''' \tag{1.246}$$

for the roll free case, and the incidence angle i_p will be zero for the panel to face fully the Sun vector in an orthogonal orientation, being understood that σ is given in Equation (1.243).

But when roll is constrained at zero, $\sigma = 0$ and the $[\sigma]$ rotation matrix is the unit matrix such that $\hat{\mathbf{R}}'''$ is now given by

$$\begin{pmatrix} R_x''' \\ R_y''' \\ R_z''' \end{pmatrix} = [\theta']\,[\psi_y]\,\hat{\mathbf{R}}_s \tag{1.247}$$

$\hat{\mathbf{N}}$ being initially along $\hat{\mathbf{e}}_{z'''}$, it is rotated by the angle ζ to be determined below, such that $\hat{\mathbf{N}}$ becomes $\hat{\mathbf{N}}'$, and from

$$\hat{\mathbf{N}}' = \begin{pmatrix} c_\zeta & 0 & s_\zeta \\ 0 & 1 & 0 \\ -s_\zeta & 0 & c_\zeta \end{pmatrix} \begin{pmatrix} 0 \\ 0 \\ 1 \end{pmatrix} = \begin{pmatrix} s_\zeta \\ 0 \\ c_\zeta \end{pmatrix} \tag{1.248}$$

it follows that

$$\hat{\mathbf{N}}'^T \hat{\mathbf{R}}''' = c_{i_p} = s_\zeta R_x''' + c_\zeta R_z''' \tag{1.249}$$

which is the equation for the incidence angle i_p. This angle can be minimized, i.e., c_{i_p} maximized, by $\frac{\partial c_{i_p}}{\partial \zeta} = 0$, yielding

$$c_\zeta = \frac{R_z'''}{\left(R_x'''^2 + R_z'''^2\right)^{1/2}} \tag{1.250}$$

$$s_\zeta = \frac{R_x'''}{\left(R_x'''^2 + R_z'''^2\right)^{1/2}} \tag{1.251}$$

Thus,

$$c_{i_p} = \left(R_x'''^2 + R_z'''^2\right)^{1/2} \tag{1.252}$$

Finally, R_x''', R_y''', R_z''' being the components of the Sun vector in the $(e_{x'''}, e_{y'''}, e_{z'''})$ system after the three Eulerian rotations, they are also the cosines of the angles between the three normals to the sides of the spacecraft and the Sun vector, yielding the incidence angles on those three sides.

1.11 Examples of Minimum-Time Transfers from LEO to GEO

One of the main challenges associated with a resuable EOTV (electric orbit transfer vehicle) consists of the degradation of the solar array power output due to radiation damage from electrons and protons during transit in the Van Allen radiation belts. Figures 1.7 and 1.8, borrowed from [6], show the electron and proton contributions to the total flux as a function of radial distance normalized with respect to the Earth radius R_e, for various latitudes.

This field model consists of a collection of 1 MEV equivalent electron and proton values defined at discrete spatial points around the Earth, with solar cell shield thickness as a parameter. These discrete points are then fit with analytic functions through a technique of multivariate regression such that the flux $\dot{N}$ is modeled as

$$\dot{N} = \exp\left[\sum_{i=0}^{4}\left(\sum_{k=0}^{4}\sum_{j=0}^{2} B_{ij} t_s^j C_{ik} \ell^k\right) S^i\right] \tag{1.253}$$

where $S = \ell n(R - 1)$ with R standing for radial distance from the center of the Earth, t_s for shield thickness, ℓ for geographic latitude and B_{ij} and C_{ik} are coefficients whose

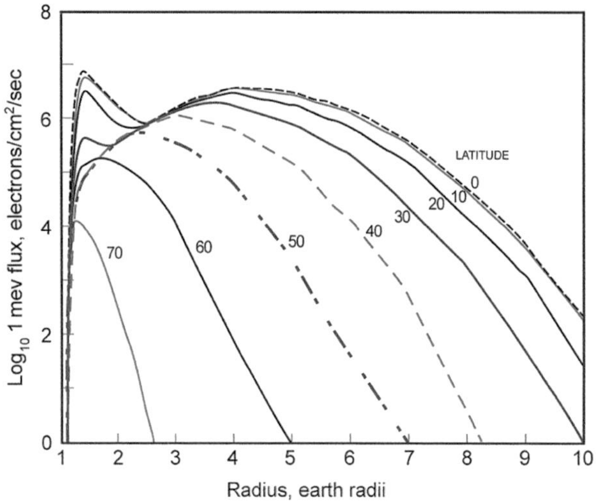

Figure 1.7 Electron contribution to total flux versus radial distance for $t_s = 10$ mils.

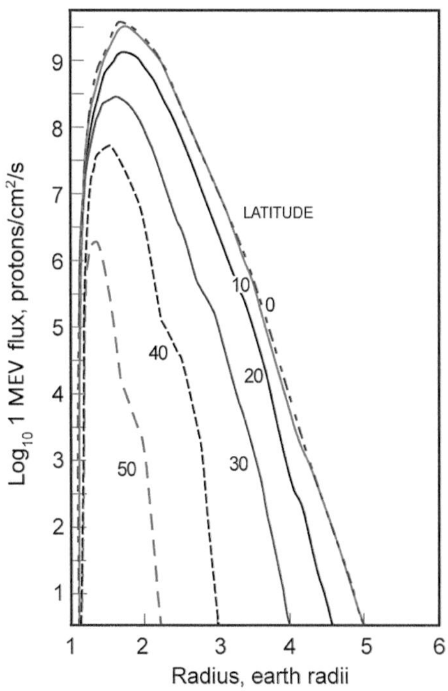

Figure 1.8 Proton contribution to total flux versus radial distance for $t_s = 10$ mils.

Table 1.1 LEO-GEO performance for various initial inclinations

i (deg)	ΔV (km/s)	T_f (days)	M_f (kg)	N (1 MEV electrons/cm^2)
28.5	5.722927	172.824	904.77	$1.53354179 \times 10^{16}$
30.0	5.829989	175.811	902.40	$1.46766247 \times 10^{16}$
35.0	6.203601	186.166	974.62	1.2556365×10^{16}
40.0	6.594618	196.894	1030.78	1.072459×10^{16}
45.0	6.993132	207.71	1087.41	0.9346549×10^{16}
50.0	7.391172	218.40	1143.37	—
55.0	7.782092	228.79	1197.76	—
60.0	8.160527	238.74	1249.86	—
65.0	8.522116	248.16	1299.16	—
70.0	8.863321	256.96	1345.23	—
75.0	9.181268	265.09	1387.79	—
80.0	9.473609	272.50	1426.60	—
85.0	9.738530	279.17	1461.51	—
90.0	9.974480	285.07	1492.39	—
95.0	10.180319	290.18	1519.18	—
100.0	10.355301	294.51	1541.83	—

products are given in Tables 4 and 5 of [6] for both the 1 MEV electron and proton contributions to the total flux.

However, this field model is not completely general since it is valid only for $R > 1.5$ R_e in the latitude region 50 deg $< \ell <$ 60 deg, and for $R > 2.5\ R_e$ in the latitude region 60 deg $< \ell <$ 75 deg. Above a latitude of 75 deg, the field is set to zero. This is then the radiation model representation in SECKSPOT.

However, inspection of Figures 1.7 and 1.8 reveals a dramatic drop in the flux values at high latitudes due to the fact that the radiation belts are thicker at the equator and rather thin near the poles.

It is then possible to evaluate the reduction in the total accumulated fluence for a spacecraft which starts its low-thrust transfer from an initially higher inclination LEO to the final GEO, as well as the penalty in propellant and transfer time associated with the more demanding inclination change.

Let us assume for the sake of the discussion that a combined EOTV–payload mass is in LEO at 225 nm circular orbit of 6350.35 kg or 14,000 lbs, and an I_{sp} of 3800 s.

The nominal transfer from 28.5 deg inclination LEO, to GEO with a 60 kW power level at 70% efficiency (effective 42 kW), without degradation, J_2 effect or shadowing, requires a transfer time of some 172 days, a ΔV of 5.722927 km/s and a propellant consumption of 904 kg (min-time solution). The total accumulated fluence is at an equivalent 1 MEV electrons/cm^2 of $1.53354179 \times 10^{16}$. Table 1.1 shows the corresponding numbers for other initial inclinations up to 100 deg. However, the fluence is not calculated by SECKSPOT due to the limitations in latitude and radial distance mentioned above, for initial inclinations beyond 45 deg.

Nevertheless, a drop of 40 percent in the fluence N is already evident when the initial LEO is at 45 deg inclination as compared to the nominal 28.5 deg. This is depicted

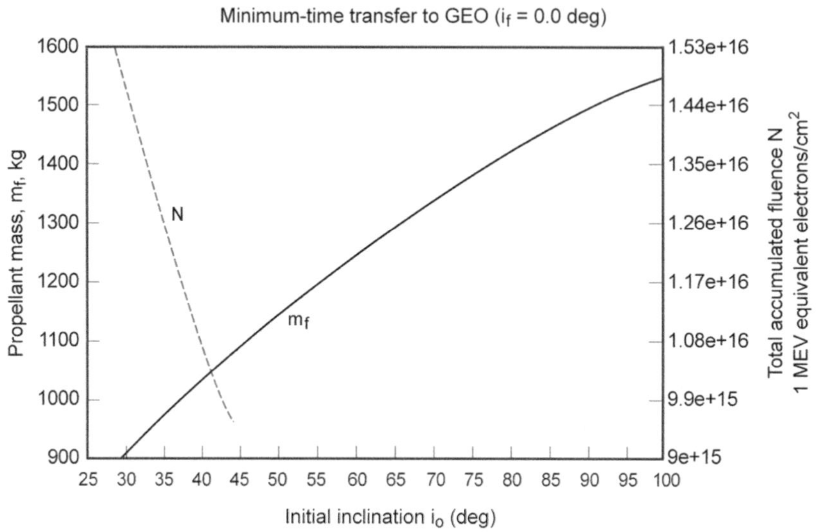

Figure 1.9 Accumulated fluence and propellant expenditure for various initial orbit inclinations.

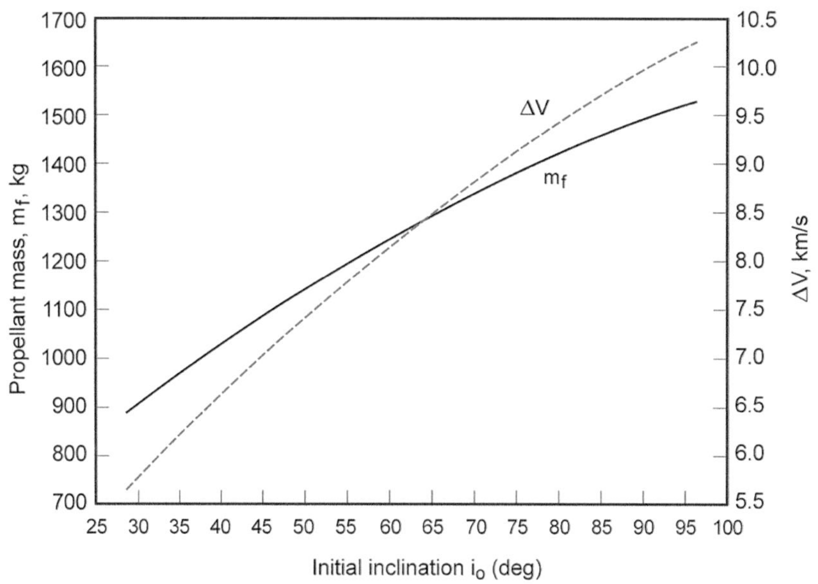

Figure 1.10 Velocity change and propellant expenditure for various initial orbit inclinations.

graphically in Figure 1.9, where propellant mass is also shown. It is interesting to observe that even if one starts from a polar orbit, the penalty in propellant is of the order of 600 kg and transfer time of the order of 100 days.

Figures 1.10 and 1.11 depict ΔV and flight time t_f as a function of initial LEO inclination. ΔV increases from 5.722 km/s to 9.974 km/s between $i = 28.5$ deg and $i = 90$ deg.

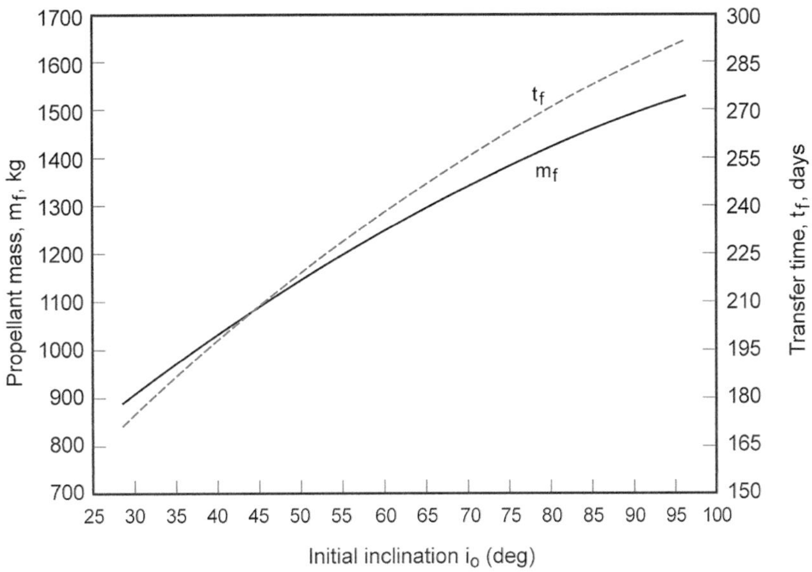

Figure 1.11 Transfer time and propellant expenditure for various initial orbit inclinations.

The 40 percent reduction in fluence mentioned above without activating the power degradation is associated with a 20 percent increase in propellant usage of 180 kg and a flight time extension of 35 days.

In Table 1.2, the values in internal units of all the initial Lagrange multipliers generated by SECKSPOT are depicted in order to show how λ_a and λ_q in particular experience important variations from one LEO to another.

1.12 A Discussion of the SECKSPOT Software Capabilities and Limitations

SECKSPOT is based on a highly sophisticated averaging theory using nonsingular orbital elements called the equinoctial elements. Despite its outstanding ability to solve the minimum-time low-thrust transfer between elliptic orbits for the unconstrained thrust direction problem as well as for some particular attitude-constrained problems, it is not capable in its present form of solving a variety of other optimization problems such as the minimum-fuel transfer, not does it optimize thrust magnitude, specific impulse or take into account a variety of other attitude and attitude rate constraints that must be considered for operational mode simulation studies.

In this section, we attempt to describe the capabilities of this outstanding program created at MIT/Draper by Sackett, Malchow and Edelbaum [2] in the early 1970s, as well as its limitations, which leads us then to consider the various analytic updates that must be worked out in order to extend its usefulness. After a thorough discussion of some of these updates, the problem of the mechanics of optimization of the thrust magnitude, as well as direction, is addressed heuristically. Finally, new software

Table 1.2 Converged values of initial multipliers

i (deg)	λ_a	λ_h	λ_k	λ_p	λ_q	λ_m
28.5	316.020	−4.1339E-9	2.5963E-10	7.1528E-9	−2.4152E4	2.3849E-3
30.0	284.387	−1.9259E-10	−1.8573E-10	−3.4152E-11	−2.4704E4	5.2614E-4
35.0	181.045	−3.2309E-9	3.8962E-9	−3.0885E-10	−2.5993E4	2.0352E-2
40.0	83.208	−9.4922E-10	−8.3180E-11	1.2081E-9	−2.6539E4	3.0894E-3
45.0	−7.414	−4.9422E-11	−8.5255E-9	−9.3241E-10	−2.647E4	7.377E-5
50.0	−90.292	−1.0511E-9	−5.2702E-9	4.8872E-9	−2.5927E4	3.4589E-3
55.0	−165.471	7.4291E-10	−3.5621E-9	3.3086E-10	−2.5008E4	1.7341E-3
60.0	−233.273	1.7086E-10	5.3481E-11	−1.5105E-10	−2.3804E4	−2.4316E-6
65.0	−294.130	−2.3455E-10	−1.9489E-11	−4.1825E-11	−2.2382E4	2.9858E-5
70.0	−348.499	2.7687E-11	−6.9494E-12	1.8738E-10	−2.0795E4	−1.0367E-5
75.0	−396.807	1.3792E-10	5.4426E-11	7.7346E-11	−1.9086E4	1.3062E-4
80.0	−439.434	−1.6156E-10	3.7394E-11	−1.0676E-10	−1.7287E4	−5.1386E-4
85.0	−476.702	2.2150E-10	−3.2567E-10	−3.9449E-11	−1.5423E4	−4.0654E-3
90.0	−508.899	−7.7040E-10	2.1624E-9	4.2685E-11	−1.3513E4	−4.2578E-4
95.0	−536.229	1.7466E-10	6.1269E-9	−3.0060E-11	−1.1574E4	−1.7823E-3
100.0	−558.826	1.6509E-8	3.9461E-8	−5.9243E-11	−9.6220E3	−7.1956E-3

requirements are addressed both within the SECKSPOT formulation environment and the more realistic mission operations environment leading to the future creation of high-fidelity programs capable of accurately simulating mission characteristics for a variety of spaceflight problems using realistic constraints, as well as supporting other functions such as navigation and guidance.

1.12.1 The Capability of the SECKSPOT Program

SECKSPOT is one of the main programs used at NASA and other research centers to carry out mission design studies involving low-thrust transfers between any given earthbound orbits. It was developed at the MIT/Draper Laboratory in the early 1970s in support of the NASA/Lewis SERT-C mission. It is based on the technique of averaging, pioneered in its applications to spaceflight by the late Theordore Edelbaum [1]. The averaging technique requires that the thrust acceleration be small, of the order of 10^{-4} g, such as in the case of the solar-electric propulsion systems using ion engines. This program generates minimum-time geocentric transfer trajectories rather quickly, because the method of averaging it uses allows the consideration of large time steps during the numerical integration of the state and adjoint equations. These equations represent averaged rates of change of the mean values of the state and costate parameters. The short-period variations in these parameters are eliminated by the averaging process which, thereby, eliminates the position element or anomaly from the dynamic system of equations. This, in effect, removes the need to take small time steps to carry out the integration of the exact dynamical equations and allows the use of time steps of the order of days, instead of seconds, to integrate the averaged or approximate dynamical equations.

Another important feature of the SECKSPOT program is the use of a set of nonsingular orbit elements called the equinoctial elements first introduced by Lagrange.

The dynamic equations in these elements have no singularities at zero eccentricity and inclination which, of course, is not the case for the equations based on the classical elements. The transformations between these two sets of elements, where the mean longitude λ is selected as the sixth orbit element, are given by

$$a = a \tag{1.254}$$

$$h = e \sin(\omega + \Omega) \tag{1.255}$$

$$k = e \cos(\omega + \Omega) \tag{1.256}$$

$$p = \tan\left(\frac{i}{2}\right) \sin \Omega \tag{1.257}$$

$$q = \tan\left(\frac{i}{2}\right) \cos \Omega \tag{1.258}$$

$$\lambda = M + \omega + \Omega \tag{1.259}$$

And conversely,

$$a = a \tag{1.260}$$

$$e = \left(h^2 + k^2\right)^{1/2} \tag{1.261}$$

$$i = 2 \tan^{-1}\left(p^2 + q^2\right)^{1/2} \tag{1.262}$$

$$\Omega = \tan^{-1}\left(\frac{p}{q}\right) \tag{1.263}$$

$$\omega = \tan^{-1}\left(\frac{h}{k}\right) - \tan^{-1}\left(\frac{p}{q}\right) \tag{1.264}$$

$$M = \lambda - \tan^{-1}\left(\frac{h}{k}\right) \tag{1.265}$$

The elements a, e, i, Ω, ω and M are the well-known classical elements, while a, h, k, p, q and λ represent the equinoctial elements, with λ standing for the mean longitude.

The derivation of the exact unaveraged differential equations in terms of the equinoctial elements is somewhat tedious and draws on the work of Broucke and Cefola [7]. These elements are, of course, slowly varying when the thrust is on, and they remain constant when the thrust is off, except of course for λ.

Furthermore, the program makes use of an analytic radiation model and can account for such effects as Earth oblateness, solar motion, shadowing, thruster startup delay, and solar cell degradation.

As in all optimal control problems using the indirect shooting method, the initial values of some or all of the adjoint variables must be first guessed and then adjusted in an iterative manner in order to solve the two-point boundary value problem, which satisfies the initial and final conditions, thereby resulting in an optimal transfer solution minimizing flight time. For this particular problem, a guess of the total transfer time is also needed to start the iteration process.

SECKSPOT uses a simple Newton iteration scheme to converge to the required initial values of the unspecified state and adjoint parameters and the optimal transfer time that lead to the satisfaction of the final conditions to within a specified tolerance. This requires the generation of a certain sensitivity matrix, which is created by varying the initial conditions one at a time and running a set of neighboring trajectories.

The program computes the optimal thrust direction time-history that achieves the optimal transfer, and it is assumed that this direction is established entirely by the spacecraft attitude rotations. The specification of this thrust vector orientation in three-dimensional space is equivalent to the specification of the roll, yaw and pitch angles where the yaw axis is parallel to the spacecraft–Earth radius vector pointed to the Earth center, the pitch axis is perpendicular to the orbit plane and directed south, and the roll axis lies in the orbit plane and is perpendicular to the Earth–spacecraft vector. The nominal attitude is such that the principal body axes are aligned with this coordinate system, the ion thrusters being mounted on the negative roll face of the spacecraft, parallel to the roll axis. It is also assumed that the solar panels can be rotated about their longitudinal axis which is along the spacecraft pitch axis.

Besides the unconstrained thrust direction case, SECKSPOT also generates optimal minimum-time geocentric transfers between any elliptic orbits for the attitude-constrained case where only yaw and roll are free, while pitch is held constant at zero. This is equivalent to saying that the thrust vector lies in a plane perpendicular to the spacecraft–Earth axis at all times such that the radial component of the acceleration vector is zero. When the roll angle is free, the solar panels can always be rotated so as to face the Sun and provide maximum power. This is achieved by rolling the spacecraft to the appropriate roll angle, which varies continuously as the vehicle moves along its orbit. For these two cases, the power is not a function of the thrust direction and is always at its maximum available level.

If both pitch and roll are held at zero while yaw is free, then the panels cannot always be oriented normal to the Sun vector such that power becomes, now, a function of thrust direction, Sun vector, and spacecraft position, complicating the analysis to a great extent. The thrust vector still lies in a plane normal to the Earth vector such that its orientation is defined by a single angle which is derived analytically from the solution of a cubic after some simplifying assumptions. These assumptions are such that the transfer is not a strictly minimum-time solution and is called, therefore, a suboptimal solution. The optimal panel orientation angle which leads to the generation of the maximum power is also derived analytically in terms of the control angle mentioned above.

The case of zero roll but free yaw and pitch is considerably more difficult since, in this case, the thrust vector orientation is specified by two control angles, and the panel orientation angle which leads to the maximum power output is now a function of these two angles. The author has derived the optimal control law for this latter case in Section 1.7, because it cannot be handled by SECKSPOT. However, the control angles cannot be obtained analytically, as for the previous cases discussed above.

The last two constrained transfers will require substantial increases in flight time when compared to the unconstrained or the zero pitch only case, because the power developed is coupled to the thrust direction and because the thrust direction is constrained. The case of zero yaw is not interesting except for coplanar transfers because no plane change is possible without yaw.

Finally, the thrust magnitude is assumed proportional to power, and both the specific impulse and efficiency are assumed to be constant in SECKSPOT. The assumption of constant specific impulse or constant exhaust velocity is valid only if the beam voltage is held constant. It is possible to optimize not only the thrust magnitude, but beam voltage and beam current as well, leading to an optimum I_{sp} or exhaust velocity during the mission, as discussed in Chapter 4.

It is possible to assume a so-called ramp law or linear variation of beam voltage versus beam current, and even define a two-dimensional surface or envelope in the beam voltage–current space in order to select the optimum combination of these two parameters during the optimization process.

The analytic radiation and power degradation model deserves a few more comments. The thruster beam power is given by

$$P = \eta P_0 D(N) \frac{1}{R_s^2} \cos\phi \tag{1.266}$$

with P_0 standing for the initial maximum array power at 1 au, η for the total power efficiency factor, $D(N)$ for the damage factor which is a function of the fluence N, with a maximum value of 1, R_s for the distance between the spacecraft and the Sun in au, and ϕ for the angle between the normal to the solar panels and the spacecraft–Sun vector. When roll is free, $\phi = 0$, and

$$P = \frac{P_0}{R_s^2} \eta D(N) \tag{1.267}$$

Because P, the beam power, is given by

$$P = \frac{fc}{2} \tag{1.268}$$

where f is the thrust magnitude, and because $f = -\dot{m}c$ with $\dot{m} < 0$, then

$$P = \frac{fc}{2} = -\frac{f^2}{2\dot{m}} \tag{1.269}$$

$$\dot{m} = -\frac{f}{c} = -\frac{f^2}{2P} = -\frac{2P}{c^2} \tag{1.270}$$

If c is assumed constant, then from Equation (1.270) it follows that

$$\Delta V = c\ell n\left(\frac{m_0}{m}\right) \tag{1.271}$$

The damage function $D(N)$ represents the power degradation as a function of accumulated fluence N, or, more precisely, the accumulated equivalent 1 MEV electron and proton fluence. It is given analytically in SECKSPOT by

$$D(N) = e^{c_1 + c_2 (\log_{10} N)^{12}} \tag{1.272}$$

The constants C_1 and C_2 are dependent on the cell thickness and base resistivity. The fluence N is derived from the flux equation which is given in terms of the spatial coordinates

$$\dot{N} = \exp\left(\sum_{1}^{5} A_i S^{i-1}\right) \tag{1.273}$$

such that $S = \ell n(R-1)$, with R the distance from the Earth center in au and

$$A_i = \sum_{j=1}^{5}\sum_{m=0}^{2} K_{i,j+5m}\ell^{j-1}t_s^m \tag{1.274}$$

Here, ℓ is, as before, geographic latitude, t_s is shield thickness and the Ks are coefficients given in Tables 4 and 5 of [6].

Equation (1.273) constitutes an additional state equation that must be solved simultaneously with the differential equations involving the five equinoctial elements and the single scalar equation for the mass given by equation (1.270).

These analytic radiation and power loss models give only an approximate account of the solar cell power degradation by exposure of the cell to the energetic trapped particles. Accurate trajectory simulations must, however, rely on detailed radiation degradation calculations based, for example, on the models of Obenschain [8]. Otherwise, much more accurate mission simulations cannot be carried out.

These detailed models use the National Space Science Data Center (NSSDC) routines to calculate the magnetic field and spectral flux in order to determine the degradation integral at each trajectory sample point, whereas the analytic models constitute data files of orbital average degradation. These analytic models minimize computation time in optimization codes such as SECKSPOT, because the equivalent 1 MEV flux values are averaged over one day in these models.

In SECKSPOT, the construction of the 1 MEV flux analytic model is carried out by first generating a space of discrete values of 1 MEV flux from particle spectral flux data and from data concerning the conversion of spectral flux to 1 MEV flux, then by fitting the discrete points thus generated with analytic expressions based on multivariate regression techniques. The geocentric coordinates of radial distance from the Earth center, longitude and latitude are converted to B (magnetic field intensity) and L (magnetic invariant associated with particle momentum along the magnetic field lines), and the choice for the magnetic field model is based on the International Geophysical Reference Field (IGRF) for the year 1978.

A series of integrations is then carried out to first eliminate the longitude by averaging because the rotation of the Earth and of the radiation belts does effectively average the fluence, then to convert the azimuth-averaged flux data to an equivalent 1 MEV flux.

1.12.2 Some Comments Concerning the Optimization of the Thrust Magnitude for Low-Thrust Power-Limited Vehicles

One of the limitations of the SECKSPOT program is its inability to generate a variable thrust transfer trajectory, which is due to the fact that it is mainly concerned with minimum-time solutions. It assumes that the thrust magnitude is proportional to the array power, even though this power is allowed to degrade due to solar cell degradation and attitude constraints which prevent the solar panels from being oriented normal to the Sun vector.

It is worthwhile considering optimal variable thrust as well as optimal beam current and voltage within a predefined envelope, as was mentioned above. Drawing on Edelbaum [9], let us expose the mechanics of variable thrust briefly.

Using position and velocity as state variables, the original second-order differential equation of motion for a thrusting vehicle in a variable gravity field

$$m\ddot{\mathbf{r}} = \dot{m}\mathbf{c} + m\mathbf{g} \tag{1.275}$$

can be reduced to two first-order equations

$$\dot{\mathbf{v}} = \frac{\mathbf{f}}{m} + \mathbf{g}\,(\mathbf{r}, t) \tag{1.276}$$

$$\dot{\mathbf{r}} = \mathbf{v} \tag{1.277}$$

where **f** is the thrust vector and **g** the acceleration of gravity. The mass obeys the general form

$$\dot{m} = \dot{m}\,(\mathbf{r}, t, \mathbf{f}) \tag{1.278}$$

A Hamiltonian H is thus formed from which the state and Lagrange multiplier differential equations are derived:

$$H = \boldsymbol{\lambda}_v \cdot \frac{\mathbf{f}}{m} + \boldsymbol{\lambda}_v \cdot \mathbf{g} + \boldsymbol{\lambda}_r \mathbf{v} + \lambda_m \dot{m} \tag{1.279}$$

$$\dot{\boldsymbol{\lambda}}_v = -\partial H/\partial \mathbf{v} = -\boldsymbol{\lambda}_r \tag{1.280}$$

$$\dot{\boldsymbol{\lambda}}_r = -\partial H/\partial \mathbf{r} = -\frac{\boldsymbol{\lambda}_v k}{m} \cdot \frac{\partial \mathbf{f}_{max}}{\partial \mathbf{r}} - \boldsymbol{\lambda}_v \cdot \frac{\partial \mathbf{g}}{\partial \mathbf{r}} - \boldsymbol{\lambda}_m \frac{\partial \dot{m}}{\partial \mathbf{r}} \tag{1.281}$$

$$\dot{\lambda}_m = -\partial H/\partial m = \frac{\boldsymbol{\lambda}_v \cdot k\mathbf{f}_{max}}{m^2} \tag{1.282}$$

Here, **f** has been replaced by $k\mathbf{f}_{\max}(\mathbf{r}, \mathrm{t})$ where k is the throttle variable with $0 \leq k \leq 1$, and $\mathbf{f}_{\max}$ is the maximum magnitude thrust vector dependent on both position and time. The control variable being the thrust vector, H is maximized by

$$\frac{\boldsymbol{\lambda}_v \cdot k\mathbf{f}_{\max}}{m} + \lambda_m \dot{m} = \text{maximum} \tag{1.283}$$

If the thrust direction is unconstrained, then $\dot{m}$ is only a function of thrust magnitude $f = |\mathbf{f}|$ and not thrust direction $\hat{\mathbf{f}} = \frac{\mathbf{f}}{|\mathbf{f}|}$. Then Equation (1.283) is maximized if **f** is aligned with $\boldsymbol{\lambda}_v$, Lawden's famous primer vector. This maximum then takes place for a unique combination of f and $\dot{m}$, because Equation (1.283) is reduced to the equation of a straight line in the $(f, -\dot{m})$ space,

$$f + \frac{\lambda_m m}{\lambda_v} \dot{m} = \text{maximum} \tag{1.284}$$

and because $f = \sqrt{-2\dot{m}P}$ represents a parabola in the same $(f, -\dot{m})$ system, because of Equation (1.270).

If $\dot{m} = \frac{-f^2}{2P}$ is replaced in Equation (1.284), then

$$f - \frac{\lambda_m m}{2\lambda_v P} f^2 = \text{maximum} \tag{1.285}$$

is solved for the optimal thrust magnitude f^* in terms of the two multipliers λ_v and λ_m, as well as the power P.

$$f^* = \frac{\lambda_v P}{\lambda_m m} \tag{1.286}$$

The $\dot{\lambda}_m$ and $\dot{m}$ equations, which are written in terms of the thrust magnitude f, reduce to

$$\dot{\lambda}_m = \frac{\lambda_v^2 P}{\lambda_m m^3} \tag{1.287}$$

$$\dot{m} = -\frac{\lambda_v^2 P}{2\lambda_m^2 m^2} \tag{1.288}$$

These last two equations can be combined and then integrated to give

$$\lambda_m m^2 = (\lambda_m)_0\, m_0^2 \tag{1.289}$$

which is a constant that can be set to P_0. Therefore, the optimal acceleration program obeys the following law:

$$a^* = \frac{f^*}{m} = \frac{P\lambda_v}{\lambda_m m^2} = \frac{P\lambda_v}{P_0} \tag{1.290}$$

where P_0 is power at time zero at the start of the transfer. The multiplier λ_v is, of course, obtained from Equations (1.280) and (1.281), such that

$$\ddot{\boldsymbol{\lambda}}_v = \frac{\boldsymbol{\lambda}_v k}{m} \cdot \frac{\partial \mathbf{f}_{max}}{\partial \mathbf{r}} + \boldsymbol{\lambda}_v \cdot \frac{\partial \mathbf{g}}{\partial \mathbf{r}} + \lambda_m \frac{\partial \dot{m}}{\partial \mathbf{r}} \tag{1.291}$$

The optimal control law, Equation (1.286), must be used in Equation (1.291) during numerical integration such that $\boldsymbol{\lambda}_v \cdot k \frac{\partial \mathbf{f}_{max}}{\partial \mathbf{r}} = \boldsymbol{\lambda}_v \cdot \frac{\partial \mathbf{f}^*}{\partial \mathbf{r}}$ with $\mathbf{f}^*$ along $\boldsymbol{\lambda}_v$; P is, of course, a function of $\mathbf{r}$ and t, such that as a function of P, the optimal $\mathbf{f}^*$ as well as λ_v, λ_m and m can be integrated numerically to yield the optimal exhaust velocity c^* and optimal I_{sp}^*, because

$$P = \frac{1}{2} f^* c^* = \frac{1}{2} f^* I_{sp}^* g \tag{1.292}$$

$$c^* = \frac{2P}{f^*} \tag{1.293}$$

Beam voltage and current must be adjusted continuously within their operating envelope and thrusters from the thruster bank turned on and off when needed to satisfy the optimal acceleration program. This sort of optimization may be needed for transfers using fixed initial and end times while minimizing fuel consumption. The position and velocity treatment was used to highlight the mechanics of the optimization; a derivation of this problem using the formulation of the SECKSPOT analysis is presented in Chapter 4.

1.12.3 Some Recommendations for Future Investigations

Limitations and Updates within the SECKSPOT Environment

There are both short- and long-term goals in analysis and software updates. The short-term goals consist of various updates to extend the applicability of SECKSPOT to carry out certain parametric studies needed to rapidly define a mission profile that would demonstrate all of the key technologies for the future development of an electric orbit transfer vehicle.

For some parametric studies where some of the three steering angles are deliberately held constant, additional propellant needed to carry out the same orbit transfer must be evaluated and compared to the RCS propellant needed to carry out the attitude variations. For example, the roll maneuvers could be completely eliminated to save RCS propellant and the transfer carried out with this zero roll attitude constraint with some main propulsion system propellant and transfer time penalties. This update has been carried out analytically for the zero roll and free pitch and yaw case in Section 1.7, and it can be programmed in SECKSPOT.

SECKSPOT provides the time history or rather the eccentric longitude history of all three steering angles. The angle rates and accelerations can be generated numerically in SECKSPOT, and analytic expressions for the rates can also be developed for incorporation in the code. This information is needed to evaluate the steering torques and, therefore, the RCS propellant requirements. In its present form, SECKSPOT does not account for other important constraints, whether they are geometric or dynamic constraints.

One important constraint could be the so-called Sun angle constraint, whereby the angle between the Sun vector and the thrust vector must never go below a certain minimum value. This is equivalent to saying that the Sun vector must be at all times outside a cone centered at the thrust vector. This geometric constraint could be factored in the optimization formulation as well as other attitude constraints such as due to thermal or sensor orientation.

Another important limitation of SECKSPOT is that it has no provisions to optimize the thrust magnitude, concerning itself only with the optimization of the thrust vector direction in three-dimensional space. It also assumes a constant exhaust velocity and specific impulse with the thrust magnitude always proportional to the array power, which is always maximized by optimal orientation to the Sun vector. The developments in Chapter 4 show how the thrust magnitude, as well as direction, can be optimized and the optimum number of thrusters could be used out of a given thruster bank configuration.

Some low-thrust software specialists are already using similar capabilities for their heliocentric transfer software. For some problems such as minimum fuel transfers for given transfer time, and rendezvous problems, this capability could be useful to implement.

Minimum fuel transfers will, in general, introduce coasting arcs during the transfer, especially if the transfer time is relaxed further from the minimum time solution. It would also appear for certain elliptic transfers in which the relative geometry of the initial and final orbits is such that only certain portions of the intermediate transfer orbits are adequate for thrusting. This important capability does not exist in SECKSPOT. Appropriate transversality conditions can be developed and included in the program to enable the solution of this problem. This type of solution is important to generate because it could result in substantial savings in propellant at the expense of accepting a slightly longer transfer time than the minimum time. Propellant trade studies require this addition.

SECKSPOT uses, as was discussed in some detail earlier in this chapter, an analytic radiation and power degradation model. The model is analytic for computational efficiency, but lacks accuracy when compared with detailed radiation degradation calculations used in maximum realism simulations. The total accumulated fluence is computed by SECKSPOT using the analytic models, but there are no constraints associated with this parameter.

It would be useful to consider the total accumulated fluence as a constraint such that the minimum-time transfer solution is constrained to satisfy a user-defined maximum fluence. These solutions could be useful in prolonging the useful lifetime of the solar panels by flying slightly different trajectories with perhaps shorter segments in the radiation belts and longer segments outside the belts. Interesting scenarios could result from this constrained optimization problem. For example, it is possible that a transfer from LEO to a HEO whose altitude is just outside the belts, and whose orbit is substantially inclined with respect to the LEO orbit, would be initially totally coplanar to quickly gain in altitude and get out of the belts before the inclination change is carried out as opposed to a transfer that does both altitude raising and inclination change simultaneously without worrying about dwell time in the radiation belts. This capability could be implemented to find out whether solar panel lifetime could be thus extended at the expense of some additional flight time and propellant expenditure for a given transfer.

Finally, one must address the problem of rendezvous using low-thrust propulsion. This includes both orbital and terminal rendezvous, and they arise from the requirement of placing a payload over a desired station longitude over the equator at geosynchronous altitude. SECKSPOT is not only a transfer program in which the location along the final orbit is free, but it is also an approximate solution-generating program because it is not of the precision integration or the high-fidelity type. This means that the final orbit will be reached with some error, which must be corrected to enter it in a precise manner. This could be done either by RCS propellant or by creating a precision integration low-thrust rendezvous program that would use the ion engines. A parametric study can be carried out to find out RCS propellant requirements for this final phase where apogee, perigee, inclination, and position errors must be corrected to insert the payload on station at GEO. If found excessive, then the rationale for the creation of precision low-thrust rendezvous software would exist.

For missions that have inadequate solar array power for continuous firing of the arcjet engines, the trajectory optimization using this low-thrust propulsion system is much harder to do than for the ion thruster's case because of the need to charge the batteries prior to any firing of the arcjets. The firing, coasting and shadowing sequences become extremely complex in an optimization context, where natural coasting arcs may be introduced additionally for some minimum-fuel transfers. Needless to say,

SECKSPOT is inadequate for tackling this constrained optimization problem, which requires a precision integration scheme and an almost totally new formulation.

High-Fidelity Optimization Simulation Software

Low-thrust trajectory optimization software that simulates optimal transfers with a high degree of realism is useful to establish the gold standard trajectory that determines without ambiguity the level of confidence that would be placed in the approximate tools such as SECKSPOT in carrying out trade analyses. They are also needed as an operational tool to be used in mission operations during actual transfers in support of EOTV missions.

References

[1] Edelbaum, T. N., Sackett, L. L. and Malchow, H. L. (1973) *Optimal Low Thrust Geocentric Transfer*, AIAA Paper 73–1074, AIAA 10th Electric Propulsion Conference, Lake Tahoe, NV.

[2] Sackett, L. L., Malchow, H. L. and Edelbaum, T. N. (1975) *Solar Electric Geocentric Transfer with Attitude Constraints: Analysis*, NASA CR 134927, Washington, DC.

[3] Sackett, L. L. and Edelbaum, T. N. (1975) *Effect of Attitude Constraints on Solar-Electric Geocentric Transfers*, AIAA Paper 75–350, AIAA 11th Electric Propulsion Conference, New Orleans, LA.

[4] Kryloff, N. and Bogoliuboff, N. (1947) *Introduction to Non-Linear Mechanics*, Princeton University Press, Princeton, NJ.

[5] Cefola, P. J. (1972) *Equinoctial Orbit Elements: Application to Artificial Satellite Orbits*, AIAA Paper 72–937, AIAA/AAS Astrodynamics Conference, Palo Alto, CA.

[6] Malchow, H. L. and Whitney, C. K. (1975) *A Radiation Model for Geocentric Trajectory Calculations*, AIAA Paper 75–351, AIAA 11th Electric Propulsion Conference, New Orleans, LA.

[7] Broucke, R. A. and Cefola, P. J. (1972) On the Equinoctial Orbital Elements, *Celestial Mechanics*, Vol. 5, 303–310.

[8] Obenschain, A. F. and Rasmussen, R. (1969) *The Solar Array Synthesis Computer Program*, NASA TMX-63709.

[9] Edelbaum, T. N. (1969) *Optimal Space Trajectories*, Analytic Mechanics Associates, Inc., New York, NY.

2 The Analysis of the Six-Element Formulation

2.1 Introduction

The theory of optimal low-thrust orbit transfer has received a great deal of attention in the astrodynamics and flight mechanics literature over the past several decades. This chapter begins with the detailed description of some fundamental analytic results obtained by Edelbaum [1] which are widely used by practitioners in the aerospace industry. The reader, after becoming familiar with the simplified transfer analysis, is invited to consider the treatment of the exact transfer problem in the subsequent sections. Drawing on the pioneering work of the Americans Broucke, Cefola and Edelbaum, who perfected the theory of orbital mechanics in terms of nonsingular orbital elements, examples of optimal orbit transfers are generated and discussed and all the relevant equations needed to develop unconstrained orbit transfer computer codes are exposed and derived.

2.2 The Edelbaum Low-Thrust Orbit Transfer Problem

A discussion of the problem of optimal low-thrust transfer between inclined circular orbits was presented by Edelbaum [1] in the early 1960s. Assuming constant acceleration and constant thrust vector yaw angle within each revolution, Edelbaum linearizes the Lagrange planetary equations of orbital motion about a circular orbit and, using the velocity as the independent variable, reduces the transfer optimization problem to a problem in the theory of maxima. The variational integral involves a single constant Lagrange multiplier because it involves a single integral constraint equation for the transfer time or velocity change, while maximizing the change in orbital inclination. The control variable being the yaw angle, the necessary condition for a stationary solution is obtained by simply setting the partial derivative of the integrand of the variational integral with respect to the control to zero. This optimum control is then used in the right-hand sides of the original equations of motion which are integrated analytically to provide expressions for the time and inclination in terms of the independent variable, the orbital velocity. Two expressions for the inclination are provided to cover the case of large inclination change transfers. This complication arises if orbital velocity is adopted as the independent variable. However, a single expression for the inclination change can be obtained that is uniformly valid throughout any desired transfer if the original Edelbaum problem is cast into a minimum-time transfer problem using the more direct formalism of optimal control theory. Edelbaum's original analysis and the formulation using optimal control theory are discussed below.

2.2.1 Edelbaum's Analysis

The full set of the Gaussian form of the Lagrange planetary equations for near-circular orbits is given by

$$\dot{a} = \frac{2af_t}{V} \tag{2.1}$$

$$\dot{e}_x = \frac{2f_t c_\alpha}{V} - \frac{f_n s_\alpha}{V} \tag{2.2}$$

$$\dot{e}_y = \frac{2f_t s_\alpha}{V} + \frac{f_n c_\alpha}{V} \tag{2.3}$$

$$\dot{i} = \frac{f_h c_\alpha}{V} \tag{2.4}$$

$$\dot{\Omega} = \frac{f_h s_\alpha}{V s_i} \tag{2.5}$$

$$\dot{\alpha} = n + \frac{2f_n}{V} - \frac{f_h s_\alpha}{V \tan i} \tag{2.6}$$

where s_α and c_α stand for $\sin\alpha$ and $\cos\alpha$ respectively, and where a stands for the orbit semimajor axis, i for inclination, Ω for the right ascension of the ascending node, $e_x = e\cos\omega$, $e_y = e\sin\omega$ with e and ω standing for orbital eccentricity and argument of perigee. Finally, $\alpha = \omega + M$ represents the mean angular position, M the mean anomaly and $n = \left(\mu/a^3\right)^{1/2}$ the orbit mean motion with μ standing for the Earth gravity constant. For near-circular orbits, $V = na = (\mu/a)^{1/2}$. The components of the thrust acceleration vector along the tangent, normal and out-of-plane directions are depicted by f_t, f_n and f_h with the normal direction oriented toward the center of attraction. If we assume only tangential and out-of-plane acceleration, and that the orbit remains circular during the transfer, Equations (2.1)–(2.6) reduce to

$$\dot{a} = \frac{2af_t}{V} \tag{2.7}$$

$$\dot{i} = \frac{f_h c_\alpha}{V} \tag{2.8}$$

$$\dot{\Omega} = \frac{f_h s_\alpha}{V s_i} \tag{2.9}$$

$$\dot{\alpha} = n - \frac{f_h s_\alpha}{V \tan i} \tag{2.10}$$

If f represents the magnitude of the acceleration vector, and β the out-of-plane or thrust yaw angle, then $f_t = fc_\beta$ and $f_h = fs_\beta$. Furthermore, $\alpha = \omega + M = \omega + \theta^* = \theta$, the angular position when $e = 0$, with $\theta = nt$ and θ^* the true anomaly. If the angle β is held piecewise constant, switching sign at the orbital antinodes, then the $f_h s_\alpha$ terms above will have a net zero contribution such that the system of differential equations further reduces to

$$\dot{a} = \frac{2af_t}{V} \tag{2.11}$$

$$\dot{i} = \frac{c_\theta f_h}{V} \tag{2.12}$$

$$\dot{\theta} = n \tag{2.13}$$

We can now average out the angular position θ in Equation (2.12) by integrating with respect to θ and by holding f, β and V constant:

$$\int_0^{2\pi} \left(\frac{di}{dt}\right) d\theta = \frac{2fs_\beta}{V} \int_{-\pi/2}^{\pi/2} c_\theta d\theta$$

$$2\pi \frac{di}{dt} = \frac{4fs_\beta}{V}$$

$$\frac{di}{dt} = \frac{2fs_\beta}{\pi V} \tag{2.14}$$

From the energy equation $V^2/2 - \mu/r = -\mu/2a$, with $r = a$, and with Equation (2.11) used to eliminate the semimajor axis,

$$dV = -\left[\frac{\mu}{2Va^2}\right] da = -fc_\beta dt$$

$$\frac{dV}{dt} = -fc_\beta \tag{2.15}$$

Equation (2.14) can also be obtained by dividing Equation (2.12) by Equation (2.13):

$$\frac{di}{d\theta} = \frac{c_\theta f_h}{Vn}$$

$$\Delta i = \frac{2f_h}{Vn} \int_{-\pi/2}^{\pi/2} c_\theta d\theta = \frac{4f_h}{Vn}$$

and because $\Delta t = 2\pi\, a/V$,

$$\frac{di}{dt} = \frac{\Delta i}{\Delta t} = \frac{2f_h}{V\pi}$$

Equations (2.14) and (2.15) can be replaced by the following set, where V is now the independent variable:

$$\frac{di}{dV} = -\frac{2\tan\beta}{\pi V} \tag{2.16}$$

$$\frac{dt}{dV} = -\frac{1}{fc_\beta} \tag{2.17}$$

Let I represent the function to be maximized,

$$I = \int_{V_0}^{V_f} \left(\frac{di}{dV}\right) dV = -\int_{V_0}^{V_f} \frac{2}{\pi V} \tan\beta dV \tag{2.18}$$

and let J represent the integral constraint given by

$$J = \int_{V_0}^{V_f} \left(\frac{dt}{dV}\right) dV = \text{const} \tag{2.19}$$

Let us adjoin J to I by way of a constant Lagrange multiplier λ such that the optimization problem is now reduced to a succession of ordinary maximum problems for each value of V between V_0 and V_f, the initial and final velocities, respectively. The necessary condition for a stationary solution of the augmented integral

$$K = I + \lambda J = \int_{V_0}^{V_f} \left[-\frac{2}{\pi V} \tan\beta - \frac{\lambda}{f c_\beta} \right] dV \tag{2.20}$$

is then simply given by

$$\frac{\partial}{\partial \beta} \left[\frac{2}{\pi V} \tan\beta + \frac{\lambda}{f c_\beta} \right] = 0 \tag{2.21}$$

The optimization problem consists therefore of the maximization of the inclination change subject to the constraint of given total transfer time because

$$\Delta V = ft \tag{2.22}$$

This constraint is equivalent to the fixed ΔV constraint for constant acceleration f. Furthermore, V_0 and V_f being given, the initial and final radii are, therefore, given too because the orbits are assumed to be circular. The acceleration being applied continuously, this problem is equivalent to minimizing the total transfer time for a given change in the inclination and velocity. This is also equivalent to minimizing the total ΔV or propellant usage because the thrust is always on and no coasting arcs are allowed. In this latter case, I and J are simply interchanged to yield the optimality condition

$$\frac{\partial}{\partial \beta} \left[\frac{1}{f c_\beta} + \lambda_i \frac{2}{\pi V} \tan\beta \right] = 0 \tag{2.23}$$

From Equation (2.21), it follows that

$$V s_\beta = -\frac{2f}{\pi \lambda} = \text{const} = V_0 s_{\beta_0} \tag{2.24}$$

$$\lambda = -\frac{2f}{\pi V_0 s_{\beta_0}} \tag{2.25}$$

The optimal β steering law given by Equation (2.24) can be used in Equation (2.17) for dV/dt in order to obtain the expression for the velocity as a function of time t or $\Delta V = ft$:

$$\frac{dV}{dt} = -f c_\beta$$

$$f dt = -\frac{dV}{c_\beta} = \frac{-\,dV}{\pm\left(1 - s_\beta^2\right)^{1/2}}$$

$$\int_0^t f dt = -\int_{V_0}^{V} \frac{V dV}{\pm\left(V^2 - V_0^2 s_{\beta_0}^2\right)^{1/2}}$$

$$\Delta V = ft = -\frac{1}{\pm}\left[\left(V^2 - V_0^2 s_{\beta_0}^2\right)^{1/2} - (\pm)\, V_0 c_{\beta_0}\right]$$

$$\Delta V = V_0 c_{\beta_0} \mp \left(V^2 - V^2 s_\beta^2\right)^{1/2} = V_0 c_{\beta_0} \mp (\pm) V c_\beta \tag{2.26}$$

$$\Delta V = V_0 c_{\beta_0} - V c_\beta \tag{2.27}$$

From Equation (2.26),

$$\Delta V - V_0 c_{\beta_0} = \mp \left(V^2 - V^2 s_\beta^2\right)^{1/2} = \mp \left(V^2 - V_0^2 s_{\beta_0}^2\right)^{1/2}$$

and after squaring,

$$V^2 = V_0^2 + \Delta V^2 - 2\Delta V\, V_0 c_{\beta_0} \tag{2.28}$$

This then represents V as a function of time because $\Delta V = ft$. The initial yaw angle β_0 must still be determined. In a similar way, Equation (2.16) for di/dV can be integrated to provide an expression for the evolution of the inclination in time,

$$\frac{di}{dV} = -\frac{2}{\pi V} \tan\beta = -\frac{2}{\pi V}\frac{s_\beta}{c_\beta} = -\frac{2}{\pi V}\frac{V s_\beta}{\left(V^2 - V^2 s_\beta^2\right)^{1/2}}$$

such that, with the use of $V s_\beta = V_0 s_{\beta_0}$,

$$\int_{i_0}^{i} di = -\frac{2}{\pi} V_0 s_{\beta_0} \int_{V_0}^{V} \frac{dV}{V\left(V^2 - V_0^2 s_{\beta_0}^2\right)^{1/2}}$$

$$\Delta i = -\frac{2}{\pi} \sin^{-1}\left(\frac{-V_0 s_{\beta_0}}{V}\right)\Bigg|_{V_0}^{V}$$

$$\Delta i = -\frac{2}{\pi}\left[\sin^{-1}\left(s_{\beta_0}\right) - \sin^{-1}\left(\frac{V_0 s_{\beta_0}}{V}\right)\right]$$

$$\Delta i = -\frac{2}{\pi}\beta_0 + \frac{2}{\pi}\sin^{-1}\left(\frac{V_0 s_{\beta_0}}{V}\right) \tag{2.29}$$

and because $V_0 s_{\beta_0} = V s_\beta$,

$$\Delta i = -\frac{2}{\pi}(\beta - \beta_0)$$

Now, because the inverse sine function in Equation (2.29) is double-valued in the interval $(0, 2\pi)$, it is necessary to write this function as

$$\sin^{-1}\left(\frac{V_0 s_{\beta_0}}{V}\right) \quad \text{if } \sin^{-1}\left(\frac{V_0 s_{\beta_0}}{V}\right) < \frac{\pi}{2}$$

and

$$\frac{\pi}{2} + \left[\frac{\pi}{2} - \sin^{-1}\left(\frac{V_0 s_{\beta_0}}{V}\right)\right] = \pi - \sin^{-1}\left(\frac{V_0 s_{\beta_0}}{V}\right) \quad \text{if } \sin^{-1}\left(\frac{V_0 s_{\beta_0}}{V}\right) > \frac{\pi}{2}$$

due to the function being symmetrical with respect to $\pi/2$. In the second of the preceding conditions, Δi can be written as

$$\Delta i = \frac{2}{\pi}\left[\pi - \sin^{-1}\left(\frac{V_0 s_{\beta_0}}{V}\right)\right] - \frac{2}{\pi}\beta_0 \tag{2.30}$$

or

$$\Delta i = 2 - \frac{2}{\pi} \sin^{-1}\left(\frac{V_0 s_{\beta_0}}{V}\right) - \frac{2}{\pi}\beta_0 \tag{2.31}$$

This is equivalent to writing Equations (2.29) and (2.31) as

$$\Delta i = \frac{2}{\pi}(\beta - \beta_0) \qquad \text{if } \beta < \frac{\pi}{2} \tag{2.32}$$

$$\Delta i = 2 - \frac{2}{\pi}(\beta + \beta_0) \qquad \text{if } \beta > \frac{\pi}{2} \tag{2.33}$$

Of course, the 2 appearing in Equation (2.33) is given in radians, and it corresponds to 114.6 deg.

Finally, from Equation (2.26),

$$\Delta V = V_0 c_{\beta_0} - \left(V^2 - V_0^2 s_{\beta_0}^2\right)^{1/2} \qquad \text{if } \Delta V - V_0 c_{\beta_0} < 0 \tag{2.34}$$

$$\Delta V = V_0 c_{\beta_0} + \left(V^2 - V_0^2 s_{\beta_0}^2\right)^{1/2} \qquad \text{if } \Delta V - V_0 c_{\beta_0} > 0 \tag{2.35}$$

From $\Delta V = V_0 c_{\beta_0} - V c_\beta$, the condition $\Delta V - V_0 c_{\beta_0} < 0$ is identical to $c_\beta > 0$ or $\beta < \pi/2$ or $\sin^{-1}\left(V_0 s_{\beta_0}/V\right) < \pi/2$, and the condition $\Delta V - V_0 c_{\beta_0} > 0$ is identical to $\beta > \pi/2$ or $\sin^{-1}\left(V_0 s_{\beta_0}/V\right) > \pi/2$, such that the Edelbaum analysis leads to the following set of equations:

1. If $\Delta V - V_0 c_{\beta_0} < 0$, then,

$$\left.\begin{aligned} V &= \left(V_0^2 - 2V_0 \Delta V c_{\beta_0} + \Delta V^2\right)^{1/2} \\ \Delta i &= \frac{2}{\pi}\sin^{-1}\left(\frac{V_0 s_{\beta_0}}{V}\right) - \frac{2}{\pi}\beta_0 = \frac{2}{\pi}(\beta - \beta_0) \end{aligned}\right\} \tag{2.36}$$

2. If $\Delta V - V_0 c_{\beta_0} > 0$, then,

$$\left.\begin{aligned} V &= \left(V_0^2 - 2V_0 \Delta V c_{\beta_0} + \Delta V^2\right)^{1/2} \\ \Delta i &= 2 - \frac{2}{\pi}\sin^{-1}\left(\frac{V_0 s_{\beta_0}}{V}\right) - \frac{2}{\pi}\beta_0 = 2 - \frac{2}{\pi}(\beta + \beta_0) \end{aligned}\right\} \tag{2.37}$$

The preceding equations show that one must monitor the condition $\Delta V - V_0 c_{\beta_0}$ and use Equations (2.36) to describe the transfer starting from time 0 and later switch to Equations (2.37) as soon as $t = \Delta V/f$ exceeds $V_0 c_{\beta_0}/f$, which will take place for large transfers, as will be shown later with an example. For large transfers, ΔV as given in Equation (2.26) could become double-valued in V such that one must use $\Delta V = V_0 c_{\beta_0} - \left(V^2 - V_0^2 s_{\beta_0}^2\right)^{1/2}$ from V_0 to $V_0 s_{\beta_0}$ and $\Delta V = V_0 c_{\beta_0} + \left(V^2 - V_0^2 s_{\beta_0}^2\right)^{1/2}$ from $V = V_0 s_{\beta_0}$ to V_f where $V_0 s_{\beta_0} < V_f$. This minimum velocity takes place when $\Delta V = V_0 c_{\beta_0}$, indicating that the orbit will grow to become larger than the final desired orbit and later shrink to that desired orbit. This will happen when larger inclination changes are required because then the orbit plane rotation will be carried out mostly at those higher intermediate altitudes. This, of course, is the result of the trade off between inclination and radius or velocity. From

$$\Delta V = V_0 c_{\beta_0} \mp \left(V^2 - V_0^2 s_{\beta_0}^2\right)^{1/2}$$

we have

$$\frac{\partial \Delta V}{\partial V} = \mp \frac{V}{\left(V^2 - V_0^2 s_{\beta_0}^2\right)^{1/2}}$$

which is equal to ∞ for $V = V_0 s_{\beta_0}$, the minimum velocity reached. The initial yaw angle β_0 can be obtained from the terminal conditions at time t_f. At $t = t_f$, $V = V_f$ and $\Delta i = \Delta i_f$. Using Equation (2.36) for Δi, we get

$$\frac{\pi}{2}\Delta i_f + \beta_0 = \sin^{-1}\left(\frac{V_0 s_{\beta_0}}{V_f}\right)$$

$$\sin\left(\beta_0 + \frac{\pi}{2}\Delta i_f\right) = \frac{V_0 s_{\beta_0}}{V_f}$$

$$\sin\beta_0 \cos\frac{\pi}{2}\Delta i_f + \cos\beta_0 \sin\frac{\pi}{2}\Delta i_f = \frac{V_0 s_{\beta_0}}{V_f}$$

Dividing both sides by c_{β_0} yields

$$\left[\cos\frac{\pi}{2}\Delta i_f - \frac{V_0}{V_f}\right]\tan\beta_0 = -\sin\frac{\pi}{2}\Delta i_f$$

$$\tan\beta_0 = \frac{\sin\frac{\pi}{2}\Delta i_f}{\frac{V_0}{V_f} - \cos\frac{\pi}{2}\Delta i_f} \tag{2.38}$$

Now, carrying out the same manipulations using Equation (2.37), we get

$$\frac{\pi}{2}\left(\Delta i_f - 2 + \frac{2}{\pi}\beta_0\right) = -\sin^{-1}\left(\frac{V_0 s_{\beta_0}}{V_f}\right)$$

$$\pi - \left(\beta_0 + \frac{\pi}{2}\Delta i_f\right) = \sin^{-1}\left(\frac{V_0 s_{\beta_0}}{V_f}\right)$$

$$\sin\left(\beta_0 + \frac{\pi}{2}\Delta i_f\right) = \frac{V_0 s_{\beta_0}}{V_f}$$

$$\sin\beta_0 \cos\frac{\pi}{2}\Delta i_f + c_{\beta_0} \sin\frac{\pi}{2}\Delta i_f = \frac{V_0 s_{\beta_0}}{V_f}$$

Dividing by c_{β_0},

$$\tan\beta_0 = \frac{\sin\frac{\pi}{2}\Delta i_f}{\frac{V_0}{V_f} - \cos\frac{\pi}{2}\Delta i_f} \tag{2.39}$$

Equations (2.38) and (2.39) indicate that β_0 is given by

$$\beta_0 = \tan^{-1}\left[\frac{\sin\frac{\pi}{2}\Delta i_f}{\frac{V_0}{V_f} - \cos\frac{\pi}{2}\Delta i_f}\right] \tag{2.40}$$

regardless of whether $\Delta V - V_0 c_{\beta_0} < 0$ or $\Delta V - V_0 c_{\beta_0} > 0$ and from $\Delta V = V_0 c_{\beta_0} - V c_\beta$ the yaw angle β at future times is given by

$$\beta = \cos^{-1}\left[\frac{V_0 c_{\beta_0} - \Delta V}{\left(V_0^2 - 2V_0 \Delta V c_{\beta_0} + \Delta V^2\right)^{1/2}}\right] \tag{2.41}$$

where $\Delta V = ft$ and where $0 \leq \beta \leqslant \pi$. This expression is better than $V s_\beta = V_0 s_{\beta_0}$, which would yield $\beta = \sin^{-1}\left(V_0 s_{\beta_0}/V\right)$ because β could, for large transfers, exceed $\pi/2$.

If the evolution of Δi as a function of time or velocity or ΔV is desired, Equation (2.36) for Δi is used until $\Delta V = V_0 c_{\beta_0}$. When $\Delta V - V_0 c_{\beta_0} > 0$, Δi as given in Equation (2.37) is used. However, in Equation (2.37), the inverse sine function will always return a β angle that is always less than $\pi/2$, and this value for β is the correct value to be used in $\Delta i = 2 - (2/\pi)(\beta + \beta_0)$. This β angle is clearly not the real yaw angle because, in this case, it would be given by $\pi - \beta$ with $\beta < \pi/2$, such that the yaw angle is now larger than $\pi/2$. If the real β angle is used in $\Delta i = 2 - (2/\pi)(\beta + \beta_0)$, we get

$$\Delta i = 2 - \frac{2}{\pi}(\pi - \beta + \beta_0) = \frac{2}{\pi}(\beta - \beta_0) \tag{2.42}$$

Equation (2.42) is universally valid for all yaw angles $0 \leq \beta < 180$ deg and should be the only one used. Equation (2.42) will effectively replace Equations (2.36) and (2.37) provided that the angle β is computed from Equation (2.41). Because the sign of $\Delta V - V_0 c_{\beta_0}$ is effectively accounted for in Equation (2.41), it will return the yaw angle β to be used in Equation (2.42) for the unambiguous evaluation of Δi. Expressions for β_0 and β can also be obtained by using the identity in Equation (2.42) because

$$\cos\frac{\pi}{2}\Delta i = c_\beta c_{\beta_0} + s_\beta s_{\beta_0} \tag{2.43}$$

$$\sin\frac{\pi}{2}\Delta i = s_\beta c_{\beta_0} - s_{\beta_0} c_\beta \tag{2.44}$$

If we multiply Equation (2.43) by VV_0 and replace $V s_\beta$ by $V_0 s_{\beta_0}$ and $V c_\beta$ by $V_0 c_{\beta_0} - \Delta V$, after regrouping terms we get,

$$c_{\beta_0} = \frac{V_0 - V\cos\frac{\pi}{2}\Delta i}{\Delta V} \tag{2.45}$$

In a similar manner, from Equation (2.43), if we replace this time $V_0 s_{\beta_0}$ by $V s_\beta$ and $V_0 c_{\beta_0}$ by $\Delta V + V c_\beta$ we get an expression for c_β:

$$c_\beta = \frac{V_0\cos\frac{\pi}{2}\Delta i - V}{\Delta V} \tag{2.46}$$

Equation (2.44) can also be written as

$$VV_0 \sin \frac{\pi}{2} \Delta i = Vs_\beta V_0 c_{\beta_0} - V_0 s_{\beta_0} Vc_\beta$$

If we replace Vs_β by $V_0 s_{\beta_0}$ and $V_0 c_{\beta_0}$ by $\Delta V + Vc_\beta$, then

$$s_{\beta_0} = \frac{V \sin \frac{\pi}{2} \Delta i}{\Delta V} \tag{2.47}$$

If, on the other hand, we replace $V_0 c_{\beta_0}$ by $\Delta V + Vc_\beta$, then the identity will yield

$$s_\beta = \frac{V_0 \sin \frac{\pi}{2} \Delta i}{\Delta V} \tag{2.48}$$

Now these expressions will readily yield

$$\tan \beta_0 = \frac{V \sin \frac{\pi}{2} \Delta i}{V_0 - V \cos \frac{\pi}{2} \Delta i} \tag{2.49}$$

$$\tan \beta = \frac{V_0 \sin \frac{\pi}{2} \Delta i}{V_0 \cos \frac{\pi}{2} \Delta i - V} \tag{2.50}$$

These last two expressions can be used to obtain the initial β_0 and current β provided that the appropriate Δi expression of Equations (2.36) or (2.37) is used according to whether $\Delta V - V_0 c_{\beta_0}$ is positive or negative. Although Equations (2.49) and (2.50) are valid for any optimal $(V, \Delta i)$ pair during the transfer, there is clearly a singularity at time 0 when $V = V_0$ and $\Delta i = 0$. The angle β_0 is best obtained by setting $V = V_f$ and $\Delta i = \Delta i_f$ in Equation (2.49). It is better to use Equation (2.41) for the control time history instead of Equation (2.50) because we do not have to switch between two Δi expressions to describe that evolution in the first case. Finally, Edelbaum's ΔV equation in terms of the velocities and inclination is obtained from the velocity equation

$$V = \left(V_0^2 - 2V_0 \Delta V c_{\beta_0} + \Delta V^2\right)^{1/2}$$

If we square this expression, replace ΔV in the product term by $V_0 c_{\beta_0} - Vc_\beta$, and then use the identity

$$c_\beta c_{\beta_0} = \frac{1}{2} c_{\beta-\beta_0} + \frac{1}{2} c_{\beta+\beta_0}$$

with $c_{\beta-\beta_0} = \cos \pi/2\ \Delta i$ from Equation (2.42), then we get with $Vs_\beta = V_0 s_{\beta_0}$:

$$V^2 = -V_0^2 + V_0^2 s_{\beta_0}^2 + VV_0 \cos \frac{\pi}{2} \Delta i + V_0 c_{\beta_0} Vc_\beta + \Delta V^2$$

However, $V_0^2 s_{\beta_0}^2 = V_0 s_{\beta_0} Vs_\beta$ and if this term is combined with $V_0 c_{\beta_0} Vc_\beta$, the result will be $VV_0 c_{\beta-\beta_0}$, which can be replaced by $VV_0 \cos \pi/2 \Delta i$.

The final result is given by

$$V^2 = -V_0^2 + 2VV_0 \cos \frac{\pi}{2} \Delta i + \Delta V^2$$

from which

$$\Delta V = \left(V_0^2 - 2VV_0 \cos \frac{\pi}{2} \Delta i + V^2\right)^{1/2} \tag{2.51}$$

This is Edelbaum's ΔV equation for constant-acceleration circle to inclined circle transfer. It is valid for any $(V, \Delta i)$ pair along the transfer provided once again that the appropriate Δi expression is used, i.e., Equations (2.36) or (2.37) according to whether $\Delta V - V_0 c_{\beta_0}$ is < 0 or > 0, respectively. As shown earlier, ΔV is double-valued in the velocity because Δi itself is double-valued in that same variable. However, Equation (2.51) is mainly used to obtain the total $\Delta V_{\rm tot}$ required to achieve a given transfer between V_0 and V_f with a relative inclination change of Δi_f. It is valid for any $0 < \Delta i_f < 114.6$ deg or $0 < \Delta i_f < 2$ rad because this is the limiting Δi in Equation (2.37). The transfer time t_f is simply obtained from

$$t_f = \frac{\Delta V_{tot}}{f} \tag{2.52}$$

2.2.2 The Formulation Using Optimal Control Theory

Let the system equations be given by Equations (2.14) and (2.15), with time as the independent variable and i and V as the state variables. The yaw angle β is the control variable:

$$\frac{di}{dt} = \frac{2}{\pi} \frac{f}{V} s_\beta \tag{2.53}$$

$$\frac{dV}{dt} = -f c_\beta \tag{2.54}$$

This problem is now cast as a minimum-time transfer problem between initial and final parameters i_0, V_0 and i_f, V_f respectively. The variational Hamiltonian is then given by

$$H = 1 + \lambda_i \left(\frac{2}{\pi} \frac{f}{V} s_\beta\right) + \lambda_V \left(-f c_\beta\right) \tag{2.55}$$

because the performance index is simply given by

$$J = \int_{t_0}^{t_f} L\, dt$$

with $L = 1$. The Euler–Lagrange differential equations are given by

$$\dot{\lambda}_V = -\frac{\partial H}{\partial V} = \frac{2}{\pi} \frac{f s_\beta}{V^2} \lambda_i \tag{2.56}$$

$$\dot{\lambda}_i = -\frac{\partial H}{\partial i} = 0 \tag{2.57}$$

Therefore λ_i is a constant. The optimality condition is given by

$$\frac{\partial H}{\partial \beta} = \lambda_i \frac{2}{\pi} \frac{f}{V} c_\beta + f \lambda_V s_\beta = 0 \tag{2.58}$$

which yields the optimal control law

$$\tan\beta = -\frac{2}{\pi}\frac{\lambda_i}{V\lambda_V} \tag{2.59}$$

There is no need to integrate $\dot{\lambda}_V$ because we can use the transversality condition $H_f = 0$ at the final time. The Hamiltonian is a constant of the motion because it is not an explicit function of time. Therefore, it is equal to zero all the time. We can therefore solve for λ_V and λ_i from

$$H = 0 = 1 + \frac{2}{\pi}\frac{f}{V}s_\beta\lambda_i - fc_\beta\lambda_V$$
$$\frac{\partial H}{\partial\beta} = 0 = \frac{2}{\pi}\frac{f}{V}c_\beta\lambda_i + fs_\beta\lambda_V$$

This results in

$$\lambda_i = -\frac{\pi s_\beta V}{2f} = \text{const} \tag{2.60}$$

$$\lambda_V = \frac{c_\beta}{f} \tag{2.61}$$

Equation (2.60) reveals that $Vs_\beta = V_0 s_{\beta_0}$ because the acceleration f is assumed to be a constant. We can now take advantage of this constancy of Vs_β in order to integrate the velocity from Equation (2.54):

$$fdt = -\frac{dV}{c_\beta} = \frac{-dV}{\pm\left(1 - s_\beta^2\right)^{1/2}}$$

$$f\int_0^t dt = -\int_{V_0}^{V}\frac{VdV}{\pm\left(V^2 - V_0^2 s_{\beta_0}^2\right)^{1/2}} = ft = \Delta V$$

This yields, as in the previous section,

$$V^2 = V_0^2 + \Delta V^2 - 2\Delta V V_0 c_{\beta_0}$$
$$V = \left(V_0^2 + f^2t^2 - 2ftV_0c_{\beta_0}\right)^{1/2} \tag{2.62}$$

We can also obtain the preceding equation without integrating dV/dt by simply writing

$$V = \frac{V_0 s_{\beta_0}}{s_\beta} = V_0 s_{\beta_0}\frac{\left(1 + \tan^2\beta\right)}{\tan\beta} \tag{2.63}$$

However, an expression for $\tan\beta$ is needed first. If we differentiate Equation (2.59),

$$\frac{d}{dt}(\tan\beta) = \frac{2}{\pi}\lambda_i\frac{\left(\dot{V}\lambda_V + V\dot{\lambda}_V\right)}{V^2\lambda_V^2} \tag{2.64}$$

Replacing $\dot{V}$ and $\dot{\lambda}_V$ by Equations (2.54) and (2.56) and using Equations (2.60) and (2.61) to eliminate λ_i and λ_V, the above derivative can be written as

$$\frac{d}{d\beta}(\tan\beta)\,\dot{\beta} = \frac{\dot{\beta}}{c_\beta^2} = \frac{fs_\beta}{Vc_\beta^2}$$

which yields

$$\dot{\beta} = \frac{f s_\beta}{V} \tag{2.65}$$

Because $V s_\beta = V_0 s_{\beta_0}$, this can also be written as

$$\dot{\beta} = \frac{f\, s_\beta^2}{V_0 s_{\beta_0}}$$

such that

$$\int_{\beta_0}^{\beta} \frac{d\beta}{s_\beta^2} = \frac{f}{V_0 s_{\beta_0}} \int_0^t dt$$

$$\cot\beta_0 - \cot\beta = \frac{ft}{V_0 s_{\beta_0}}$$

and finally, the control law

$$\tan\beta = \frac{V_0 s_{\beta_0}}{V_0 c_{\beta_0} - ft} \tag{2.66}$$

Going back to Equation (2.63) and replacing $\tan\beta$ with the above expression results in

$$V = \left(V_0^2 + f^2 t^2 - 2V_0 c_{\beta_0} ft\right)^{1/2} \tag{2.67}$$

Now the inclination time history can be obtained by direct integration of Equation (2.53) by using the expression for V in Equation (2.67),

$$\frac{di}{dt} = \frac{2}{\pi}\frac{f}{V^2} V s_\beta = \frac{2}{\pi}\frac{f}{V^2} V_0 s_{\beta_0}$$

$$\int_0^i di = \frac{2}{\pi} V_0 s_{\beta_0} f \int_0^t \frac{dt}{\left(V_0^2 + f^2 t^2 - 2V_0 c_{\beta_0} f\right) t}$$

which yields

$$\Delta i = \frac{2}{\pi}\left[\tan^{-1}\left(\frac{ft - V_0 c_{\beta_0}}{V_0 s_{\beta_0}}\right) - \tan^{-1}\left(-\cot\beta_0\right)\right]$$

Because $\tan^{-1} x = -\tan^{-1}(-x)$ and $\tan^{-1}(\cot x) = \pi/2 - \tan^{-1} x$, the final result can be written as

$$\Delta i = \frac{2}{\pi}\left[\tan^{-1}\left(\frac{ft - V_0 c_{\beta_0}}{V_0 s_{\beta_0}}\right) + \frac{\pi}{2} - \beta_0\right] \tag{2.68}$$

This formula is uniformly valid for all t unlike the formulation of the previous section, which resulted in a set of two expressions for Δi because Δi was double-valued in the velocity. This simplification is achieved because time is selected as the independent variable instead of the velocity. If we integrate Equation (2.56) for $\dot{\lambda}_V$ and use Equation (2.60) to eliminate the constant λ_i, then

$$\dot{\lambda}_V = -\frac{V_0^2 s_{\beta_0}^2}{V^3} = -V_0^2 s_{\beta_0}^2 \left(V_0^2 + f^2 t^2 - 2V_0 c_{\beta_0} ft\right)^{-3/2}$$

which, upon integration, yields

$$\lambda_V = \frac{V_0 c_{\beta_0} - ft}{fV} \tag{2.69}$$

with $(\lambda_V)_0 = c_{\beta_0}/f$ and, in view of Equation (2.61),

$$c_\beta = \frac{V_0 c_{\beta_0} - ft}{V} \tag{2.70}$$

From the definition of the influence functions λ_i and λ_V, we have

$$(\lambda_V)_0 = \frac{c_{\beta_0}}{f} = \frac{\partial t_f}{\partial V_0}$$

$$\delta t_f = \frac{c_{\beta_0}}{f}\delta V_0 \tag{2.71}$$

$$(\lambda_i)_0 = -\frac{\pi V_0 s_{\beta_0}}{2f} = \frac{\partial t_f}{\partial (\Delta i)_0}$$

$$\delta t_f = -\frac{\pi V_0 s_{\beta_0}}{2f}\delta(\Delta i)_0 \tag{2.72}$$

Equations (2.71) and (2.72) show how the total transfer time will vary for small variations in the initial velocity and inclination. Let us obtain an expression for ΔV in terms of β_0, V_0 and Δi. From Equation (2.68), we have

$$\frac{ft - V_0 c_{\beta_0}}{V_0 s_{\beta_0}} = \tan\left(\frac{\pi}{2}\Delta i + \beta_0 - \frac{\pi}{2}\right) = \frac{-1}{\tan\left(\frac{\pi}{2}\Delta i + \beta_0\right)}$$

and because $ft = \Delta V$

$$\Delta V = V_0 c_{\beta_0} - \frac{V_0 s_{\beta_0}}{\tan\left(\frac{\pi}{2}\Delta i + \beta_0\right)} \tag{2.73}$$

This expression can be used to evaluate the total ΔV for the desired transfer once β_0, the initial yaw angle, is known. To obtain β_0, let us first observe that during the integration of $\dot{V}$ in Equation (2.54), the same intermediate results shown in the previous section and resulting from that particular integration are valid here, too. They are

$$\Delta V = ft = V_0 c_{\beta_0} - V c_\beta$$

$$\Delta V = V_0 c_{\beta_0} \pm \left(V^2 - V_0^2 s_{\beta_0}^2\right)^{1/2} \tag{2.74}$$

From Δi in Equation (2.68), and using the control law for $\tan\beta$ given in Equation (2.66), we get

$$\Delta i = \frac{2}{\pi}(\beta - \beta_0) \tag{2.75}$$

We can now obtain β_0 by using the identity in Equation (2.74) and writing Δi as

$$\Delta i = \frac{2}{\pi}\left\{\tan^{-1}\left[\frac{\pm\left(V^2 - V_0^2 s_{\beta_0}^2\right)^{1/2}}{V_0 s_{\beta_0}}\right] + \frac{\pi}{2} - \beta_0\right\}$$

Because $\tan^{-1} x = \pm\cos^{-1}\left[1/(x^2+1)^{1/2}\right]$ according to whether $x > 0$ or $x < 0$, the preceding expression for Δi can be cast as

$$\Delta i = \frac{2}{\pi}\left[\pm\cos^{-1}\left(\frac{V_0 s_{\beta_0}}{V}\right) + \frac{\pi}{2} - \beta_0\right]$$

or

$$\cos\left\{\mp\left[\frac{\pi}{2} - \left(\beta_0 + \frac{\pi}{2}\Delta i\right)\right]\right\} = \frac{V_0 s_{\beta_0}}{V}$$

or, because $\cos(-x) = \cos x$,

$$\sin\left(\beta_0 + \frac{\pi}{2}\Delta i\right) = \frac{V_0 s_{\beta_0}}{V}$$

which, after expansion and division by c_{β_0}, yields

$$\tan\beta_0 = \frac{\sin\frac{\pi}{2}\Delta i}{\frac{V_0}{V} - \cos\frac{\pi}{2}\Delta i} \tag{2.76}$$

For given V_0, V_f and $(\Delta i)_f$, β_0 can be obtained from Equation (2.76), which then allows us to describe V and Δi, as well as β as a function of time, from Equations (2.67), (2.68) and (2.66), respectively. Equation (2.70) shows that as Δi approaches the 2 rad value or $\Delta i = 114.59$ deg, $\sin\pi/2\,\Delta i$ will approach zero so that $\beta_0 \to 0$, indicating that the initial phase of the transfer will be coplanar. The ΔV equation of Edelbaum given by Equation (2.51) will then approach $\Delta V = V_0 + V_f$, which is the sum of the initial and final velocities. Given that V_0 represents the ΔV needed to transfer from V_0 to ∞ or escape, and V_f represents the ΔV needed to transfer from ∞ back to V_f, the transfer is initially coplanar until escape. At infinity, $V_\infty = 0$ and the inclination change is achieved at zero cost, after which the return leg to V_f is also coplanar, resulting in $\Delta V = V_0 + V_f$. This is shown from

$$\Delta V_1 = \left(V_0^2 - 2V_0V_\infty + V_\infty^2\right)^{1/2} = V_0$$

$$\Delta V_2 = \left(V_\infty^2 - 2V_\infty V_f + V_f^2\right)^{1/2} = V_f$$

$$\Delta V = \Delta V_1 + \Delta V_2 = V_0 + V_f \tag{2.77}$$

For any inclination larger than $\Delta i = 114.59$ deg, the cost of the orbit rotation is zero and the ΔV remains stationary at $V_0 + V_f$. If the transfer is purely coplanar, i.e., $\Delta i = 0$, then

$$\Delta V = \left|V_0 - V_f\right| \tag{2.78}$$

the difference of the boundary velocities. For given V_0 and V_f, ΔV reaches a maximum if we set

$$\frac{\partial\Delta V}{\partial\Delta i} = 0$$

where $\Delta V = \left(V_0^2 - 2V_fV_0\cos\pi/2\,\Delta i + V_f^2\right)^{1/2}$. This results in $\sin\pi/2\,\Delta i = 0$ or $\Delta i = 114.59$ deg as discussed earlier. Therefore, Edelbaum's Equation (2.51) is to be used for $0 \leq \Delta i \leq 114.59$ deg only. For $\Delta i > 114.59$ deg, Equation (2.77) must be used instead, as the use of Equation (2.51) in this case will yield the wrong ΔV. Furthermore, from Equation (2.74), as Δi approaches 114.59 deg from below, ΔV must approach $V_0 + V_f$, which implies that $\beta_0 \to 0$, as shown earlier, and $\beta_f \to 180$ deg. The ΔV in Equation (2.73) also approaches $V_0 + V_f$ because, with $\beta_0 = \varepsilon$, $s_{\beta_0} \sim \beta_0$, $c_{\beta_0} \sim 1$, $s_{\beta_f} \sim -s_{\beta_0} \sim -\beta_0$ and $\pi/2\,\Delta i \sim 180$ deg.

$$\Delta V \cong V_0 - \frac{V_f s_{\beta_f}}{\tan(180 + \varepsilon)} \cong V_0 + \frac{V_f \varepsilon}{\varepsilon} = V_0 + V_f$$

2.2.3 The Algorithm of the Edelbaum Transfer Problem

In summary, we have the following algorithm. First, one computes V_0 and V_f from the knowledge of initial and final semimajor axes a_0 and a_f, $V_0 = (\mu/a_0)^{1/2}$, $V_f = (\mu/a_f)^{1/2}$. Given $\mu = 398{,}601.3\,\text{km}^3/\text{s}^2$, Earth's gravity constant, Δi, the total inclination change desired, and f, the low-thrust acceleration, one computes β_0 from Equation (2.76) and the total ΔV_{tot} from Equation (2.73) such that the transfer time is known from $t_f = \Delta V_{\text{tot}}/f$. The variation with time of the various variables of interest is obtained from

$$\begin{aligned}
\Delta V &= ft \\
\beta &= \tan^{-1}\left(\frac{V_0 s_{\beta_0}}{V_0 c_{\beta_0} - ft}\right) \\
V &= \left(V_0^2 - 2V_0 ft\, c_{\beta_0} + f^2 t^2\right)^{1/2} \\
\lambda_V &= \frac{c_\beta}{f} \\
\Delta i &= \frac{2}{\pi}\left[\tan^{-1}\left(\frac{ft - V_0 c_{\beta_0}}{V_0 s_{\beta_0}}\right) + \frac{\pi}{2} - \beta_0\right]
\end{aligned}$$

λ_i is, of course, constant and given by $-\pi V_0 s_{\beta_0}/(2f)$. The total $(\Delta i)_f$ is obtained from $|i_0 - i_f|$ where i_0 and i_f are the initial and final inclination respectively. If $i_f > i_0$, the current inclination i is given by $i = i_0 + \Delta i$. If $i_f < i_0$, then $i = i_0 - \Delta i$. This is needed because we assumed $\Delta i > 0$ so that $\beta > \beta_0$, too.

An example of a constant acceleration LEO to GEO transfer is shown in Figures 2.1 and 2.2. The transfer is from $a_0 = 7000$ km, $i_0 = 28.5$ deg to $a_f = 42{,}166$ km, $i_f = 0.0$ deg with $f = 3.5 \times 10^{-7}\,\text{km/s}^2$. The total transfer time $t_f = 191.26259$ days with corresponding $\Delta V_{\text{tot}} = 5.78378$ km/s. The thrust yaw angle increases from its initial value $\beta_0 = 21.98$ deg to $\beta_f = 66.75$ deg as it is more efficient to rotate the orbit plane at higher altitudes. Figures 2.3 and 2.4 show the variation of the thrust yaw angle β, semimajor axis a, velocity V and inclination i as a function of time for a low-thrust transfer between $a_0 = 7000$ km, $i_0 = 90.0$ deg and $a_f = 42{,}166$ km, $i_f = 0.0$ deg orbit. The acceleration is still constant at $f = 3.5 \times 10^{-7}\,\text{km/s}^2$. Starting from its initial value at $\beta_0 = 10.92$ deg, β stays almost stationary for the first 100 days before surging to the final value $\beta_f = 152.29$ deg. It goes through 90 deg at day 245, where it starts to decelerate the vehicle because the orbit radius has exceeded the desired final altitude

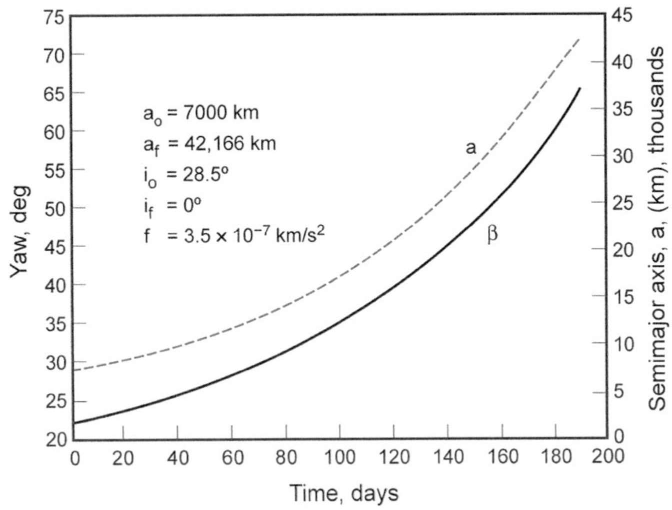

Figure 2.1 Optimal thrust yaw profile and semimajor axis variation for a low-acceleration LEO to GEO transfer.

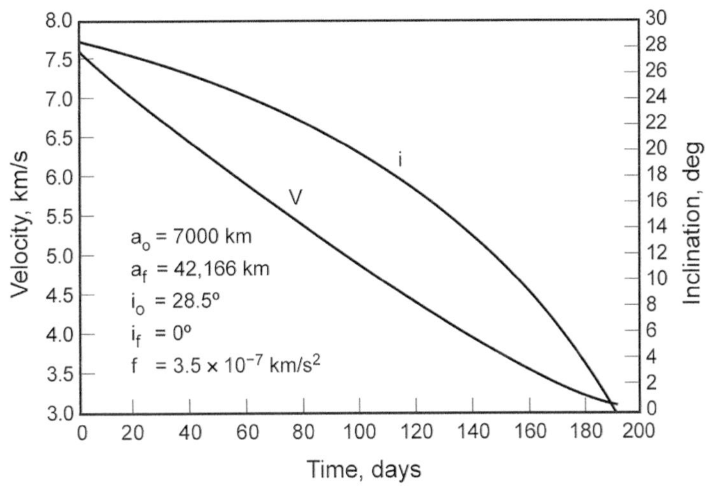

Figure 2.2 Orbital velocity and inclination variations for a low-acceleration LEO to GEO.

and, thus, must be shrunk until the final transfer time of $t_f = 335$ days for a total $\Delta V_{\text{tot}} = 10.13$ km/s. This is shown in Figure 2.4, where the intermediate velocity is much less than the final desired velocity of 3.07 km/s at GEO. Figures 2.5 and 2.6 are for $i_0 = 114.591$ deg, the uppermost limit of Edelbaum's theory. The β angle stays at zero, indicating a coplanar transfer up to $t = 250$ days, where $V \cong 0$, and the semimajor axis a at infinity, before flipping to 180 deg for the return leg to GEO. The inclination change is carried out instantaneously at infinity with zero cost. The total transfer time is $t_f = 351$ days with a ΔV of 10.61 km/s. This algorithm is valid regardless of whether the transfer is to a higher orbit or a lower orbit, irrespective of the direction of the inclination change.

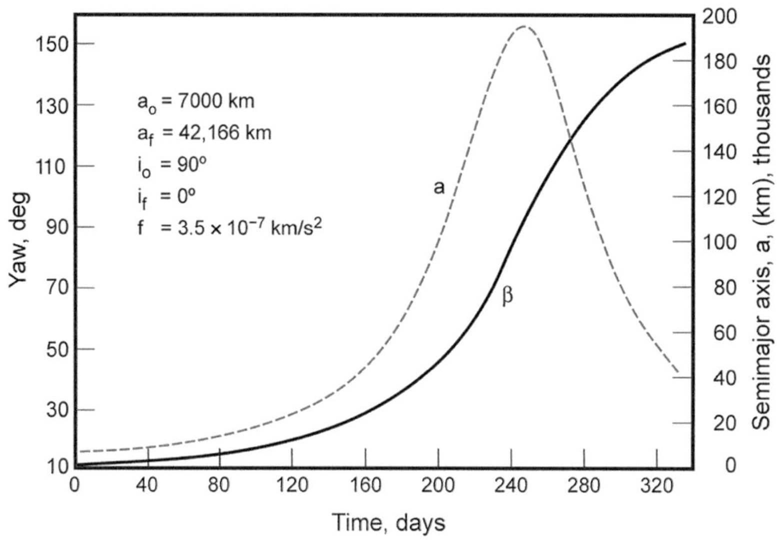

Figure 2.3 Optimal thrust yaw profile and semimajor axis variation for a large inclination change transfer.

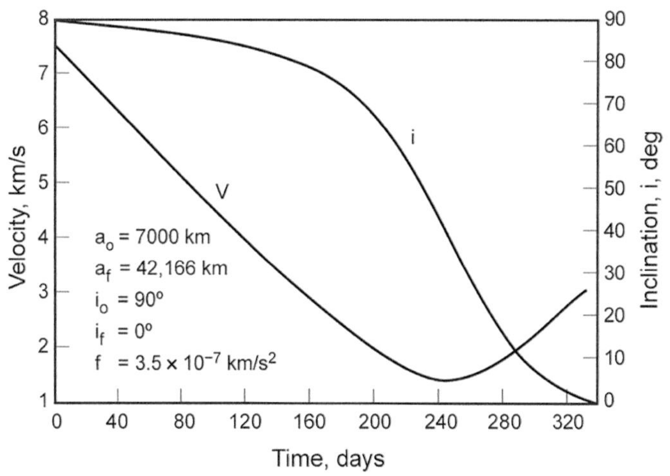

Figure 2.4 Evolution of velocity and inclination during optimal transfer for a large inclination change case.

2.3 The Full Six-State Formulation Using Nonsingular Equinoctial Orbit Elements

The mathematical theory of orbital mechanics in terms of the nonsingular equinoctial orbit elements has benefited from the contribution of Broucke and Cefola [2], who took advantage of the well-established results concerning the Lagrange and Poisson brackets of classical elements which are found, for example, in [3] and [4], in

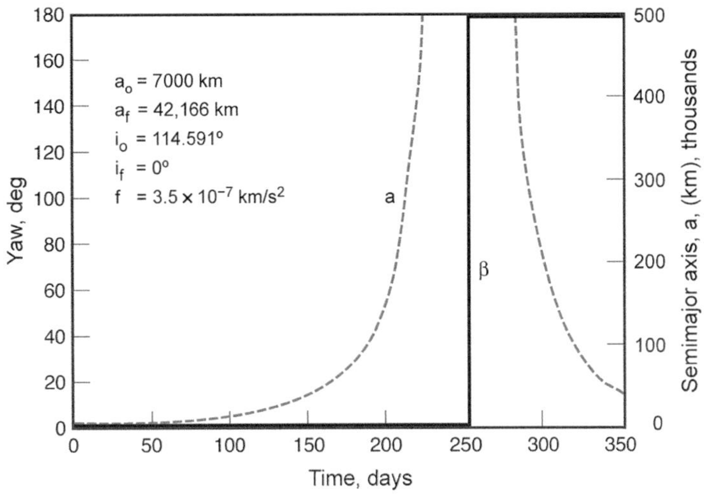

Figure 2.5 Optimal thrust yaw profile and semimajor axis variation for the limiting case of Edelbaum's theory.

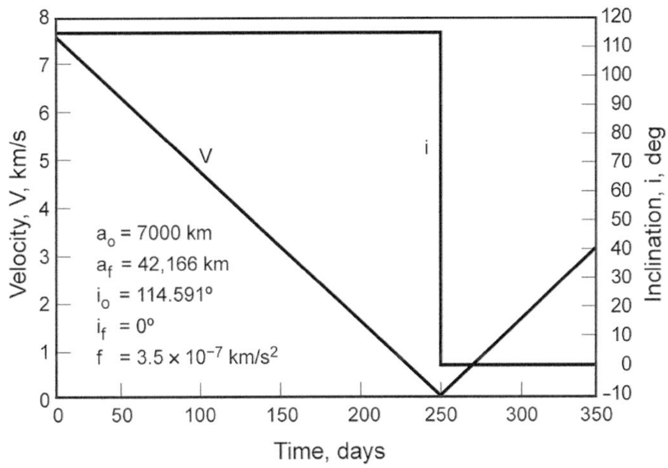

Figure 2.6 Velocity and inclination profiles for the limiting case of Edelbaum's theory.

order to transform these brackets in terms of the nonsingular elements. Cefola later developed the single-averaged variation of parameters equations for these elements [5], which were applied by Edelbaum, Sackett and Malchow [6] to the problem of optimal low-thrust transfer. Further applications of nonsingular orbit prediction and orbit-transfer optimization problems appeared in [7] and [8]. The variation of parameters perturbation equations based on the nonsingular equinoctial orbit elements are free from singularities for zero eccentricity and 0 and 90 deg inclination orbits. This fact, as well as many additional properties of the equinoctial elements, are derived in a systematic way in [2]. The matrizant, or state transition matrix corresponding to these elements, is based on the partial derivatives of the position and velocity vectors with respect to the equinoctial elements, as well as the inverse partial derivatives, meaning the partial derivatives of the equinoctial elements with respect to the position and velocity vectors. These partials were derived in [2] in terms of the classical elements, and transformed later in [5] and [7] in terms of the equinoctial elements. The applications of these elements to general and special perturbations are also discussed in the above-mentioned references. The position and velocity components in the direct equinoctial frame $\hat{\mathbf{f}}$, $\hat{\mathbf{g}}$, $\hat{\mathbf{w}}$ of Figure 2.7, to be described later, are given in terms of the equinoctial elements and the eccentric longitude F, which is related to the mean longitude λ by way of Kepler's equation. The equinoctial elements are defined in terms of the classical elements a, e, i, Ω, ω, M. Cefola et al. [5] and [7] consider the eccentric longitude F to be a function of λ, h and k, where h and k are the eccentricity vector components along $\hat{\mathbf{f}}$ and $\hat{\mathbf{g}}$, such that the partials of the position and velocity vectors with respect to the equinoctial elements are derived accordingly, F being defined by $F = E + \omega + \Omega$, where E is the classical eccentric anomaly. Given that the partials of the elements with respect to the velocity are obtained from the partials of the position with respect to the elements as well as the Poisson brackets,

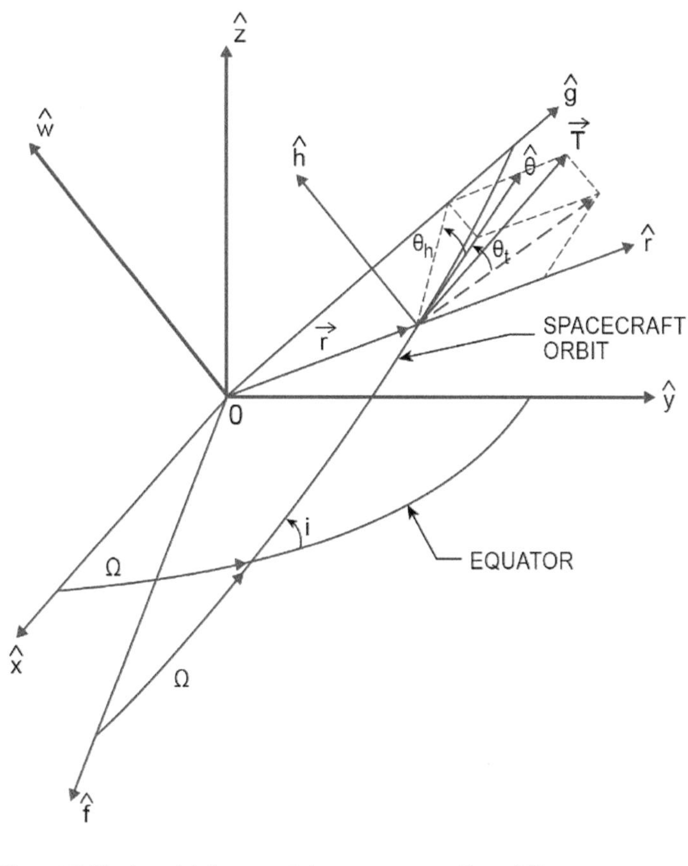

Figure 2.7 Equinoctial frame and thrust geometry (from [5]).

$$\frac{\partial a_\alpha}{\partial \dot{\mathbf{r}}} = -\sum_{\beta=1}^{6} \left(a_\alpha, a_\beta\right) \frac{\partial \mathbf{r}}{\partial a_\beta}$$

where a_α and a_β are the generic elements and (a_α, a_β) are the Poisson brackets of the equinoctial elements, then the partials $\partial a_\alpha / \partial \dot{\mathbf{r}}$ are obtained in a form that accounts for the dependence of F on λ, h and k. It will be shown later how the preceding equation is used to integrate the state differential equations forward in time. In problems of optimal transfer and rendezvous, the Euler–Lagrange differential equations, which must be integrated simultaneously with the state differential equations for the solution of the two-point boundary value problem, involve the partial derivatives of $\partial a_\alpha / \partial \dot{\mathbf{r}}$ with respect to the equinoctial elements. The velocity partials for the first five elements were derived for the case of the optimal transfer problem in [6], by assuming that F is an independent orbit element. However, we will assume in what follows that F is dependent on λ, h and k. This assumption is consistent with Broucke and Cefola's, such that we can provide the full set of the governing equations used in the solution of spaceflight optimization problems that are consistent with the original nonsingular state differential equations used in orbit prediction.

2.3.1 Orbital Mechanics in Terms of Equinoctial Elements

The equinoctial orbit elements were introduced in order to avoid the singularities associated with the use of the classical elements, a, e, i, Ω, ω and M. They are defined with respect to these elements by

$$a = a \tag{2.79}$$

$$h = e \sin(\omega + \Omega) \tag{2.80}$$

$$k = e \cos(\omega + \Omega) \tag{2.81}$$

$$p = \tan\left(\frac{i}{2}\right) \sin\Omega \tag{2.82}$$

$$q = \tan\left(\frac{i}{2}\right) \cos\Omega \tag{2.83}$$

$$\lambda = M + \omega + \Omega \tag{2.84}$$

The mean longitude λ defined in terms of the mean anomaly M can be replaced by the eccentric longitude F or the true longitude L defined respectively in terms of the eccentric anomaly E and the true anomaly θ^* such that

$$F = E + \omega + \Omega \tag{2.85}$$

$$L = \theta^* + \omega + \Omega \tag{2.86}$$

The inverse transformation is readily obtained from the preceding definitions:

$$a = a \tag{2.87}$$

$$e = \left(h^2 + k^2\right)^{1/2} \tag{2.88}$$

$$i = 2\tan^{-1}\left(p^2 + q^2\right)^{1/2} \tag{2.89}$$

$$\Omega = \tan^{-1}\left(\frac{p}{q}\right) \tag{2.90}$$

$$\omega = \tan^{-1}\left(\frac{h}{k}\right) - \tan^{-1}\left(\frac{p}{q}\right) \tag{2.91}$$

$$M = \lambda - \tan^{-1}\left(\frac{h}{k}\right) \tag{2.92}$$

The last expression can be replaced by either of the two following ones:

$$E = F - \tan^{-1}\left(\frac{h}{k}\right) \tag{2.93}$$

$$\theta^* = L - \tan^{-1}\left(\frac{h}{k}\right) \tag{2.94}$$

This set of equinoctial elements is called the direct equinoctial elements. They do not exhibit any singularity for $e = 0$ and $i = 0$ deg, 90 deg. However, they are singular for $i = 180$ deg. The equinoctial frame $\hat{\mathbf{f}}$, $\hat{\mathbf{g}}$, $\hat{\mathbf{w}}$ is defined in Figure 2.7 respective to the inertial frame $\hat{\mathbf{x}}$, $\hat{\mathbf{y}}$, $\hat{\mathbf{z}}$. The direction of the unit vector $\hat{\mathbf{f}}$ is such that both $\hat{\mathbf{f}}$ and $\hat{\mathbf{g}}$ are in the orbit plane and $\hat{\mathbf{f}}$ is obtained through a clockwise rotation of an angle Ω from the direction of the ascending node. The three rotations Ω, $-i$, $-\Omega$ transform the $\hat{\mathbf{f}}$, $\hat{\mathbf{g}}$, $\hat{\mathbf{w}}$ system into the $\hat{\mathbf{x}}$, $\hat{\mathbf{y}}$, $\hat{\mathbf{z}}$ system as shown in Equation (2.95).

It is then possible to obtain expressions relating the various elements of the transformation matrix in Equation (2.97) in terms of the equinoctial elements p and q such that

$$\begin{pmatrix}\hat{\mathbf{x}}\\ \hat{\mathbf{y}}\\ \hat{\mathbf{z}}\end{pmatrix} = \begin{pmatrix} c_\Omega c_{-\Omega} - s_\Omega c_i s_{-\Omega} & -c_\Omega s_{-\Omega} - s_\Omega c_i c_{-\Omega} & s_\Omega s_i \\ s_\Omega c_{-\Omega} + c_\Omega c_i s_{-\Omega} & -s_\Omega s_{-\Omega} + c_\Omega c_i c_{-\Omega} & -c_\Omega s_i \\ s_i s_{-\Omega} & s_i c_{-\Omega} & c_i \end{pmatrix}\begin{pmatrix}\hat{\mathbf{f}}\\ \hat{\mathbf{g}}\\ \hat{\mathbf{w}}\end{pmatrix} \tag{2.95}$$

which reduces to

$$\begin{pmatrix}\hat{\mathbf{x}}\\ \hat{\mathbf{y}}\\ \hat{\mathbf{z}}\end{pmatrix} = \begin{pmatrix} c_\Omega^2 + c_i s_\Omega^2 & c_\Omega s_\Omega - s_\Omega c_i c_\Omega & s_\Omega s_i \\ s_\Omega c_\Omega - s_\Omega c_\Omega c_i & s_\Omega^2 + c_i c_\Omega^2 & -c_\Omega s_i \\ -s_\Omega s_i & s_i c_\Omega & c_i \end{pmatrix}\begin{pmatrix}\hat{\mathbf{f}}\\ \hat{\mathbf{g}}\\ \hat{\mathbf{w}}\end{pmatrix} \tag{2.96}$$

The inverse transformation is given by

$$\begin{pmatrix}\hat{\mathbf{f}}\\ \hat{\mathbf{g}}\\ \hat{\mathbf{w}}\end{pmatrix} = \begin{pmatrix} c_\Omega^2 + c_i s_\Omega^2 & s_\Omega c_\Omega - s_\Omega c_\Omega c_i & -s_\Omega s_i \\ s_\Omega c_\Omega - s_\Omega c_\Omega c_i & s_\Omega^2 + c_i c_\Omega^2 & s_i c_\Omega \\ s_\Omega s_i & -c_\Omega s_i & c_i \end{pmatrix}\begin{pmatrix}\hat{\mathbf{x}}\\ \hat{\mathbf{y}}\\ \hat{\mathbf{z}}\end{pmatrix} \tag{2.97}$$

From the definitions of the equinoctial elements, the following relationships can be established:

$$c_{2\Omega} = \frac{q^2 - p^2}{p^2 + q^2}$$

$$s_\Omega = \frac{p}{\left(p^2 + q^2\right)^{1/2}}$$

$$
\begin{aligned}
c_\Omega &= \frac{q}{\left(p^2+q^2\right)^{1/2}} \\
s_{i/2}^2 &= \frac{p^2+q^2}{1+p^2+q^2} \\
c_{i/2}^2 &= \frac{1}{1+p^2+q^2} \\
s_i &= \frac{2\left(p^2+q^2\right)^{1/2}}{1+p^2+q^2} \\
c_i &= \frac{1-p^2-q^2}{1+p^2+q^2} \\
(1+c_i)^2 &= \frac{4}{\left(1+p^2+q^2\right)^2}
\end{aligned}
$$

It is then possible to obtain expressions relating the various elements of the transformation matrix in Equation (2.97) in terms of the equinoctial elements p and q such that

$$
\begin{aligned}
c_\Omega^2 + c_\text{i}\, s_\Omega^2 &= \frac{1-p^2+q^2}{1+p^2+q^2} \\
s_\Omega c_\Omega - s_\Omega c_\Omega c_\text{i} &= \frac{2pq}{1+p^2+q^2} \\
s_\Omega s_\text{i} &= \frac{2p}{1+p^2+q^2} \\
s_\Omega^2 + c_\text{i}\, c_\Omega^2 &= \frac{1+p^2-q^2}{1+p^2+q^2} \\
s_\text{i} c_\Omega &= \frac{2q}{1+p^2+q^2}
\end{aligned}
$$

From Equation (2.97), the components of the $\hat{\mathbf{f}}$, $\hat{\mathbf{g}}$, $\hat{\mathbf{w}}$ unit vectors in the inertial $\hat{\mathbf{x}}$, $\hat{\mathbf{y}}$, $\hat{\mathbf{z}}$ frame are now readily obtained in terms of the direct equinoctial elements

$$
\hat{\mathbf{f}} = \frac{1}{\left(1+p^2+q^2\right)} \begin{pmatrix} 1-p^2+q^2 \\ 2pq \\ -2p \end{pmatrix} \tag{2.98}
$$

$$
\hat{\mathbf{g}} = \frac{1}{\left(1+p^2+q^2\right)} \begin{pmatrix} 2pq \\ 1+p^2-q^2 \\ 2q \end{pmatrix} \tag{2.99}
$$

$$
\hat{\mathbf{w}} = \frac{1}{\left(1+p^2+q^2\right)} \begin{pmatrix} 2p \\ -2q \\ 1-p^2-q^2 \end{pmatrix} \tag{2.100}
$$

The derivation of Kepler's equation in terms of the equinoctial elements is carried out by using the definitions of e as well as h and k in the classical Kepler equation,

$$M = E - e\sin E$$
$$\lambda - (\omega + \Omega) = F - (\omega + \Omega) - e\sin[F - (\omega + \Omega)]$$
$$\lambda = F - e\left(s_F\frac{k}{e} - c_F\frac{h}{e}\right)$$
$$\lambda = F - k\sin F + h\cos F \tag{2.101}$$

The orbit equation $r = a(1 - e\cos E)$ can be written in terms of the equinoctial elements by using $F = E + \omega + \Omega$ and the definitions in Equations (2.80), (2.81) and (2.88), which relate e to h and k:

$$r = a\{1 - e\cos[F - (\omega + \Omega)]\}$$
$$r = a(1 - k\cos F - h\sin F) \tag{2.102}$$

The radius vector r being contained in the orbit plane, its components X_1 and Y_1 along $\hat{\mathbf{f}}$ and $\hat{\mathbf{g}}$ must be derived. Then, $\dot{\mathbf{r}} = \dot{X}_1\,\hat{\mathbf{f}} + \dot{Y}_1\,\hat{\mathbf{g}}$ can also be evaluated. To do this, let us observe that $X_1 = r\cos(\theta^* + \omega + \Omega)$ and $Y_1 = r\sin(\theta^* + \omega + \Omega)$. From classical orbital mechanics, the following relations $rc_{\theta^*} = a(\cos E - e)$ and $rs_{\theta^*} = a(1 - e^2)^{1/2}\sin E$ can now be used before the definitions in Equations (2.80), (2.81), (2.85) and (2.88); then

$$X_1 = rc_{\theta^*}\,c_{\omega+\Omega} - rs_{\theta^*}\,s_{\omega+\Omega}$$
$$= a\left\{\frac{k^2 + h^2\left(1 - h^2 - k^2\right)^{1/2}}{\left(h^2 + k^2\right)}c_F + \frac{hk\left[1 - \left(1 - h^2 - k^2\right)^{1/2}\right]}{\left(h^2 + k^2\right)}s_F - k\right\}$$

This intermediary result is written in order to introduce the quantity β, which is not to be confused with the thrust yaw angle β of the Edelbaum theory described earlier:

$$\beta = \frac{1 - \left(1 - h^2 - k^2\right)^{1/2}}{\left(h^2 + k^2\right)}$$
$$\beta = \frac{1}{1 + \left(1 - h^2 - k^2\right)^{1/2}} \tag{2.103}$$

We observe that

$$\frac{k^2 + h^2\left(1 - h^2 - k^2\right)^{1/2}}{\left(h^2 + k^2\right)} = 1 - h^2\beta$$

and find that X_1 takes the final form

$$X_1 = a\left[\left(1 - h^2\beta\right)\cos F + hk\beta\sin F - k\right] \tag{2.104}$$

The same manipulations are carried out for the definition of Y_1, which reduces to

$$Y_1 = a\left[hk\beta\cos F + \left(1 - k^2\beta\right)\sin F - h\right] \tag{2.105}$$

The velocity components $\dot{X}_1$ and $\dot{Y}_1$ can be obtained directly from Equations (2.104) and (2.105) by holding a, h and k, and therefore also β, constant and varying only F.

$$\dot{X}_1 = a\left[-\left(1 - h^2\beta\right) s_F\dot{F} + hk\beta c_F\dot{F}\right]$$
$$\dot{Y}_1 = a\left[-hk\beta s_F\dot{F} + \left(1 - k^2\beta\right) c_F\dot{F}\right]$$

From Equation (2.101) and with t_0 designating time at epoch

$$\lambda = n\,(t - t_0) + \lambda_0 = F - ks_F + hc_F$$
$$\dot{\lambda} = n = \dot{F} - kc_F\dot{F} - hs_F\dot{F}$$
$$\dot{F} = \frac{n}{(1 - kc_F - hs_F)}$$

and, in view of Equation (2.102), the orbit equation

$$\dot{F} = \frac{na}{r} \tag{2.106}$$

The velocity components can then be cast into the following form:

$$\dot{X}_1 = \frac{a^2 n}{r}\left[hk\beta c_F - \left(1 - h^2\beta\right) s_F\right] \tag{2.107}$$
$$\dot{Y}_1 = \frac{na^2}{r}\left[\left(1 - k^2\beta\right) c_F - hk\beta s_F\right] \tag{2.108}$$

Given $\mathbf{r}$ and $\dot{\mathbf{r}}$ in the inertial $\hat{\mathbf{x}}$, $\hat{\mathbf{y}}$, $\hat{\mathbf{z}}$ system, the equinoctial elements can be obtained by first evaluating a and $\mathbf{e}$ with the semimajor axis obtained from the energy equation

$$a = \left(\frac{2}{|\mathbf{r}|} - \frac{|\dot{\mathbf{r}}|^2}{\mu}\right)^{-1} \tag{2.109}$$
$$\mathbf{e} = -\frac{\mathbf{r}}{|\mathbf{r}|} - \frac{(\mathbf{r} \times \dot{\mathbf{r}}) \times \dot{\mathbf{r}}}{\mu} \tag{2.110}$$

Because $\hat{\mathbf{w}}$ is along the angular momentum vector $\mathbf{h} = \mathbf{r} \times \dot{\mathbf{r}}$, we have

$$\hat{\mathbf{w}} = \frac{\mathbf{r} \times \dot{\mathbf{r}}}{|\mathbf{r} \times \dot{\mathbf{r}}|} = \begin{pmatrix} w_x \\ w_y \\ w_z \end{pmatrix} \tag{2.111}$$

From the definition of $\hat{\mathbf{w}}$ given by Equation (2.100), it follows that p and q can be determined from Equation (2.111) such that

$$p = w_x/(1 + w_z) \tag{2.112}$$
$$q = -w_y/(1 + w_z) \tag{2.113}$$

Because k and h are the components of the eccentricity vector $\mathbf{e}$ along $\hat{\mathbf{f}}$ and $\hat{\mathbf{g}}$, we have

$$k = \mathbf{e} \bullet \hat{\mathbf{f}} \tag{2.114}$$
$$h = \mathbf{e} \bullet \hat{\mathbf{g}} \tag{2.115}$$

Finally, the components of **r** along $\hat{\mathbf{f}}$ and $\hat{\mathbf{g}}$ are obtained from

$$X_1 = \mathbf{r} \bullet \hat{\mathbf{f}} \tag{2.116}$$

$$Y_1 = \mathbf{r} \bullet \hat{\mathbf{g}} \tag{2.117}$$

These quantities are needed to evaluate the mean longitude λ, which itself requires the evaluation of the eccentric longitude F because $\lambda = F - ks_F + hc_F$. The quantities s_F and c_F are obtained by solving the system of Equations (2.104) and (2.105), which express X_1 and Y_1 in terms of h, k and F. This results in

$$s_F = h + \frac{Y_1\left(1 - h^2\beta\right) - hk\beta X_1}{a\left(1 - h^2 - k^2\right)^{1/2}}$$

$$c_F = \frac{\left(1 - k^2\beta\right)X_1 - hk\beta Y_1}{a\left(1 - h^2 - k^2\right)^{1/2}} + k$$

X_1 and Y_1 are first obtained from Equations (2.116) and (2.117). Finally, the partials of X_1, Y_1, $\dot{X}_1$ and $\dot{Y}_1$ with respect to h and k, which will be needed later, are evaluated from Equations (2.104), (2.105), (2.107) and (2.108). The following partials must, however, first be derived from

$$\beta = \frac{1}{1 + \left(1 - h^2 - k^2\right)^{1/2}}$$

with the observation that $(1 - h^2 - k^2)^{1/2} = (1 - \beta)/\beta$. They are

$$\frac{\partial\beta}{\partial h} = \frac{h\beta^3}{(1 - \beta)}$$

$$\frac{\partial\beta}{\partial k} = \frac{k\beta^3}{(1 - \beta)}$$

From $\lambda = F - ks_F + hc_F = M + \omega + \Omega$, it follows that

$$\begin{aligned}\frac{\partial\lambda}{\partial h} = 0 &= \frac{\partial F}{\partial h} - kc_F\frac{\partial F}{\partial h} + c_F - hs_F\frac{\partial F}{\partial h}\\ &= \frac{\partial F}{\partial h}\left[1 - kc_F - hs_F\right] + c_F = \frac{\partial F}{\partial h}\left(\frac{r}{a}\right) + c_F\end{aligned}$$

$$\frac{\partial F}{\partial h} = -\frac{a}{r}c_F \tag{2.118}$$

Similarly, from $\frac{\partial\lambda}{\partial k} = 0$,

$$\frac{\partial F}{\partial k} = \frac{a}{r}s_F \tag{2.119}$$

which is then used in Equation (2.104), together with $\partial\beta/\partial h$ developed earlier, to yield, after some manipulations,

$$\frac{\partial X_1}{\partial h} = a\left[-\left(hc_F - ks_F\right)\left(\beta + \frac{h^2\beta^3}{(1 - \beta)}\right) - \frac{a}{r}c_F\left(h\beta - s_F\right)\right] \tag{2.120}$$

In a similar way, and as in [5],

$$\frac{\partial X_1}{\partial k} = -a\left[(hc_F - ks_F)\frac{hk\beta^3}{(1-\beta)} + 1 + \frac{a}{r}s_F(s_F - h\beta)\right] \tag{2.121}$$

$$\frac{\partial Y_1}{\partial h} = a\left[(hc_F - ks_F)\frac{hk\beta^3}{(1-\beta)} - 1 + \frac{a}{r}c_F(k\beta - c_F)\right] \tag{2.122}$$

$$\frac{\partial Y_1}{\partial k} = a\left[(hc_F - ks_F)\left(\beta + \frac{k^2\beta^3}{(1-\beta)}\right) + \frac{a}{r}s_F(c_F - k\beta)\right] \tag{2.123}$$

Letting $\mathbf{z} = (a\ h\ k\ p\ q\ \lambda_0)^T$ represent the orbit state vector at time t, with λ_0 the mean longitude at epoch corresponding to the mean anomaly M_0, and given that $\mathbf{z} = f(\mathbf{r}, \dot{\mathbf{r}})$, we have

$$\dot{\mathbf{z}} = \frac{\partial \mathbf{z}}{\partial \mathbf{r}}\dot{\mathbf{r}} + \frac{\partial \mathbf{z}}{\partial \dot{\mathbf{r}}}\ddot{\mathbf{r}} \tag{2.124}$$

where

$$\ddot{\mathbf{r}} = \frac{\mathbf{T}}{m} + \mathbf{g} = \frac{\mathbf{T}}{m} - \frac{\mu}{r^3}\mathbf{r} \tag{2.125}$$

and where $\mathbf{T}$ and m stand for the thrust vector and spacecraft mass, respectively. The partial derivatives of the equinoctial elements with respect to the position and velocity vectors, the so-called inverse partials, are related to the partials of the position and velocity vectors with respect to the equinoctial elements using the Poisson brackets (a_α, a_β) of equinoctial elements:

$$\frac{\partial a_\alpha}{\partial \mathbf{r}} = \sum_{\beta=1}^{6}(a_\alpha, a_\beta)\frac{\partial \dot{\mathbf{r}}}{\partial a_\beta} \tag{2.126}$$

$$\frac{\partial a_\alpha}{\partial \dot{\mathbf{r}}} = -\sum_{\beta=1}^{6}(a_\alpha, a_\beta)\frac{\partial \mathbf{r}}{\partial a_\beta} \tag{2.127}$$

If we write Equation (2.124) one component at a time, then, using Equation (2.125),

$$\begin{aligned}
\dot{z} &= \frac{\partial z}{\partial \mathbf{r}}\dot{\mathbf{r}} + \frac{\partial z}{\partial \dot{\mathbf{r}}}\ddot{\mathbf{r}} \\
&= \sum_\beta (a_\alpha, a_\beta)\frac{\partial \dot{\mathbf{r}}}{\partial a_\beta}\dot{\mathbf{r}} - \sum_\beta (a_\alpha, a_\beta)\frac{\partial \mathbf{r}}{\partial a_\beta}\ddot{\mathbf{r}} \\
&= \sum_\beta (a_\alpha, a_\beta)\frac{\partial \dot{\mathbf{r}}}{\partial a_\beta}\dot{\mathbf{r}} + \sum_\beta (a_\alpha, a_\beta)\frac{\partial \mathbf{r}}{\partial a_\beta}\left(\frac{\mu}{r^3}\right)\mathbf{r} + \frac{\partial z}{\partial \dot{\mathbf{r}}}\frac{\mathbf{T}}{m} \\
&= \sum_\beta (a_\alpha, a_\beta)\left(\frac{\partial \dot{\mathbf{r}}}{\partial a_\beta}\dot{\mathbf{r}} + \frac{\mu}{r^3}\frac{\partial \mathbf{r}}{\partial a_\beta}\mathbf{r}\right) + \frac{\partial z}{\partial \dot{\mathbf{r}}}\frac{\mathbf{T}}{m}
\end{aligned} \tag{2.128}$$

The term in brackets can be written as

$$\frac{1}{2}\frac{\partial\left(\dot{\mathbf{r}}^2\right)}{\partial z} + \frac{\mu}{r^3}\frac{1}{2}\frac{\partial\left(\mathbf{r}^2\right)}{\partial z} = \frac{1}{2}\left(\frac{\partial v^2}{\partial z} + \frac{\mu}{r^3}\frac{\partial r^2}{\partial z}\right) \tag{2.129}$$

However, $|\dot{\mathbf{r}}| = v$ and, from the energy equation,

$$\frac{1}{2}v^2 - \frac{\mu}{r} = \text{C}$$

We have

$$\frac{1}{2}\frac{\partial v^2}{\partial z} = -\frac{\mu}{r^2}\frac{\partial r}{\partial z}$$

And, in view of $\partial r^2/\partial z = 2r\ \partial r/\partial z$, the bracket in Equation (2.129) cancels out such that

$$\dot{z} = \frac{\partial z}{\partial \dot{\mathbf{r}}}\frac{\mathbf{T}}{m}$$

If $\hat{\mathbf{u}}$ represents a unit vector in the direction of the thrust,

$$\mathbf{T} = T\ \hat{\mathbf{u}} \tag{2.130}$$

and writing z as a vector, the variation of parameters equations, where both $\partial \mathbf{z}/\partial \mathbf{r}$ and $\hat{\mathbf{u}}$ are expressed in the direct equinoctial frame,

$$\dot{\mathbf{z}} = \frac{\partial \mathbf{z}}{\partial \dot{\mathbf{r}}}\frac{T}{m}\hat{\mathbf{u}} = \frac{\partial \mathbf{z}}{\partial \dot{\mathbf{r}}}f\hat{\mathbf{u}} \tag{2.131}$$

2.3.2 Partials of the Equinoctial Orbit Elements with Respect to Velocity

The first-order differential equations in the equinoctial orbit elements involve the partials of these elements with respect to the velocity vector $\dot{\mathbf{r}}$. These partials are computed using the Poisson brackets of the equinoctial elements (a_α, a_β). However, the partials of the position vector $\mathbf{r}$ with respect to the equinoctial elements must be computed first. Furthermore, the Poisson brackets of the equinoctial elements can be obtained from the Poisson brackets of the classical elements, which are found in texts such as [3] and [4]. These brackets have been derived in [2] and [5]. These "velocity partials" are given by Equation (2.127). First, let us derive the $\partial \mathbf{r}/\partial a_\beta$ partials. From the definition of $\mathbf{r} = X_1\hat{\mathbf{f}} + Y_1\hat{\mathbf{g}}$, it follows that

$$\frac{\partial \mathbf{r}}{\partial a} = \frac{\partial X_1}{\partial a}\hat{\mathbf{f}} + \frac{\partial Y_1}{\partial a}\hat{\mathbf{g}}$$

Refer back to the definition of λ to arrive at

$$\begin{aligned}\frac{\partial \lambda}{\partial a} &= \frac{\partial \lambda}{\partial F}\frac{\partial F}{\partial a} = -\frac{3}{2a}nt \\ \frac{\partial F}{\partial a} &= -\frac{3}{2}\frac{n}{r}t\end{aligned} \tag{2.132}$$

Then, using the definitions of $\dot{X}_1$ and $\dot{Y}_1$ in Equations (2.107) and (2.108) to obtain

$$\begin{aligned}\frac{\partial X_1}{\partial a} &= \left[\left(1 - h^2\beta\right)c_F + hk\beta s_F - k\right] + a\left[-\left(1 - h^2\beta\right)s_F + hk\beta c_F\right]\frac{\partial F}{\partial a} \\ &= \frac{X_1}{a} + \frac{r\dot{X}_1}{na}\frac{\partial F}{\partial a} = \frac{X_1}{a} - \frac{3}{2}\frac{t}{a}\dot{X}_1\end{aligned}$$

In a similar way,

$$\frac{\partial Y_1}{\partial a} = \frac{Y_1}{a} - \frac{3}{2}\frac{t}{a}\dot{Y}_1$$

and, therefore,

$$\frac{\partial \mathbf{r}}{\partial a} = \left(\frac{X_1}{a} - \frac{3}{2}\frac{t}{a}\dot{X}_1\right)\hat{\mathbf{f}} + \left(\frac{Y_1}{a} - \frac{3}{2}\frac{t}{a}\dot{Y}_1\right)\hat{\mathbf{g}}$$

The partials $\partial\mathbf{r}/\partial h$ and $\partial\mathbf{r}/\partial k$ are written in a straightforward manner

$$\frac{\partial \mathbf{r}}{\partial h} = \frac{\partial X_1}{\partial h}\hat{\mathbf{f}} + \frac{\partial Y_1}{\partial h}\hat{\mathbf{g}} \tag{2.133}$$

$$\frac{\partial \mathbf{r}}{\partial k} = \frac{\partial X_1}{\partial k}\hat{\mathbf{f}} + \frac{\partial Y_1}{\partial k}\hat{\mathbf{g}} \tag{2.134}$$

Equations (2.120)–(2.123) must, of course, be used in Equations (2.133) and (2.134). Now, by way of some manipulations,

$$\frac{\partial \mathbf{r}}{\partial \lambda} = \frac{\partial \mathbf{r}}{\partial F}\frac{\partial F}{\partial \lambda} = \left(\frac{\partial X_1}{\partial F}\hat{\mathbf{f}} + \frac{\partial Y_1}{\partial F}\hat{\mathbf{g}}\right)\frac{\partial F}{\partial \lambda}$$

$$\frac{\partial \mathbf{r}}{\partial \lambda} = \frac{\dot{X}_1}{n}\hat{\mathbf{f}} + \frac{\dot{Y}_1}{n}\hat{\mathbf{g}} = \frac{\dot{\mathbf{r}}}{n} \tag{2.135}$$

The remaining partials $\partial\mathbf{r}/\partial p$ and $\partial\mathbf{r}/\partial q$ are derived by observing first that X_1 and Y_1 are not functions of either p or q but that the components of the unit vectors $\hat{\mathbf{f}}$ and $\hat{\mathbf{g}}$ are such functions; therefore,

$$\frac{\partial \mathbf{r}}{\partial p} = X_1\frac{\partial \hat{\mathbf{f}}}{\partial p} + Y_1\frac{\partial \hat{\mathbf{g}}}{\partial p}$$

From Equations (2.98)–(2.100), it follows that

$$\frac{\partial \mathbf{r}}{\partial p} = \frac{2X_1}{\left(1+p^2+q^2\right)^2}\begin{pmatrix} -2p - 2pq^2 \\ 2q - q\left(1+p^2-q^2\right) \\ -2q^2 - \left(1-p^2-q^2\right) \end{pmatrix} + \frac{2Y_1}{\left(1+p^2+q^2\right)^2}\begin{pmatrix} q\left(1+q^2-p^2\right) \\ 2pq^2 \\ -2pq \end{pmatrix}$$

$$\frac{\partial \mathbf{r}}{\partial p} = \frac{2}{\left(1+p^2+q^2\right)}\left[q\left(Y_1\hat{\mathbf{f}} - X_1\hat{\mathbf{g}}\right) - X_1\hat{\mathbf{w}}\right] \tag{2.136}$$

In a similar way,

$$\frac{\partial \mathbf{r}}{\partial q} = X_1\frac{\partial \hat{\mathbf{f}}}{\partial q} + Y_1\frac{\partial \hat{\mathbf{g}}}{\partial q} = \frac{2}{\left(1+p^2+q^2\right)}\left[p\left(X_1\hat{\mathbf{g}} - Y_1\hat{\mathbf{f}}\right) + Y_1\hat{\mathbf{w}}\right] \tag{2.137}$$

The Poisson brackets of equinoctial elements are very tedious to derive unless one uses the transformation equation that converts the Poisson brackets of classical elements into the brackets of equinoctial elements. The transformation equation is given in [2], and, using the nomenclature introduced by Broucke and Cefola, we write

$$\left[\left(p_\alpha, p_\beta\right)\right] = \left[\frac{\partial p_\alpha}{\partial a_\lambda}\right]\left[\left(a_\lambda, a_\mu\right)\right]\left[\frac{\partial p_\beta}{\partial a_\mu}\right]^T \tag{2.138}$$

The various matrices appearing in Equation (2.138) are given by

$$[(p_\alpha, p_\beta)] = \begin{bmatrix} (a, a) & (a, h) & (a, k) & (a, \lambda_0) & (a, p) & (a, q) \\ (h, a) & (h, h) & (h, k) & (h, \lambda_0) & (h, p) & (h, q) \\ (k, a) & (k, h) & (k, k) & (k, \lambda_0) & (k, p) & (k, q) \\ (\lambda_0, a) & (\lambda_0, h) & (\lambda_0, k) & (\lambda_0, \lambda_0) & (\lambda_0, p) & (\lambda_0, q) \\ (p, a) & (p, h) & (p, k) & (p, \lambda_0) & (p, p) & (p, q) \\ (q, a) & (q, h) & (q, k) & (q, \lambda_0) & (q, p) & (q, q) \end{bmatrix}$$

$$[(a_\lambda, a_\mu)] = \begin{bmatrix} (a,a) & (a, e) & (a, i) & (a, \Omega) & (a, \omega) & (a, M_0) \\ (e,a) & (e,e) & (e, i) & (e, \Omega) & (e,\omega) & (e, M_0) \\ (i, a) & (i,e) & (i,i) & (i,\Omega) & (i,\omega) & (i, M_0) \\ (\Omega, a) & (\Omega, e) & (\Omega, i) & (\Omega, \Omega) & (\Omega,\omega) & (\Omega, M_0) \\ (\omega, a) & (\omega, e) & (\omega, i) & (\omega, \Omega) & (\omega,\omega) & (\omega, M_0) \\ (M_0, a) & (M_0, e) & (M_0, i) & (M_0, \Omega) & (M_0,\omega) & (M_0, M_0) \end{bmatrix}$$

$$\left[\frac{\partial p_\alpha}{\partial a_\lambda}\right] = \begin{bmatrix} \frac{\partial a}{\partial a} & \frac{\partial a}{\partial e} & \frac{\partial a}{\partial i} & \frac{\partial a}{\partial \Omega} & \frac{\partial a}{\partial \omega} & \frac{\partial a}{\partial M_0} \\ \frac{\partial h}{\partial a} & \frac{\partial h}{\partial e} & \frac{\partial h}{\partial i} & \frac{\partial h}{\partial \Omega} & \frac{\partial h}{\partial \omega} & \frac{\partial h}{\partial M_0} \\ \frac{\partial k}{\partial a} & \frac{\partial k}{\partial e} & \frac{\partial k}{\partial i} & \frac{\partial k}{\partial \Omega} & \frac{\partial k}{\partial \omega} & \frac{\partial k}{\partial M_0} \\ \frac{\partial \lambda_0}{\partial a} & \frac{\partial \lambda_0}{\partial e} & \frac{\partial \lambda_0}{\partial i} & \frac{\partial \lambda_0}{\partial \Omega} & \frac{\partial \lambda_0}{\partial \omega} & \frac{\partial \lambda_0}{\partial M_0} \\ \frac{\partial p}{\partial a} & \frac{\partial p}{\partial e} & \frac{\partial p}{\partial i} & \frac{\partial p}{\partial \Omega} & \frac{\partial p}{\partial \omega} & \frac{\partial p}{\partial M_0} \\ \frac{\partial q}{\partial a} & \frac{\partial q}{\partial e} & \frac{\partial q}{\partial i} & \frac{\partial q}{\partial \Omega} & \frac{\partial q}{\partial \omega} & \frac{\partial q}{\partial M_0} \end{bmatrix}$$

$$\left[\frac{\partial p_\beta}{\partial a_\mu}\right]^T = \begin{bmatrix} \frac{\partial a}{\partial a} & \frac{\partial h}{\partial a} & \frac{\partial k}{\partial a} & \frac{\partial \lambda_0}{\partial a} & \frac{\partial p}{\partial a} & \frac{\partial q}{\partial a} \\ \frac{\partial a}{\partial e} & \frac{\partial h}{\partial e} & \frac{\partial k}{\partial e} & \frac{\partial \lambda_0}{\partial e} & \frac{\partial p}{\partial e} & \frac{\partial q}{\partial e} \\ \frac{\partial a}{\partial i} & \frac{\partial h}{\partial i} & \frac{\partial k}{\partial i} & \frac{\partial \lambda_0}{\partial i} & \frac{\partial p}{\partial i} & \frac{\partial q}{\partial i} \\ \frac{\partial a}{\partial \Omega} & \frac{\partial h}{\partial \Omega} & \frac{\partial k}{\partial \Omega} & \frac{\partial \lambda_0}{\partial \Omega} & \frac{\partial p}{\partial \Omega} & \frac{\partial q}{\partial \Omega} \\ \frac{\partial a}{\partial \omega} & \frac{\partial h}{\partial \omega} & \frac{\partial k}{\partial \omega} & \frac{\partial \lambda_0}{\partial \omega} & \frac{\partial p}{\partial \omega} & \frac{\partial q}{\partial \omega} \\ \frac{\partial a}{\partial M_0} & \frac{\partial h}{\partial M_0} & \frac{\partial k}{\partial M_0} & \frac{\partial \lambda_0}{\partial M_0} & \frac{\partial p}{\partial M_0} & \frac{\partial q}{\partial M_0} \end{bmatrix}$$

The elements of the matrix $\left[\partial p_\alpha / \partial a_\lambda\right]$ can be calculated from the definitions of the equinoctial elements that appear in Equations (2.79)–(2.84):

$$\left[\frac{\partial p_\alpha}{\partial a_\lambda}\right] = \begin{bmatrix} 1 & 0 & 0 & 0 & 0 & 0 \\ 0 & s_{\omega+\Omega} & 0 & ec_{\omega+\Omega} & ec_{\omega+\Omega} & 0 \\ 0 & c_{\omega+\Omega} & 0 & -es_{\omega+\Omega} & -es_{\omega+\Omega} & 0 \\ 0 & 0 & 0 & 1 & 1 & 1 \\ 0 & 0 & \dfrac{s_\Omega}{2\cos^2\frac{i}{2}} & \tan\dfrac{i}{2}c_\Omega & 0 & 0 \\ 0 & 0 & \dfrac{c_\Omega}{2\cos^2\frac{i}{2}} & -\tan\dfrac{i}{2}s_\Omega & 0 & 0 \end{bmatrix}$$

which converts to

$$\left[\frac{\partial p_\alpha}{\partial a_\lambda}\right] = \begin{bmatrix} 1 & 0 & 0 & 0 & 0 & 0 \\ 0 & h/\left(h^2+k^2\right)^{1/2} & 0 & k & k & 0 \\ 0 & k/\left(h^2+k^2\right)^{1/2} & 0 & -h & -h & 0 \\ 0 & 0 & 0 & 1 & 1 & 1 \\ 0 & 0 & \dfrac{p\left(1+p^2+q^2\right)}{2\left(p^2+q^2\right)^{1/2}} & q & 0 & 0 \\ 0 & 0 & \dfrac{q\left(1+p^2+q^2\right)}{2\left(p^2+q^2\right)^{1/2}} & -p & 0 & 0 \end{bmatrix}$$

The Poisson brackets in classical elements, though difficult to derive, are readily available in advanced books on astrodynamics such as [3] and [4]. The matrix of the Poisson brackets in the classical elements is found to be, with $p' = a\left(1-e^2\right)$, the orbit parameter:

$$\left[\left(a_\lambda, a_\mu\right)\right] = P_{cl} = \begin{bmatrix} 0 & 0 & 0 & 0 & 0 & -2\left(\dfrac{a}{\mu}\right)^{1/2} \\ & 0 & 0 & 0 & \dfrac{\left(1-e^2\right)^{1/2}}{e\left(\mu a\right)^{1/2}} & \dfrac{-\left(1-e^2\right)}{e\left(\mu a\right)^{1/2}} \\ & & 0 & \dfrac{1}{\left(\mu p'\right)^{1/2}s_i} & \dfrac{-c_i}{\left(\mu p'\right)^{1/2}s_i} & 0 \\ & & & 0 & 0 & 0 \\ -\mathbf{Sym} & & & & 0 & 0 \\ & & & & & 0 \end{bmatrix} \tag{2.139}$$

Carrying out the algebra, and using the identities introduced earlier, as well as the following identities, we obtain

$$\begin{aligned} s_i c_\Omega &= 2q/\left(1+p^2+q^2\right) \\ s_\Omega s_i &= 2p/\left(1+p^2+q^2\right) \\ -c_\Omega s_i &= -2q/\left(1+p^2+q^2\right) \end{aligned}$$

The Poisson brackets of equinoctial elements are derived to result in

$$(a,\lambda_0) = -2/(na)$$

$$(\lambda_0,h) = \frac{-h\left(1-h^2-k^2\right)^{1/2}}{na^2\left[1+\left(1-h^2-k^2\right)^{1/2}\right]}$$

$$(\lambda_0,k) = \frac{-k\left(1-h^2-k^2\right)^{1/2}}{na^2\left[1+\left(1-h^2-k^2\right)^{1/2}\right]}$$

$$(\lambda_0,p) = \frac{-p\left(1+p^2+q^2\right)}{2na^2\left(1-h^2-k^2\right)^{1/2}}$$

$$(\lambda_0,q) = \frac{-q\left(1+p^2+q^2\right)}{2na^2\left(1-h^2-k^2\right)^{1/2}}$$

$$(h,k) = \frac{-\left(1-h^2-k^2\right)^{1/2}}{na^2}$$

$$(h,p) = \frac{-kp\left(1+p^2+q^2\right)}{2na^2\left(1-h^2-k^2\right)^{1/2}}$$

$$(h,q) = \frac{-kq\left(1+p^2+q^2\right)}{2na^2\left(1-h^2-k^2\right)^{1/2}}$$

$$(k,p) = \frac{hp\left(1+p^2+q^2\right)}{2na^2\left(1-h^2-k^2\right)^{1/2}}$$

$$(k,q) = \frac{hq\left(1+p^2+q^2\right)}{2na^2\left(1-h^2-k^2\right)^{1/2}}$$

$$(p,q) = \frac{-\left(1+p^2+q^2\right)^2}{4na^2\left(1-h^2-k^2\right)^{1/2}}$$

The property that emanates from the definition of the brackets, namely $(\alpha,\alpha)=0$ and $(\alpha,\beta)=-(\beta,\alpha)$, is responsible for the antisymmetric nature of the foregoing matrices. Going back to our original nomenclature for $\left(a_\alpha,a_\beta\right)$ and having developed all the Poisson brackets $\left(a_\alpha,a_\beta\right)$ and all the partials $\partial r/\partial a_\beta$, we can now derive the partials $\partial \mathbf{z}/\partial \dot{\mathbf{r}}$. These partials are

$$\frac{\partial a}{\partial \dot{\mathbf{r}}} = \frac{2}{n^2 a}\left(\dot{X}_1\hat{\mathbf{f}} + \dot{Y}_1\hat{\mathbf{g}}\right) = M_{11}\hat{\mathbf{f}} + M_{12}\hat{\mathbf{g}} + M_{13}\hat{\mathbf{w}} \tag{2.140}$$

$$\frac{\partial h}{\partial \dot{\mathbf{r}}} = \frac{\left(1-h^2-k^2\right)^{1/2}}{na^2}\left[\left(\frac{\partial X_1}{\partial k} - \frac{h\beta}{n}\dot{X}_1\right)\hat{\mathbf{f}} + \left(\frac{\partial Y_1}{\partial k} - \frac{h\beta}{n}\dot{Y}_1\right)\hat{\mathbf{g}}\right]$$
$$+ \frac{k\left(qY_1 - pX_1\right)}{na^2\left(1-h^2-k^2\right)^{1/2}}\hat{\mathbf{w}} = M_{21}\hat{\mathbf{f}} + M_{22}\hat{\mathbf{g}} + M_{23}\hat{\mathbf{w}} \tag{2.141}$$

$$\frac{\partial k}{\partial \dot{\mathbf{r}}} = -\frac{\left(1-h^2-k^2\right)^{1/2}}{na^2}\left[\left(\frac{\partial X_1}{\partial h} + k\beta\frac{\dot{X}_1}{n}\right)\hat{\mathbf{f}} + \left(\frac{\partial Y_1}{\partial k} + k\beta\frac{\dot{Y}_1}{n}\right)\hat{\mathbf{g}}\right]$$
$$- \frac{h\left(qY_1 - pX_1\right)}{na^2\left(1-h^2-k^2\right)^{1/2}}\hat{\mathbf{w}} = M_{31}\hat{\mathbf{f}} + M_{32}\hat{\mathbf{g}} + M_{33}\hat{\mathbf{w}} \tag{2.142}$$

$$\frac{\partial p}{\partial \dot{\mathbf{r}}} = \frac{\left(1+p^2+q^2\right) Y_1}{2na^2\left(1-h^2-k^2\right)^{1/2}} \hat{\mathbf{w}} = M_{41}\hat{\mathbf{f}} + M_{42}\hat{\mathbf{g}} + M_{43}\hat{\mathbf{w}} \tag{2.143}$$

$$\frac{\partial q}{\partial \dot{\mathbf{r}}} = \frac{\left(1+p^2+q^2\right) X_1}{2na^2\left(1-h^2-k^2\right)^{1/2}} \hat{\mathbf{w}} = M_{51}\hat{\mathbf{f}} + M_{52}\hat{\mathbf{g}} + M_{53}\hat{\mathbf{w}} \tag{2.144}$$

$$\begin{aligned}\frac{\partial \lambda_0}{\partial \dot{\mathbf{r}}} &= \frac{1}{na^2}\left[-2X_1 + 3\dot{X}_1 t + \left(1-h^2-k^2\right)^{1/2}\left(h\beta\frac{\partial X_1}{\partial h} + k\beta\frac{\partial X_1}{\partial k}\right)\right]\hat{\mathbf{f}} \\ &+ \frac{1}{na^2}\left[-2Y_1 + 3\dot{Y}_1 t + \left(1-h^2-k^2\right)^{1/2}\left(h\beta\frac{\partial Y_1}{\partial h} + k\beta\frac{\partial Y_1}{\partial k}\right)\right]\hat{\mathbf{g}} \\ &+ \frac{\left(qY_1 - pX_1\right)}{na^2\left(1-h^2-k^2\right)^{1/2}}\hat{\mathbf{w}} = M_{61}\hat{\mathbf{f}} + M_{62}\hat{\mathbf{g}} + M_{63}\hat{\mathbf{w}}\end{aligned} \tag{2.145}$$

Let us now use $\partial\lambda/\partial\dot{\mathbf{r}}$ instead of $\partial\lambda_0/\partial\dot{\mathbf{r}}$ as we did above. There is a simple relationship that links the two partials because $\lambda = \lambda_0 + n(t - t_0)$, such that

$$\frac{\partial \lambda}{\partial \dot{\mathbf{r}}} = \frac{\partial \lambda_0}{\partial \dot{\mathbf{r}}} + t\frac{\partial n}{\partial a}\frac{\partial a}{\partial \dot{\mathbf{r}}}$$

and because $\partial n/\partial a = -3n/(2a)$ and $\partial a/\partial \dot{\mathbf{r}} = 2\dot{\mathbf{r}}/\left(n^2 a\right)$, then

$$\frac{\partial \lambda}{\partial \dot{\mathbf{r}}} = \frac{\partial \lambda_0}{\partial \dot{\mathbf{r}}} - \frac{3}{na^2}t\dot{\mathbf{r}} \tag{2.146}$$

The matrix M is then a 6×3 matrix such that M_{61}, M_{62}, and M_{63} are as in Equation (2.145) without the t terms.

$$M = \begin{bmatrix} (\partial a/\partial \dot{\mathbf{r}})^T \\ (\partial h/\partial \dot{\mathbf{r}})^T \\ (\partial k/\partial \dot{\mathbf{r}})^T \\ (\partial p/\partial \dot{\mathbf{r}})^T \\ (\partial q/\partial \dot{\mathbf{r}})^T \\ (\partial \lambda/\partial \dot{\mathbf{r}})^T \end{bmatrix} = \begin{bmatrix} M_{11} & M_{12} & M_{13} \\ M_{21} & M_{22} & M_{23} \\ M_{31} & M_{32} & M_{33} \\ M_{41} & M_{42} & M_{43} \\ M_{51} & M_{52} & M_{53} \\ M_{61} & M_{62} & M_{63} \end{bmatrix} \tag{2.147}$$

The equations of motion are therefore given by

$$\dot{a} = (\partial a/\partial \dot{\mathbf{r}})^T \bullet \hat{\mathbf{u}}f \tag{2.148}$$
$$\dot{h} = (\partial h/\partial \dot{\mathbf{r}})^T \bullet \hat{\mathbf{u}}f \tag{2.149}$$
$$\dot{k} = (\partial k/\partial \dot{\mathbf{r}})^T \bullet \hat{\mathbf{u}}f \tag{2.150}$$
$$\dot{p} = (\partial p/\partial \dot{\mathbf{r}})^T \bullet \hat{\mathbf{u}}f \tag{2.151}$$
$$\dot{q} = (\partial q/\partial \dot{\mathbf{r}})^T \bullet \hat{\mathbf{u}}f \tag{2.152}$$
$$\dot{\lambda} = n + (\partial \lambda/\partial \dot{\mathbf{r}})^T \bullet \hat{\mathbf{u}}f \tag{2.153}$$

2.3.3 The Euler–Lagrange Differential Equations

The Euler–Lagrange or adjoint differential equations for the set a, h, k, p, q and λ will now be derived. The necessary condition for optimality consists of maximizing the Hamiltonian of the system given by

$$H = \boldsymbol{\lambda}_z^T \dot{\mathbf{z}} = \boldsymbol{\lambda}_z^T M(\mathbf{z}, F) f\hat{\mathbf{u}} + \lambda_\lambda n \tag{2.154}$$

where the seventh differential equation for the mass flow rate has been neglected. The Euler–Lagrange equations are then given by

$$\dot{\boldsymbol{\lambda}}_z = -\frac{\partial H}{\partial \mathbf{z}} = -\boldsymbol{\lambda}_z^T \frac{\partial M}{\partial \mathbf{z}} f\hat{\mathbf{u}} - \lambda_\lambda \frac{\partial n}{\partial \mathbf{z}} \tag{2.155}$$

The optimal thrust direction $\hat{\mathbf{u}}$ is chosen such that it is at all times parallel to $\boldsymbol{\lambda}_z^T M(\mathbf{z}, F)$ in order to maximize the Hamiltonian. The matrix M for the set (a, h, k, p, q, λ) is as given by Equation (2.147). The complete set of the partials $\partial M/\partial \mathbf{z}$ is shown in Appendix 3.1 of Chapter 3. These analytic partials are used in Equation (2.155) for the numerical integration of the $\dot{\boldsymbol{\lambda}}_z$ rates. The only nonzero $\partial n/\partial \mathbf{z}$ partial is given by

$$\frac{\partial n}{\partial a} = -\frac{3n}{2a} = -\frac{3}{2}\mu^{1/2} a^{-5/2}$$

The following partials are used to generate the $\partial M/\partial \mathbf{z}$ partials of Appendix 3.1. Here, F must be considered to be independent of a but not of h, k and λ because λ is being integrated and is therefore independent of a, h, k, p and q such that $\partial\lambda/\partial a$ is now equal to zero. From Kepler's equation, this implies that $\partial F/\partial a = 0$. Thus,

$$\begin{aligned}
\frac{\partial r}{\partial a} &= \frac{r}{a} \\
\frac{\partial r}{\partial h} &= \frac{a^2}{r}(h - s_F) \\
\frac{\partial r}{\partial k} &= \frac{a^2}{r}(k - c_F) \\
\frac{\partial r}{\partial F} &= a(k s_F - h c_F) \\
\frac{\partial F}{\partial a} &= 0 \\
\frac{\partial F}{\partial h} &= -\frac{a}{r} c_F \\
\frac{\partial F}{\partial k} &= \frac{a}{r} s_F \\
\frac{\partial F}{\partial \lambda} &= \frac{a}{r}
\end{aligned}$$

It is also true that $\partial\lambda/\partial F = r/a$. The components u_f, u_g, u_w of the unit vector $\hat{\mathbf{u}}$ in the equinoctial frame are obtained from

$$\hat{\mathbf{u}} = \left(\boldsymbol{\lambda}_z^T M\right)^T / \left|\boldsymbol{\lambda}_z^T M\right| \tag{2.156}$$

From Figure 2.7, the thrust vector $\mathbf{T} = T\hat{\mathbf{u}}$ is defined by the thrust pitch and thrust yaw angles θ_t and θ_h respectively, in the rotating $\hat{\mathbf{r}}$, $\hat{\boldsymbol{\theta}}$, $\hat{\mathbf{h}}$ frame; $\hat{\mathbf{r}}$ is a unit vector along the instantaneous position vector $\mathbf{r}$, with $\hat{\boldsymbol{\theta}}$ in the orbit plane and along the direction of motion, and $\hat{\mathbf{h}}$ along the angular momentum vector. Therefore,

$$\theta_t = \tan^{-1}(u_r/u_\theta) \tag{2.157}$$

$$\theta_h = \tan^{-1}(u_w/u_\theta) \tag{2.158}$$

with u_r, u_θ and u_h, representing the components of the unit vector $\hat{\mathbf{u}}$ along the $\hat{\mathbf{r}}$, $\hat{\boldsymbol{\theta}}$ and $\hat{\mathbf{h}}$ directions. We have

$$\hat{\mathbf{r}} = \frac{X_1}{|\mathbf{r}|}\hat{\mathbf{f}} + \frac{Y_1}{|\mathbf{r}|}\hat{\mathbf{g}} \tag{2.159}$$

$$\hat{\boldsymbol{\theta}} = \hat{\mathbf{h}} \times \hat{\mathbf{r}} = -\frac{Y_1}{|\mathbf{r}|}\hat{\mathbf{f}} + \frac{X_1}{|\mathbf{r}|}\hat{\mathbf{g}} \tag{2.160}$$

such that

$$u_r = \frac{X_1}{r}u_f + \frac{Y_1}{r}u_g \tag{2.161}$$

$$u_\theta = -\frac{Y_1}{r}u_f + \frac{X_1}{r}u_g \tag{2.162}$$

$$u_h = u_w \tag{2.163}$$

where

$$r = |\mathbf{r}| = \left(X_1^2 + Y_1^2\right)^{1/2}$$

2.4 Orbit Transfer with Continuous Constant Acceleration

This section presents examples of precision integrated, optimized low Earth orbit (LEO) to geostationary Earth orbit (GEO) minimum-time transfer and compares them to the solutions obtained by way of the averaging technique. A 10^{-2} g acceleration applied in a constant and continuous manner is taken as an example in order to generate fast sub-day transfers that could be flown with nuclear thermal propulsion upper stages. The six-state formulation used here allows the user to generate optimal transfers that first start from a given fixed location on the initial orbit while optimizing the arrival point on the target or final orbit. The analysis is further extended to optimize both departure and arrival points in order to obtain the overall minimum-time free–free solution. This requires the vanishing of the Lagrange multiplier adjoint to the mean longitude at both initial and final times with fixed initial time and optimized final time. These fast, few-revolution, five-state transfers are sensitive to initial and final orbital position, thereby necessitating the use of the full six-state dynamics. These exact results are then compared to the approximate solutions obtained using averaged dynamics with robust and fast convergence characteristics. These examples determine that the ΔVs or transfer time solutions compare rather well, even for these short-duration transfers, but that the element time histories, and especially the eccentricity, are poorly simulated by the approximate solutions. Furthermore, due to the nature of the averaging technique, the sensitivity of the solution to orbital position is totally removed such that the precision-integrated solution must be used instead for accurate guidance.

2.4.1 Minimum-Time Transfer from Fixed Initial State with Continuous Constant Acceleration

We minimize total transfer time by maximizing the performance index, $J = \int_{t_0}^{t_f} L dt = -\int_{t_0}^{t_f} dt = -\left(t_f - t_0\right)$. For fixed t_0, the minimization of t_f or maximization of $-t_f$ gives rise to the transversality condition $H_f = 1$ for $H = \boldsymbol{\lambda}_z^T \dot{\mathbf{z}}$ because $H_f = 0$ for the augmented Hamiltonian $H = -1 + \boldsymbol{\lambda}_z^T \dot{\mathbf{z}}$. We now use the full six-element formulation to solve five-state orbit transfer problems by first starting from a fixed initial state and optimizing the final arrival point on the terminal orbit. Given $(a)_0$, $(e)_0$, $(i)_0$, $(\Omega)_0$, $(\omega)_0$ and $(M)_0$ at time $t_0 = 0$ or, equivalently, $(a)_0$, $(h)_0$, $(k)_0$, $(p)_0$, $(q)_0$, $(\lambda)_0$, the initial values of the Lagrange multipliers, namely $(\lambda_a)_0$, $(\lambda_h)_0$, $(\lambda_k)_0$, $(\lambda_p)_0$, $(\lambda_q)_0$ and $(\lambda_\lambda)_0$ are guessed, and the dynamic equations $\dot{\mathbf{z}} = (\partial \mathbf{z}/\partial \dot{\mathbf{r}}) \bullet \hat{\mathbf{u}} f$ as well as the adjoint Equation (2.155) are integrated forward to the guessed transfer time t_f by using the optimal control $\hat{\mathbf{u}} = \left[\boldsymbol{\lambda}_z^T M(\mathbf{z}, F)\right]^T / \left|\boldsymbol{\lambda}_z^T M(\mathbf{z}, F)\right|$. An iterative scheme is used next in order to adjust the initial values of the six multipliers as well as t_f such that the five terminal state parameters, a_f, h_f, k_f, p_f and q_f are matched and $(\lambda_\lambda)_f = 0$, $H_f = 1$ are satisfied. This is done by minimizing the following objective function:

$$
\begin{aligned}
F' &= w_1 \left(a - a_f\right)^2 + w_2 \left(h - h_f\right)^2 + w_3 \left(k - k_f\right)^2 + w_4 \left(p - p_f\right)^2 \\
&\quad + w_5 \left(q - q_f\right)^2 + w_6 \left(\lambda_\lambda - \lambda_{\lambda_f}\right)^2 + w_7 \left(H - H_f\right)^2
\end{aligned}
$$

or

$$
F' = \sum_{i=1}^{5} w_i \left(z_i - z_{i_f}\right)^2 + w_6 \left(\lambda_\lambda - 0\right)^2 + w_7 \left(H - 1\right)^2
$$

with w_i standing for certain weights that can be adjusted in order to favor the rapid convergence of some elements relative to others and alleviate, to some extent, certain sensitivity and scaling problems associated with the use of a given optimizer. For this purpose, we make use of the minimization algorithm UNCMIN of [9], which is designed for the unconstrained minimization of a real-valued function $F'(\mathbf{x})$ of n variables denoted by the vector $\mathbf{x}$. This subroutine is based on a general descent method and uses a quasi-Newton algorithm. In the Newton method, the step $\mathbf{p}''$ is computed from the solution of a set of n linear equations known as the Newton equations:

$$
\nabla^2 F'(\mathbf{x})\, \mathbf{p}'' = -\nabla F'(\mathbf{x}) \tag{2.164}
$$

Therefore, the solution is updated by using

$$
\mathbf{x}_{k+1} = \mathbf{x}_k + \mathbf{p}'' = \mathbf{x}_k - \left[\nabla^2 F'(\mathbf{x}_k)\right]^{-1} \nabla F'(\mathbf{x}_k) \tag{2.165}
$$

Here, $\nabla F'(\mathbf{x})$ denotes the gradient of F' at $\mathbf{x}$, while $\nabla^2 F'(\mathbf{x})$, the constant matrix of the second partial derivatives of F' at $\mathbf{x}$, represents the Hessian matrix. The Newton direction given by $\mathbf{p}''$ is guaranteed to be a descent direction only if $\left[\nabla^2 F'\right]^{-1}$ is positive definite, i.e., $\mathbf{z}^T \left[\nabla^2 F'\right]^{-1} \mathbf{z} > 0$ for all $\mathbf{z} \neq 0$ because, in that case for small ε,

$$
\begin{aligned}
F'\left(\mathbf{x} + \varepsilon \mathbf{p}''\right) &= F'(\mathbf{x}) + \varepsilon \nabla F'^T \mathbf{p}'' + O\left(\varepsilon^2\right) \\
&= F'(\mathbf{x}) - \varepsilon \nabla F'^T \left[\nabla^2 F'\right]^{-1} \nabla F' + O\left(\varepsilon^2\right)
\end{aligned} \tag{2.166}
$$

such that $F'\left(\mathbf{x} + \varepsilon \mathbf{p}''\right) < F'(\mathbf{x})$. This is equivalent to requiring the linear term in Newton's quadratic approximation for the function F' to be negative. This is, of course, the second term in the Taylor series expansion for F' at $\mathbf{x} + \mathbf{p}''$. The algorithm UNCMIN builds a secant approximation B_k to the Hessian as the function is being minimized such that, at $\mathbf{x}_k$,

$$B_k \mathbf{p}'' = -\nabla F'_k \tag{2.167}$$

with the matrix B_k positive definite because there is no guarantee that $\left[\nabla^2 F'\right]^{-1}$ will always be positive definite for each $\mathbf{x}$. Once the descent direction is established, a line search on α' is established such that $F'\left(\mathbf{x}_k + \alpha'\mathbf{p}''\right) < F'_k$ and the solution updated via $\mathbf{x}_{k+1} = \mathbf{x}_k + \alpha'\mathbf{p}''$. The approximate Hessian B_k is updated next using $\mathbf{x}_{k+1}$ and the gradient $\nabla F'\left(\mathbf{x}_{k+1}\right)$. For example, if F' is quadratic [9],

$$B_{k+1}\left(\mathbf{x}_{k+1} - \mathbf{x}_k\right) = \nabla F'_{k+1} - \nabla F'_k \tag{2.168}$$

Only the gradient values are needed for the update of the approximate Hessian, which is achieved by finite differencing such that the user must provide only the function F' itself. Let the initial orbit be given by

$$a_0 = 7000\,\text{km}, \quad e_0 = 0, \quad i_0 = 28.5\,\text{deg},$$
$$\Omega_0 = 0\,\text{deg}, \quad \omega_0 = 0\,\text{deg}, \quad M_0 = -220\,\text{deg}$$

and let $f = 9.8 \times 10^{-5}\,\text{km/s}^2$ or roughly $10^{-2}\,g$. The final orbit is given by

$$a_f = 42{,}000\,\text{km}, \quad e_f = 10^{-3}, \quad i_f = 1\,\text{deg}$$
$$\Omega_f = 0\,\text{deg}, \quad \omega_f = 0\,\text{deg}$$

with M_f free. The following solution is obtained:

$$\begin{aligned}
(\lambda_a)_0 &= 1.253187761\,\text{s/km} \\
(\lambda_h)_0 &= 3.809805448 \times 10^2\,\text{s} \\
(\lambda_k)_0 &= -9.407287686 \times 10^3\,\text{s} \\
\left(\lambda_p\right)_0 &= -2.118635388 \times 10^3\,\text{s} \\
\left(\lambda_q\right)_0 &= -1.747000680 \times 10^4\,\text{s} \\
(\lambda_\lambda)_0 &= 5.232351705 \times 10^2\,\text{s/rad} \\
t_f &= 58{,}623.992\,\text{s}
\end{aligned}$$

The achieved orbit parameters at $t = t_f$ are

$$a_f = 42{,}000.007\,\text{km}, \quad e_f = 1.00025 \times 10^{-3},$$
$$i_f = 1.000010\,\text{deg}, \quad \Omega f = 359.999909\,\text{deg},$$
$$\omega_f = 1.970582 \times 10^{-2}\,\text{deg}, \quad M_f = 43.984113\,\text{deg}$$

with

$$H_f = 1.002755, \qquad (\lambda_\lambda)_f = -7.761098 \times 10^{-3}\,\text{s/rad}.$$

The value of $\lambda_f = 19.617565$ rad is such that with $\lambda_0 = -3.839724$ rad a total travel of $\Delta\lambda/(2\pi) = \left(\lambda_f - \lambda_0\right)/(2\pi) = 23.457289/(2\pi) = 3.733$ revolutions around the Earth is accomplished. The initial values of the equinoctial elements are given by

$$a_0 = 7000\,\text{km}, \quad h_0 = 0, \quad k_0 = 0, \quad p_0 = 0,$$
$$q_0 = 0.2539676465, \quad \lambda_0 = -3.839724354\,\text{rad}$$

Figure 2.8 shows the variations of the classical elements a and e as a function of time during the 16.284442 h transfer. The eccentricity reaches a peak of around 0.4 with most of the orbit rotation taking place in the final 2 h of the transfer. Figure 2.9 depicts

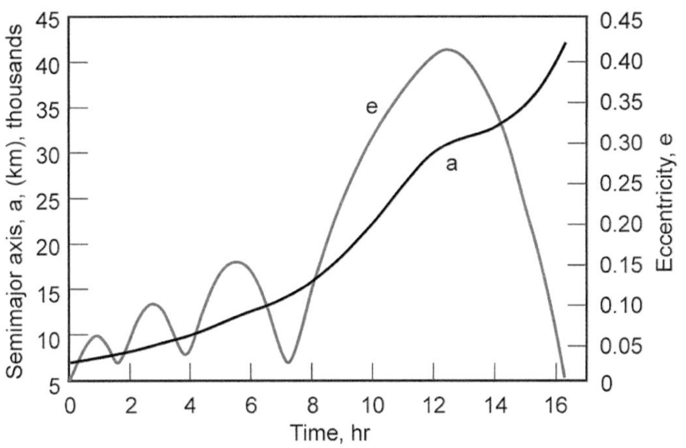

Figure 2.8 Variation of semimajor axis and eccentricity for a LEO to a near-GEO transfer with initial $M_0 = -220$ deg.
Source: [10].

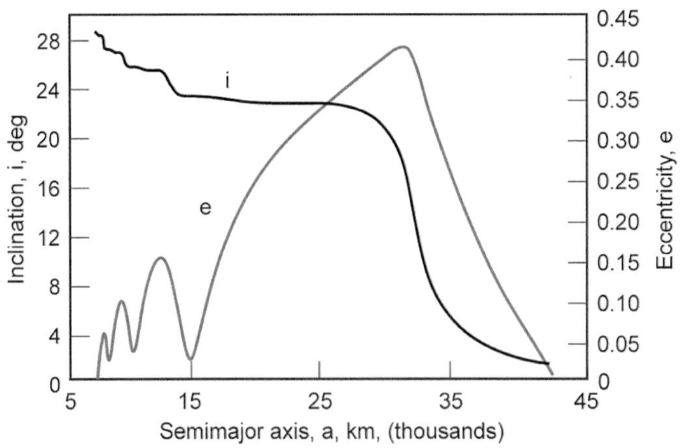

Figure 2.9 Eccentricity and inclination versus semimajor axis for a LEO to a near-GEO transfer with initial $M_0 = -220$ deg.
Source: [10].

e and i as a function of the semimajor axis which is monotonically increasing, as shown in Figure 2.8. The final position given by λ_f or M_f with $M_f = 43.984113$ deg is the optimal arrival point on the target orbit that results in the minimum-time transfer for the given or fixed initial $M_0 = -220$ deg. The total ΔV is given by

$$\Delta V = f t_f = 5.745151 \text{ km/s}$$

We now solve the same transfer problem except that $M_0 = -300$ deg instead of -220 deg on the initial orbit at LEO. This will show whether or not there exists any sensitivity related to the departure location on the initial orbit. The solution is now given by

$$\begin{aligned}
(\lambda_a)_0 &= 7.130443217 \text{ s/km} \\
(\lambda_h)_0 &= -6.614582933 \times 10^2 \text{ s} \\
(\lambda_k)_0 &= 9.143744201 \times 10^3 \text{ s} \\
(\lambda_p)_0 &= 1.526341606 \times 10^3 \text{ s} \\
(\lambda_q)_0 &= -2.416719191 \times 10^4 \text{ s} \\
(\lambda_\lambda)_0 &= -3.774461310 \times 10^2 \text{ s/rad} \\
t_f &= 58158.819 \text{ s}
\end{aligned}$$

This transfer requires a flight time of some 465.173 s less than the previous transfer or some 7.75 min. The achieved parameters are

$$\begin{aligned}
a_f &= 42000.012 \text{ km}, & e_f &= 9.9972 \times 10^{-4}, \\
i_f &= 1.000017 \text{ deg}, & \Omega_f &= 359.999324 \text{ deg}, \\
\omega_f &= 359.975014 \text{ deg}, & M_f &= 227.221692 \text{ deg},
\end{aligned}$$

with

$$\begin{aligned}
H_f &= 1.000312805 \\
(\lambda_\lambda)_f &= -8.895998 \times 10^{-3} \text{ s/rad}
\end{aligned}$$

and

$$\begin{aligned}
\lambda_0 &= -5.235987 \text{ rad} \\
\lambda_f &= 16.531689 \text{ rad}
\end{aligned}$$

such that the total angular travel consists of $(\lambda_f - \lambda_0)/(2\pi) = 21.767676/(2\pi) = 3.464$ revolutions around the Earth. Figure 2.10 shows the variations of the classical orbit elements a and e during the optimal transfer with smaller oscillation amplitudes in the eccentricity during the first revolutions. Figure 2.11 shows e and i as a function of the semimajor axis with essentially linear buildup and decay for e. The total ΔV is given by

$$\Delta V = f t_f = 5.699564 \text{ km/s}$$

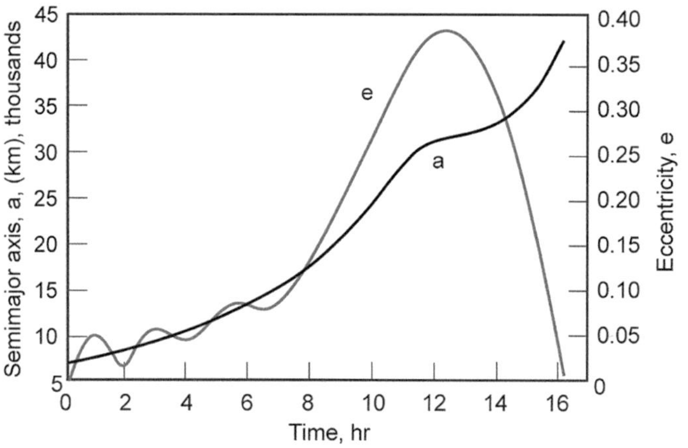

Figure 2.10 Time history of semimajor axis and eccentricity for a LEO to a near-GEO transfer with initial $M_0 = -300$ deg.
Source: [10].

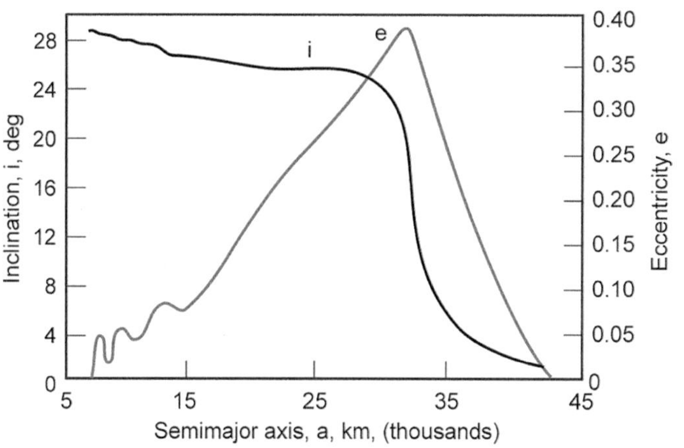

Figure 2.11 Eccentricity and inclination versus semimajor axis for a LEO to a near-GEO transfer with initial $M_0 = -300$ deg.
Source: [10].

2.4.2 Minimum-Time Transfer with Optimized Departure and Arrival Locations

In this problem, we are given $(a)_0$, $(e)_0$, $(i)_0$, $(\Omega)_0$, $(\omega)_0$, or equivalently $(a)_0$, $(h)_0$, $(k)_0$, $(p)_0$ and $(q)_0$. The guessed parameters are $(\lambda_a)_0$, $(\lambda_h)_0$, $(\lambda_k)_0$, $(\lambda_p)_0$, $(\lambda_q)_0$ and the initial mean longitude $(\lambda)_0$ as well as the transfer time t_f. The precision integration of the dynamic and adjoint systems in Equations (2.148) through (2.153) and Equation (2.155) is carried out using $(\lambda_\lambda)_0 = 0$.

The boundary conditions at the unknown final time are given by a_f, h_f, k_f, p_f, q_f, $(\lambda_\lambda)_f = 0$, and $H_f = 1$ for a minimum-time solution. The optimal thrust direction is still given by

$$\hat{\mathbf{u}} = \left[\boldsymbol{\lambda}_z^T M(\mathbf{z}, F)\right]^T / \left|\boldsymbol{\lambda}_z^T M(\mathbf{z}, F)\right|$$

The objective function to minimize is still given by F' above. We have essentially traded $(\lambda)_0$, given with $(\lambda_\lambda)_0 = 0$, because small variations in $(\lambda)_0$ must be such that the performance index $J = -t_f$ is stationary, indicating zero sensitivity to initial location. The same argument still holds true for the arrival point on the target orbit. The solution of the optimal free–free transfer is given by

$$\begin{aligned}(\lambda_a)_0 &= 4.853916817\,\mathrm{s/km}\\ (\lambda_h)_0 &= 5.621721342 \times 10^2\ \mathrm{s}\\ (\lambda_k)_0 &= -9.554222267 \times 10^3\ \mathrm{s}\\ (\lambda_p)_0 &= 1.904458303 \times 10^1\ \mathrm{s}\\ (\lambda_q)_0 &= -2.344660077 \times 10^4\ \mathrm{s}\\ (\lambda)_0 &= -2.274699655\,\mathrm{rad}\end{aligned}$$

corresponding to $M_0 = -130.330689$ deg, and the overall minimum-time $t_f = 58{,}089.469$ s, with corresponding $\Delta V = 5.692767$ km/s. The total angular travel is obtained from $(\lambda)_0$ and $(\lambda)_f = 19.65526752$ rad or 3.4899 revolutions around the Earth. The final achieved parameters are

$$\begin{aligned}a_f &= 41{,}999.979\,\mathrm{km}\\ e_f &= 1.00757 \times 10^{-3}\\ i_f &= 1.000962\,\mathrm{deg}\\ \Omega_f &= 0.025218\,\mathrm{deg}\\ \omega_f &= 0.026815\,\mathrm{deg}\\ M_f &= 46.111840\,\mathrm{deg}\end{aligned}$$

with

$$H_f = 1.042050399$$

and

$$(\lambda_\lambda)_f = -5.1171240 \times 10^{-3}\ \mathrm{s/rad}$$

The Hamiltonian is constant because the dynamic equations are not explicit functions of time. Furthermore, we can multiply all the initial values of the multipliers by a common constant factor in order to get $H = 1$ exactly, if we so desire. The solution is of course unchanged because of the common scaling used. This also means that we can arbitrarily select one of the λ and reduce the order of the integration by 1 because H, a positive constant, is in effect a first integral of the motion. The initial values of the multipliers that scale the Hamiltonian to the unit value are

$$\begin{aligned}(\lambda_a)_0 &= 4.658044200\,\mathrm{s/km}\\ (\lambda_h)_0 &= 5.394865112 \times 10^2\ \mathrm{s}\\ (\lambda_k)_0 &= -9.168675792\,\mathrm{s}\end{aligned}$$

$$(\lambda_p)_0 = 1.827606712 \times 10^1 \text{ s}$$
$$(\lambda_q)_0 = -2.250044796 \times 10^4 \text{ s}$$

Figure 2.12 shows the variations of e and i as a function of the semimajor axis. Most of the inclination change is taking place at or near maximum eccentricity. The thrust pitch and yaw programs are shown in Figure 2.13. The yaw profile, as expected, changes sign every one-half revolution during the first three orbits, rotating the orbit slowly, but the pitch angle stays near zero for maximum energy buildup. The sharp buildup in the yaw angle is responsible for most of the orbit rotation, which takes place as soon as eccentricity reaches its maximum value, as observed above. Figure 2.14 shows r versus θ, the angular position with $\theta = \omega + \theta^*$. As soon as the orbit reaches the proper energy level, it is able then to transfer directly to the GEO altitude on a highly eccentric orbit. The last portion of the trajectory consists of a circularization maneuver in order to cancel the eccentricity and enter the GEO orbit. The pitch angle θ_t becomes negative during the last portion of this phase as shown in Figure 2.13 for final eccentricity control. The θ_t is such that $-90\,\text{deg} < \theta_t < 90\,\text{deg}$ at all times, indicating that the thrust is never applied in a breaking mode. This is not always the case, especially for larger inclination changes or larger orbit transfers. As another example, the solution for the optimal transfer from

$$a_0 = 7000\,\text{km}, \quad e_0 = 0, \quad i_0 = 28.5\,\text{deg},$$
$$\Omega_0 = 0\,\text{deg}, \quad \omega_0 = 0\,\text{deg}$$

to a circular equatorial orbit given by

$$a_f = 42{,}000\,\text{km}, \quad e_f = 0, \quad i_f = 0\,\text{deg},$$
$$\Omega_f = 0\,\text{deg}, \quad \omega_f = 0\,\text{deg}$$

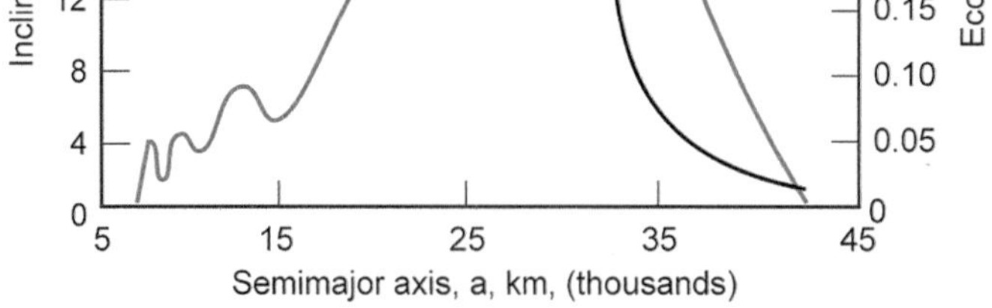

Figure 2.12 Eccentricity and inclination versus semimajor axis for absolute minimum-time LEO to near-GEO transfer.

Source: [10].

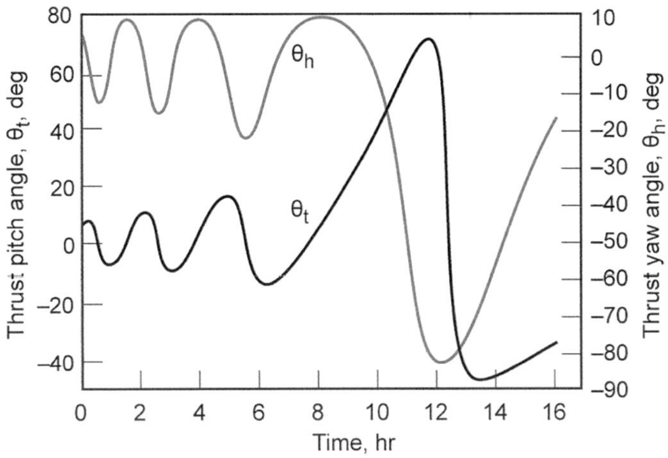

Figure 2.13 Optimal thrust pitch and yaw programs for absolute minimum-time LEO to near-GEO transfer.
Source: [10].

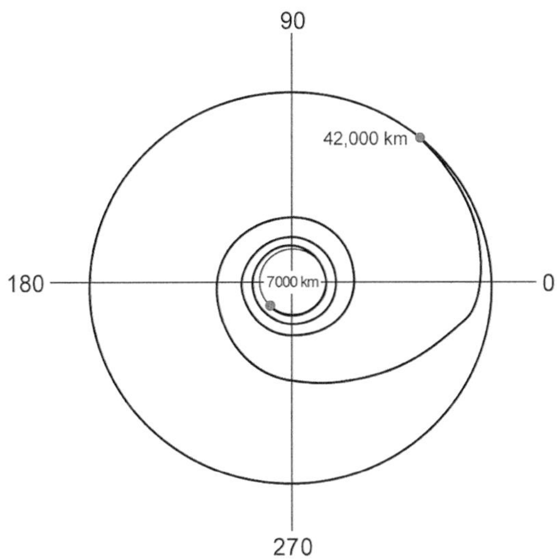

Figure 2.14 Polar plot of absolute minimum-time LEO to near-GEO transfer.
Source: [10].

with $f = 9.8 \times 10^{-5}$ km/s^2 is given by

$$
\begin{aligned}
(\lambda_a)_0 &= 4.841918234\,\text{s/km} \\
(\lambda_h)_0 &= 5.966487872 \times 10^2\,\text{s} \\
(\lambda_k)_0 &= -9.728162329 \times 10^3\,\text{s} \\
(\lambda_p)_0 &= -4.967618705\,\text{s}
\end{aligned}
$$

$$(\lambda_q)_0 = -2.518203349 \times 10^4 \text{ s}$$
$$(\lambda)_0 = -2.332590176 \text{ rad corresponding to } M_0 = -133.647572 \text{ deg}$$

with

$$t_f = 58{,}324.803 \text{ s with corresponding } \Delta V = 5.715830 \text{ km/s}$$

The first example used in Section 2.2.3 with Edelbaum's theory consisted of a transfer from $a_0 = 7000$ km, $i_0 = 28.5$ deg to $a_f = 42{,}166$ km, $i_f = 0$ deg with a low acceleration of $f = 3.5 \times 10^{-7}$ km/s^2. The optimal transfer for that same example with

$$e_0 = 0, \quad \Omega_0 = 0 \text{ deg}, \quad \omega_0 = 0 \text{ deg}, \quad e_f = 0$$
$$\Omega_f = 0 \text{ deg}, \quad \omega_f = 0 \text{ deg} \quad \text{and} \quad f = 9.8 \times 10^{-5} \text{ km/s}^2$$

is given by

$$(\lambda_a)_0 = 4.596409022 \text{ s/km}$$
$$(\lambda_h)_0 = 5.516332308 \times 10^2 \text{ s}$$
$$(\lambda_k)_0 = -9.212474616 \times 10^3 \text{ s}$$
$$(\lambda_p)_0 = -1.856853566 \times 10^2 \text{ s}$$
$$(\lambda_q)_0 = -2.358291542 \times 10^4 \text{ s}$$
$$(\lambda)_0 = -2.330144473 \text{ rad corresponding to } M_0 = -133.507443 \text{ deg}$$
$$t_f = 58{,}413.638 \text{ s with corresponding } \Delta V = 5.724536 \text{ km/s.}$$

The ΔV_{tot} from Edelbaum's formula, being independent of the acceleration level, yielded $\Delta V_{\text{tot}} = 5.77584$ km/s or some 50 m/s more than the exact solution, which is quite accurate even for this relatively high-acceleration example.

2.4.3 Continuous Constant Acceleration Orbit Transfer Using Averaging

Following [6], an averaged Hamiltonian $\tilde{H}$ is formed first, from which a first-order approximation to the state and costate is derived by holding these quantities constant over the averaging interval and varying only F, the eccentric longitude on a conic orbit.

$$H = \frac{1}{T_0}\int_0^{T_0} H dt = \frac{1}{T_0}\int_{-\pi}^{\pi} \frac{H dt}{\dot{F}(\tilde{\mathbf{z}}, F)} \tag{2.169}$$

where T_0, the orbital period at time t, is given by $T_0 = 2\pi/\tilde{n}$, with $\tilde{n} = \mu^{1/2}\,\tilde{a}^{-3/2}$ and $\tilde{a}$ indicating the averaged value of the semimajor axis at time t, and so forth. From Kepler's equation, $\tilde{\lambda} = F - \tilde{k}s_F + \tilde{h}c_F = \tilde{n}t$, we have

$$\dot{F} = \frac{2\pi}{T_0\left(1 - \tilde{k}c_F - \tilde{h}s_F\right)} \tag{2.170}$$

such that, with $s = (1/T_0)/\dot{F} = (1/(2\pi))(1 - \tilde{k}c_F - \tilde{h}s_F)$ the equations for the approximation for the state and costate variables are given by [6]:

$$\dot{\tilde{\mathbf{z}}} = \left(\frac{\partial \tilde{H}}{\partial \tilde{\boldsymbol{\lambda}}_z}\right)^T = \int_{-\pi}^{\pi} \left(\frac{\partial H}{\partial \tilde{\boldsymbol{\lambda}}_z}\right)^T s\left(\tilde{\mathbf{z}}, F\right) dF \tag{2.171}$$

$$\dot{\tilde{\boldsymbol{\lambda}}}_z = -\left(\frac{\partial \tilde{H}}{\partial \tilde{\mathbf{z}}}\right)^T = -\int_{-\pi}^{\pi} \left[\left(\frac{\partial H}{\partial \tilde{\mathbf{z}}}\right)^T s\left(\tilde{\mathbf{z}}, F\right) + H\left(\frac{\partial s}{\partial \tilde{\mathbf{z}}}\right)^T\right] \mathrm{d}F \tag{2.172}$$

The partials $\partial s/\partial \tilde{\mathbf{z}}$ are given by

$$\frac{\partial s}{\partial \tilde{a}} = 0, \frac{\partial s}{\partial \tilde{h}} = -\frac{s_F}{2\pi}, \frac{\partial s}{\partial \tilde{k}} = -\frac{c_F}{2\pi}, \frac{\partial s}{\partial \tilde{p}} = 0, \frac{\partial s}{\partial \tilde{q}} = 0$$

and

$$\frac{\partial s}{\partial \tilde{\lambda}} = \frac{\left(\tilde{k}s_F - \tilde{h}c_F\right)}{\left(1 - \tilde{k}c_F - \tilde{h}s_F\right)2\pi}$$

because

$$\frac{\partial s}{\partial \tilde{\lambda}} = \left(\frac{\partial s}{\partial F}\right)\left(\frac{\partial F}{\partial \tilde{\lambda}}\right)$$

and

$$\tilde{\lambda} = F - \tilde{k}s_F + \tilde{h}c_F$$

The optimal trajectory is such that H is constant throughout. This means that, when thrust is applied continuously during every complete revolution, the contribution from the $H\left(\partial s/\partial \tilde{\mathbf{z}}\right)^T$ term is equal to zero because the $\partial s/\partial \tilde{\mathbf{z}}$ terms will give a zero net contribution. For a continuous constant acceleration f, the averaged equations are given by

$$\dot{\tilde{a}} = \frac{1}{2\pi}f \int_{-\pi}^{\pi} \left(\frac{\partial a}{\partial \dot{\mathbf{r}}}\right)^T \bullet \hat{\mathbf{u}} \left(1 - \tilde{k}c_F - \tilde{h}s_F\right) dF \tag{2.173}$$

$$\dot{\tilde{h}} = \frac{1}{2\pi}f \int_{-\pi}^{\pi} \left(\frac{\partial h}{\partial \dot{\mathbf{r}}}\right)^T \bullet \hat{\mathbf{u}} \left(1 - \tilde{k}c_F - \tilde{h}s_F\right) dF \tag{2.174}$$

$$\dot{\tilde{k}} = \frac{1}{2\pi}f \int_{-\pi}^{\pi} \left(\frac{\partial k}{\partial \dot{\mathbf{r}}}\right)^T \bullet \hat{\mathbf{u}} \left(1 - \tilde{k}c_F - \tilde{h}s_F\right) dF \tag{2.175}$$

$$\dot{\tilde{p}} = \frac{1}{2\pi}f \int_{-\pi}^{\pi} \left(\frac{\partial p}{\partial \dot{\mathbf{r}}}\right)^T \bullet \hat{\mathbf{u}} \left(1 - \tilde{k}c_F - \tilde{h}s_F\right) dF \tag{2.176}$$

$$\dot{\tilde{q}} = \frac{1}{2\pi}f \int_{-\pi}^{\pi} \left(\frac{\partial q}{\partial \dot{\mathbf{r}}}\right)^T \bullet \hat{\mathbf{u}} \left(1 - \tilde{k}c_F - \tilde{h}s_F\right) dF \tag{2.177}$$

$$\begin{aligned} \dot{\tilde{\lambda}} &= \frac{1}{2\pi}f \int_{-\pi}^{\pi} \left(\frac{\partial \lambda}{\partial \dot{\mathbf{r}}}\right)^T \bullet \hat{\mathbf{u}} \left(1 - \tilde{k}c_F - \tilde{h}s_F\right) dF \\ &+ \frac{1}{2\pi} \int_{-\pi}^{\pi} \tilde{n} \left(1 - \tilde{k}c_F - \tilde{h}s_F\right) dF \end{aligned} \tag{2.178}$$

$$\dot{\tilde{\lambda}}_a = \frac{1}{2\pi} f \int_{-\pi}^{\pi} -\tilde{\boldsymbol{\lambda}}_z^T \frac{\partial M}{\partial \tilde{a}} \bullet \hat{\mathbf{u}} \left(1 - \tilde{k} c_F - \tilde{h} s_F\right) dF + \frac{1}{2\pi} \int_{-\pi}^{\pi} -\tilde{\lambda}_\lambda \frac{\partial \tilde{n}}{\partial \tilde{a}} \left(1 - \tilde{k} c_F - \tilde{h} s_F\right) dF \tag{2.179}$$

$$\dot{\tilde{\lambda}}_h = \frac{1}{2\pi} f \int_{-\pi}^{\pi} -\tilde{\boldsymbol{\lambda}}_z^T \frac{\partial M}{\partial \tilde{h}} \bullet \hat{\mathbf{u}} \left(1 - \tilde{k} c_F - \tilde{h} s_F\right) dF \tag{2.180}$$

$$\dot{\tilde{\lambda}}_k = \frac{1}{2\pi} f \int_{-\pi}^{\pi} -\tilde{\boldsymbol{\lambda}}_z^T \frac{\partial M}{\partial \tilde{k}} \bullet \hat{\mathbf{u}} \left(1 - \tilde{k} c_F - \tilde{h} s_F\right) dF \tag{2.181}$$

$$\dot{\tilde{\lambda}}_p = \frac{1}{2\pi} f \int_{-\pi}^{\pi} -\tilde{\boldsymbol{\lambda}}_z^T \frac{\partial M}{\partial \tilde{p}} \bullet \hat{\mathbf{u}} \left(1 - \tilde{k} c_F - \tilde{h} s_F\right) dF \tag{2.182}$$

$$\dot{\tilde{\lambda}}_q = \frac{1}{2\pi} f \int_{-\pi}^{\pi} -\tilde{\boldsymbol{\lambda}}_z^T \frac{\partial M}{\partial \tilde{q}} \bullet \hat{\mathbf{u}} \left(1 - \tilde{k} c_F - \tilde{h} s_F\right) dF \tag{2.183}$$

$$\dot{\tilde{\lambda}}_\lambda = \frac{1}{2\pi} f \int_{-\pi}^{\pi} -\tilde{\boldsymbol{\lambda}}_z^T \frac{\partial M}{\partial \tilde{\lambda}} \bullet \hat{\mathbf{u}} \left(1 - \tilde{k} c_F - \tilde{h} s_F\right) dF \tag{2.184}$$

The thrust direction $\hat{\mathbf{u}}$ is a function of $\tilde{\mathbf{z}}$ and F as well as $\tilde{\boldsymbol{\lambda}}_z$, and the second term in Equation (2.178) is equal to zero. An eighth-order Gauss–Legendre quadrature is used to evaluate the averaged rates given by Equations (2.173) through (2.184), which then are used by a variable step size integrator in order to integrate trajectories from given initial conditions, with step sizes spanning several orbital revolutions. The two-point boundary value problem is identical to the one used in Section 2.4.2, except that all variables as well as the Hamiltonian have averaged values. The averaged Hamiltonian is obtained from the integrated variables:

$$\tilde{H} = \tilde{\lambda}_a \, \dot{\tilde{a}} + \tilde{\lambda}_h \, \dot{\tilde{h}} + \tilde{\lambda}_k \, \dot{\tilde{k}} + \tilde{\lambda}_p \, \dot{\tilde{p}} + \tilde{\lambda}_q \, \dot{\tilde{q}} + \tilde{\lambda}_\lambda \, \dot{\tilde{\lambda}}$$

The same example is used with the optimized $(\lambda)_0$ or equivalently

$$(M)_0 = -130.330689 \text{ deg}$$

yielding the following solution given by

$$\left(\tilde{\lambda}_a\right)_0 = 0.597196796 \times 10^{-4} \text{ day/km}$$
$$\left(\tilde{\lambda}_h\right)_0 = -0.810808977 \times 10^{-11} \text{ day}$$
$$\left(\tilde{\lambda}_k\right)_0 = 0.502638066 \times 10^{-3} \text{ day}$$
$$\left(\tilde{\lambda}_p\right)_0 = -0.103393201 \times 10^{-11} \text{ day}$$
$$\left(\tilde{\lambda}_q\right)_0 = -0.967633761 \text{ day}$$
$$\left(\tilde{\lambda}_\lambda\right)_0 = 0 \text{ day/rad}$$
$$t_f = 0.6566499466 \text{ days (or 56,734.555 s)}$$

The achieved final orbit is given by

$$\tilde{a}_f = 42{,}000.000\,\text{km}, \qquad \tilde{e}_f = 9.999999 \times 10^{-4},$$
$$\tilde{i}_f = 1.000000\,\text{deg}, \qquad \tilde{\Omega}_f = 360.000000\,\text{deg},$$
$$\tilde{\omega}_f = 359.999998\,\text{deg}, \qquad \tilde{M}_f = 126.928280\,\text{deg}$$

with $\tilde{\lambda}_f = 21.064872\,\text{rad}$, $\left(\tilde{\lambda}_\lambda\right)_f = 0\,\text{day/rad}$ and $\tilde{H}_f = 1.000000166$, indicating a converged solution. Because $\left(\tilde{\lambda}\right)_0 = -2.274699\,\text{rad}$, the transfer requires $\left(\tilde{\lambda}_f - \tilde{\lambda}_0\right)/(2\pi) = 3.7146$ revolutions of the Earth. This information is available because we also have integrated the $\dot{\tilde{\lambda}}$ equation. The multiplier $\tilde{\lambda}_\lambda$ stays constant at zero throughout the transfer as it should, due to the averaging procedure, which eliminates all sensitivity to the position along the orbit. This is verified by starting from $\left(\tilde{\lambda}\right)_0 = 0\,\text{rad}$ or $\tilde{M}_0 = 0$ deg and generating the exact duplicate trajectory, except that $\left(\tilde{\lambda}\right)_f = 23.33957196\,\text{rad}$ or $\tilde{M}_f = 257.258952\,\text{deg}$ such that the transfer still requires 3.7146 revolutions of the Earth. This averaged transfer requires some 1354.914 s less time than the precision integrated exact transfer of the preceding section, which required $t_f = 58{,}089.469\,\text{s}$. Therefore, the averaged transfer solution is optimistic by some 22.581 min requiring a ΔV of 5.559986 km/s, which is rather accurate. In order to compare the various λ, we first must convert the units from days to seconds, such that

$$\left(\tilde{\lambda}_a\right)_0 = 5.159780317\,\text{s/km}$$
$$\left(\tilde{\lambda}_h\right)_0 = -7.00538956 \times 10^{-7}\,\text{s}$$
$$\left(\tilde{\lambda}_k\right)_0 = 4.342792890 \times 10^{1}\,\text{s}$$
$$\left(\tilde{\lambda}_p\right)_0 = -8.93317257 \times 10^{-8}\,\text{s}$$
$$\left(\tilde{\lambda}_q\right)_0 = -8.360355695 \times 10^{4}\,\text{s}$$
$$\left(\tilde{\lambda}_\lambda\right)_0 = 0\,\text{s/rad}$$

The exact and averaged plots for the semimajor axis, inclination and mean anomaly are shown in Figures 2.15, 2.16 and 2.17, respectively, while Figures 2.18 and 2.19 depict the equinoctial elements q and λ. On the other hand, the averaged classical elements $\tilde{\Omega}$ and $\tilde{\omega}$ stay at zero, while the eccentricity $\tilde{e}$ slowly increases from 0 to 10^{-3} at the final time. The $\tilde{\lambda}_a$ multiplier is shown in Figure 2.20 with its exact counterpart, while $\tilde{\lambda}_q$ is constant at -8.360354382×10^4 s, $\tilde{\lambda}_\lambda = 0$ throughout, $\tilde{\lambda}_h$ and $\tilde{\lambda}_p$ at essentially constant and zero, and $\tilde{\lambda}_k$ varying between 45 and -12 s, which is very small when compared to the exact λ_k curve, which oscillates at around 10^4 s and therefore essentially is negligibly small. The value of $\tilde{\lambda}_q$ is about four times larger than the exact solution, as is seen in Figure 2.21. Finally, Figures 2.22 and 2.23 show the exact and the averaged radial distance, r and $\tilde{r}$, and the exact and averaged angular position $\theta = \omega + \theta^*$, and $\tilde{\theta} = \tilde{\omega} + \tilde{\theta}^*$ with θ^* standing for the true anomaly. The effect of averaging is to

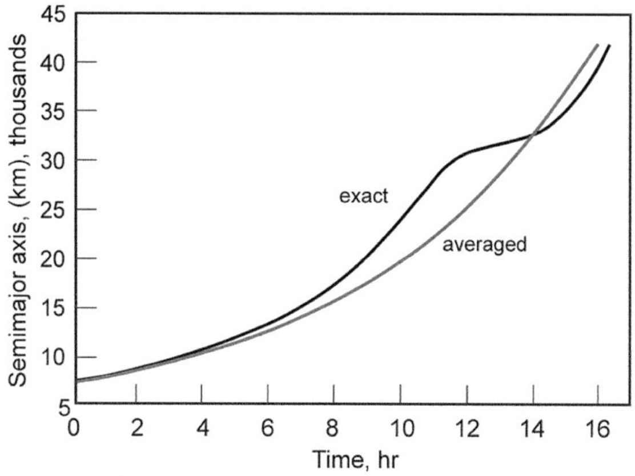

Figure 2.15 Evolution of exact and averaged solutions for the semimajor axis for a LEO to near-GEO transfer.
Source: [10].

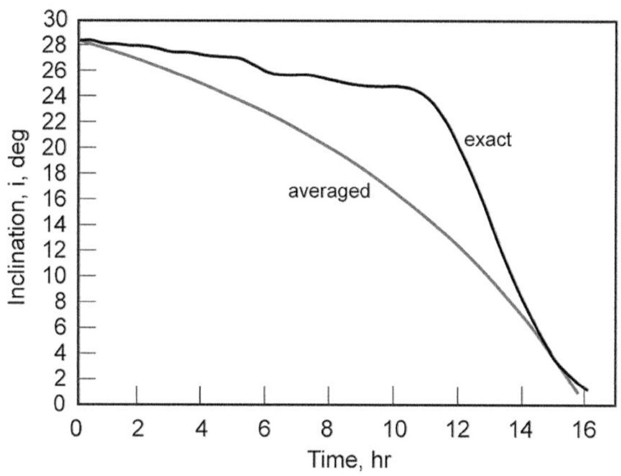

Figure 2.16 Evolution of exact and averaged solutions for the inclination for a LEO to near-GEO transfer.
Source: [10].

smooth out all the fluctuations seen in the exact curves. The main shortcoming of the averaging of the dynamics is the inability to describe the important eccentricity buildup seen in the exact solution. However, the ΔV evaluation or equivalent transfer time is quite accurate even for this difficult, short-duration, large-transfer problem. If we now use an acceleration f of 3.5×10^{-7} km/s^2, this averaged transfer will be achieved in the minimum time of 183.861985 days or $t_f = 1.588567551 \times 10^7$ s.

This corresponds to a ΔV of 5.559986 km/s, which is identical to the ΔV required by the fast transfer using averaging. This shows that the averaged transfer time is exactly inversely proportional to the acceleration, such that $\Delta V = ft_f$ remains constant,

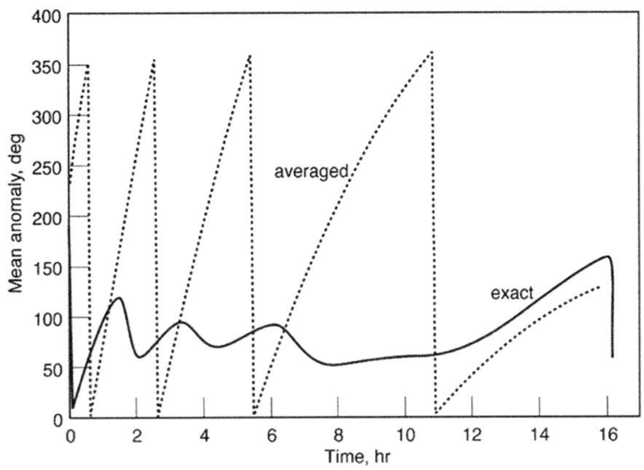

Figure 2.17 Exact and averaged mean anomaly.

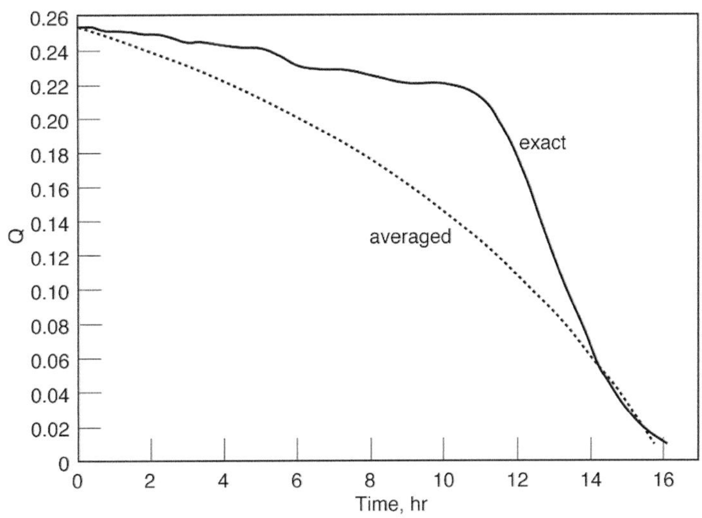

Figure 2.18 Exact and averaged equinoctial q.

regardless of the duration of the transfer. Going back to the example of Figure 2.1 in Section 2.2.3, consisting of a LEO to GEO transfer with

$$\tilde{a}_0 = 7000\,\text{km}, \quad \tilde{i}_0 = 28.5\,\text{deg}, \quad \tilde{e}_0 = 0, \quad \tilde{\Omega}_0 = 0\,\text{deg}$$
$$\tilde{\omega}_0 = 0\,\text{deg}, \quad \tilde{a}_f = 42{,}166\,\text{km}, \quad \tilde{e}_f = 0,$$
$$\tilde{i}_f = 0\,\text{deg} \quad \tilde{\Omega}_f = 0\,\text{deg} \quad \tilde{\omega}_f = 0\,\text{deg}$$

and using $f = 9.8 \times 10^{-5}\,\text{km/s}^2$, the optimal solution using averaging is given by

$$\left(\tilde{\lambda}_a\right)_0 = 0.595400888 \times 10^{-4}\,\text{day/km}$$
$$\left(\tilde{\lambda}_h\right)_0 = 0.205831846 \times 10^{-11}\,\text{day}$$

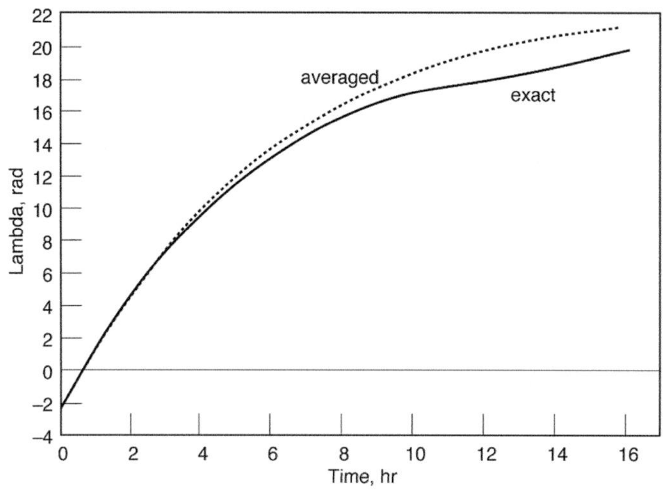

Figure 2.19 Exact and averaged mean longitude.

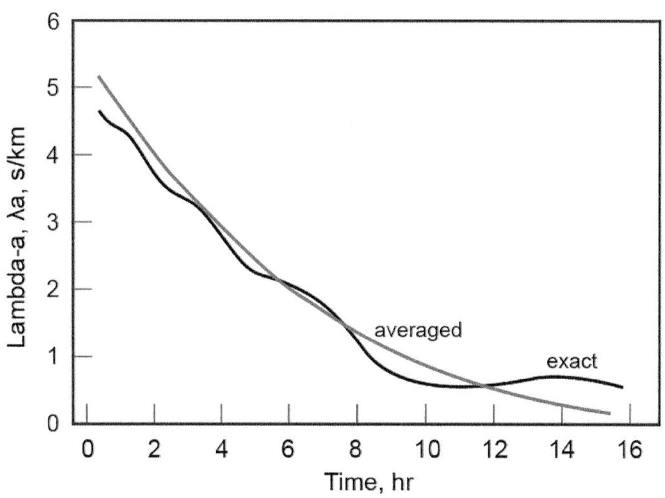

Figure 2.20 Comparison between exact and averaged solutions for λ_a for a LEO to near-GEO transfer.
Source: [10].

$$\left(\tilde{\lambda}_k\right)_0 = -0.129577414 \times 10^{-13}\,\text{day}$$
$$\left(\tilde{\lambda}_p\right)_0 = -0.111281493 \times 10^{-11}\,\text{day}$$
$$\left(\tilde{\lambda}_q\right)_0 = -0.986509406\,\text{day}$$
$$\left(\tilde{\lambda}_\lambda\right)_0 = 0\,\text{day/rad}$$
$$t_f = 57{,}505.17220\,\text{s}$$

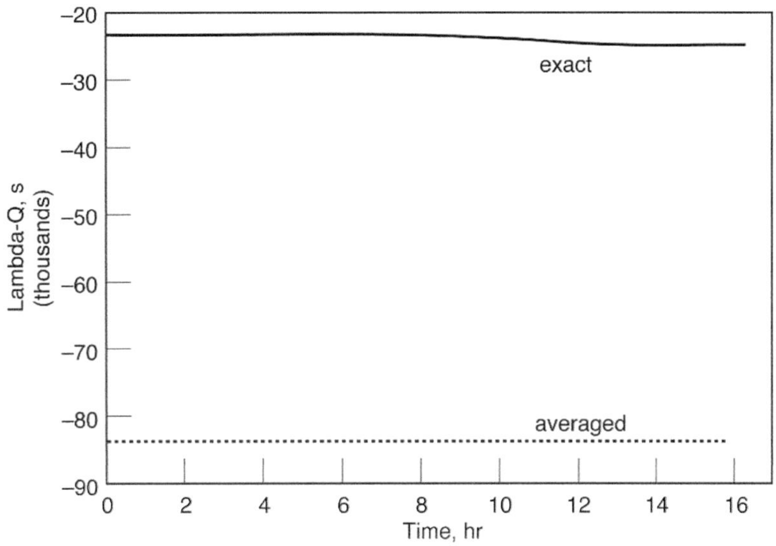

Figure 2.21 Exact and averaged λ_q time-history.

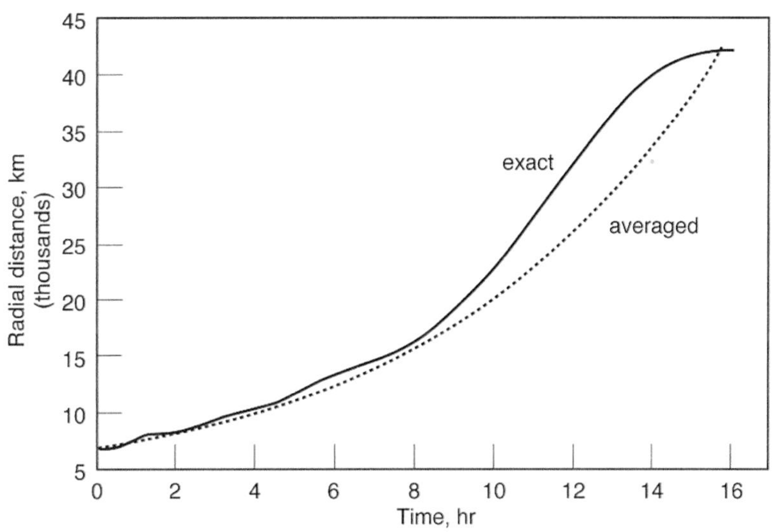

Figure 2.22 Exact and averaged radial distance.

with corresponding $\Delta V = 5.635506\,\text{km/s}$. The exact solution requires a ΔV of 5.724536 km/s and Edelbaum's $\Delta V_{\text{tot}} = 5.78378\,\text{km/s}$. The solution of this problem for $f = 3.5 \times 10^{-7}\,\text{km/s}^2$ with the present averaging method is given by

$$\left(\tilde{\lambda}_a\right)_0 = 0.166712237 \times 10^{-1}\,\text{day/km}$$

$$\left(\tilde{\lambda}_h\right)_0 = 0.109835994 \times 10^{-6}\,\text{day}$$

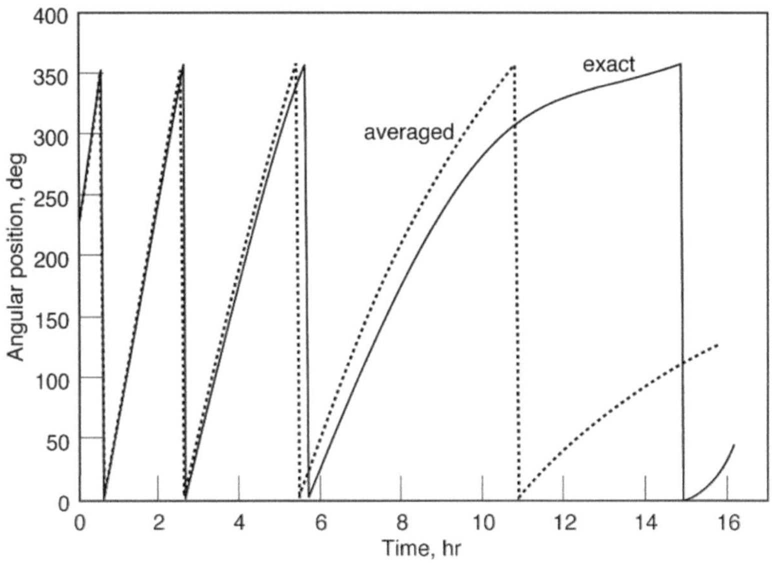

Figure 2.23 Exact and averaged angular position.

$$\left(\tilde{\lambda}_k\right)_0 = -0.269552075 \times 10^{-5} \text{ day}$$
$$\left(\tilde{\lambda}_p\right)_0 = 0.375375682 \times 10^{-6} \text{ day}$$
$$\left(\tilde{\lambda}_q\right)_0 = -2.762226146 \times 10^{2} \text{ day}$$
$$\left(\tilde{\lambda}_\lambda\right)_0 = 0 \text{ day/rad}$$
$$t_f = 1.610144782 \times 10^{7} \text{ s}$$

with corresponding $\Delta V = 5.635506$ km/s. This is only some 148 m/s less than Edelbaum's ΔV. The exact solution for this case would require very long integration times, and it has not been attempted. However, it is expected that the exact ΔV will be very close to the value of 5.635506 km/s because the transfer orbit will remain nearly circular throughout the exact transfer; this is the case with the averaged solution, too, which also uses a variable thrust yaw and pitch profile within each revolution during the ascent. Gravity losses are inherently more important when the orbit departs further from the circular shape, as is the case with the higher acceleration of $f = 9.8 \times 10^{-5}$ km/s^2. The averaged solution fails to account effectively for the gravity losses because it provides an optimistic ΔV with respect to the exact solution, whereas Edelbaum's ΔV_{tot} equation is overall more pessimistic, reflecting the losses associated with the use of a nonoptimal thrust profile with fixed yaw angle within each revolution. This more than compensates for its inherent optimistic characteristics since it assumes a perfectly circular orbit with no gravity losses.

References

[1] Edelbaum, T. N. (1961) Propulsion Requirements for Controllable Satellites, *ARS Journal*, Vol. 31, 1079–1089.

[2] Broucke, R. A. and Cefola, P. J. (1972) On the Equinoctial Orbit Elements, *Celestial Mechanics*, Vol. 5, 303–310.

[3] Herrick, S. (1972) *Astrodynamics*, Vol. II, Van Nostrand Reinhold, London.

[4] Plummer, H. C. (1960) *An Introductory Treatise on Dynamical Astronomy*, Dover, New York.

[5] Cefola, P. J. (1972) *Equinoctial Orbit Elements: Application to Artificial Satellite Orbits*, AIAA Paper 72–937, AIAA/AAS Astrodynamics Conference, Palo Alto, CA.

[6] Edelbaum, T. N., Sackett, L. L. and Malchow, H. L. (1973) *Optimal Low Thrust Geocentric Transfer*, AIAA Paper 73–1074, AIAA 10th Electric Propulsion Conference, Lake Tahoe, NV.

[7] Cefola, P. J., Long, A. C. and Holloway, G. Jr. (1974) *The Long-Term Prediction of Artificial Satellite Orbits*, AIAA Paper 74–170, AIAA 12th Aerospace Sciences Meeting, Washington, DC.

[8] Sackett, L. L. and Edelbaum, T. N. (1975) *Effect of Attitude Constraints on Solar-Electric Geocentric Transfers*, AIAA Paper 75–350, AIAA 11th Electric Propulsion Conference, New Orleans, LA.

[9] Kahaner, D., Moler, C. and Nash, S. (1989) *Numerical Methods and Software*, Prentice Hall, Englewood Cliffs, NJ.

[10] Kechichian, J. A. (1994) *Optimal LEO–GEO Intermediate Acceleration Orbit Transfer*, AAS Paper 94–125, AAS/AIAA Spaceflight Mechanics Meeting, Cocoa Beach, FL.

3 Optimal Low-Thrust Rendezvous Using Equinoctial Orbit Elements

3.1 Introduction to the Minimum-Time Rendezvous Problem

Analytic solutions for the low-thrust rendezvous problem in the vicinity of the initial circular or elliptical orbit were obtained in the early 1960s by Gobetz and Edelbaum [1, 2]. Marec and Vinh [3] have presented both analytic and numerical techniques to solve the more general low-thrust problem with rigorous treatment of the analysis of the conjugate point.

In the foregoing, the problem of minimum-time low-thrust orbit rendezvous using continuous constant acceleration is analyzed. The variation of parameter equations are written in terms of a set of nonsingular equinoctial orbit elements which are free of singularities for zero eccentricity, and 0 and 90 deg inclination. This orbit theory was developed early in the 1970s by Broucke and Cefola and later extended by Edelbaum and his colleagues to solve five-state orbit transfer problems [4–7]. State and adjoint equations are integrated numerically by applying the thrust vector along a direction that maximizes the variational Hamiltonian at each instant of time. The two-point boundary value problem is solved via a Newton–Raphson scheme after guessing the initial values of the Lagrange multipliers as well as the total flight time. The transversality condition for the Hamiltonian at the final time is also used to carry out the 7×7 search.

Equinoctial elements have previously been used to carry out minimum-time orbit transfers by considering only the five slowly varying elements. The inclusion of the sixth element representing the fast variable will allow extending the applicability of the nonsingular formulation to problems of orbital rendezvous.

The full set of six-state adjoint and state equations using the mean longitude as the fast variable is then applied to solve the optimal rendezvous problem. It is assumed that there exists enough control authority to prevent the use of coasting arcs, and that the relative geometry between the two orbits is such that a direct rendezvous trajectory is possible. This can be achieved by waiting in the initial orbit until the proper orbital configuration is reached for direct trajectory rendezvous initiation. If coasting arcs are allowed, the optimal initial waiting period will be obtained from the optimization process. Nevertheless, in our simpler problem, the thrust is on continuously from time zero, and as an example we consider a rendezvous problem in near-circular condition to show the effectiveness of using the equinoctial elements as opposed to using the classical elements which would undoubtedly introduce undesirable singularities during the numerical integration of the various exact equations of motion.

3.2 The Differential Equations in Terms of the Equinoctial Elements

The basic set of the six-state differential equations of motion have been developed in [4, 6, 7]. They are derived in Chapter 2 and are repeated in this section for completeness.

The equinoctial elements in terms of the classical elements are given by

$$a = a \tag{3.1}$$

$$h = e \sin(\omega + \Omega) \tag{3.2}$$

$$k = e \cos(\omega + \Omega) \tag{3.3}$$

$$p = \tan\left(\frac{i}{2}\right) \sin\Omega \tag{3.4}$$

$$q = \tan\left(\frac{i}{2}\right) \cos\Omega \tag{3.5}$$

$$\lambda = M + \omega + \Omega \tag{3.6}$$

The last expression that defines the mean longitude can be replaced by

$$F = E + \omega + \Omega \tag{3.7}$$

or

$$L = \theta^* + \omega + \Omega \tag{3.8}$$

defining the eccentric longitude and true longitude, respectively. Here θ^* represents the true anomaly. The inverse transformation is given by

$$a = a \tag{3.9}$$

$$e = \left(h^2 + k^2\right)^{1/2} \tag{3.10}$$

$$i = 2\tan^{-1}\left(p^2 + q^2\right)^{1/2} \tag{3.11}$$

$$\Omega = \tan^{-1}\left(\frac{p}{q}\right) \tag{3.12}$$

$$\omega = \tan^{-1}\left(\frac{h}{k}\right) - \tan^{-1}\left(\frac{p}{q}\right) \tag{3.13}$$

$$M = \lambda - \tan^{-1}\left(\frac{h}{k}\right) \tag{3.14}$$

If F or L are used as the fast variable defining the sixth equinoctial element instead of λ, then the last expression in Equation (3.14) should be replaced by

$$E = F - \tan^{-1}\left(\frac{h}{k}\right) \tag{3.15}$$

or

$$\theta^* = L - \tan^{-1}\left(\frac{h}{k}\right) \tag{3.16}$$

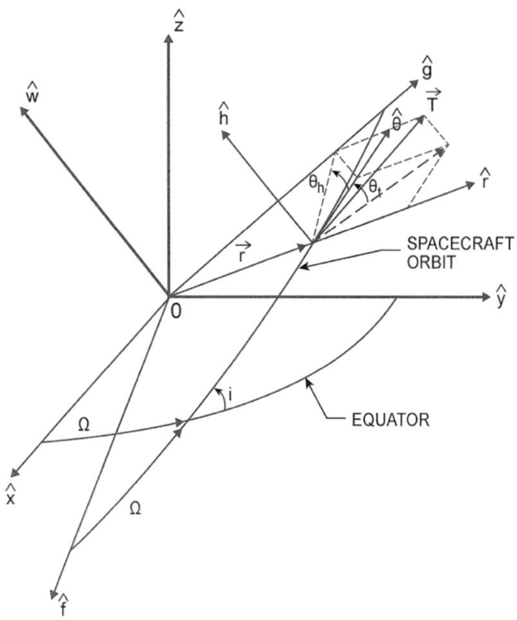

Figure 3.1 Equinoctial frame and thrust geometry.

The direct equinoctial frame is shown in Figure 3.1 with unit vectors $\hat{\mathbf{f}}$, $\hat{\mathbf{g}}$, $\hat{\mathbf{w}}$, such that $\hat{\mathbf{f}}$ and $\hat{\mathbf{g}}$ are contained in the instantaneous orbit plane with the direction of $\hat{\mathbf{f}}$ obtained through a clockwise rotation of an angle Ω from the direction of the ascending node. The position vector and velocity $\dot{\mathbf{r}}$ are such that

$$\mathbf{r} = X_1 \hat{\mathbf{f}} + Y_1 \hat{\mathbf{g}} \tag{3.17}$$

$$\dot{\mathbf{r}} = \dot{X}_1 \hat{\mathbf{f}} + \dot{Y}_1 \hat{\mathbf{g}} \tag{3.18}$$

where the components are written in terms of the equinoctial elements and the eccentric longitude F, with s_F and c_F standing for $\sin F$ and $\cos F$, etc.

$$X_1 = a\left[\left(1 - h^2\beta\right) c_F + hk\beta s_F - k\right] \tag{3.19}$$

$$Y_1 = a\left[hk\beta c_F + \left(1 - k^2\beta\right) s_F - h\right] \tag{3.20}$$

$$\dot{X}_1 = a^2 n r^{-1}\left[hk\beta c_F - \left(1 - h^2\beta\right) s_F\right] \tag{3.21}$$

$$\dot{Y}_1 = a^2 n r^{-1}\left[\left(1 - k^2\beta\right) c_F - hk\beta s_F\right] \tag{3.22}$$

with β given by Equation (3.23), and where $n = \mu^{1/2} a^{-3/2}$ represents the orbit mean motion; μ is the Earth gravity constant

$$\beta = \frac{1}{(1 + G)} \tag{3.23}$$

and where

$$G = \left(1 - h^2 - k^2\right)^{1/2}$$

Here $r = a\,(1 - kc_F - hs_F)$. Letting $\mathbf{z} = (a,\ h,\ k,\ p,\ q,\ \lambda)^{\mathrm{T}}$ represent the orbit state vector at time t, and letting $\mathbf{f}$ and m stand for the thrust vector and spacecraft mass, respectively, with $\mathbf{f} = f\hat{\mathbf{u}}$ where $\hat{\mathbf{u}}$ is a unit vector in the direction of the thrust, the variation of parameters equations for the thrust perturbation are given by Equation (3.24) with the addition of an n term to $\dot{\lambda}$

$$\dot{\mathbf{z}} = \left(\frac{\partial \mathbf{z}}{\partial \dot{\mathbf{r}}}\right)\hat{\mathbf{u}} f_t \tag{3.24}$$

where $f_t = f/m$ represents the instantaneous perturbation acceleration, with both $\partial \mathbf{z}/\partial \dot{\mathbf{r}}$ and $\hat{\mathbf{u}}$ expressed in the direct equinoctial frame, and where with $K = 1 + p^2 + q^2$,

$$\frac{\partial a}{\partial \dot{\mathbf{r}}} = 2a^{-1}n^{-2}\left(\dot{X}_1\hat{\mathbf{f}} + \dot{Y}_1\hat{\mathbf{g}}\right) = M_{11}\hat{\mathbf{f}} + M_{12}\hat{\mathbf{g}} + M_{13}\hat{\mathbf{w}} \tag{3.25}$$

$$\begin{aligned}\frac{\partial h}{\partial \dot{\mathbf{r}}} &= Gn^{-1}a^{-2}\left[\left(\frac{\partial X_1}{\partial k} - h\beta\frac{\dot{X}_1}{n}\right)\hat{\mathbf{f}} + \left(\frac{\partial Y_1}{\partial k} - h\beta\frac{\dot{Y}_1}{n}\right)\hat{\mathbf{g}}\right] \\ &\quad + k\,(qY_1 - pX_1)\,n^{-1}a^{-2}G^{-1}\hat{\mathbf{w}} = M_{21}\hat{\mathbf{f}} + M_{22}\hat{\mathbf{g}} + M_{23}\hat{\mathbf{w}}\end{aligned} \tag{3.26}$$

$$\begin{aligned}\frac{\partial k}{\partial \dot{\mathbf{r}}} &= -Gn^{-1}a^{-2}\left[\left(\frac{\partial X_1}{\partial h} + k\beta\frac{\dot{X}_1}{n}\right)\hat{\mathbf{f}} + \left(\frac{\partial Y_1}{\partial h} + k\beta\frac{\dot{Y}_1}{n}\right)\hat{\mathbf{g}}\right] \\ &\quad - h\,(qY_1 - pX_1)\,n^{-1}a^{-2}G^{-1}\hat{\mathbf{w}} = M_{31}\hat{\mathbf{f}} + M_{32}\hat{\mathbf{g}} + M_{33}\hat{\mathbf{w}}\end{aligned} \tag{3.27}$$

$$\frac{\partial p}{\partial \dot{\mathbf{r}}} = KY_1\frac{n^{-1}a^{-2}G^{-1}}{2}\hat{\mathbf{w}} = M_{41}\hat{\mathbf{f}} + M_{42}\hat{\mathbf{g}} + M_{43}\hat{\mathbf{w}} \tag{3.28}$$

$$\frac{\partial q}{\partial \dot{\mathbf{r}}} = KX_1\frac{n^{-1}a^{-2}G^{-1}}{2}\hat{\mathbf{w}} = M_{51}\hat{\mathbf{f}} + M_{52}\hat{\mathbf{g}} + M_{53}\hat{\mathbf{w}} \tag{3.29}$$

$$\begin{aligned}\frac{\partial \lambda}{\partial \dot{\mathbf{r}}} &= n^{-1}a^{-2}\left[-2X_1 + G\left(h\beta\frac{\partial X_1}{\partial h} + k\beta\frac{\partial X_1}{\partial k}\right)\right]\hat{\mathbf{f}} \\ &\quad + n^{-1}a^{-2}\left[-2Y_1 + G\left(h\beta\frac{\partial Y_1}{\partial h} + k\beta\frac{\partial Y_1}{\partial k}\right)\right]\hat{\mathbf{g}} \\ &\quad + n^{-1}a^{-2}G^{-1}\,(qY_1 - pX_1)\,\hat{\mathbf{w}} = M_{61}\hat{\mathbf{f}} + M_{62}\hat{\mathbf{g}} + M_{63}\hat{\mathbf{w}}\end{aligned} \tag{3.30}$$

$$\frac{\partial X_1}{\partial h} = a\left[-(hc_F - ks_F)\left(\beta + \frac{h^2\beta^3}{(1-\beta)}\right) - \frac{a}{r}c_F\,(h\beta - s_F)\right] \tag{3.31}$$

$$\frac{\partial X_1}{\partial k} = -a\left[-(hc_F - ks_F)\frac{hk\beta^3}{(1-\beta)} + 1 + \frac{a}{r}s_F\,(s_F - h\beta)\right] \tag{3.32}$$

$$\frac{\partial Y_1}{\partial h} = a\left[(hc_F - ks_F)\frac{hk\beta^3}{(1-\beta)} - 1 + \frac{a}{r}c_F\,(k\beta - c_F)\right] \tag{3.33}$$

$$\frac{\partial Y_1}{\partial k} = a\left[(hc_F - ks_F)\left(\beta + \frac{k^2\beta^3}{(1-\beta)}\right) + \frac{a}{r}s_F\,(c_F - k\beta)\right] \tag{3.34}$$

The system of differential equations for the state variables is given by

$$\dot{a} = \left(\frac{\partial a}{\partial \dot{\mathbf{r}}}\right)^T \cdot \hat{\mathbf{u}} f_t \tag{3.35}$$

$$\dot{h} = \left(\frac{\partial h}{\partial \dot{\mathbf{r}}}\right)^T \cdot \hat{\mathbf{u}} f_t \tag{3.36}$$

$$\dot{k} = \left(\frac{\partial k}{\partial \dot{\mathbf{r}}}\right)^T \cdot \hat{\mathbf{u}} f_t \tag{3.37}$$

$$\dot{p} = \left(\frac{\partial p}{\partial \dot{\mathbf{r}}}\right)^T \cdot \hat{\mathbf{u}} f_t \tag{3.38}$$

$$\dot{q} = \left(\frac{\partial q}{\partial \dot{\mathbf{r}}}\right)^T \cdot \hat{\mathbf{u}} f_t \tag{3.39}$$

$$\dot{\lambda} = n + \left(\frac{\partial \lambda}{\partial \dot{\mathbf{r}}}\right)^T \cdot \hat{\mathbf{u}} f_t \tag{3.40}$$

where n, the mean motion is oscillating too, because it is dependent on the semimajor axis through Equation (3.35). Because we are integrating the mean longitude λ as the sixth element, and because X_1, Y_1, $\dot{X}_1$, $\dot{Y}_1$, r, as well as the partials $\partial X_1/\partial h$ through $\partial Y_1/\partial k$ are given in terms of the eccentric longitude F, it is necessary at each integration step to solve Kepler's equation

$$\lambda = F - k s_F + h c_F \tag{3.41}$$

for F, through a Newton–Raphson iteration scheme.

The intermediate results for the partial derivatives of the position vector with respect to the six elements used to generate the partials in Equation (3.24) are given by

$$\frac{\partial \mathbf{r}}{\partial a} = \frac{\partial X_1}{\partial a}\hat{\mathbf{f}} + \frac{\partial Y_1}{\partial a}\hat{\mathbf{g}} = \left(\frac{X_1}{a} - \frac{3}{2}\frac{t}{a}\dot{X}_1\right)\hat{\mathbf{f}} + \left(\frac{Y_1}{a} - \frac{3}{2}\frac{t}{a}\dot{Y}_1\right)\hat{\mathbf{g}}$$

$$\frac{\partial \mathbf{r}}{\partial h} = \frac{\partial X_1}{\partial h}\hat{\mathbf{f}} + \frac{\partial Y_1}{\partial h}\hat{\mathbf{g}}$$

$$\frac{\partial \mathbf{r}}{\partial k} = \frac{\partial X_1}{\partial k}\hat{\mathbf{f}} + \frac{\partial Y_1}{\partial k}\hat{\mathbf{g}}$$

$$\frac{\partial \mathbf{r}}{\partial \lambda} = \frac{\dot{X}_1}{n}\hat{\mathbf{f}} + \frac{\dot{Y}_1}{n}\hat{\mathbf{g}} = \frac{\dot{\mathbf{r}}}{n}$$

$$\frac{\partial \mathbf{r}}{\partial p} = X_1\frac{\partial \hat{\mathbf{f}}}{\partial p} + Y_1\frac{\partial \hat{\mathbf{g}}}{\partial p} = \frac{2}{K}\left[q\left(Y_1\hat{\mathbf{f}} - X_1\hat{\mathbf{g}}\right) - X_1\hat{\mathbf{w}}\right]$$

$$\frac{\partial \mathbf{r}}{\partial q} = X_1\frac{\partial \hat{\mathbf{f}}}{\partial q} + Y_1\frac{\partial \hat{\mathbf{g}}}{\partial q} = \frac{2}{K}\left[p\left(X_1\hat{\mathbf{g}} - Y_1\hat{\mathbf{f}}\right) + Y_1\hat{\mathbf{w}}\right]$$

3.3 The Euler–Lagrange Differential Equations

The Euler–Lagrange or adjoint differential equations for the set a, h, k, p, q and λ have been derived in [8] and [9] and in Chapter 2 of this book, and are shown here for the purpose of the discussion. The necessary condition for optimality, which consists of maximizing the Hamiltonian of the system, can be found in [10–12].

The Hamiltonian of the system is given by

$$H = \boldsymbol{\lambda}_z^T \dot{\mathbf{z}} = \boldsymbol{\lambda}_z^T M(\mathbf{z}, F) f_t \hat{\mathbf{u}} + \lambda_\lambda n \tag{3.42}$$

where the seventh differential equation for the mass flow rate has been neglected. The Euler–Lagrange equations are given by

$$\dot{\boldsymbol{\lambda}}_z = -\frac{\partial H}{\partial \mathbf{z}} = -\boldsymbol{\lambda}_z^T \frac{\partial M}{\partial \mathbf{z}} f_t \hat{\mathbf{u}} - \lambda_\lambda \frac{\partial n}{\partial \mathbf{z}} \tag{3.43}$$

The optimal thrust direction $\hat{\mathbf{u}}$ is chosen such that it is at all times parallel to $\boldsymbol{\lambda}_z^T M(\mathbf{z}, F) f_t$ in order to maximize the Hamiltonian. The matrix M for the set (a, h, k, p, q, λ) is as given by Equations (3.25)–(3.30). The partials $\partial M/\partial \mathbf{z}$ appeared in [8] and are also shown here in Appendix 3.1. If we let u_f, u_g, u_w represent the components of $\hat{\mathbf{u}}$ in the equinoctial frame, then

$$\begin{aligned} H = \big[&\lambda_a (M_{11} u_f + M_{12} u_g + M_{13} u_w) + \lambda_h (M_{21} u_f + M_{22} u_g + M_{23} u_w) \\ &+ \lambda_k (M_{31} u_f + M_{32} u_g + M_{33} u_w) + \lambda_p (M_{41} u_f + M_{42} u_g + M_{43} u_w) \\ &+ \lambda_q (M_{51} u_f + M_{52} u_g + M_{53} u_w) + \lambda_\lambda (M_{61} u_f + M_{62} u_g + M_{63} u_w)\big] \\ &\times f_t + \lambda_\lambda n \end{aligned} \tag{3.44}$$

The Euler–Lagrange equations are now written as

$$\begin{aligned} \dot{\lambda}_a = -\frac{\partial H}{\partial a} = \Big[&-\lambda_a \left(\frac{\partial M_{11}}{\partial a} u_f + \frac{\partial M_{12}}{\partial a} u_g + \frac{\partial M_{13}}{\partial a} u_w\right) \\ &- \lambda_h \left(\frac{\partial M_{21}}{\partial a} u_f + \frac{\partial M_{22}}{\partial a} u_g + \frac{\partial M_{23}}{\partial a} u_w\right) \\ &- \lambda_k \left(\frac{\partial M_{31}}{\partial a} u_f + \frac{\partial M_{32}}{\partial a} u_g + \frac{\partial M_{33}}{\partial a} u_w\right) - \lambda_p \\ &\times \left(\frac{\partial M_{41}}{\partial a} u_f + \frac{\partial M_{42}}{\partial a} u_g + \frac{\partial M_{43}}{\partial a} u_w\right) \\ &- \lambda_q \left(\frac{\partial M_{51}}{\partial a} u_f + \frac{\partial M_{52}}{\partial a} u_g + \frac{\partial M_{53}}{\partial a} u_w\right) \\ &-\lambda_\lambda \left(\frac{\partial M_{61}}{\partial a} u_f + \frac{\partial M_{62}}{\partial a} u_g + \frac{\partial M_{63}}{\partial a} u_w\right)\Big] f_t - \lambda_\lambda \frac{\partial n}{\partial a} \end{aligned} \tag{3.45}$$

where

$$\frac{\partial n}{\partial a} = -\frac{3n}{2a} = -\frac{3}{2}\mu^{1/2} a^{-5/2}$$

$$\begin{aligned} \dot{\lambda}_h = -\frac{\partial H}{\partial h} = \Big[&-\lambda_a \left(\frac{\partial M_{11}}{\partial h} u_f + \frac{\partial M_{12}}{\partial h} u_g + \frac{\partial M_{13}}{\partial h} u_w\right) \\ &-\lambda_h \left(\frac{\partial M_{21}}{\partial h} u_f + \frac{\partial M_{22}}{\partial h} u_g + \frac{\partial M_{23}}{\partial h} u_w\right) - \lambda_k \left(\frac{\partial M_{31}}{\partial h} u_f + \frac{\partial M_{32}}{\partial h} u_g + \frac{\partial M_{33}}{\partial h} u_w\right) \end{aligned}$$

$$- \lambda_p \left(\frac{\partial M_{41}}{\partial h} u_f + \frac{\partial M_{42}}{\partial h} u_g + \frac{\partial M_{43}}{\partial h} u_w \right) - \lambda_q \left(\frac{\partial M_{51}}{\partial h} u_f + \frac{\partial M_{52}}{\partial h} u_g + \frac{\partial M_{53}}{\partial h} u_w \right)$$
$$- \lambda_\lambda \left(\frac{\partial M_{61}}{\partial h} u_f + \frac{\partial M_{62}}{\partial h} u_g + \frac{\partial M_{63}}{\partial h} u_w \right) \bigg] f_t \tag{3.46}$$

$$\dot{\lambda}_k = -\frac{\partial H}{\partial k} = \bigg[-\lambda_a \left(\frac{\partial M_{11}}{\partial k} u_f + \frac{\partial M_{12}}{\partial k} u_g + \frac{\partial M_{13}}{\partial k} u_w \right)$$
$$- \lambda_h \left(\frac{\partial M_{21}}{\partial k} u_f + \frac{\partial M_{22}}{\partial k} u_g + \frac{\partial M_{23}}{\partial k} u_w \right) - \lambda_k \left(\frac{\partial M_{31}}{\partial k} u_f + \frac{\partial M_{32}}{\partial k} u_g + \frac{\partial M_{33}}{\partial k} u_w \right)$$
$$- \lambda_p \left(\frac{\partial M_{41}}{\partial k} u_f + \frac{\partial M_{42}}{\partial k} u_g + \frac{\partial M_{43}}{\partial k} u_w \right) - \lambda_q \left(\frac{\partial M_{51}}{\partial k} u_f + \frac{\partial M_{52}}{\partial k} u_g + \frac{\partial M_{53}}{\partial k} u_w \right)$$
$$- \lambda_\lambda \left(\frac{\partial M_{61}}{\partial k} u_f + \frac{\partial M_{62}}{\partial k} u_g + \frac{\partial M_{63}}{\partial k} u_w \right) \bigg] f_t \tag{3.47}$$

$$\dot{\lambda}_p = -\frac{\partial H}{\partial p} = \bigg[-\lambda_a \left(\frac{\partial M_{11}}{\partial p} u_f + \frac{\partial M_{12}}{\partial p} u_g + \frac{\partial M_{13}}{\partial p} u_w \right)$$
$$- \lambda_h \left(\frac{\partial M_{21}}{\partial p} u_f + \frac{\partial M_{22}}{\partial p} u_g + \frac{\partial M_{23}}{\partial p} u_w \right) - \lambda_k \left(\frac{\partial M_{31}}{\partial p} u_f + \frac{\partial M_{32}}{\partial p} u_g + \frac{\partial M_{33}}{\partial p} u_w \right)$$
$$- \lambda_p \left(\frac{\partial M_{41}}{\partial p} u_f + \frac{\partial M_{42}}{\partial p} u_g + \frac{\partial M_{43}}{\partial p} u_w \right) - \lambda_q \left(\frac{\partial M_{51}}{\partial p} u_f + \frac{\partial M_{52}}{\partial p} u_g + \frac{\partial M_{53}}{\partial p} u_w \right)$$
$$- \lambda_\lambda \left(\frac{\partial M_{61}}{\partial p} u_f + \frac{\partial M_{62}}{\partial p} u_g + \frac{\partial M_{63}}{\partial p} u_w \right) \bigg] f_t \tag{3.48}$$

$$\dot{\lambda}_q = -\frac{\partial H}{\partial q} = \bigg[-\lambda_a \left(\frac{\partial M_{11}}{\partial q} u_f + \frac{\partial M_{12}}{\partial q} u_g + \frac{\partial M_{13}}{\partial q} u_w \right)$$
$$- \lambda_h \left(\frac{\partial M_{21}}{\partial q} u_f + \frac{\partial M_{22}}{\partial q} u_g + \frac{\partial M_{23}}{\partial q} u_w \right) - \lambda_k \left(\frac{\partial M_{31}}{\partial q} u_f + \frac{\partial M_{32}}{\partial q} u_g + \frac{\partial M_{33}}{\partial q} u_w \right)$$
$$- \lambda_p \left(\frac{\partial M_{41}}{\partial q} u_f + \frac{\partial M_{42}}{\partial q} u_g + \frac{\partial M_{43}}{\partial q} u_w \right) - \lambda_q \left(\frac{\partial M_{51}}{\partial q} u_f + \frac{\partial M_{52}}{\partial q} u_g + \frac{\partial M_{53}}{\partial q} u_w \right)$$
$$- \lambda_\lambda \left(\frac{\partial M_{61}}{\partial q} u_f + \frac{\partial M_{62}}{\partial q} u_g + \frac{\partial M_{63}}{\partial q} u_w \right) \bigg] f_t \tag{3.49}$$

$$\dot{\lambda}_\lambda = -\frac{\partial H}{\partial \lambda} = \bigg[-\lambda_a \left(\frac{\partial M_{11}}{\partial \lambda} u_f + \frac{\partial M_{12}}{\partial \lambda} u_g + \frac{\partial M_{13}}{\partial \lambda} u_w \right)$$
$$- \lambda_h \left(\frac{\partial M_{21}}{\partial \lambda} u_f + \frac{\partial M_{22}}{\partial \lambda} u_g + \frac{\partial M_{23}}{\partial \lambda} u_w \right) - \lambda_k \left(\frac{\partial M_{31}}{\partial \lambda} u_f + \frac{\partial M_{32}}{\partial \lambda} u_g + \frac{\partial M_{33}}{\partial \lambda} u_w \right)$$
$$- \lambda_p \left(\frac{\partial M_{41}}{\partial \lambda} u_f + \frac{\partial M_{42}}{\partial \lambda} u_g + \frac{\partial M_{43}}{\partial \lambda} u_w \right) - \lambda_q \left(\frac{\partial M_{51}}{\partial \lambda} u_f + \frac{\partial M_{52}}{\partial \lambda} u_g + \frac{\partial M_{53}}{\partial \lambda} u_w \right)$$
$$- \lambda_\lambda \left(\frac{\partial M_{61}}{\partial \lambda} u_f + \frac{\partial M_{62}}{\partial \lambda} u_g + \frac{\partial M_{63}}{\partial \lambda} u_w \right) \bigg] f_t \tag{3.50}$$

The following partials are used to generate the $\partial M/\partial \mathbf{z}$ partials of Appendix 3.1. Here F must be considered to be independent of a but not of h, k and λ, because $\partial\lambda/\partial a$ is now equal to zero:

$$\frac{\partial r}{\partial a} = \frac{r}{a}, \quad \frac{\partial r}{\partial h} = \frac{a^2}{r}(h - s_F), \quad \frac{\partial r}{\partial k} = \frac{a^2}{r}(k - c_F), \quad \frac{\partial r}{\partial F} = a(ks_F - hc_F), \quad \frac{\partial F}{\partial a} = 0,$$
$$\frac{\partial F}{\partial h} = -\frac{a}{r}c_F, \quad \frac{\partial F}{\partial k} = \frac{a}{r}s_F, \quad \frac{\partial F}{\partial \lambda} = \frac{r}{a}$$

It is also true that $\partial\lambda/\partial F = r/a$. The components of the unit vector $\hat{\mathbf{u}}$ are obtained from

$$\hat{\mathbf{u}} = \frac{\left(\boldsymbol{\lambda}_z^T M\right)^T}{\left|\boldsymbol{\lambda}_z^T M\right|} \tag{3.51}$$

From Figure 3.1, the thrust vector $\mathbf{T} = \mathbf{f} = f\hat{\mathbf{u}}$ is defined by the thrust pitch and thrust yaw angles, θ_t and θ_h, respectively, in the rotating $\hat{\mathbf{r}}, \hat{\boldsymbol{\theta}}, \hat{\mathbf{h}}$ frame; $\hat{\mathbf{r}}$ is a unit vector along the instantaneous position vector $\mathbf{r}$, with $\hat{\boldsymbol{\theta}}$ in the orbit plane and along the direction of motion, and $\hat{\mathbf{h}}$ along the angular momentum vector. Therefore

$$\theta_t = \tan^{-1}\left(\frac{u_r}{u_\theta}\right) \tag{3.52}$$

$$\theta_h = \tan^{-1}\left(\frac{u_w}{u_\theta}\right) \tag{3.53}$$

with u_r, u_θ and u_h representing the components of the vector $\hat{\mathbf{u}}$ along the $\hat{\mathbf{r}}$, $\hat{\boldsymbol{\theta}}$ and $\hat{\mathbf{h}}$ directions. We have

$$\hat{\mathbf{r}} = \frac{X_1}{|\mathbf{r}|}\hat{\mathbf{f}} + \frac{Y_1}{|\mathbf{r}|}\hat{\mathbf{g}} \tag{3.54}$$

$$\hat{\boldsymbol{\theta}} = \hat{\mathbf{h}} \times \hat{\mathbf{r}} = -\frac{Y_1}{|\mathbf{r}|}\hat{\mathbf{f}} + \frac{Y_1}{|\mathbf{r}|}\hat{\mathbf{g}} \tag{3.55}$$

such that

$$u_r = \frac{X_1}{r}u_f + \frac{Y_1}{r}u_g \tag{3.56}$$

$$u_\theta = -\frac{Y_1}{r}u_f + \frac{X_1}{r}u_g \tag{3.57}$$

$$u_h = u_w \tag{3.58}$$

where

$$r = |\mathbf{r}| = \left(X_1^2 + Y_1^2\right)^{1/2}$$

3.4 Numerical Results

Because we are minimizing the total rendezvous time, i.e.,

$$J = \int_{t_0}^{t_f} dt = \left(t_f - t_0\right) \tag{3.59}$$

this is equivalent to maximizing

$$J = \int_{t_0}^{t_f} L dt = -\int_{t_0}^{t_f} dt = -\left(t_f - t_0\right) \tag{3.60}$$

Therefore $L = -1$, such that the transversality condition at the unknown final time t_f, namely, $H_f = 0$ for the augmented Hamiltonian $H = -1 + \boldsymbol{\lambda}_z^T \dot{\mathbf{z}}$ is equivalent to $H_f = 1$ for our Hamiltonian $H = \boldsymbol{\lambda}_z^T \dot{\mathbf{z}}$. The two-point boundary value problem consists of guessing the initial values of the six Lagrange multipliers $(\lambda_a)_0$, $(\lambda_h)_0$, $(\lambda_k)_0$, $(\lambda_p)_0$, $(\lambda_q)_0$, $(\lambda_\lambda)_0$, as well as the rendezvous time t_f (we assume here $t_0 = 0$) and integrating numerically the 12 differential equations of motion, namely Equations (3.35)–(3.40) and Equations (3.45)–(3.50) using the optimal control given in Equation (3.51), such that the six state parameters at time t_f namely $(a)_f$, $(h)_f$, $(k)_f$, $(p)_f$, $(q)_f$, $(\lambda)_f$ and $H_f = 1$ are satisfied. This consists therefore of a 7×7 search with given initial parameters $(a)_0$, $(h)_0$, $(k)_0$, $(p)_0$, $(q)_0$, $(\lambda)_0$.

For this purpose, we make use of the minimization algorithm UNCMIN of [13], which is designed for the unconstrained minimization of a real-valued function $F(\mathbf{x})$ of n variables denoted by the vector $\mathbf{x}$. This subroutine is based on a general descent method and uses a quasi-Newton algorithm. In the Newton method, the step $\mathbf{p}$ is computed from the solution of a set of n linear equations known as the Newton equations

$$\nabla^2 F(\mathbf{x})\mathbf{p} = -\nabla F(\mathbf{x}) \tag{3.61}$$

Therefore, the solution is updated by using

$$\mathbf{x}_{k+1} = \mathbf{x}_k + \mathbf{p} = \mathbf{x}_k - \left[\nabla^2 F(\mathbf{x}_k)\right]^{-1} \nabla F(\mathbf{x}_k) \tag{3.62}$$

Here, $\nabla^2 F(\mathbf{x})$ denotes the gradient of F at $\mathbf{x}$ while $\nabla^2 F(\mathbf{x})$, the constant matrix of the second partial derivatives of F at $\mathbf{x}$ represents the Hessian matrix.

The Newton direction given by $\mathbf{p}$ is guaranteed to be a descent direction only if $\left[\nabla^2 F\right]^{-1}$ is positive definite, i.e., $\mathbf{z}^T \left[\nabla^2 F\right]^{-1} \mathbf{z} > 0$ for all $\mathbf{z} \neq 0$, because then, for small ε,

$$F(\mathbf{x} + \varepsilon\mathbf{p}) = F(\mathbf{x}) + \varepsilon \nabla F^T \mathbf{p} + O\left(\varepsilon^2\right) = F(\mathbf{x}) - \varepsilon \nabla F^T \left[\nabla^2 F\right]^{-1} \nabla F + O\left(\varepsilon^2\right) \tag{3.63}$$

such that $F(\mathbf{x} + \varepsilon\mathbf{p}) < F(\mathbf{x})$. This is equivalent to requiring the linear term in Newton's quadratic approximation for the function F to be negative. This is, of course, the second term in the Taylor series expansion for F at $\mathbf{x} + \mathbf{p}$.

The algorithm UNCMIN builds a secant approximation B_k to the Hessian as the function is being minimized such that at $\mathbf{x}_k$

$$B_k \mathbf{p} = -\nabla F_k \tag{3.64}$$

with the matrix B_k positive definite because there is no guarantee that $\left[\nabla^2 F\right]^{-1}$ will always be positive definite for each $\mathbf{x}$. Once the descent direction is established, a line search on $\mathbf{x}$ is established, such that $F(\mathbf{x}_k + \alpha\mathbf{p}) < F_k$ and the solution updated via $\mathbf{x}_{k+1} = \mathbf{x}_k + \alpha\mathbf{p}$. The approximate Hessian B_k is updated next using $\mathbf{x}_{k+1}$ and the gradient $\nabla F(\mathbf{x}_{k+1})$. For example, if F is quadratic [13],

$$B_{k+1}\left(\mathbf{x}_{k+1} - \mathbf{x}_k\right) = \nabla F_{k+1} - F_k \tag{3.65}$$

Only gradient values are needed for the update of the approximate Hessian, which is achieved by finite differencing such that the user must provide only the function F itself.

The function to be minimized is chosen as

$$F = \frac{|a - a_T|}{a_T} + |h - h_T| + |k - k_T| + |p - p_T| + |q - q_T| + \frac{|\lambda - \lambda_T|}{2\pi} + |H - 1| \quad (3.66)$$

where all the values are at the final unknown time t_f, and where the subscript T is used for the target parameters that must be matched. These are, of course, the same as $(a)_f$, $(h)_f$, $(k)_f$, $(p)_f$, $(q)_f$ and $(\lambda)_f$. Starting from $a_0 = 42,000$ km, $e_0 = 0$, $i_0 = 28.5$ deg, $\Omega_0 = 30$ deg, $\omega_0 = 10$ deg, $M_0 = 0$ deg, and using a constant acceleration $f_t = 3.5 \times 10^{-7}$ km/s^2, we seek the minimum-time solution that achieves the final target parameters $a_f = 42,767.073$ km, $e_f = 1.64459 \times 10^{-4}$, $i_f = 28.343$ deg, $\Omega_f = 29.999$ deg, $\omega_f = 247.299$ deg and $M_f = 120.905$ deg.

The value of F is minimized at $F = 0.38989 \times 10^{-5}$ with the solution given as $(\lambda_a)_0 = 7347.174908$ s/km, $(\lambda_h)_0 = -6.994059338 \times 10^5$ s, $(\lambda_k)_0 = 8.246922768 \times 10^5$ s, $(\lambda_p)_0 = -3.631161518 \times 10^8$ s, $(\lambda_q)_0 = -6.198272749 \times 10^8$ s, and $(\lambda_\lambda)_0 = -1.107002405 \times 10^6$ s/rad and $t_f = 86,402.453$ s. Figure 3.2 depicts the evolution of the semimajor axis and eccentricity during the 24 h transfer, while Figure 3.3 shows the evolution of the inclination and the right ascension of the ascending node. The relative minima in the Ω curve correspond to the stationary points of the i curve with the maxima corresponding to the largest negative gradients of the inclination. Figure 3.4 shows ω and M, which exhibit large gradients as soon as the eccentricity reaches the near-circular condition. Figure 3.5 shows the time history of the Hamiltonian, which is a constant in this formulation due to the independence of the accessory variable, namely the eccentric longitude on the semimajor axis. Figure 3.6 shows the control angles θ_t and θ_h that achieve the desired orbit transfer. Finally, Figures 3.7–3.9 show the evolution of the Lagrange multipliers during the transfer.

The evolution of the equinoctial elements h, k, p, q and the radial distance r is depicted in Figures 3.10, 3.11 and 3.12.

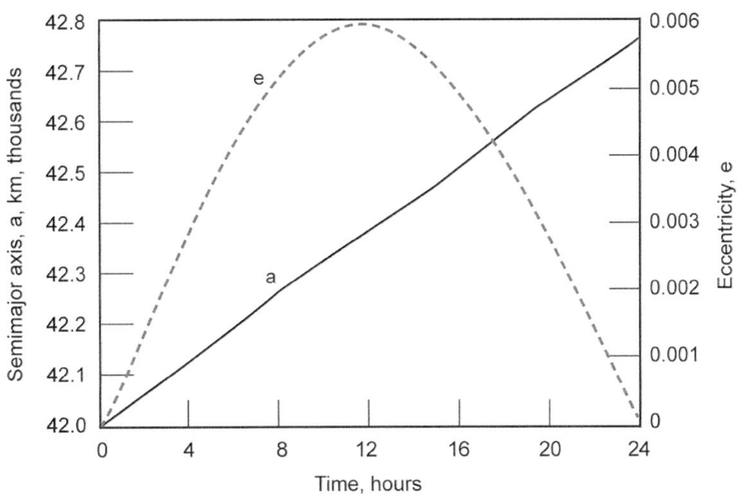

Figure 3.2 Evolution of semimajor axis and eccentricity.

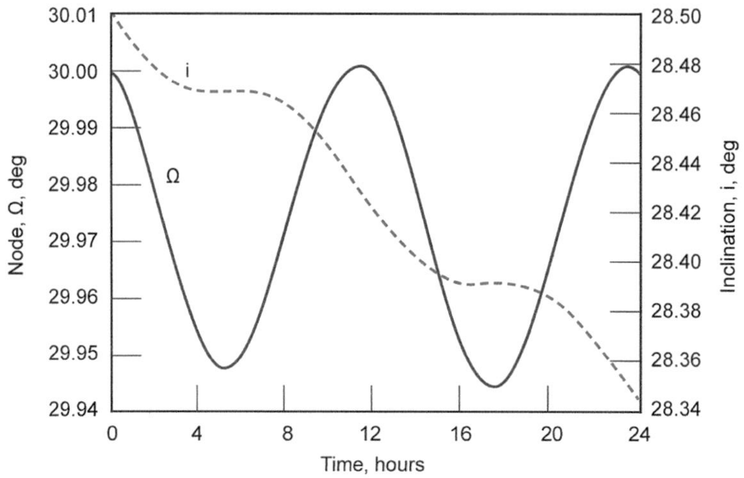

Figure 3.3 Evolution of orbit inclination and node.

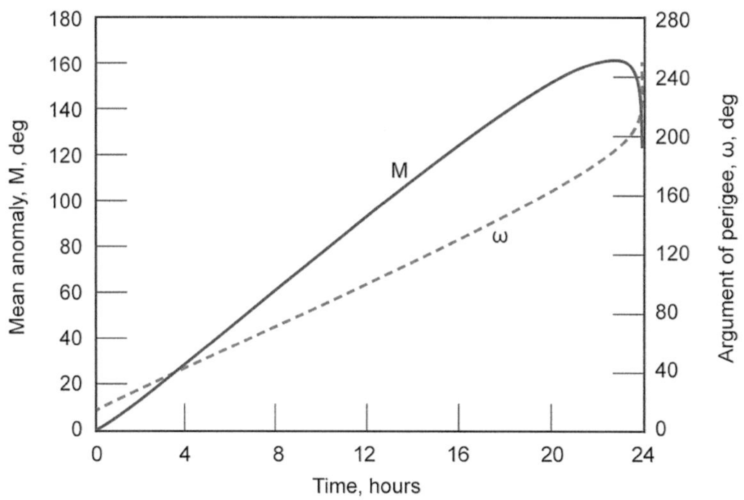

Figure 3.4 Evolution of mean anomaly and argument of perigee.

3.5 Summary of the Minimum-Time Rendezvous Problem

The minimum-time low-thrust orbit rendezvous using a set of nonsingular elements has been investigated and a simple illustrative example has been shown. This theory can be extended in order to allow the appearance of initial and intermediate time coasting arcs to solve minimum-fuel time-fixed rendezvous problems as well. Further improvements would necessitate the consideration of the J_2 zonal harmonic and possibly the acceleration due to air drag for LEO applications. Although the present analysis presents exact solutions using precision integration, long-duration rendezvous applications can effectively make use of the averaging technique for rapid calculation.

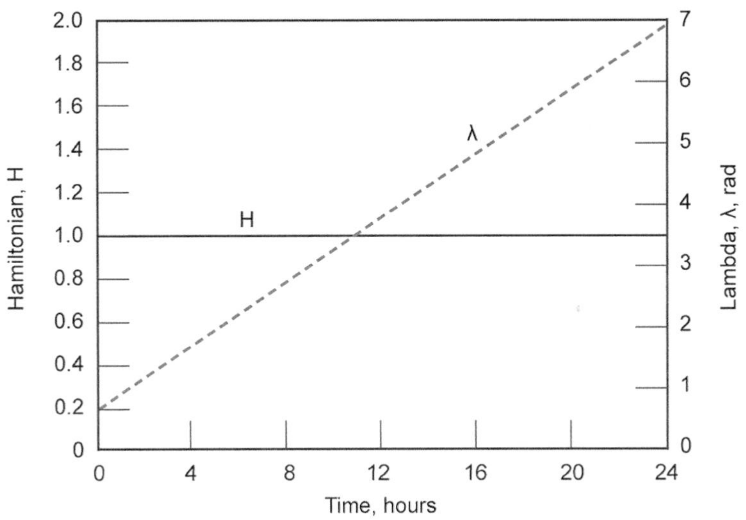

Figure 3.5 Evolution of optimal Hamiltonian during transfer.

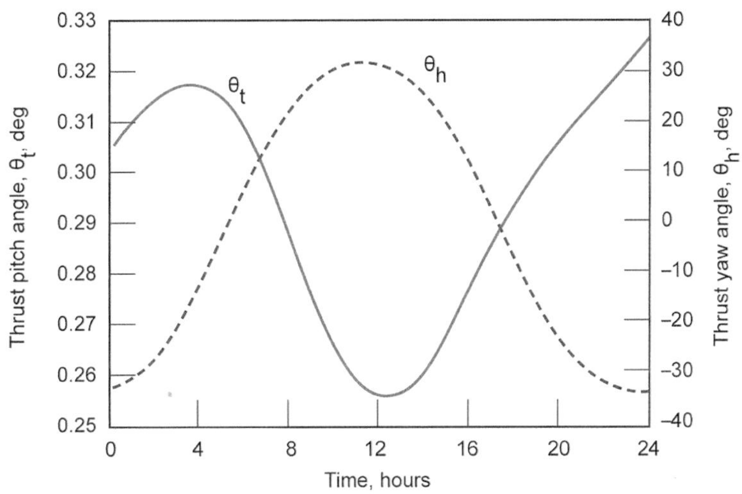

Figure 3.6 Optimal thrust pitch and yaw control angles during transfer.

3.6 Minimum-Fuel Time-Fixed Rendezvous Using Constant Low-Thrust

The problem of minimum-fuel time-fixed rendezvous using a set of nonsingular orbit elements is analyzed next. Constant thrust is applied during the powered phases of the trajectory with the thrust vector orientation optimized to minimize fuel expenditure. The selection of mean longitude as the fast orbit element results in a simpler dynamic system not involving time explicitly. The Hamiltonian is therefore constant throughout the powered and unpowered phases of the trajectory with simpler boundary conditions to satisfy. Exact trajectories are generated by simultaneously integrating a system of 14

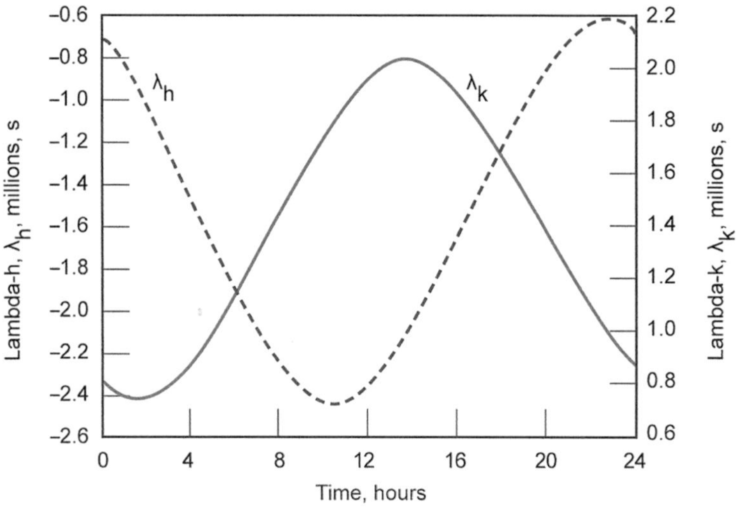

Figure 3.7 Time history of λ_h and λ_k.

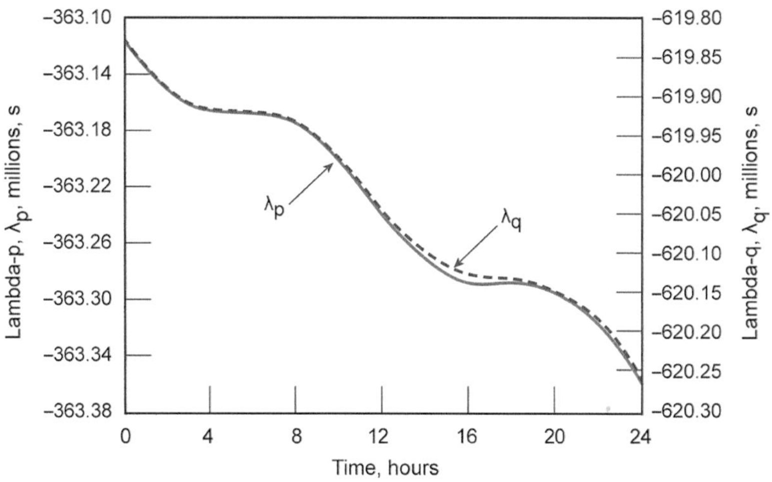

Figure 3.8 Time history of λ_p and λ_q.

differential equations for the state and adjoint variables involving the orbit elements and the spacecraft mass.

3.7 Introduction to the Minimum-Fuel Rendezvous Problem

The problem of minimum-fuel time-fixed rendezvous using constant low thrust is presented based on the use of a set of nonsingular orbit elements where the current mean longitude represents the fast element; [8] shows the state and adjoint equations for this particular set. The spacecraft mass is considered as the seventh state variable such that

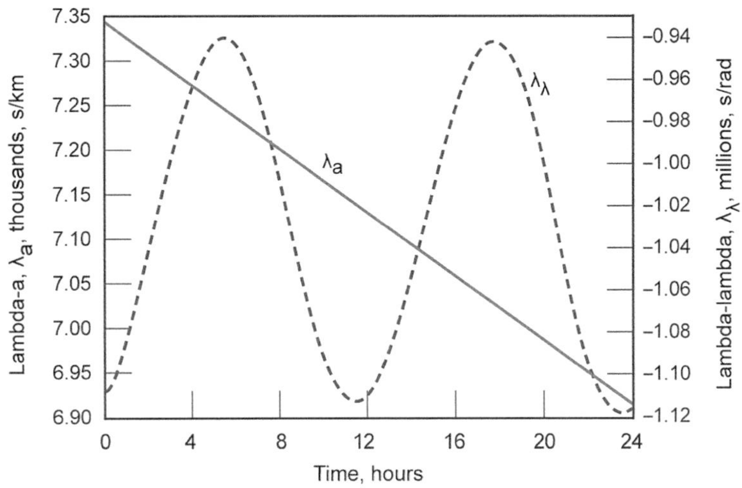

Figure 3.9 Time history of λ_a and λ_λ.

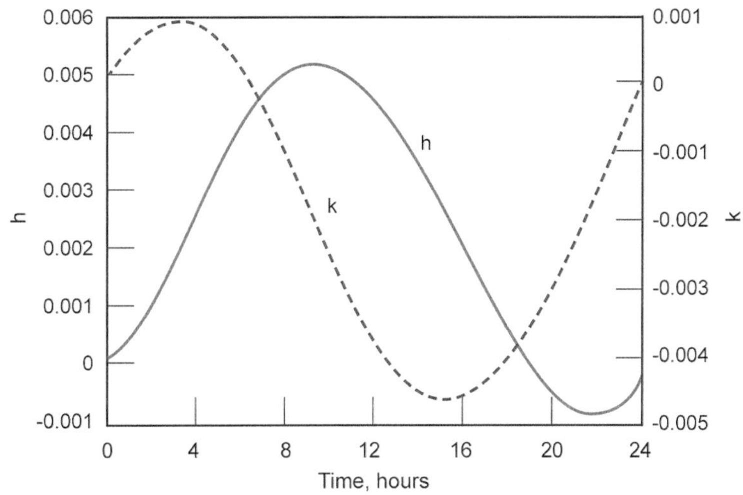

Figure 3.10 Time history of h and k.

a system of 14 differential equations for the state and adjoints is used to integrate an initial orbit state vector.

The mass of the spacecraft at the known final time is maximized by selecting the thrust direction that maximizes the variational Hamiltonian at each instant of time.

Using only constant thrust, the Hamiltonian appears as a linear function of the power which multiplies a certain switching function. The switching function is given in terms of the adjoints to the mass and velocity variables as well as the mass variable itself. Bang-Bang control for the power is used according to the sign of the switching function such that the Hamiltonian is always maximized. This results in an optimal trajectory made up of thrust arcs as well as coast arcs with power on and off respectively. The

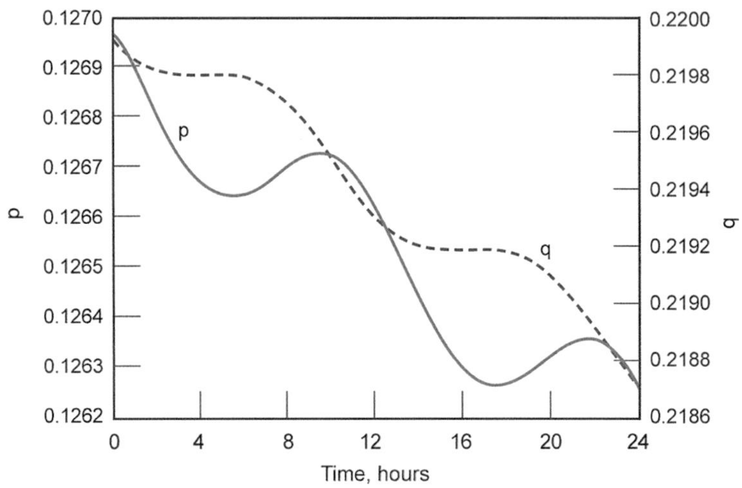

Figure 3.11 Time history of p and q.

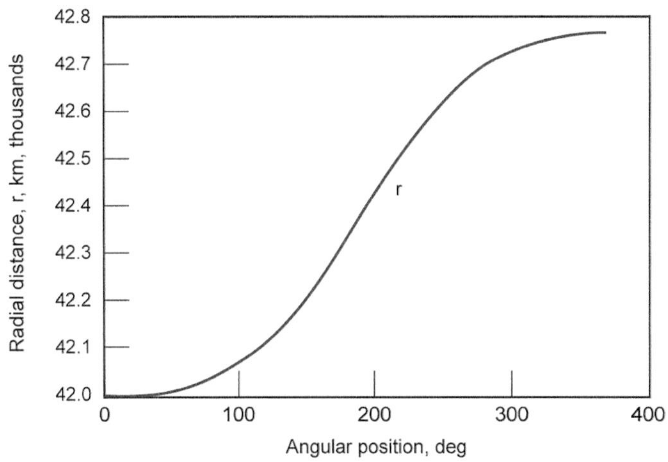

Figure 3.12 Radial distance r versus spacecraft angular position.

two-point boundary value problem is solved by guessing the initial values of the seven Lagrange multipliers such that the six orbit elements at the prescribed final time are satisfied and the adjoint to the mass variable at the same time driven to the unit value. The problem of minimum-time orbit rendezvous neglected the mass equation in the previous sections in order to create a well posed problem because otherwise instantaneous fuel dumps could be required to obtain a minimum-time solution. When the final mass is being maximized as in the present analysis, this undesirable outcome will be prevented from taking place. The propellant used by low-thrust systems is a small fraction of the total vehicle mass. However, for large transfers or long-duration rendezvous where the final time is considerably relaxed from the minimum-time solution, considerable savings in propellant are realized by the introduction of the coast arcs where it is not efficient to thrust.

Examples of minimum-fuel trajectories are included in the following sections to support the analysis. The limiting case in which the thrust is continuously on duplicates the constant acceleration minimum-time solution obtained in the previous sections.

3.8 The Dynamic Equations for the Seven State Variables

Using the current mean longitude λ as the fast element, the six orbit state variables are defined as before in terms of the classical elements by

$$a = a \tag{3.67}$$

$$h = e\sin(\omega + \Omega) \tag{3.68}$$

$$k = e\cos(\omega + \Omega) \tag{3.69}$$

$$p = \tan\left(\frac{i}{2}\right)\sin\Omega \tag{3.70}$$

$$q = \tan\left(\frac{i}{2}\right)\cos\Omega \tag{3.71}$$

$$\lambda = M + \omega + \Omega \tag{3.72}$$

Position and velocity are expressed in the direct equinoctial frame with unit vectors $\hat{\mathbf{f}}, \hat{\mathbf{g}}, \hat{\mathbf{w}}$ as before

$$\mathbf{r} = X_1\hat{\mathbf{f}} + Y_1\hat{\mathbf{g}} \tag{3.73}$$

$$\dot{\mathbf{r}} = \dot{X}_1\hat{\mathbf{f}} + \dot{Y}_1\hat{\mathbf{g}} \tag{3.74}$$

where the components are written in terms of the equinoctial elements and the eccentric longitude F, with s_F and c_F standing for $\sin F$ and $\cos F$, etc.

$$X_1 = a\left[\left(1 - h^2\beta\right)c_F + hk\beta s_F - k\right] \tag{3.75}$$

$$Y_1 = a\left[hk\beta c_F + \left(1 - k^2\beta\right)s_F - h\right] \tag{3.76}$$

$$\dot{X}_1 = a^2 n r^{-1}\left[hk\beta c_F - \left(1 - h^2\beta\right)s_F\right] \tag{3.77}$$

$$\dot{Y}_1 = a^2 n r^{-1}\left[\left(1 - k^2\beta\right)c_F - hk\beta s_F\right] \tag{3.78}$$

with β given below and with $n = \mu^{1/2}a^{-3/2}$ the orbit mean motion

$$\beta = \frac{1}{(1+G)}$$

with

$$G = (1 - h^2 - k^2)^{1/2}$$

Here $r = \left(X_1^2 + Y_1^2\right)^{1/2} = a\left(1 - kc_F - hs_F\right)$ represents the radial distance. Letting $\mathbf{z} = (a, h, k, p, q, \lambda)^T$ represent the orbit state vector at time t, and letting $\mathbf{f}$ and m stand for the thrust vector and spacecraft mass respectively with $\mathbf{f} = f\hat{\mathbf{u}}$ where $\hat{\mathbf{u}}$ is a unit vector in the direction of the thrust, the variation of parameters equations for the thrust perturbation are given by Equation (3.79) for the first slowly varying elements

$$\dot{\mathbf{z}} = 2\left(\frac{\partial \mathbf{z}}{\partial \dot{\mathbf{r}}}\right)\hat{\mathbf{u}}\frac{P}{mc} \tag{3.79}$$

and by

$$\dot{\lambda} = n + 2\left(\frac{\partial \lambda}{\partial \dot{\mathbf{r}}}\right)^T \hat{\mathbf{u}}\frac{P}{mc} \tag{3.80}$$

where $2P/(mc)$ represents the instantaneous acceleration, with both $\partial \mathbf{z}/\partial \dot{\mathbf{r}}$ and $\hat{\mathbf{u}}$ expressed in the direct equinoctial frame. Here, P represents the exhaust stream power given in terms of the thrust magnitude f and exhaust velocity c:

$$P = \frac{1}{2}fc \tag{3.81}$$

The partial derivatives in Equations (3.79) and (3.80) are written below as in [8] and the previous sections:

$$\frac{\partial a}{\partial \dot{\mathbf{r}}} = 2a^{-1}n^{-2}\left(\dot{X}_1\hat{\mathbf{f}} + \dot{Y}\hat{\mathbf{g}}\right) = M_{11}\hat{\mathbf{f}} + M_{12}\hat{\mathbf{g}} + M_{13}\hat{\mathbf{w}} \tag{3.82}$$

$$\begin{aligned}\frac{\partial h}{\partial \dot{\mathbf{r}}} &= Gn^{-1}a^{-2}\left[\left(\frac{\partial X_1}{\partial k} - h\beta\frac{\dot{X}_1}{n}\right)\hat{\mathbf{f}} + \left(\frac{\partial Y_1}{\partial k} - h\beta\frac{\dot{Y}_1}{n}\right)\hat{\mathbf{g}}\right]\\ &\quad + k\left(qY_1 - pX_1\right)n^{-1}a^{-2}G^{-1}\hat{\mathbf{w}} = M_{21}\hat{\mathbf{f}} + M_{22}\hat{\mathbf{g}} + M_{23}\hat{\mathbf{w}}\end{aligned} \tag{3.83}$$

$$\begin{aligned}\frac{\partial k}{\partial \dot{\mathbf{r}}} &= -Gn^{-1}a^{-2}\left[\left(\frac{\partial X_1}{\partial h} + k\beta\frac{\dot{X}_1}{n}\right)\hat{\mathbf{f}} + \left(\frac{\partial Y_1}{\partial h} + k\beta\frac{\dot{Y}_1}{n}\right)\hat{\mathbf{g}}\right]\\ &\quad - h(qY_1 - pX_1)n^{-1}a^{-2}G^{-1}\hat{\mathbf{w}} = M_{31}\hat{\mathbf{f}} + M_{32}\hat{\mathbf{g}} + M_{33}\hat{\mathbf{w}}\end{aligned} \tag{3.84}$$

$$\frac{\partial p}{\partial \dot{\mathbf{r}}} = KY_1\frac{n^{-1}a^{-2}G^{-1}}{2}\hat{\mathbf{w}} = M_{41}\hat{\mathbf{f}} + M_{42}\hat{\mathbf{g}} + M_{43}\hat{\mathbf{w}} \tag{3.85}$$

$$\frac{\partial q}{\partial \dot{\mathbf{r}}} = KX_1\frac{n^{-1}a^{-2}G^{-1}}{2}\hat{\mathbf{w}} = M_{51}\hat{\mathbf{f}} + M_{52}\hat{\mathbf{g}} + M_{53}\hat{\mathbf{w}} \tag{3.86}$$

$$\begin{aligned}\frac{\partial \lambda}{\partial \dot{\mathbf{r}}} &= n^{-1}a^{-2}\left[-2X_1 + G\left(h\beta\frac{\partial X_1}{\partial h} + k\beta\frac{\partial X_1}{\partial k}\right)\right]\hat{\mathbf{f}}\\ &\quad + n^{-1}a^{-2}\left[-2Y_1 + G\left(h\beta\frac{\partial Y_1}{\partial h} + k\beta\frac{\partial Y_1}{\partial k}\right)\right]\hat{\mathbf{g}}\\ &\quad + n^{-1}a^{-2}G^{-1}(qY_1 - pX_1)\hat{\mathbf{w}} = M_{61}\hat{\mathbf{f}} + M_{62}\hat{\mathbf{g}} + M_{63}\hat{\mathbf{w}}\end{aligned} \tag{3.87}$$

with $K = 1 + p^2 + q^2$ and

$$\frac{\partial X_1}{\partial h} = a\left[-(hc_F - ks_F)\left(\beta + \frac{h^2\beta^3}{(1-\beta)}\right) - \frac{a}{r}c_F(h\beta - s_F)\right]$$

$$\frac{\partial X_1}{\partial k} = -a\left[(hc_F - ks_F)\frac{hk\beta^3}{(1-\beta)} + 1 + \frac{a}{r}s_F(s_F - h\beta)\right]$$

$$\frac{\partial Y_1}{\partial h} = a\left[(hc_F - ks_F)\frac{hk\beta^3}{(1-\beta)} - 1 + \frac{a}{r}c_F(k\beta - c_F)\right]$$

$$\frac{\partial Y_1}{\partial k} = a\left[(hc_F - ks_F)\left(\beta + \frac{k^2\beta^3}{(1-\beta)}\right) + \frac{a}{r}s_F(c_F - k\beta)\right]$$

The vehicle mass m is governed by

$$\dot{m} = -\frac{2P}{c^2} \tag{3.88}$$

such that Equations (3.79), (3.80) and (3.88) describe the evolution of the seven state variables namely, a, h, k, p, q, λ and m. These equations can be integrated numerically between the initial time t_0 and the fixed final time t_f for a given constant power P and exhaust velocity c and initial conditions $(a)_0$, $(h)_0$, $(k)_0$, $(p)_0$, $(q)_0$, $(\lambda)_0$ and $(m)_0$. The eccentric longitude F appearing in the right-hand side of the dynamic equations is evaluated at each integration step from Kepler's equation

$$\lambda = F - ks_F + hc_F$$

Because P and c are constants, $\dot{m}$ is also constant during powered flight and is identically equal to zero during coast. Equation (3.88) can therefore be integrated analytically and this formulation is a constant thrust magnitude formulation for a variable-mass vehicle. The thrust acceleration is therefore also variable.

3.9 The Euler–Lagrange Differential Equations

The Hamiltonian of the system is given by

$$H = \boldsymbol{\lambda}_z^T\dot{\mathbf{z}} + \lambda_m\dot{m} = 2\boldsymbol{\lambda}_z^T M(\mathbf{z}, F)\,\hat{\mathbf{u}}\frac{P}{mc} - 2\lambda_m\frac{P}{c^2} + \lambda_\lambda n \tag{3.89}$$

The Euler–Lagrange equations are then given by

$$\dot{\boldsymbol{\lambda}}_z = -\frac{\partial H}{\partial \mathbf{z}} = -2\boldsymbol{\lambda}_z^T\frac{\partial M}{\partial \mathbf{z}}\hat{\mathbf{u}}\frac{P}{mc} - \lambda_\lambda\frac{\partial n}{\partial \mathbf{z}} \tag{3.90}$$

$$\dot{\lambda}_m = -\frac{\partial H}{\partial m} = 2\boldsymbol{\lambda}_z^T M\hat{\mathbf{u}}\frac{P}{m^2c} \tag{3.91}$$

The optimal thrust direction $\hat{\mathbf{u}}$ is chosen such that it is at all times parallel to $\boldsymbol{\lambda}_z^T M(\mathbf{z}F)$ in order to maximize the Hamiltonian. The matrix M for the set (a, h, k, p, q, λ) is as given by Equations (3.82) through (3.87). The partials $\partial M/\partial \mathbf{z}$ appear in Appendix 3.1. The Hamiltonian can be written as

$$H = \frac{2P}{mc}\left(\boldsymbol{\lambda}_z^T M\hat{\mathbf{u}} - \frac{m\lambda_m}{c}\right) + \lambda_\lambda n = \frac{2P}{mc}S + \lambda_\lambda n \tag{3.92}$$

where S represents the switching function:

$$S = \boldsymbol{\lambda}_z^T M\hat{\mathbf{u}} - \frac{m\lambda_m}{c} \tag{3.93}$$

This is equivalent to $\lambda_v - (m\lambda_m)/c$ when the Cartesian formulation $(\mathbf{r}, \mathbf{v})$ is used instead of the orbit elements formulation adopted here. If $S > 0$, then we use $P = P_{\max}$ and if $S < 0$ then $P = 0$, indicating coast. Singular arcs possible when $S = 0$ over a finite period of time and using intermediate power levels $0 < P < P_{\max}$ are not considered here because we assumed f and c constant. If we let u_f, u_g, u_w represent the components of $\hat{\mathbf{u}}$ in the equinoctial frame, then

$$\begin{aligned} H = \big[&\lambda_a \left(M_{11}u_f + M_{12}u_g + M_{13}u_w\right) + \lambda_h \left(M_{21}u_f + M_{22}u_g + M_{23}u_w\right) \\ &+ \lambda_k \left(M_{31}u_f + M_{32}u_g + M_{33}u_w\right) + \lambda_p \left(M_{41}u_f + M_{42}u_g + M_{43}u_w\right) \\ &+\lambda_q \left(M_{51}u_f + M_{52}u_g + M_{53}u_w\right) + \lambda_\lambda \left(M_{61}u_f + M_{62}u_g + M_{63}u_w\right)\big] \\ &\times \frac{2P}{mc} + \lambda_\lambda n - \frac{2P}{c^2}\lambda_m \end{aligned} \tag{3.94}$$

The Euler–Lagrange equations are now written as

$$\begin{aligned} \dot{\lambda}_a = -\frac{\partial H}{\partial a} = \Bigg[&-\lambda_a \left(\frac{\partial M_{11}}{\partial a}u_f + \frac{\partial M_{12}}{\partial a}u_g + \frac{\partial M_{13}}{\partial a}u_w\right) \\ &-\lambda_h \left(\frac{\partial M_{21}}{\partial a}u_f + \frac{\partial M_{22}}{\partial a}u_g + \frac{\partial M_{23}}{\partial a}u_w\right) - \lambda_k \left(\frac{\partial M_{31}}{\partial a}u_f + \frac{\partial M_{32}}{\partial a}u_g + \frac{\partial M_{33}}{\partial a}u_w\right) \\ &- \lambda_p \left(\frac{\partial M_{41}}{\partial a}u_f + \frac{\partial M_{42}}{\partial a}u_g + \frac{\partial M_{43}}{\partial a}u_w\right) - \lambda_q \left(\frac{\partial M_{51}}{\partial a}u_f + \frac{\partial M_{52}}{\partial a}u_g + \frac{\partial M_{53}}{\partial a}u_w\right) \\ &-\lambda_\lambda \left(\frac{\partial M_{61}}{\partial a}u_f + \frac{\partial M_{62}}{\partial a}u_g + \frac{\partial M_{63}}{\partial a}u_w\right)\Bigg] \frac{2P}{mc} - \lambda_\lambda \frac{\partial n}{\partial a} \end{aligned} \tag{3.95}$$

where

$$\frac{\partial n}{\partial a} = -\frac{3n}{2a} = -\frac{3}{2}\mu^{1/2}a^{-5/2}$$

$$\begin{aligned} \dot{\lambda}_h = -\frac{\partial H}{\partial h} = \Bigg[&-\lambda_a \left(\frac{\partial M_{11}}{\partial h}u_f + \frac{\partial M_{12}}{\partial h}u_g + \frac{\partial M_{13}}{\partial h}u_w\right) \\ &-\lambda_h \left(\frac{\partial M_{21}}{\partial h}u_f + \frac{\partial M_{22}}{\partial h}u_g + \frac{\partial M_{23}}{\partial h}u_w\right) - \lambda_k \left(\frac{\partial M_{31}}{\partial h}u_f + \frac{\partial M_{32}}{\partial h}u_g + \frac{\partial M_{33}}{\partial h}u_w\right) \\ &- \lambda_p \left(\frac{\partial M_{41}}{\partial h}u_f + \frac{\partial M_{42}}{\partial h}u_g + \frac{\partial M_{43}}{\partial h}u_w\right) - \lambda_q \left(\frac{\partial M_{51}}{\partial h}u_f + \frac{\partial M_{52}}{\partial h}u_g + \frac{\partial M_{53}}{\partial h}u_w\right) \\ &-\lambda_\lambda \left(\frac{\partial M_{61}}{\partial h}u_f + \frac{\partial M_{62}}{\partial h}u_g + \frac{\partial M_{63}}{\partial h}u_w\right)\Bigg] \frac{2P}{mc} \end{aligned} \tag{3.96}$$

$$\begin{aligned} \dot{\lambda}_k = -\frac{\partial H}{\partial k} = \Bigg[&-\lambda_a \left(\frac{\partial M_{11}}{\partial k}u_f + \frac{\partial M_{12}}{\partial k}u_g + \frac{\partial M_{13}}{\partial k}u_w\right) \\ &- \lambda_h \left(\frac{\partial M_{21}}{\partial k}u_f + \frac{\partial M_{22}}{\partial k}u_g + \frac{\partial M_{23}}{\partial k}u_w\right) - \lambda_k \left(\frac{\partial M_{31}}{\partial k}u_f + \frac{\partial M_{32}}{\partial k}u_g + \frac{\partial M_{33}}{\partial k}u_w\right) \\ &- \lambda_p \left(\frac{\partial M_{41}}{\partial k}u_f + \frac{\partial M_{42}}{\partial k}u_g + \frac{\partial M_{43}}{\partial k}u_w\right) - \lambda_q \left(\frac{\partial M_{51}}{\partial k}u_f + \frac{\partial M_{52}}{\partial k}u_g + \frac{\partial M_{53}}{\partial k}u_w\right) \\ &-\lambda_\lambda \left(\frac{\partial M_{61}}{\partial k}u_f + \frac{\partial M_{62}}{\partial k}u_g + \frac{\partial M_{63}}{\partial k}u_w\right)\Bigg] \frac{2P}{mc} \end{aligned} \tag{3.97}$$

$$\begin{aligned}\dot{\lambda}_p = -\frac{\partial H}{\partial p} = \Bigg[&-\lambda_a\left(\frac{\partial M_{11}}{\partial p}u_f + \frac{\partial M_{12}}{\partial p}u_g + \frac{\partial M_{13}}{\partial p}u_w\right)\\ &- \lambda_h\left(\frac{\partial M_{21}}{\partial p}u_f + \frac{\partial M_{22}}{\partial p}u_g + \frac{\partial M_{23}}{\partial p}u_w\right) - \lambda_k\left(\frac{\partial M_{31}}{\partial p}u_f + \frac{\partial M_{32}}{\partial p}u_g + \frac{\partial M_{33}}{\partial p}u_w\right)\\ &- \lambda_p\left(\frac{\partial M_{41}}{\partial p}u_f + \frac{\partial M_{42}}{\partial p}u_g + \frac{\partial M_{43}}{\partial p}u_w\right) - \lambda_q\left(\frac{\partial M_{51}}{\partial p}u_f + \frac{\partial M_{52}}{\partial p}u_g + \frac{\partial M_{53}}{\partial p}u_w\right)\\ &-\lambda_\lambda\left(\frac{\partial M_{61}}{\partial p}u_f + \frac{\partial M_{62}}{\partial p}u_g + \frac{\partial M_{63}}{\partial p}u_w\right)\Bigg]\frac{2P}{mc}\end{aligned} \tag{3.98}$$

$$\begin{aligned}\dot{\lambda}_q = -\frac{\partial H}{\partial q} = \Bigg[&-\lambda_a\left(\frac{\partial M_{11}}{\partial q}u_f + \frac{\partial M_{12}}{\partial q}u_g + \frac{\partial M_{13}}{\partial q}u_w\right)\\ &- \lambda_h\left(\frac{\partial M_{21}}{\partial q}u_f + \frac{\partial M_{22}}{\partial q}u_g + \frac{\partial M_{23}}{\partial q}u_w\right) - \lambda_k\left(\frac{\partial M_{31}}{\partial q}u_f + \frac{\partial M_{32}}{\partial q}u_g + \frac{\partial M_{33}}{\partial q}u_w\right)\\ &- \lambda_p\left(\frac{\partial M_{41}}{\partial q}u_f + \frac{\partial M_{42}}{\partial q}u_g + \frac{\partial M_{43}}{\partial q}u_w\right) - \lambda_q\left(\frac{\partial M_{51}}{\partial q}u_f + \frac{\partial M_{52}}{\partial q}u_g + \frac{\partial M_{53}}{\partial q}u_w\right)\\ &-\lambda_\lambda\left(\frac{\partial M_{61}}{\partial q}u_f + \frac{\partial M_{62}}{\partial q}u_g + \frac{\partial M_{63}}{\partial q}u_w\right)\Bigg]\frac{2P}{mc}\end{aligned} \tag{3.99}$$

$$\begin{aligned}\dot{\lambda}_\lambda = -\frac{\partial H}{\partial \lambda} = \Bigg[&-\lambda_a\left(\frac{\partial M_{11}}{\partial \lambda}u_f + \frac{\partial M_{12}}{\partial \lambda}u_g + \frac{\partial M_{13}}{\partial \lambda}u_w\right)\\ &- \lambda_h\left(\frac{\partial M_{21}}{\partial \lambda}u_f + \frac{\partial M_{22}}{\partial \lambda}u_g + \frac{\partial M_{23}}{\partial \lambda}u_w\right) - \lambda_k\left(\frac{\partial M_{31}}{\partial \lambda}u_f + \frac{\partial M_{32}}{\partial \lambda}u_g + \frac{\partial M_{33}}{\partial \lambda}u_w\right)\\ &- \lambda_p\left(\frac{\partial M_{41}}{\partial \lambda}u_f + \frac{\partial M_{42}}{\partial \lambda}u_g + \frac{\partial M_{43}}{\partial \lambda}u_w\right) - \lambda_q\left(\frac{\partial M_{51}}{\partial \lambda}u_f + \frac{\partial M_{52}}{\partial \lambda}u_g + \frac{\partial M_{53}}{\partial \lambda}u_w\right)\\ &-\lambda_\lambda\left(\frac{\partial M_{61}}{\partial \lambda}u_f + \frac{\partial M_{62}}{\partial \lambda}u_g + \frac{\partial M_{63}}{\partial \lambda}u_w\right)\Bigg]\frac{2P}{mc}\end{aligned} \tag{3.100}$$

$$\begin{aligned}\dot{\lambda}_m = -\frac{\partial H}{\partial m} = \big[&\lambda_a\left(M_{11}u_f + M_{12}u_g + M_{13}u_w\right)\\ &+ \lambda_h\left(M_{21}u_f + M_{22}u_g + M_{23}u_w\right)\\ &+ \lambda_k\left(M_{31}u_f + M_{32}u_g + M_{33}u_w\right)\\ &+ \lambda_p\left(M_{41}u_f + M_{42}u_g + M_{43}u_w\right)\\ &+ \lambda_q\left(M_{51}u_f + M_{52}u_g + M_{53}u_w\right)\\ &+\lambda_\lambda\left(M_{61}u_f + M_{62}u_g + M_{63}u_w\right)\big]\frac{2P}{m^2c}\end{aligned} \tag{3.101}$$

The switching function S is written in expanded form as

$$\begin{aligned}S = &\lambda_a\left(M_{11}u_f + M_{12}u_g + M_{13}u_w\right) + \lambda_h\left(M_{21}u_f + M_{22}u_g + M_{23}u_w\right)\\ &+ \lambda_k\left(M_{31}u_f + M_{32}u_g + M_{33}u_w\right) + \lambda_p\left(M_{41}u_f + M_{42}u_g + M_{43}u_w\right)\\ &+ \lambda_q\left(M_{51}u_f + M_{52}u_g + M_{53}u_w\right) + \lambda_\lambda\left(M_{61}u_f + M_{62}u_g + M_{63}u_w\right) - \frac{\lambda_\lambda m}{c}\end{aligned} \tag{3.102}$$

The following partials are used in the generation of the $\partial M/\partial \mathbf{z}$ partials of Appendix 3.1. Because λ is being integrated, F is independent of a because $\partial\lambda/\partial a$ yields $\partial F/\partial a = 0$ from Kepler's equation:

$$\frac{\partial r}{\partial a} = \frac{r}{a}, \quad \frac{\partial r}{\partial h} = \frac{a^2}{r}(h - s_F), \quad \frac{\partial r}{\partial k} = \frac{a^2}{r}(k - c_F), \quad \frac{\partial r}{\partial F} = a(ks_F - hc_F),$$
$$\frac{\partial F}{\partial a} = 0, \quad \frac{\partial F}{\partial h} = -\frac{a}{r}c_F, \quad \frac{\partial F}{\partial k} = \frac{a}{r}s_F, \quad \frac{\partial F}{\partial \lambda} = \frac{a}{r}$$

3.10 Examples

The minimization of fuel expenditure is equivalent to the maximization of the seventh component of the state vector at the fixed terminal time t_f. Therefore the performance index is given by $J = \phi = m_f$ and the maximum principle is applied by selecting the control $\hat{\mathbf{u}}$ which is the thrust orientation such that the Hamiltonian is maximized at each instant of time:

$$\hat{\mathbf{u}} = \frac{\left(\boldsymbol{\lambda}_z^T M\right)^T}{\left|\boldsymbol{\lambda}_z^T M\right|} \tag{3.103}$$

Furthermore, when the current mean longitude λ is selected as the fast element, the equations of motion are not explicit functions of time such that the system is autonomous and therefore the Hamiltonian is constant throughout the optimal trajectory, which is made up of powered as well as coast arcs. Now given initial parameters $(a)_0$, $(h)_0$, $(k)_0$, $(p)_0$, $(q)_0$, $(\lambda)_0$, $(m)_0$, one guesses the initial values of the Lagrange multipliers $(\lambda_a)_0$, $(\lambda_h)_0$, $(\lambda_k)_0$, $(\lambda_p)_0$, $(\lambda_q)_0$, $(\lambda_\lambda)_0$, $(\lambda_m)_0$, and integrates from t_0 to t_f Equations (3.79), (3.80) and (3.88) for the states as well as Equations (3.95) through (3.101) for the adjoints by selecting the optimal control in Equation (3.103). The power is on and off according to whether the sign of the switching function S in Equation (3.102) is positive or negative. An iteration is carried out on the initial values of the multipliers until the desired terminal state is reached, namely $(a)_f$, $(h)_f$, $(k)_f$, $(p)_f$, $(q)_f$, $(\lambda)_f$ and $(\lambda_m)_f = (\partial\phi/\partial m)_{t_f} = 1$ for a minimum fuel solution. In practice an objective function F' is driven to zero to within a small tolerance with

$$F' = \frac{\left|a - (a)_f\right|}{(a)_f} + \left|h - (h)_f\right| + \left|k - (k)_f\right| + \left|p - (p)_f\right| + \left|q - (q)_f\right| + \frac{\left|\lambda - (\lambda)_f\right|}{2\pi} + \left|\lambda_m - 1\right| \tag{3.104}$$

where a, h, k, p, q, λ and λ_m are evaluated at the fixed terminal time t_f. As a first example, let $a_0 = 42{,}000$ km, $e_0 = 0$, $i_0 = 28.5$ deg, $\Omega_0 = 30$ deg, $\omega_0 = 10$ deg, $M_0 = 0$ deg, $m_0 = 6000$ kg, $I_{sp} = 3800$ s and $P = 39,112.3$ W. The target parameters are $(a)_f = 42{,}767.073$ km, $(e)_f = 1.64459 \times 10^{-4}$, $(i)_f = 28.343$ deg, $(\Omega)_f = 29.999$ deg, $(\omega)_f = 247.299$ deg, $(M)_f = 120.905$ deg, the same as for the example used in the previous section. In that section, the mass equation was neglected and a constant continuous acceleration $f_t = 3.5 \times 10^{-7}$ km/s^2 was used. Because we are now using the mass equation, as well as power P and exhaust velocity $c = I_{sp}g$ with $g = 9.80665$ m/s^2, the acceleration of gravity, and I_{sp} the specific impulse of the propulsion system, and because the minimum-time solution required $t_f = 86{,}402.453$ s, the value of the power $P = 39{,}112.3$ W used here is an average value such that the average acceleration is also $f_t = 3.5 \times 10^{-7}$ km/s^2. Using first $P = 40{,}000$W, a mass flow rate $\dot{m} = (-2P)/(c^2) = [-(2 \times 40{,}000)]/(3800 \times 9.80665)^2 = -5.760782 \times 10^{-5}$kg/s is computed which gives a fuel expenditure Δm of 4.977 kg during $t_f = 86{,}402.453$ s. This means that the initial mass of 6000 kg is dropped to 5999.022 kg at the end of the rendezvous maneuver such that the average mass is 5997.511 kg and with $I_{sp} = 3800$ s held constant.

$$\frac{2P}{mc} = \frac{2P}{5997.511 \times (3800 \times 9.80665)} = 3.5 \times 10^{-4}\text{m/s}^2 \tag{3.105}$$

requires $P = 39{,}112.3$ W. The minimum-time solution is recovered with $(\lambda_a)_0 = 4.362055989 \times 10^{-4}$ kg/m, $(\lambda_h)_0 = -4.405792619 \times 10^1$ kg, $(\lambda_k)_0 = 5.409257264 \times 10^1$ kg, $(\lambda_p)_0 = -2.199834264 \times 10^4$ kg, $(\lambda_q)_0 = -3.634568409 \times 10^4$ kg, $(\lambda_\lambda)_0 = -8.591412279 \times 10^1$ kg/rad, $(\lambda_m)_0 = 0.9358240874$ kg/kg with the final state parameters of $a = 42{,}767.052$ km, $e = 1.64475 \times 10^{-4}$, $i = 28.343$ deg, $\Omega = 29.995$ deg, $\omega = 247.336$ deg, $M = 120.873$ deg and $\lambda_m = 1.000000090$, indicating convergence to a locally optimal solution.

Figures 3.13 through 3.16 depict the evolution of λ_a, λ_λ, λ_h, λ_k, λ_p, λ_q, λ_m and the switching function S, which remains positive throughout, indicating no coast. These Lagrange multipliers are almost exact duplicates of those in the previous sections on the minimum-time transfer except for a scaling factor, and this is due to the fact that this minimum-fuel solution is also the minimum-time solution because the thrust is continuously on. This is expected because we used the fixed terminal time t_f of the minimum-time solution and an equivalent constant acceleration.

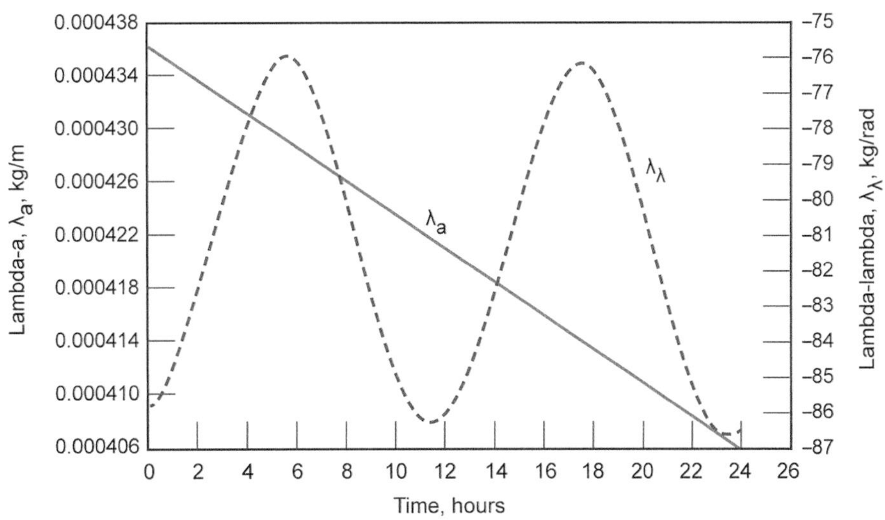

Figure 3.13 Evolution of λ_a and λ_λ for continuous burn case.

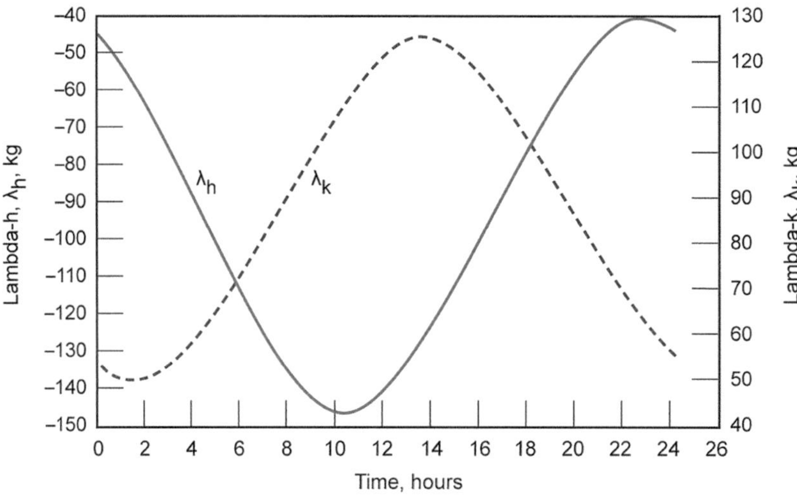

Figure 3.14 Evolution of λ_h and λ_k for continuous burn case.

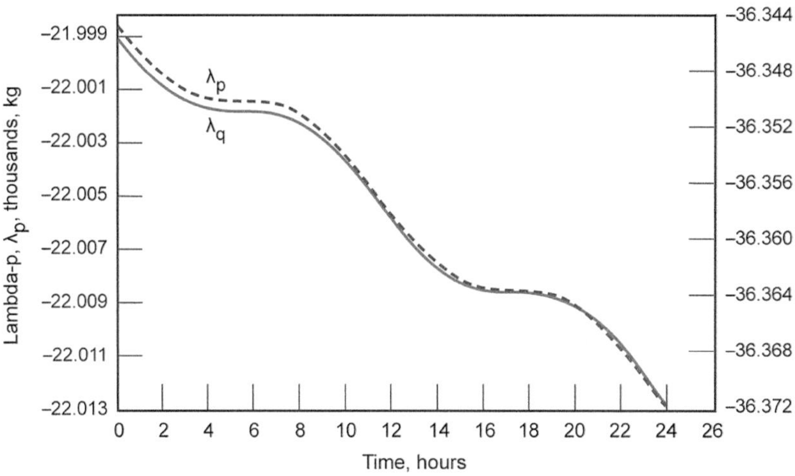

Figure 3.15 Evolution of λ_p and λ_q for continuous burn case.

In our second example, we let $t_f = 175{,}100$ s, $(e)_f = 9.0 \times 10^{-4}$, $(\omega)_f = 195$ deg, $(M)_f = 173.1$ deg, leaving otherwise all other parameters unchanged, and iterate numerically to $(\lambda_a)_0 = 1.450621648 \times 10^{-4}$ kg/m, $(\lambda_h)_0 = -8.061568896 \times 10^2$ kg, $(\lambda_k)_0 = 4.482581471 \times 10^2$ kg, $(\lambda_p)_0 = -1.973556477 \times 10^2$ kg, $(\lambda_q)_0 = -9.445686274 \times 10^2$ kg, $(\lambda_\lambda)_0 = -4.366236008 \times 10^2$ kg/rad, $(\lambda_m)_0 = 0.9841195033$ kg/kg leading to $a = 42{,}768.260$ km, $e = 9.257 \times 10^{-4}$, $i = 28.341$ deg, $\Omega = 30.065$ deg, $\omega = 195.181$ deg, $M = 172.958$ deg, $\lambda_m = 1.0000742$ at the final time for a locally optimal solution. There exists now a coast arc whose duration is over an hour such that the rendezvous trajectory is composed of burn–coast–burn arcs. It is necessary to stop and restart the integrations at the boundaries between the thrust and coast arcs for an accurate integration. Also, the discontinuities in the adjoints

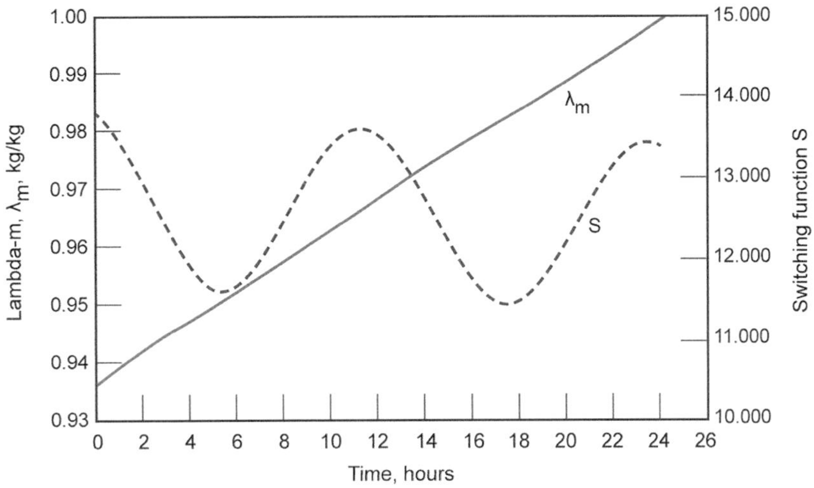

Figure 3.16 Evolution of λ_m and the switching function S for continuous burn case.

at those boundaries need to be considered as additional search variables for accurate results.

The orbit grows to become larger than the final orbit and the yawing of the thrust vector is delayed for a more efficient orbit plane rotation.

Dicussion of orbit transfers with coasting periods as well as time-fixed minimum-fuel transfers appears in [14–20]. Many such contributions by Betts, Kluever and others are already referenced in Chapters 5, 7 and 8, using the direct methods of collocation. However, it is beyond the scope of this book to carry out detailed descriptions of these contributions. The books by Conway [15], *Spacecraft Trajectory Optimization*, and Betts [14], *Practical Methods for Optimal Control Using Nonlinear Programming*, are important sources for direct method applications, and recent papers by Tarzi [17], Mazzini [18], Lundberg [16] and Graham and Rao [19, 20] also address eclipsing. Many of these researchers, such as Tarzi, use equinoctial elements and averaging for rapid solution generation [17], while Mazzini makes use of a form of Hill's variables, albeit singular for $i = 0$, but with averaging [18]. Lundberg [16] adopts precision integration for exact solutions with eclipsing by handling the discontinuities in the dynamics due to the eclipse-dependent thrust jumps. The semimajor axis is often replaced by the semilatus rectum as the first element of the equinoctial set to allow the consideration of escape trajectories as well.

3.11 Conclusion

The general solution of the minimum-fuel time-fixed rendezvous problem has been analyzed based on the use of a set of nonsingular orbit elements and by direct application of optimal control theory. The rendezvous time being larger than the minimum-time solution, one or more coasting arcs are naturally introduced in order to minimize fuel use. The boundary conditions are simplified by the adoption of the mean longitude as the fast element or sixth orbital variable.

Appendix 3.1 The Partials of the *M* Matrix

The partial derivatives of M with respect to h are

$$\frac{\partial M_{11}}{\partial h} = \frac{2}{n^2 a}\frac{\partial \dot{X}_1}{\partial h} \tag{A3.1}$$

$$\frac{\partial M_{12}}{\partial h} = \frac{2}{n^2 a}\frac{\partial \dot{Y}_1}{\partial h} \tag{A3.2}$$

$$\frac{\partial M_{13}}{\partial h} = 0 \tag{A3.3}$$

$$\begin{aligned}\frac{\partial M_{21}}{\partial h} &= \frac{-h}{na^2(1-h^2-k^2)^{1/2}}\left(\frac{\partial X_1}{\partial k} - \frac{h\beta}{n}\dot{X}_1\right) \\ &+ \frac{(1-h^2-k^2)^{1/2}}{na^2}\left[\frac{\partial^2 X_1}{\partial h \partial k} - \frac{\dot{X}_1}{n}\left(\beta + \frac{h^2\beta^3}{1-\beta}\right) - \frac{h\beta}{n}\frac{\partial \dot{X}_1}{\partial h}\right]\end{aligned} \tag{A3.4}$$

In a similar way,

$$\begin{aligned}\frac{\partial M_{22}}{\partial h} &= \frac{-h}{na^2(1-h^2-k^2)^{1/2}}\left(\frac{\partial Y_1}{\partial k} - \frac{h\beta}{n}\dot{Y}_1\right) \\ &+ \frac{(1-h^2-k^2)^{1/2}}{na^2}\left[\frac{\partial^2 Y_1}{\partial h \partial k} - \frac{\dot{Y}_1}{n}\left(\beta + \frac{h^2\beta^3}{1-\beta}\right) - \frac{h\beta}{n}\frac{\partial \dot{Y}_1}{\partial h}\right]\end{aligned} \tag{A3.5}$$

$$\frac{\partial M_{23}}{\partial h} = \frac{hk(1-h^2-k^2)^{-3/2}}{na^2}(qY_1 - pX_1) + \frac{k\left(q\frac{\partial Y_1}{\partial h} - p\frac{\partial X_1}{\partial h}\right)}{na^2(1-h^2-k^2)^{1/2}} \tag{A3.6}$$

$$\begin{aligned}\frac{\partial M_{31}}{\partial h} &= \frac{h}{na^2(1-h^2-k^2)^{1/2}}\left(\frac{\partial X_1}{\partial h} + k\beta\frac{\dot{X}_1}{n}\right) \\ &- \frac{(1-h^2-k^2)^{1/2}}{na^2}\left(\frac{\partial^2 X_1}{\partial h^2} + \frac{hk\beta^3}{1-\beta}\frac{\dot{X}_1}{n} + \frac{k\beta}{n}\frac{\partial \dot{X}_1}{\partial h}\right)\end{aligned} \tag{A3.7}$$

$$\begin{aligned}\frac{\partial M_{32}}{\partial h} &= \frac{h}{na^2(1-h^2-k^2)^{1/2}}\left(\frac{\partial Y_1}{\partial h} + k\beta\frac{\dot{Y}_1}{n}\right) \\ &- \frac{(1-h^2-k^2)^{1/2}}{na^2}\left(\frac{\partial^2 Y_1}{\partial h^2} + \frac{hk\beta^3}{1-\beta}\frac{\dot{Y}_1}{n} + \frac{k\beta}{n}\frac{\partial \dot{Y}_1}{\partial h}\right)\end{aligned} \tag{A3.8}$$

$$\begin{aligned}\frac{\partial M_{33}}{\partial h} &= \frac{-1}{na^2(1-h^2-k^2)^{1/2}}\left[(qY_1 - pX_1) + h\left(q\frac{\partial Y_1}{\partial h} - p\frac{\partial X_1}{\partial h}\right)\right] \\ &- \frac{h^2(qY_1 - pX_1)}{na^2(1-h^2-k^2)^{3/2}}\end{aligned} \tag{A3.9}$$

$$\frac{\partial M_{43}}{\partial h} = \frac{(1+p^2+q^2)}{2na^2(1-h^2-k^2)^{1/2}}\left[\frac{\partial Y_1}{\partial h} + \frac{hY_1}{(1-h^2-k^2)}\right] \tag{A3.10}$$

$$\frac{\partial M_{53}}{\partial h} = \frac{(1+p^2+q^2)}{2na^2(1-h^2-k^2)^{1/2}}\left[\frac{\partial X_1}{\partial h} + \frac{hX_1}{(1-h^2-k^2)}\right] \tag{A3.11}$$

The following partials are identically equal to zero, namely

$$\frac{\partial M_{41}}{\partial h} = \frac{\partial M_{42}}{\partial h} = \frac{\partial M_{51}}{\partial h} = \frac{\partial M_{52}}{\partial h} = 0$$

$$\begin{aligned}\frac{\partial M_{61}}{\partial h} = \frac{1}{na^2}\Bigg\{&-2\frac{\partial X_1}{\partial h} - h\beta(1-h^2-k^2)^{-1/2}\left(h\frac{\partial X_1}{\partial h} + k\frac{\partial X_1}{\partial k}\right) + (1-h^2-k^2)^{1/2} \\ &\times\left[\left(\beta + \frac{h^2\beta^3}{1-\beta}\right)\frac{\partial X_1}{\partial h} + \frac{hk\beta^3}{1-\beta}\frac{\partial X_1}{\partial k} + \beta\left(h\frac{\partial^2 X_1}{\partial h^2} + k\frac{\partial^2 X_1}{\partial h\partial k}\right)\right]\Bigg\}\end{aligned} \tag{A3.12}$$

$$\begin{aligned}\frac{\partial M_{62}}{\partial h} = \frac{1}{na^2}\Bigg\{&-2\frac{\partial Y_1}{\partial h} - h\beta(1-h^2-k^2)^{-1/2}\left(h\frac{\partial Y_1}{\partial h} + k\frac{\partial Y_1}{\partial k}\right) + (1-h^2-k^2)^{1/2} \\ &\times\left[\left(\beta + \frac{h^2\beta^3}{1-\beta}\right)\frac{\partial Y_1}{\partial h} + \frac{hk\beta^3}{1-\beta}\frac{\partial Y_1}{\partial k} + \beta\left(h\frac{\partial^2 Y_1}{\partial h^2} + k\frac{\partial^2 Y_1}{\partial h\partial k}\right)\right]\Bigg\}\end{aligned} \tag{A3.13}$$

$$\frac{\partial M_{63}}{\partial h} = \frac{(1-h^2-k^2)^{-1/2}}{na^2}\left[\left(q\frac{\partial Y_1}{\partial h} - p\frac{\partial X_1}{\partial h}\right) + h(1-h^2-k^2)^{-1}(qY_1 - pX_1)\right] \tag{A3.14}$$

The partial derivatives of M with respect to k are

$$\frac{\partial M_{11}}{\partial k} = \frac{2}{n^2 a}\frac{\partial \dot{X}_1}{\partial k} \tag{A3.15}$$

$$\frac{\partial M_{12}}{\partial k} = \frac{2}{n^2 a}\frac{\partial \dot{Y}_1}{\partial k} \tag{A3.16}$$

$$\frac{\partial M_{13}}{\partial k} = 0 \tag{A3.17}$$

$$\begin{aligned}\frac{\partial M_{21}}{\partial k} &= \frac{-k}{na^2(1-h^2-k^2)^{1/2}}\left(\frac{\partial X_1}{\partial k} - \frac{h\beta}{n}\dot{X}_1\right) \\ &+ \frac{(1-h^2-k^2)^{1/2}}{na^2}\left[\frac{\partial^2 X_1}{\partial k^2} - \frac{hk\beta^3}{n(1-\beta)}\dot{X}_1 - \frac{h\beta}{n}\frac{\partial \dot{X}_1}{\partial k}\right]\end{aligned} \tag{A3.18}$$

$$\begin{aligned}\frac{\partial M_{22}}{\partial k} &= \frac{-k}{na^2(1-h^2-k^2)^{1/2}}\left(\frac{\partial Y_1}{\partial k} - \frac{h\beta}{n}\dot{Y}_1\right) \\ &+ \frac{(1-h^2-k^2)^{1/2}}{na^2}\left[\frac{\partial^2 Y_1}{\partial k^2} - \frac{hk\beta^3}{n(1-\beta)}\dot{Y}_1 - \frac{h\beta}{n}\frac{\partial \dot{Y}_1}{\partial k}\right]\end{aligned} \tag{A3.19}$$

$$\frac{\partial M_{23}}{\partial k} = \frac{(qY_1 - pX_1)}{na^2(1-h^2-k^2)^{1/2}} + \frac{1}{na^2(1-h^2-k^2)^{1/2}} \times \left[k\left(q\frac{\partial Y_1}{\partial k} - p\frac{\partial X_1}{\partial k}\right) + \frac{k^2(qY_1 - pX_1)}{(1-h^2-k^2)}\right] \tag{A3.20}$$

$$\frac{\partial M_{31}}{\partial k} = \frac{k}{na^2(1-h^2-k^2)^{1/2}}\left(\frac{\partial X_1}{\partial h} + k\beta\frac{\dot{X}_1}{n}\right) - \frac{(1-h^2-k^2)^{1/2}}{na^2} \times \left[\frac{\partial^2 X_1}{\partial k \partial h} + \left(\beta + \frac{k^2\beta^3}{1-\beta}\right)\frac{\dot{X}_1}{n} + \frac{k\beta}{n}\frac{\partial \dot{X}_1}{\partial k}\right] \tag{A3.21}$$

$$\frac{\partial M_{32}}{\partial k} = \frac{k}{na^2(1-h^2-k^2)^{1/2}}\left(\frac{\partial Y_1}{\partial h} + k\beta\frac{\dot{Y}_1}{n}\right) - \frac{(1-h^2-k^2)^{1/2}}{na^2} \times \left[\frac{\partial^2 Y_1}{\partial k \partial h} + \left(\beta + \frac{k^2\beta^3}{1-\beta}\right)\frac{\dot{Y}_1}{n} + \frac{k\beta}{n}\frac{\partial \dot{Y}_1}{\partial k}\right] \tag{A3.22}$$

$$\frac{\partial M_{33}}{\partial k} = \frac{-h}{na^2(1-h^2-k^2)^{1/2}}\left(q\frac{\partial Y_1}{\partial k} - p\frac{\partial X_1}{\partial k}\right) - \frac{hk}{na^2(1-h^2-k^2)^{3/2}}(qY_1 - pX_1) \tag{A3.23}$$

$$\frac{\partial M_{43}}{\partial k} = \frac{(1+p^2+q^2)}{2na^2(1-h^2-k^2)^{1/2}}\frac{\partial Y_1}{\partial k} + \frac{k(1+p^2+q^2)}{2na^2(1-h^2-k^2)^{3/2}}Y_1 \tag{A3.24}$$

$$\frac{\partial M_{53}}{\partial k} = \frac{(1+p^2+q^2)}{2na^2(1-h^2-k^2)^{1/2}}\frac{\partial X_1}{\partial k} + \frac{k(1+p^2+q^2)}{2na^2(1-h^2-k^2)^{3/2}}X_1 \tag{A3.25}$$

$$\frac{\partial M_{41}}{\partial k} = \frac{\partial M_{42}}{\partial k} = \frac{\partial M_{51}}{\partial k} = \frac{\partial M_{52}}{\partial k} = 0$$

$$\frac{\partial M_{61}}{\partial k} = \frac{1}{na^2}\left\{-2\frac{\partial X_1}{\partial k} - k\beta(1-h^2-k^2)^{-1/2}\left(h\frac{\partial X_1}{\partial h} + k\frac{\partial X_1}{\partial k}\right) + (1-h^2-k^2)^{1/2} \times \left[\left(\beta + \frac{k^2\beta^3}{1-\beta}\right)\frac{\partial X_1}{\partial k} + \frac{hk\beta^3}{1-\beta}\frac{\partial X_1}{\partial h} + \beta\left(h\frac{\partial^2 X_1}{\partial k \partial h} + k\frac{\partial^2 X_1}{\partial k^2}\right)\right]\right\} \tag{A3.26}$$

$$\frac{\partial M_{62}}{\partial k} = \frac{1}{na^2}\left\{-2\frac{\partial Y_1}{\partial k} - k\beta(1-h^2-k^2)^{-1/2}\left(h\frac{\partial Y_1}{\partial h} + k\frac{\partial Y_1}{\partial k}\right) + (1-h^2-k^2)^{1/2} \times \left[\left(\beta + \frac{k^2\beta^3}{1-\beta}\right)\frac{\partial Y_1}{\partial k} + \frac{hk\beta^3}{1-\beta}\frac{\partial Y_1}{\partial h} + \beta\left(h\frac{\partial^2 Y_1}{\partial k \partial h} + k\frac{\partial^2 Y_1}{\partial k^2}\right)\right]\right\} \tag{A3.27}$$

$$\frac{\partial M_{63}}{\partial k} = \frac{(1-h^2-k^2)^{-1/2}}{na^2}\left[\left(q\frac{\partial Y_1}{\partial k} - p\frac{\partial X_1}{\partial k}\right) + k(1-h^2-k^2)^{-1}(qY_1 - pX_1)\right] \tag{A3.28}$$

The partial derivatives of M with respect to p are given by the foregoing, with the only nonzero partials being

$$\frac{\partial M_{23}}{\partial p} = \frac{-kX_1}{na^2(1-h^2-k^2)^{1/2}} \tag{A3.29}$$

$$\frac{\partial M_{33}}{\partial p} = \frac{hX_1}{na^2(1-h^2-k^2)^{1/2}} \tag{A3.30}$$

$$\frac{\partial M_{43}}{\partial p} = \frac{pY_1}{na^2(1-h^2-k^2)^{1/2}} \tag{A3.31}$$

$$\frac{\partial M_{53}}{\partial p} = \frac{pX_1}{na^2(1-h^2-k^2)^{1/2}} \tag{A3.32}$$

$$\frac{\partial M_{63}}{\partial p} = \frac{-X_1}{na^2(1-h^2-k^2)^{1/2}} \tag{A3.33}$$

The partial derivatives of M with respect to q are given below, with the only nonzero partials being

$$\frac{\partial M_{23}}{\partial q} = \frac{kY_1}{na^2(1-h^2-k^2)^{1/2}} \tag{A3.34}$$

$$\frac{\partial M_{33}}{\partial q} = \frac{-hY_1}{na^2(1-h^2-k^2)^{1/2}} \tag{A3.35}$$

$$\frac{\partial M_{43}}{\partial q} = \frac{qY_1}{na^2(1-h^2-k^2)^{1/2}} \tag{A3.36}$$

$$\frac{\partial M_{53}}{\partial q} = \frac{qX_1}{na^2(1-h^2-k^2)^{1/2}} \tag{A3.37}$$

$$\frac{\partial M_{63}}{\partial q} = \frac{Y_1}{na^2(1-h^2-k^2)^{1/2}} \tag{A3.38}$$

The partial derivatives of $\dot{X}_1$ with respect to h and k are

$$\begin{aligned}\frac{\partial \dot{X}_1}{\partial h} &= \frac{a}{r}\dot{X}_1\left[s_F + \frac{a}{r}c_F\left(ks_F - hc_F\right)\right] + \frac{na^2}{r}\left\{h\beta s_F + \left(kc_F + hs_F\right)\left(\beta + \frac{h^2\beta^3}{1-\beta}\right)\right.\\ &\left.+\frac{a}{r}c_F\left[hk\beta s_F + \left(1 - h^2\beta\right)c_F\right]\right\}\end{aligned} \tag{A3.39}$$

$$\begin{aligned}\frac{\partial \dot{X}_1}{\partial k} &= -\frac{\dot{X}_1}{r}a\left[-c_F + \frac{a}{r}s_F\left(ks_F - hc_F\right)\right] + \frac{na^2}{r}\\ &\times\left\{\frac{hk\beta^3}{1-\beta}\left(kc_F + hs_F\right) + h\beta c_F - \frac{a}{r}s_F\left[hk\beta s_F + \left(1 - h^2\beta\right)c_F\right]\right\}\end{aligned} \tag{A3.40}$$

The partial derivatives of $\dot{Y}_1$ with respect to h and k are

$$\begin{aligned}\frac{\partial \dot{Y}_1}{\partial h} &= -\frac{\dot{Y}_1}{r}a\left[-s_F - \frac{a}{r}c_F\left(ks_F - hc_F\right)\right] + \frac{na^2}{r}\\ &\times\left\{\frac{-hk\beta^3}{1-\beta}\left(kc_F + hs_F\right) - k\beta s_F + \left[hk\beta c_F + \left(1 - k^2\beta\right)s_F\right]\frac{a}{r}c_F\right\}\end{aligned} \tag{A3.41}$$

$$\begin{aligned}\frac{\partial \dot{Y}_1}{\partial k} &= \frac{-\dot{Y}_1}{r}a\left[-c_F + \frac{a}{r}s_F\left(ks_F - hc_F\right)\right] + \frac{na^2}{r}\\ &\times\left\{-\left(\beta + \frac{k^2\beta^3}{1-\beta}\right)\left(kc_F + hs_F\right) - k\beta c_F - \frac{a}{r}s_F\left[hk\beta c_F + \left(1 - k^2\beta\right)s_F\right]\right\}\end{aligned} \tag{A3.42}$$

The second partial derivatives of X_1 and Y_1 with respect to h and k are

$$\frac{\partial^2 X_1}{\partial h^2} = a\left\{-\frac{2a}{r}c_F\left(\beta+\frac{h^2\beta^3}{1-\beta}\right)-\frac{h\beta^3}{1-\beta}(hc_F-ks_F)\left[3+\frac{h^2\beta^2(3-2\beta)}{(1-\beta)^2}\right]\right.$$
$$\left.+\frac{a^2}{r^2}c_F(h\beta-s_F)\left[-s_F+\frac{a}{r}(h-s_F)\right]-\frac{a^2}{r^2}c_F^3\right\} \quad \text{(A3.43)}$$

$$\frac{\partial^2 X_1}{\partial k^2} = -a\left\{-\frac{2a}{r}s_F\frac{hk\beta^3}{(1-\beta)}+(hc_F-ks_F)\left[1+\frac{k^2\beta^2(3-2\beta)}{(1-\beta)^2}\right]\frac{h\beta^3}{(1-\beta)}\right.$$
$$\left.+\frac{a^2}{r^2}s_F(h\beta-s_F)\left[-c_F+\frac{a}{r}(k-c_F)\right]+\frac{a^2}{r^2}c_Fs_F^2\right\} \quad \text{(A3.44)}$$

$$\frac{\partial^2 X_1}{\partial h\partial k} = -a\left\{\frac{a}{r}c_F\frac{hk\beta^3}{(1-\beta)}+(hc_F-ks_F)\left[1+\frac{h^2\beta^2(3-2\beta)}{(1-\beta)^2}\right]\frac{k\beta^3}{(1-\beta)}\right.$$
$$\left.+(s_F-h\beta)\left[\frac{a}{r}\left(s_F^2-hs_F\right)-c_F^2\right]\frac{a^2}{r^2}-\frac{a^2}{r^2}s_Fc_F^2-\frac{a}{r}s_F\left(\beta+\frac{h^2\beta^3}{1-\beta}\right)\right\}$$
$$\text{(A3.45)}$$

$$\frac{\partial^2 X_1}{\partial k\partial h} = a\left\{\frac{a}{r}s_F\left(\beta+\frac{h^2\beta^3}{(1-\beta)}\right)-(hc_F-ks_F)\left[1+\frac{h^2\beta^2(3-2\beta)}{(1-\beta)^2}\right]\frac{k\beta^3}{(1-\beta)}\right.$$
$$\left.+\frac{a^2}{r^2}\left[\frac{a}{r}\left(kc_F-c_F^2\right)+s_F^2\right](h\beta-s_F)-\frac{a}{r}c_F\frac{hk\beta^3}{(1-\beta)}+\frac{a^2}{r^2}c_F^2s_F\right\} \quad \text{(A3.46)}$$

$$\frac{\partial^2 Y_1}{\partial h^2} = a\left\{\frac{2a}{r}c_F\frac{hk\beta^3}{(1-\beta)}+(hc_F-ks_F)\frac{k\beta^3}{(1-\beta)}\left[1+\frac{h^2\beta^2(3-2\beta)}{(1-\beta)^2}\right]+\frac{a^2}{r^2}c_F\right.$$
$$\left.\times\left[-\frac{a}{r}(h-s_F)+s_F\right](k\beta-c_F)-\frac{a^2}{r^2}s_Fc_F^2\right\} \quad \text{(A3.47)}$$

$$\frac{\partial^2 Y_1}{\partial k^2} = a\left\{\frac{-2a}{r}s_F\left(\beta+\frac{k^2\beta^3}{1-\beta}\right)+(hc_F-ks_F)\left[3+\frac{k^2\beta^2(3-2\beta)}{(1-\beta)^2}\right]\frac{k\beta^3}{(1-\beta)}+\frac{a^2}{r^2}s_F\right.$$
$$\left.\times\left[-\frac{a}{r}(k-c_F)+c_F\right](c_F-k\beta)-\frac{a^2}{r^2}s_F^3\right\} \quad \text{(A3.48)}$$

$$\frac{\partial^2 Y_1}{\partial h\partial k} = a\left\{\frac{a}{r}c_F\left(\beta+\frac{k^2\beta^3}{1-\beta}\right)+(hc_F-ks_F)\frac{h\beta^3}{(1-\beta)}\left[1+\frac{k^2\beta^2(3-2\beta)}{(1-\beta)^2}\right]\right.$$
$$\left.-\frac{a^2}{r^2}\left[\frac{a}{r}s_F(h-s_F)+c_F^2\right](c_F-k\beta)+\frac{a^2}{r^2}c_Fs_F^2-\frac{a}{r}s_F\frac{hk\beta^3}{(1-\beta)}\right\} \quad \text{(A3.49)}$$

$$\frac{\partial^2 Y_1}{\partial k\partial h} = a\left\{-\frac{a}{r}s_F\frac{hk\beta^3}{(1-\beta)}+(hc_F-ks_F)\frac{h\beta^3}{(1-\beta)}\right.$$
$$\times\left[1+\frac{k^2\beta^2(3-2\beta)}{(1-\beta)^2}\right]-\frac{a^2}{r^2}\left[\frac{a}{r}c_F(k-c_F)+s_F^2\right]$$
$$\left.\times(k\beta-c_F)+\frac{a}{r}c_F\left(\beta+\frac{k^2\beta^3}{1-\beta}\right)+\frac{a^2}{r^2}c_Fs_F^2\right\}. \quad \text{(A3.50)}$$

Next, the accessory partials $\partial^2 X_1/\partial a \partial k$, $\partial^2 X_1/\partial a \partial h$, $\partial^2 Y_1/\partial a \partial k$ and $\partial^2 Y_1/\partial a \partial h$ are generated with $\partial F/\partial a = 0$. In all the following partials, $\partial X_1/\partial a = X_1/a$ and $\partial Y_1/\partial a = Y_1/a$

$$\frac{\partial^2 X_1}{\partial a \partial k} = \frac{1}{a}\frac{\partial X_1}{\partial k} \tag{A3.51}$$

$$\frac{\partial^2 X_1}{\partial a \partial h} = \frac{1}{a}\frac{\partial X_1}{\partial h} \tag{A3.52}$$

$$\frac{\partial^2 Y_1}{\partial a \partial k} = \frac{1}{a}\frac{\partial Y_1}{\partial k} \tag{A3.53}$$

$$\frac{\partial^2 Y_1}{\partial a \partial h} = \frac{1}{a}\frac{\partial Y_1}{\partial h} \tag{A3.54}$$

The partial derivatives of M with respect to a are

$$\frac{\partial M_{11}}{\partial a} = \frac{4}{n^2a^2}\dot{X}_1 + \frac{2}{n^2a}\frac{\partial \dot{X}_1}{\partial a} \tag{A3.55}$$

$$\frac{\partial M_{12}}{\partial a} = \frac{4}{n^2a^2}\dot{Y}_1 + \frac{2}{n^2a}\frac{\partial \dot{Y}_1}{\partial a} \tag{A3.56}$$

$$\frac{\partial M_{13}}{\partial a} = 0 \tag{A3.57}$$

$$\frac{\partial M_{21}}{\partial a} = \frac{(1-h^2-k^2)^{1/2}}{na^2}\left[-\frac{1}{2a}\frac{\partial X_1}{\partial k} + \frac{\partial^2 X_1}{\partial a \partial k} - \frac{h\beta}{na}\dot{X}_1 - \frac{h\beta}{n}\frac{\partial \dot{X}_1}{\partial a}\right] \tag{A3.58}$$

$$\frac{\partial M_{22}}{\partial a} = \frac{(1-h^2-k^2)^{1/2}}{na^2}\left[-\frac{1}{2a}\frac{\partial Y_1}{\partial k} + \frac{\partial^2 Y_1}{\partial a \partial k} - \frac{h\beta}{na}\dot{Y}_1 - \frac{h\beta}{n}\frac{\partial \dot{Y}_1}{\partial a}\right] \tag{A3.59}$$

$$\frac{\partial M_{23}}{\partial a} = \frac{k}{na^2(1-h^2-k^2)^{1/2}}\left[-\frac{1}{2a}(qY_1 - pX_1) + q\frac{\partial Y_1}{\partial a} - p\frac{\partial X_1}{\partial a}\right] \tag{A3.60}$$

$$\frac{\partial M_{31}}{\partial a} = -\frac{(1-h^2-k^2)^{1/2}}{na^2}\left[-\frac{1}{2a}\frac{\partial X_1}{\partial h} + \frac{\partial^2 X_1}{\partial a \partial h} + \frac{k\beta}{na}\dot{X}_1 + \frac{k\beta}{n}\frac{\partial \dot{X}_1}{\partial a}\right] \tag{A3.61}$$

$$\frac{\partial M_{32}}{\partial a} = -\frac{(1-h^2-k^2)^{1/2}}{na^2}\left[-\frac{1}{2a}\frac{\partial Y_1}{\partial h} + \frac{\partial^2 Y_1}{\partial a \partial h} + \frac{k\beta}{na}\dot{Y}_1 + \frac{k\beta}{n}\frac{\partial \dot{Y}_1}{\partial a}\right] \tag{A3.62}$$

$$\frac{\partial M_{33}}{\partial a} = \frac{-h}{na^2(1-h^2-k^2)^{1/2}}\left[-\frac{1}{2a}(qY_1 - pX_1) + q\frac{\partial Y_1}{\partial a} - p\frac{\partial X_1}{\partial a}\right] \tag{A3.63}$$

$$\frac{\partial M_{41}}{\partial a} = 0 \tag{A3.64}$$

$$\frac{\partial M_{42}}{\partial a} = 0 \tag{A3.65}$$

$$\frac{\partial M_{43}}{\partial a} = \frac{(1+p^2+q^2)}{2na^2(1-h^2-k^2)^{1/2}}\left(-\frac{1}{2a}Y_1 + \frac{\partial Y_1}{\partial a}\right) \tag{A3.66}$$

$$\frac{\partial M_{51}}{\partial a} = 0 \tag{A3.67}$$

$$\frac{\partial M_{52}}{\partial a} = 0 \tag{A3.68}$$

$$\frac{\partial M_{53}}{\partial a} = \frac{(1+p^2+q^2)}{2na^2(1-h^2-k^2)^{1/2}}\left(-\frac{1}{2a}X_1 + \frac{\partial X_1}{\partial a}\right) \tag{A3.69}$$

$$\frac{\partial M_{61}}{\partial a} = \frac{-M_{61}}{2a} + \frac{1}{na^2}\left[-2\frac{\partial X_1}{\partial a} + (1-h^2-k^2)^{1/2}\left(h\beta\frac{\partial^2 X_1}{\partial a\partial h} + k\beta\frac{\partial^2 X_1}{\partial a\partial k}\right)\right] \tag{A3.70}$$

$$\frac{\partial M_{62}}{\partial a} = -\frac{M_{62}}{2a} + \frac{1}{na^2}\left[-2\frac{\partial Y_1}{\partial a} + (1-h^2-k^2)^{1/2}\left(h\beta\frac{\partial^2 Y_1}{\partial a\partial h} + k\beta\frac{\partial^2 Y_1}{\partial a\partial k}\right)\right] \tag{A3.71}$$

$$\frac{\partial M_{63}}{\partial a} = -\frac{M_{63}}{2a} + \frac{1}{na^2}\left[\left(q\frac{\partial Y_1}{\partial a} - p\frac{\partial X_1}{\partial a}\right)(1-h^2-k^2)^{-1/2}\right] \tag{A3.72}$$

$$\frac{\partial \dot{X}_1}{\partial a} = -\frac{1}{2}\frac{na}{r}\left[hk\beta c_F - \left(1-h^2\beta\right)s_F\right] \tag{A3.73}$$

$$\frac{\partial \dot{Y}_1}{\partial a} = \frac{1}{2}\frac{na}{r}\left[hk\beta s_F - \left(1-k^2\beta\right)c_F\right] \tag{A3.74}$$

The partial derivatives of M with respect to λ are

$$\frac{\partial M_{11}}{\partial \lambda} = \frac{2}{n^2 r}\frac{\partial \dot{X}_1}{\partial F} \tag{A3.75}$$

$$\frac{\partial M_{12}}{\partial \lambda} = \frac{2}{n^2 r}\frac{\partial \dot{Y}_1}{\partial F} \tag{A3.76}$$

$$\frac{\partial M_{13}}{\partial \lambda} = 0 \tag{A3.77}$$

$$\frac{\partial M_{21}}{\partial \lambda} = \frac{(1-h^2-k^2)^{1/2}}{nar}\left(\frac{\partial^2 X_1}{\partial F\partial k} - \frac{h\beta}{n}\frac{\partial \dot{X}_1}{\partial F}\right) \tag{A3.78}$$

$$\frac{\partial M_{22}}{\partial \lambda} = \frac{(1-h^2-k^2)^{1/2}}{nar}\left(\frac{\partial^2 Y_1}{\partial F\partial k} - \frac{h\beta}{n}\frac{\partial \dot{Y}_1}{\partial F}\right) \tag{A3.79}$$

$$\frac{\partial M_{23}}{\partial \lambda} = \frac{k\left(q\frac{\partial Y_1}{\partial F} - p\frac{\partial X_1}{\partial F}\right)}{nar(1-h^2-k^2)^{1/2}} \tag{A3.80}$$

$$\frac{\partial M_{31}}{\partial \lambda} = -\frac{(1-h^2-k^2)^{1/2}}{nar}\left(\frac{\partial^2 X_1}{\partial F\partial h} + \frac{k\beta}{n}\frac{\partial \dot{X}_1}{\partial F}\right) \tag{A3.81}$$

$$\frac{\partial M_{32}}{\partial \lambda} = -\frac{(1-h^2-k^2)^{1/2}}{nar}\left(\frac{\partial^2 Y_1}{\partial F\partial h} + \frac{k\beta}{n}\frac{\partial \dot{Y}_1}{\partial F}\right) \tag{A3.82}$$

$$\frac{\partial M_{33}}{\partial \lambda} = \frac{-h\left(q\frac{\partial Y_1}{\partial F} - p\frac{\partial X_1}{\partial F}\right)}{nar(1-h^2-k^2)^{1/2}} \tag{A3.83}$$

$$\frac{\partial M_{41}}{\partial \lambda} = \frac{\partial M_{42}}{\partial \lambda} = 0 \tag{A3.84}$$

$$\frac{\partial M_{43}}{\partial \lambda} = \frac{(1+p^2+q^2)}{2nar(1-h^2-k^2)^{1/2}}\frac{\partial Y_1}{\partial F} \tag{A3.85}$$

$$\frac{\partial M_{51}}{\partial \lambda} = \frac{\partial M_{52}}{\partial \lambda} = 0 \tag{A3.86}$$

$$\frac{\partial M_{53}}{\partial \lambda} = \frac{(1+p^2+q^2)}{2nar(1-h^2-k^2)^{1/2}}\frac{\partial X_1}{\partial F} \tag{A3.87}$$

$$\frac{\partial M_{61}}{\partial \lambda} = \frac{1}{nar}\left[-2\frac{\partial X_1}{\partial F} + (1-h^2-k^2)^{1/2}\left(h\beta\frac{\partial^2 X_1}{\partial F \partial h} + k\beta\frac{\partial^2 X_1}{\partial F \partial k}\right)\right] \tag{A3.88}$$

$$\frac{\partial M_{62}}{\partial \lambda} = \frac{1}{nar}\left[-2\frac{\partial Y_1}{\partial F} + (1-h^2-k^2)^{1/2}\left(h\beta\frac{\partial^2 Y_1}{\partial F \partial h} + k\beta\frac{\partial^2 Y_1}{\partial F \partial k}\right)\right] \tag{A3.89}$$

$$\frac{\partial M_{63}}{\partial \lambda} = \frac{\left(q\frac{\partial Y_1}{\partial F} - p\frac{\partial X_1}{\partial F}\right)}{nar(1-h^2-k^2)^{1/2}} \tag{A3.90}$$

The auxiliary partials are

$$\frac{\partial X_1}{\partial F} = a\left[hk\beta c_F - \left(1-h^2\beta\right)s_F\right] \tag{A3.91}$$

$$\frac{\partial Y_1}{\partial F} = a\left[-hk\beta s_F + \left(1-k^2\beta\right)c_F\right] \tag{A3.92}$$

$$\frac{\partial \dot{X}_1}{\partial F} = -\frac{a}{r}(ks_F - hc_F)\dot{X}_1 + \frac{a^2 n}{r}\left[-hk\beta s_F - \left(1-h^2\beta\right)c_F\right] \tag{A3.93}$$

$$\frac{\partial \dot{Y}_1}{\partial F} = -\frac{a}{r}(ks_F - hc_F)\dot{Y}_1 + \frac{a^2 n}{r}\left[-hk\beta c_F - \left(1-k^2\beta\right)s_F\right] \tag{A3.94}$$

$$\frac{\partial^2 X_1}{\partial F \partial h} = a\left[(hs_F + kc_F)\left(\beta + \frac{h^2\beta^3}{1-\beta}\right) + \frac{a^2}{r^2}(h\beta - s_F)(s_F - h) + \frac{a}{r}c_F^2\right] \tag{A3.95}$$

$$\frac{\partial^2 X_1}{\partial F \partial k} = -a\left[-(hs_F + kc_F)\frac{hk\beta^3}{1-\beta} + \frac{a^2}{r^2}(s_F - h\beta)(c_F - k) + \frac{a}{r}s_F c_F\right] \tag{A3.96}$$

$$\frac{\partial^2 Y_1}{\partial F \partial h} = a\left[-(hs_F + kc_F)\frac{hk\beta^3}{1-\beta} - \frac{a^2}{r^2}(k\beta - c_F)(s_F - h) + \frac{a}{r}s_F c_F\right] \tag{A3.97}$$

$$\frac{\partial^2 Y_1}{\partial F \partial k} = a\left[-(hs_F + kc_F)\left(\beta + \frac{k^2\beta^3}{1-\beta}\right) + \frac{a^2}{r^2}(c_F - k\beta)(c_F - k) - \frac{a}{r}s_F^2\right] \tag{A3.98}$$

References

[1] Gobetz, F. W. (1965) A Linear Theory of Optimum Low-Thrust Rendezvous Trajectories, *JAS*, Vol. 12, 69–76.

[2] Edelbaum, T. N. (1964) Optimum Low-Thrust Rendezvous and Station Keeping, *AIAA Journal*, Vol. 2, 1196–1201.

[3] Marec, J.-P. and Vinh, N. X. (1977) Optimum Low-Thrust Limited-Power Transfers Between Arbitrary Elliptical Orbits, *Acta Astronautica*, Vol. 4, 511–540.

[4] Broucke, R. A. and Cefola, P. J. (1972) On the Equinoctial Orbit Elements, *Celestial Mechanics*, Vol. 5, 303–310.

[5] Cefola, P. (1972) *Equinoctial Orbit Elements: Application to Artificial Satellite Orbits*, AIAA Paper 72–937, AIAA/AAS Astrodynamics Conference, Palo Alto, CA.

[6] Edelbaum, T. N., Sackett, L. L. and Malchow, H. L. (1973) *Optimal Low Thrust Geocentric Transfer*, AIAA Paper 73–1074, AIAA 10th Electric Propulsion Conference, Lake Tahoe, NV.

[7] Cefola, P., Long, A. and Holloway, G. (1974) *The Long-Term Prediction of Artificial Satellite Orbits*, AIAA Paper 74–170, AIAA 12th Aerospace Sciences Meeting, Washington, DC.

[8] Kechichian, J. (1990) *Equinoctial Orbit Elements: Application to Optimal Transfer Problems*, AIAA Paper 90–2976, AIAA/AAS Astrodynamics Conference, Portland, OR.
[9] Kechichian, J. (1991) *Trajectory Optimization with a Modified Set of Equinoctial Orbit Elements*, AAS/AIAA Paper 91–524, Astrodynamics Specialist Conference, Durango, CO.
[10] Bryson, A. E. and Ho, Y.-C. (1969) *Applied Optimal Control*, Ginn and Company, Waltham, MA.
[11] Marec, J.-P. (1979) *Optimal Space Trajectories*, Elsevier, Amsterdam.
[12] Vinh, N. X. (1981) *Optimal Trajectories in Atmospheric Flight*, Elsevier, Amsterdam.
[13] Kahaner, D., Moler, C. and Nash, S. (1989) *Numerical Methods and Software*, Prentice-Hall, Englewood Cliffs, NJ.
[14] Betts, J. T. (2001) *Practical Methods for Optimal Control Using Nonlinear Programming*, SIAM, Philadelphia, PA.
[15] Conway, B. A. (ed.) (2010) *Spacecraft Trajectory Optimization*, Cambridge University Press, Cambridge
[16] Lundberg, B. N. (2010) *An Adjoint Jump Shooting Method for Parametric Low Thrust Orbit Transfer Problems with Eclipsing*, Proceedings of the Fourth International Conference on Astrodynamics Tools and Techniques (WPP-308), ESAC, Madrid, Spain
[17] Tarzi, Z., Speyer, J. and Wirz, R. (2013) Fuel Optimum Low-Thrust Elliptic Transfer Using Numerical Averaging, *Acta Astronautica*, 86, 95–118.
[18] Mazzini, L. (2014) Finite Thrust Orbital Transfers, *Acta Astronautica*, Vol. 100, 107–128.
[19] Graham, K. F. and Rao, A. V. (2015) Minimum-Time Trajectory Optimization of Many Revolution Low-Thrust Earth-Orbit Transfers, *Journal of Spacecraft and Rockets*, Vol. 52, 711–727.
[20] Graham, K. F. and Rao, A. V. (2016) Minimum-Time Trajectory Optimization of Low-Thrust Earth-Orbit Transfers with Eclipsing, *Journal of Spacecraft and Rockets*, Vol. 53, 289–303.

4 Optimal Low-Thrust Transfer Using Variable Bounded Thrust

4.1 Introduction

The problem of minimum-fuel time-fixed orbit transfer and rendezvous using continuous low thrust, bounded from above and below, is analyzed here. The formulation is based on the use of the variation of parameters equations, which are written in terms of a set of nonsingular equinoctial elements, where the mean longitude represents the fast element. The consideration of maximum and minimum bounds on the thrust magnitude with constant power is equivalent to constraining the specific impulse of the propulsion system, such that this particular variable is optimized during the transfer. The thrust vector orientation is also optimized in order to minimize fuel consumption. Exact nonlinear dynamics for state vector propagation and constraint inequalities on the thrust magnitude are considered in the formulation of this low-thrust transfer problem.

Specific impulse or I_{sp} is no longer considered to be a fixed quantity during the transfer, and is now allowed to vary between user-defined minimum and maximum bounds, such that both thrust magnitude and direction are optimized to yield the overall minimum-fuel solution. Assuming that power remains constant, the thrust magnitude is inversely proportional to the specific impulse, which is continuously adjusted by varying the beam voltage. We first revisit the fundamentals of flight mechanics and low-thrust propulsion developed in [1–3], and derive the equivalent expression for the optimal controls for the thrust magnitude unconstrained case, using equinoctial elements instead of the usual Cartesian coordinates. The necessary conditions for optimality for the thrust-bounded case are derived following [4] and [5], and the problem of low-thrust transfer and rendezvous of [6–8] and Chapter 3 of this book extended to the case of continuously varying I_{sp}. An example of minimum-fuel time-fixed rendezvous with different I_{sp} bounds is solved numerically and compared with earlier results generated with constant thrust.

As the fixed transfer time approaches infinity, the solutions will approach the optimal multi-impulse chemical transfer solution with an infinite number of negligibly small thrust arcs replacing each impulse. In this case, the thrust magnitude will approach zero, and I_{sp} will approach infinity, or in practice, a lower and upper bound respectively. However, for finite transfer times, both the lower and upper bounds on the thrust magnitude will effectively act as constraints to shape the transfer trajectory, in order to minimize the fuel expenditure.

4.2 The Optimization of the Thrust Magnitude

From rocket propulsion fundamentals, and using Newton's law for a variable-mass body, the equation of motion of a rocket-powered vehicle is given by

$$m\ddot{\mathbf{r}} = \dot{m}\mathbf{c} + m\mathbf{g} \tag{4.1}$$

where $\mathbf{r}$ is the vehicle position vector, $\mathbf{g}$ is the acceleration of gravity, $\mathbf{c}$ is the exhaust velocity, and $\dot{m} < 0$ is the rate at which mass is expelled from the engine. The thrust vector $\mathbf{f} = \dot{m}\mathbf{c}$ is directed opposite the exhaust velocity vector, such that the acceleration can be written as

$$\mathbf{a} = \frac{\mathbf{f}}{m} = \ddot{\mathbf{r}} - \mathbf{g} \tag{4.2}$$

For a solar electric-powered ion thruster, the exhaust stream or beam power can be written as

$$P_{\mathrm{B}} = \frac{fc}{2} \tag{4.3}$$

and, because $f = -\dot{m}c$, it can be written as

$$P_{\mathrm{B}} = \frac{f^2}{2\dot{m}} \tag{4.4}$$

It can also be expressed in terms of the beam voltage V_{B} and the beam current I_{B} as

$$P_{\mathrm{B}} = V_{\mathrm{B}} I_{\mathrm{B}} \tag{4.5}$$

Equation (4.3) can be derived from the following two expressions based on electrostatics considerations

$$V_{\mathrm{B}} e = \frac{m_i}{2} c^2 \tag{4.6}$$

$$I_{\mathrm{B}} = \left(\frac{e}{m_i}\right)(-\dot{m}) \tag{4.7}$$

The first expression states that if a particle of mass m_i and charge e, and negligible initial velocity, passes through a potential difference V_{B} it will acquire a kinetic energy of $1/2m_i c^2$, where c is the exhaust velocity. The second expression is the definition of the current such that the beam power P_{B} is written as

$$P_{\mathrm{B}} = I_{\mathrm{B}} V_{\mathrm{B}} = \frac{1}{2}(-\dot{m})\, c^2$$

and, because $f = -\dot{m}c$, then

$$P_{\mathrm{B}} = \frac{1}{2} fc$$

Alternatively, from Equation (4.6), an expression relating the I_{sp} to the beam voltage can be obtained, because, with $c = I_{\mathrm{sp}} g$

$$c = \left[2\left(\frac{e}{m_i}\right) V_{\mathrm{B}}\right]^{1/2}$$

$$I_{\mathrm{sp}} = \frac{1}{g}\left[2\left(\frac{e}{m_i}\right) V_{\mathrm{B}}\right]^{1/2} \tag{4.8}$$

For a given beam power P_B, the thrust versus mass flow rate curve of an ion rocket is parabolic, because, from Equation (4.4)

$$f = \sqrt{-2\dot{m}P_B} \tag{4.9}$$

This behavior is very different from that of a constant exhaust velocity rocket, because in the latter case the curve is linear

$$f = -\dot{m}c \tag{4.10}$$

Conversely, the mass flow rate expressions for both types of vehicle are

$$\dot{m} = -\frac{f}{c} \tag{4.11}$$

$$\dot{m} = \frac{-f^2}{2P_B} \tag{4.12}$$

It is therefore advantageous from a propellant consumption point of view to have high exhaust velocity or high power.

Furthermore, Equation (4.9) shows that the same level of thrust can be achieved by different combinations of $\dot{m}$ and P_B because, if a lower power level is selected, an appropriate increase in $\dot{m}$ will maintain f constant.

However, from Equation (4.12), and for a given f, it is seen that $\dot{m}$ is at a minimum if P_B is chosen at its maximum level.

This means that at each instant of time the selection of P_B at its maximum value, namely $P_{B_{max}}$, will achieve the required thrust for minimum $\dot{m}$ or minimum fuel consumption. In other words, from all the possible ways of flying a required trajectory, meaning an acceleration time history $\mathbf{a}(t) = \ddot{\mathbf{r}}(t) - \mathbf{g}(t)$, the selection of $P_{B_{max}}$ is the only one that results in minimum fuel expenditure [3].

Ion rockets must therefore always operate at $P_{B_{max}}$. Then, it is clear from Equation (4.3), with $P_B = P_{B_{max}}$, that as the thrust is decreased, I_{sp} will increase and vice versa. Because the I_{sp} is dependent on the beam voltage V_B, it can be obtained by adjusting V_B, provided that I_B is also adjusted to obtain the required $P_{B_{max}}$ power, because $P_B = I_B V_B$. All these fundamental ideas are encapsulated in Figures 4.1, 4.2 and 4.3. In Figure 4.1, the thrust versus mass flow rate curve for a low-thrust rocket with constant exhaust velocity is depicted. Because c cannot be varied in this type of rocket, the thruster operates at f_{max} only [3].

Figure 4.2 corresponds to the variable c case, with $f = (-2\dot{m}P)^{1/2}$ for a given P. The power levels below P_{max} result in Region II, with Region I being completely inaccessible, because it corresponds to $P > P_{max}$, which is impossible.

As pointed out by Marec [3], it is not optimal to operate these types of thrusters at a power level $P_1 < P_{max}$, because for the same thrust f_1, the operation at P_{max} depicted by point 2 results in the minimum mass flow rate or propellant expenditure.

In Figure 4.3, and for a given P_{max}, any operating point A corresponds to a unique combination of beam voltage V_B and beam current I_B. If V_B is increased, the I_{sp} or c will also increase according to Equation (4.8), and, because power is held at P_{max} = constant, the thrust will decrease accordingly, because $P = 1/2fc$. Conversely, decreasing I_{sp} or V_B will result in increasing thrust and I_B.

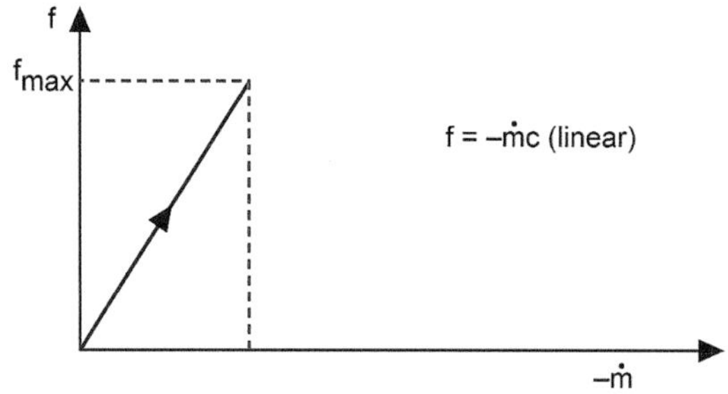

Figure 4.1 Thrust versus mass flow rate for a constant exhaust velocity.
Source: [3].

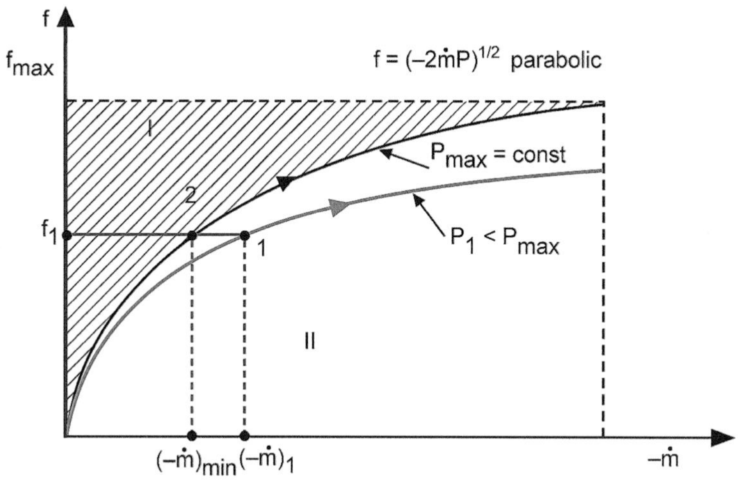

Figure 4.2 Selection of optimal thrust.
Source: [3].

In Figure 4.4, the unreachable Region I is extended further by the inclusion of the boundary 00′, which corresponds to the equation $f = -\dot{m}c_{\max}$, where $c_{\max}$ is the maximum exhaust velocity achieved by the rocket. This boundary is necessary to prevent the exhaust velocity or the I_{sp} from growing to very large values, as the thrust is decreased toward its minimum value. This minimum is conveniently defined at point 0, such that the operating arc is the arc 03 on the $P_{\max}$ parabola.

Edelbaum's analysis of the optimization of the thrust magnitude using the Cartesian formulation is first shown for the sake of completeness. This analysis is next translated to the formulation using the nonsingular equinoctial orbit elements, which is then extended to the case of constrained thrust magnitude. Following Edelbaum and letting **r** and **v** stand for the spacecraft position and velocity vectors, the second-order differential equation of motion is reduced to the following first-order form:

$$\dot{\mathbf{v}} = \frac{\mathbf{f}}{m} + \mathbf{g}(\mathbf{r}, t) \tag{4.13}$$

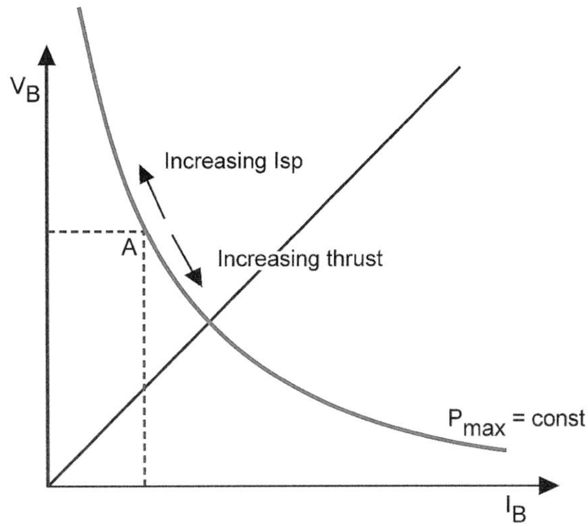

Figure 4.3 Beam voltage and current variation for a given power $P_{\max}$.

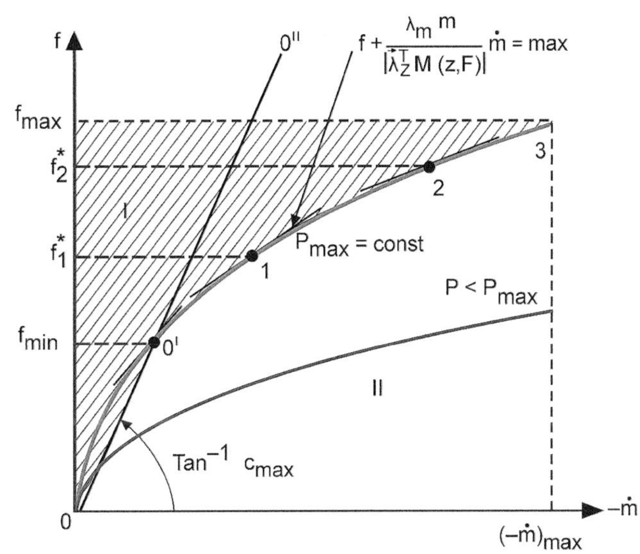

Figure 4.4 Thrust versus mass flow rate for a variable exhaust velocity rocket.
Sources: [1–3].

$$\dot{\mathbf{r}} = \mathbf{v} \tag{4.14}$$

with **f** and **g** representing the thrust and acceleration of gravity vectors, respectively. The mass flow rate obeys the general form

$$\dot{m} = \dot{m}(\mathbf{r}, t, \mathbf{f}) \tag{4.15}$$

The Hamiltonian of such a system is given by

$$H = \boldsymbol{\lambda}_v \cdot \frac{\mathbf{f}}{m}(\mathbf{r}, t) + \boldsymbol{\lambda}_v \cdot \mathbf{g}(\mathbf{r}, t) + \boldsymbol{\lambda}_r \cdot \mathbf{v} + \lambda_m \dot{m}(\mathbf{r}, t, \mathbf{f}) \tag{4.16}$$

The Euler–Lagrange equations are therefore

$$\dot{\boldsymbol{\lambda}}_v = -\frac{\partial H}{\partial \mathbf{v}} = -\boldsymbol{\lambda}_r \tag{4.17}$$

$$\dot{\boldsymbol{\lambda}}_r = -\frac{\partial H}{\partial \mathbf{r}} = -\frac{\boldsymbol{\lambda}_v}{m}\cdot\frac{\partial \mathbf{f}}{\partial \mathbf{r}} - \boldsymbol{\lambda}_v\cdot\frac{\partial \mathbf{g}}{\partial \mathbf{r}} - \lambda_m\frac{\partial \dot{m}}{\partial \mathbf{r}} \tag{4.18}$$

$$\dot{\lambda}_m = -\frac{\partial H}{\partial m} = \frac{\boldsymbol{\lambda}_v\cdot\mathbf{f}}{m^2} \tag{4.19}$$

The application of Pontryagin's maximum principle requires that the control vector, namely the thrust vector **f**, must be selected in such a manner as to maximize H at each instant of time. Because **f** appears in $\dot{m}$, this necessary condition translates into

$$\frac{\boldsymbol{\lambda}_v\cdot\mathbf{f}(\mathbf{r},t)}{m} + \lambda_m\dot{m}(\mathbf{r},t,\mathbf{f}) = maximum \tag{4.20}$$

This can be simplified further if $\dot{m}$ is a function of thrust magnitude but not direction. This is the case of the unconstrained transfer with free yaw, pitch and roll. In these cases, $\dot{m} = \dot{m}(\mathbf{r},t)$, such that the above condition reduces to

$$\frac{\boldsymbol{\lambda}_v\cdot\mathbf{f}(\mathbf{r},t)}{m} + \lambda_m\dot{m}(\mathbf{r},t) = maximum \tag{4.21}$$

This requires that the thrust vector **f** remains aligned with $\boldsymbol{\lambda}_v$ at all times, because $\boldsymbol{\lambda}_v\cdot\mathbf{f}$ is then maximum. This results in

$$\frac{\lambda_v f}{m} + \lambda_m\dot{m} = maximum \tag{4.22}$$

From Equation (4.12), $\dot{m}$ can be replaced by $-f^2/2P$, which reduces the expression in Equation (4.22) to

$$f - \frac{\lambda_m m}{2\lambda_v P}f^2 = max \tag{4.23}$$

Maximizing this expression with respect to the thrust magnitude f results in

$$f^* = \frac{\lambda_v P}{\lambda_m m} \tag{4.24}$$

and from Equation (4.3), the optimal I^*_{sp} is then

$$I^*_{\mathrm{sp}} = \frac{2P}{gf^*} = \frac{2m\lambda_m}{g\lambda_v} \tag{4.25}$$

Let us now carry out the optimization of f for the equinoctial formulation, using the equinoctial orbit elements represented by the vector $\mathbf{z} = (ahkpq\lambda)^{\mathrm{T}}$ instead of the **r** and **v** formulation used above.

These nonsingular elements are defined as before in terms of the classical elements by the following relationships, namely $a = a$, $h = e\sin(\omega+\Omega)$, $k = e\cos(\omega+\Omega)$, $p = \tan(i/2)\sin\Omega$, $q = \tan(i/2)\cos\Omega$ and $\lambda = M+\omega+\Omega$, with λ standing for the mean longitude. Letting $\hat{\mathbf{u}}$ represent a unit vector in the direction of the thrust, the variation of parameters equations for the thrust perturbation are given by Equation (4.26) with the addition of an n term to $\dot{\lambda}$. Here, n stands for the orbit mean motion at the current time t:

$$\dot{\mathbf{z}} = \left(\frac{\partial \mathbf{z}}{\partial \dot{\mathbf{r}}}\right) \hat{\mathbf{u}} f_t = M \hat{\mathbf{u}} f_t \tag{4.26}$$

$f_t = f/m$ represents the instantaneous perturbation acceleration with $\partial \mathbf{z}/\partial \dot{\mathbf{r}}$ and $\hat{\mathbf{u}}$ expressed in the direct equinoctial frame defined in [7] and Chapter 3 of this book. The M matrix above is given as before by

$$\frac{\partial a}{\partial \dot{\mathbf{r}}} = 2a^{-1} n^{-2} \left(\dot{X}_1 \hat{\mathbf{f}} + \dot{Y}_1 \hat{\mathbf{g}}\right) = M_{11} \hat{\mathbf{f}} + M_{12} \hat{\mathbf{g}} + M_{13} \hat{\mathbf{w}} \tag{4.27}$$

$$\begin{aligned} \frac{\partial h}{\partial \dot{\mathbf{r}}} &= G n^{-1} a^{-2} \left[\left(\frac{\partial X_1}{\partial k} - h\beta \frac{\dot{X}_1}{n}\right) \hat{\mathbf{f}} + \left(\frac{\partial Y_1}{\partial k} - h\beta \frac{\dot{Y}_1}{n}\right) \hat{\mathbf{g}}\right] \\ &\quad + k\left(qY_1 - pX_1\right) n^{-1} a^{-2} G^{-1} \hat{\mathbf{w}} \\ &= M_{21} \hat{\mathbf{f}} + M_{22} \hat{\mathbf{g}} + M_{23} \hat{\mathbf{w}} \end{aligned} \tag{4.28}$$

$$\begin{aligned} \frac{\partial k}{\partial \dot{\mathbf{r}}} &= -G n^{-1} a^{-2} \left[\left(\frac{\partial X_1}{\partial h} + k\beta \frac{\dot{X}_1}{n}\right) \hat{\mathbf{f}} + \left(\frac{\partial Y_1}{\partial h} + k\beta \frac{\dot{Y}_1}{n}\right) \hat{\mathbf{g}}\right] \\ &\quad - h\left(qY_1 - pX_1\right) n^{-1} a^{-2} G^{-1} \hat{\mathbf{w}} \\ &= M_{31} \hat{\mathbf{f}} + M_{32} \hat{\mathbf{g}} + M_{33} \hat{\mathbf{w}} \end{aligned} \tag{4.29}$$

$$\frac{\partial p}{\partial \dot{\mathbf{r}}} = K Y_1 \frac{n^{-1} a^{-2} G^{-1}}{2} \hat{\mathbf{w}} = M_{41} \hat{\mathbf{f}} + M_{42} \hat{\mathbf{g}} + M_{43} \hat{\mathbf{w}} \tag{4.30}$$

$$\frac{\partial q}{\partial \dot{\mathbf{r}}} = K X_1 \frac{n^{-1} a^{-2} G^{-1}}{2} \hat{\mathbf{w}} = M_{51} \hat{\mathbf{f}} + M_{52} \hat{\mathbf{g}} + M_{53} \hat{\mathbf{w}} \tag{4.31}$$

$$\begin{aligned} \frac{\partial \lambda}{\partial \dot{\mathbf{r}}} &= n^{-1} a^{-2} \left[-2X_1 + G\left(h\beta \frac{\partial X_1}{\partial h} + k\beta \frac{\dot{X}_1}{\partial k}\right) \hat{\mathbf{f}}\right] \\ &\quad + n^{-1} a^{-2} \left[-2Y_1 + G\left(h\beta \frac{\partial Y_1}{\partial h} + k\beta \frac{\partial Y_1}{\partial k}\right)\right] \hat{\mathbf{g}} + n^{-1} a^{-2} G^{-1} \left(qY_1 - pX_1\right) \hat{\mathbf{w}} \\ &= M_{61} \hat{\mathbf{f}} + M_{62} \hat{\mathbf{g}} + M_{63} \hat{\mathbf{w}} \end{aligned} \tag{4.32}$$

where $\beta = 1/(1+G)$, $G = (1-h^2-k^2)^{1/2}$, $K = 1+p^2+q^2$, $r = a(1-kc_F-hs_F)$ with $(\hat{\mathbf{f}}, \hat{\mathbf{g}}, \hat{\mathbf{w}})$ representing the equinoctial frame and with the position and velocity vectors given by

$$\mathbf{r} = X_1 \hat{\mathbf{f}} + Y_1 \hat{\mathbf{g}} \tag{4.33}$$

$$\dot{\mathbf{r}} = \dot{X}_1 \hat{\mathbf{f}} + \dot{Y}_1 \hat{\mathbf{g}} \tag{4.34}$$

and where

$$X_1 = a\left[\left(1 - h^2\beta\right) c_F + hk\beta s_F - k\right] \tag{4.35}$$

$$Y_1 = a\left[hk\beta c_F + \left(1 - k^2\beta\right) s_F - h\right] \tag{4.36}$$

$$\dot{X}_1 = a^2 n r^{-1} \left[hk\beta c_F - \left(1 - h^2\beta\right) s_F \right] \tag{4.37}$$

$$\dot{Y}_1 = a^2 n r^{-1} \left[\left(1 - k^2\beta\right) c_F - hk\beta s_F \right] \tag{4.38}$$

with F representing the eccentric longitude, which is obtained from Kepler's equation by iteration

$$\lambda = F - k s_F + h c_F \tag{4.39}$$

Finally,

$$\frac{\partial X_1}{\partial h} = a \left[-\left(h c_F - k s_F\right) \left(\beta + \frac{h^2 \beta^3}{(1-\beta)} \right) - \frac{a}{r} c_F \left(h\beta - s_F\right) \right] \tag{4.40}$$

$$\frac{\partial X_1}{\partial k} = -a \left[\left(h c_F - k s_F\right) \frac{hk\beta^3}{(1-\beta)} + 1 + \frac{a}{r} s_F \left(s_F - h\beta\right) \right] \tag{4.41}$$

$$\frac{\partial Y_1}{\partial h} = a \left[\left(h c_F - k s_F\right) \frac{hk\beta^3}{(1-\beta)} - 1 + \frac{a}{r} c_F \left(k\beta - c_F\right) \right] \tag{4.42}$$

$$\frac{\partial Y_1}{\partial k} = a \left[\left(h c_F - k s_F\right) \left(\beta + \frac{k^2 \beta^3}{(1-\beta)} \right) + \frac{a}{r} s_F \left(c_F - k\beta\right) \right] \tag{4.43}$$

The equation for the mass is as before given by

$$\dot{m} = -\frac{f^2}{2P} = -\frac{2P}{c^2} = -\frac{f}{c}$$

The system equations can now be written as

$$\dot{\mathbf{z}} = \frac{f}{m} M(\mathbf{z}, F)\, \hat{\mathbf{u}} \tag{4.44}$$

$$\dot{m} = -\frac{f^2}{2P(\mathbf{z}, F, \hat{\mathbf{u}})} \tag{4.45}$$

Here, P is assumed to be a function of the vehicle orientation. Let us further assume that the thrust magnitude is not constrained, such that the Hamiltonian of the system above is written as

$$H = \boldsymbol{\lambda}_z^T \cdot \dot{\mathbf{z}} + \lambda_m \dot{m} = \boldsymbol{\lambda}_z^T \frac{f}{m} M(\mathbf{z}, F)\hat{\mathbf{u}} - \lambda_m \frac{f^2}{2P(\mathbf{z}, F, \hat{\mathbf{u}})} + \lambda_\lambda n \tag{4.46}$$

If P is not a function of $\hat{\mathbf{u}}$, then

$$H = \boldsymbol{\lambda}_z^T \cdot \frac{f}{m} M(\mathbf{z}, F)\hat{\mathbf{u}} - \lambda_m \frac{f^2}{2P(\mathbf{z}, F)} + \lambda_\lambda n \tag{4.47}$$

H is maximized if $\hat{\mathbf{u}}$ is chosen parallel to $\boldsymbol{\lambda}_z^T (f/m) M(\mathbf{z}, F)$ or

$$\hat{\mathbf{u}} = \frac{\boldsymbol{\lambda}_z^T \frac{f}{m} M(\mathbf{z}, F)}{\left| \boldsymbol{\lambda}_z^T \frac{f}{m} M(\mathbf{z}, F) \right|} \tag{4.48}$$

Then, H is reduced to

$$H = \left| \boldsymbol{\lambda}_z^T \frac{f}{m} M(\mathbf{z}, F) \right| - \lambda_m \frac{f^2}{2P(\mathbf{z}, F)} + \lambda_\lambda n \tag{4.49}$$

and the optimal thrust magnitude is obtained from the optimality condition $\partial H/\partial f = 0$, which results in

$$f^* = \frac{\left|\boldsymbol{\lambda}_z^T M(\mathbf{z}, F)\right| P(\mathbf{z}, F)}{\lambda_m m} \tag{4.50}$$

This is equivalent to the maximization of

$$f + \frac{m\lambda_m}{\left|\boldsymbol{\lambda}_z^T M(\mathbf{z}, F)\right|}\dot{m} = max \tag{4.51}$$

because $\dot{m}$ is equal to $-f^2/2P$ in Equation (4.49). Equation (4.50) is identical to Equation (4.24), where the $\mathbf{r}$, $\mathbf{v}$ formulation was used. Here, $|\boldsymbol{\lambda}_z^T M(\mathbf{z}, F)|$ replaces λ_v, because the state variables are the equinoctial elements. Equation (4.51) is the equation of the straight line in the $(f, -\dot{m})$ system, and the maximum takes place when the line is tangent to the $P_{\max}$ curve (Figure 4.4), such as at point 1. At a later time, the optimal operating point is at 2, with f_2^* representing the optimal time-varying thrust magnitude [1, 3].

The corresponding mass flow rate at each instant of time is, of course, obtained from $\dot{m} = -f^2/2P$, and the optimal $I_{\rm sp}^*$ from

$$I_{\rm sp}^* = \frac{2m\lambda_m}{g\left|\boldsymbol{\lambda}_z^T M(\mathbf{z}, F)\right|} \tag{4.52}$$

The Euler–Lagrange differential equations are given by

$$\dot{\boldsymbol{\lambda}}_z^T = -\frac{\partial H}{\partial \mathbf{z}} = -\boldsymbol{\lambda}_z^T \frac{\partial M}{\partial \mathbf{z}} \frac{f}{m}\hat{\mathbf{u}} - \lambda_\lambda \frac{\partial n}{\partial \mathbf{z}} \tag{4.53}$$

$$\dot{\lambda}_m = -\frac{\partial H}{\partial m} = \boldsymbol{\lambda}_z^T M \frac{f}{m^2}\hat{\mathbf{u}} \tag{4.54}$$

where it is assumed that the power P is independent of the state vector $\mathbf{z}$, being considered constant throughout the transfer. The $\partial M/\partial \mathbf{z}$ partials have been derived in [7] and in Chapter 3 of this book, and are not repeated here.

For the optimization of the thrust orientation, the maximum principle is used to get

$$\hat{\mathbf{u}} = \frac{\left(\boldsymbol{\lambda}_z^T M\right)^T}{\left|\boldsymbol{\lambda}_z^T M\right|} \tag{4.55}$$

and thereby the three components u_f, u_g and u_w, which in turn provide the values of u_r, u_θ and u_h in the rotating $\hat{\mathbf{r}}, \hat{\boldsymbol{\theta}}, \hat{\mathbf{h}}$ frame. The instantaneous pitch and yaw angles are obtained from $\theta_t = \tan^{-1}(u_r/u_\theta)$ and $\theta_h = \tan^{-1}(u_h/u_\theta)$, as in [7] and in Chapter 3 of this book.

If the thrust magnitude is unconstrained, its optimal value is obtained from the optimality condition

$$\frac{\partial H}{\partial f} = 0 = \boldsymbol{\lambda}_z^T \frac{M}{m}\hat{\mathbf{u}} - \frac{f}{P}\lambda_m \tag{4.56}$$

which yields the optimal thrust f^*:

$$f^* = \frac{P\left(\boldsymbol{\lambda}_z^T M\hat{\mathbf{u}}\right)^T}{m\lambda_m} \tag{4.57}$$

The optimal value of the specific impulse is then given by

$$I_{sp}^* = \frac{2P}{f^* g} \tag{4.58}$$

because $c = 2P/f = I_{sp} g$. Now the optimal acceleration program is obtained from

$$f_t^* = \frac{f^*}{m} = \frac{\boldsymbol{\lambda}_z^T M \hat{\mathbf{u}}}{m^2 \lambda_m} P \tag{4.59}$$

If the thrust magnitude is bounded from above and below, then

$$f_{\min} < f < f_{\max} \tag{4.60}$$

such that we have the following inequality constraints on the control variable f

$$C_1 = f - f_{\max} \leq 0 \tag{4.61}$$

$$C_2 = -f + f_{\min} \leq 0 \tag{4.62}$$

These constraints can be adjoined to the original Hamiltonian by way of Lagrange multipliers, μ_1 and μ_2 such that

$$H = \frac{f}{m} \boldsymbol{\lambda}_z^T M \hat{\mathbf{u}} - \frac{f^2}{2P} \lambda_m + \lambda_\lambda n + \mu_1 (f - f_{\max}) + \mu_2 (-f + f_{\min}) \tag{4.63}$$

The necessary condition on H is

$$H_f = \frac{\partial H}{\partial f} = \boldsymbol{\lambda}_z^T \frac{M}{m} \hat{\mathbf{u}} - \frac{f}{P} \lambda_m + \mu_1 - \mu_2 = 0 \tag{4.64}$$

The multipliers μ_1 and μ_2 are such that $\mu_1 > 0$ when $C_1 = 0$ or $f = f_{\max}$, and $\mu_1 = 0$ when $C_1 < 0$ or $f < f_{\max}$, and similarly $\mu_2 > 0$ when $C_2 = 0$ or $f = f_{\min}$, and $\mu_2 = 0$ when $C_2 < 0$ or $f > f_{\min}$. When $f_{\min} < f < f_{\max}$ assumes an intermediate value, $\mu_1 = \mu_2 = 0$, and $H_f = 0$ reduces to $\boldsymbol{\lambda}_z^T (M/m)\hat{\mathbf{u}} - (f/P)\lambda_m = 0$, yielding the optimal control given by Equation (4.57). The values of μ_1 and μ_2 are readily obtained from

$$\mu_1 = \frac{f_{\max}}{P} \lambda_m - \frac{\boldsymbol{\lambda}_z^T M \hat{\mathbf{u}}}{m} \tag{4.65}$$

$$\mu_2 = -\frac{f_{\min}}{P} \lambda_m + \frac{\boldsymbol{\lambda}_z^T M \hat{\mathbf{u}}}{m} \tag{4.66}$$

The Lagrange multipliers are still given by Equations (4.53) and (4.54), and the optimal f^* is selected by monitoring the value of $\boldsymbol{\lambda}_z^T M \hat{\mathbf{u}} P/(m\lambda_m)$. If it is less than $f_{\min}$, then we use $f^* = f_{\min}$, and if it is larger than $f_{\max}$, then we use, $f = f_{\max}$, and finally, if it is intermediate between $f_{\min}$ and $f_{\max}$ then we use f^* as given by Equation (4.57), which is the presently calculated value of $\boldsymbol{\lambda}_z^T M \hat{\mathbf{u}} P/(m\lambda_m)$.

We can also use the simpler Hamiltonian H^* without adjoining the constraints, namely

$$H^* = \boldsymbol{\lambda}_z^T \dot{\mathbf{z}} + \lambda_\lambda \dot{m} = \frac{f}{m} \left(\boldsymbol{\lambda}_z^T M \hat{\mathbf{u}} - \frac{m \lambda_m}{c} \right) + \lambda_\lambda n \tag{4.67}$$

This is equivalent to Equation (4.68) because $c = 2P/f$ is a function of the control f:

$$H^* = \frac{f}{m}\boldsymbol{\lambda}_z^T M\hat{\mathbf{u}} - \frac{f^2}{2P}\lambda_m + \lambda_\lambda n \tag{4.68}$$

The optimality condition yields, with

$$\frac{\partial H^*}{\partial f} = H_f^* = \boldsymbol{\lambda}_z^T \frac{M}{m}\hat{\mathbf{u}} - \frac{f}{P}\lambda_m \tag{4.69}$$

the following relation

$$\delta H^* = H_f^* \delta f \leq 0 \tag{4.70}$$

because this is equivalent to

$$\delta J = \int_{t_0}^{t_f} H_f^* \delta f dt = \int_{t_0}^{t_f} \delta H^* dt \leq 0$$

for the control f to be maximizing for all admissible values of δf. δJ is the variation in J, the performance index, due to variations in f, for fixed $\mathbf{z}(t_0)$. The optimal control is selected by monitoring the value of H_f^*. If H_f^* as calculated by Equation (4.69) is positive, then we use $f = f_{\max}$, and if H_f^* is negative we use $f = f_{\min}$ and, finally, if $H_f^* = 0$, then we use f^* from Equation (4.69):

$$H_f^* > 0 \Rightarrow \delta f < 0 \Rightarrow f = f_{\max} \tag{4.71}$$

$$H_f^* < 0 \Rightarrow \delta f > 0 \Rightarrow f = f_{\min} \tag{4.72}$$

$$H_f^* = 0 \Rightarrow f = f^* = \boldsymbol{\lambda}_z^T \frac{M\hat{\mathbf{u}}}{m\lambda_m P} \tag{4.73}$$

In practice, the last condition for $H_f^* = 0$ is replaced by $|H_f^*| < \varepsilon$, where, ε is a small number, say, 10^{-10}. The Euler–Lagrange equations are still given by Equations (4.53) and (4.54).

4.3 A Simple Example of Rendezvous in Near-Circular Orbit

Let us revisit our rendezvous example of [7], also shown in the previous chapter, in which a minimum-time solution with a continuous constant acceleration was generated. The mass equation was not used in [7] because the acceleration vector assumed effectively the role of the control, and it was then sufficient to optimize its direction to achieve the minimum-time solution. Under these assumptions, this solution was also equivalent to the minimum-fuel solution as was demonstrated in [8] and also discussed in the previous chapter, in which the mass equation was used, and the minimum-fuel time-fixed problem formulated using the same set of equinoctial nonsingular orbit elements. For a given initial mass $m_0 = 6000$ kg and a constant $I_{sp} = 3800$ s, an equivalent power of $P = 39{,}112.3$ W was used in order to obtain an average acceleration of $f_t = 3.5 \times 10^{-7}$ km/s^2, equal to the value of the constant acceleration used in [7] and Chapter 3 of this book. We now attempt to solve the same problem by allowing the I_{sp} to vary in order to improve the fuel consumption. As in [8] and Chapter 3 of this book, we maximize the value of the mass at the fixed final time t_f, such that the performance index $J = \phi = m_f$, with the optimal thrust direction given by Equation (4.55) obtained directly from the maximum principle. Because the equations of motion given by Equations (4.44) and (4.45) are not explicit functions of time, the Hamiltonian H in Equation (4.46) or H^* in Equation (4.68) is constant throughout the transfer. Given initial state parameters $(a)_0$, $(h)_0$, $(k)_0$, $(p)_0$, $(q)_0$, $(\lambda)_0$ and $(m)_0$, the initial values of the seven Lagrange multipliers are guessed, namely $(\lambda_a)_0$, $(\lambda_h)_0$, $(\lambda_k)_0$, $(\lambda_p)_0$, $(\lambda_q)_0$, $(\lambda_\lambda)_0$, $(\lambda_m)_0$, and the state adjoint equations given by Equations (4.44), (4.45) and (4.53), (4.54) integrated forward from $t_0 = 0$ to $t_f = 86{,}402.453$ s, by using the optimal thrust direction $\hat{\mathbf{u}}$ in Equation (4.55), and the thrust magnitude from Equations (4.71), (4.72) and (4.73). The initial values of the multipliers are iterated until the desired terminal state given by $(a)_f$, $(h)_f$, $(k)_f$, $(p)_f$, $(q)_f$, $(\lambda)_f$ and $(m)_f = (\partial\phi/\partial m)_{t_f} = 1$ are satisfied. This is achieved by minimizing the following objective function:

$$\begin{aligned} F' &= w_1\left[a-(a)_f\right]^2 + w_2\left[h-(h)_f\right]^2 + w_3\left[k-(k)_f\right]^2 + w_4\left[p-(p)_f\right]^2 \\ &\quad + w_5\left[q-(q)_f\right]^2 + w_6\left[\lambda-(\lambda)_f\right]^2 + w_7\left[\lambda_m - 1\right]^2 \end{aligned} \tag{4.74}$$

where w_j are certain weights, and where the values of a, h, k, p, q, λ and λ_m are evaluated at the fixed terminal time t_f. Let $a_0 = 42{,}000{,}000$ m, $e_0 = 0$, $i_0 = 28.5$ deg, $\Omega_0 = 30$ deg, $\omega_0 = 10$ deg, $M_0 = 0$ deg, $m_0 = 6000$ kg, $P_0 = 40{,}000$ W, and $a_f = 42{,}767{,}073$ m, $e_f = 1.64459 \times 10^{-4}$, $i_f = 28.343$ deg, $\Omega_f = 29.999$ deg, $\omega_f = 247.299$ deg, $M_f = 120.905$ deg, and $t_f = 86{,}402.453$ s. Because the power is assumed to remain constant, let us select $(I_{sp})_{min} = 3000$ s and $(I_{sp})_{max} = 4000$ s, such that

$$f_{min} = \frac{2P}{\left(I_{sp}\right)_{max} g} = \frac{2 \times 40{,}000}{4000 \times 9.80665} = 2.039432\,\text{N}$$

$$f_{max} = \frac{2P}{\left(I_{sp}\right)_{min} g} = 2.719243\,\text{N}$$

The values of f_{max} and f_{min} are used as input to the optimizer, and they correspond to initial accelerations of $(f_t)_{max} = f_{max}/m_0 = 2.719243/6000 = 4.532071 \times 10^{-4}$ m/s^2, and $(f_t)_{min} = f_{min}/m_0 = 2.039432/6000 = 3.399053 \times 10^{-4}$ m/s^2, or an average of 3.965562×10^{-4} m/s^2, or 3.965562×10^{-7} km/s^2, which is slightly higher than the constant $f_t = 3.5 \times 10^{-7}$ km/s^2 used in [7] and duplicated in [8] and the previous chapter, by adjusting the power accordingly. Here we leave the power at 40,000 W, so that the I_{sp} bounds are exactly at the $(I_{sp})_{min}$ and $(I_{sp})_{max}$ values. Furthermore, the acceleration being now variable, it will be optimized to achieve minimum-fuel consumption. The solution is given by $(\lambda_a)_0 = 1.05651899 \times 10^{-5}$, $(\lambda_h)_0 = -3.102865136$, $(\lambda_k)_0 = 3.077180522$, $(\lambda_p)_0 = -5.066278561 \times 10^{-2}$, $(\lambda_q)_0 = -8.672939167 \times 10^{-2}$, $(\lambda_\lambda)_0 = -2.160048598$, $(\lambda_m)_0 = 0.9984527097$, where the various λs have units that are the inverse of the corresponding element rates, namely s/m for λ_a, s/rad for λ_λ and s/kg for λ_m, with the remaining λ units being seconds because h, k, p and q are unitless. This is needed because the Hamiltonian is unitless.

The initial values of the elements are $a_0 = 4.2 \times 10^7$ m, $h_0 = 0$, $k_0 = 0$, $p_0 = 1.269838 \times 10^{-1}$, $q_0 = 2.199424 \times 10^{-1}$, $\lambda_0 = 6.981317 \times 10^{-1}$ rad, and

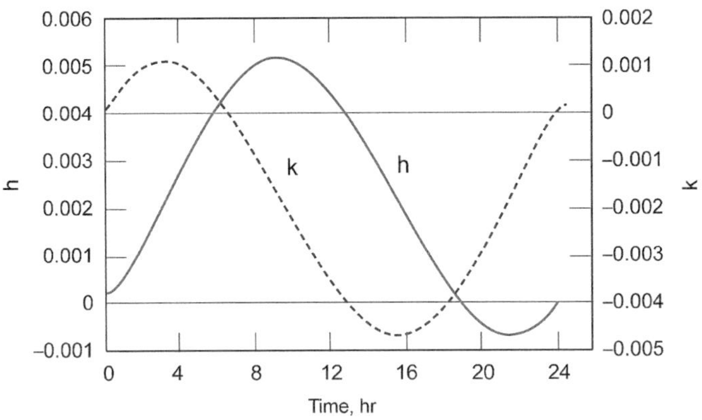

Figure 4.5 Variation of h and k on an optimal trajectory.

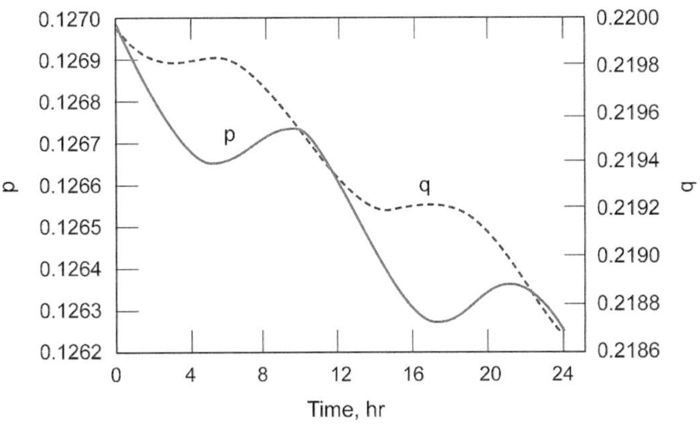

Figure 4.6 Variation of p and q on an optimal trajectory.

$m_0 = 6000$ kg. The final achieved parameters are $a = 42,767,073$ m, $e = 1.64459 \times 10^{-4}$, $i = 28.343$ deg, $\Omega = 29.999$ deg, $\omega = 247.299$ deg, $M = 120.905$ deg, $m_f = 5995.251893$ kg and $\lambda_m = 1.000000000$, indicating a local optimum. The Hamiltonian stays constant at $H = -0.959 \times 10^{-4}$ or essentially near zero. Figures 4.5–4.7 show the variations of h, k, p, q, H, and λ as a function of time, while Figures 4.8–4.10 depict the variations of the classical orbit elements. Figures 4.11–4.14 show how λ_a, λ_λ, λ_h, λ_k, λ_p, λ_q, m and λ_m vary during the optimal transfer. The terminal mass $m_f = 5995.251893$ kg is as expected, higher than the final mass of 5995.022 kg obtained with a constant $I_{sp} = 3800$ s in [8] and Chapter 3 of this book, showing the benefit of using a variable I_{sp}, and resulting in a decrease in fuel consumption of 0.229 kg. Figure 4.15 shows the optimal thrust orientation given by the pitch and yaw angles θ_t and θ_h, while Figure 4.16 depicts the acceleration program as well as the I_{sp} as a function of time. There are two arcs of constant maximum I_{sp} in this particular example.

Now if we relax the I_{sp} bounds or, equivalently, the thrust bounds, we could generate a trajectory that uses only intermediate thrust levels. Let $(I_{sp})_{min} = 1000$ s, and $(I_{sp})_{max} =$ 10,000 s, or equivalently $(f)_{min} = 8.157729 \times 10^{-1}$ N, and $(f)_{max} = 8.157729$ N

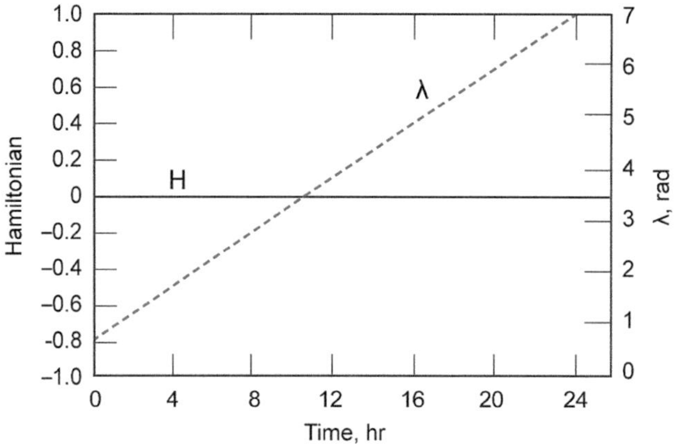

Figure 4.7 Variation of λ and H on optimal Hamiltonian.

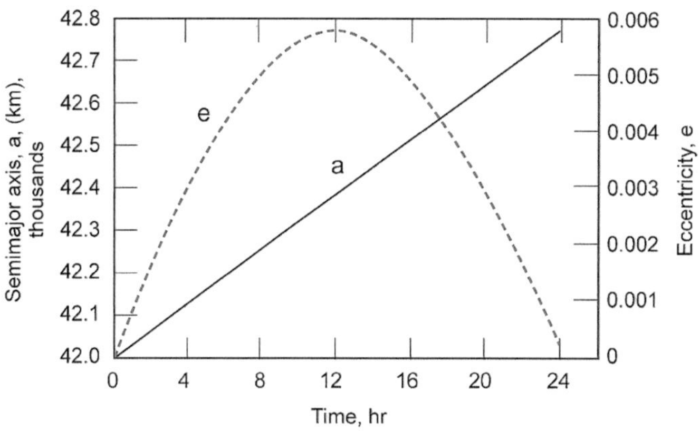

Figure 4.8 Evolution of a and e on an optimal rendezvous trajectory.

for our example at hand. The solution is given by $(\lambda_a)_0 = 1.087325882 \times 10^{-5}$, $(\lambda_h)_0 = -4.548851268$, $(\lambda_k)_0 = 4.130745710$, $(\lambda_p)_0 = -5.080206079 \times 10^2$, $(\lambda_q)_0 = -8.719703291 \times 10^2$, $(\lambda_\lambda)_0 = -2.630702853$, $(\lambda_m)_0 = 0.9984192335$, with $m_f = 5995.255825$ kg, which is only slightly higher than the value obtained in the example above. Figure 4.17 shows the acceleration program as well as the I_{sp} time history, which remains at intermediate values never reaching either the minimum or maximum bounds. This solution represents therefore the thrust magnitude unconstrained case providing the overall minimum-fuel expenditure. As a final example, let $(I_{sp})_{min} = 3700$ s and $(I_{sp})_{max} = 4000$ s, such that the transfer must now use a smaller I_{sp} bandwidth than our first example. These bounds correspond to $(f)_{min} = 2.039432$ N, and $(f)_{max} = 2.204791$ N, and the solution is given by $(\lambda_a)_0 = 1.058971845 \times 10^{-5}$, $(\lambda_h)_0 = -3.140806779$, $(\lambda_k)_0 = 3.401400619$, $(\lambda_p)_0 = -5.071540433 \times 10^2$, $(\lambda_q)_0 = -8.679605832 \times 10^2$, $(\lambda_\lambda)_0 = -2.307546419$, $(\lambda_m)_0 = 0.9984514728$, with the final mass at $m_f = 5995.251875$ kg. This value is slightly less than the corresponding

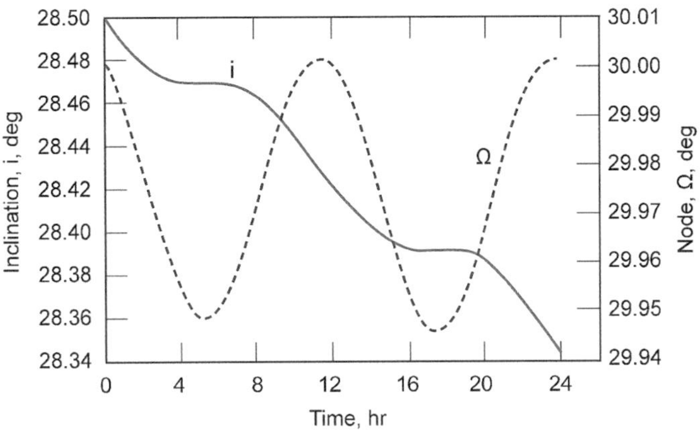

Figure 4.9 Evolution of i and Ω on an optimal rendezvous trajectory.

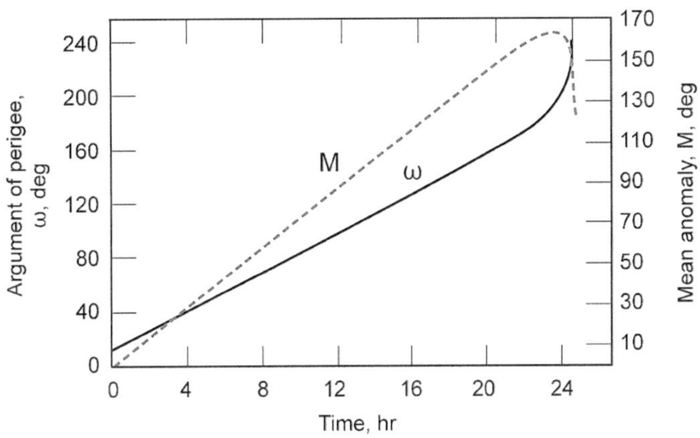

Figure 4.10 Evolution of M and ω on an optimal rendezvous trajectory.

value obtained for the absolute minimum-fuel transfer of 5995.255825 kg with unconstrained thrust magnitude, and also slightly less than the value of 5995.251893 kg of the first example. This is because, in this last example, both minimum I_{sp} and maximum I_{sp} thrust arcs are generated as shown in Figure 4.18 due to the narrowing of the I_{sp} bandwidth, which further constrains the transfer. These examples validate the analytical developments and verify that the code is error-free, because the correct trends are clearly appearing, even though the savings in fuel are very small for this particular rendezvous trajectory. The final state is met with a very small tolerance in order to produce the exact trajectories for a meaningful comparison. Substantial savings in fuel are possible especially with the higher values of specific power, or power to mass ratio, for highly elliptical orbit transfers and for larger changes in the energy and angular momentum vector.

Furthermore, because the thrust is continuously on, or otherwise the control is continuous, it is much easier to generate a series of transfers with increasing I_{sp} bandwidths,

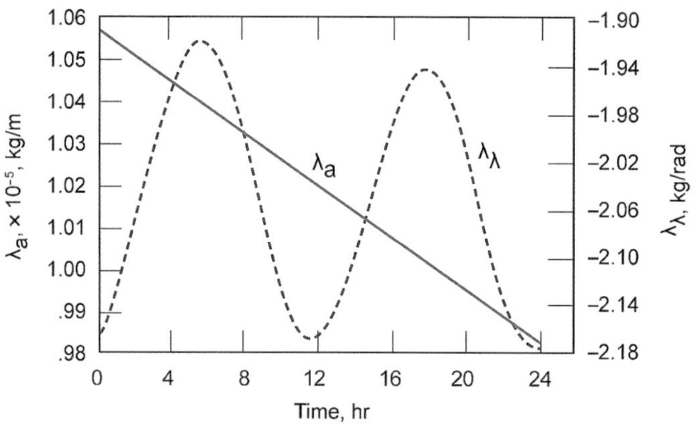

Figure 4.11 λ_a and λ_λ versus time.

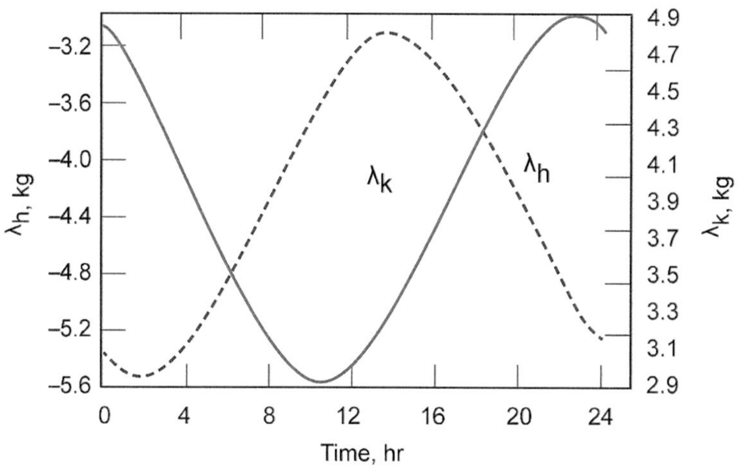

Figure 4.12 λ_h and λ_k versus time.

such that, in the limit, the discontinuous thrust case can be produced with full thrust cut-off arcs, which correspond to infinite I_{sp} or, in practice, negligible thrust. These coast arcs are beneficial for certain transfer geometries, especially, if the fixed transfer time is further relaxed from its minimum-time value obtained with continuous thrust. Finally, when a given engine design is used with specified I_{sp} bounds, the thrust magnitude unconstrained solution may not be possible to fly due to the engine limitations, which in turn will necessitate the generation of the optimal I_{sp}-constrained solution as shown in these examples.

4.4 Conclusion

The use of the variation of parameters equations based on a set of nonsingular equinoctial orbit elements is extended to the important case of minimum-fuel time-fixed transfer

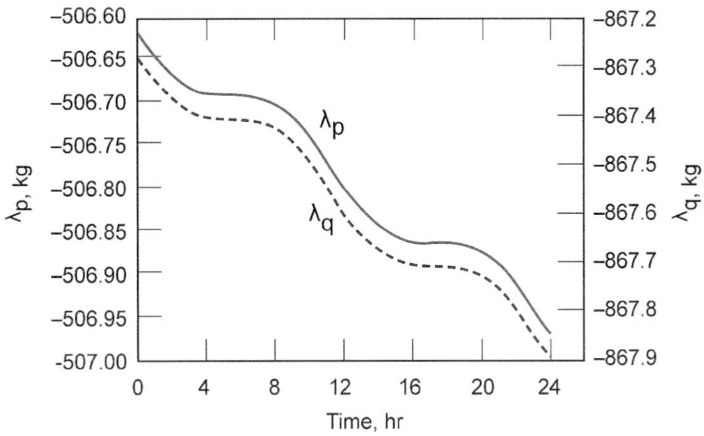

Figure 4.13 λ_p and λ_q versus time.

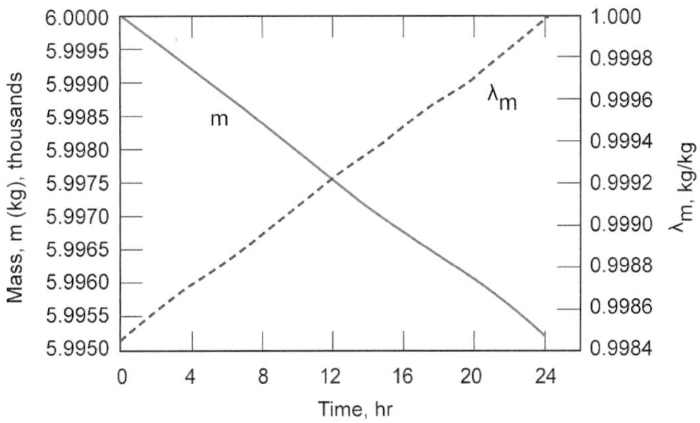

Figure 4.14 λ_m and m versus time.

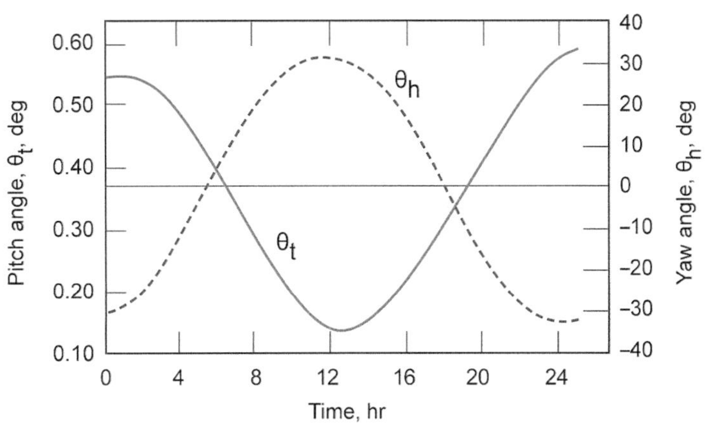

Figure 4.15 Optimal thrust pitch and yaw program.

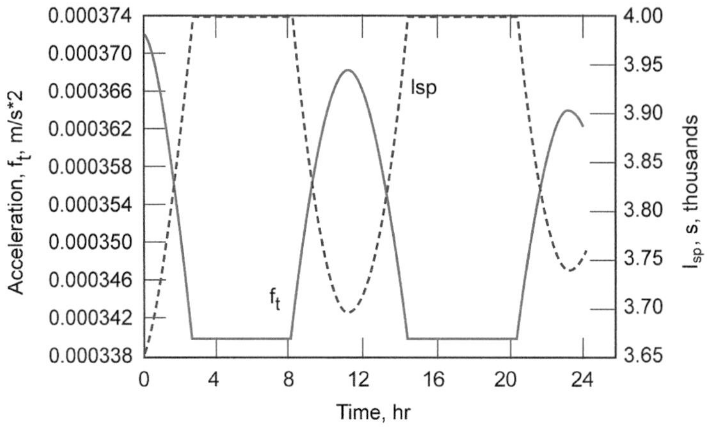

Figure 4.16 Optimal acceleration and I_{sp} program with $(I_{sp})_{max} = 4000$ s and $(I_{sp})_{min} = 3000$ s.

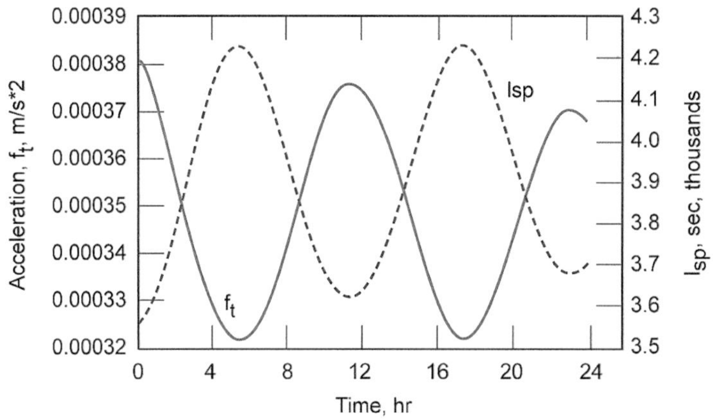

Figure 4.17 Optimal acceleration and I_{sp} program with $(I_{sp})_{max} = 10,000$ s and $(I_{sp})_{min} = 1000$ s.

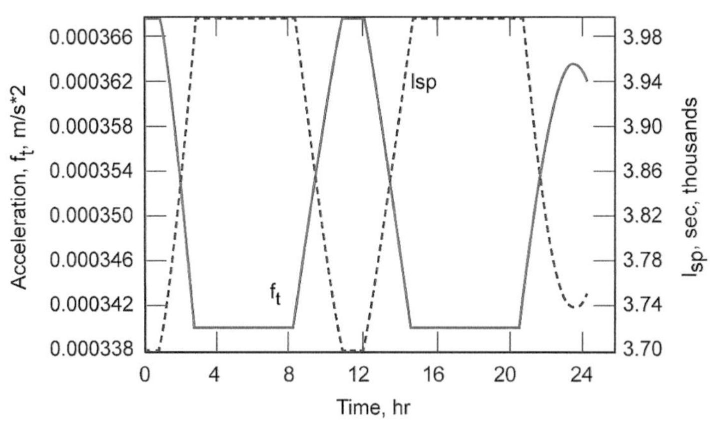

Figure 4.18 Optimal acceleration and I_{sp} program with $(I_{sp})_{max} = 4000$ s and $(I_{sp})_{min} = 3700$ s.

and rendezvous, where both thrust magnitude and orientation are simultaneously optimized. Furthermore, minimum and maximum bounds on the thrust magnitude, or equivalently I_{sp}, are considered in order to allow the throttling of the propulsion system, which in turn can result in substantial savings in propellant consumption, especially when the fixed flight time is further relaxed. The limiting case of coasting arcs can be approached by increasing the maximum I_{sp} bound, such that the discontinuous control transfer problem can later be solved after the proper thrust–coast sequence is essentially guessed using the simpler and more robust continuous control algorithm developed in this chapter.

References

[1] Edelbaum, T. N. (1969) *Optimal Space Trajectories*, Analytic Mechanics Associates Inc., Jericho, NY.

[2] Hill, P. G. and Peterson, C. R. (1970) *Mechanics and Thermodynamics of Propulsion*, Addison-Wesley, New York, NY.

[3] Marec, J.-P. (1979) *Optimal Space Trajectories*, Elsevier, Amsterdam.

[4] Vinh, N. X. (1981) *Optimal Trajectories in Atmospheric Flight*, Elsevier, Amsterdam.

[5] Bryson, A. E. and Ho, Y.-C. (1969) *Applied Optimal Control*, Ginn, Waltham, MA.

[6] Edelbaum, T. N., Sackett, L. L. and Malchow, H. L. (1973) *Optimal Low Thrust Geocentric Transfer*, AIAA Paper 73–1074, AIAA 10th Electric Propulsion Conference, Lake Tahoe, NV.

[7] Kechichian, J. A. (1992) *Optimal Low-Thrust Rendezvous Using Equinoctial Orbit Elements*, IAF Paper 92–0014, 43rd Congress of the International Astronautical Federation, Washington, DC.

[8] Kechichian, J. A. (1993) *Minimum-Fuel Time-Fixed Rendezvous Using Constant Low-Thrust*, AAS Paper 93–130, AAS/AIAA Spaceflight Mechanics Meeting, Pasadena, CA.

5 Minimum-Time Low-Thrust Rendezvous and Transfer Using Epoch Mean Longitude Formulation

5.1 Introduction

The mathematics of minimum-time rendezvous and transfer using a set of nonsingular orbit elements involving the mean longitude at epoch is shown to represent a basis from which the current time formulations are derived. The state and adjoint differential equations are explicit functions of time in this formulation that involves natural orbital elements which stay constant if no perturbation is applied. The optimal Hamiltonian also varies in time; however, the function that defines the transversality condition at the end time in minimum-time problems is shown to remain constant during the optimal transfer, effectively replacing the numerical check for the constant Hamiltonian in accepting a converged trajectory as being truly optimal. The mapping matrices for the Lagrange multipliers of this as well as several other formulations are also shown explicitly. A numerical example is depicted that uses the shooting method coupled with a quasi-Newton scheme to solve the two-point boundary value problem and generate the numerically integrated trajectory.

The use of the nonsingular equinoctial orbit elements in solving the problem of optimal orbit transfer using low-thrust acceleration was initiated by Edelbaum et al. [1]. The full set of the state and adjoint differential equations for the case in which the current mean longitude is chosen as the fast element appeared in [2] and [3], and Chapter 3 of this book, by extending the analysis of Cefola et al. [4, 5]. The low-thrust transfer and rendezvous problem using continuous constant acceleration with the current mean longitude as the fast variable [6, and Chapter 3] is solved here using the epoch mean longitude as the sixth state variable instead, this element being a natural orbit element that stays constant in the absence of any perturbation acceleration. This epoch formulation is fundamental in the sense that the current time formulations are directly derived from it. Its development is of considerable academic as well as practical engineering interest inasmuch as it constitutes a basis for the generation of nonsingular trajectory optimization software that are calculus of variations based and which use the equinoctial elements as coordinates. The transversality condition at the unknown final time needed in the numerical solution of the two-point boundary value problem for minimum-time transfers is more complicated than for the case in which the current mean longitude is selected as the fast variable. Nevertheless, the seven-parameter search is established to satisfy the boundary conditions and generate converged optimal transfer and rendezvous trajectories, duplicating previously generated results for mutual validation.

Table 5.1 Classical and equinoctial elements transformations

a	a
$h = e\sin(\omega + \Omega)$	$e = (h^2 + k^2)^{1/2}$
$k = e\cos(\omega + \Omega)$	$i = 2\tan^{-1}(p^2 + q^2)^{1/2}$
$p = \tan(i/2)\sin\Omega$	$\Omega = \tan^{-1}(p/q)$
$q = \tan(i/2)\cos\Omega$	$\omega = \tan^{-1}(h/k) - \tan^{-1}(p/q)$
$\lambda_0 = M_0 + \omega + \Omega$	$M_0 = \lambda_0 - \tan^{-1}(h/k)$

5.2 The Equations of Motion for the Epoch Mean Longitude Formulation

This section provides a general description of the equations of motion for the fundamental set of equinoctial elements by first showing how the partial derivatives of these elements, with respect to the position and velocity vectors, are derived from the knowledge of the Poisson brackets of the elements, and thereby constructing the required set of the variation of parameters equations. This general analysis is shown here for completeness, with many of the important expressions for the components of the position and velocity vectors in the direct equinoctial reference frame, with reference to the original contributions in [4, 5, 7]. Let Table 5.1 define the correspondence between the fundamental classical and equinoctial element sets.

Now, letting $\mathbf{z} = (ahk\,pq\,\lambda_0)^T$ represent the state vector at time t, and due to $\mathbf{z} = f(\mathbf{r}, \dot{\mathbf{r}})$,

$$\dot{\mathbf{z}} = \frac{\partial \mathbf{z}}{\partial \mathbf{r}}\dot{\mathbf{r}} + \frac{\partial \mathbf{z}}{\partial \dot{\mathbf{r}}}\ddot{\mathbf{r}} \tag{5.1}$$

where

$$\ddot{\mathbf{r}} = \frac{\mathbf{f}}{m} + \mathbf{g} = \frac{\mathbf{f}}{m} - \frac{\mu}{r^3}\mathbf{r} \tag{5.2}$$

The partial derivatives of the equinoctial elements with respect to $\mathbf{r}$ and $\dot{\mathbf{r}}$ are related to the partials of $\mathbf{r}$ and $\dot{\mathbf{r}}$ with respect to the elements $\mathbf{z}$ through the Poisson brackets (a_α, a_β) of the elements by

$$\frac{\partial a_\alpha}{\partial \mathbf{r}} = \sum_{\beta=1}^{6} (a_\alpha, a_\beta)\frac{\partial \dot{\mathbf{r}}}{\partial a_\beta} \tag{5.3}$$

$$\frac{\partial a_\alpha}{\partial \dot{\mathbf{r}}} = -\sum_{\beta=1}^{6} (a_\alpha, a_\beta)\frac{\partial \mathbf{r}}{\partial a_\beta} \tag{5.4}$$

The Poisson brackets P are related [8] to the Lagrange brackets L by $P = -L^{-1}$ and $P^T = L^{-1}$ with

$$L = \left(\frac{\partial \mathbf{r}}{\partial \mathbf{z}}\right)^T \frac{\partial \dot{\mathbf{r}}}{\partial \mathbf{z}} - \left(\frac{\partial \dot{\mathbf{r}}}{\partial \mathbf{z}}\right)^T \frac{\partial \mathbf{r}}{\partial \mathbf{z}} \tag{5.5}$$

$$P = \frac{\partial \mathbf{z}}{\partial \mathbf{r}}\left(\frac{\partial \mathbf{z}}{\partial \dot{\mathbf{r}}}\right)^T - \frac{\partial \mathbf{z}}{\partial \dot{\mathbf{r}}}\left(\frac{\partial \mathbf{z}}{\partial \mathbf{r}}\right)^T \tag{5.6}$$

with the Lagrange brackets given by $\left[a_\alpha, a_\beta\right] = \frac{\partial \mathbf{r}^T}{\partial a_\alpha}\frac{\partial \dot{\mathbf{r}}}{\partial a_\beta} - \frac{\partial \mathbf{r}^T}{\partial a_\beta}\frac{\partial \dot{\mathbf{r}}}{\partial a_\alpha}$ and the Poisson brackets given by $\left(a_\alpha, a_\beta\right) = \frac{\partial a_\alpha}{\partial \mathbf{r}}\left(\frac{\partial a_\beta}{\partial \dot{\mathbf{r}}}\right)^T - \frac{\partial a_\alpha}{\partial \dot{\mathbf{r}}}\left(\frac{\partial a_\beta}{\partial \mathbf{r}}\right)^T$. The various matrices of interest are given in terms of $\mathbf{r} = (x, y, z)$ and $\dot{\mathbf{r}} = (\dot{x}, \dot{y}, \dot{z})$, such that, for example,

$$[a, a] = \left(\frac{\partial x}{\partial a}\frac{\partial \dot{x}}{\partial a} - \frac{\partial x}{\partial a}\frac{\partial \dot{x}}{\partial a}\right) + \left(\frac{\partial y}{\partial a}\frac{\partial \dot{y}}{\partial a} - \frac{\partial y}{\partial a}\frac{\partial \dot{y}}{\partial a}\right) + \left(\frac{\partial z}{\partial a}\frac{\partial \dot{z}}{\partial a} - \frac{\partial z}{\partial a}\frac{\partial \dot{z}}{\partial a}\right) = 0$$

$$(a, a) = \left(\frac{\partial a}{\partial x}\frac{\partial a}{\partial \dot{x}} - \frac{\partial a}{\partial x}\frac{\partial a}{\partial \dot{x}}\right) + \left(\frac{\partial a}{\partial y}\frac{\partial a}{\partial \dot{y}} - \frac{\partial a}{\partial y}\frac{\partial a}{\partial \dot{y}}\right) + \left(\frac{\partial a}{\partial z}\frac{\partial a}{\partial \dot{z}} - \frac{\partial a}{\partial z}\frac{\partial a}{\partial \dot{z}}\right) = 0$$

The various brackets can also be written as the sum of three Jacobian determinants [8, 9]:

$$\left[a_\alpha, a_\beta\right] = \frac{\partial (x, \dot{x})}{\partial \left(a_\alpha, a_\beta\right)} + \frac{\partial (y, \dot{y})}{\partial \left(a_\alpha, a_\beta\right)} + \frac{\partial (z, \dot{z})}{\partial \left(a_\alpha, a_\beta\right)} \tag{5.7}$$

$$\left(a_\alpha, a_\beta\right) = \frac{\partial \left(a_\alpha, a_\beta\right)}{\partial (x, \dot{x})} + \frac{\partial \left(a_\alpha, a_\beta\right)}{\partial (y, \dot{y})} + \frac{\partial \left(a_\alpha, a_\beta\right)}{\partial (z, \dot{z})} \tag{5.8}$$

where, for example,

$$\frac{\partial (x, \dot{x})}{\partial \left(a_\alpha, a_\beta\right)} = \begin{vmatrix} \frac{\partial x}{\partial a_\alpha} & \frac{\partial x}{\partial a_\beta} \\ \frac{\partial \dot{x}}{\partial a_\alpha} & \frac{\partial \dot{x}}{\partial a_\beta} \end{vmatrix} = \frac{\partial x}{\partial a_\alpha}\frac{\partial \dot{x}}{\partial a_\beta} - \frac{\partial x}{\partial a_\beta}\frac{\partial \dot{x}}{\partial a_\alpha} \tag{5.9}$$

The matrix of the Poisson brackets P of the elements a, h, k, p, q, λ_0 can be obtained in a more direct manner [7] by a simple transformation applied to the matrix of the Poisson brackets of the classical elements $a, e, i, \Omega, \omega, M_0$. This mapping is shown in great detail [2, 10] as

$$P = \frac{1}{na^2}\begin{bmatrix} 0 & 0 & 0 & -2a & 0 & 0 \\ & 0 & -G & \frac{hG}{F'} & \frac{-kpK}{2G} & \frac{-kqK}{2G} \\ & & 0 & \frac{kG}{F'} & \frac{hpK}{2G} & \frac{hqK}{2G} \\ & & & 0 & \frac{-pK}{2G} & \frac{-qK}{2G} \\ & & & & 0 & \frac{-K^2}{4G} \\ -Sym & & & & & 0 \end{bmatrix}$$

$$= \begin{bmatrix} (a,a) & (a,h) & (a,k) & (a,\lambda_0) & (a,p) & (a,q) \\ (h,a) & (h,h) & (h,k) & (h,\lambda_0) & (h,p) & (h,q) \\ (k,a) & (k,h) & (k,k) & (k,\lambda_0) & (k,p) & (k,q) \\ (\lambda_0,a) & (\lambda_0,h) & (\lambda_0,k) & (\lambda_0,\lambda_0) & (\lambda_0,p) & (\lambda_0,q) \\ (p,a) & (p,h) & (p,k) & (p,\lambda_0) & (p,p) & (p,q) \\ (q,a) & (q,h) & (q,k) & (q,\lambda_0) & (q,p) & (q,q) \end{bmatrix} \tag{5.10}$$

Now using the property $\dfrac{\partial \begin{pmatrix} \mathbf{r} \\ \dot{\mathbf{r}} \end{pmatrix} \Big/ \partial \mathbf{z}}{\partial \mathbf{z} \Big/ \partial \begin{pmatrix} \mathbf{r} \\ \dot{\mathbf{r}} \end{pmatrix}} = I_{(6\times6)}$ or in expanded form

$$\begin{pmatrix} \dfrac{\partial \mathbf{r}}{\partial \mathbf{z}} \\ \dfrac{\partial \dot{\mathbf{r}}}{\partial \mathbf{z}} \end{pmatrix} \begin{pmatrix} \dfrac{\partial \mathbf{z}}{\partial \mathbf{r}} \dfrac{\partial \mathbf{z}}{\partial \dot{\mathbf{r}}} \end{pmatrix} = \begin{bmatrix} \dfrac{\partial \mathbf{r}}{\partial \mathbf{z}}\dfrac{\partial \mathbf{z}}{\partial \mathbf{r}} & \dfrac{\partial \mathbf{r}}{\partial \mathbf{z}}\dfrac{\partial \mathbf{z}}{\partial \dot{\mathbf{r}}} \\ \dfrac{\partial \dot{\mathbf{r}}}{\partial \mathbf{z}}\dfrac{\partial \mathbf{z}}{\partial \mathbf{r}} & \dfrac{\partial \dot{\mathbf{r}}}{\partial \mathbf{z}}\dfrac{\partial \mathbf{z}}{\partial \dot{\mathbf{r}}} \end{bmatrix} = \begin{pmatrix} I_{(3\times3)} & 0_{(3\times3)} \\ 0_{(3\times3)} & I_{(3\times3)} \end{pmatrix} \tag{5.11}$$

and post-multiplying matrix P in Equation (5.6) by $\left(\frac{\partial \dot{\mathbf{r}}}{\partial \mathbf{z}}\right)^T$ and $\left(\frac{\partial \mathbf{r}}{\partial \mathbf{z}}\right)^T$ respectively will yield $P\left(\frac{\partial \dot{\mathbf{r}}}{\partial \mathbf{z}}\right)^T = \frac{\partial \mathbf{z}}{\partial \mathbf{r}}$ and $P\left(\frac{\partial \mathbf{r}}{\partial \mathbf{z}}\right)^T = -\frac{\partial \mathbf{z}}{\partial \dot{\mathbf{r}}}$, or in other words, the identities in Equations (5.3) and (5.4). This is true because $\left(\frac{\partial \mathbf{z}}{\partial \mathbf{r}}\right)^T\left(\frac{\partial \dot{\mathbf{r}}}{\partial \mathbf{z}}\right)^T = 0_{(3\times3)}$ and $\left(\frac{\partial \mathbf{z}}{\partial \dot{\mathbf{r}}}\right)^T\left(\frac{\partial \mathbf{r}}{\partial \mathbf{z}}\right)^T = 0_{(3\times3)}$. Going back to Equation (5.1), writing one component at a time, and making use of Equation (5.2),

$$\begin{aligned} \dot{z} &= \frac{\partial z}{\partial \mathbf{r}}\dot{\mathbf{r}} + \frac{\partial z}{\partial \dot{\mathbf{r}}}\ddot{\mathbf{r}} \\ &= \sum_\beta \left(a_\alpha, a_\beta\right) \frac{\partial \dot{\mathbf{r}}}{\partial a_\beta}\dot{\mathbf{r}} - \sum_\beta \left(a_\alpha, a_\beta\right) \frac{\partial \mathbf{r}}{\partial a_\beta}\ddot{\mathbf{r}} \\ &= \sum_\beta \left(a_\alpha, a_\beta\right) \frac{\partial \dot{\mathbf{r}}}{\partial a_\beta}\dot{\mathbf{r}} + \sum_\beta \left(a_\alpha, a_\beta\right) \frac{\partial \mathbf{r}}{\partial a_\beta}\left(\frac{\mu}{r^3}\right)\mathbf{r} + \frac{\partial z}{\partial \dot{\mathbf{r}}}\frac{\mathbf{f}}{m} \\ &= \sum_\beta \left(a_\alpha, a_\beta\right) \left(\frac{\partial \dot{\mathbf{r}}}{\partial a_\beta}\dot{\mathbf{r}} + \frac{\mu}{r^3}\frac{\partial \mathbf{r}}{\partial a_\beta}\mathbf{r}\right) + \frac{\partial z}{\partial \dot{\mathbf{r}}}\frac{\mathbf{f}}{m} \end{aligned} \tag{5.12}$$

The term in brackets can be written as

$$\frac{1}{2}\frac{\partial\left(\dot{\mathbf{r}}^2\right)}{\partial z} + \frac{\mu}{r^3}\frac{1}{2}\frac{\partial\left(\mathbf{r}^2\right)}{\partial z} = \frac{1}{2}\left(\frac{\partial v^2}{\partial z} + \frac{\mu}{r^3}\frac{\partial r^2}{\partial z}\right) \tag{5.13}$$

because $|\dot{\mathbf{r}}| = v$, and using the energy equation $\frac{1}{2}v^2 - \frac{\mu}{r} = C$ which yields $\frac{1}{2}\frac{\partial v^2}{\partial z} = -\frac{\mu}{r^2}\frac{\partial r}{\partial z}$, and in view of $\partial r^2/\partial z = 2r\partial r/\partial z$, the brackets in Equation (5.13) cancel out such that $\dot{z} = \frac{\partial z}{\partial \dot{\mathbf{r}}}\frac{\mathbf{f}}{m}$ and in vector form the variation of parameters equations

$$\dot{\mathbf{z}} = \frac{\partial \mathbf{z}}{\partial \dot{\mathbf{r}}}\frac{f}{m}\hat{\mathbf{u}} = \frac{\partial \mathbf{z}}{\partial \dot{\mathbf{r}}}f_t\hat{\mathbf{u}} \tag{5.14}$$

From Equation (5.4), and the known Poisson brackets in Equation (5.10), the $\dot{\mathbf{z}}$ differential equations above can be obtained. If both $\hat{\mathbf{u}}$ and $\frac{\partial \mathbf{z}}{\partial \dot{\mathbf{r}}}$ are expressed in the direct equinoctial orbital frame with unit vectors $\hat{\mathbf{f}}, \hat{\mathbf{g}}, \hat{\mathbf{w}}$ shown in Figure 5.1 such that $\hat{\mathbf{f}}$ and $\hat{\mathbf{g}}$ are contained in the instantaneous orbit plane with the direction of $\hat{\mathbf{f}}$ obtained through a clockwise rotation of an angle Ω from the direction of the ascending node, then $\mathbf{r}$ and $\dot{\mathbf{r}}$ are such that [4, 5, 7] $\mathbf{r} = X_1\hat{\mathbf{f}} + Y_1\hat{\mathbf{g}}$ and $\dot{\mathbf{r}} = \dot{X}_1\hat{\mathbf{f}} + \dot{Y}_1\hat{\mathbf{g}}$, where the components X_1, Y_1, $\dot{X}_1$, $\dot{Y}_1$ are written in terms of the equinoctial elements and F, and also shown on p. 804 of [11] and Chapter 1 of this book.

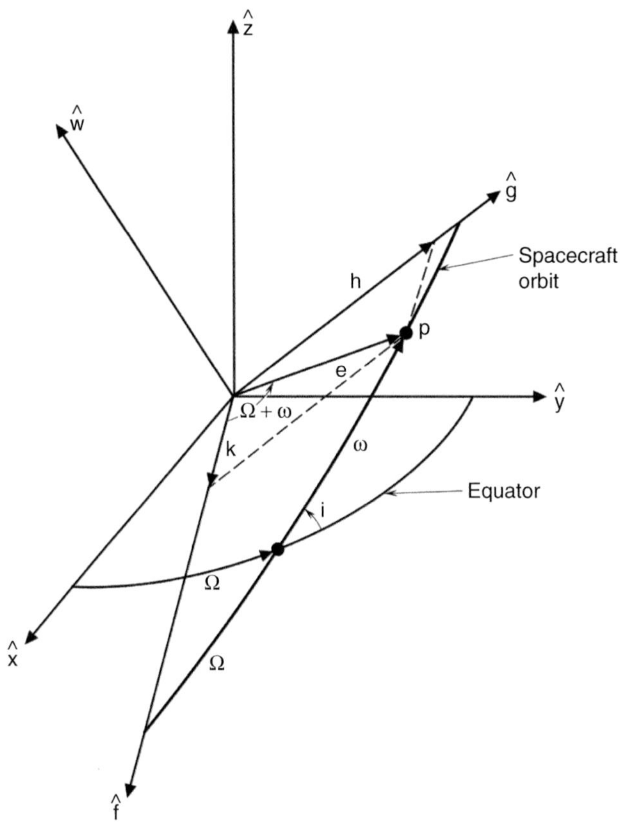

Figure 5.1 Direct equinoctial reference frame.

The following position vector partials are used in generating the equations of motion in Equation (5.14):

$$\frac{\partial \mathbf{r}}{\partial a} = \frac{\partial X_1}{\partial a}\hat{\mathbf{f}} + \frac{\partial Y_1}{\partial a}\hat{\mathbf{g}} = \left(\frac{X_1}{a} - \frac{3}{2}\frac{t}{a}\dot{X}_1\right)\hat{\mathbf{f}} + \left(\frac{Y_1}{a} - \frac{3}{2}\frac{t}{a}\dot{Y}_1\right)\hat{\mathbf{g}}$$

$$\frac{\partial \mathbf{r}}{\partial h} = \frac{\partial X_1}{\partial h}\hat{\mathbf{f}} + \frac{\partial Y_1}{\partial h}\hat{\mathbf{g}}$$

$$\frac{\partial \mathbf{r}}{\partial k} = \frac{\partial X_1}{\partial k}\hat{\mathbf{f}} + \frac{\partial Y_1}{\partial k}\hat{\mathbf{g}}$$

$$\frac{\partial \mathbf{r}}{\partial \lambda_0} = \frac{\partial \mathbf{r}}{\partial F}\frac{\partial F}{\partial \lambda}\frac{\partial \lambda}{\partial \lambda_0} = \frac{\dot{\mathbf{r}}}{n}$$

because $\partial \mathbf{r}/\partial \lambda = (\dot{X}_1/n)\hat{\mathbf{f}} + (\dot{Y}_1/n)\hat{\mathbf{g}} = \dot{\mathbf{r}}/n$. Finally,

$$\frac{\partial \mathbf{r}}{\partial p} = X_1\frac{\partial \hat{\mathbf{f}}}{\partial p} + Y_1\frac{\partial \hat{\mathbf{g}}}{\partial p} = \frac{2}{K}\left[q\left(Y_1\hat{\mathbf{f}} - X_1\hat{\mathbf{g}}\right) - X_1\hat{\mathbf{w}}\right]$$

$$\frac{\partial \mathbf{r}}{\partial q} = X_1\frac{\partial \hat{\mathbf{f}}}{\partial q} + Y_1\frac{\partial \hat{\mathbf{g}}}{\partial q} = \frac{2}{K}\left[p\left(X_1\hat{\mathbf{g}} - Y_1\hat{\mathbf{f}}\right) - Y_1\hat{\mathbf{w}}\right]$$

These expressions make use of the following partial derivatives of F with respect to the elements λ_0, a, h and k, namely $\frac{\partial F}{\partial \lambda_0} = \frac{a}{r}$, $\frac{\partial F}{\partial a} = -\frac{3}{2}\frac{nt}{r}$, $\frac{\partial F}{\partial h} = -\frac{a}{r}c_F$ and $\frac{\partial F}{\partial k} = \frac{a}{r}s_F$. They are easily derived from $\lambda = \lambda_0 + nt = F - ks_F + hc_F$, which yields

$$dF = \left[d\lambda_0 + ndt - \frac{3}{2}\frac{nt}{a}da + s_F dk - c_F dh\right]\frac{a}{r}$$

Also, the partial $\frac{\partial \mathbf{r}}{\partial \lambda_0}$ can be directly obtained from $\frac{\partial \mathbf{r}}{\partial \lambda_0} = \frac{\partial \mathbf{r}}{\partial F}\frac{\partial F}{\partial \lambda_0} = \left(\frac{\partial X_1}{\partial F}\hat{\mathbf{f}} + \frac{\partial Y_1}{\partial F}\hat{\mathbf{g}}\right)\frac{a}{r} = \left(\frac{r}{a}\frac{\dot{X}_1}{n}\hat{\mathbf{f}} + \frac{r}{a}\frac{\dot{Y}_1}{n}\hat{\mathbf{g}}\right)\frac{a}{r} = \frac{\dot{\mathbf{r}}}{n}$. The partials $\frac{\partial \mathbf{r}}{\partial p}$ and $\frac{\partial \mathbf{r}}{\partial q}$ are obtained by observing that X_1 and Y_1 are not functions of p and q, but that the unit vectors $\hat{\mathbf{f}}$ and $\hat{\mathbf{g}}$ are such functions because $\hat{\mathbf{f}}, \hat{\mathbf{g}}, \hat{\mathbf{w}}$ are given in the inertial $\hat{\mathbf{x}}, \hat{\mathbf{y}}, \hat{\mathbf{z}}$ frame by

$$\hat{\mathbf{f}} = \frac{1}{1+p^2+q^2}\begin{pmatrix} 1-p^2+q^2 \\ 2pq \\ -2p \end{pmatrix}$$

$$\hat{\mathbf{g}} = \frac{1}{1+p^2+q^2}\begin{pmatrix} 2pq \\ 1+p^2-q^2 \\ 2q \end{pmatrix}$$

$$\hat{\mathbf{w}} = \frac{1}{1+p^2+q^2}\begin{pmatrix} 2p \\ -2q \\ 1-p^2-q^2 \end{pmatrix}$$

Therefore,

$$\frac{\partial \mathbf{r}}{\partial p} = \frac{2X_1}{\left(1+p^2+q^2\right)^2}\begin{pmatrix} -2p-2pq^2 \\ 2q - q\left(1+p^2-q^2\right) \\ -2q^2 - \left(1-p^2-q^2\right) \end{pmatrix} + \frac{2Y_1}{\left(1+p^2+q^2\right)^2}\begin{pmatrix} q\left(1-p^2+q^2\right) \\ 2pq^2 \\ -2pq \end{pmatrix}$$

which yields the form written above. Similar manipulations yield the expression for $\frac{\partial \mathbf{r}}{\partial q}$. Also, the partials $\partial X_1/\partial h$, $\partial X_1/\partial k$, $\partial Y_1/\partial h$ and $\partial Y_1/\partial k$ are generated by using $\frac{\partial F}{\partial h} = -\frac{a}{r}c_F$, $\frac{\partial F}{\partial k} = \frac{a}{r}s_F$ and $\frac{\partial \beta}{\partial h} = \frac{h\beta^3}{1-\beta}$, $\frac{\partial \beta}{\partial k} = \frac{k\beta^3}{1-\beta}$. They are shown explicitly in [4] and [5] and also on p. 804 of [11] and Chapter 2 of this book.

Carrying out the algebra results in the generation of the partials of the five slowly varying elements a, h, k, p and q with respect to the velocity vector, namely $\partial a/\partial \dot{\mathbf{r}}$, $\partial h/\partial \dot{\mathbf{r}}$, $\partial k/\partial \dot{\mathbf{r}}$, $\partial p/\partial \dot{\mathbf{r}}$ and $\partial q/\partial \dot{\mathbf{r}}$ shown, for example, in Equation (11) of [11] and Chapter 2 of this book, and by $\partial \lambda_0/\partial \dot{\mathbf{r}}$ given below:

$$\begin{aligned}\frac{\partial \lambda_0}{\partial \dot{\mathbf{r}}} &= n^{-1}a^{-2}\left[-2X_1 + 3\dot{X}_1 t + G\left(h\beta\frac{\partial X_1}{\partial h} + k\beta\frac{\partial X_1}{\partial k}\right)\right]\hat{\mathbf{f}} \\ &+ n^{-1}a^{-2}\left[-2Y_1 + 3\dot{Y}_1 t + G\left(h\beta\frac{\partial Y_1}{\partial h} + k\beta\frac{\partial Y_1}{\partial k}\right)\right]\hat{\mathbf{g}} \\ &+ n^{-1}a^{-2}G^{-1}\left(qY_1 - pX_1\right)\hat{\mathbf{w}} = M_{61}^0\hat{\mathbf{f}} + M_{62}^0\hat{\mathbf{g}} + M_{63}^0\hat{\mathbf{w}} \qquad (5.15)\end{aligned}$$

The magnitude of $\mathbf{r}$ is simply obtained from the orbit equation, $r = a(1 - kc_F - hs_F)$. The system of differential equations to integrate is then given by $\dot{a} = \left(\frac{\partial a}{\partial \dot{\mathbf{r}}}\right)^T \bullet \hat{\mathbf{u}} f_t$, $\dot{h} = \left(\frac{\partial h}{\partial \dot{\mathbf{r}}}\right)^T \bullet \hat{\mathbf{u}} f_t$, $\dot{k} = \left(\frac{\partial k}{\partial \dot{\mathbf{r}}}\right)^T \bullet \hat{\mathbf{u}} f_t$, $\dot{p} = \left(\frac{\partial p}{\partial \dot{\mathbf{r}}}\right)^T \bullet \hat{\mathbf{u}} f_t$, $\dot{q} = \left(\frac{\partial q}{\partial \dot{\mathbf{r}}}\right)^T \bullet \hat{\mathbf{u}} f_t$ and $\dot{\lambda}_0 = \left(\frac{\partial \lambda_0}{\partial \dot{\mathbf{r}}}\right)^T \bullet \hat{u} f_t$ where n is osculating too because it is dependent on the semimajor axis. Because we are integrating the epoch mean longitude λ_0 as the sixth element and due to the fact that X_1, Y_1, $\dot{X}_1$, $\dot{Y}_1$, r as well as the partials $\partial X_1/\partial h$ through $\partial Y_1/\partial k$ are given in terms of the eccentric longitude F, it is necessary at each integration step to solve Kepler's equation after evaluating the current mean longitude λ from $\lambda = \lambda_0 + nt$:

$$\lambda = F - ks_F + hc_F \tag{5.16}$$

for F, through a Newton–Raphson iteration scheme. The error tolerance is set to 10^{-15}.

5.3 The Variational Hamiltonian

This section provides an explicit form of the differential equations for the adjoint variables which in turn show how these variables remain truly constant in the absence of perturbations for this particular set. The assumptions needed in generating these equations are also discussed, leading to the form of the partial derivatives of the eccentric longitude accessory variable with respect to the four elements upon which it is dependent, namely the elements λ_0, h, k and a, the semimajor axis.

The Hamiltonian of the system is given by

$$H = \boldsymbol{\lambda}_z^T \dot{\mathbf{z}} = \boldsymbol{\lambda}_z^T M(\mathbf{z}, F) f_t \hat{\mathbf{u}} \tag{5.17}$$

where the seventh differential equation for the mass flow rate has been neglected due to the small expenditure of propellant during the transfer for a spacecraft with a low-thrust propulsion system. The Euler–Lagrange equations are given by

$$\dot{\boldsymbol{\lambda}}_z = -\frac{\partial H}{\partial \mathbf{z}} = -\boldsymbol{\lambda}_z^T \frac{\partial M}{\partial \mathbf{z}} f_t \hat{\mathbf{u}} \tag{5.18}$$

The optimal thrust direction $\hat{\mathbf{u}}$ is chosen such that it is at all times parallel to $\boldsymbol{\lambda}_z^T M(\mathbf{z}, F)$ in order to maximize the Hamiltonian. The matrix M for the set $(a, h, k, p, q, \lambda_0)$ is as given by Equation (11) of [11] and by Equation (5.15) of this chapter. The partials $\partial M/\partial \mathbf{z}$ are shown in Appendix 5.1. If we let u_i represent the components of $\hat{\mathbf{u}}$ in the equinoctial frame, then with $u_1 = u_f$, $u_2 = u_g$ and $u_3 = u_w$ and using the summation notation:

$$\begin{aligned} H = \Big[& \lambda_a (M_{1i} u_i) + \lambda_h (M_{2i} u_i) + \lambda_k (M_{3i} u_i) + \lambda_p (M_{4i} u_i) \\ & + \lambda_q (M_{5i} u_i) + \lambda_{\lambda_0} \left(M_{6i}^0 u_i \right) \Big] f_t \end{aligned} \tag{5.19}$$

The corresponding Lagrange equations are simply given by the partial derivatives of the above Hamiltonian with respect to the six elements of interest:

$$\dot{\lambda}_a = -\left(\lambda_a\,\lambda_h\,\lambda_k\,\lambda_p\,\lambda_q\,\lambda_{\lambda_0}\right)\frac{\partial M}{\partial a}f_t\hat{\mathbf{u}} = -\frac{\partial H}{\partial a}$$
$$= \left[-\lambda_a\left(\frac{\partial M_{1i}}{\partial a}u_i\right) - \lambda_h\left(\frac{\partial M_{2i}}{\partial a}u_i\right) - \lambda_k\left(\frac{\partial M_{3i}}{\partial a}u_i\right) - \lambda_p\left(\frac{\partial M_{4i}}{\partial a}u_i\right)\right.$$
$$\left. - \lambda_q\left(\frac{\partial M_{5i}}{\partial a}u_i\right) - \lambda_{\lambda_0}\left(\frac{\partial M^0_{6i}}{\partial a}u_i\right)\right]f_t \tag{5.20}$$

$$\dot{\lambda}_h = -\frac{\partial H}{\partial h} = \left[-\lambda_a\left(\frac{\partial M_{1i}}{\partial h}u_i\right) - \lambda_h\left(\frac{\partial M_{2i}}{\partial h}u_i\right) - \lambda_k\left(\frac{\partial M_{3i}}{\partial h}u_i\right)\right.$$
$$\left. -\lambda_p\left(\frac{\partial M_{4i}}{\partial h}u_i\right) - \lambda_q\left(\frac{\partial M_{5i}}{\partial h}u_i\right) - \lambda_{\lambda_0}\left(\frac{\partial M^0_{6i}}{\partial h}u_i\right)\right]f_t \tag{5.21}$$

$$\dot{\lambda}_k = -\frac{\partial H}{\partial k} = \left[-\lambda_a\left(\frac{\partial M_{1i}}{\partial k}u_i\right) - \lambda_h\left(\frac{\partial M_{2i}}{\partial k}u_i\right) - \lambda_k\left(\frac{\partial M_{3i}}{\partial k}u_i\right)\right.$$
$$\left. -\lambda_p\left(\frac{\partial M_{4i}}{\partial k}u_i\right) - \lambda_q\left(\frac{\partial M_{5i}}{\partial k}u_i\right) - \lambda_{\lambda_0}\left(\frac{\partial M^0_{6i}}{\partial k}u_i\right)\right]f_t \tag{5.22}$$

$$\dot{\lambda}_p = -\frac{\partial H}{\partial p} = \left[-\lambda_a\left(\frac{\partial M_{1i}}{\partial p}u_i\right) - \lambda_h\left(\frac{\partial M_{2i}}{\partial p}u_i\right) - \lambda_k\left(\frac{\partial M_{3i}}{\partial p}u_i\right)\right.$$
$$\left. -\lambda_p\left(\frac{\partial M_{4i}}{\partial p}u_i\right) - \lambda_q\left(\frac{\partial M_{5i}}{\partial p}u_i\right) - \lambda_{\lambda_0}\left(\frac{\partial M^0_{6i}}{\partial p}u_i\right)\right]f_t \tag{5.23}$$

$$\dot{\lambda}_q = -\frac{\partial H}{\partial q} = \left[-\lambda_a\left(\frac{\partial M_{1i}}{\partial q}u_i\right) - \lambda_h\left(\frac{\partial M_{2i}}{\partial q}u_i\right) - \lambda_k\left(\frac{\partial M_{3i}}{\partial q}u_i\right) - \lambda_p\left(\frac{\partial M_{4i}}{\partial q}u_i\right)\right.$$
$$\left. -\lambda_q\left(\frac{\partial M_{5i}}{\partial q}u_i\right) - \lambda_{\lambda_0}\left(\frac{\partial M^0_{6i}}{\partial q}u_i\right)\right]f_t \tag{5.24}$$

$$\dot{\lambda}_{\lambda_0} = -\frac{\partial H}{\partial \lambda_0} = \left[-\lambda_a\left(\frac{\partial M_{1i}}{\partial \lambda_0}u_i\right) - \lambda_h\left(\frac{\partial M_{2i}}{\partial \lambda_0}u_i\right) - \lambda_k\left(\frac{\partial M_{3i}}{\partial \lambda_0}u_i\right) - \lambda_p\left(\frac{\partial M_{4i}}{\partial \lambda_0}u_i\right)\right.$$
$$\left. -\lambda_q\left(\frac{\partial M_{5i}}{\partial \lambda_0}u_i\right) - \lambda_{\lambda_0}\left(\frac{\partial M^0_{6i}}{\partial \lambda_0}u_i\right)\right]f_t \tag{5.25}$$

All partials of the matrix M with respect to the elements appearing in the above differential equations are shown in Appendix 5.1. The following partials are used to generate the $\partial M/\partial \mathbf{z}$ partials of Appendix 5.1. In this formulation, F is a function of λ_0, h, k and a, because from Kepler's equation $\lambda_0 + nt = F - ks_F + hc_F$, we have the partials in Table 5.2.

The partials of the first five rows of the M matrix with respect to h, k, p and q are identical to those in [11], and also shown in Appendix 3.1, and they are not repeated here. Because F is dependent on a, all the partials with respect to a involve the explicit appearance of terms in t. In minimum-fuel problems, all the state and adjoint variables remain constant during the coasting arc when the present λ_0 formulation is used. The switching

Table 5.2 Partials of F and r w.r.t elements

$\partial F/\partial \lambda_0 = \frac{a}{r}$	$\partial r/\partial F = a(ks_F - hc_F)$
$\partial F/\partial a = -\frac{3}{2}\frac{nt}{r}$	$\partial r/\partial a = \frac{r}{a} - \frac{3}{2}\frac{nat}{r}(ks_F - hc_F)$
$\partial F/\partial h = -\frac{a}{r}c_F$	$\partial r/\partial h = \frac{a^2}{r}(h - s_F)$
$\partial F/\partial k = -\frac{a}{r}s_F$	$\partial r/\partial k = \frac{a^2}{r}(k - c_F)$

function which determines the locations where the thrust must be either turned on or turned off becomes a function of the current anomaly on the conic orbit corresponding to the coast arc. The numerical interpolation needed to find the zeros of the switching function that correspond to the beginning of the powered arcs is thus easily carried out with the present λ_0 formulation. Also, in the context of averaging, the averaged rate $\tilde{\dot{\lambda}}_a$ involves only one Gauss–Legendre quadrature as opposed to two quadratures for the λ formulation of [11] and Chapter 3.

5.4 Canonical Transformations

One of the main purposes of deriving the various formulations corresponding to different orbit element sets is to mutually validate the mathematics of all the corresponding differential equations, and the underlying assumptions used in their derivations. These formulations are thus coded and run to generate duplicate transfers and the various state and adjoint variables' evolutions during the given transfer plotted for comparison. Canonical transformations are of great use in relating mathematically the various adjoint variables for further verification purposes and also for selecting the initial conditions for some or all of the adjoint variables for a given set once the solution corresponding to another set is known. This allows for straightforward trajectory runs without the need to solve the two-point boundary value problem repeatedly in order to validate a given formulation. To this end, the mappings between all the equinoctial sets are shown explicitly in this section, beginning with the classical and equinoctial pair with the current mean anomaly and current mean longitude as sixth state variable respectively. The canonicity condition for this particular transformation is given by

$$\begin{aligned}\lambda_a da + \lambda_h dh + \lambda_k dk + \lambda_p dp + \lambda_q dq + \lambda_\lambda d\lambda - H^\lambda dt \\ = \lambda_a da + \lambda_e de + \lambda_i di + \lambda_\Omega d\Omega + \lambda_\omega d\omega + \lambda_M dM - H^M dt \qquad (5.26)\end{aligned}$$

Examples of canonical transformations can be found in [12]. From the inverse transformation given in Table 5.1, and $M = \lambda - \tan^{-1}\left(\frac{h}{k}\right)$, we have respectively

$$de = \frac{h}{e}dh + \frac{k}{e}dk$$

$$di = \frac{2c_{i/2}^2}{\tan\frac{i}{2}}pdp + \frac{2c_{i/2}^2}{\tan\frac{i}{2}}qdq$$

$$d\Omega = c_{\Omega}^{2}\frac{(qdp - pdq)}{q^2}$$
$$d\omega = \frac{(kdh - hdk)}{e^2} - \frac{(qdp - pdq)}{\tan^2\frac{i}{2}}$$
$$dM = d\lambda - \frac{(kdh - hdk)}{e^2}$$

Inserting these identities in Equation (5.26) yields

$$\lambda_a = \lambda_a \tag{5.27}$$
$$\lambda_h = \lambda_e s_{\omega+\Omega} + \frac{\lambda_\omega}{e}c_{\omega+\Omega} - \frac{\lambda_M}{e}c_{\omega+\Omega} \tag{5.28}$$
$$\lambda_k = \lambda_e c_{\omega+\Omega} - \frac{\lambda_\omega}{e}s_{\omega+\Omega} + \frac{\lambda_M}{e}s_{\omega+\Omega} \tag{5.29}$$
$$\lambda_p = 2\lambda_i s_\Omega c_{i/2}^2 + \lambda_\Omega\frac{c_\Omega}{\tan\frac{i}{2}} - \lambda_\omega\frac{c_\Omega}{\tan\frac{i}{2}} \tag{5.30}$$
$$\lambda_q = 2\lambda_i c_\Omega c_{i/2}^2 - \lambda_\Omega\frac{s_\Omega}{\tan\frac{i}{2}} + \lambda_\omega\frac{s_\Omega}{\tan\frac{i}{2}} \tag{5.31}$$
$$\lambda_\lambda = \lambda_M \tag{5.32}$$
$$H^M = H^\lambda \tag{5.33}$$

Here H^M stands for the Hamiltonian of the classical system and H^λ for the equinoctial system. Now if we consider the transformation between the set (a, h, k, p, q, λ) and the set $(a, h, k, p, q, \lambda_0)$ and using

$$d\lambda = d\lambda_0 + tdn + ndt \tag{5.34}$$

because $\lambda = \lambda_0 + nt$ and with $dn = -\frac{3}{2}\frac{n}{a}da$, the canonicity condition requires

$$\lambda_a^\lambda da + \lambda_\lambda^\lambda d\lambda - H^\lambda dt = \lambda_a^{\lambda_0} da + \lambda_{\lambda_0}^{\lambda_0} d\lambda_0 - H^{\lambda_0} dt \tag{5.35}$$

where H^{λ_0} is given by Equation (5.19). The superscript λ_0 is used in the right-hand side of the above relation to indicate that it is the one corresponding to the $(a, h, k, p, q, \lambda_0)$ set. This yields

$$\lambda_\lambda^\lambda = \lambda_{\lambda_0}^{\lambda_0} \tag{5.36}$$
$$H^{\lambda_0} = H^\lambda - n\lambda_\lambda^\lambda = H^\lambda - n\lambda_{\lambda_0}^{\lambda_0} \tag{5.37}$$
$$\lambda_a^\lambda = \lambda_a^{\lambda_0} + \frac{3}{2}\frac{nt}{a}\lambda_\lambda^\lambda = \lambda_a^{\lambda_0} + \frac{3}{2}\frac{nt}{a}\lambda_{\lambda_0}^{\lambda_0} \tag{5.38}$$

It is also true that $\lambda_{\lambda_0}^{\lambda_0} = \lambda_\lambda^\lambda = \lambda_M = \lambda_{M_0}$ such that Equations (5.28) through (5.31) are still valid with λ_M identically replaced by λ_{M_0} in the first two expressions. Now using the superscripts F and F_0 for the (a, h, k, p, q, F) and (a, h, k, p, q, F_0) formulations of [13] and [14] respectively, we have from [13] and Chapter 6:

$$\lambda_a^F = \lambda_a^\lambda$$
$$\lambda_h^F = \lambda_h^\lambda + \lambda_\lambda^\lambda c_F$$

$$\begin{aligned}
\lambda_k^F &= \lambda_k^\lambda - \lambda_\lambda^\lambda s_F \\
\lambda_p^F &= \lambda_p^\lambda \\
\lambda_q^F &= \lambda_q^\lambda \\
\lambda_F^F &= \frac{r}{a}\lambda_\lambda^\lambda \\
H^F &= H^\lambda
\end{aligned} \tag{5.39}$$

and from [14], and Chapter 7:

$$\begin{aligned}
\lambda_a^{F_0} &= \lambda_a^F - \frac{3}{2}\frac{nt}{a}\frac{a}{r}\lambda_F^F \\
\lambda_h^{F_0} &= \lambda_h^F - \frac{a}{r}c_F\lambda_F^F \\
\lambda_k^{F_0} &= \lambda_k^F + \frac{a}{r}s_F\lambda_F^F \\
\lambda_p^{F_0} &= \lambda_p^F \\
\lambda_q^{F_0} &= \lambda_q^F \\
\lambda_{F_0}^{F_0} &= \frac{a}{r}\frac{r_0}{a_0}\lambda_F^F \\
H^{F_0} &= H^F - n\frac{a}{r}\lambda_F^F
\end{aligned} \tag{5.40}$$

Furthermore, from [15] and the present chapter using the $(a, h, k, p, q, \lambda_0)$ formulation, we have:

$$\begin{aligned}
\lambda_a^{\lambda_0} &= \lambda_a^\lambda - \frac{3}{2}\frac{nt}{a}\lambda_\lambda^\lambda \\
\lambda_h^{\lambda_0} &= \lambda_h^\lambda \\
\lambda_k^{\lambda_0} &= \lambda_k^\lambda \\
\lambda_p^{\lambda_0} &= \lambda_p^\lambda \\
\lambda_q^{\lambda_0} &= \lambda_q^\lambda \\
\lambda_{\lambda_0}^{\lambda_0} &= \lambda_\lambda^\lambda \\
H^{\lambda_0} &= H^\lambda - n\lambda_\lambda^\lambda
\end{aligned} \tag{5.41}$$

The expressions in Equations (5.39) and (5.41) combine to yield

$$\begin{aligned}
\lambda_a^F &= \lambda_a^{\lambda_0} + \frac{3}{2}\frac{nt}{a}\lambda_{\lambda_0}^{\lambda_0} \\
\lambda_h^F &= \lambda_h^{\lambda_0} + \lambda_{\lambda_0}^{\lambda_0}c_F \\
\lambda_k^F &= \lambda_k^{\lambda_0} - \lambda_{\lambda_0}^{\lambda_0}s_F \\
\lambda_p^F &= \lambda_p^{\lambda_0} \\
\lambda_q^F &= \lambda_q^{\lambda_0} \\
\lambda_F^F &= \frac{r}{a}\lambda_{\lambda_0}^{\lambda_0} \\
H^F &= H^{\lambda_0} + n\lambda_{\lambda_0}^{\lambda_0}
\end{aligned} \tag{5.42}$$

which are then combined with Equations (5.40) to yield

$$\begin{aligned}
\lambda_a^{F_0} &= \lambda_a^{\lambda_0}\\
\lambda_h^{F_0} &= \lambda_h^{\lambda_0}\\
\lambda_k^{F_0} &= \lambda_k^{\lambda_0}\\
\lambda_p^{F_0} &= \lambda_p^{\lambda_0}\\
\lambda_q^{F_0} &= \lambda_q^{\lambda_0}\\
\lambda_{F_0}^{F_0} &= \frac{r_0}{a_0}\lambda_{\lambda_0}^{\lambda_0}\\
H^{F_0} &= H^{\lambda_0}
\end{aligned} \tag{5.43}$$

In the matrix form, the following mappings are obtained:

$$\begin{pmatrix}\lambda_a^\lambda\\ \lambda_h^\lambda\\ \lambda_k^\lambda\\ \lambda_p^\lambda\\ \lambda_q^\lambda\\ \lambda_\lambda^\lambda\end{pmatrix} = \begin{pmatrix} 1 & 0 & 0 & 0 & 0 & 0\\ 0 & s_{\omega+\Omega} & 0 & 0 & \dfrac{c_{\omega+\Omega}}{e} & \dfrac{-c_{\omega+\Omega}}{e}\\ 0 & c_{\omega+\Omega} & 0 & 0 & \dfrac{-s_{\omega+\Omega}}{e} & \dfrac{s_{\omega+\Omega}}{e}\\ 0 & 0 & 2s_\Omega c_{i/2}^2 & \dfrac{c_\Omega}{\tan(i/2)} & \dfrac{-c_\Omega}{\tan(i/2)} & 0\\ 0 & 0 & 2c_\Omega c_{i/2}^2 & \dfrac{-s_\Omega}{\tan(i/2)} & \dfrac{s_\Omega}{\tan(i/2)} & 0\\ 0 & 0 & 0 & 0 & 0 & 1\end{pmatrix}\begin{pmatrix}\lambda_a\\ \lambda_e\\ \lambda_i\\ \lambda_\Omega\\ \lambda_\omega\\ \lambda_M\end{pmatrix} \tag{5.44}$$

with $H^\lambda = H^M$.

Equations (5.41) and (5.44) combine to yield

$$\begin{pmatrix}\lambda_a^{\lambda_0}\\ \lambda_h^{\lambda_0}\\ \lambda_k^{\lambda_0}\\ \lambda_p^{\lambda_0}\\ \lambda_q^{\lambda_0}\\ \lambda_{\lambda_0}^{\lambda_0}\end{pmatrix} = \begin{pmatrix} 1 & 0 & 0 & 0 & 0 & -\dfrac{3}{2}\dfrac{nt}{a}\\ 0 & s_{\omega+\Omega} & 0 & 0 & \dfrac{c_{\omega+\Omega}}{e} & \dfrac{-c_{\omega+\Omega}}{e}\\ 0 & c_{\omega+\Omega} & 0 & 0 & \dfrac{-s_{\omega+\Omega}}{e} & \dfrac{s_{\omega+\Omega}}{e}\\ 0 & 0 & 2s_\Omega c_{i/2}^2 & \dfrac{c_\Omega}{\tan(i/2)} & \dfrac{-c_\Omega}{\tan(i/2)} & 0\\ 0 & 0 & 2c_\Omega c_{i/2}^2 & \dfrac{-s_\Omega}{\tan(i/2)} & \dfrac{s_\Omega}{\tan(i/2)} & 0\\ 0 & 0 & 0 & 0 & 0 & 1\end{pmatrix}\begin{pmatrix}\lambda_a\\ \lambda_e\\ \lambda_i\\ \lambda_\Omega\\ \lambda_\omega\\ \lambda_M\end{pmatrix}$$

with $H^{\lambda_0} = H^\lambda - n\lambda_\lambda^\lambda = H^M - n\lambda_M$.

Equations (5.39) and (5.44) combine to yield

$$\begin{pmatrix}\lambda_a^F\\ \lambda_h^F\\ \lambda_k^F\\ \lambda_p^F\\ \lambda_q^F\\ \lambda_\lambda^F\end{pmatrix} = \begin{pmatrix} 1 & 0 & 0 & 0 & 0 & 0\\ 0 & s_{\omega+\Omega} & 0 & 0 & \dfrac{c_{\omega+\Omega}}{e} & \dfrac{-c_{\omega+\Omega}}{e} + c_F\\ 0 & c_{\omega+\Omega} & 0 & 0 & \dfrac{-s_{\omega+\Omega}}{e} & \dfrac{s_{\omega+\Omega}}{e} - s_F\\ 0 & 0 & 2s_\Omega c_{i/2}^2 & \dfrac{c_\Omega}{\tan(i/2)} & \dfrac{-c_\Omega}{\tan(i/2)} & 0\\ 0 & 0 & 2c_\Omega c_{i/2}^2 & \dfrac{-s_\Omega}{\tan(i/2)} & \dfrac{s_\Omega}{\tan(i/2)} & 0\\ 0 & 0 & 0 & 0 & 0 & \dfrac{r}{a}\end{pmatrix}\begin{pmatrix}\lambda_a\\ \lambda_e\\ \lambda_i\\ \lambda_\Omega\\ \lambda_\omega\\ \lambda_M\end{pmatrix}$$

with $H^F = H^\lambda = H^M$.

Equations (5.39), (5.40) and (5.44) combine to yield

$$\begin{pmatrix}\lambda_a^{F_0}\\ \lambda_h^{F_0}\\ \lambda_k^{F_0}\\ \lambda_p^{F_0}\\ \lambda_q^{F_0}\\ \lambda_{\lambda_0}^{F_0}\end{pmatrix} = \begin{pmatrix} 1 & 0 & 0 & 0 & 0 & -\frac{3}{2}\frac{nt}{a}\\ 0 & s_{\omega+\Omega} & 0 & 0 & \frac{c_{\omega+\Omega}}{e} & \frac{-c_{\omega+\Omega}}{e}\\ 0 & c_{\omega+\Omega} & 0 & 0 & \frac{-s_{\omega+\Omega}}{e} & \frac{s_{\omega+\Omega}}{e}\\ 0 & 0 & 2s_\Omega c_{i/2}^2 & \frac{c_\Omega}{\tan(i/2)} & \frac{-c_\Omega}{\tan(i/2)} & 0\\ 0 & 0 & 2c_\Omega c_{i/2}^2 & \frac{-s_\Omega}{\tan(i/2)} & \frac{s_\Omega}{\tan(i/2)} & 0\\ 0 & 0 & 0 & 0 & 0 & \frac{r_0}{a_0}\end{pmatrix}\begin{pmatrix}\lambda_a\\ \lambda_e\\ \lambda_i\\ \lambda_\Omega\\ \lambda_\omega\\ \lambda_M\end{pmatrix}$$

with $H^{F_0} = H^F - n\frac{a}{r}\lambda_F^F = H^F - n\lambda_\lambda^\lambda = H^F - n\lambda_M = H^\lambda - n\lambda_M = H^{\lambda_0}$ and $H^{F_0} = H^M - n\lambda_M$ with $\frac{r_0}{a_0} = 1 - k_0 c_{F_0} - h_0 s_{F_0}$. The preceding mapping matrices can also be written in terms of the equinoctial variables and associated quantities by observing that $(p^2+q^2)^{1/2} = (K-1)^{1/2}$, $p^2+q^2 = \tan^2(i/2)$, $s_\Omega = p/(p^2+q^2)^{1/2}$, $c_\Omega = q/(p^2+q^2)^{1/2}$, $c_{i/2}^2 = \frac{1}{1+p^2+q^2}$, such that we have the identities in Table 5.3.

These transformations lead to

$$\begin{pmatrix}\lambda_a^{\lambda_0}\\ \lambda_h^{\lambda_0}\\ \lambda_k^{\lambda_0}\\ \lambda_p^{\lambda_0}\\ \lambda_q^{\lambda_0}\\ \lambda_{\lambda_0}^{\lambda_0}\end{pmatrix} = \begin{pmatrix} 1 & 0 & 0 & 0 & 0 & -\frac{3}{2}\frac{nt}{a}\\ 0 & \frac{h\beta}{(2\beta-1)^{1/2}} & 0 & 0 & \frac{k\beta^2}{2\beta-1} & \frac{-k\beta^2}{2\beta-1}\\ 0 & \frac{k\beta}{(2\beta-1)^{1/2}} & 0 & 0 & \frac{-h\beta^2}{2\beta-1} & \frac{h\beta^2}{2\beta-1}\\ 0 & 0 & \frac{2p}{K(K-1)^{1/2}} & \frac{q}{K-1} & \frac{-q}{K-1} & 0\\ 0 & 0 & \frac{2q}{K(K-1)^{1/2}} & \frac{-p}{K-1} & \frac{p}{K-1} & 0\\ 0 & 0 & 0 & 0 & 0 & 1\end{pmatrix}\begin{pmatrix}\lambda_a\\ \lambda_e\\ \lambda_i\\ \lambda_\Omega\\ \lambda_\omega\\ \lambda_M\end{pmatrix} \tag{5.45}$$

$$\begin{pmatrix}\lambda_a^{F_0}\\ \lambda_h^{F_0}\\ \lambda_k^{F_0}\\ \lambda_p^{F_0}\\ \lambda_q^{F_0}\\ \lambda_{F_0}^{F_0}\end{pmatrix} = \begin{pmatrix} 1 & 0 & 0 & 0 & 0 & -\frac{3}{2}\frac{nt}{a}\\ 0 & \frac{h\beta}{(2\beta-1)^{1/2}} & 0 & 0 & \frac{k\beta^2}{2\beta-1} & \frac{-k\beta^2}{2\beta-1}\\ 0 & \frac{k\beta}{(2\beta-1)^{1/2}} & 0 & 0 & \frac{-h\beta^2}{2\beta-1} & \frac{h\beta^2}{2\beta-1}\\ 0 & 0 & \frac{2p}{K(K-1)^{1/2}} & \frac{q}{K-1} & \frac{-q}{K-1} & 0\\ 0 & 0 & \frac{2q}{K(K-1)^{1/2}} & \frac{-p}{K-1} & \frac{p}{K-1} & 0\\ 0 & 0 & 0 & 0 & 0 & 1-k_0c_{F_0}-h_0s_{F_0}\end{pmatrix}\begin{pmatrix}\lambda_a\\ \lambda_e\\ \lambda_i\\ \lambda_\Omega\\ \lambda_\omega\\ \lambda_M\end{pmatrix} \tag{5.46}$$

The last element of the above 6×6 matrix is equal to r_0/a_0 with a_0 fixed and r_0 osculating because F_0 is itself osculating. Both h_0 and k_0 are also fixed and for an initial circular orbit, $e_0 = 0$ such that h_0 and k_0 are both equal to 0 and therefore $r_0/a_0 = 1$ will not osculate but remain constant. Furthermore,

Table 5.3 Identities for the transformation of the mapping matrices

$s_{\omega+\Omega} = \dfrac{h}{e} = \dfrac{h\beta}{(2\beta-1)^{1/2}}$	$2s_\Omega c_{i/2}^2 = \dfrac{2p}{K(K-1)^{1/2}}$
$c_{\omega+\Omega} = \dfrac{k}{e} = \dfrac{k\beta}{(2\beta-1)^{1/2}}$	$2c_\Omega c_{i/2}^2 = \dfrac{2q}{K(K-1)^{1/2}}$
$\dfrac{c_{\omega+\Omega}}{e} = \dfrac{k\beta^2}{2\beta-1}$	$\dfrac{s_\Omega}{\tan(i/2)} = \dfrac{p}{K-1}$
$\dfrac{s_{\omega+\Omega}}{e} = \dfrac{h\beta^2}{2\beta-1}$	$\dfrac{c_\Omega}{\tan(i/2)} = \dfrac{q}{K-1}$

$$\begin{pmatrix} \lambda_a^F \\ \lambda_h^F \\ \lambda_k^F \\ \lambda_p^F \\ \lambda_q^F \\ \lambda_F^F \end{pmatrix} = \begin{pmatrix} 1 & 0 & 0 & 0 & 0 & 0 \\ 0 & \dfrac{h\beta}{(2\beta-1)^{1/2}} & 0 & 0 & \dfrac{k\beta^2}{2\beta-1} & \dfrac{-k\beta^2}{2\beta-1} + c_F \\ 0 & \dfrac{k\beta}{(2\beta-1)^{1/2}} & 0 & 0 & \dfrac{-h\beta^2}{2\beta-1} & \dfrac{h\beta^2}{2\beta-1} - s_F \\ 0 & 0 & \dfrac{2p}{K(K-1)^{1/2}} & \dfrac{q}{K-1} & \dfrac{-q}{K-1} & 0 \\ 0 & 0 & \dfrac{2q}{K(K-1)^{1/2}} & \dfrac{-p}{K-1} & \dfrac{p}{K-1} & 0 \\ 0 & 0 & 0 & 0 & 0 & 1 - kc_F - hs_F \end{pmatrix} \begin{pmatrix} \lambda_a \\ \lambda_e \\ \lambda_i \\ \lambda_\Omega \\ \lambda_\omega \\ \lambda_M \end{pmatrix} \tag{5.47}$$

Here, $1 - kc_F - hs_F = \frac{r}{a}$. Finally,

$$\begin{pmatrix} \lambda_a^\lambda \\ \lambda_h^\lambda \\ \lambda_k^\lambda \\ \lambda_p^\lambda \\ \lambda_q^\lambda \\ \lambda_\lambda^\lambda \end{pmatrix} = \begin{pmatrix} 1 & 0 & 0 & 0 & 0 & 0 \\ 0 & \dfrac{h\beta}{(2\beta-1)^{1/2}} & 0 & 0 & \dfrac{k\beta^2}{2\beta-1} & \dfrac{-k\beta^2}{2\beta-1} \\ 0 & \dfrac{k\beta}{(2\beta-1)^{1/2}} & 0 & 0 & \dfrac{-h\beta^2}{2\beta-1} & \dfrac{h\beta^2}{2\beta-1} \\ 0 & 0 & \dfrac{2p}{K(K-1)^{1/2}} & \dfrac{q}{K-1} & \dfrac{-q}{K-1} & 0 \\ 0 & 0 & \dfrac{2q}{K(K-1)^{1/2}} & \dfrac{-p}{K-1} & \dfrac{p}{K-1} & 0 \\ 0 & 0 & 0 & 0 & 0 & 1 \end{pmatrix} \begin{pmatrix} \lambda_a \\ \lambda_e \\ \lambda_i \\ \lambda_\Omega \\ \lambda_\omega \\ \lambda_M \end{pmatrix} \tag{5.48}$$

The transformation matrix between the multipliers of the (a, h, k, p, q, L) and (a, h, k, p, q, λ) sets have been derived in [16] and [17] and in Chapters 8 and 9, such that

$$\begin{pmatrix} \lambda_a^\lambda \\ \lambda_h^\lambda \\ \lambda_k^\lambda \\ \lambda_p^\lambda \\ \lambda_q^\lambda \\ \lambda_\lambda^\lambda \end{pmatrix} = \begin{pmatrix} 1 & 0 & 0 & 0 & 0 & 0 \\ 0 & 1 & 0 & 0 & 0 & \left(\dfrac{\partial L}{\partial h}\right)_{\text{tot}} \\ 0 & 0 & 1 & 0 & 0 & \left(\dfrac{\partial L}{\partial k}\right)_{\text{tot}} \\ 0 & 0 & 0 & 1 & 0 & 0 \\ 0 & 0 & 0 & 0 & 1 & 0 \\ 0 & 0 & 0 & 0 & 0 & \dfrac{a^2 G}{r^2} \end{pmatrix} \begin{pmatrix} \lambda_a^L \\ \lambda_h^L \\ \lambda_k^L \\ \lambda_p^L \\ \lambda_q^L \\ \lambda_L^L \end{pmatrix}$$

where $\left(\frac{\partial L}{\partial h}\right)_{\text{tot}}$ and $\left(\frac{\partial L}{\partial k}\right)_{\text{tot}}$ are given either in terms of the true longitude L or the eccentric longitude F by

$$\left(\frac{\partial L}{\partial h}\right)_{\text{tot}} = \frac{\left(a^2/r^2\right) G\left[\left(1-h^2\beta\right) s_L - hk\beta c_L\right] - s_L - 2\,(a/r)\,h}{hc_L - ks_L}$$
$$= \frac{\left(-a^2/r^2\right)(s_F - h)\,G^2 + Y_1/a + 2h}{G\,(ks_F - hc_F)}$$
$$\left(\frac{\partial L}{\partial k}\right)_{\text{tot}} = \frac{\left(a^2/r^2\right) G\left[\left(1-k^2\beta\right) c_L - hk\beta s_L\right] - c_L - 2\,(a/r)\,k}{hc_L - ks_L}$$
$$= \frac{\left(-a^2/r^2\right)(c_F - k)\,G^2 + X_1/a + 2k}{G\,(ks_F - hc_F)}$$

The inverse transformation is given by

$$\begin{pmatrix} \lambda_a^L \\ \lambda_h^L \\ \lambda_k^L \\ \lambda_p^L \\ \lambda_q^L \\ \lambda_L^L \end{pmatrix} = \begin{pmatrix} 1 & 0 & 0 & 0 & 0 & 0 \\ 0 & 1 & 0 & 0 & 0 & \dfrac{-r^2}{a^2 G}\left(\dfrac{\partial L}{\partial h}\right)_{\text{tot}} \\ 0 & 0 & 1 & 0 & 0 & \dfrac{-r^2}{a^2 G}\left(\dfrac{\partial L}{\partial k}\right)_{\text{tot}} \\ 0 & 0 & 0 & 1 & 0 & 0 \\ 0 & 0 & 0 & 0 & 1 & 0 \\ 0 & 0 & 0 & 0 & 0 & \dfrac{r^2}{a^2 G} \end{pmatrix} \begin{pmatrix} \lambda_a^\lambda \\ \lambda_h^\lambda \\ \lambda_k^\lambda \\ \lambda_p^\lambda \\ \lambda_q^\lambda \\ \lambda_\lambda^\lambda \end{pmatrix} \tag{5.49}$$

The mapping between $(\lambda_a^L, \lambda_h^L, \lambda_k^L, \lambda_p^L, \lambda_q^L, \lambda_L^L)$ and $(\lambda_a, \lambda_e, \lambda_i, \lambda_\Omega, \lambda_\omega, \lambda_M)$ is obtained from Equations (5.48) and (5.49).

5.5 Boundary Conditions for Minimum-Time Rendezvous and Example of a Free–Free Minimum-Time Transfer

In problems of minimum-time orbital rendezvous, rendezvous time is minimized by maximizing the performance index:

$$J = \int_{t_0}^{t_f} L dt = -\int_{t_0}^{t_f} dt = -\left(t_f - t_0\right)$$

with $L = -1$. The two-point boundary value problem consists of guessing the initial values of the Lagrange multipliers $\left(\lambda_a^{\lambda_0}\right)_0, \left(\lambda_h^{\lambda_0}\right)_0, \left(\lambda_k^{\lambda_0}\right)_0, \left(\lambda_p^{\lambda_0}\right)_0, \left(\lambda_q^{\lambda_0}\right)_0, \left(\lambda_{\lambda_0}^{\lambda_0}\right)_0$ and the rendezvous time t_f and with given initial parameters a_0, h_0, k_0, p_0, q_0, $(\lambda_0)_0$ and integrating the equations of motion and Equations (5.20) through (5.25) using the optimal thrust direction $\hat{\mathbf{u}} = \left(\boldsymbol{\lambda}_z^T M\right)^T / |\boldsymbol{\lambda}_z^T M|$, such that the final state given by a_f, h_f, k_f, p_f, q_f, λ_f, and a particular transversality condition are matched. The matching of $\lambda_f = (\lambda_0)_f + n_f t_f$ acts as a constraint because λ_f is a function of the state variables

λ_0 and a. Following [18] and letting ν represent a constant multiplier, the transversality condition at the unknown final time t_f is given by

$$\left(\frac{\partial \Phi}{\partial t} + L + \boldsymbol{\lambda}_z^T \dot{\mathbf{z}}\right)_{t_f} = 0 \tag{5.50}$$

where $\Phi = \nu\left[(\lambda_0)_f + n_f t_f - \lambda_f\right] = \nu\psi$. We must also require that $\left(\lambda_{\lambda_0}^{\lambda_0}\right)_f = \left(\nu\frac{\partial\psi}{\partial\lambda_0}\right)_{t_f}$, which yields the trivial result $\nu = \left(\lambda_{\lambda_0}^{\lambda_0}\right)_f$. Now, Equation (5.50) is equivalent to $\left(\frac{\partial\Phi}{\partial t} + \boldsymbol{\lambda}_z^T\dot{\mathbf{z}}\right)_{t_f} = 1$ and because $(\partial\Phi/\partial t)_{t_f} = \nu n_f = \left(\lambda_{\lambda_0}^{\lambda_0}\right)_f n_f$, and because H^{λ_0} is defined as $H^{\lambda_0} = \boldsymbol{\lambda}_z^T\dot{\mathbf{z}}$, we have for the transversality condition at the final time t_f

$$T_r = \left(\lambda_{\lambda_0}^{\lambda_0}\right)_f n_f + H_f^{\lambda_0} = 1 \tag{5.51}$$

When the set (a, h, k, p, q, λ) is used as in [6] and Chapter 3, the augmented Hamiltonian $H^\lambda = -1 + \boldsymbol{\lambda}_z^T\dot{\mathbf{z}}$ is not explicitly a function of time, such that the system is autonomous and the transversality condition is simply given by $H_f^\lambda = 0$ [6], or $H_f^\lambda = 1$ for the unaugmented Hamiltonian $H^\lambda = \boldsymbol{\lambda}_z^T\dot{\mathbf{z}}$ used there. However, because $(\partial\lambda/\partial\dot{\mathbf{r}}) = (\partial\lambda_0/\partial\dot{\mathbf{r}}) - 3t\dot{\mathbf{r}}n^{-1}a^{-2}$ and $\dot{\lambda} = n + (\partial\lambda/\partial\dot{\mathbf{r}})\hat{\mathbf{u}}f_t$, and in view of Equation (5.38), one recovers $H^\lambda = H^{\lambda_0} + n\lambda_{\lambda_0}^{\lambda_0}$ of Equation (5.37) where H^{λ_0} is the Hamiltonian used here. Because $H^\lambda = 1$ throughout the optimal trajectory [6, and chapter 3] the quantity defined by $n\lambda_{\lambda_0}^{\lambda_0} + H^{\lambda_0} = 1$ remains constant throughout the transfer because it is equivalent to $H^\lambda = 1$. This is especially useful in accepting a converged trajectory as a locally optimal trajectory. For the rendezvous problem, the objective function to minimize is taken as follows, with the appropriate weights attached to each term if so desired:

$$F' = (a - a_f)^2 + (h - h_f)^2 + (k - k_f)^2 + (p - p_f)^2 + (q - q_f)^2 + (\lambda - \lambda_f)^2 + (T_r - 1)^2$$

We now duplicate the converged minimum-time transfer example of [16] and Chapters 2 and 8 of this book by running a direct trajectory using the present formulation and starting from the low orbit given in Table 5.4. Because of the relationships given in Equation (5.41), the values of the multipliers at time zero must be identical to the values generated in [16] and Chapter 8 with the λ formulation. Using $\left(\lambda_a^{\lambda_0}\right)_0 = 4.675229762$ s/km, $\left(\lambda_h^{\lambda_0}\right)_0 = 5.413413947 \times 10^2$ s, $\left(\lambda_k^{\lambda_0}\right)_0 = -9.202702084 \times 10^3$ s, $\left(\lambda_p^{\lambda_0}\right)_0 = 1.778011878 \times 10$ s, $\left(\lambda_q^{\lambda_0}\right)_0 = -2.258455855 \times 10^4$ s, $\left(\lambda_{\lambda_0}^{\lambda_0}\right)_0 = 0$ s/rad and the constant acceleration $f_t = 9.8 \times 10^{-5}$ km/s^2, the equations of motion as well as the adjoint equations are integrated simultaneously using the optimal control $\hat{\mathbf{u}}$ from $t = 0$ to $t_f = 58{,}089.90058$ s, and with the relative and absolute error controls set to 10^{-9}.

The achieved final conditions are shown in Table 5.4 with $(H^{\lambda_0})_f = 1.003704076$. These parameters are very close to the target parameters and to the optimized final location $\lambda_f = M_f + \omega_f + \Omega_f = 19.655283$ rad found in [16] and Chapter 8 of this book, and $\lambda_f = 19.655267$ rad of Chapter 2. We can now scale these initial multipliers to get

Table 5.4 Transfer parameters and solution

Orbit	Initial	Target	Achieved	Solution	Initial values
a, km	7000	42,000	41,999.992	$\lambda_a^{\lambda_0}$, s/km	4.657973438
e	0	10^{-3}	9.983×10^{-4}	$\lambda_h^{\lambda_0}$, s	5.393432977×10^2
i, deg	28.5	1	0.999797	$\lambda_k^{\lambda_0}$, s	-9.168734810×10^3
Ω, deg	0	0	0.000332	$\lambda_p^{\lambda_0}$, s	1.771449217×10
ω, deg	0	0	359.994533	$\lambda_q^{\lambda_0}$, s	-2.250119870×10^4
M_0, M, deg	$M_0 = -130.3331648$ (optimized)	Free	$M = 46.169872$ (optimized)	$\lambda_{\lambda_0}^{\lambda_0}$, s/rad	0.0

$(H^{\lambda_0})_f = 1$, resulting in the solution shown in Table 5.4. A run using these scaled values results in $(H^{\lambda_0})_f = 0.999999393$, which is close to 1. Both [6] and [16] formulations use the λ formulation, meaning that λ, the mean longitude, is selected as the sixth state variable, the difference being that unlike [6], which uses the eccentric longitude F as the accessory variable that defines the radial distance, [16] uses the true longitude L instead. After correcting an error in the form of h instead of k in Equation (A-96) of the appendix of [6, 10, 11], which should read as in Equation (A5.48) of Appendix 5.1 here and Equation (A3.96) of Appendix 3.1, a direct run is carried out here using the formulation of [6] and Chapter 3 of this book, yielding $\left(\lambda_\lambda^\lambda\right)_f = 1.63163 \times 10^{-5}$ s/rad, which is close to the value found in [16] as $\left(\lambda_\lambda^\lambda\right)_f = -1.8239 \times 10^{-4}$ s/rad. These values are close to the theoretical value of zero needed for the optimization of the final arrival point on the high orbit. The present λ_0 formulation run yields $\left(\lambda_{\lambda_0}^{\lambda_0}\right)_f = 8.4844 \times 10^{-3}$ s/rad at the final time, with the quantity $\left(\lambda_{\lambda_0}^{\lambda_0}\right) n + H^{\lambda_0} = 1.000000015$ constant throughout the transfer and the theoretical transversality condition for the minimum-time transfer $\left(H^{\lambda_0}\right)_f = 1$, closely matched by $\left(H^{\lambda_0}\right)_f = 0.999999393$ as mentioned earlier. Figures 5.2 and 5.3 show the variations of the elements h, k, p, q on the optimal transfer trajectory, while Figures 5.4 through 5.7 depict the evolutions of all six multipliers and compare them with those corresponding to the λ formulation of [6] and Chapter 3. All the multipliers have identical values for this transfer example, which requires $\left(\lambda_{\lambda_0}^{\lambda_0}\right)_0 = \left(\lambda_\lambda^\lambda\right)_0 = \left(\lambda_{\lambda_0}^{\lambda_0}\right)_f = \left(\lambda_\lambda^\lambda\right)_f = 0$, and therefore from Equation (5.41), $\left(\lambda_a^{\lambda_0}\right)_0 = \left(\lambda_a^\lambda\right)_0$ and $\left(\lambda_a^{\lambda_0}\right)_f = \left(\lambda_a^\lambda\right)_f$ and different λ_a profiles. In Figure 5.8, the Hamiltonian of the λ formulation H^λ remains constant throughout, while H^{λ_0} varies but ends at $\left(H^{\lambda_0}\right)_f = 1$ at the final time. The thrust pitch and yaw profiles are perfectly matched in Figure 5.9 with both λ and λ_0 formulations, and finally, the variations of a, λ, and the epoch mean longitude λ_0 are shown in Figure 5.10.

5.6 Conclusion

The minimum-time rendezvous and transfer problem between given initial and final general elliptic or circular orbits and based on the application of a continuous constant

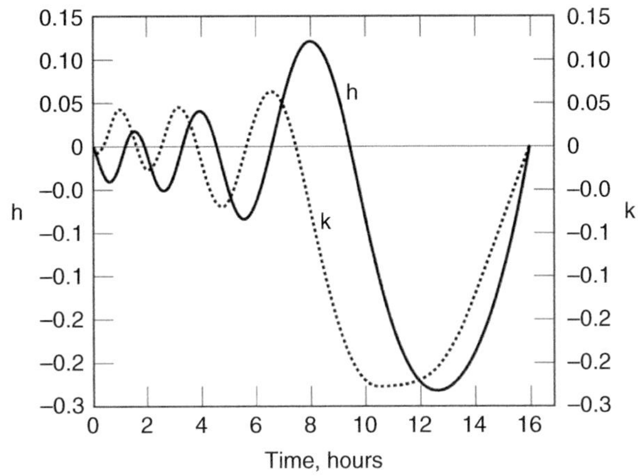

Figure 5.2 Time histories of equinoctial elements h and k during minimum-time transfer.

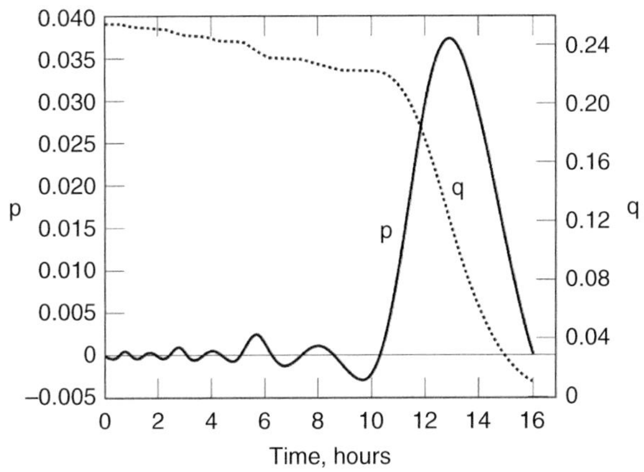

Figure 5.3 Time histories of equinoctial elements p and q during minimum-time transfer.

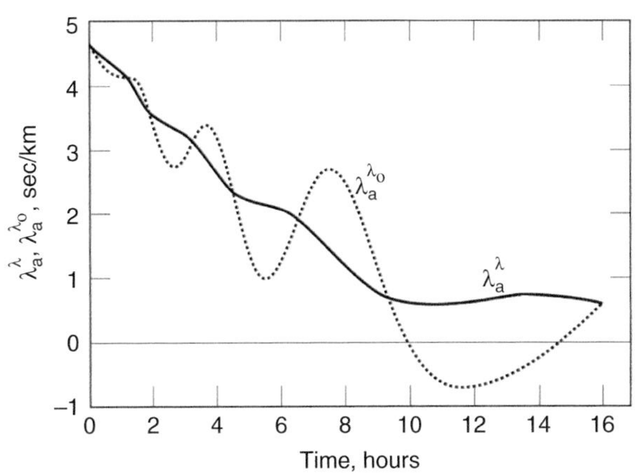

Figure 5.4 Evolution of $\lambda_a^{\lambda_0}$ and λ_a^{λ} multipliers during optimal transfer.

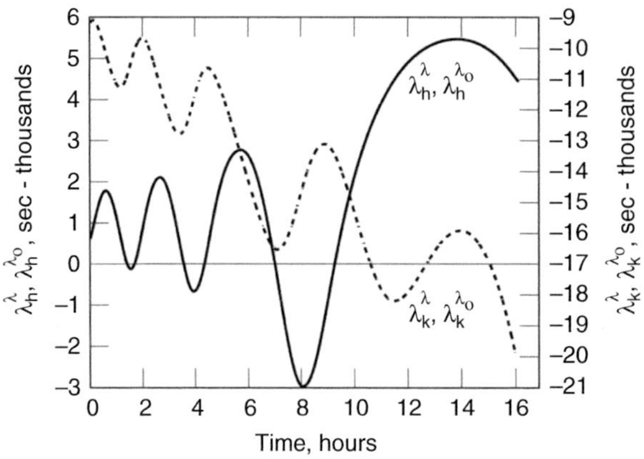

Figure 5.5 Evolution of $\lambda_h^{\lambda_0}$, λ_h^{λ} and $\lambda_k^{\lambda_0}$, λ_k^{λ} multipliers during optimal transfer.

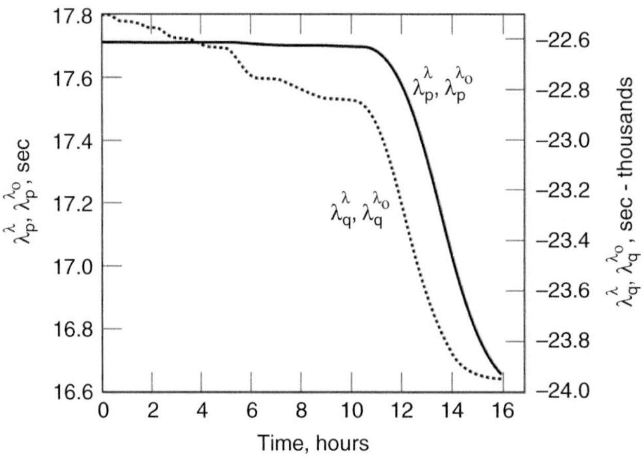

Figure 5.6 Evolution of $\lambda_p^{\lambda_0}$, λ_p^{λ} and $\lambda_q^{\lambda_0}$, λ_q^{λ} multipliers during optimal transfer.

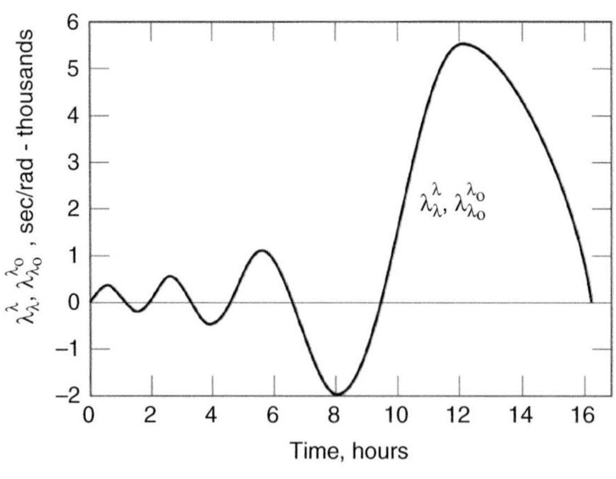

Figure 5.7 Evolution of $\lambda_{\lambda_0}^{\lambda_0}$ and $\lambda_{\lambda}^{\lambda}$ multipliers during optimal transfer.

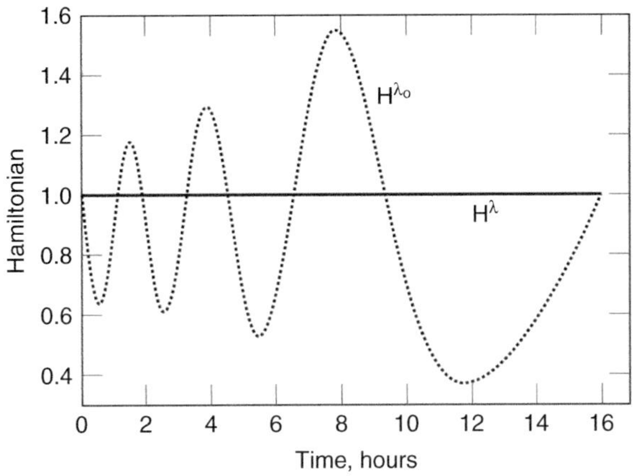

Figure 5.8 Variation of H^{λ_0} and constancy of H^{λ} during optimal transfer.

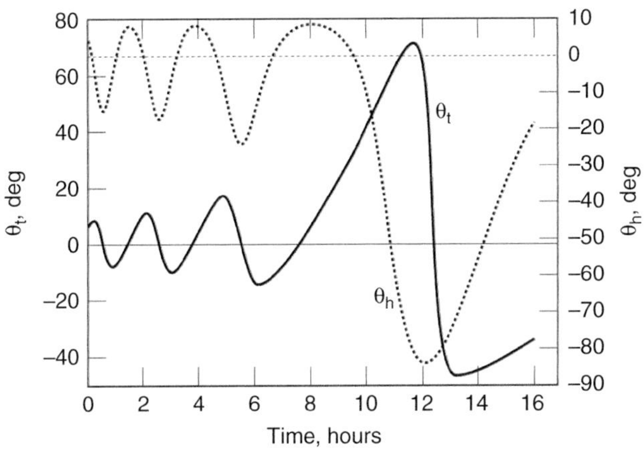

Figure 5.9 Optimal-thrust pitch and yaw profiles during minimum-time transfer.

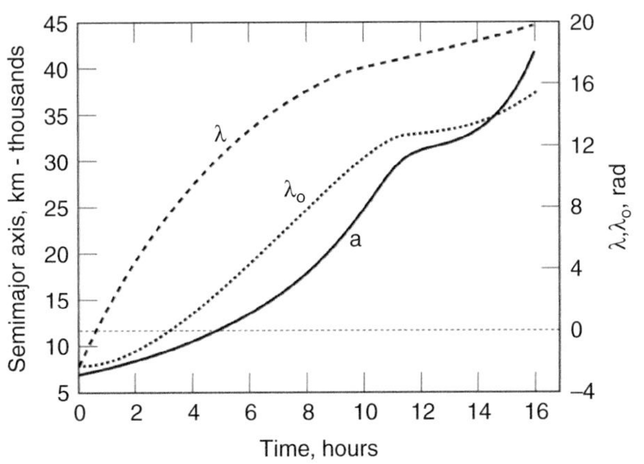

Figure 5.10 Semimajor axis change and variation of λ and λ_0 during optimal transfer.

low-thrust acceleration has been shown for a particular choice of the fast variable selected as a fundamental epoch element. Besides providing added insight into the mathematics of nonsingular trajectory optimization and mutual validation and verification of the other formulations developed, the analysis further reveals that additional simplifications in computing averaged transfers as well as certain aspects of minimum-fuel transfers are possible using this fundamental set. Further comparisons are necessary between the epoch and current time formulations, in order to find out whether certain formulations lead to overall numerically more robust orbit transfer optimization codes. The results presented here are closely related to those published earlier in [6], and those in Chapter 3, which used the current mean longitude as the fast orbital element instead. The mathematical correspondence between the two Hamiltonians and the various multipliers is also established and verified by way of a numerical example. Furthermore, the transformation matrices relating the Lagrange multipliers of the classical elements to the multipliers corresponding to different sets of equinoctial elements are also fully derived.

Appendix 5.1 The Nonzero Partials of Matrix *M*

A5.1.1 The Partial Derivatives of *M* with Respect to *h*

The partial derivatives $\partial M_{11}/\partial h$, $\partial M_{12}/\partial h$, $\partial M_{21}/\partial h$, $\partial M_{22}/\partial h$, $\partial M_{23}/\partial h$, $\partial M_{31}/\partial h$, $\partial M_{32}/\partial h$, $\partial M_{33}/\partial h$, $\partial M_{43}/\partial h$ and $\partial M_{53}/\partial h$ are identical to the partials given in Equations (A1), (A2) and (A4) through (A11) of [11], and also derived in Chapter 3, and therefore they are not repeated here.

$$\frac{\partial M_{61}^0}{\partial h} = n^{-1}a^{-2}\left\{-2\frac{\partial X_1}{\partial h} + 3\frac{\partial \dot{X}_1}{\partial h}t - h\beta G^{-1}\left(h\frac{\partial X_1}{\partial h} + k\frac{\partial X_1}{\partial k}\right)\right.$$
$$\left. + G\left[\left(\beta + \frac{h^2\beta^3}{1-\beta}\right)\frac{\partial X_1}{\partial h} + \frac{hk\beta^3}{1-\beta}\frac{\partial X_1}{\partial k} + \beta\left(h\frac{\partial^2 X_1}{\partial h^2} + k\frac{\partial^2 X_1}{\partial h \partial k}\right)\right]\right\} \quad \text{(A5.1)}$$

$$\frac{\partial M_{62}^0}{\partial h} = n^{-1}a^{-2}\left\{-2\frac{\partial Y_1}{\partial h} + 3\frac{\partial \dot{Y}_1}{\partial h}t - h\beta G^{-1}\left(h\frac{\partial Y_1}{\partial h} + k\frac{\partial Y_1}{\partial k}\right)\right.$$
$$\left. + G\left[\left(\beta + \frac{h^2\beta^3}{1-\beta}\right)\frac{\partial Y_1}{\partial h} + \frac{hk\beta^3}{1-\beta}\frac{\partial Y_1}{\partial k} + \beta\left(h\frac{\partial^2 Y_1}{\partial h^2} + k\frac{\partial^2 Y_1}{\partial h \partial k}\right)\right]\right\} \quad \text{(A5.2)}$$

$$\frac{\partial M_{63}^0}{\partial h} = \frac{G^{-1}}{na^2}\left[\left(q\frac{\partial Y_1}{\partial h} - p\frac{\partial X_1}{\partial h}\right) + hG^{-2}\left(qY_1 - pX_1\right)\right] \quad \text{(A5.3)}$$

A5.1.2 The Partial Derivatives of *M* with Respect to *k*

The partial derivatives $\partial M_{11}/\partial k$, $\partial M_{12}/\partial k$, $\partial M_{21}/\partial k$, $\partial M_{22}/\partial k$, $\partial M_{23}/\partial k$, $\partial M_{31}/\partial k$, $\partial M_{32}/\partial k$, $\partial M_{33}/\partial k$, $\partial M_{43}/\partial k$ and $\partial M_{53}/\partial k$ are identical to the partials given in Equations (A15), (A16) and (A18) through (A25) of [11], and also derived in Chapter 3, and they also are not repeated here.

$$\frac{\partial M_{61}^0}{\partial k} = n^{-1}a^{-2}\left\{-2\frac{\partial X_1}{\partial k} + 3\frac{\partial \dot{X}_1}{\partial k}t - k\beta G^{-1}\left(h\frac{\partial X_1}{\partial h} + k\frac{\partial X_1}{\partial k}\right)\right.$$
$$\left. + G\left[\left(\beta + \frac{k^2\beta^3}{1-\beta}\right)\frac{\partial X_1}{\partial k} + \frac{hk\beta^3}{1-\beta}\frac{\partial X_1}{\partial h} + \beta\left(h\frac{\partial^2 X_1}{\partial k\partial h} + k\frac{\partial^2 X_1}{\partial k^2}\right)\right]\right\} \tag{A5.4}$$

$$\frac{\partial M_{62}^0}{\partial k} = n^{-1}a^{-2}\left\{-2\frac{\partial Y_1}{\partial k} + 3\frac{\partial \dot{Y}_1}{\partial k}t - k\beta G^{-1}\left(h\frac{\partial Y_1}{\partial h} + k\frac{\partial Y_1}{\partial k}\right)\right.$$
$$\left. + G\left[\left(\beta + \frac{k^2\beta^3}{1-\beta}\right)\frac{\partial Y_1}{\partial k} + \frac{hk\beta^3}{1-\beta}\frac{\partial Y_1}{\partial h} + \beta\left(h\frac{\partial^2 Y_1}{\partial k\partial h} + k\frac{\partial^2 Y_1}{\partial k^2}\right)\right]\right\} \tag{A5.5}$$

$$\frac{\partial M_{63}^0}{\partial k} = \frac{G^{-1}}{na^2}\left[\left(q\frac{\partial Y_1}{\partial k} - p\frac{\partial X_1}{\partial k}\right) + kG^{-2}\left(qY_1 - pX_1\right)\right] \tag{A5.6}$$

A5.1.3 The Partial Derivatives of M with Respect to p

The partial derivatives $\partial M_{23}/\partial p$, $\partial M_{33}/\partial p$, $\partial M_{43}/\partial p$ and $\partial M_{53}/\partial p$ are identical to the partials in Equations (A29) through (A32) of [11], and also derived in Chapter 3, respectively, while

$$\frac{\partial M_{63}^0}{\partial p} = \frac{-X_1}{na^2G} \tag{A5.7}$$

A5.1.4 The Partial Derivatives of M with Respect to q

The partial derivatives $\partial M_{23}/\partial q$, $\partial M_{33}/\partial q$, $\partial M_{43}/\partial q$ and $\partial M_{53}/\partial q$ are identical to the partials in Equations (A34) through (A37) of [11] and also derived in Chapter 3, and

$$\frac{\partial M_{63}^0}{\partial q} = \frac{Y_1}{na^2G} \tag{A5.8}$$

The partial derivatives of $\dot{X}_1$ with respect to h and k, namely $\partial\dot{X}_1/\partial h$ and $\partial\dot{X}_1/\partial k$, are identical to those in Equations (A39) and (A40) of [11], and also derived in Chapter 3, while the partials of $\dot{Y}_1$ with respect to h and k, namely $\partial\dot{Y}_1/\partial h$ and $\partial\dot{Y}_1/\partial k$, are identical to Equations (A41) and (A42) of the appendix of [11] and also derived in Chapter 3. The second partials of X_1 and Y_1 with respect to h and k or $\partial^2X_1/\partial h^2$, $\partial^2X_1/\partial k^2$, $\partial^2X_1/\partial h\partial k$, $\partial^2X_1/\partial k\partial h$ and $\partial^2Y_1/\partial h^2$, $\partial^2Y_1/\partial k^2$, $\partial^2Y_1/\partial h\partial k$, $\partial^2Y_1/\partial k\partial h$ are identical to Equations (A43) through (A50) of [11] and also derived in Chapter 3. It can be shown that $\frac{\partial^2X_1}{\partial h\partial k} = \frac{\partial^2X_1}{\partial k\partial h}$ and $\frac{\partial^2Y_1}{\partial h\partial k} = \frac{\partial^2Y_1}{\partial k\partial h}$. Next, the accessory partials $\partial^2X_1/\partial a\partial k$, $\partial^2X_1/\partial a\partial h$, $\partial^2Y_1/\partial a\partial k$ and $\partial^2Y_1/\partial a\partial h$ are

$$\frac{\partial^2X_1}{\partial a\partial k} = \frac{1}{a}\frac{\partial X_1}{\partial k} - \frac{3}{2}\frac{nat}{r}\left[\left(hs_F + kc_F\right)\frac{hk\beta^3}{(1-\beta)}\right.$$
$$\left. + \frac{a^2}{r^2}\left(k - c_F\right)\left(s_F - h\beta\right) - \frac{a}{r}s_Fc_F\right] \tag{A5.9}$$

$$\frac{\partial^2 X_1}{\partial a \partial h} = \frac{1}{a}\frac{\partial X_1}{\partial h} - \frac{3}{2}\frac{nat}{r}\left\{(hs_F + kc_F)\left[\beta + \frac{h^2\beta^3}{(1-\beta)}\right]\right.$$
$$\left. - \frac{a^2}{r^2}(h - s_F)(h\beta - s_F) + \frac{a}{r}c_F^2\right\} \quad \text{(A5.10)}$$

and also,

$$\frac{\partial^2 Y_1}{\partial a \partial k} = \frac{1}{a}\frac{\partial Y_1}{\partial k} + \frac{3}{2}\frac{nat}{r}\left\{(hs_F + kc_F)\left[\beta + \frac{k^2\beta^3}{(1-\beta)}\right]\right.$$
$$\left. + \frac{a^2}{r^2}(k - c_F)(c_F - k\beta) + \frac{a}{r}s_F^2\right\} \quad \text{(A5.11)}$$

$$\frac{\partial^2 Y_1}{\partial a \partial h} = \frac{1}{a}\frac{\partial Y_1}{\partial h} + \frac{3}{2}\frac{nat}{r}\left[(hs_F + kc_F)\frac{hk\beta^3}{(1-\beta)}\right.$$
$$\left. - \frac{a^2}{r^2}(h - s_F)(k\beta - c_F) - \frac{a}{r}s_F c_F\right] \quad \text{(A5.12)}$$

A5.1.5 The Partial Derivatives of M with Respect to a

These partial derivatives are written in terms of the first-order partials of X_1, $\dot{X}_1$ and Y_1, $\dot{Y}_1$ with respect to a as well as $\partial^2 X_1/\partial a\partial h$, $\partial^2 X_1/\partial a\partial k$, $\partial^2 Y_1/\partial a\partial h$ and $\partial^2 Y_1/\partial a\partial k$ shown below and in Equations (A5.9) through (A5.12).

$$\frac{\partial M_{11}}{\partial a} = \frac{4}{n^2a^2}\dot{X}_1 + \frac{2}{n^2 a}\frac{\partial \dot{X}_1}{\partial a} \quad \text{(A5.13)}$$

$$\frac{\partial M_{12}}{\partial a} = \frac{4}{n^2a^2}\dot{Y}_1 + \frac{2}{n^2 a}\frac{\partial \dot{Y}_1}{\partial a} \quad \text{(A5.14)}$$

$$\frac{\partial M_{21}}{\partial a} = \frac{G}{na^2}\left[-\frac{1}{2a}\frac{\partial X_1}{\partial k} + \frac{\partial^2 X_1}{\partial a \partial k} - \frac{h\beta}{na}\dot{X}_1 - \frac{h\beta}{n}\frac{\partial \dot{X}_1}{\partial a}\right] \quad \text{(A5.15)}$$

$$\frac{\partial M_{22}}{\partial a} = \frac{G}{na^2}\left[-\frac{1}{2a}\frac{\partial Y_1}{\partial k} + \frac{\partial^2 Y_1}{\partial a \partial k} - \frac{h\beta}{na}\dot{Y}_1 - \frac{h\beta}{n}\frac{\partial \dot{Y}_1}{\partial a}\right] \quad \text{(A5.16)}$$

$$\frac{\partial M_{23}}{\partial a} = \frac{k}{na^2G}\left[-\frac{1}{2a}(qY_1 - pX_1) + q\frac{\partial Y_1}{\partial a} - p\frac{\partial X_1}{\partial a}\right] \quad \text{(A5.17)}$$

$$\frac{\partial M_{31}}{\partial a} = -\frac{G}{na^2}\left[-\frac{1}{2a}\frac{\partial X_1}{\partial h} + \frac{\partial^2 X_1}{\partial a \partial h} + \frac{k\beta}{na}\dot{X}_1 + \frac{k\beta}{n}\frac{\partial \dot{X}_1}{\partial a}\right] \quad \text{(A5.18)}$$

$$\frac{\partial M_{32}}{\partial a} = -\frac{G}{na^2}\left[-\frac{1}{2a}\frac{\partial Y_1}{\partial h} + \frac{\partial^2 Y_1}{\partial a \partial h} + \frac{k\beta}{na}\dot{Y}_1 + \frac{k\beta}{n}\frac{\partial \dot{Y}_1}{\partial a}\right] \quad \text{(A5.19)}$$

$$\frac{\partial M_{33}}{\partial a} = \frac{-h}{na^2G}\left[-\frac{1}{2a}(qY_1 - pX_1) + q\frac{\partial Y_1}{\partial a} - p\frac{\partial X_1}{\partial a}\right] \quad \text{(A5.20)}$$

$$\frac{\partial M_{43}}{\partial a} = \frac{K}{2na^2G}\left(-\frac{1}{2a}Y_1 + \frac{\partial Y_1}{\partial a}\right) \quad \text{(A5.21)}$$

$$\frac{\partial M_{53}}{\partial a} = \frac{K}{2na^2G}\left(-\frac{1}{2a}X_1 + \frac{\partial X_1}{\partial a}\right) \quad \text{(A5.22)}$$

$$\frac{\partial M_{61}^{0}}{\partial a} = -\frac{M_{61}^{0}}{2a} + n^{-1}a^{-2}\left[-2\frac{\partial X_1}{\partial a} + 3\frac{\partial \dot{X}_1}{\partial a}t + G\left(h\beta\frac{\partial^2 X_1}{\partial a \partial h} + k\beta\frac{\partial^2 X_1}{\partial a \partial k}\right)\right] \tag{A5.23}$$

$$\frac{\partial M_{62}^{0}}{\partial a} = -\frac{M_{62}^{0}}{2a} + n^{-1}a^{-2}\left[-2\frac{\partial Y_1}{\partial a} + 3\frac{\partial \dot{Y}_1}{\partial a}t + G\left(h\beta\frac{\partial^2 Y_1}{\partial a \partial h} + k\beta\frac{\partial^2 Y_1}{\partial a \partial k}\right)\right] \tag{A5.24}$$

$$\frac{\partial M_{63}^{0}}{\partial a} = -\frac{M_{63}^{0}}{2a} + \frac{1}{na^2}\left[\left(q\frac{\partial Y_1}{\partial a} + p\frac{\partial X_1}{\partial a}\right)G^{-1}\right] \tag{A5.25}$$

with

$$\frac{\partial \dot{X}_1}{\partial a} = -\frac{1}{2}\frac{na}{r}\left[1 - \frac{3na^2 t}{r^2}\left(ks_F - hc_F\right)\right]$$
$$\times\left[hk\beta c_F - \left(1 - h^2\beta\right)s_F\right] + \frac{3}{2}\frac{n^2a^2t}{r^2}\left[hk\beta s_F + \left(1 - h^2\beta\right)c_F\right] \tag{A5.26}$$

$$\frac{\partial \dot{Y}_1}{\partial a} = \frac{1}{2}\frac{na}{r}\left[1 - \frac{3na^2 t}{r^2}\left(ks_F - hc_F\right)\right]$$
$$\times\left[hk\beta s_F - \left(1 - k^2\beta\right)c_F\right] + \frac{3}{2}\frac{n^2a^2t}{r^2}\left[hk\beta c_F + \left(1 - k^2\beta\right)s_F\right] \tag{A5.27}$$

$$\frac{\partial X_1}{\partial a} = \frac{X_1}{a} - \frac{3}{2}\frac{t}{a}\dot{X}_1 \tag{A5.28}$$

$$\frac{\partial Y_1}{\partial a} = \frac{Y_1}{a} - \frac{3}{2}\frac{t}{a}\dot{Y}_1 \tag{A5.29}$$

A5.1.6 The Partial Derivatives of M with respect to λ_0

$$\frac{\partial M_{11}}{\partial \lambda_0} = \frac{\partial M_{11}}{\partial \lambda}\frac{\partial \lambda}{\partial \lambda_0} = \frac{\partial M_{11}}{\partial \lambda} = \frac{2}{n^2 r}\frac{\partial \dot{X}_1}{\partial F} \tag{A5.30}$$

$$\frac{\partial M_{12}}{\partial \lambda_0} = \frac{2}{n^2 r}\frac{\partial \dot{Y}_1}{\partial F} \tag{A5.31}$$

$$\frac{\partial M_{21}}{\partial \lambda_0} = \frac{G}{nar}\left(\frac{\partial^2 X_1}{\partial F \partial k} - \frac{h\beta}{n}\frac{\partial \dot{X}_1}{\partial F}\right) \tag{A5.32}$$

$$\frac{\partial M_{22}}{\partial \lambda_0} = \frac{G}{nar}\left(\frac{\partial^2 Y_1}{\partial F \partial k} - \frac{h\beta}{n}\frac{\partial \dot{Y}_1}{\partial F}\right) \tag{A5.33}$$

$$\frac{\partial M_{23}}{\partial \lambda_0} = \frac{k\left(q\frac{\partial Y_1}{\partial F} - p\frac{\partial X_1}{\partial F}\right)}{narG} \tag{A5.34}$$

$$\frac{\partial M_{31}}{\partial \lambda_0} = -\frac{G}{nar}\left(\frac{\partial^2 X_1}{\partial F \partial h} + \frac{k\beta}{n}\frac{\partial \dot{X}_1}{\partial F}\right) \tag{A5.35}$$

$$\frac{\partial M_{32}}{\partial \lambda_0} = -\frac{G}{nar}\left(\frac{\partial^2 Y_1}{\partial F \partial h} + \frac{k\beta}{n}\frac{\partial \dot{Y}_1}{\partial F}\right) \tag{A5.36}$$

$$\frac{\partial M_{33}}{\partial \lambda_0} = \frac{-h\left(q\frac{\partial Y_1}{\partial F} - p\frac{\partial X_1}{\partial F}\right)}{narG} \tag{A5.37}$$

$$\frac{\partial M_{43}}{\partial \lambda_0} = \frac{K}{2narG} \frac{\partial Y_1}{\partial F} \tag{A5.38}$$

$$\frac{\partial M_{53}}{\partial \lambda_0} = \frac{K}{2narG} \frac{\partial X_1}{\partial F} \tag{A5.39}$$

$$\frac{\partial M_{61}^0}{\partial \lambda_0} = n^{-1}a^{-1}r^{-1}\left[-2\frac{\partial X_1}{\partial F} + 3\frac{\partial \dot{X}_1}{\partial F}t + G\left(h\beta\frac{\partial^2 X_1}{\partial F \partial h} + k\beta\frac{\partial^2 X_1}{\partial F \partial k}\right)\right] \tag{A5.40}$$

$$\frac{\partial M_{62}^0}{\partial \lambda_0} = n^{-1}a^{-1}r^{-1}\left[-2\frac{\partial Y_1}{\partial F} + 3\frac{\partial \dot{Y}_1}{\partial F}t + G\left(h\beta\frac{\partial^2 Y_1}{\partial F \partial h} + k\beta\frac{\partial^2 Y_1}{\partial F \partial k}\right)\right] \tag{A5.41}$$

$$\frac{\partial M_{63}^0}{\partial \lambda_0} = \frac{\left(q\frac{\partial Y_1}{\partial F} - p\frac{\partial X_1}{\partial F}\right)}{narG} \tag{A5.42}$$

The auxiliary partials are

$$\frac{\partial X_1}{\partial F} = a\left[hk\beta c_F - \left(1 - h^2\beta\right)s_F\right] \tag{A5.43}$$

$$\frac{\partial Y_1}{\partial F} = a\left[-hk\beta s_F + \left(1 - k^2\beta\right)c_F\right] \tag{A5.44}$$

$$\frac{\partial \dot{X}_1}{\partial F} = -\frac{a}{r}\left(ks_F - hc_F\right)\dot{X}_1 + \frac{a^2 n}{r}\left[-hk\beta s_F - \left(1 - h^2\beta\right)c_F\right] \tag{A5.45}$$

$$\frac{\partial \dot{Y}_1}{\partial F} = -\frac{a}{r}\left(ks_F - hc_F\right)\dot{Y}_1 + \frac{a^2 n}{r}\left[-hk\beta c_F - \left(1 - k^2\beta\right)s_F\right] \tag{A5.46}$$

$$\frac{\partial^2 X_1}{\partial F \partial h} = a\left[\left(hs_F + kc_F\right)\left(\beta + \frac{h^2\beta^3}{1-\beta}\right) + \frac{a^2}{r^2}\left(h\beta - s_F\right)\left(s_F - h\right) + \frac{a}{r}c_F^2\right] \tag{A5.47}$$

$$\frac{\partial^2 X_1}{\partial F \partial k} = -a\left[-\left(hs_F + kc_F\right)\frac{hk\beta^3}{1-\beta} + \frac{a^2}{r^2}\left(s_F - h\beta\right)\left(c_F - k\right) + \frac{a}{r}s_F c_F\right] \tag{A5.48}$$

$$\frac{\partial^2 Y_1}{\partial F \partial h} = a\left[-\left(hs_F + kc_F\right)\frac{hk\beta^3}{1-\beta} - \frac{a^2}{r^2}\left(k\beta - c_F\right)\left(s_F - h\right) + \frac{a}{r}s_F c_F\right] \tag{A5.49}$$

$$\frac{\partial^2 Y_1}{\partial F \partial k} = a\left[-\left(hs_F + kc_F\right)\left(\beta + \frac{k^2\beta^3}{1-\beta}\right) + \frac{a^2}{r^2}\left(c_F - k\beta\right)\left(c_F - k\right) - \frac{a}{r}s_F^2\right] \tag{A5.50}$$

References

[1] Edelbaum, T. N., Sackett, L. L. and Malchow, H. L. (1973) *Optimal Low Thrust Geocentric Transfer*, AIAA Paper 73–1074, AIAA 10th Electric Propulsion Conference, Lake Tahoe, NV.

[2] Kechichian, J. A. (1990) *Equinoctial Orbit Elements: Application to Optimal Transfer Problems*, AIAA Paper 90–2976, AIAA/AAS Astrodynamics Conference, Portland, OR.

[3] Kechichian, J. A. (1991) *Trajectory Optimization with a Modified Set of Equinoctial Orbit Elements*, AAS Paper 91–524, AAS/AIAA Astrodynamics Specialist Conference, Durango, CO.

[4] Cefola, P. (1972) *Equinoctial Orbit Elements: Application to Artificial Satellite Orbits*, AIAA Paper 72–937, AIAA/AAS Astrodynamics Conference, Palo Alto, CA.

[5] Cefola, P., Long, A. and Holloway, G. (1974) *The Long-Term Prediction of Artificial Satellite Orbits*, AIAA Paper, 74–170, AIAA 12th Aerospace Sciences Meeting, Washington, DC.

[6] Kechichian, J. A. (1996) Optimal Low-Thrust Rendezvous Using Equinoctial Orbit Elements, *Acta Astronautica*, Vol. 38, No. 1, 1–14.

[7] Broucke, R. A. and Cefola, P. J. (1972) On the Equinoctial Orbit Elements, *Celestial Mechanics*, Vol. 5, 303–310.

[8] Battin, R. H. (1987) *An Introduction to the Mathematics and Methods of Astrodynamics*, AIAA, New York, NY, 471–514.

[9] Brouwer, D. and Clemence, G. M. (1961) *Methods of Celestial Mechanics*, Academic Press, New York, NY, 273–307.

[10] Kechichian, J. A. (1996) Optimal Low-Thrust Orbit Transfer, *Orbital Mechanics*, 2nd ed., edited by V. A. Chobotov, AIAA, Reston, VA, Ch. 14.

[11] Kechichian, J. A. (1997) Optimal Low-Earth Orbit Geostationary-Earth Orbit Intermediate Acceleration Orbit Transfer, *Journal of Guidance, Control, and Dynamics*, Vol. 20, No. 4, 803–811.

[12] Marec, J.-P. (1979) *Optimal Space Trajectories*, Elsevier, Amsterdam.

[13] Kechichian, J. A. (1998) Trajectory Optimization Using Eccentric Longitude Formulation, *Journal of Spacecraft and Rockets*, Vol. 35, No. 3, 317–326.

[14] Kechichian, J. A. (1994) *Low-Thrust Trajectory Optimization Based on Epoch Eccentric Longitude Formulation*, AAS Paper 94–124, AAS/AIAA Spaceflight Mechanics Meeting, Cocoa Beach, FL.

[15] Kechichian, J. A. (1993) *Minimum-Time Low-Thrust Rendezvous Using Epoch Mean Longitude Formulation*, AAS Paper 93–129, AAS/AIAA Spaceflight Mechanics Meeting, Pasadena, CA.

[16] Kechichian, J. A. (1997) Mechanics of Trajectory Optimization Using Nonsingular Variational Equations in Polar Coordinates, *Journal of Guidance, Control, and Dynamics*, Vol. 20, No. 4, 812–818.

[17] Kechichian, J. A. (1997) Trajectory Optimization Using Nonsingular Orbital Elements and True Longitude, *Journal of Guidance, Control, and Dynamics*, Vol. 20, No. 5, 1003–1009.

[18] Bryson, A. E. and Ho, Y.-C. (1969) *Applied Optimal Control*, Ginn and Company, Waltham, MA.

6 Trajectory Optimization Using Eccentric Longitude Formulation

6.1 Introduction

The problem of optimal transfer using a set of nonsingular orbit elements where the eccentric longitude represents the sixth state variable is extended and completed by fully considering the six-state dynamics. Because the eccentric longitude appears in the right-hand sides of the dynamic equations, the use of this particular formulation removes the need for solving the transcendental Kepler equation at each integration step, easing thereby to some extent the numerical computations. Furthermore, because the eccentric longitude is being integrated directly, it effectively becomes an independent orbital element such that the adjoint differential equations are derived by assuming that this longitude is independent of the other elements. The variational Hamiltonian is constant throughout the transfer with the boundary conditions given simply in terms of the elements. An example of a general minimum-time transfer using continuous low thrust is generated, duplicating a previous result using the mean longitude formulation, in order to validate the mathematical derivations.

Equinoctial orbit elements have been used in [1–3] to solve optimal low-thrust transfer problems by considering only the five slowly varying orbit elements. The dynamic equations based on these elements are singularity-free for both zero eccentricity and zero inclination, and are valid for any elliptical orbit because they are exact nonlinear equations that describe the exact motion in a central force field, and are perturbed by the thrust acceleration. In problems of low-thrust transfer, the adoption of orbital elements allows for the technique of averaging to be applied because these elements vary slowly during an entire orbit. When precision integration is used instead to obtain exact solutions, the Cartesian formulation, which is also singularity-free, can also be used to generate optimal trajectories. However, with the boundary conditions being given in terms of the elements, it is more natural to use an element-based formulation even though the Cartesian formulation is mathematically very simple. In an element-based formulation, the equations of motion contain an accessory variable in the right-hand sides, namely the eccentric anomaly, if classical elements are considered, or eccentric longitude when equinoctial variables are considered instead. These equations can also be written in terms of the true anomaly or eccentric anomaly respectively as the accessory variable.

When all six elements are considered, the differential equation for the fast element or sixth independent variable can be written as the time rate of change of the mean longitude, the eccentric longitude or the true longitude; the accessory variable appearing in the right-hand sides of these dynamic equations being the eccentric longitude or the true longitude but not the mean longitude. The reason why the mean longitude cannot be used as the accessory variable is due to the fact that the position vector cannot be written explicitly in terms of the mean longitude. If we select the mean longitude as the sixth independent variable, then we must solve for the eccentric longitude through Kepler's equation at each integration step in order to evaluate the right-hand sides of all six differential equations. This additional effort in solving Kepler's equation is clearly not needed if the eccentric longitude is selected as the sixth variable instead. The particular choice of the sixth variable has important implications for how the differential equations for the adjoints are written [4–12]. This observation holds true even if only the first five dynamic equations for the slowly varying elements are considered, and the sixth equation neglected. There is no ambiguity if all six equations are considered, provided that the dependence of the eccentric longitude on the other elements is accounted for when the mean longitude is selected as the sixth variable, and considered independent of the other elements when the eccentric longitude itself is selected as the sixth variable. The formulation for the case in which the eccentric longitude is selected as the sixth state variable while considering only the differential equations for the first five slowly varying elements [1, 2] is extended in this paper by considering all six differential equations for an exact description of the orbit motion. The full six-element formulation for the case in which the mean longitude represents the fast variable was developed in [7], and in Chapter 3 of this book.

The next two sections show all the pertinent derivations with a duplication of the general example of [7] and Chapter 3 of this book in order to validate the mathematics of this version.

6.2 Equations of Motion in Terms of the Eccentric Longitude

The equations of motion in terms of the eccentric longitude for the thrust perturbation are given by [5]:

$$\dot{a} = \left(\frac{\partial a}{\partial \dot{\mathbf{r}}}\right)^T \bullet \hat{\mathbf{u}} f_t \tag{6.1}$$

$$\dot{h} = \left(\frac{\partial h}{\partial \dot{\mathbf{r}}}\right)^T \bullet \hat{\mathbf{u}} f_t \tag{6.2}$$

$$\dot{k} = \left(\frac{\partial k}{\partial \dot{\mathbf{r}}}\right)^T \bullet \hat{\mathbf{u}} f_t \tag{6.3}$$

$$\dot{p} = \left(\frac{\partial p}{\partial \dot{\mathbf{r}}}\right)^T \bullet \hat{\mathbf{u}} f_t \tag{6.4}$$

$$\dot{q} = \left(\frac{\partial q}{\partial \dot{\mathbf{r}}}\right)^T \bullet \hat{\mathbf{u}} f_t \tag{6.5}$$

$$\dot{F} = \frac{na}{r} + \left(\frac{\partial F}{\partial \dot{\mathbf{r}}}\right)^T \bullet \hat{\mathbf{u}} f_t \tag{6.6}$$

The thrust vector $\mathbf{f} = f\hat{\mathbf{u}}$ is measured in the direct equinoctial frame $(\hat{\mathbf{f}}, \hat{\mathbf{g}}, \hat{\mathbf{w}})$ with $\hat{\mathbf{f}}$ and $\hat{\mathbf{g}}$ contained in the instantaneous orbit plane and with $\hat{\mathbf{f}}$ obtained through a clockwise rotation of an angle Ω from the direction of the ascending node. The position and velocity vectors $\mathbf{r}$ and $\dot{\mathbf{r}}$ are written as $\mathbf{r} = X_1\hat{\mathbf{f}} + Y_1\hat{\mathbf{g}}$, and $\dot{\mathbf{r}} = \dot{X}_1\hat{\mathbf{f}} + \dot{Y}_1\hat{\mathbf{g}}$. The various components appearing above are given in terms of the equinoctial orbit elements as well as the eccentric longitude F, by

$$X_1 = a\left[\left(1 - h^2\beta\right)c_F + hk\beta s_F - k\right] \tag{6.7}$$

$$Y_1 = a\left[hk\beta c_F + \left(1 - k^2\beta\right)s_F - h\right] \tag{6.8}$$

$$\dot{X}_1 = a^2nr^{-1}\left[hk\beta c_F - \left(1 - h^2\beta\right)s_F\right] \tag{6.9}$$

$$\dot{Y}_1 = a^2nr^{-1}\left[\left(1 - k^2\beta\right)c_F - hk\beta s_F\right] \tag{6.10}$$

Here, $\beta = 1/(1+G)$, $n = \mu^{1/2}a^{-3/2}$, the mean motion, and $r = a(1 - kc_F - hs_F)$.

The partials appearing in Equations (6.1)–(6.6) are given by

$$\frac{\partial a}{\partial \dot{\mathbf{r}}} = 2a^{-1}n^{-2}(\dot{X}_1\hat{\mathbf{f}} + \dot{Y}_1\hat{\mathbf{g}}) = M_{11}\hat{\mathbf{f}} + M_{12}\hat{\mathbf{g}} + M_{13}\hat{\mathbf{w}} \tag{6.11}$$

$$\begin{aligned}\frac{\partial h}{\partial \dot{\mathbf{r}}} &= Gn^{-1}a^{-2}\left[\left(\frac{\partial X_1}{\partial k} - h\beta\frac{\dot{X}_1}{n}\right)\hat{\mathbf{f}} + \left(\frac{\partial Y_1}{\partial k} - h\beta\frac{\dot{Y}_1}{n}\right)\hat{\mathbf{g}}\right] \\ &\quad + k(qY_1 - pX_1)n^{-1}a^{-2}G^{-1}\hat{\mathbf{w}} = M_{21}\hat{\mathbf{f}} + M_{22}\hat{\mathbf{g}} + M_{23}\hat{\mathbf{w}}\end{aligned} \tag{6.12}$$

$$\begin{aligned}\frac{\partial k}{\partial \dot{\mathbf{r}}} &= -Gn^{-1}a^{-2}\left[\left(\frac{\partial X_1}{\partial h} + k\beta\frac{\dot{X}_1}{n}\right)\hat{\mathbf{f}} + \left(\frac{\partial Y_1}{\partial h} + k\beta\frac{\dot{Y}_1}{n}\right)\hat{\mathbf{g}}\right] \\ &\quad - h(qY_1 - pX_1)n^{-1}a^{-2}G^{-1}\hat{\mathbf{w}} = M_{31}\hat{\mathbf{f}} + M_{32}\hat{\mathbf{g}} + M_{33}\hat{\mathbf{w}}\end{aligned} \tag{6.13}$$

$$\frac{\partial p}{\partial \dot{\mathbf{r}}} = KY_1\frac{n^{-1}a^{-2}G^{-1}}{2}\hat{\mathbf{w}} = M_{41}\hat{\mathbf{f}} + M_{42}\hat{\mathbf{g}} + M_{43}\hat{\mathbf{w}} \tag{6.14}$$

$$\frac{\partial q}{\partial \dot{\mathbf{r}}} = KX_1\frac{n^{-1}a^{-2}G^{-1}}{2}\hat{\mathbf{w}} = M_{51}\hat{\mathbf{f}} + M_{52}\hat{\mathbf{g}} + M_{53}\hat{\mathbf{w}} \tag{6.15}$$

$$\frac{\partial F}{\partial \dot{\mathbf{r}}} = \frac{a}{r}\left[\frac{\partial \lambda}{\partial \dot{\mathbf{r}}} + s_F\frac{\partial k}{\partial \dot{\mathbf{r}}} - c_F\frac{\partial h}{\partial \dot{\mathbf{r}}}\right] = M_{61}^F\hat{\mathbf{f}} + M_{62}^F\hat{\mathbf{g}} + M_{63}^F\hat{\mathbf{w}} \tag{6.16}$$

with both Equations (6.6) and (6.16) derived later in this section, and where

$$\begin{aligned}\frac{\partial \lambda}{\partial \dot{\mathbf{r}}} &= n^{-1}a^{-2}\left[-2X_1 + G\left(h\beta\frac{\partial X_1}{\partial h} + k\beta\frac{\partial X_1}{\partial k}\right)\right]\hat{\mathbf{f}} \\ &\quad + n^{-1}a^{-2}\left[-2Y_1 + G\left(h\beta\frac{\partial Y_1}{\partial h} + k\beta\frac{\partial Y_1}{\partial k}\right)\right]\hat{\mathbf{g}} \\ &\quad + n^{-1}a^{-2}G^{-1}(qY_1 - pX_1)\hat{\mathbf{w}} = M_{61}\hat{\mathbf{f}} + M_{62}\hat{\mathbf{g}} + M_{63}\hat{\mathbf{w}}\end{aligned} \tag{6.17}$$

with

$$\frac{\partial X_1}{\partial h} = a\left[-(hc_F - ks_F)\left(\beta + \frac{h^2\beta^3}{(1-\beta)}\right) - \frac{a}{r}c_F(h\beta - s_F)\right] \tag{6.18}$$

$$\frac{\partial X_1}{\partial k} = -a\left[(hc_F - ks_F)\frac{hk\beta^3}{(1-\beta)} + 1 + \frac{a}{r}s_F(s_F - h\beta)\right] \tag{6.19}$$

$$\frac{\partial Y_1}{\partial h} = a\left[(hc_F - ks_F)\frac{hk\beta^3}{(1-\beta)} - 1 + \frac{a}{r}c_F(k\beta - c_F)\right] \tag{6.20}$$

$$\frac{\partial Y_1}{\partial k} = a\left[(hc_F - ks_F)\left(\beta + \frac{k^2\beta^3}{(1-\beta)}\right) + \frac{a}{r}s_F(c_F - k\beta)\right] \tag{6.21}$$

The derivations that lead to these equations are given in great detail in [4] and [5] and Chapter 2 of this book. The partial $\partial F/\partial \dot{\mathbf{r}}$ is written directly in terms of the partials $\partial\lambda/\partial\dot{\mathbf{r}}$, $\partial h/\partial\dot{\mathbf{r}}$ and $\partial k/\partial\dot{\mathbf{r}}$. Because the partials of matrix $M$with respect to the elements were developed in [7] and Chapter 3 of this book using λ as the fast element, it is straightforward to use these partials with minor modifications in order to generate the partials of $\partial F/\partial\dot{\mathbf{r}}$ with respect to the elements needed in the Euler–Lagrange equations. It is for this reason that $\partial F/\partial\dot{\mathbf{r}}$ is not written explicitly and left as in Equation (6.16). The explicit form of $\partial F/\partial\dot{\mathbf{r}}$ is written as

$$\begin{aligned}\frac{\partial F}{\partial \dot{\mathbf{r}}} = \frac{1}{nar}&\left[\left\{-2X_1 + G\left(h\beta - s_F\right)\frac{\partial X_1}{\partial h} + G\left(k\beta - c_F\right)\frac{\partial X_1}{\partial k} - \beta G\left(ks_F - hc_F\right)\frac{\dot{X}_1}{n}\right\}\hat{\mathbf{f}}\right.\\ &+\left\{-2Y_1 + G\left(h\beta - s_F\right)\frac{\partial Y_1}{\partial h} + G\left(k\beta - c_F\right)\frac{\partial Y_1}{\partial k} - \beta G\left(ks_F - hc_F\right)\frac{\dot{Y}_1}{n}\right\}\hat{\mathbf{g}}\\ &\left.+\frac{r}{aG}(qY_1 - pX_1)\hat{\mathbf{w}}\right]\end{aligned}$$

The basic equations of motion can be written in a compact form as

$$\dot{\mathbf{z}} = M\,\hat{\mathbf{u}}f_t + \begin{pmatrix}0 & 0 & 0 & 0 & 0 & \frac{na}{r}\end{pmatrix}^T \tag{6.22}$$

Here, $\mathbf{z} = (a \quad h \quad k \quad q \quad p \quad F)^T$ is the state vector and the 6×3 matrix M is given by

$$M = \begin{bmatrix}\left(\frac{\partial a}{\partial\dot{\mathbf{r}}}\right)^T\\ \left(\frac{\partial h}{\partial\dot{\mathbf{r}}}\right)^T\\ \left(\frac{\partial k}{\partial\dot{\mathbf{r}}}\right)^T\\ \left(\frac{\partial p}{\partial\dot{\mathbf{r}}}\right)^T\\ \left(\frac{\partial q}{\partial\dot{\mathbf{r}}}\right)^T\\ \left(\frac{\partial F}{\partial\dot{\mathbf{r}}}\right)^T\end{bmatrix} = \begin{bmatrix}M_{11} & M_{12} & M_{13}\\ M_{21} & M_{22} & M_{23}\\ M_{31} & M_{32} & M_{33}\\ M_{41} & M_{42} & M_{43}\\ M_{51} & M_{52} & M_{53}\\ M^F_{61} & M^F_{62} & M^F_{63}\end{bmatrix}$$

These are the variation of parameters equations for this particular set of state variables, namely a, h, k, p, q, F, and they are obtained directly from the corresponding equations of the fundamental set a, h, k, p, q, λ_0, where λ_0 represents the mean longitude at epoch. The matrix of the Poisson brackets of the fundamental equinoctial set is obtained by applying the following transformation equation to the matrix of the Poisson brackets of the fundamental classical set $a, e, i, \Omega, \omega, M_0$:

$$[p_\alpha, p_\beta] = \left[\frac{\partial p_\alpha}{\partial a_\lambda}\right][(a_\lambda, a_\mu)]\left[\frac{\partial p_\beta}{\partial a_\mu}\right]^T$$

where

$$[(a_\lambda, a_\mu)] = P_{cl} = \begin{bmatrix} 0 & 0 & 0 & 0 & 0 & -2\left(\frac{a}{\mu}\right)^{1/2} \\ & 0 & 0 & 0 & \frac{(1-e^2)^{1/2}}{e(\mu a)^{1/2}} & \frac{-(1-e^2)}{e(\mu a)^{1/2}} \\ & & 0 & \frac{1}{(\mu p')^{1/2} s_i} & \frac{-c_i}{(\mu p')^{1/2} s_i} & 0 \\ & & & 0 & 0 & 0 \\ & & & & 0 & 0 \\ -Sym & & & & & 0 \end{bmatrix}$$

$$\left[\frac{\partial p_\alpha}{\partial a_\lambda}\right] = \begin{bmatrix} 1 & 0 & 0 & 0 & 0 & 0 \\ 0 & s_{\omega+\Omega} & 0 & ec_{\omega+\Omega} & ec_{\omega+\Omega} & 0 \\ 0 & c_{\omega+\Omega} & 0 & -es_{\omega+\Omega} & -es_{\omega+\Omega} & 0 \\ 0 & 0 & 0 & 1 & 1 & 1 \\ 0 & 0 & \frac{s_\Omega}{2\cos^2\frac{i}{2}} & \tan\frac{i}{2}c_\Omega & 0 & 0 \\ 0 & 0 & \frac{c_\Omega}{2\cos^2\frac{i}{2}} & -\tan\frac{i}{2}s_\Omega & 0 & 0 \end{bmatrix}$$

Converting the elements of the above two matrices from the classical elements to the equinoctial elements and carrying out the transformation yields

$$[(p_\alpha, p_\beta)] = \begin{bmatrix} (a,a) & (a,h) & (a,k) & (a,\lambda_0) & (a,p) & (a,q) \\ (h,a) & (h,h) & (h,k) & (h,\lambda_0) & (h,p) & (h,q) \\ (k,a) & (k,h) & (k,k) & (k,\lambda_0) & (k,p) & (k,q) \\ (\lambda_0,a) & (\lambda_0,h) & (\lambda_0,k) & (\lambda_0,\lambda_0) & (\lambda_0,p) & (\lambda_0,q) \\ (p,a) & (p,h) & (p,k) & (p,\lambda_0) & (p,p) & (p,q) \\ (q,a) & (q,h) & (q,k) & (q,\lambda_0) & (q,p) & (q,q) \end{bmatrix}$$

$$= \frac{1}{na^2}\begin{bmatrix} 0 & 0 & 0 & -2a & 0 & 0 \\ & 0 & -G & h\beta G & \frac{-kpK}{2G} & \frac{-kqK}{2G} \\ & & 0 & k\beta G & \frac{hpK}{2G} & \frac{hqK}{2G} \\ & & & 0 & \frac{-pK}{2G} & \frac{-qK}{2G} \\ -Sym & & & & 0 & \frac{-K^2}{4G} \\ & & & & & 0 \end{bmatrix}$$

The partials of the equinoctial elements of the fundamental set with respect to the velocity vector $\dot{\mathbf{r}}$ are obtained from $\frac{\partial p_\alpha}{\partial \dot{\mathbf{r}}} = -\sum_{\beta=1}^{6}(p_\alpha, p_\beta)\frac{\partial \mathbf{r}}{\partial p_\beta}$, such that for λ_0,

$$\frac{\partial \lambda_0}{\partial \mathbf{r}} = -(\lambda_0,\, a)\frac{\partial \mathbf{r}}{\partial a} - (\lambda_0,\, h)\frac{\partial \mathbf{r}}{\partial \mathbf{h}} - (\lambda_0,\, k)\frac{\partial \mathbf{r}}{\partial k} - (\lambda_0,\, p)\frac{\partial \mathbf{r}}{\partial p} - (\lambda_0,\, q)\frac{\partial \mathbf{r}}{\partial q} - (\lambda_0,\, \lambda_0)\frac{\partial \mathbf{r}}{\partial \lambda_0}$$

The partials of $\mathbf{r}$ with respect to the elements are generated by allowing for the variation of F with respect to the elements, with $\partial F/\partial a = -\frac{3}{2}\frac{n}{r}t$, $\partial F/\partial h = -\frac{a}{r}c_F$, $\partial F/\partial k = \frac{a}{r}s_F$, $\frac{\partial F}{\partial \lambda} = \frac{a}{r}$, with the latter partial used in $\frac{\partial r}{\partial \lambda} = \frac{\partial r}{\partial F}\frac{\partial F}{\partial \lambda}$ such that $\frac{\partial r}{\partial \lambda_0} = \frac{\partial r}{\partial \lambda}\frac{\partial \lambda}{\partial \lambda_0}$ with $\partial\lambda/\partial\lambda_0 = 1$. These relevant partials of $\mathbf{r}$ with respect to the elements are shown in [4] and [11] and Chapters 2 and 5 of this book with all the details of their derivations. They are:

$$\frac{\partial \mathbf{r}}{\partial a} = \left(\frac{X_1}{a} - \frac{3}{2}\frac{t}{a}\dot{X}_1\right)\hat{\mathbf{f}} + \left(\frac{Y_1}{a} - \frac{3}{2}\frac{t}{a}\dot{Y}_1\right)\hat{\mathbf{g}}$$
$$\frac{\partial \mathbf{r}}{\partial h} = \frac{\partial X_1}{\partial h}\hat{\mathbf{f}} + \frac{\partial Y_1}{\partial h}\hat{\mathbf{g}}$$
$$\frac{\partial \mathbf{r}}{\partial k} = \frac{\partial X_1}{\partial k}\hat{\mathbf{f}} + \frac{\partial Y_1}{\partial k}\hat{\mathbf{g}}$$
$$\frac{\partial \mathbf{r}}{\partial p} = \frac{2}{K}\left[q\left(Y_1\hat{\mathbf{f}} - X_1\hat{\mathbf{g}}\right) - X_1\hat{\mathbf{w}}\right]$$
$$\frac{\partial \mathbf{r}}{\partial q} = \frac{2}{K}\left[p\left(X_1\hat{\mathbf{g}} - Y_1\hat{\mathbf{f}}\right) + Y_1\hat{\mathbf{w}}\right]$$
$$\frac{\partial \mathbf{r}}{\partial \lambda_0} = \frac{\dot{X}_1}{n}\hat{\mathbf{f}} + \frac{\dot{Y}_1}{n}\hat{\mathbf{g}}$$

Thus,

$$\frac{\partial \lambda_0}{\partial \dot{\mathbf{r}}} = \frac{1}{na^2}\left\{\left[-2X_1 + 3\dot{X}_1 t + G\beta\left(h\frac{\partial X_1}{\partial h} + k\frac{\partial X_1}{\partial k}\right)\right]\hat{\mathbf{f}} + \left[-2Y_1 + 3\dot{Y}_1 t + G\beta\left(h\frac{\partial Y_1}{\partial h} + k\frac{\partial Y_1}{\partial k}\right)\right]\hat{\mathbf{g}} + \frac{1}{G}\left(qY_1 - pX_1\right)\hat{\mathbf{w}}\right\}$$

The explicit appearance of time above is due to the selection of the epoch variable, namely the epoch mean longitude λ_0, which is eliminated by replacing the variable λ_0 by λ. From $\lambda = \lambda_0 + nt$, $\frac{\partial \lambda}{\partial \dot{\mathbf{r}}} = \frac{\partial \lambda_0}{\partial \dot{\mathbf{r}}} - \frac{3}{na^2}t\dot{\mathbf{r}}$. Thus the partials of the elements with respect to the velocity vector for the set a, h, k, p, q, λ are given by Equations (6.11)–(6.15), which remain unchanged and Equation (6.17), which replaces the partial $\partial\lambda_0/\partial\dot{\mathbf{r}}$ shown above. The differential equation for the sixth variable is thus changed from $\dot{\lambda}_0 = \left(\frac{\partial \lambda_0}{\partial \dot{\mathbf{r}}}\right)^T \hat{\mathbf{u}}f_t$ to $\dot{\lambda} = n + \left(\frac{\partial \lambda}{\partial \dot{\mathbf{r}}}\right)^T \hat{\mathbf{u}}f_t$ when λ_0 is replaced by λ, whereas the first five Equations (6.1)–(6.5) remain identical. Equation (6.6) is obtained from Kepler's equation $\lambda = F - ks_F + hc_F$ which yields $\dot{\lambda} = (1 - kc_F - hs_F)\dot{F} - s_F\dot{k} + c_F\dot{h}$ such that $\dot{F} = \frac{a}{r}\left[n + \left(\frac{\partial \lambda}{\partial \dot{\mathbf{r}}}\right)^T \hat{\mathbf{u}}f_t\right] + \frac{a}{r}\left[s_F\left(\frac{\partial k}{\partial \dot{\mathbf{r}}}\right)^T - c_F\left(\frac{\partial h}{\partial \dot{\mathbf{r}}}\right)^T\right]\hat{\mathbf{u}}f_t$ which is further written as $\dot{F} = \frac{na}{r} + \left(\frac{\partial F}{\partial \dot{\mathbf{r}}}\right)^T \hat{\mathbf{u}}f_t$, because from $\lambda = F - ks_F + hc_F$, we have $\frac{\partial \lambda}{\partial \dot{\mathbf{r}}} = (1 - kc_F - hs_F)\frac{\partial F}{\partial \dot{\mathbf{r}}} - s_F\frac{\partial k}{\partial \dot{\mathbf{r}}} + c_F\frac{\partial h}{\partial \dot{\mathbf{r}}}$, and therefore $\frac{\partial F}{\partial \dot{\mathbf{r}}} = \frac{a}{r}\left[\frac{\partial \lambda}{\partial \dot{\mathbf{r}}} + s_F\frac{\partial k}{\partial \dot{\mathbf{r}}} - c_F\frac{\partial h}{\partial \dot{\mathbf{r}}}\right]$.

The Hamiltonian corresponding to this set (a, h, k, p, q, F) can now be written in compact form as

$$H = \boldsymbol{\lambda}_z^T\dot{\mathbf{z}} = \boldsymbol{\lambda}_z^T M(\mathbf{z}, F)f_t\hat{\mathbf{u}} + \lambda_F\frac{na}{r} \tag{6.23}$$

This Hamiltonian is maximized by making $\hat{\mathbf{u}}$ be parallel to $\boldsymbol{\lambda}_z^T M(\mathbf{z}, F)$. The Euler–Lagrange equations for the set (a, h, k, p, q, F) can now be written in a straightforward manner after making use of the following partials, namely $\frac{\partial r}{\partial a} = \frac{r}{a}$, $\frac{\partial r}{\partial h} = -as_F$, $\frac{\partial r}{\partial k} = -ac_F$, $\frac{\partial r}{\partial F} = a(ks_F - hc_F)$ and $\frac{\partial F}{\partial a} = 0$, $\frac{\partial F}{\partial h} = 0$, $\frac{\partial F}{\partial k} = 0$. The last three partials are identically equal to zero because F is being integrated and must therefore be considered an orbital element such that it is now independent of the other in-plane elements, a, h, and k. Likewise, from $r = a(1 - kc_F - hs_F)$, the partials $\partial r/\partial h$ and $\partial r/\partial k$ do not account for the variation of F with respect to h and k. This dependence had to be accounted for when the system of dynamic equations were originally generated. In taking the partial derivatives of H with respect to the six elements, the eccentric longitude F which appears in the right-hand side of all six differential equations in the quantities $X_1, Y_1, \dot{X}_1, \dot{Y}_1, \partial X_1/\partial h, \partial X_1/\partial k, \partial Y_1/\partial h, \partial Y_1/\partial k$, as well as r and the trigonometric functions s_F and c_F, must be held fixed because it is being integrated directly, becoming independent of the other five state variables a, h, k, p, and q. Now the Euler–Lagrange equations for the adjoints are given by

$$\dot{\boldsymbol{\lambda}}_z = -\frac{\partial H}{\partial \mathbf{z}} = -\boldsymbol{\lambda}_z^T \frac{\partial M}{\partial \mathbf{z}} f_t \hat{\mathbf{u}} - \lambda_F \frac{\partial \left(\frac{na}{r}\right)}{\partial \mathbf{z}} \tag{6.24}$$

Explicitly, and using the summation notation with $i = 1, 2, 3$ and $u_1 = u_f$, $u_2 = u_g$ and $u_3 = u_w$, we have the following differential equations:

$$\dot{\lambda}_a = -\frac{\partial H}{\partial a} = \left[-\lambda_a \frac{\partial M_{1i}}{\partial a} u_i - \lambda_h \frac{\partial M_{2i}}{\partial a} u_i - \lambda_k \frac{\partial M_{3i}}{\partial a} u_i - \lambda_p \frac{\partial M_{4i}}{\partial a} u_i \right.$$
$$\left. -\lambda_q \frac{\partial M_{5i}}{\partial a} u_i - \lambda_F \frac{\partial M_{6i}^F}{\partial a} u_i \right] f_t - \lambda_F \frac{\partial}{\partial a}\left(\frac{na}{r}\right) \tag{6.25}$$

$$\dot{\lambda}_h = -\frac{\partial H}{\partial h} = \left[-\lambda_a \frac{\partial M_{1i}}{\partial h} u_i - \lambda_h \frac{\partial M_{2i}}{\partial h} u_i - \lambda_k \frac{\partial M_{3i}}{\partial h} u_i - \lambda_p \frac{\partial M_{4i}}{\partial h} u_i \right.$$
$$\left. -\lambda_q \frac{\partial M_{5i}}{\partial h} u_i - \lambda_F \frac{\partial M_{6i}^F}{\partial h} u_i \right] f_t - \lambda_F \frac{\partial}{\partial h}\left(\frac{na}{r}\right) \tag{6.26}$$

$$\dot{\lambda}_k = -\frac{\partial H}{\partial k} = \left[-\lambda_a \frac{\partial M_{1i}}{\partial k} u_i - \lambda_h \frac{\partial M_{2i}}{\partial k} u_i - \lambda_k \frac{\partial M_{3i}}{\partial k} u_i - \lambda_p \frac{\partial M_{4i}}{\partial k} u_i \right.$$
$$\left. -\lambda_q \frac{\partial M_{5i}}{\partial k} u_i - \lambda_F \frac{\partial M_{6i}^F}{\partial k} u_i \right] f_t - \lambda_F \frac{\partial}{\partial k}\left(\frac{na}{r}\right) \tag{6.27}$$

$$\dot{\lambda}_p = -\frac{\partial H}{\partial p} = \left[-\lambda_a \frac{\partial M_{1i}}{\partial p} u_i - \lambda_h \frac{\partial M_{2i}}{\partial p} u_i - \lambda_k \frac{\partial M_{3i}}{\partial p} u_i - \lambda_p \frac{\partial M_{4i}}{\partial p} u_i \right.$$
$$\left. -\lambda_q \frac{\partial M_{5i}}{\partial p} u_i - \lambda_F \frac{\partial M_{6i}^F}{\partial p} u_i \right] f_t \tag{6.28}$$

$$\dot{\lambda}_q = -\frac{\partial H}{\partial q} = \left[-\lambda_a \frac{\partial M_{1i}}{\partial q} u_i - \lambda_h \frac{\partial M_{2i}}{\partial q} u_i - \lambda_k \frac{\partial M_{3i}}{\partial q} u_i - \lambda_p \frac{\partial M_{4i}}{\partial q} u_i \right.$$
$$\left. -\lambda_q \frac{\partial M_{5i}}{\partial q} u_i - \lambda_F \frac{\partial M_{6i}^F}{\partial q} u_i \right] f_t \tag{6.29}$$

$$\dot{\lambda}_F = -\frac{\partial H}{\partial F} = \left[-\lambda_a \frac{\partial M_{1i}}{\partial F} u_i - \lambda_h \frac{\partial M_{2i}}{\partial F} u_i - \lambda_k \frac{\partial M_{3i}}{\partial F} u_i - \lambda_p \frac{\partial M_{4i}}{\partial F} u_i \right.$$
$$\left. -\lambda_q \frac{\partial M_{5i}}{\partial F} u_i - \lambda_F \frac{\partial M_{6i}^F}{\partial F} u_i \right] f_t - \lambda_F \frac{\partial}{\partial F}\left(\frac{na}{r}\right) \tag{6.30}$$

However, the $\partial M/\partial \mathbf{z}$ partials must be derived with the assumption that $\partial F/\partial h = \partial F/\partial k = \partial F/\partial a = 0$, consistent with assuming F as an independent variable. For the first five rows of the M matrix, these partials were derived [6], and are included in Appendix 6.1 for completeness. This formulation is the only one that results in mixed second-order partials for both X_1 and Y_1, namely $\partial^2 X_1/\partial h \partial k \neq \partial^2 X_1/\partial k \partial h$, $\partial^2 Y_1/\partial h \partial k \neq \partial^2 Y_1/\partial k \partial h$ not being identical, as can be seen from Equations (A6.35), (A6.36), (A6.39) and (A6.40). This lack of symmetry is due to the independence of F with respect to both h and k. This dependence can be restored if the Hamiltonian in Equation (6.23) is written without the $\lambda_F na/r$ term, such that Equation (6.6) is written as $\dot{F}' = (\partial F/\partial \dot{\mathbf{r}})^T \bullet \hat{\mathbf{u}} f_t$ with F' representing the sixth orbital element, which remains constant in the absence of any thrust perturbation. The eccentric longitude F will then be obtained by a separate integration such that $\dot{F} = \dot{F}' + na/r$. When only the first five elements are considered, the Euler–Lagrange equations can be written either as in [7] or as developed in this chapter, the results being identical when averaging is used because then the dynamic and adjoint equations are effectively decoupled from the orbital position.

However, when the Hamiltonian is written as in Equation (6.23), we must accept that F is independent of a, h and k when deriving the adjoint equations. There is no ambiguity when the full six differential equations, mainly Equations (6.1)–(6.6) are used as in the present chapter within the context of precision integration. The mixed second-order partials of X_1 and Y_1 are not identical even though X_1 and Y_1 are continuous in h and k. Although X_1 and Y_1 are continuous in these two variables, the first-order partials $\partial X_1/\partial h$, $\partial X_1/\partial k$, $\partial Y_1/\partial h$ and $\partial Y_1/\partial k$ appearing explicitly in the elements of the M matrix and giving rise to the second-order partials in $\partial M/\partial \mathbf{z}$ are combined with other terms as is clear from Equations (6.12), (6.13), (6.16) and (6.17), such that the totality of the $\hat{\mathbf{f}}$ and $\hat{\mathbf{g}}$ components of the pertinent elements of M must be considered as unique functions of the elements after replacement of the $\partial X_1/\partial h$, $\partial X_1/\partial k$, $\partial Y_1/\partial h$ and $\partial Y_1/\partial k$ partials by their functional representation given in Equations (6.18)–(6.21). We should therefore revisit the assumptions made in this regard in [6] and not worry about the explicit presence of these first-order partials in M because Equations (6.12), (6.13), (6.16) and (6.17) are shown as they are for convenience and compactness in writing. During the generation of the $\partial M/\partial \mathbf{z}$ partials in the Appendices, care must be taken in writing the partials of X_1 and Y_1 with respect to h and k whenever X_1 and Y_1 appear explicitly in the matrix M. As stated earlier, these partials must not allow for the variation of F with respect to h and k, such that

$$\left(\frac{\partial X_1}{\partial h}\right)_F = a\left[-(hc_F - ks_F)\left(\beta + \frac{h^2\beta^3}{1-\beta}\right) - h\beta c_F\right]$$
$$\left(\frac{\partial X_1}{\partial k}\right)_F = -a\left[(hc_F - ks_F)\frac{hk\beta^3}{1-\beta} + 1 - h\beta s_F\right]$$

$$\left(\frac{\partial Y_1}{\partial h}\right)_F = a\left[(hc_F - ks_F)\frac{hk\beta^3}{1-\beta} - 1 + k\beta c_F\right]$$
$$\left(\frac{\partial Y_1}{\partial k}\right)_F = a\left[(hc_F - ks_F)\left(\beta + \frac{k^2\beta^3}{1-\beta}\right) - k\beta s_F\right]$$

This distinction between $\partial X_1/\partial h$ and $(\partial X_1/\partial h)_F$, as well as the other three pairs of partials, was not properly documented in the original conference paper in [13]. An error in Equation (131) of [13] in the form of h instead of k that appears in the second term there has also been corrected in the corresponding Equation (A6.85) of Appendix 6.2. The numerical results presented in this chapter and based on the exact six-state dynamics validate these assertions. For the sixth row of the M matrix, the nonzero partials are given in Appendix 6.2.

6.3 Numerical Example

Let us now try to duplicate the transfer trajectory used in [12] and Chapter 8 of this book. The minimum-time transfer solution must satisfy the transversality condition at the unknown final time $H_f = 0$ for the Hamiltonian $H = -1 + \boldsymbol{\lambda}_z^T \dot{\mathbf{z}}$ or equivalently $H_f = 1$ for our Hamiltonian $H = \boldsymbol{\lambda}_z^T \dot{\mathbf{z}}$. The initial values of the five Lagrange multipliers $(\lambda_a)_0$, $(\lambda_h)_0$, $(\lambda_k)_0$, $(\lambda_p)_0$, $(\lambda_q)_0$ are guessed as well as the total transfer time and the initial value of the eccentric longitude, and the shooting method is used to integrate from the initial given state with $(\lambda_F)_0 = 0$ in an iterative manner until the required final state is matched. In practice, the following objective function is minimized:

$$F_{\text{obj}} = w_1\left(a_f - a_T\right)^2 + w_2\left(h_f - h_T\right)^2 + w_3\left(k_f - k_T\right)^2 + w_4\left(p_f - p_T\right)^2 + w_5\left(q_f - q_T\right)^2 + w_6\left[(\lambda_F)_f - 0\right]^2 + w_7\left(H_f - 1\right)^2$$

where the w_js are weights associated to each element and where the f subscript represents the values of the equinoctial elements, the multiplier λ_F and H at the current final time of that particular iteration. The subscript T represents the targeted values that the iterator must match closely for a converged solution.

Using the F superscript for the present F formulation and the λ superscript for the λ formulation of [7] and Chapter 3 of this book, a canonical transformation between the set (a, h, k, p, q, λ) and the set (a, h, k, p, q, F) will require that the following condition hold true:

$$\lambda_a^\lambda da + \lambda_h^\lambda dh + \lambda_k^\lambda dk + \lambda_p^\lambda dp + \lambda_q^\lambda dq + \lambda_\lambda^\lambda d\lambda + H^\lambda dt = \lambda_a^F da + \lambda_h^F dh + \lambda_k^F dk + \lambda_p^F dp + \lambda_q^F dq + \lambda_F^F dF + H^F dt \tag{6.31}$$

Because the dynamic system is autonomous in both formulations, $H^\lambda = H^F = 1$ throughout the transfer, and from Kepler's equation $\lambda = F - ks_F + hc_F$, we have $d\lambda = (1 - kc_F - hs_F)\, dF - s_F dk + c_F dh$, or

$$d\lambda = \frac{r}{a} dF - s_F dk + c_F dh \tag{6.32}$$

which when used in Equation (6.31) yields the following six relationships:

$$\lambda_a^F = \lambda_a^\lambda \tag{6.33}$$
$$\lambda_h^F = \lambda_h^\lambda + \lambda_\lambda^\lambda c_F \tag{6.34}$$

$$\lambda_k^F = \lambda_k^\lambda - \lambda_\lambda^\lambda s_F \tag{6.35}$$

$$\lambda_p^F = \lambda_p^\lambda \tag{6.36}$$

$$\lambda_q^F = \lambda_q^\lambda \tag{6.37}$$

$$\lambda_F^F = \frac{r}{a}\lambda_\lambda^\lambda \tag{6.38}$$

Both initial and final eccentric longitudes are free, and therefore they are optimized because $(\lambda_F)_0 = 0$ is imposed, and $(\lambda_F)_f = 0$ is iterated on. We can also verify that the derivations presented in this chapter are indeed correct and error-free by running a direct trajectory using the solution shown in [12] and Chapter 8 of this book. Using $a_0 = 7000$ km, $e_0 = 0$, $i_0 = 28.5$ deg, $\Omega_0 = 0$ deg, $\omega_0 = 0$ deg, and the target conditions $a_f = 42{,}000$ km, $e_f = 10^{-3}$, $i_f = 1$ deg, $\Omega_f = \omega_f = 0$ deg, and the acceleration $f_t = 9.8 \times 10^{-5}$ km/s^2, the λ formulation resulted in the following solution [12, and Chapter 8] given by $(\lambda_a)_0 = 4.675229762$ s/km, $(\lambda_h)_0 = 5.413413947 \times 10^2$ s, $(\lambda_k)_0 = -9.202702084 \times 10^3$ s, $(\lambda_p)_0 = 1.778011878 \times 10^1$ s, $(\lambda_q)_0 = -2.258455855 \times 10^4$ s, with an optimized initial mean longitude $(\lambda)_0 = -2.274742851$ rad corresponding to an initial mean anomaly $M_0 = -130.3331648$ deg, and the minimum time of $t_f = 58{,}089.90058$ s. Because $(\lambda_\lambda)_0 = 0$ was used at time zero to start the integration, and because $(r)_0 = a_0$ due to the initial circular condition, Equations (6.33)–(6.38) show that we can select λ_a, λ_h, λ_k, λ_p, λ_q and $\lambda_F = 0$ at the initial time to be identical to the values shown earlier. Furthermore $(F)_0 = (\lambda)_0 = (M)_0$ because $e_0 = 0$ and $\Omega_0 = \omega_0 = 0$ deg in this example. The direct integration to t_f results in the final parameters $a_f = 41{,}999.991$ km, $e_f = 9.9833 \times 10^{-4}$, $i_f = 0.999797$ deg, $\Omega_f = 3.435 \times 10^{-4}$ deg, $\omega_f = 359.992909$ deg, $F_f = 46.206027$ deg with $(\lambda_F)_f = 9.089 \times 10^{-3}$ s/rad, duplicating the minimum-time transfer example of [12] and Chapter 8 of this book. This F_f value corresponds to $M_f = 46.171481$ deg, which is close to $M_f = 46.146408$ deg obtained in [12] and Chapter 8 of this book, while the locations on the final orbit are given by $\lambda_f = 19.655282$ rad and $\lambda_f = 19.655283$ rad, respectively. Figure 6.1 shows how F and M vary during the short optimal flight starting from the same initial value of 229.666835 deg or -130.333164 deg, with the eccentric longitude cycling between 0 and 360 deg. Figures 6.2–6.7 show the evolution of the various multipliers and compare

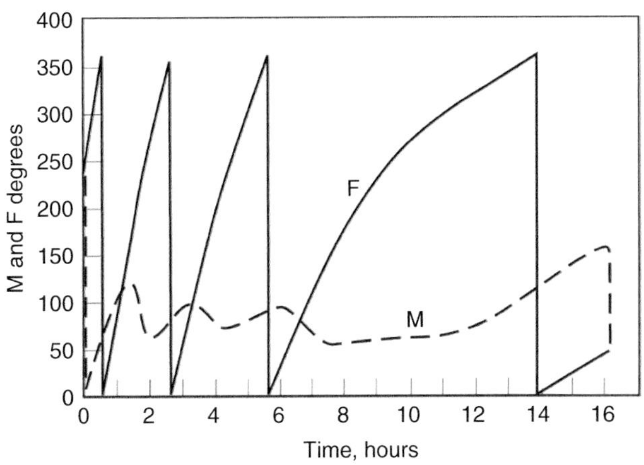

Figure 6.1 Mean anomaly and eccentric longitude time histories during optimal transfer.

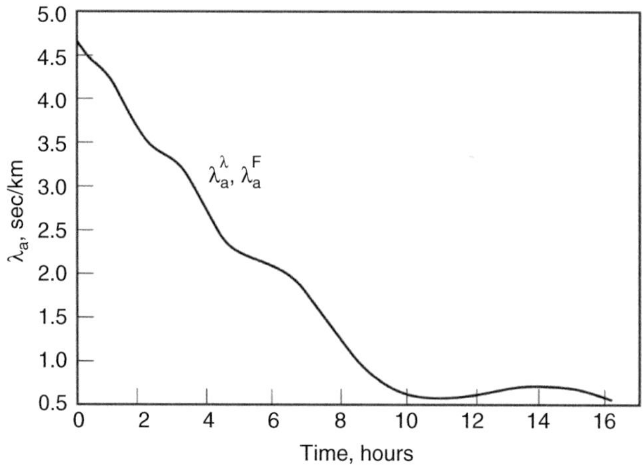

Figure 6.2 Evolution of λ_a^λ and λ_a^F multipliers during optimal transfer.

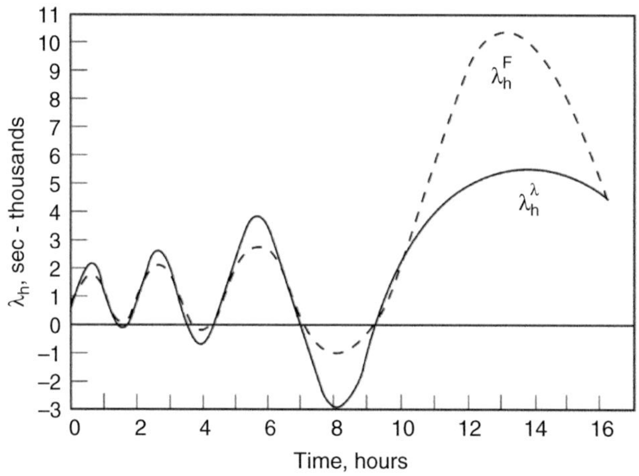

Figure 6.3 Evolution of λ_h^λ and λ_h^F multipliers during optimal transfer.

them to the corresponding multipliers of the λ formulation of [12] and Chapter 8 of this book with identical λ_a, λ_p and λ_q, as predicted by Equations (6.33), (6.36) and (6.37). Both λ_λ and λ_F start and end at zero as required for the solution of the free–free transfer because both initial and final longitudes are free and are therefore optimized to yield the absolute minimum-time transfer. Finally, Figures 6.8 and 6.9 depict the variations of the Earth-centered inertial Cartesian position components, with Figure 6.9 showing clearly how the orbit initially inclined at 28.5 deg rotates near the end of its run to match the 1 deg target. These expressions clearly show why the in-plane adjoints exhibit different plots with the exception of the various λ_a. Because our example transfer remains near-circular throughout except near the end, r/a stays close to the unit value such that the λ_F and λ_λ curves are nearly identical. This near-identity of the λ_F and λ_λ curves will not appear when the orbits are elliptical because then r/a will effectively modulate the λ_λ curve to generate the λ_F curve.

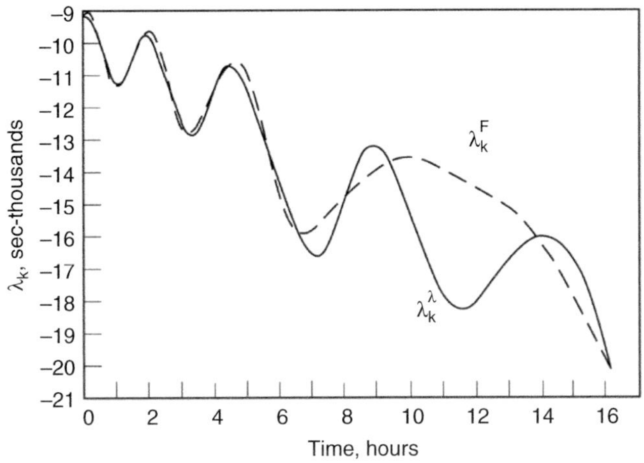

Figure 6.4 Evolution of λ_k^λ and λ_k^F multipliers during optimal transfer.

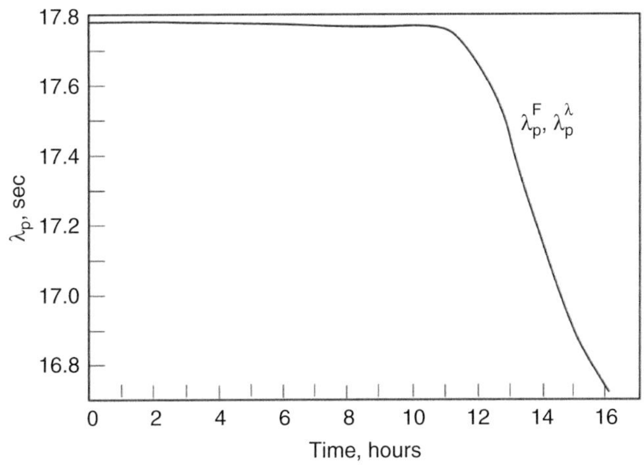

Figure 6.5 Evolution of λ_p^λ and λ_p^F multipliers during optimal transfer.

6.4 Conclusion

The mathematics of trajectory optimization for low-thrust orbit transfer and rendezvous are derived for the particular case in which the eccentric longitude represents the sixth state variable. The use of this nonsingular set removes the need for solving the Kepler transcendental equation at each integration step because the eccentric longitude is now directly integrated. An example of minimum-time transfer using continuous constant acceleration obtained with a different formulation in which the mean longitude represents the sixth state variable instead, is duplicated by way of this formulation. The correspondence between the adjoints to the six elements of both formulations is also established, and the result used to set the initial values of the six Lagrange multipliers

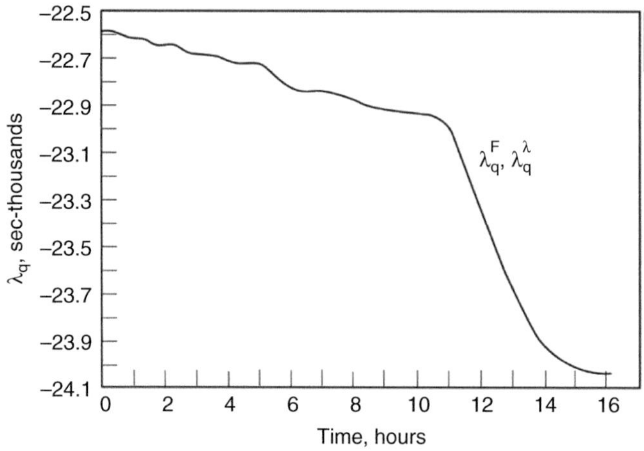

Figure 6.6 Evolution of λ_q^λ and λ_q^F multipliers during optimal transfer.

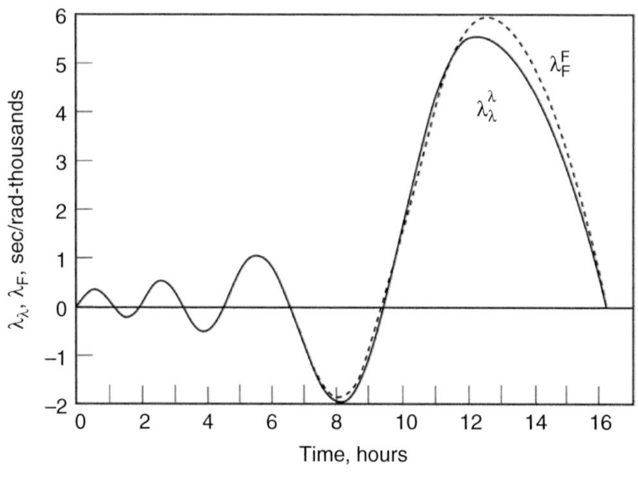

Figure 6.7 Evolution of λ_λ^λ and λ_F^F multipliers during optimal transfer.

corresponding to the present formulation in order to duplicate a direct transfer trajectory for validation and verification of the various mathematical derivations.

Appendix 6.1 The $\partial M/\partial z$ Partial Derivatives

The nonzero partials of the first five rows of matrix M:

$$\frac{\partial M_{11}}{\partial h} = \frac{2}{n^2 a}\frac{\partial \dot{X}_1}{\partial h} \tag{A6.1}$$

$$\frac{\partial M_{12}}{\partial h} = \frac{2}{n^2 a}\frac{\partial \dot{Y}_1}{\partial h} \tag{A6.2}$$

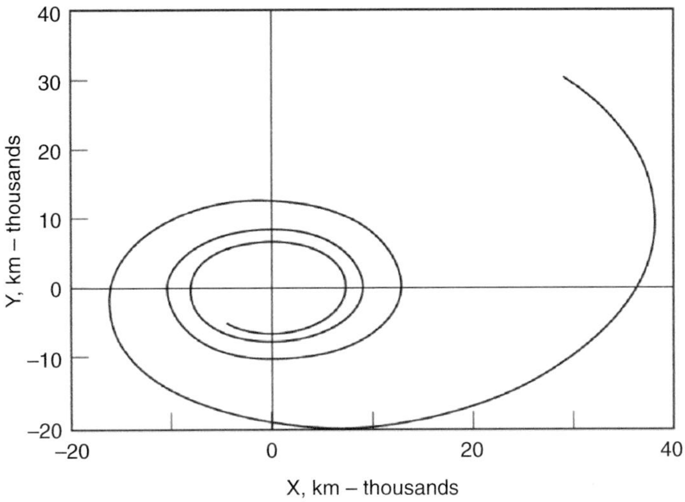

Figure 6.8 Inertial Cartesian coordinates Y versus X during optimal transfer.

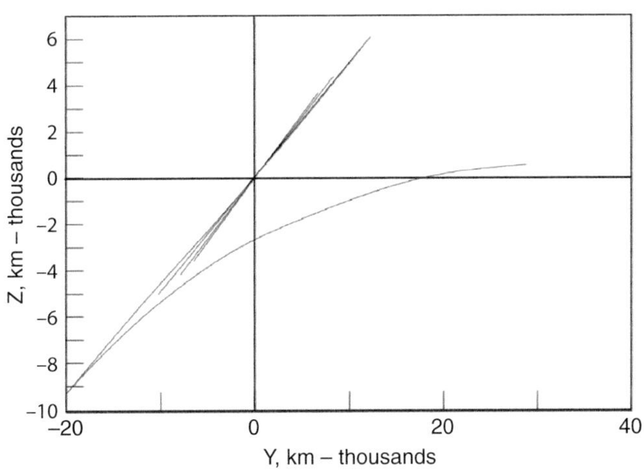

Figure 6.9 Inertial Cartesian coordinates Z versus Y during optimal transfer.

$$\frac{\partial M_{21}}{\partial h} = \frac{-h}{na^2G}\left(\frac{\partial X_1}{\partial k} - \frac{h\beta}{n}\dot{X}_1\right) + \frac{G}{na^2}\left[\frac{\partial^2 X_1}{\partial h \partial k} - \frac{\dot{X}_1}{n}\left(\beta + \frac{h^2\beta^3}{1-\beta}\right) - \frac{h\beta}{n}\frac{\partial \dot{X}_1}{\partial h}\right] \tag{A6.3}$$

In a similar way,

$$\frac{\partial M_{22}}{\partial h} = \frac{-h}{na^2G}\left(\frac{\partial Y_1}{\partial k} - \frac{h\beta}{n}\dot{Y}_1\right) + \frac{G}{na^2}\left[\frac{\partial^2 Y_1}{\partial h \partial k} - \frac{\dot{Y}_1}{n}\left(\beta + \frac{h^2\beta^3}{1-\beta}\right) - \frac{h\beta}{n}\frac{\partial \dot{Y}_1}{\partial h}\right] \tag{A6.4}$$

$$\frac{\partial M_{23}}{\partial h} = \frac{hkG^{-3}}{na^2}(qY_1 - pX_1) + \frac{k\left[q\left(\frac{\partial Y_1}{\partial h}\right)_F - p\left(\frac{\partial X_1}{\partial h}\right)_F\right]}{na^2G} \tag{A6.5}$$

$$\frac{\partial M_{31}}{\partial h} = \frac{h}{na^2G}\left(\frac{\partial X_1}{\partial h} + k\beta\frac{\dot{X}_1}{n}\right) - \frac{G}{na^2}\left[\frac{\partial^2 X_1}{\partial h^2} + \frac{hk\beta^3}{1-\beta}\frac{\dot{X}_1}{n} + \frac{k\beta}{n}\frac{\partial \dot{X}_1}{\partial h}\right] \tag{A6.6}$$

$$\frac{\partial M_{32}}{\partial h} = \frac{h}{na^2G}\left(\frac{\partial Y_1}{\partial h} + k\beta\frac{\dot{Y}_1}{n}\right) - \frac{G}{na^2}\left[\frac{\partial^2 Y_1}{\partial h^2} + \frac{hk\beta^3}{1-\beta}\frac{\dot{Y}_1}{n} + \frac{k\beta}{n}\frac{\partial \dot{Y}_1}{\partial h}\right] \tag{A6.7}$$

$$\frac{\partial M_{33}}{\partial h} = \frac{-1}{na^2G}\left\{(qY_1 - pX_1) + h\left[q\left(\frac{\partial Y_1}{\partial h}\right)_F - p\left(\frac{\partial X_1}{\partial h}\right)_F\right]\right\} - \frac{h^2(qY_1 - pX_1)}{na^2G^3} \tag{A6.8}$$

$$\frac{\partial M_{43}}{\partial h} = \frac{K}{2na^2G}\left[\left(\frac{\partial Y_1}{\partial h}\right)_F + \frac{hY_1}{G^2}\right] \tag{A6.9}$$

$$\frac{\partial M_{53}}{\partial h} = \frac{K}{2na^2G}\left[\left(\frac{\partial X_1}{\partial h}\right)_F + \frac{hX_1}{G^2}\right] \tag{A6.10}$$

$$\frac{\partial M_{11}}{\partial k} = \frac{2}{n^2a}\frac{\partial \dot{X}_1}{\partial k} \tag{A6.11}$$

$$\frac{\partial M_{12}}{\partial k} = \frac{2}{n^2a}\frac{\partial \dot{Y}_1}{\partial k} \tag{A6.12}$$

$$\frac{\partial M_{21}}{\partial k} = \frac{-k}{na^2G}\left(\frac{\partial X_1}{\partial k} - \frac{h\beta}{n}\dot{X}_1\right) + \frac{G}{na^2}\left[\frac{\partial^2 X_1}{\partial k^2} - \frac{hk\beta^3}{n(1-\beta)}\dot{X}_1 - \frac{h\beta}{n}\frac{\partial \dot{X}_1}{\partial k}\right] \tag{A6.13}$$

$$\frac{\partial M_{22}}{\partial k} = \frac{-k}{na^2G}\left(\frac{\partial Y_1}{\partial k} - \frac{h\beta}{n}\dot{Y}_1\right) + \frac{G}{na^2}\left[\frac{\partial^2 Y_1}{\partial k^2} - \frac{hk\beta^3}{n(1-\beta)}\dot{Y}_1 - \frac{h\beta}{n}\frac{\partial \dot{Y}_1}{\partial k}\right] \tag{A6.14}$$

$$\begin{aligned}\frac{\partial M_{23}}{\partial k} &= \frac{(qY_1 - pX_1)}{na^2G} + \frac{1}{na^2G}\\ &\quad\times\left\{k\left[q\left(\frac{\partial Y_1}{\partial k}\right)_F - p\left(\frac{\partial X_1}{\partial k}\right)_F\right] + \frac{k^2\,(qY_1 - pX_1)}{G^2}\right\}\end{aligned} \tag{A6.15}$$

$$\begin{aligned}\frac{\partial M_{31}}{\partial k} &= \frac{k}{na^2G}\left(\frac{\partial X_1}{\partial h} + k\beta\frac{\dot{X}_1}{n}\right)\\ &\quad - \frac{G}{na^2}\left[\frac{\partial^2 X_1}{\partial k\partial h} + \left(\beta + \frac{k^2\beta^3}{1-\beta}\right)\frac{\dot{X}_1}{n} + \frac{k\beta}{n}\frac{\partial \dot{X}_1}{\partial k}\right]\end{aligned} \tag{A6.16}$$

$$\begin{aligned}\frac{\partial M_{32}}{\partial k} &= \frac{k}{na^2G}\left(\frac{\partial Y_1}{\partial h} + k\beta\frac{\dot{Y}_1}{n}\right)\\ &\quad - \frac{G}{na^2}\left[\frac{\partial^2 Y_1}{\partial k\partial h} + \left(\beta + \frac{k^2\beta^3}{1-\beta}\right)\frac{\dot{Y}_1}{n} + \frac{k\beta}{n}\frac{\partial \dot{Y}_1}{\partial k}\right]\end{aligned} \tag{A6.17}$$

$$\frac{\partial M_{33}}{\partial k} = \frac{-h}{na^2G}\left[q\left(\frac{\partial Y_1}{\partial k}\right)_F - p\left(\frac{\partial X_1}{\partial k}\right)_F\right] - \frac{hk}{na^2G^3}(qY_1 - pX_1) \tag{A6.18}$$

$$\frac{\partial M_{43}}{\partial k} = \frac{K}{2na^2G}\left(\frac{\partial Y_1}{\partial k}\right)_F + \frac{kK}{2na^2G^3}Y_1 \tag{A6.19}$$

$$\frac{\partial M_{53}}{\partial k} = \frac{K}{2na^2G}\left(\frac{\partial X_1}{\partial k}\right)_F + \frac{kK}{2na^2G^3}X_1 \tag{A6.20}$$

The only nonzero partials with respect to p are

$$\frac{\partial M_{23}}{\partial p} = \frac{-kX_1}{na^2G} \tag{A6.21}$$

$$\frac{\partial M_{33}}{\partial p} = \frac{hX_1}{na^2G} \tag{A6.22}$$

$$\frac{\partial M_{43}}{\partial p} = \frac{pY_1}{na^2G} \tag{A6.23}$$

$$\frac{\partial M_{53}}{\partial p} = \frac{pX_1}{na^2G} \tag{A6.24}$$

The only nonzero partials with respect to q are

$$\frac{\partial M_{23}}{\partial q} = \frac{kY_1}{na^2G} \tag{A6.25}$$

$$\frac{\partial M_{33}}{\partial q} = \frac{-hY_1}{na^2G} \tag{A6.26}$$

$$\frac{\partial M_{43}}{\partial q} = \frac{qY_1}{na^2G} \tag{A6.27}$$

$$\frac{\partial M_{53}}{\partial q} = \frac{qX_1}{na^2G} \tag{A6.28}$$

The accessory partials appearing in the definition of the preceding partial derivatives are obtained from Equations (6.9), (6.10) and (6.18)–(6.21) by assuming that $\partial F/\partial h = \partial F/\partial k = 0$:

$$\frac{\partial \dot{X}_1}{\partial h} = \frac{a}{r}\dot{X}_1 s_F + \frac{na^2}{r}\left[h\beta s_F + (kc_F + hs_F)\left(\beta + \frac{h^2\beta^3}{(1-\beta)}\right)\right] \tag{A6.29}$$

$$\frac{\partial \dot{X}_1}{\partial k} = \frac{a}{r}\dot{X}_1 c_F + \frac{na^2}{r}\left[h\beta c_F + (kc_F + hs_F)\frac{hk\beta^3}{(1-\beta)}\right] \tag{A6.30}$$

$$\frac{\partial \dot{Y}_1}{\partial h} = \frac{a}{r}\dot{Y}_1 s_F + \frac{na^2}{r}\left[-k\beta s_F - (kc_F + hs_F)\frac{hk\beta^3}{(1-\beta)}\right] \tag{A6.31}$$

$$\frac{\partial \dot{Y}_1}{\partial k} = \frac{a}{r}\dot{Y}_1 c_F + \frac{na^2}{r}\left[-k\beta c_F - (kc_F + hs_F)\left(\beta + \frac{k^2\beta^3}{(1-\beta)}\right)\right] \tag{A6.32}$$

$$\begin{aligned}\frac{\partial^2 X_1}{\partial h^2} = a\Bigg[&-\left(1+\frac{a}{r}\right)c_F\left(\beta + \frac{h^2\beta^3}{(1-\beta)}\right) - \frac{h\beta^3}{(1-\beta)}(hc_F - ks_F)\\ &\times\left(3 + \frac{h^2\beta^2(3-2\beta)}{(1-\beta)^2}\right) - \frac{a^2}{r^2}s_F c_F(h\beta - s_F)\Bigg]\end{aligned} \tag{A6.33}$$

$$\begin{aligned}\frac{\partial^2 X_1}{\partial k^2} = -a\Bigg[&-s_F\left(1+\frac{a}{r}\right)\frac{hk\beta^3}{(1-\beta)} + \frac{h\beta^3}{(1-\beta)}(hc_F - ks_F)\\ &\times\left(1 + \frac{k^2\beta^2(3-2\beta)}{(1-\beta)^2}\right) - \frac{a^2}{r^2}s_F c_F(h\beta - s_F)\Bigg]\end{aligned} \tag{A6.34}$$

$$\frac{\partial^2 X_1}{\partial h \partial k} = -a\left[c_F \frac{hk\beta^3}{(1-\beta)} + \frac{k\beta^3}{(1-\beta)}(hc_F - ks_F)\left(1 + \frac{h^2\beta^2(3-2\beta)}{(1-\beta)^2}\right)\right.$$
$$\left. + \frac{a^2}{r^2} s_F^2 (s_F - h\beta) - \frac{a}{r} s_F \left(\beta + \frac{h^2\beta^3}{(1-\beta)}\right)\right] \tag{A6.35}$$

$$\frac{\partial^2 X_1}{\partial k \partial h} = a\left[s_F\left(\beta + \frac{h^2\beta^3}{(1-\beta)}\right) - \frac{k\beta^3}{(1-\beta)}(hc_F - ks_F)\left(1 + \frac{h^2\beta^2(3-2\beta)}{(1-\beta)^2}\right)\right.$$
$$\left. - \frac{a^2}{r^2} c_F^2 (h\beta - s_F) - \frac{a}{r} c_F \frac{hk\beta^3}{(1-\beta)}\right] \tag{A6.36}$$

$$\frac{\partial^2 Y_1}{\partial h^2} = a\left[c_F\left(1 + \frac{a}{r}\right)\frac{hk\beta^3}{(1-\beta)} + \frac{k\beta^3}{(1-\beta)}(hc_F - ks_F)\left(1 + \frac{h^2\beta^2(3-2\beta)}{(1-\beta)^2}\right)\right.$$
$$\left. + \frac{a^2}{r^2} s_F c_F (k\beta - c_F)\right] \tag{A6.37}$$

$$\frac{\partial^2 Y_1}{\partial k^2} = a\left[-s_F\left(1 + \frac{a}{r}\right)\left(\beta + \frac{k^2\beta^3}{(1-\beta)}\right) + \frac{k\beta^3}{(1-\beta)}(hc_F - ks_F)\left(3 + \frac{k^2\beta^2(3-2\beta)}{(1-\beta)^2}\right)\right.$$
$$\left. + \frac{a^2}{r^2} s_F c_F (c_F - k\beta)\right] \tag{A6.38}$$

$$\frac{\partial^2 Y_1}{\partial h \partial k} = a\left[c_F\left(\beta + \frac{k^2\beta^3}{(1-\beta)}\right) + \frac{h\beta^3}{(1-\beta)}(hc_F - ks_F)\left(1 + \frac{k^2\beta^2(3-2\beta)}{(1-\beta)^2}\right)\right.$$
$$\left. + \frac{a^2}{r^2} s_F^2 (c_F - k\beta) - \frac{a}{r} s_F \frac{hk\beta^3}{(1-\beta)}\right] \tag{A6.39}$$

$$\frac{\partial^2 Y_1}{\partial k \partial h} = a\left[-s_F \frac{hk\beta^3}{(1-\beta)} + \frac{h\beta^3}{(1-\beta)}(hc_F - ks_F)\left(1 + \frac{k^2\beta^2(3-2\beta)}{(1-\beta)^2}\right)\right.$$
$$\left. + \frac{a^2}{r^2} c_F^2 (k\beta - c_F) + \frac{a}{r} c_F \left(\beta + \frac{k^2\beta^3}{(1-\beta)}\right)\right] \tag{A6.40}$$

$$\frac{\partial^2 X_1}{\partial a \partial h} = \frac{1}{a}\frac{\partial X_1}{\partial h} \tag{A6.41}$$
$$\frac{\partial^2 X_1}{\partial a \partial k} = \frac{1}{a}\frac{\partial X_1}{\partial k} \tag{A6.42}$$
$$\frac{\partial^2 Y_1}{\partial a \partial h} = \frac{1}{a}\frac{\partial Y_1}{\partial h} \tag{A6.43}$$
$$\frac{\partial^2 Y_1}{\partial a \partial k} = \frac{1}{a}\frac{\partial Y_1}{\partial k} \tag{A6.44}$$

where the $\partial X_1/\partial h$, $\partial X_1/\partial k$, $\partial Y_1/\partial h$, $\partial Y_1/\partial k$ partials just appearing are given by Equations (6.18)–(6.21). The last four second-order partials are needed to define the $\partial M/\partial a$ partials in Equation (A6.49), produced by assuming $\partial F/\partial a = 0$. We have

$$\frac{\partial X_1}{\partial a} = \frac{X_1}{a} \tag{A6.45}$$
$$\frac{\partial Y_1}{\partial a} = \frac{Y_1}{a} \tag{A6.46}$$

$$\frac{\partial \dot{X}_1}{\partial a} = -\frac{na}{2r}[hk\beta c_F - (1 - h^2\beta)s_F] = -\frac{\dot{X}_1}{2a} \tag{A6.47}$$

$$\frac{\partial \dot{Y}_1}{\partial a} = -\frac{na}{2r}[hk\beta s_F - (1 - k^2\beta)c_F] = -\frac{\dot{Y}_1}{2a} \tag{A6.48}$$

This, together with the second-order partials of Equations (A6.41) through (A6.44) give

$$\frac{\partial M}{\partial a} = \frac{1}{2a}\begin{bmatrix} 3 & 0 & 0 & 0 & 0 \\ & 1 & 0 & 0 & 0 \\ & & 1 & 0 & 0 \\ & & & 1 & 0 \\ 0 & & & & 1 \end{bmatrix} M \tag{A6.49}$$

Appendix 6.2 The $\partial M^F/\partial z$ Partial Derivatives

The partials of the sixth row of the matrix M:

$$\frac{\partial M_{61}^F}{\partial a} = \frac{a}{r}\left[\frac{\partial M_{61}}{\partial a} + s_F\frac{\partial M_{31}}{\partial a} - c_F\frac{\partial M_{21}}{\partial a}\right] \tag{A6.50}$$

$$\frac{\partial M_{62}^F}{\partial a} = \frac{a}{r}\left[\frac{\partial M_{62}}{\partial a} + s_F\frac{\partial M_{32}}{\partial a} - c_F\frac{\partial M_{22}}{\partial a}\right] \tag{A6.51}$$

$$\frac{\partial M_{63}^F}{\partial a} = \frac{a}{r}\left[\frac{\partial M_{63}}{\partial a} + s_F\frac{\partial M_{33}}{\partial a} - c_F\frac{\partial M_{23}}{\partial a}\right] \tag{A6.52}$$

$$\frac{\partial}{\partial a}\left(\frac{na}{r}\right) = -\frac{3}{2}\frac{\mu^{1/2}a^{-3/2}}{r} \tag{A6.53}$$

$$\begin{aligned}\frac{\partial M_{61}^F}{\partial h} &= \frac{a^2}{r^2}s_F(M_{61} + s_F M_{31} - c_F M_{21}) \\ &+ \frac{a}{r}\left[\frac{\partial M_{61}}{\partial h} + s_F\frac{\partial M_{31}}{\partial h} - c_F\frac{\partial M_{21}}{\partial h}\right]\end{aligned} \tag{A6.54}$$

$$\begin{aligned}\frac{\partial M_{62}^F}{\partial h} &= \frac{a^2}{r^2}s_F(M_{62} + s_F M_{32} - c_F M_{22}) \\ &+ \frac{a}{r}\left[\frac{\partial M_{62}}{\partial h} + s_F\frac{\partial M_{32}}{\partial h} - c_F\frac{\partial M_{22}}{\partial h}\right]\end{aligned} \tag{A6.55}$$

$$\begin{aligned}\frac{\partial M_{63}^F}{\partial h} &= \frac{a^2}{r^2}s_F(M_{63} + s_F M_{33} - c_F M_{23}) \\ &+ \frac{a}{r}\left[\frac{\partial M_{63}}{\partial h} + s_F\frac{\partial M_{33}}{\partial h} - c_F\frac{\partial M_{23}}{\partial h}\right]\end{aligned} \tag{A6.56}$$

$$\frac{\partial}{\partial h}\left(\frac{na}{r}\right) = \frac{na^2}{r^2}s_F \tag{A6.57}$$

$$\begin{aligned}\frac{\partial M_{61}^F}{\partial k} &= \frac{a^2}{r^2}c_F(M_{61} + s_F M_{31} - c_F M_{21}) \\ &+ \frac{a}{r}\left[\frac{\partial M_{61}}{\partial k} + s_F\frac{\partial M_{31}}{\partial k} - c_F\frac{\partial M_{21}}{\partial k}\right]\end{aligned} \tag{A6.58}$$

$$\frac{\partial M_{62}^F}{\partial k} = \frac{a^2}{r^2} c_F (M_{62} + s_F M_{32} - c_F M_{22}) + \frac{a}{r}\left[\frac{\partial M_{62}}{\partial k} + s_F \frac{\partial M_{32}}{\partial k} - c_F \frac{\partial M_{22}}{\partial k}\right] \tag{A6.59}$$

$$\frac{\partial M_{63}^F}{\partial k} = \frac{a^2}{r^2} c_F (M_{63} + s_F M_{33} - c_F M_{23}) + \frac{a}{r}\left[\frac{\partial M_{63}}{\partial k} + s_F \frac{\partial M_{33}}{\partial k} - c_F \frac{\partial M_{23}}{\partial k}\right] \tag{A6.60}$$

$$\frac{\partial}{\partial k}\left(\frac{na}{r}\right) = \frac{na^2}{r^2} c_F \tag{A6.61}$$

$$\frac{\partial M_{63}^F}{\partial p} = \frac{a}{r}\left(\frac{\partial M_{63}}{\partial p} + s_F \frac{\partial M_{33}}{\partial p} - c_F \frac{\partial M_{23}}{\partial p}\right) \tag{A6.62}$$

$$\frac{\partial M_{63}^F}{\partial q} = \frac{a}{r}\left(\frac{\partial M_{63}}{\partial q} + s_F \frac{\partial M_{33}}{\partial q} - c_F \frac{\partial M_{23}}{\partial q}\right) \tag{A6.63}$$

$$\frac{\partial M_{61}^F}{\partial F} = -\frac{a^2}{r^2}(k s_F - h c_F)(M_{61} + s_F M_{31} - c_F M_{21}) + \frac{a}{r}\left[\frac{\partial M_{61}}{\partial F} + s_F \frac{\partial M_{31}}{\partial F} - c_F \frac{\partial M_{21}}{\partial F} + c_F M_{31} + s_F M_{21}\right] \tag{A6.64}$$

$$\frac{\partial M_{62}^F}{\partial F} = -\frac{a^2}{r^2}(k s_F - h c_F)(M_{62} + s_F M_{32} - c_F M_{22}) + \frac{a}{r}\left[\frac{\partial M_{62}}{\partial F} + s_F \frac{\partial M_{32}}{\partial F} - c_F \frac{\partial M_{22}}{\partial F} + c_F M_{32} + s_F M_{22}\right] \tag{A6.65}$$

$$\frac{\partial M_{63}^F}{\partial F} = -\frac{a^2}{r^2}(k s_F - h c_F)(M_{63} + s_F M_{33} - c_F M_{23}) + \frac{a}{r}\left[\frac{\partial M_{63}}{\partial F} + s_F \frac{\partial M_{33}}{\partial F} - c_F \frac{\partial M_{23}}{\partial F} + c_F M_{33} + s_F M_{23}\right] \tag{A6.66}$$

Finally, the partial derivatives of M with respect to F which are not identically equal to zero are given by

$$\frac{\partial M_{11}}{\partial F} = \frac{2}{n^2 a}\frac{\partial \dot{X}_1}{\partial F} \tag{A6.67}$$

$$\frac{\partial M_{12}}{\partial F} = \frac{2}{n^2 a}\frac{\partial \dot{Y}_1}{\partial F} \tag{A6.68}$$

$$\frac{\partial M_{21}}{\partial F} = \frac{G}{na^2}\left(\frac{\partial^2 X_1}{\partial F \partial k} - \frac{h\beta}{n}\frac{\partial \dot{X}_1}{\partial F}\right) \tag{A6.69}$$

$$\frac{\partial M_{22}}{\partial F} = \frac{G}{na^2}\left(\frac{\partial^2 Y_1}{\partial F \partial k} - \frac{h\beta}{n}\frac{\partial \dot{Y}_1}{\partial F}\right) \tag{A6.70}$$

$$\frac{\partial M_{23}}{\partial F} = \frac{k\left(q\frac{\partial Y_1}{\partial F} - p\frac{\partial X_1}{\partial F}\right)}{na^2 G} \tag{A6.71}$$

$$\frac{\partial M_{31}}{\partial F} = -\frac{G}{na^2}\left(\frac{\partial^2 X_1}{\partial F \partial h} + \frac{k\beta}{n}\frac{\partial \dot{X}_1}{\partial F}\right) \tag{A6.72}$$

$$\frac{\partial M_{32}}{\partial F} = -\frac{G}{na^2}\left(\frac{\partial^2 Y_1}{\partial F \partial h} + \frac{k\beta}{n}\frac{\partial \dot{Y}_1}{\partial F}\right) \tag{A6.73}$$

$$\frac{\partial M_{33}}{\partial F} = \frac{-h\left(q\frac{\partial Y_1}{\partial F} - p\frac{\partial X_1}{\partial F}\right)}{na^2 G} \tag{A6.74}$$

$$\frac{\partial M_{43}}{\partial F} = \frac{K}{2na^2 G}\frac{\partial Y_1}{\partial F} \tag{A6.75}$$

$$\frac{\partial M_{53}}{\partial F} = \frac{K}{2na^2 G}\frac{\partial X_1}{\partial F} \tag{A6.76}$$

$$\frac{\partial M_{61}}{\partial F} = \frac{1}{na^2}\left[-2\frac{\partial X_1}{\partial F} + G\left(h\beta\frac{\partial^2 X_1}{\partial F \partial h} + k\beta\frac{\partial^2 X_1}{\partial F \partial k}\right)\right] \tag{A6.77}$$

$$\frac{\partial M_{62}}{\partial F} = \frac{1}{na^2}\left[-2\frac{\partial Y_1}{\partial F} + G\left(h\beta\frac{\partial^2 Y_1}{\partial F \partial h} + k\beta\frac{\partial^2 Y_1}{\partial F \partial k}\right)\right] \tag{A6.78}$$

$$\frac{\partial M_{63}}{\partial F} = \frac{\left(q\frac{\partial Y_1}{\partial F} - p\frac{\partial X_1}{\partial F}\right)}{na^2 G} \tag{A6.79}$$

The auxiliary partials are

$$\frac{\partial X_1}{\partial F} = a[hk\beta c_F - (1 - h^2\beta)s_F] \tag{A6.80}$$

$$\frac{\partial Y_1}{\partial F} = a[-hk\beta s_F + (1 - k^2\beta)c_F] \tag{A6.81}$$

$$\frac{\partial \dot{X}_1}{\partial F} = -\frac{a}{r}(ks_F - hc_F)\dot{X}_1 + \frac{a^2 n}{r}\left[-hk\beta s_F - (1 - h^2\beta)c_F\right] \tag{A6.82}$$

$$\frac{\partial \dot{Y}_1}{\partial F} = -\frac{a}{r}(ks_F - hc_F)\dot{Y}_1 + \frac{a^2 n}{r}\left[-hk\beta c_F - (1 - k^2\beta)s_F\right] \tag{A6.83}$$

$$\frac{\partial^2 X_1}{\partial F \partial h} = a\left[(hs_F + kc_F)\left(\beta + \frac{h^2\beta^3}{1-\beta}\right) + \frac{a^2}{r^2}(h\beta - s_F)(s_F - h) + \frac{a}{r}c_F^2\right] \tag{A6.84}$$

$$\frac{\partial^2 X_1}{\partial F \partial k} = -a\left[-(hs_F + kc_F)\frac{hk\beta^3}{1-\beta} + \frac{a^2}{r^2}(s_F - h\beta)(c_F - k) + \frac{a}{r}s_F c_F\right] \tag{A6.85}$$

$$\frac{\partial^2 Y_1}{\partial F \partial h} = a\left[-(hs_F + kc_F)\frac{hk\beta^3}{1-\beta} - \frac{a^2}{r^2}(k\beta - c_F)(s_F - h) + \frac{a}{r}s_F c_F\right] \tag{A6.86}$$

$$\frac{\partial^2 Y_1}{\partial F \partial k} = a\left[-(hs_F + kc_F)\left(\beta + \frac{k^2\beta^3}{1-\beta}\right) + \frac{a^2}{r^2}(c_F - k\beta)(c_F - k) - \frac{a}{r}s_F^2\right] \tag{A6.87}$$

$$\frac{\partial}{\partial F}\left(\frac{na}{r}\right) = -n\frac{a^2}{r^2}(ks_F - hc_F) \tag{A6.88}$$

Finally,

$$\frac{\partial M_{61}}{\partial a} = \frac{-M_{61}}{2a} + \frac{1}{na^2}\left[-2\frac{\partial X_1}{\partial a} + G\left(h\beta\frac{\partial^2 X_1}{\partial a \partial h} + k\beta\frac{\partial^2 X_1}{\partial a \partial k}\right)\right] \tag{A6.89}$$

$$\frac{\partial M_{62}}{\partial a} = -\frac{M_{62}}{2a} + \frac{1}{na^2}\left[-2\frac{\partial Y_1}{\partial a} + G\left(h\beta\frac{\partial^2 Y_1}{\partial a \partial h} + k\beta\frac{\partial^2 Y_1}{\partial a \partial k}\right)\right] \tag{A6.90}$$

$$\frac{\partial M_{63}}{\partial a} = -\frac{M_{63}}{2a} + \frac{1}{na^2}\left[\left(q\frac{\partial Y_1}{\partial a} - p\frac{\partial X_1}{\partial a}\right)G^{-1}\right] \tag{A6.91}$$

$$\frac{\partial M_{61}}{\partial h} = \frac{1}{na^2}\left\{-2\left(\frac{\partial X_1}{\partial h}\right)_F - h\beta G^{-1}\left(h\frac{\partial X_1}{\partial h} + k\frac{\partial X_1}{\partial k}\right) + G\right.$$
$$\left.\times\left[\left(\beta + \frac{h^2\beta^3}{1-\beta}\right)\frac{\partial X_1}{\partial h} + \frac{hk\beta^3}{1-\beta}\frac{\partial X_1}{\partial k} + \beta\left(h\frac{\partial^2 X_1}{\partial h^2} + k\frac{\partial^2 X_1}{\partial h \partial k}\right)\right]\right\} \quad \text{(A6.92)}$$

$$\frac{\partial M_{62}}{\partial h} = \frac{1}{na^2}\left\{-2\left(\frac{\partial Y_1}{\partial h}\right)_F - h\beta G^{-1}\left(h\frac{\partial Y_1}{\partial h} + k\frac{\partial Y_1}{\partial k}\right) + G\right.$$
$$\left.\times\left[\left(\beta + \frac{h^2\beta^3}{1-\beta}\right)\frac{\partial Y_1}{\partial h} + \frac{hk\beta^3}{1-\beta}\frac{\partial Y_1}{\partial k} + \beta\left(h\frac{\partial^2 X_1}{\partial h^2} + k\frac{\partial^2 X_1}{\partial h \partial k}\right)\right]\right\} \quad \text{(A6.93)}$$

$$\frac{\partial M_{63}}{\partial h} = \frac{G^{-1}}{na^2}\left\{\left[q\left(\frac{\partial Y_1}{\partial h}\right)_F - p\left(\frac{\partial X_1}{\partial h}\right)_F\right] + hG^{-2}(qY_1 - pX_1)\right\} \quad \text{(A6.94)}$$

$$\frac{\partial M_{61}}{\partial k} = \frac{1}{na^2}\left\{-2\left(\frac{\partial X_1}{\partial k}\right)_F - k\beta G^{-1}\left(h\frac{\partial X_1}{\partial h} + k\frac{\partial X_1}{\partial k}\right) + G\right.$$
$$\left.\times\left[\left(\beta + \frac{k^2\beta^3}{1-\beta}\right)\frac{\partial X_1}{\partial k} + \frac{hk\beta^3}{1-\beta}\frac{\partial X_1}{\partial h} + \beta\left(h\frac{\partial^2 X_1}{\partial k \partial h} + k\frac{\partial^2 X_1}{\partial k^2}\right)\right]\right\} \quad \text{(A6.95)}$$

$$\frac{\partial M_{62}}{\partial k} = \frac{1}{na^2}\left\{-2\left(\frac{\partial Y_1}{\partial k}\right)_F - k\beta G^{-1}\left(h\frac{\partial Y_1}{\partial h} + k\frac{\partial Y_1}{\partial k}\right) + G\right.$$
$$\left.\times\left[\left(\beta + \frac{k^2\beta^3}{1-\beta}\right)\frac{\partial Y_1}{\partial k} + \frac{hk\beta^3}{1-\beta}\frac{\partial Y_1}{\partial h} + \beta\left(h\frac{\partial^2 Y_1}{\partial k \partial h} + k\frac{\partial^2 Y_1}{\partial k^2}\right)\right]\right\} \quad \text{(A6.96)}$$

$$\frac{\partial M_{63}}{\partial k} = \frac{G^{-1}}{na^2}\left\{\left[q\left(\frac{\partial Y_1}{\partial k}\right)_F - p\left(\frac{\partial X_1}{\partial k}\right)_F\right] + kG^{-2}(qY_1 - pX_1)\right\} \quad \text{(A6.97)}$$

$$\frac{\partial M_{63}}{\partial p} = \frac{-X_1}{na^2 G} \quad \text{(A6.98)}$$

$$\frac{\partial M_{63}}{\partial q} = \frac{Y_1}{na^2 G} \quad \text{(A6.99)}$$

References

[1] Edelbaum, T. N., Sackett, L. L. and Malchow, H. L. (1973) *Optimal Low Thrust Geocentric Transfer*, AIAA Paper 73–1074, AIAA 10th Electric Propulsion Conference, Lake Tahoe, NV.

[2] Sackett, L. L., Malchow, H. L. and Edelbaum, T. N. (1975) *Solar Electric Geocentric Transfer with Attitude Constraints: Analysis*, Draper Laboratory Report R901, Cambridge, MA.

[3] Sackett, L. L., Malchow, H. L. and Edelbaum, T. N. (1975) *Solar Electric Geocentric Transfer with Attitude Constraints: Program Manual*, Draper Laboratory Report R902, Cambridge, MA.

[4] Kechichian, J. A. (1990) *Equinoctial Orbit Elements: Application to Optimal Transfer Problems*, AIAA Paper 90–2976, AIAA/AAS Astrodynamics Conference, Portland, OR.

[5] Kechichian, J. A. (1991) *Trajectory Optimization with a Modified Set of Equinoctial Orbit Elements*, AAS Paper 91–524, AAS/AIAA Astrodynamics Specialist Conference, Durango, CO.

[6] Yee, T. and Kechichian, J. A. (1992) *On the Dynamic Modeling in Optimal Low-Thrust Orbit Transfer*, AAS Paper 92–177, AAS/AIAA Spaceflight Mechanics Meeting, Colorado Springs, CO.

[7] Kechichian, J. A. (1996) Optimal Low-Thrust Rendezvous Using Equinoctial Orbit Elements, *Acta Astronautica*, Vol. 38, No. 1, 1–14.

[8] Kechichian, J. A. (1993) *Minimum-Time Low-Thrust Rendezvous Using Epoch Mean Longitude Formulation*, AAS Paper 93–129, AAS/AIAA Spaceflight Mechanics Meeting, Pasadena, CA.

[9] Kechichian, J. A. (1993) *Minimum-Fuel Time-Fixed Rendezvous Using Constant Low-Thrust*, AAS Paper 93–130, AAS/AIAA Spaceflight Mechanics Meeting, Pasadena, CA.

[10] Kechichian, J. A. (1994) *Low-Thrust Trajectory Optimization Based on Epoch Eccentric Longitude Formulation*, AAS Paper 94–124, AAS/AIAA Spaceflight Mechanics Meeting, Cocoa Beach, FL.

[11] Kechichian, J. A. (1996) Optimal Low-Thrust Orbit Transfer, *Orbital Mechanics*, 2nd ed., edited by V. A. Chobotov, AIAA, Reston, VA, 327–402.

[12] Kechichian, J. A. (1997) Mechanics of Trajectory Optimization Using Nonsingular Variational Equations in Polar Coordinates, *Journal of Guidance, Control, and Dynamics*, Vol. 20, No. 4, 812–818.

[13] Kechichian, J. A. (1993) *Trajectory Optimization Using Eccentric Longitude Formulation*, AAS Paper 93–664, AAS/AIAA Astrodynamics Specialist Conference, Victoria, BC.

7 Low-Thrust Trajectory Optimization Based on Epoch Eccentric Longitude Formulation

7.1 Introduction

The mathematics of trajectory optimization based on the use of nonsingular orbit elements involving the eccentric longitude at epoch as the sixth element are fully derived in this chapter. The epoch eccentric longitude and epoch mean longitude are fundamental nonsingular orbit elements that stay constant in the absence of perturbations. These formulations constitute the basis from which the current time formulations are derived and therefore are also important from a theoretical point of view. They are also beneficial in minimum-fuel problems during the natural coasting parts of the trajectory where the adjoint equations need not be propagated by either analytical or numerical integration. The state and adjoint differential equations are explicit functions of time in this formulation that involves natural orbit elements, with the optimal Hamiltonian also varying in time. The mathematics of this epoch formulation provides added insight into the problem of trajectory optimization by relating the various assumptions used in generating the differential equations for the adjoint variables that correspond to various sets of orbital elements. Furthermore, the function that defines the transversality condition at the end time in minimum-time problems is shown to remain constant during the optimal transfer, providing a further check in accepting a converged trajectory as truly optimal. This formulation is also related to the one that uses the current eccentric longitude as the sixth state variable and the mathematical relationship between the Hamiltonian and the Lagrange multipliers of these two formulations is also shown. A pair of continuous constant acceleration minimum-time transfer examples are duplicated using this new formulation in order to further validate the mathematical derivations.

The use of the classical and modern optimization methods in solving low-thrust optimal transfer and rendezvous problems in Earth orbit [1–4], interplanetary flight [5–7], and even within the context of the restricted three-body problem [8, 9], has been adopted by many contributors with a preference for analytic methods. Scheel and Conway [2] solved the minimum-time problem by using a direct-transcription approach, while Kluever and Pierson [3] adopted a combined direct/indirect approach to solve minimum-fuel Earth–Moon low-thrust transfer trajectories involving coast arcs and using the restricted circular three-body model. In problems involving two active spacecraft, Coverstone-Carroll and Prussing [4] solved the cooperative power-limited linearized rendezvous problem analytically by further enforcing propellant constraints for a more realistic simulation. The purpose of this present formulation is not to solve a specific transfer problem or to compare the advantages and disadvantages of selecting the indirect method as opposed to the more commonly used direct method involving collocation and nonsingular programming techniques. Its purpose is rather to provide theoretical insight into the mathematical formulations of optimal transfer and to present the analysis and derivations that lead to the formulation of the optimal transfer problem using a set of fundamental nonsingular orbit elements in which the sixth element is an epoch element in the form of the epoch eccentric longitude. The epoch formulations are truly fundamental formulations and form the basis from which the current time formulations are readily derived. The state and adjoint variables, in addition, remain constant in the absence of any perturbation when epoch formulations are used. In an effort to mutually validate the various formulations using various sets of equinoctial elements and in particular the various sets of the adjoint differential equations that are derived with different assumptions, this chapter derives the epoch eccentric longitude formulation, also showing how it is related mathematically to the corresponding current time eccentric longitude formulation. Simple examples of transfer using continuous constant acceleration are used to compare the time histories of the Lagrange multipliers of both formulations. Furthermore, the constancy of the Hamiltonian in the current time formulation is replaced by the constancy of a certain function that defines the transversality condition for minimum-time problems in this epoch formulation. The Hamiltonian of this epoch formulation is not constant, but time varying, such that the numerical check, in accepting a converged solution, must rely now on the constancy of the above-mentioned transversality function. A minor drawback of the epoch formulations consist of the appearance of time, which is the independent variable, explicitly in the state and adjoint equations. This explicit appearance of time is possibly a nuisance only for applications involving planetary motions in which the state vector is integrated over a very long span. Furthermore, the epoch formulations cannot avoid the added computations needed to solve for the eccentric longitude from Kepler's equation through iteration. This avoidance is possible only by considering either the current time eccentric longitude or the current time true longitude as the sixth or fast variable and by expressing the right-hand side of each of the six state equations in terms of F or L respectively. Numerical comparisons within the context of averaging between epoch and current time formulations have not yet been carried out in order to determine whether some formulations are numerically more robust than others. Additional comparisons are necessary when solving minimum-fuel problems involving coast arcs and/or when considering additional perturbations such as due to J_2 or third-body gravity, as well as natural eclipsing during a given transfer. For Earth-orbiting applications, nonsingular elements are used in [10–14] instead of the classical elements because they do not exhibit any singularities for zero eccentricity and zero inclination. The full set of governing equations to solve precision integrated optimized transfer problems for the current mean longitude [12, 15], epoch mean longitude [13], current eccentric longitude [14] and true longitude [16] have been derived. The derivation of the equations of motion for the epoch eccentric longitude formulation is presented here by direct transformation of the Poisson brackets matrix that corresponds to the epoch mean longitude formulation. As stated above, this particular orbit element is considered a natural element, because it is perturbation-dependent only and therefore stays constant in the absence of any thrust perturbation, like the first five elements. Precision integration is used to solve the corresponding two point boundary value problem by the iterative shooting method and a quasi-Newton scheme. The mathematical derivations are validated by duplicating two examples of minimum-time transfer generated with other nonsingular formulations.

7.2 System and Adjoint Differential Equations for the Epoch Eccentric Longitude Formulation

The most straightforward way of deriving the equations of motion for the set (a, h, k, p, q, F_0) is by mapping the matrix of the Poisson brackets that corresponds to the set $(a, h, k, p, q, \lambda_0)$, which is itself mapped from the matrix of the fundamental classical set $(a, e, i, \Omega, \omega, M_0)$. Once the elements of the Poisson matrix are obtained, the partial derivatives of the elements with respect to the velocity vector are derived, after establishing the partial derivatives of the position vector with respect to the same elements. The former partials provide the required equations of motion after multiplication by the thrust acceleration vector itself. The differential equations of the adjoint variables are derived next from the Hamiltonian after establishing the functional dependence of r and F with respect to the element set at hand; the r and F variables appear in the right-hand sides of the differential equations for the state variables. Finally, a canonical transformation shows how the various multipliers and the Hamiltonian of the sets (a, h, k, p, q, F) and (a, h, k, p, q, F_0) are related. The plots of these multipliers and Hamiltonians are compared in the next section using a common transfer example, further verifying the corresponding mathematical relationships.

Given initial values of the classical orbit elements at time zero, namely, $a_0, e_0, i_0, \Omega_0, \omega_0, M_0$, the eccentric anomaly E_0 is first obtained from Kepler's equation by iteration:

$$M_0 = E_0 - e_0 \sin E_0 \tag{7.1}$$

such that the equinoctial elements can now be evaluated. The equations of motion as well as the adjoint system of differential equations using λ as the sixth state variable were shown earlier [12, 17, 18, and the preceding chapters] with $\dot{a} = \left(\frac{\partial a}{\partial \dot{\mathbf{r}}}\right)^T \bullet \hat{\mathbf{u}} f_t$, $\dot{h} = \left(\frac{\partial h}{\partial \dot{\mathbf{r}}}\right)^T \bullet \hat{\mathbf{u}} f_t$, $\dot{k} = \left(\frac{\partial k}{\partial \dot{\mathbf{r}}}\right)^T \bullet \hat{\mathbf{u}} f_t$, $\dot{p} = \left(\frac{\partial p}{\partial \dot{\mathbf{r}}}\right)^T \bullet \hat{\mathbf{u}} f_t$, $\dot{q} = \left(\frac{\partial q}{\partial \dot{\mathbf{r}}}\right)^T \bullet \hat{\mathbf{u}} f_t$, and $\dot{\lambda} = n + \left(\frac{\partial \lambda}{\partial \dot{\mathbf{r}}}\right)^T \bullet \hat{\mathbf{u}} f_t$ given in [17] and Chapter 3 of this book. The velocity partials or the partials of the equinoctial elements with respect to the velocity vector $\dot{\mathbf{r}}$ constitute the rows of the so-called 6×3 M matrix such that they are given directly in terms of the elements, the components of the velocity vector $\dot{\mathbf{r}} = \dot{X}_1 \hat{\mathbf{f}} + \dot{Y}_1 \hat{\mathbf{g}}$, namely $\dot{X}_1$ and $\dot{Y}_1$, and the components of the position vector $\mathbf{r} = X_1 \hat{\mathbf{f}} + Y_1 \hat{\mathbf{g}}$, and X_1 and Y_1 as well as their partial derivatives with respect to h and k.

The position and velocity vector components are written in terms of the equinoctial elements, as in [17] and Chapter 3 of this book, which also shows the partial derivatives $\partial X_1/\partial h$, $\partial X_1/\partial k$ and $\partial Y_1/\partial h$, $\partial Y_1/\partial k$ as a function of the elements, with $\beta = 1/(1+G)$ and $r = a(1 - kc_F - hs_F)$. These partials are derived by varying F and β with respect to h and k with $\partial\beta/\partial h = h\beta^3/(1-\beta)$, $\partial\beta/\partial k = k\beta^3/(1-\beta)$, and $\partial F/\partial h = -\frac{a}{r}c_F$, $\partial F/\partial k = \frac{a}{r}s_F$. The expressions for X_1 and Y_1 are obtained by direct transformation of the expressions given in terms of the classical elements $X_1 = r\cos(\theta^* + \omega + \Omega)$, $Y_1 = r\sin(\theta^* + \omega + \Omega)$, and using the identities $rc_{\theta^*} = a(c_E - e)$ and $rs_{\theta^*} = a(1-e^2)^{1/2}s_E$, and the definitions $h = es_{\omega+\Omega}$, $k = ec_{\omega+\Omega}$, $F = E + \omega + \Omega$, $e = \left(h^2+k^2\right)^{1/2}$. Letting $\beta = \left[1-(1-h^2-k^2)^{1/2}\right]\Big/\left(h^2+k^2\right) = \left[1+\left(1-h^2-k^2\right)^{1/2}\right]^{-1}$, and observing that $\left[k^2 + h^2\left(1-h^2-k^2\right)^{1/2}\right]\Big/\left(h^2+k^2\right) = 1 - h^2\beta$, the expression for X_1 is obtained. Similar manipulations lead to the expression for Y_1. Holding a, h, k and therefore β constant, and varying only F as a function of time, the expressions for the velocity components $\dot{X}_1$ and $\dot{Y}_1$ are obtained directly from X_1 and Y_1 after making use of $\dot{\lambda} = n = \dot{F} - kc_F\dot{F} - hs_F\dot{F}$, which yields $\dot{F} = na/r$. The epoch mean longitude λ_0 was used as the sixth state variable [13, and Chapter 5] instead of λ, and this required replacing the equation for the fast variable λ, namely $\dot{\lambda} = n + \left(\frac{\partial\lambda}{\partial\dot{\mathbf{r}}}\right)\hat{\mathbf{u}}f_t$ by

$$\dot{\lambda}_0 = \left(\frac{\partial\lambda_0}{\partial\dot{\mathbf{r}}}\right)^T \bullet \hat{\mathbf{u}}f_t \tag{7.2}$$

where

$$\begin{aligned}\frac{\partial\lambda_0}{\partial\dot{\mathbf{r}}} &= n^{-1}a^{-2}\left[-2X_1 + 3\dot{X}_1 t + G\left(h\beta\frac{\partial X_1}{\partial h} + k\beta\frac{\partial X_1}{\partial k}\right)\right]\hat{\mathbf{f}} \\ &+ n^{-1}a^{-2}\left[-2Y_1 + 3\dot{Y}_1 t + G\left(h\beta\frac{\partial Y_1}{\partial h} + k\beta\frac{\partial Y_1}{\partial k}\right)\right]\hat{\mathbf{g}} \\ &+ n^{-1}a^{-2}G^{-1}\left(qY_1 - pX_1\right)\hat{\mathbf{w}} = M_{61}^0\hat{\mathbf{f}} + M_{62}^0\hat{\mathbf{g}} + M_{63}^0\hat{\mathbf{w}}\end{aligned} \tag{7.3}$$

When Equation (7.2) is integrated numerically, we must evaluate λ from $\lambda = \lambda_0 + nt$, where n is osculating, and solve for F by iteration from Kepler's equation $\lambda = F - ks_F + hc_F$. This calculation is needed because F appears in the right-hand sides of the dynamic as well as adjoint equations, as will be shown. The need for this particular iteration was removed by using the eccentric longitude F as the sixth state variable [10, 14, and Chapter 6]. In this case, Equation (7.2), or the $\dot{\lambda}$ differential equation, is replaced by

$$\dot{F} = \frac{na}{r} + \left(\frac{\partial F}{\partial\dot{\mathbf{r}}}\right)^T \bullet \hat{\mathbf{u}}f_t \tag{7.4}$$

where $\partial F/\partial\dot{\mathbf{r}}$ is given by

$$\frac{\partial F}{\partial\dot{\mathbf{r}}} = \frac{a}{r}\left[\frac{\partial\lambda}{\partial\dot{\mathbf{r}}} + s_F\frac{\partial k}{\partial\dot{\mathbf{r}}} - c_F\frac{\partial h}{\partial\dot{\mathbf{r}}}\right] = M_{61}^F\hat{\mathbf{f}} + M_{62}^F\hat{\mathbf{g}} + M_{63}^F\hat{\mathbf{w}} \tag{7.5}$$

or in explicit form

$$\begin{aligned}\frac{\partial F}{\partial\dot{\mathbf{r}}} = \frac{1}{nar}\Bigg[&\left\{-2X_1 + G(h\beta - s_F)\frac{\partial X_1}{\partial h} + G(k\beta - c_F)\frac{\partial X_1}{\partial k} - \beta G(ks_F - hc_F)\frac{\dot{X}_1}{n}\right\}\hat{\mathbf{f}} \\ &+ \left\{-2Y_1 + G(h\beta - s_F)\frac{\partial Y_1}{\partial h} + G(k\beta - c_F)\frac{\partial Y_1}{\partial k} - \beta G\left(ks_F - hc_F\right)\frac{\dot{Y}_1}{n}\right\}\hat{\mathbf{g}} \\ &+ \frac{r}{aG}(qY_1 - pX_1)\hat{\mathbf{w}}\Bigg]\end{aligned} \tag{7.6}$$

The adjoint differential equations were developed for the above three formulations in [12–14] and Chapters 3, 5, and 6 of this book, where the sixth state variable is chosen respectively as the mean longitude, the epoch mean longitude and the eccentric longitude. All three formulations make use of the same second-order partial derivative of X_1 that appears in the description of the adjoint equations. This particular partial, shown in Equation (A7.48) of Appendix 7.1, was written erroneously in [12, 13, 17, 18] as

$$\frac{\partial^2 X_1}{\partial F \partial k} = -a\left[-(hs_F + kc_F)\frac{hk\beta^3}{1-\beta} + \frac{a^2}{r^2}(s_F - h\beta)(c_F - h) + \frac{a}{r}s_F c_F\right]$$

which had the unfortunate effect of introducing a small error in some of the examples shown there. In this chapter, we consider F_0 as the fast variable, or rather sixth state variable, such that the $\dot{\lambda}$ equation as well as Equations (7.2) and (7.4) are now replaced by

$$\dot{F}_0 = \left(\frac{\partial F_0}{\partial \dot{\mathbf{r}}}\right)^T \bullet \hat{\mathbf{u}} f_t \tag{7.7}$$

where

$$\begin{aligned}\frac{\partial F_0}{\partial \dot{\mathbf{r}}} &= n^{-1}a^{-2}\frac{a_0}{r_0}\left[-2X_1 + 3\dot{X}_1 t + G\left(h\beta\frac{\partial X_1}{\partial h} + k\beta\frac{\partial X_1}{\partial k}\right)\right]\hat{\mathbf{f}} \\ &+ n^{-1}a^{-2}\frac{a_0}{r_0}\left[-2Y_1 + 3\dot{Y}_1 t + G\left(h\beta\frac{\partial Y_1}{\partial h} + k\beta\frac{\partial Y_1}{\partial k}\right)\right]\hat{\mathbf{g}} \\ &+ n^{-1}a^{-2}G^{-1}\frac{a_0}{r_0}(qY_1 - pX_1)\,\hat{\mathbf{w}}\end{aligned} \tag{7.8}$$

This particular form was derived [11] by also showing how $\partial\lambda_0/\partial\dot{\mathbf{r}}$ and $\partial F_0/\partial\dot{\mathbf{r}}$ are related. From Kepler's equation evaluated at epoch, we have

$$\lambda_0 = F_0 - k_0 s_{F_0} + h_0 c_{F_0} \tag{7.9}$$

$$\frac{\partial \lambda_0}{\partial \dot{\mathbf{r}}} = \frac{\partial F_0}{\partial \dot{\mathbf{r}}}\left[1 - k_0 c_{F_0} - h_0 s_{F_0}\right] = \frac{r_0}{a_0}\frac{\partial F_0}{\partial \dot{\mathbf{r}}} \tag{7.10}$$

Equation (7.10) shows that

$$\frac{\partial F_0}{\partial \dot{\mathbf{r}}} = \frac{a_0}{r_0}\frac{\partial \lambda_0}{\partial \dot{\mathbf{r}}} = \frac{a_0}{r_0}M^0_{61}\hat{\mathbf{f}} + \frac{a_0}{r_0}M^0_{62}\hat{\mathbf{g}} + \frac{a_0}{r_0}M^0_{63}\hat{\mathbf{w}} \tag{7.11}$$

The Poisson brackets of the fundamental equinoctial elements set $(a, h, k, \lambda_0, p, q)$ are obtained by direct transformation of the brackets of the fundamental classical set $(a, e, i, M_0, \Omega, \omega)$. The transformed matrix is given by

$$[(a_\lambda, a_\mu)] = \frac{1}{na^2}\begin{bmatrix} 0 & 0 & 0 & -2a & 0 & 0 \\ & 0 & -G & \dfrac{hG}{F'} & \dfrac{-kpK}{2G} & \dfrac{-kqK}{2G} \\ & & 0 & \dfrac{kG}{F'} & \dfrac{hpK}{2G} & \dfrac{hqK}{2G} \\ & & & 0 & \dfrac{-pK}{2G} & \dfrac{-qK}{2G} \\ & & & & 0 & \dfrac{-K^2}{4G} \\ -Sym & & & & & 0 \end{bmatrix}$$

with $F' = 1 + G = 1/\beta$. The Poisson brackets of the set (a, h, k, F_0, p, q) are obtained by applying the transformation equation $\left[\left(p_\alpha, p_\beta\right)\right] = \left[\frac{\partial p_\alpha}{\partial a_\lambda}\right]\left[a_\lambda, a_\mu\right]\left[\frac{\partial p_\beta}{\partial a_\mu}\right]^T$ where the mapping matrix $\left[\frac{\partial p_\alpha}{\partial a_\lambda}\right]$ is given by

$$\left[\frac{\partial p_\alpha}{\partial a_\lambda}\right] = \begin{bmatrix} \frac{\partial a}{\partial a} & \frac{\partial a}{\partial h} & \frac{\partial a}{\partial k} & \frac{\partial a}{\partial \lambda_0} & \frac{\partial a}{\partial p} & \frac{\partial a}{\partial q} \\ \frac{\partial h}{\partial a} & \frac{\partial h}{\partial h} & \frac{\partial h}{\partial k} & \frac{\partial h}{\partial \lambda_0} & \frac{\partial h}{\partial p} & \frac{\partial h}{\partial q} \\ \frac{\partial k}{\partial a} & \frac{\partial k}{\partial h} & \frac{\partial k}{\partial k} & \frac{\partial k}{\partial \lambda_0} & \frac{\partial k}{\partial p} & \frac{\partial k}{\partial q} \\ \frac{\partial F_0}{\partial a} & \frac{\partial F_0}{\partial h} & \frac{\partial F_0}{\partial k} & \frac{\partial F_0}{\partial \lambda_0} & \frac{\partial F_0}{\partial p} & \frac{\partial F_0}{\partial q} \\ \frac{\partial p}{\partial a} & \frac{\partial p}{\partial h} & \frac{\partial p}{\partial k} & \frac{\partial p}{\partial \lambda_0} & \frac{\partial p}{\partial p} & \frac{\partial p}{\partial q} \\ \frac{\partial q}{\partial a} & \frac{\partial q}{\partial h} & \frac{\partial q}{\partial k} & \frac{\partial q}{\partial \lambda_0} & \frac{\partial q}{\partial p} & \frac{\partial q}{\partial q} \end{bmatrix}$$

From Kepler's equation at epoch, $\lambda_0 = F_0 - k_0 s_{F_0} + h_0 c_{F_0}$, $\frac{\partial \lambda_0}{\partial \lambda_0} = 1 = \frac{\partial F_0}{\partial \lambda_0}\left(1 - k_0 c_{F_0} - h_0 s_{F_0}\right)$ such that $\frac{\partial F_0}{\partial \lambda_0} = \frac{a_0}{r_0}$. Because F_0 is now an orbital element and is therefore independent of the elements a h, k, p and q, its partials with respect to these five elements are equal to zero, and the above matrix is then purely diagonal:

$$\left[\frac{\partial p_\alpha}{\partial a_\lambda}\right] = \begin{bmatrix} 1 & 0 & 0 & 0 & 0 & 0 \\ & 1 & 0 & 0 & 0 & 0 \\ & & 1 & 0 & 0 & 0 \\ & & & \frac{a_0}{r_0} & 0 & 0 \\ & & & & 1 & 0 \\ & & 0 & & & 1 \end{bmatrix}$$

Because $\left[\frac{\partial p_\beta}{\partial a_\mu}\right]^T = \left[\frac{\partial p_\alpha}{\partial a_\lambda}\right]^T$, the matrix of the Poisson brackets of the set (a, h, k, F_0, p, q) reduces to

$$\left[\left(p_\alpha, p_\beta\right)\right] = \frac{1}{na^2} \begin{bmatrix} 0 & 0 & 0 & -2a\frac{a_0}{r_0} & 0 & 0 \\ & 0 & -G & Gh\beta\frac{a_0}{r_0} & \frac{-kpK}{2G} & \frac{-kqK}{2G} \\ & & 0 & Gk\beta\frac{a_0}{r_0} & \frac{hpK}{2G} & \frac{hqK}{2G} \\ & & & 0 & \frac{-pK}{2G}\frac{a_0}{r_0} & \frac{-qK}{2G}\frac{a_0}{r_0} \\ & & & & 0 & \frac{-K^2}{4G} \\ -Sym & & & & & 0 \end{bmatrix}$$

The partials of the elements of the set (a, h, k, F_0, p, q) with respect to the velocity vector $\dot{\mathbf{r}}$ can now be written as

$$\frac{\partial a}{\partial \dot{\mathbf{r}}} = -(a, F_0)\frac{\partial \mathbf{r}}{\partial F_0}$$
$$\frac{\partial h}{\partial \dot{\mathbf{r}}} = -(h,k)\frac{\partial \mathbf{r}}{\partial k} - (h,F_0)\frac{\partial \mathbf{r}}{\partial F_0} - (h,p)\frac{\partial \mathbf{r}}{\partial p} - (h,q)\frac{\partial \mathbf{r}}{\partial q}$$
$$\frac{\partial k}{\partial \dot{\mathbf{r}}} = -(k,h)\frac{\partial \mathbf{r}}{\partial h} - (k,F_0)\frac{\partial \mathbf{r}}{\partial F_0} - (k,p)\frac{\partial \mathbf{r}}{\partial p} - (k,q)\frac{\partial \mathbf{r}}{\partial q}$$
$$\frac{\partial F_0}{\partial \dot{\mathbf{r}}} = -(F_0,a)\frac{\partial \mathbf{r}}{\partial a} - (F_0,h)\frac{\partial \mathbf{r}}{\partial h} - (F_0,k)\frac{\partial \mathbf{r}}{\partial k} - (F_0,p)\frac{\partial \mathbf{r}}{\partial p} - (F_0,q)\frac{\partial \mathbf{r}}{\partial q}$$
$$\frac{\partial p}{\partial \dot{\mathbf{r}}} = -(p,h)\frac{\partial \mathbf{r}}{\partial h} - (p,k)\frac{\partial \mathbf{r}}{\partial k} - (p,F_0)\frac{\partial \mathbf{r}}{\partial F_0} - (p,q)\frac{\partial \mathbf{r}}{\partial q}$$
$$\frac{\partial q}{\partial \dot{\mathbf{r}}} = -(q,h)\frac{\partial \mathbf{r}}{\partial h} - (q,k)\frac{\partial \mathbf{r}}{\partial k} - (q,F_0)\frac{\partial \mathbf{r}}{\partial F_0} - (q,p)\frac{\partial \mathbf{r}}{\partial p}$$

The partials of $\mathbf{r}$ with respect to the elements are obtained by allowing for the variation of F with respect to the same elements, with $\partial F/\partial a = -\frac{3}{2}\frac{n}{r}t$, $\partial F/\partial h = -\frac{a}{r}c_F$, $\partial F/\partial k = \frac{a}{r}s_F$. From $\lambda = F - ks_F + hc_F$, partial $\partial\lambda/\partial F = r/a$ such that from $\partial\lambda/\partial a = (\partial\lambda/\partial F)(\partial F/\partial a) = \frac{\partial n}{\partial a}t = -\frac{3}{2}\frac{n}{a}t$, it follows that $\partial\lambda/\partial a = -\frac{3}{2}\frac{n}{r}t$. In a similar way, from $\partial\lambda/\partial h = 0 = \partial F/\partial h - kc_F\partial F/\partial h + c_F - hs_F\partial F/\partial h$, it follows that $\partial F/\partial h = -\frac{a}{r}c_F$, and in an equivalent way, from $\partial\lambda/\partial k = 0$, it follows that $\partial F/\partial k = \frac{a}{r}s_F$. Finally, from $\partial\mathbf{r}/\partial\lambda = (\partial\mathbf{r}/\partial F)(\partial F/\partial\lambda) = \left(\frac{\partial X_1}{\partial F}\hat{f} + \frac{\partial Y_1}{\partial F}\hat{g}\right)\frac{\partial F}{\partial\lambda}$ and because from $\partial\lambda/\partial\lambda = 1 = \frac{\partial F}{\partial\lambda}(1 - kc_F - hs_F) = \frac{\partial F}{\partial\lambda}\frac{r}{a}$, we have $\partial F/\partial\lambda = a/r$, the partial $\partial\mathbf{r}/\partial\lambda$ reduces to $\frac{\partial\mathbf{r}}{\partial\lambda} = \frac{\dot{X}_1}{n}\hat{\mathbf{f}} + \frac{\dot{Y}_1}{n}\hat{\mathbf{g}} = \frac{\dot{\mathbf{r}}}{n}$. This expression is now used in $\frac{\partial\mathbf{r}}{\partial F_0} = \frac{\partial\mathbf{r}}{\partial\lambda_0}\frac{\partial\lambda_0}{\partial F_0} = \frac{\partial\mathbf{r}}{\partial\lambda}\frac{\partial\lambda}{\partial\lambda_0}\frac{\partial\lambda_0}{\partial F_0}\frac{\partial\mathbf{r}}{\partial F_0} = \frac{\partial\mathbf{r}}{\partial\lambda_0}\frac{\partial\lambda_0}{\partial F_0} = \frac{\partial\mathbf{r}}{\partial\lambda}\frac{\partial\lambda}{\partial\lambda_0}\frac{\partial\lambda_0}{\partial F_0}$ because $\partial\lambda/\partial\lambda_0 = 1$ and from $\lambda_0 = F_0 - k_0 s_{F_0} + h_0 c_{F_0}$, $\frac{\partial\lambda_0}{\partial F_0} = \frac{r_0}{a_0}$. Therefore

$$\frac{\partial\mathbf{r}}{\partial a} = \frac{\partial X_1}{\partial a}\hat{\mathbf{f}} + \frac{\partial Y_1}{\partial a}\hat{\mathbf{g}} = \left(\frac{X_1}{a} - \frac{3}{2}\frac{t}{a}\dot{X}_1\right)\hat{\mathbf{f}} + \left(\frac{Y_1}{a} - \frac{3}{2}\frac{t}{a}\dot{Y}_1\right)\hat{\mathbf{g}}$$
$$\frac{\partial\mathbf{r}}{\partial h} = \frac{\partial X_1}{\partial h}\hat{\mathbf{f}} + \frac{\partial Y_1}{\partial h}\hat{\mathbf{g}}$$
$$\frac{\partial\mathbf{r}}{\partial k} = \frac{\partial X_1}{\partial k}\hat{\mathbf{f}} + \frac{\partial Y_1}{\partial k}\hat{\mathbf{g}}$$
$$\frac{\partial\mathbf{r}}{\partial F_0} = \frac{r_0}{a_0}\left(\frac{\dot{X}_1}{n}\hat{\mathbf{f}} + \frac{\dot{Y}_1}{n}\hat{\mathbf{g}}\right)$$

where the partials $\partial X_1/\partial h$, $\partial X_1/\partial k$, $\partial Y_1/\partial h$, $\partial Y_1/\partial k$ are given in [17] and Chapter 3 of this book as stated earlier. The transformation from the inertial $(\hat{\mathbf{x}}, \hat{\mathbf{y}}, \hat{\mathbf{z}})$ frame to the equinoctial
$(\hat{\mathbf{f}}, \hat{\mathbf{g}}, \hat{\mathbf{w}})$ frame is given by

$$\begin{pmatrix}\hat{\mathbf{f}}\\ \hat{\mathbf{g}}\\ \hat{\mathbf{w}}\end{pmatrix} = \begin{pmatrix} c_\Omega^2 + c_i s_\Omega^2 & s_\Omega c_\Omega - s_\Omega c_\Omega c_i & -s_\Omega s_i \\ s_\Omega c_\Omega - s_\Omega c_\Omega c_i & s_\Omega^2 + c_i c_\Omega^2 & s_i c_\Omega \\ s_\Omega s_i & -c_\Omega s_i & c_i \end{pmatrix}\begin{pmatrix}\hat{\mathbf{x}}\\ \hat{\mathbf{y}}\\ \hat{\mathbf{z}}\end{pmatrix}$$
$$= \frac{1}{K}\begin{pmatrix} 1 - p^2 + q^2 & 2pq & -2p \\ 2pq & 1 + p^2 - q^2 & 2q \\ 2p & -2q & 1 - p^2 - q^2 \end{pmatrix}\begin{pmatrix}\hat{\mathbf{x}}\\ \hat{\mathbf{y}}\\ \hat{\mathbf{z}}\end{pmatrix}$$

The partials of the unit vectors $\hat{\mathbf{f}}$ and $\hat{\mathbf{g}}$ with respect to p and q are readily obtained to yield

$$\frac{\partial \mathbf{r}}{\partial p} = X_1 \frac{\partial \hat{\mathbf{f}}}{\partial p} + Y_1 \frac{\partial \hat{\mathbf{g}}}{\partial p} = \frac{2}{K}\left[q\left(Y_1\hat{\mathbf{f}} - X_1\hat{\mathbf{g}}\right) - X_1\hat{\mathbf{w}}\right]$$

$$\frac{\partial \mathbf{r}}{\partial q} = X_1 \frac{\partial \hat{\mathbf{f}}}{\partial q} + Y_1 \frac{\partial \hat{\mathbf{g}}}{\partial q} = \frac{2}{K}\left[p\left(X_1\hat{\mathbf{g}} - Y_1\hat{\mathbf{f}}\right) + Y_1\hat{\mathbf{w}}\right]$$

The partials of the elements a, h, k, p, q and F_0 with respect to the velocity vector take the form given in Equations (7) through (11) of [17] and Equation (7.8) given here. Because we already have shown an expression for $\partial\lambda_0/\partial\dot{\mathbf{r}}$, we shall keep $\partial F_0/\partial\dot{\mathbf{r}}$, as given in Equation (7.11) here, in order to make use of most of the partial derivatives of $\partial\lambda_0/\partial\dot{\mathbf{r}}$ with respect to the elements developed in [13] and Chapter 5 of this book. This will save us some additional coding effort. Now, once E_0 has been computed from Equation (7.1), we get $F_0 = E_0 + \omega_0 + \Omega_0$, $h_0 = e_0 \sin(\omega_0 + \Omega_0)$ and $k_0 = e_0 \cos(\omega_0 + \Omega_0)$. The values of h_0, k_0 and therefore $e_0 = \left(h_0^2 + k_0^2\right)^{1/2}$ remain constant. However, F_0 is being integrated and therefore E_0 will vary according to $E_0 = F_0 - \tan^{-1}(h_0/k_0)$. Now λ_0 is evaluated from $\lambda_0 = E_0 - e_0 \sin E_0 + \tan^{-1}(h_0/k_0)$ because, in the definition of $\lambda = M + \omega + \Omega$, the equation $M = E - e \sin E$ is used as well as the inverse transformation relations for ω and Ω, namely $\omega = \tan^{-1}(h/k) - \tan^{-1}(p/q)$ and $\Omega = \tan^{-1}(p/q)$. The current mean longitude is now evaluated from $\lambda = \lambda_0 + nt$, where n is also osculating, and F is solved for through iteration from Kepler's equation $\lambda = F - ks_F + hc_F$. Although a_0 remains constant, $r_0 = a_0\left(1 - k_0 c_{F_0} - h_0 s_{F_0}\right)$ is a function of F_0 such that the factor a_0/r_0 in Equation (7.11) is a function of only one element, namely F_0. The following partial is then needed to evaluate $\dot{\lambda}_{F_0}$, the differential equation for the adjoint to F_0:

$$\frac{\partial}{\partial F_0}\left(\frac{a_0}{r_0}\right) = \left(\frac{a_0}{r_0}\right)^2 \left(h_0 c_{F_0} - k_0 s_{F_0}\right) \tag{7.12}$$

The Hamiltonian for the F_0 formulation is given by

$$\begin{aligned} H = \Big[& \lambda_a \left(M_{11}u_f + M_{12}u_g + M_{13}u_w\right) + \lambda_h \left(M_{21}u_f + M_{22}u_g + M_{23}u_w\right) \\ & + \lambda_k \left(M_{31}u_f + M_{32}u_g + M_{33}u_w\right) + \lambda_p \left(M_{41}u_f + M_{42}u_g + M_{43}u_w\right) \\ & + \lambda_q \left(M_{51}u_f + M_{52}u_g + M_{53}u_w\right) \\ & + \lambda_{F_0}\left(\frac{a_0}{r_0}M_{61}^0 u_f + \frac{a_0}{r_0}M_{62}^0 u_g + \frac{a_0}{r_0}M_{63}^0 u_w\right)\Big] f_t \end{aligned} \tag{7.13}$$

or, in compact form, by

$$H = \boldsymbol{\lambda}_z^T \dot{\mathbf{z}} = \boldsymbol{\lambda}_z^T M^{F_0}(\mathbf{z}, F) f_t \hat{\mathbf{u}} \tag{7.14}$$

where $z = (a\ h\ k\ p\ q\ F_0)^T$ represents the state vector and where the 6×3 matrix M^{F_0} is now given by

$$M^{F_0} = \begin{pmatrix} \left(\frac{\partial a}{\partial \dot{\mathbf{r}}}\right)^T \\ \left(\frac{\partial h}{\partial \dot{\mathbf{r}}}\right)^T \\ \left(\frac{\partial k}{\partial \dot{\mathbf{r}}}\right)^T \\ \left(\frac{\partial p}{\partial \dot{\mathbf{r}}}\right)^T \\ \left(\frac{\partial q}{\partial \dot{\mathbf{r}}}\right)^T \\ \left(\frac{\partial F_0}{\partial \dot{\mathbf{r}}}\right)^T \end{pmatrix} = \begin{pmatrix} M_{11} & M_{12} & M_{13} \\ M_{21} & M_{22} & M_{23} \\ M_{31} & M_{32} & M_{33} \\ M_{41} & M_{42} & M_{43} \\ M_{51} & M_{52} & M_{53} \\ \frac{a_0}{r_0} M_{61}^0 & \frac{a_0}{r_0} M_{62}^0 & \frac{a_0}{r_0} M_{63}^0 \end{pmatrix} \tag{7.15}$$

The Euler–Lagrange or adjoint differential equations are written as

$$\dot{\boldsymbol{\lambda}}_z = -\frac{\partial H}{\partial \mathbf{z}} = -\boldsymbol{\lambda}_z^T \frac{\partial M^{F_0}}{\partial \mathbf{z}} f_t \hat{\mathbf{u}} \tag{7.16}$$

Considering continuous constant acceleration, the thrust direction $\hat{\mathbf{u}} = (u_f, u_g, u_w)$ is optimized by selecting $\hat{\mathbf{u}}$ parallel to the vector $\boldsymbol{\lambda}_z^T M^{F_0}(\mathbf{z}, F)$. The adjoint equations are given in expanded form by

$$\begin{aligned} \dot{\lambda}_a = & -(\lambda_a\, \lambda_h\, \lambda_k\, \lambda_p\, \lambda_q\, \lambda_{F_0}) \frac{\partial M^{F_0}}{\partial a} f_t \hat{\mathbf{u}} = -\frac{\partial H}{\partial a} = \left[-\lambda_a \left(\frac{\partial M_{1i}}{\partial a} u_i \right) \right. \\ & - \lambda_h \left(\frac{\partial M_{2i}}{\partial a} u_i \right) - \lambda_k \left(\frac{\partial M_{3i}}{\partial a} u_i \right) \\ & - \lambda_p \left(\frac{\partial M_{4i}}{\partial a} u_i \right) - \lambda_q \left(\frac{\partial M_{5i}}{\partial a} u_i \right) \\ & \left. - \lambda_{F_0} \left(\frac{a_0}{r_0} \frac{\partial M_{61}^0}{\partial a} u_f + \frac{a_0}{r_0} \frac{\partial M_{62}^0}{\partial a} u_g + \frac{a_0}{r_0} \frac{\partial M_{63}^0}{\partial a} u_w \right) \right] f_t \end{aligned} \tag{7.17}$$

$$\begin{aligned} \dot{\lambda}_h = -\frac{\partial H}{\partial h} = & \left[-\lambda_a \left(\frac{\partial M_{1i}}{\partial h} u_i \right) - \lambda_h \left(\frac{\partial M_{2i}}{\partial h} u_i \right) \right. \\ & - \lambda_k \left(\frac{\partial M_{3i}}{\partial h} u_i \right) - \lambda_p \left(\frac{\partial M_{4i}}{\partial h} u_i \right) \\ & \left. - \lambda_q \left(\frac{\partial M_{5i}}{\partial h} u_i \right) - \lambda_{F_0} \left(\frac{a_0}{r_0} \frac{\partial M_{61}^0}{\partial h} u_f + \frac{a_0}{r_0} \frac{\partial M_{62}^0}{\partial h} u_g + \frac{a_0}{r_0} \frac{\partial M_{63}^0}{\partial h} u_w \right) \right] f_t \end{aligned} \tag{7.18}$$

$$\begin{aligned} \dot{\lambda}_k = -\frac{\partial H}{\partial k} = & \left[-\lambda_a \left(\frac{\partial M_{1i}}{\partial k} u_i \right) - \lambda_h \left(\frac{\partial M_{2i}}{\partial k} u_i \right) \right. \\ & - \lambda_k \left(\frac{\partial M_{3i}}{\partial k} u_i \right) - \lambda_p \left(\frac{\partial M_{4i}}{\partial k} u_i \right) \\ & \left. - \lambda_q \left(\frac{\partial M_{5i}}{\partial k} u_i \right) - \lambda_{F_0} \left(\frac{a_0}{r_0} \frac{\partial M_{61}^0}{\partial k} u_f + \frac{a_0}{r_0} \frac{\partial M_{62}^0}{\partial k} u_g + \frac{a_0}{r_0} \frac{\partial M_{63}^0}{\partial k} u_w \right) \right] f_t \end{aligned} \tag{7.19}$$

$$\dot{\lambda}_p = -\frac{\partial H}{\partial p} = \left[-\lambda_a \left(\frac{\partial M_{1i}}{\partial p} u_i \right) - \lambda_h \left(\frac{\partial M_{2i}}{\partial p} u_i \right) - \lambda_k \left(\frac{\partial M_{3i}}{\partial p} u_i \right) - \lambda_p \left(\frac{\partial M_{4i}}{\partial p} u_i \right) \right.$$
$$\left. -\lambda_q \left(\frac{\partial M_{5i}}{\partial p} u_i \right) - \lambda_{F_0} \left(\frac{a_0}{r_0} \frac{\partial M^0_{61}}{\partial p} u_f + \frac{a_0}{r_0} \frac{\partial M^0_{62}}{\partial p} u_g + \frac{a_0}{r_0} \frac{\partial M^0_{63}}{\partial p} u_w \right) \right] f_t \tag{7.20}$$

$$\dot{\lambda}_q = -\frac{\partial H}{\partial q} = \left[-\lambda_a \left(\frac{\partial M_{1i}}{\partial q} u_i \right) - \lambda_h \left(\frac{\partial M_{2i}}{\partial q} u_i \right) \right.$$
$$- \lambda_k \left(\frac{\partial M_{3i}}{\partial q} u_i \right) - \lambda_p \left(\frac{\partial M_{4i}}{\partial q} u_i \right)$$
$$\left. -\lambda_q \left(\frac{\partial M_{5i}}{\partial q} u_i \right) - \lambda_{F_0} \left(\frac{a_0}{r_0} \frac{\partial M^0_{61}}{\partial q} u_f + \frac{a_0}{r_0} \frac{\partial M^0_{62}}{\partial q} u_g + \frac{a_0}{r_0} \frac{\partial M^0_{63}}{\partial q} u_w \right) \right] f_t \tag{7.21}$$

$$\dot{\lambda}_{F_0} = -\frac{\partial H}{\partial F_0} = \left[-\lambda_a \left(\frac{\partial M_{1i}}{\partial F_0} u_i \right) - \lambda_h \left(\frac{\partial M_{2i}}{\partial F_0} u_i \right) \right.$$
$$- \lambda_k \left(\frac{\partial M_{3i}}{\partial F_0} u_i \right) - \lambda_p \left(\frac{\partial M_{4i}}{\partial F_0} u_i \right) - \lambda_q \left(\frac{\partial M_{5i}}{\partial F_0} u_i \right)$$
$$\left. -\lambda_{F_0} \left(\frac{\partial \left(\frac{a_0}{r_0} M^0_{61} \right)}{\partial F_0} u_f + \frac{\partial \left(\frac{a_0}{r_0} M^0_{62} \right)}{\partial F_0} u_g + \frac{\partial \left(\frac{a_0}{r_0} M^0_{63} \right)}{\partial F_0} u_w \right) \right] f_t \tag{7.22}$$

The last three partials in Equation (7.22) can be written as

$$\begin{aligned} \frac{\partial}{\partial F_0} \left(\frac{a_0}{r_0} M^0_{61} \right) &= \frac{a_0}{r_0} \frac{\partial M^0_{61}}{\partial F_0} + M^0_{61} \frac{\partial}{\partial F_0} \left(\frac{a_0}{r_0} \right) \\ &= \frac{a_0}{r_0} \frac{r_0}{a_0} \frac{\partial M^0_{61}}{\partial \lambda_0} + M^0_{61} \frac{\partial}{\partial F_0} \left(\frac{a_0}{r_0} \right) \\ &= \frac{\partial M^0_{61}}{\partial \lambda_0} + \left(\frac{a_0}{r_0} \right)^2 \left(h_0 c_{F_0} - k_0 s_{F_0} \right) M^0_{61} \end{aligned} \tag{7.23}$$

$$\frac{\partial}{\partial F_0} \left(\frac{a_0}{r_0} M^0_{62} \right) = \frac{\partial M^0_{62}}{\partial \lambda_0} + \left(\frac{a_0}{r_0} \right)^2 \left(h_0 c_{F_0} - k_0 s_{F_0} \right) M^0_{62} \tag{7.24}$$

$$\frac{\partial}{\partial F_0} \left(\frac{a_0}{r_0} M^0_{63} \right) = \frac{\partial M^0_{63}}{\partial \lambda_0} + \left(\frac{a_0}{r_0} \right)^2 \left(h_0 c_{F_0} - k_0 s_{F_0} \right) M^0_{63} \tag{7.25}$$

where use is made of Equation (7.12) and of the following identities

$$\frac{\partial M_{ij}}{\partial F_0} = \frac{\partial M_{ij}}{\partial \lambda_0} \frac{\partial \lambda_0}{\partial F_0} = \frac{r_0}{a_0} \frac{\partial M_{ij}}{\partial \lambda_0} \qquad i = 1,6;\, j = 1,3 \tag{7.26}$$

because from $\lambda_0 = F_0 - k_0 s_{F_0} + h_0 c_{F_0}$ we have $\partial \lambda_0 / \partial F_0 = 1 - k_0 c_{F_0} - h_0 s_{F_0} = r_0/a_0$. It is also true that $\partial F_0 / \partial \lambda_0 = a_0/r_0$. Now the following partial derivatives must be used when the derivations of the $\partial M^{F_0} / \partial \mathbf{z}$ partial derivatives shown in Appendix 7.1 are carried out, namely:

$$\frac{\partial r}{\partial a} = \frac{r}{a} - \frac{3}{2} \frac{nat}{r} \left(k s_F - h c_F \right) \tag{7.27}$$

$$\frac{\partial r}{\partial h} = \frac{a^2}{r} \left(h - s_F \right) \tag{7.28}$$

$$\frac{\partial r}{\partial k} = \frac{a^2}{r}(k - c_F) \tag{7.29}$$

$$\frac{\partial r}{\partial F} = a(ks_F - hc_F) \tag{7.30}$$

$$\frac{\partial F}{\partial a} = -\frac{3}{2}\frac{nt}{r} \tag{7.31}$$

$$\frac{\partial F}{\partial h} = -\frac{a}{r}c_F \tag{7.32}$$

$$\frac{\partial F}{\partial k} = \frac{a}{r}s_F \tag{7.33}$$

$$\frac{\partial F}{\partial F_0} = \frac{r_0}{a_0}\frac{a}{r} = \frac{r_0}{a_0}\frac{\partial F}{\partial \lambda_0} \tag{7.34}$$

These expressions can be shown to be true from $r = a(1 - kc_F - hs_F)$ by first writing the differential

$$dr = (1 - kc_F - hs_F)\,da - ac_F dk - as_F dh + a(ks_F - hc_F)\,dF \tag{7.35}$$

From $\lambda = \lambda_0 + nt = F - ks_F + hc_F$, it follows that

$$d\lambda_0 + ndt + tdn = (1 - kc_F - hs_F)\,dF - s_F dk + c_F dh$$

and because $dn = -(3/2)n/a\,da$,

$$dF = \left[d\lambda_0 + ndt - \frac{3}{2}\frac{nt}{a}da + s_F dk - c_F dh\right]\frac{a}{r} \tag{7.36}$$

However, from Kepler's equation at epoch with h_0 and k_0 fixed, $\lambda_0 = F_0 - k_0 s_{F_0} + h_0 c_{F_0}$,

$$d\lambda_0 = \left(1 - k_0 c_{F_0} - h_0 s_{F_0}\right)dF_0 = \frac{r_0}{a_0}dF_0 \tag{7.37}$$

which, when replaced in Equation (7.36), yields

$$dF = \left[\frac{r_0}{a_0}dF_0 + ndt - \frac{3}{2}\frac{nt}{a}da + s_F dk - c_F dh\right]\frac{a}{r} \tag{7.38}$$

The partials in Equations (7.31) through (7.34) are now readily obtained from Equation (7.38). If we now use dF of Equation (7.38) in Equation (7.35) for dr, the partials in Equations (7.27) through (7.30) are also obtained. The last partial in Equation (7.34) also can be obtained from

$$\lambda = \lambda_0 + nt = F - ks_F + hc_F = F_0 - k_0 s_{F_0} + h_0 c_{F_0} + nt$$

such that

$$\frac{\partial F}{\partial F_0} - kc_F\frac{\partial F}{\partial F_0} - hs_F\frac{\partial F}{\partial F_0} = \frac{\partial F_0}{\partial F_0} - k_0 c_{F_0} - h_0 s_{F_0}$$

or

$$(1 - kc_F - hs_F)\frac{\partial F}{\partial F_0} = 1 - k_0 c_{F_0} - h_0 s_{F_0} = \frac{r_0}{a_0}$$

and finally $\frac{\partial F}{\partial F_0} = \frac{r_0}{a_0}\frac{a}{r}$. If we now use the canonicity condition for the transformation between $(a\ h\ k\ p\ q\ F)$ and $(a\ h\ k\ p\ q\ F_0)$, we can show how the various multipliers used in the F formulation [14, and Chapter 6] relate to the multipliers used in the F_0 formulation. Using F and F_0 superscripts, the canonicity condition is given by

$$\lambda_a^F da + \lambda_h^F dh + \lambda_k^F dk + \lambda_p^F dp + \lambda_q^F dq + \lambda_F dF - H^F dt$$
$$= \lambda_a^{F_0} da + \lambda_h^{F_0} dh + \lambda_k^{F_0} dk + \lambda_p^{F_0} dp + \lambda_q^{F_0} dq + \lambda_{F_0} dF_0 - H^{F_0} dt \qquad (7.39)$$

Replacing dF from Equation (7.38) in the above condition and identifying identical terms yields

$$\lambda_a^{F_0} = \lambda_a^F - \frac{3}{2}\frac{nt}{a}\frac{a}{r}\lambda_F \qquad (7.40)$$
$$\lambda_h^{F_0} = \lambda_h^F - \frac{a}{r} c_F \lambda_F \qquad (7.41)$$
$$\lambda_k^{F_0} = \lambda_k^F + \frac{a}{r} s_F \lambda_F \qquad (7.42)$$
$$\lambda_p^{F_0} = \lambda_p^F \qquad (7.43)$$
$$\lambda_q^{F_0} = \lambda_q^F \qquad (7.44)$$
$$\lambda_{F_0} = \frac{a}{r}\frac{r_0}{a_0}\lambda_F \qquad (7.45)$$
$$H^{F_0} = H^F - \frac{na}{r}\lambda_F \qquad (7.46)$$

7.3 Transversality Condition for Minimum-Time Rendezvous and Examples of Free–Free Minimum-Time Transfer

This section shows how the transversality condition at the final unknown time for minimum-time problems can effectively replace the Hamiltonian constancy check corresponding to the current longitude formulations in accepting a converged transfer trajectory as being truly optimal. The function that defines the transversality condition is thus shown to remain constant in this new formulation, even though the Hamiltonian is no longer constant. The optimality check is thus still possible by verifying the constancy of this function instead.

As in [13] and [14] and Chapters 5 and 6 of this book, rendezvous time is minimized by maximizing

$$J = -\int_{t_0}^{t_f} dt = -\left(t_f - t_0\right) \qquad (7.47)$$

The initial values of the six Lagrange multipliers $(\lambda_a)_0$, $(\lambda_h)_0$, $(\lambda_k)_0$, $\left(\lambda_p\right)_0$, $\left(\lambda_q\right)_0$, as well as the transfer time t_f, are guessed; starting from initial state variables a_0, h_0, k_0, p_0, q_0, $(F_0)_0$, the state and adjoint equations are integrated forward until $t = t_f$, using the optimal control

$$\hat{\mathbf{u}} = \frac{\left(\boldsymbol{\lambda}_z^T M^{F_0}\right)^T}{\left|\boldsymbol{\lambda}_z^T M^{F_0}\right|} \qquad (7.48)$$

A quasi-Newton scheme is used to iterate on the initial values of the multipliers as well as the transfer time, such that the desired terminal state given by a_f, h_f, k_f, p_f, q_f, F_f is reached and the transversality condition at the unknown final time is satisfied:

Table 7.1 Transfer parameters

Orbit	Initial	Target	Achieved	Solution	Initial values
a, km	7000	42,000	42,000.003	$(\lambda_a)_0$, s/km	4.675229762
e	0	10^{-3}	9.986×10^{-4}	$(\lambda_h)_0$, s	541.3413947
i, deg	28.5	1	0.999809	$(\lambda_k)_0$, s	−9202.702084
Ω, deg	0	0	1.155×10^{-4}	$(\lambda_p)_0$, s	17.78011878
ω, deg	0	0	1.776×10^{-2}	$(\lambda_q)_0$, s	−22,584.55855
M_0, M, deg	$M_0 =$ −130.3331648 (optimized)	Free	$M = 46.147008$	$(\lambda_{F_0})_0$, s/rad	0

$$\left(\frac{\partial \Phi}{\partial t} + \boldsymbol{\lambda}_z^T \dot{\mathbf{z}}\right)_{t_f} = 1 \tag{7.49}$$

with

$$\Phi = \nu\psi = \nu\left[(F_0)_f - k_0 s_{(F_0)_f} + h_0 c_{(F_0)_f} + n_f t_f - \lambda_f\right]$$

because we want to match $\lambda_f = (\lambda_0)_f + n_f t_f$, which is a function of the state variables a and F_0. The constant Lagrange multiplier ν is obtained from the additional condition $\left(\lambda_{F_0}\right)_f = \left(\nu \frac{\partial \psi}{\partial F_0}\right)_{t_f}$, which yields $\left(\lambda_{F_0}\right)_f = \nu\left(1 - k_0 c_{(F_0)_f} - h_0 s_{(F_0)_f}\right) = \frac{\nu (r_0)_f}{a_0}$, such that

$$\nu = \frac{a_0}{(r_0)_f}\left(\lambda_{F_0}\right)_f \tag{7.50}$$

Because $H = \boldsymbol{\lambda}_z^T \dot{\mathbf{z}}$, the transversality condition in Equation (7.49) yields $\nu \frac{\partial \psi}{\partial t} + H_f = 1$, or

$$T_r = \frac{a_0 n_f}{(r_0)_f}\left(\lambda_{F_0}\right)_f + H_f = 1 \tag{7.51}$$

From Equations (7.45) and (7.46), the above condition yields $(H^F)_f = 1$, the transversality condition of the F formulation. We can now duplicate the minimum-time transfer example of [15] and [16] and Chapter 5 of this book by generating a trajectory from the given initial orbit shown in Table 7.1. The initial values of the Lagrange multipliers are identical to the ones corresponding to the eccentric longitude formulation, as can be seen in Equations (7.40) through (7.45), because $(\lambda_F)_0 = 0$ for an optimized initial location. This equivalence holds true also between the formulations using the eccentric and mean longitudes respectively [14] and Chapter 6 of this book, such that the iterated values of the multipliers shown in [15] and [16] and Chapter 6 of this book and in Table 7.1 can be used to start the numerical integration. The constant acceleration $f_t = 9.8 \times 10^{-5}$ km/s^2 is applied from $t_0 = 0$ to $t_f = 58{,}089.9005$ s to yield the final conditions shown in Table 7.1, with $H_f = 1.003704684$, closely matching the desired target. Note that $\lambda_f = M_f + \omega_f + \Omega_f$ yields 19.655285 rad as desired.

The initial values of the multipliers can be scaled to provide $H_f = 1$ exactly, yielding $(\lambda_a)_0 = 4.657973438$ s/km, $(\lambda_h)_0 = 5.393432977 \times 10^2$ s, $(\lambda_k)_0 = -9.168734810 \times 10^3$ s, $(\lambda_p)_0 = 1.771449217 \times 10^1$ s, $(\lambda_q)_0 = -2.250119870 \times 10^4$ s and $(\lambda_{F0})_0 =$

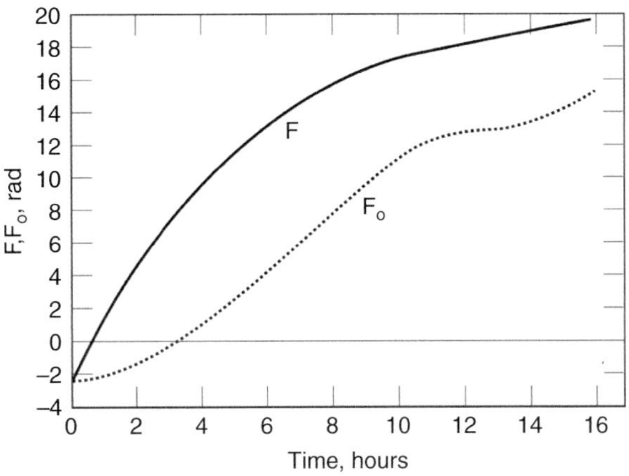

Figure 7.1 Eccentric longitude and epoch eccentric longitude time histories during optimal transfer.

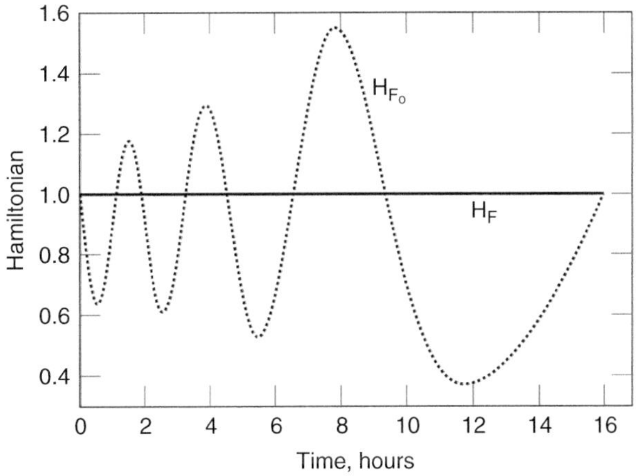

Figure 7.2 Evolution of Hamiltonians H^F and H^{F_0} during optimal transfer.

0 s/rad with $H_f = 0.999999998$. It can also be observed that $\frac{a_0}{r_0} n \left(\lambda_{F_0}\right) + H$ remains constant throughout the transfer, thereby providing a numerical check for accepting a converged trajectory as a truly optimal solution because now H is not constant during the integration. Figure 7.1 shows the evolution of F and F_0 in time whereas the Hamiltonians H^F of the F formulation and H^{F_0} of the present F_0 formulation are depicted in Figure 7.2, with $H^F = 1$ throughout and H^{F_0} varying and reaching the value of 1 at the end time. The multipliers' histories are shown in Figures 7.3–7.7 for both the F and F_0 formulations, starting from the same initial values and also ending with equal values in compliance with Equations (7.40) through (7.45) due to $(\lambda_F)_f = \left(\lambda_{F_0}\right)_f = 0$ as verified in Figure 7.7. The trajectory runs yield $(\lambda_F)_f = 8.478 \times 10^{-3}$ s/rad and $\left(\lambda_{F_0}\right)_f = 1.631 \times 10^{-5}$ s/rad, both close to the theoretical value of 0. Furthermore,

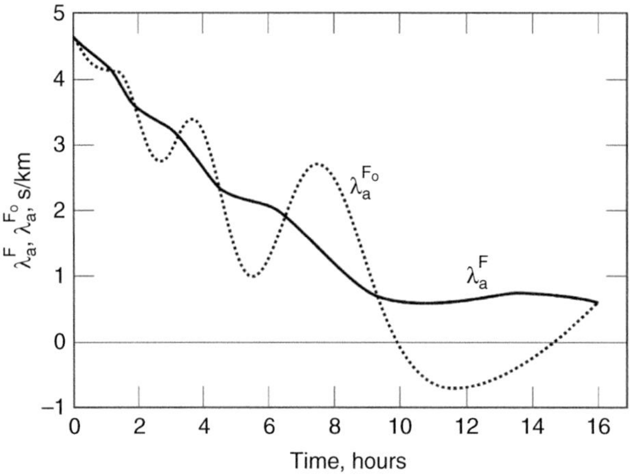

Figure 7.3 Evolution of λ_a^F and $\lambda_a^{F_0}$ multipliers during optimal transfer.

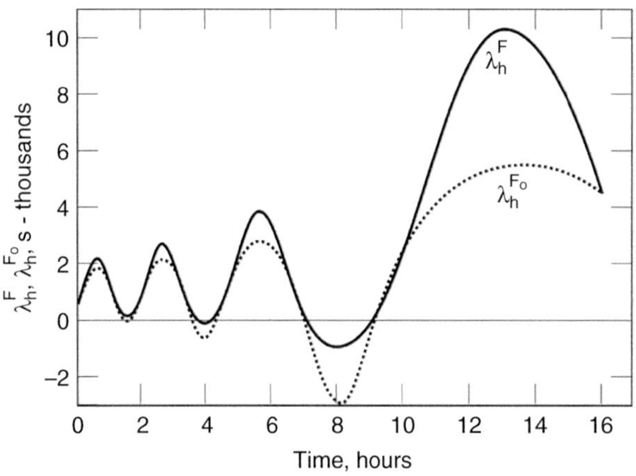

Figure 7.4 Evolution of λ_h^F and $\lambda_h^{F_0}$ multipliers during optimal transfer.

$H^F = 1.000000015$ throughout the transfer, and the quantity $\frac{a_0}{r_0} n\lambda_{F_0} + H = 1.00000000$ remains constant with the relative and absolute integration error controls set to 10^{-9}. The correspondence between the F_0 and the epoch mean longitude formulation λ_0 of [13] and Chapter 5 of this book can be shown to be $\lambda_a^{F_0} = \lambda_a^{\lambda_0}$, $\lambda_h^{F_0} = \lambda_h^{\lambda_0}$, $\lambda_k^{F_0} = \lambda_k^{\lambda_0}$, $\lambda_p^{F_0} = \lambda_p^{\lambda_0}$, $\lambda_q^{F_0} = \lambda_q^{\lambda_0}$, $\lambda_{F_0}^{F_0} = \frac{r_0}{a_0} \lambda_{\lambda_0}^{\lambda_0}$ and $H^{F_0} = H^{\lambda_0}$. For our initial circular orbit example, $h_0 = k_0 = 0$ such that r_0/a_0 remains constant and equal to 1 and, therefore, $\lambda_{F_0}^{F_0} = \lambda_{\lambda_0}^{\lambda_0}$ throughout the transfer. A trajectory is generated using the same initial values for the scaled multipliers as the earlier ones, and starting from the same optimized initial location with $M_0 = -130.3331648$ deg using the software of [13] and Chapter 5 of this book. The evolutions of the various multipliers are shown in Figures 7.8 through 7.11, with a perfect match between these λ_0 and F_0 epoch formulations. As

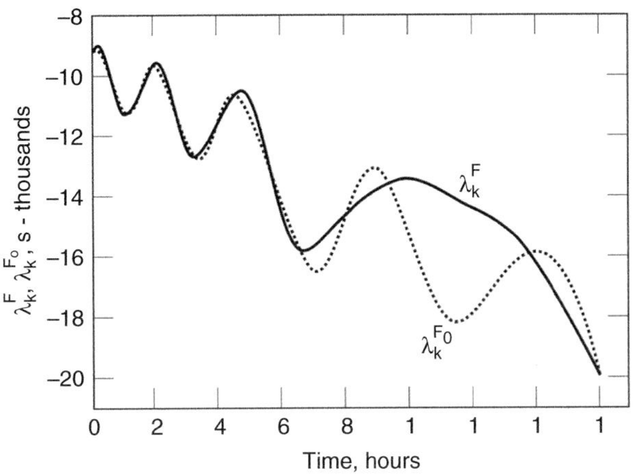

Figure 7.5 Evolution of λ_k^F and $\lambda_k^{F_0}$ multipliers during optimal transfer.

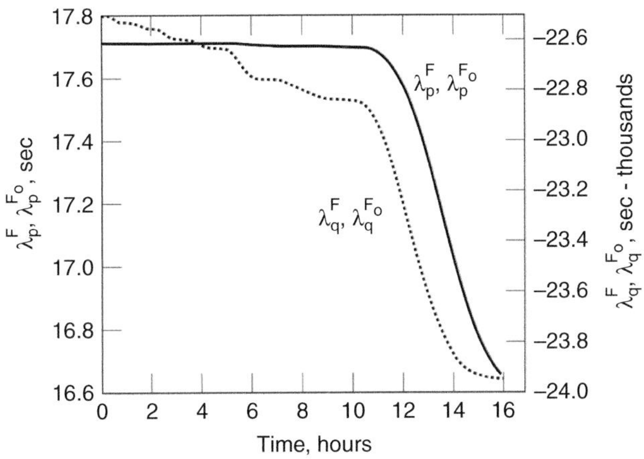

Figure 7.6 Evolution of λ_p^F, $\lambda_p^{F_0}$ and λ_q^F, $\lambda_q^{F_0}$ multipliers during optimal transfer.

shown in Figure 7.11, both Hamiltonians vary identically and both λ_0 and F_0 state variables exhibit the same variations. This can also be seen from Kepler's equation at epoch $\lambda_0 = F_0 - k_0 s_{F_0} + h_0 c_{F_0}$ with $h_0 = k_0 = 0$. It can be seen that the differential equations for the multipliers, Equations (7.17) through (7.22), become identical to the corresponding equations of the λ_0 formulation as soon as $e_0 = 0$. Then $r_0/a_0 = 1$ and with $h_0 = k_0 = 0$, Equations (7.23) through (7.25) yield $\frac{\partial}{\partial F_0}\left(\frac{a_0}{r_0} M_{6i}^0\right) = \frac{\partial M_{6i}^0}{\partial \lambda_0}$, such that using the expression $\lambda_{F_0}^{F_0} = \frac{r_0}{a_0} \lambda_{\lambda_0}^{\lambda_0}$ relating $\lambda_{F_0}^{F_0}$ and $\lambda_{\lambda_0}^{\lambda_0}$, all six pairs of the adjoint differential equations of the F_0 and λ_0 formulations become identical. For this reason, and to further validate the mathematical derivations, we now solve the same minimum-time transfer example except that the initial value of the eccentricity is nonzero, namely $e_0 = 0.05$, such that $h_0 = 0$ and $k_0 = 0.05$, and use the software of [12] and Chapter

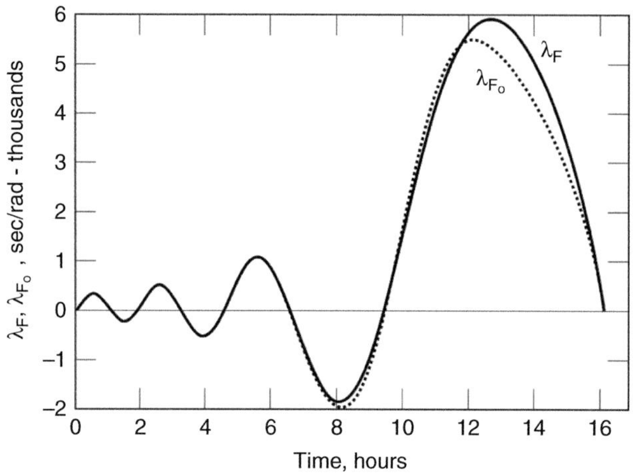

Figure 7.7 Evolution of λ_F and λ_{F_0} multipliers during optimal transfer.

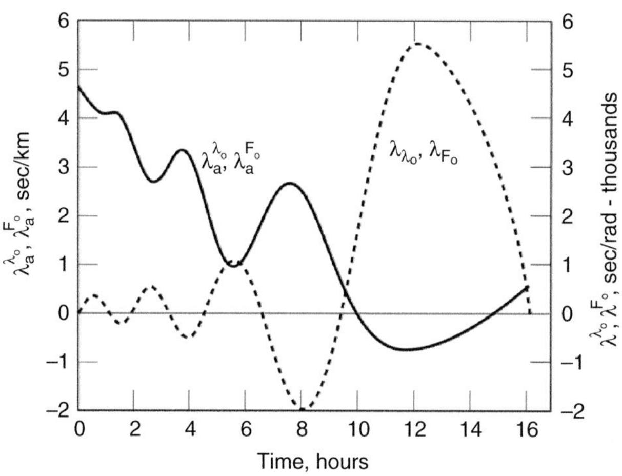

Figure 7.8 Evolution of λ_{F_0}, λ_{λ_0}, $\lambda_a^{F_0}$ and $\lambda_a^{\lambda_0}$ multipliers during optimal transfer.

3 of this book to determine the initial values of the multipliers, the optimized departure mean anomaly, and the transfer time, as $(\lambda_a)_0 = 4.719028797$ s/km, $(\lambda_h)_0 = 5.490984314 \times 10^2$ s, $(\lambda_k)_0 = -9.651612930 \times 10^3$ s, $(\lambda_p)_0 = -9.379552824 \times 10^1$ s, $(\lambda_q)_0 = -2.270946531 \times 10^4$ s, $M_0 = -130.5130449$ deg, and $t_f = 58{,}560.2470$ s. The two-point boundary value problem is solved by enforcing the transversality condition $H_f = 1$, which also stays constant in this λ formulation to within nine decimal places during the integration, indicating thereby a converged trajectory.

This solution is run using both the present F_0 formulation and the epoch mean longitude λ_0 formulation to verify the satisfaction of the target conditions which are closely matched with $a_f = 42{,}000.012$ km, $e_f = 1.00035 \times 10^{-3}$, $i_f = 1.000015$ deg, $\Omega_f = 359.999777$ deg, $\omega_f = 2.3283 \times 10^{-2}$ deg, $M_f = 46.091379$ deg and $(\lambda_{F_0})_f = -9.1906 \times 10^{-3}$ s/rad using the F_0 formulation, and with $a_f = 41{,}999.999$ km,

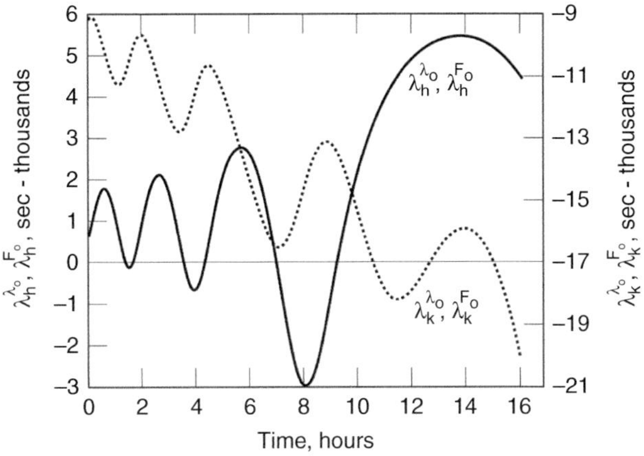

Figure 7.9 Evolution of $\lambda_h^{F_0}$, $\lambda_h^{\lambda_0}$, $\lambda_k^{F_0}$ and $\lambda_k^{\lambda_0}$ multipliers during optimal transfer.

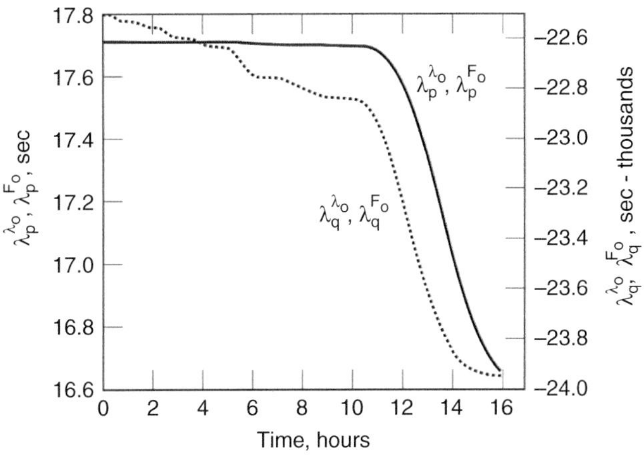

Figure 7.10 Evolution of $\lambda_p^{F_0}$, $\lambda_p^{\lambda_0}$, $\lambda_q^{F_0}$ and $\lambda_q^{\lambda_0}$ multipliers during optimal transfer.

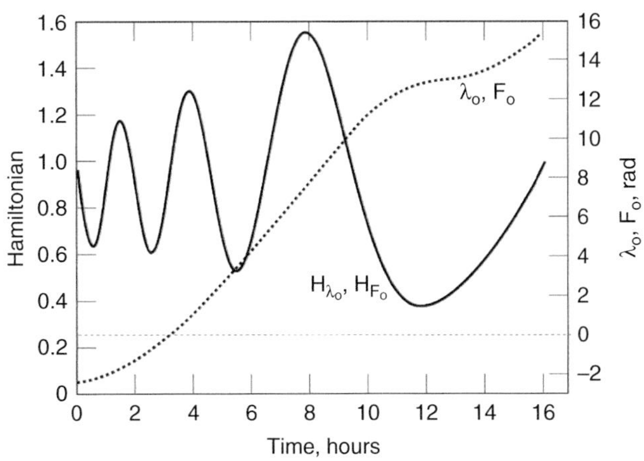

Figure 7.11 Evolution of λ_0, F_0 and Hamiltonians H^{λ_0} and H^{F_0} during optimal transfer.

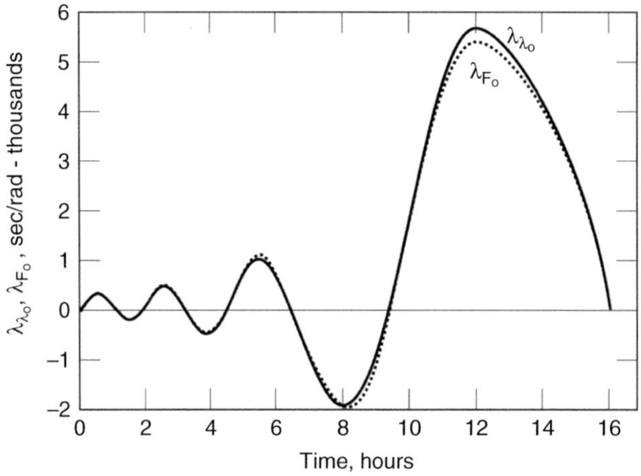

Figure 7.12 Evolution of λ_{F_0} and λ_{λ_0} multipliers for transfer with initial eccentricity $e_0 = 0.05$.

$e_f = 1.000031$ deg, $i_f = 1.000002$ deg, $\Omega_f = 8.72 \times 10^{-6}$ deg, $\omega_f = 359.999704$ deg, $M_f = 46.114676$ deg and $\left(\lambda_{\lambda_0}\right)_f = -1.683 \times 10^{-5}$ s/rad using the λ_0 formulation. Figure 7.12 shows how the λ_{F_0} and λ_{λ_0} profiles are no longer identical because e_0 is not equal to zero.

7.4 Conclusion

The mechanics of trajectory optimization based on the use of the epoch eccentric longitude has been presented as a fundamental formulation from which other formulations, based on various element sets, are mathematically derived. Besides the theoretical insight that this formulation provides into the derivations of the various adjoint differential equations and their underlying assumptions, the analysis further reveals that the adjoint equations need not be integrated during the coasting phases in minimum-fuel problems. Also, the function that defines the transversality condition at the end time in minimum-time problems can effectively be used as a numerical check in accepting a converged trajectory as a truly optimal trajectory because it stays constant during the whole transfer. Further, comparisons between the averaged solutions of this formulation and the current time formulations developed earlier are necessary to evaluate the robustness of the convergence process as well as the computational effort in generating optimized transfers for a variety of transfer geometries.

Appendix 7.1 Partial Derivatives of M^{F_0} Matrix with Respect to Orbit Elements

A7.1.1 The Nonzero Partial Derivatives of M^{F_0} with Respect to h

The partial derivatives $\partial M_{11}/\partial h$, $\partial M_{12}/\partial h$, $\partial M_{21}/\partial h$, $\partial M_{22}/\partial h$, $\partial M_{23}/\partial h$, $\partial M_{31}/\partial h$, $\partial M_{32}/\partial h$, $\partial M_{33}/\partial h$, $\partial M_{43}/\partial h$ and $\partial M_{53}/\partial h$ are identical to Equations (A1), (A2) and (A4) through (A11) of [17] and of Chapter 2 of this book, whereas

$$\frac{\partial M_{61}^0}{\partial h} = n^{-1}a^{-2}\left\{-2\frac{\partial X_1}{\partial h} + 3\frac{\partial \dot{X}_1}{\partial h}t - h\beta G^{-1}\left(h\frac{\partial X_1}{\partial h} + k\frac{\partial X_1}{\partial k}\right)\right.$$
$$\left. + G\left[\left(\beta + \frac{h^2\beta^3}{1-\beta}\right)\frac{\partial X_1}{\partial h} + \frac{hk\beta^3}{1-\beta}\frac{\partial X_1}{\partial k} + \beta\left(h\frac{\partial^2 X_1}{\partial h^2} + k\frac{\partial^2 X_1}{\partial h \partial k}\right)\right]\right\} \quad \text{(A7.1)}$$

$$\frac{\partial M_{62}^0}{\partial h} = n^{-1}a^{-2}\left\{-2\frac{\partial Y_1}{\partial h} + 3\frac{\partial \dot{Y}_1}{\partial h}t - h\beta G^{-1}\left(h\frac{\partial Y_1}{\partial h} + k\frac{\partial Y_1}{\partial k}\right)\right.$$
$$\left. + G\left[\left(\beta + \frac{h^2\beta^3}{1-\beta}\right)\frac{\partial Y_1}{\partial h} + \frac{hk\beta^3}{1-\beta}\frac{\partial Y_1}{\partial k} + \beta\left(h\frac{\partial^2 Y_1}{\partial h^2} + k\frac{\partial^2 Y_1}{\partial h \partial k}\right)\right]\right\} \quad \text{(A7.2)}$$

$$\frac{\partial M_{63}^0}{\partial h} = \frac{G^{-1}}{na^2}\left[\left(q\frac{\partial Y_1}{\partial h} - p\frac{\partial X_1}{\partial h}\right) + hG^{-2}\left(qY_1 - pX_1\right)\right] \quad \text{(A7.3)}$$

A7.1.2 The Nonzero Partial Derivatives of M^{F_0} with Respect to k

The partial derivatives $\partial M_{11}/\partial k$, $\partial M_{12}/\partial k$, $\partial M_{21}/\partial k$, $\partial M_{22}/\partial k$, $\partial M_{23}/\partial k$, $\partial M_{31}/\partial k$, $\partial M_{32}/\partial k$, $\partial M_{33}/\partial k$, $\partial M_{43}/\partial k$ and $\partial M_{53}/\partial k$ are identical to Equations (A15), (A16) and (A18) through (A25) of [17] and of Chapter 2 of this book, whereas,

$$\frac{\partial M_{61}^0}{\partial k} = n^{-1}a^{-2}\left\{-2\frac{\partial X_1}{\partial k} + 3\frac{\partial \dot{X}_1}{\partial k}t - h\beta G^{-1}\left(h\frac{\partial X_1}{\partial h} + k\frac{\partial X_1}{\partial k}\right)\right.$$
$$\left. + G\left[\left(\beta + \frac{k^2\beta^3}{1-\beta}\right)\frac{\partial X_1}{\partial k} + \frac{hk\beta^3}{1-\beta}\frac{\partial X_1}{\partial h} + \beta\left(h\frac{\partial^2 X_1}{\partial k \partial h} + k\frac{\partial^2 X_1}{\partial k^2}\right)\right]\right\} \quad \text{(A7.4)}$$

$$\frac{\partial M_{62}^0}{\partial k} = n^{-1}a^{-2}\left\{-2\frac{\partial Y_1}{\partial k} + 3\frac{\partial \dot{Y}_1}{\partial k}t - k\beta G^{-1}\left(h\frac{\partial Y_1}{\partial h} + k\frac{\partial Y_1}{\partial k}\right)\right.$$
$$\left. + G\left[\left(\beta + \frac{k^2\beta^3}{1-\beta}\right)\frac{\partial Y_1}{\partial k} + \frac{hk\beta^3}{1-\beta}\frac{\partial Y_1}{\partial h} + \beta\left(h\frac{\partial^2 Y_1}{\partial k \partial h} + k\frac{\partial^2 Y_1}{\partial k^2}\right)\right]\right\} \quad \text{(A7.5)}$$

$$\frac{\partial M_{63}^0}{\partial k} = \frac{G^{-1}}{na^2}\left[\left(q\frac{\partial Y_1}{\partial k} - p\frac{\partial X_1}{\partial k}\right) + kG^{-2}\left(qY_1 - pX_1\right)\right] \quad \text{(A7.6)}$$

A7.1.3 The Nonzero Partial Derivatives of M^{F_0} with Respect to p

The partial derivatives, $\partial M_{23}/\partial p$, $\partial M_{33}/\partial p$, $\partial M_{43}/\partial p$ and $\partial M_{53}/\partial p$ are identical to Equations (A29) through (A32) of [17] and of Chapter 2 of this book, whereas

$$\frac{\partial M_{63}^0}{\partial p} = \frac{-X_1}{na^2 G} \quad \text{(A7.7)}$$

A7.1.4 The Nonzero Partial Derivatives of M^{F_0} with Respect to q

The partial derivatives $\partial M_{23}/\partial q$, $\partial M_{33}/\partial q$, $\partial M_{43}/\partial q$ and $\partial M_{53}/\partial q$ are identical to Equations (A34) through (A37) of [17] and of Chapter 2 of this book, whereas

$$\frac{\partial M_{63}^0}{\partial q} = \frac{Y_1}{na^2 G} \quad \text{(A7.8)}$$

The partial derivatives of $\dot{X}_1$ with respect to h and k, namely $\partial\dot{X}_1/\partial h$ and $\partial\dot{X}_1/\partial k$, are identical to Equations (A39) and (A40) of [17] and of Chapter 2 of this book, whereas the partials $\partial\dot{Y}_1/\partial h$ and $\partial\dot{Y}_1/\partial k$ are identical to Equations (A41) and (A42) of the same and are not repeated here. The second partials of X_1 and Y_1 with respect to h and k, namely $\partial^2 X_1/\partial h^2$, $\partial^2 X_1/\partial k^2$, $\partial^2 X_1/\partial h\partial k$ and $\partial^2 X_1/\partial k\partial h$, and $\partial^2 Y_1/\partial h^2$, $\partial^2 Y_1/\partial k^2$, $\partial^2 Y_1/\partial h\partial k$ and $\partial^2 Y_1/\partial k\partial h$, are identical to Equations (A43) through (A46) and Equations (A47) through (A50), respectively, of [17] and also in Chapter 2 of this book. It can be shown that $\frac{\partial^2 X_1}{\partial h\partial k} = \frac{\partial^2 X_1}{\partial k\partial h}$ and $\frac{\partial^2 Y_1}{\partial h\partial k} = \frac{\partial^2 Y_1}{\partial k\partial h}$. Next, the accessory partials $\partial^2 X_1/\partial a\partial k$, $\partial^2 X_1/\partial a\partial h$, $\partial^2 Y_1/\partial a\partial k$ and $\partial^2 Y_1/\partial a\partial h$ are

$$\frac{\partial^2 X_1}{\partial a\partial k} = \frac{1}{a}\frac{\partial X_1}{\partial k} - \frac{3}{2}\frac{nat}{r}\left[(hs_F + kc_F)\frac{hk\beta^3}{(1-\beta)} + \frac{a^2}{r^2}(k - c_F)(s_F - h\beta) - \frac{a}{r}s_F c_F\right] \tag{A7.9}$$

$$\frac{\partial^2 X_1}{\partial a\partial h} = \frac{1}{a}\frac{\partial X_1}{\partial h} - \frac{3}{2}\frac{nat}{r}\left\{(hs_F + kc_F)\left[\beta + \frac{h^2\beta^3}{(1-\beta)}\right] - \frac{a^2}{r^2}(h - s_F)(h\beta - s_F) + \frac{a}{r}c_F^2\right\} \tag{A7.10}$$

and also

$$\frac{\partial^2 Y_1}{\partial a\partial k} = \frac{1}{a}\frac{\partial Y_1}{\partial k} + \frac{3}{2}\frac{nat}{r}\left\{(hs_F + kc_F)\left[\beta + \frac{k^2\beta^3}{(1-\beta)}\right] + \frac{a^2}{r^2}(k - c_F)(c_F - k\beta) + \frac{a}{r}s_F^2\right\} \tag{A7.11}$$

$$\frac{\partial^2 Y_1}{\partial a\partial h} = \frac{1}{a}\frac{\partial Y_1}{\partial h} + \frac{3}{2}\frac{nat}{r}\left[(hs_F + kc_F)\frac{hk\beta^3}{(1-\beta)} - \frac{a^2}{r^2}(h - s_F)(k\beta - c_F) - \frac{a}{r}s_F c_F\right] \tag{A7.12}$$

A7.1.5 The Nonzero Partial Derivatives of M^{F_0} with Respect to a

$$\frac{\partial M_{11}}{\partial a} = \frac{4}{n^2a^2}\dot{X}_1 + \frac{2}{n^2a}\frac{\partial\dot{X}_1}{\partial a} \tag{A7.13}$$

$$\frac{\partial M_{12}}{\partial a} = \frac{4}{n^2a^2}\dot{Y}_1 + \frac{2}{n^2a}\frac{\partial\dot{Y}_1}{\partial a} \tag{A7.14}$$

$$\frac{\partial M_{21}}{\partial a} = \frac{G}{na^2}\left[-\frac{1}{2a}\frac{\partial X_1}{\partial k} + \frac{\partial^2 X_1}{\partial a\partial k} - \frac{h\beta}{na}\dot{X}_1 - \frac{h\beta}{n}\frac{\partial\dot{X}_1}{\partial a}\right] \tag{A7.15}$$

$$\frac{\partial M_{22}}{\partial a} = \frac{G}{na^2}\left[-\frac{1}{2a}\frac{\partial Y_1}{\partial k} + \frac{\partial^2 Y_1}{\partial a\partial k} - \frac{h\beta}{na}\dot{Y}_1 - \frac{h\beta}{n}\frac{\partial\dot{Y}_1}{\partial a}\right] \tag{A7.16}$$

$$\frac{\partial M_{23}}{\partial a} = \frac{k}{na^2G}\left[-\frac{1}{2a}(qY_1 - pX_1) + q\frac{\partial Y_1}{\partial a} - p\frac{\partial X_1}{\partial a}\right] \tag{A7.17}$$

$$\frac{\partial M_{31}}{\partial a} = -\frac{G}{na^2}\left[-\frac{1}{2a}\frac{\partial X_1}{\partial h} + \frac{\partial^2 X_1}{\partial a\partial h} + \frac{k\beta}{na}\dot{X}_1 + \frac{k\beta}{n}\frac{\partial\dot{X}_1}{\partial a}\right] \tag{A7.18}$$

$$\frac{\partial M_{32}}{\partial a} = -\frac{G}{na^2}\left[-\frac{1}{2a}\frac{\partial Y_1}{\partial h} + \frac{\partial^2 Y_1}{\partial a \partial h} + \frac{k\beta}{na}\dot{Y}_1 + \frac{k\beta}{n}\frac{\partial \dot{Y}_1}{\partial a}\right] \tag{A7.19}$$

$$\frac{\partial M_{33}}{\partial a} = \frac{-h}{na^2 G}\left[-\frac{1}{2a}(qY_1 - pX_1) + q\frac{\partial Y_1}{\partial a} - p\frac{\partial X_1}{\partial a}\right] \tag{A7.20}$$

$$\frac{\partial M_{43}}{\partial a} = \frac{K}{2na^2 G}\left(-\frac{1}{2a}Y_1 + \frac{\partial Y_1}{\partial a}\right) \tag{A7.21}$$

$$\frac{\partial M_{53}}{\partial a} = \frac{K}{2na^2 G}\left(-\frac{1}{2a}X_1 + \frac{\partial X_1}{\partial a}\right) \tag{A7.22}$$

$$\frac{\partial M_{61}^0}{\partial a} = -\frac{M_{61}^0}{2a} + n^{-1}a^{-2}\left[-2\frac{\partial X_1}{\partial a} + 3\frac{\partial \dot{X}_1}{\partial a}t + G\left(h\beta\frac{\partial^2 X_1}{\partial a \partial h} + k\beta\frac{\partial^2 X_1}{\partial a \partial k}\right)\right] \tag{A7.23}$$

$$\frac{\partial M_{62}^0}{\partial a} = -\frac{M_{62}^0}{2a} + n^{-1}a^{-2}\left[-2\frac{\partial Y_1}{\partial a} + 3\frac{\partial \dot{Y}_1}{\partial a}t + G\left(h\beta\frac{\partial^2 Y_1}{\partial a \partial h} + k\beta\frac{\partial^2 Y_1}{\partial a \partial k}\right)\right] \tag{A7.24}$$

$$\frac{\partial M_{63}^0}{\partial a} = -\frac{M_{63}^0}{2a} + \frac{1}{na^2}\left[\left(q\frac{\partial Y_1}{\partial a} + p\frac{\partial X_1}{\partial a}\right)G^{-1}\right] \tag{A7.25}$$

with

$$\begin{aligned}\frac{\partial \dot{X}_1}{\partial a} &= -\frac{1}{2}\frac{na}{r}\left[1 - \frac{3na^2 t}{r^2}(ks_F - hc_F)\right] \\ &\quad \times \left[hk\beta c_F - \left(1 - h^2\beta\right)s_F\right] + \frac{3}{2}\frac{n^2a^2t}{r^2}\left[hk\beta s_F + \left(1 - h^2\beta\right)c_F\right]\end{aligned} \tag{A7.26}$$

$$\begin{aligned}\frac{\partial \dot{Y}_1}{\partial a} &= \frac{1}{2}\frac{na}{r}\left[1 - \frac{3na^2 t}{r^2}(ks_F - hc_F)\right] \\ &\quad \times \left[hk\beta s_F - \left(1 - k^2\beta\right)c_F\right] + \frac{3}{2}\frac{n^2a^2t}{r^2}\left[hk\beta c_F + \left(1 - k^2\beta\right)s_F\right]\end{aligned} \tag{A7.27}$$

$$\frac{\partial X_1}{\partial a} = \frac{X_1}{a} - \frac{3}{2}\frac{t}{a}\dot{X}_1 \tag{A7.28}$$

$$\frac{\partial Y_1}{\partial a} = \frac{Y_1}{a} - \frac{3}{2}\frac{t}{a}\dot{Y}_1 \tag{A7.29}$$

A7.1.6 The Nonzero Partial Derivatives of M^{F_0} with Respect to F_0

$$\frac{\partial M_{11}}{\partial F_0} = \frac{\partial M_{11}}{\partial \lambda}\frac{\partial \lambda}{\partial \lambda_0}\frac{\partial \lambda_0}{\partial F_0} = \frac{\partial M_{11}}{\partial \lambda}\frac{r_0}{a_0} = \frac{r_0}{a_0}\frac{2}{n^2 r}\frac{\partial \dot{X}_1}{\partial F} \tag{A7.30}$$

$$\frac{\partial M_{12}}{\partial F_0} = \frac{r_0}{a_0}\frac{2}{n^2 r}\frac{\partial \dot{Y}_1}{\partial F} \tag{A7.31}$$

$$\frac{\partial M_{21}}{\partial F_0} = \frac{r_0}{a_0}\frac{G}{nar}\left(\frac{\partial^2 X_1}{\partial F \partial k} - \frac{h\beta}{n}\frac{\partial \dot{X}_1}{\partial F}\right) \tag{A7.32}$$

$$\frac{\partial M_{22}}{\partial F_0} = \frac{r_0}{a_0}\frac{G}{nar}\left(\frac{\partial^2 Y_1}{\partial F \partial k} - \frac{h\beta}{n}\frac{\partial \dot{Y}_1}{\partial F}\right) \tag{A7.33}$$

$$\frac{\partial M_{23}}{\partial F_0} = \frac{r_0}{a_0}\frac{k\left(q\frac{\partial Y_1}{\partial F} - p\frac{\partial X_1}{\partial F}\right)}{narG} \tag{A7.34}$$

$$\frac{\partial M_{31}}{\partial F_0} = -\frac{r_0}{a_0}\frac{G}{nar}\left(\frac{\partial^2 X_1}{\partial F \partial h} + \frac{k\beta}{n}\frac{\partial \dot{X}_1}{\partial F}\right) \tag{A7.35}$$

$$\frac{\partial M_{32}}{\partial F_0} = -\frac{r_0}{a_0}\frac{G}{nar}\left(\frac{\partial^2 Y_1}{\partial F \partial h} + \frac{k\beta}{n}\frac{\partial \dot{Y}_1}{\partial F}\right) \tag{A7.36}$$

$$\frac{\partial M_{33}}{\partial F_0} = \frac{r_0}{a_0}\frac{-h\left(q\frac{\partial Y_1}{\partial F} - p\frac{\partial X_1}{\partial F}\right)}{narG} \tag{A7.37}$$

$$\frac{\partial M_{43}}{\partial F_0} = \frac{r_0}{a_0}\frac{K}{2narG}\frac{\partial Y_1}{\partial F} \tag{A7.38}$$

$$\frac{\partial M_{53}}{\partial F_0} = \frac{r_0}{a_0}\frac{K}{2narG}\frac{\partial X_1}{\partial F} \tag{A7.39}$$

$$\frac{\partial M_{61}^0}{\partial F_0} = \frac{r_0}{a_0}n^{-1}a^{-1}r^{-1}\left[-2\frac{\partial X_1}{\partial F} + 3\frac{\partial \dot{X}_1}{\partial F}t + G\left(h\beta\frac{\partial^2 X_1}{\partial F \partial h} + k\beta\frac{\partial^2 X_1}{\partial F \partial k}\right)\right] \tag{A7.40}$$

$$\frac{\partial M_{62}^0}{\partial F_0} = \frac{r_0}{a_0}n^{-1}a^{-1}r^{-1}\left[-2\frac{\partial Y_1}{\partial F} + 3\frac{\partial \dot{Y}_1}{\partial F}t + G\left(h\beta\frac{\partial^2 Y_1}{\partial F \partial h} + k\beta\frac{\partial^2 Y_1}{\partial F \partial k}\right)\right] \tag{A7.41}$$

$$\frac{\partial M_{63}^0}{\partial F_0} = \frac{r_0}{a_0}\frac{\left(q\frac{\partial Y_1}{\partial F} - p\frac{\partial X_1}{\partial F}\right)}{narG} \tag{A7.42}$$

with $\partial M_{ij}/\partial \lambda_0 = a_0/r_0 \, \partial M_{ij}/\partial F_0$. The auxiliary partials are

$$\frac{\partial X_1}{\partial F} = a\left[hk\beta c_F - \left(1 - h^2\beta\right)s_F\right] \tag{A7.43}$$

$$\frac{\partial Y_1}{\partial F} = a\left[-hk\beta s_F + \left(1 - k^2\beta\right)c_F\right] \tag{A7.44}$$

$$\frac{\partial \dot{X}_1}{\partial F} = -\frac{a}{r}(ks_F - hc_F)\dot{X}_1 + \frac{a^2 n}{r}\left[-hk\beta s_F - \left(1 - h^2\beta\right)c_F\right] \tag{A7.45}$$

$$\frac{\partial \dot{Y}_1}{\partial F} = -\frac{a}{r}(ks_F - hc_F)\dot{Y}_1 + \frac{a^2 n}{r}\left[-hk\beta c_F - \left(1 - k^2\beta\right)s_F\right] \tag{A7.46}$$

$$\frac{\partial^2 X_1}{\partial F \partial h} = a\left[(hs_F + kc_F)\left(\beta + \frac{h^2\beta^3}{1-\beta}\right) + \frac{a^2}{r^2}(h\beta - s_F)(s_F - h) + \frac{a}{r}c_F^2\right] \tag{A7.47}$$

$$\frac{\partial^2 X_1}{\partial F \partial k} = -a\left[-(hs_F + kc_F)\frac{hk\beta^3}{1-\beta} + \frac{a^2}{r^2}(s_F - h\beta)(c_F - k) + \frac{a}{r}s_F c_F\right] \tag{A7.48}$$

$$\frac{\partial^2 Y_1}{\partial F \partial h} = a\left[-(hs_F + kc_F)\frac{hk\beta^3}{1-\beta} - \frac{a^2}{r^2}(k\beta - c_F)(s_F - h) + \frac{a}{r}s_F c_F\right] \tag{A7.49}$$

$$\frac{\partial^2 Y_1}{\partial F \partial k} = a\left[-(hs_F + kc_F)\left(\beta + \frac{k^2\beta^3}{1-\beta}\right) + \frac{a^2}{r^2}(c_F - k\beta)(c_F - k) - \frac{a}{r}s_F^2\right] \tag{A7.50}$$

References

[1] Marec, J.-P. (1979) *Optimal Space Trajectories*, Elsevier Scientific, Amsterdam.

[2] Scheel, W. A. and Conway, B. A. (1994) Optimization of Very-Low-Thrust, Many-Revolution Spacecraft Trajectories, *Journal of Guidance, Control, and Dynamics*, Vol. 17, No. 6, 1185–1192.

[3] Kluever, C. A. and Pierson, B. L. (1994) Optimal Low-Thrust Earth–Moon Transfers with a Switching Function Structure, *Journal of the Astronautical Sciences*, Vol. 42, No. 3, 269–283.

[4] Coverstone-Carroll, V. and Prussing, J. E. (1995) Optimal Cooperative Power-Limited Rendezvous with Propellant Constraints, *Journal of the Astronautical Sciences*, Vol. 43, No. 3, 289–305.

[5] Pastrone, D., Casalino, L. and Colasurdo, G. (1996) Indirect Optimization Method for Round-Trip Mars Trajectories, *Advances in the Astronautical Sciences*, Vol. 90, 85–99.

[6] Casalino, L., Colasurdo, G. and Pastrone, D. (1996) *Optimal Three-Dimensional Trajectories for Manned Mars Missions*, AIAA Paper 96-3612, Astrodynamics Conference, San Diego, CA.

[7] Casalino, L., Colasurdo, G. and Pastrone, D. (1998) *Optimal Low-Thrust Trajectories using Flyby*, AAS Paper 98–200, AAS/AIAA Space Flight Mechanics Meeting, Monterey, CA.

[8] Ocampo, C. (1996) *Trajectory Optimization for Distant Earth Satellites and Satellite Constellations*, Ph.D. Thesis, Dept. of Aerospace Engineering Sciences, University of Colorado, Boulder, CO.

[9] Starchville, T. F. and Melton, R. G. (1998) *Optimal Low-Thrust Transfers to Halo Orbits About the L_2 Libration Point in the Earth-Moon System (Elliptical Problem)*, AAS Paper 98–205, AAS/AIAA Space Flight Mechanics Meeting, Monterey, CA.

[10] Edelbaum, T. N., Sackett, L. L. and Malchow, H. L. (1973) *Optimal Low-Thrust Geocentric Transfer*, AIAA Paper 73–1074, AIAA 10th Electric Propulsion Conference, Lake Tahoe, NV.

[11] Kechichian, J. (1991) *Trajectory Optimization with a Modified Set of Equinoctial Orbit Elements*, AAS Paper 91–524, AAS/AIAA Astrodynamics Specialist Conference, Durango, CO.

[12] Kechichian, J. (1992) *Optimal Low-Thrust Rendezvous Using Equinoctial Orbit Elements*, IAF paper 92–0014, 43rd Congress of the International Astronautical Federation, Washington, DC.

[13] Kechichian, J. (1993) *Minimum-Time Low-Thrust Rendezvous using Epoch Mean Longitude Formulation*, AAS Paper 93–129, AAS/AIAA Spaceflight Mechanics Meeting, Pasadena, CA.

[14] Kechichian, J. (1998) Trajectory Optimization Using Eccentric Longitude Formulation, *Journal of Spacecraft and Rockets*, Vol. 35, No. 3, 317–326.

[15] Kechichian, J. A. (1997) Mechanics of Trajectory Optimization Using Nonsingular Variational Equations in Polar Coordinates, *Journal of Guidance, Control, and Dynamics*, Vol. 20, No. 4, 812–818.

[16] Kechichian, J. A. (1997) Trajectory Optimization Using Nonsingular Orbital Elements and True Longitude, *Journal of Guidance, Control, and Dynamics*, Vol. 20, No. 5, 1003–1009.

[17] Kechichian, J. A. (1997) Optimal Low-Earth-Orbit–Geostationary-Earth-Orbit Intermediate Acceleration Orbit Transfer, *Journal of Guidance, Control, and Dynamics*, Vol. 20, No. 4, 803–811.

[18] Kechichian, J. A. (1996) Optimal Low-Thrust Orbit Transfer, *Orbital Mechanics*, edited by Chobotov, V. A., 2nd ed., AIAA, Reston, VA, 327–402.

8 Mechanics of Trajectory Optimization Using Nonsingular Variational Equations in Polar Coordinates

8.1 Introduction

The complete set of the state and adjoint differential equations for the nonsingular equinoctial orbit elements expressed in polar coordinates and written in terms of the true longitude is presented in this chapter. Previous formulations adopted the equinoctial frame as the orbital frame for component resolution of the various perturbation accelerations. The consideration of the position dependency of the J_2 acceleration in the context of precision integrated orbital transfer trajectories led us to adopt here the more convenient polar frame coupled with the use of the true longitude as the accessory variable needed in the description of the variational equations, because it is more difficult to generate the contribution of the J_2 perturbation to the adjoint equations if the variational equations are left in terms of the eccentric longitude.

The consideration of the nonsingular equinoctial orbit elements has been of great benefit in trajectory propagation and optimization. [1] through [7] made use of the corresponding variational equations resolved in the equinoctial orbital coordinate system. In [8] through [12], modern numerical methods based on the techniques of collocation and nonlinear programming are applied to the solution of low-thrust Earth-bound and interplanetary trajectories. These direct methods have introduced robustness characteristics to the optimization process and were first used with great success by launch vehicle trajectory optimization and performance analysts. In particular, Betts [9, 10] uses a variety of equinoctial element sets within the framework of the direct method. Following the more traditional approach based on the calculus of variations and the indirect shooting method, this chapter develops the differential equations for the adjoint or multiplier variables to be numerically integrated simultaneously with the state differential equations to solve the two-point boundary value orbit transfer problem. As noted by Betts, the elements of the matrix of partial derivatives of the Hamiltonian with respect to the equinoctial elements derived here in analytic form can be used to develop a certain analytic Jacobian matrix in connection with nonlinear programming (NLP) solvers used with direct methods, thereby saving considerable computation time with respect to the adoption of a numerically determined Jacobian. The variational equations are usually expressed in terms of the eccentric longitude, which can either be left as an accessory variable or be adopted as the fast or sixth orbital element. In [3], the averaged J_2 perturbation effect was accounted for in the design of the optimal transfer trajectory. However, when the exact or rather position-dependent J_2 effect must be considered in the context of precision integrated transfer solutions, the J_2 perturbation acceleration components become dependent on the eccentric longitude. Because these components are given in the rotating Euler–Hill or polar frame, it is much simpler to use this frame as the orbital frame instead of the equinoctial frame because otherwise the transformed expressions for these components become more complicated. These expressions are still quite complicated if expressed in terms of the eccentric longitude even if the polar frame is used. This is important because we must take the partial derivatives of these expressions with respect to the elements to generate the adjoint differential equations. This task is made easier by expressing the variational equations for the J_2 perturbation in terms of the true longitude, and to be completely consistent, the equations for the thrust perturbation will also be expressed in polar coordinates and in terms of the true longitude. This chapter shows the mathematical derivations that lead to the generation of the full set of the dynamic and adjoint differential equations in polar coordinates in terms of the true longitude with the mean longitude as the sixth element. The dynamic equations, which are also found in [13], are directly obtained from the equations for the classical elements in the Gaussian form. The J_2 perturbation equations will appear in the following chapters, as well as a slightly simpler set, which uses the true longitude as the sixth element itself. We duplicate an example of an optimal transfer generated earlier with a previously validated formulation in order to validate the mathematics of this new version.

8.2 Dynamic and Adjoint Differential Equations in Polar Coordinates

The variational equations for the classical elements in the Gaussian form with the components of the disturbing acceleration in the radial, transverse and out-of-plane directions are given by

$$\dot{a} = \frac{2a^2}{j}\left(es_{\theta^*}f_r + \frac{b}{r}f_\theta\right) \tag{8.1}$$

$$\dot{e} = \frac{1}{j}\left[bs_{\theta^*}f_r + \{(b+r)c_{\theta^*} + re\}f_\theta\right] \tag{8.2}$$

$$\dot{i} = \frac{rc_\theta}{j}f_h \tag{8.3}$$

$$\dot{\Omega} = \frac{rs_\theta}{js_i}f_h \tag{8.4}$$

$$\dot{\omega} = \frac{1}{je}\left[-bc_{\theta^*}f_r + (b+r)s_{\theta^*}f_\theta\right] - \frac{rs_\theta c_i}{js_i}f_h \tag{8.5}$$

$$\dot{M} = n + \frac{\left(1-e^2\right)^{1/2}}{je}\left[(bc_{\theta^*} - 2re)f_r - (b+r)s_{\theta^*}f_\theta\right] \tag{8.6}$$

where $n = \mu^{1/2}a^{-3/2}$ is the orbit mean motion, with $\mu = 398{,}601.3\,\mathrm{km^3/s^2}$, the Earth gravity constant; a, e, i, Ω, ω and M are the classical orbit elements; $b = a\,(1-e^2)$ is the orbit parameter; $j = \left[\mu a\left(1-e^2\right)\right]^{1/2}$ is the orbit angular momentum; r is the radial distance defined by $r = b/(1+ec_{\theta^*})$; θ is the angular position defined by $\theta = \omega + \theta^*$, with θ^* the true anomaly; and f_r, f_θ and f_h are the components of the disturbing acceleration vector resolved along the rotating coordinate axes $\hat{\mathbf{r}}$, $\hat{\boldsymbol{\theta}}$ and $\hat{\mathbf{h}}$, whose directions have been defined earlier. This set of equations, which is singular at zero eccentricity and zero inclination, can be converted to a nonsingular set involving the equinoctial elements defined as $a = a$, $h = e\sin(\omega + \Omega)$, $k = e\cos(\omega + \Omega)$, $p = \tan\left(\frac{i}{2}\right)\sin\Omega$, $q = \tan\left(\frac{i}{2}\right)\cos\Omega$ and $\lambda = M + \omega + \Omega$, with λ the mean longitude. Let $\bar{\omega} = \omega + \Omega$ be the longitude of pericenter, with $F = E + \bar{\omega}$ and $L = \theta^* + \bar{\omega}$ the eccentric and true longitudes, respectively, E being the eccentric anomaly. We have with s_{θ^*} and c_{θ^*} standing for $\sin\theta^*$ and $\cos\theta^*$, etc.:

$$
\begin{aligned}
s_{\theta^*} &= s_L\, c_{\bar{\omega}} - s_{\bar{\omega}}\, c_L \\
c_{\theta^*} &= c_L\, c_{\bar{\omega}} + s_L\, s_{\bar{\omega}}
\end{aligned}
$$

and because L can also be written as $L = \Omega + \theta$,

$$
\begin{aligned}
s_\theta &= s_L c_\Omega - c_L s_\Omega \\
c_\theta &= c_L c_\Omega + s_L s_\Omega
\end{aligned}
$$

In Equations (8.1) through (8.5), we have used j and b for the angular momentum and orbit parameter, respectively, to avoid confusion with the symbols used for the equinoctial elements h and p. The radial distance r can be written as

$$r = a\,(1 - kc_F - hs_F) \tag{8.7}$$

or in terms of the true longitude as

$$r = \frac{a\left(1 - h^2 - k^2\right)}{1 + hs_L + kc_L} \tag{8.8}$$

We can also write

$$
\begin{aligned}
j &= na^2\left(1 - h^2 - k^2\right)^{1/2} \\
\frac{r}{j} &= \frac{\left(1 - h^2 - k^2\right)^{1/2}}{na\,(1 + hs_L + kc_L)} \\
1 + \frac{b}{r} &= 2 + hs_L + kc_L
\end{aligned}
$$

and with

$$\beta = \frac{1}{1 + \left(1 - h^2 - k^2\right)^{1/2}} \tag{8.9}$$

we have

$$\left(1 - h^2 - k^2\right)^{1/2} = \frac{1 - \beta}{\beta} \tag{8.10}$$

$$h^2 + k^2 = \frac{2\beta - 1}{\beta^2} \tag{8.11}$$

Also, $\partial\beta/\partial h = h\beta^3/(1 - \beta)$ and $\partial\beta/\partial k = k\beta^3/(1 - \beta)$. Finally, Kepler's equation is written in terms of the eccentric longitude F as

$$\lambda = F - ks_F + hc_F \tag{8.12}$$

Using the identities $c_i = c_{i/2}^2 - s_{i/2}^2$ and $s_i = 2s_{i/2}c_{i/2}$, the equations of motion, Equations (8.1) through (8.6), can now be transformed to the following nonsingular set:

$$\dot{a} = \frac{2}{n\left(1-h^2-k^2\right)^{1/2}}\left[(ks_L - hc_L)f_r + (1 + hs_L + kc_L)f_\theta\right] \tag{8.13}$$

$$\begin{aligned}\dot{h} = &\frac{\left(1-h^2-k^2\right)^{1/2}}{na\,(1+hs_L+kc_L)}\left[-(1+hs_L+kc_L)\,c_L f_r\right.\\ &\left.+\{h + (2 + hs_L + kc_L)\,s_L\}f_\theta - k(pc_L - qs_L)f_h\right]\end{aligned} \tag{8.14}$$

$$\begin{aligned}\dot{k} = &\frac{\left(1-h^2-k^2\right)^{1/2}}{na\,(1+hs_L+kc_L)}\left[(1+hs_L+kc_L)\,s_L f_r\right.\\ &\left.+\{k + (2 + hs_L + kc_L)\,c_L\}f_\theta + h(pc_L - qs_L)f_h\right]\end{aligned} \tag{8.15}$$

$$\dot{p} = \frac{\left(1-h^2-k^2\right)^{1/2}}{2na(1+hs_L+kc_L)}\left(1+p^2+q^2\right)s_L f_h \tag{8.16}$$

$$\dot{q} = \frac{\left(1-h^2-k^2\right)^{1/2}}{2na(1+hs_L+kc_L)}(1+p^2+q^2)c_L f_h \tag{8.17}$$

$$\begin{aligned}\dot{\lambda} = n - &\frac{\left(1-h^2-k^2\right)^{1/2}}{na\,(1+hs_L+kc_L)}\left[\left\{\beta\,(1+hs_L+kc_L)\,(hs_L+kc_L) + 2\left(1-h^2-k^2\right)^{1/2}\right\}f_r\right.\\ &\left.+\beta\,(2+hs_L+kc_L)\,(hc_L - ks_L)f_\theta + (pc_L - qs_L)f_h\right]\end{aligned} \tag{8.18}$$

These equations are essentially identical to those found in [13]. It is clear that as λ is being integrated, the eccentric longitude must first be solved from Kepler's equation, after which the true longitude is obtained from the following expressions:

$$c_L = \frac{a}{r}\left[\left(1-h^2\beta\right)c_F + hk\beta s_F - k\right] \tag{8.19}$$

$$s_L = \frac{a}{r}\left[hk\beta c_F + \left(1-k^2\beta\right)s_F - h\right] \tag{8.20}$$

These equations are obtained from the components of the radius vector along the equinoctial directions $\hat{\mathbf{f}}$ and $\hat{\mathbf{g}}$, such that

$$X_1 = rc_L = a\left[\left(1-h^2\beta\right)c_F + hk\beta s_F - k\right] \tag{8.21}$$

$$Y_1 = rs_L = a\left[hk\beta c_F + \left(1-k^2\beta\right)s_F - h\right] \tag{8.22}$$

The $\left(\hat{\mathbf{f}}, \hat{\mathbf{g}}, \hat{\mathbf{w}}\right)$ equinoctial coordinate frame is defined in terms of the classical orbit elements inasmuch as $\hat{\mathbf{f}}$ and $\hat{\mathbf{g}}$ are contained in the instantaneous orbit plane with $\hat{\mathbf{f}}$ obtained through a clockwise rotation of an angle Ω from the direction of the ascending node. Furthermore, h and k are the components of the eccentricity vector $\mathbf{e}$ along $\hat{\mathbf{f}}$ and $\hat{\mathbf{g}}$, with the elements p and q defining the matrix that rotates the orbit frame $\left(\hat{\mathbf{f}}, \hat{\mathbf{g}}, \hat{\mathbf{w}}\right)$ into the inertial frame $(\hat{\mathbf{x}}, \hat{\mathbf{y}}, \hat{\mathbf{z}})$. Equations (8.13) through (8.18) are written in polar coordinates $\hat{\mathbf{r}}$, $\hat{\boldsymbol{\theta}}$, $\hat{\mathbf{h}}$ instead of the $\hat{\mathbf{f}}$, $\hat{\mathbf{g}}$, $\hat{\mathbf{w}}$ equinoctial coordinates. The transformation between these two orbital frames is simply given by

$$\begin{pmatrix} \hat{\mathbf{r}} \\ \hat{\boldsymbol{\theta}} \\ \hat{\mathbf{h}} \end{pmatrix} = \begin{pmatrix} X_1/r & Y_1/r & 0 \\ -Y_1/r & X_1/r & 0 \\ 0 & 0 & 1 \end{pmatrix} \begin{pmatrix} \hat{\mathbf{f}} \\ \hat{\mathbf{g}} \\ \hat{\mathbf{w}} \end{pmatrix} \tag{8.23}$$

The equinoctial frame is needed here only to define the true longitude L through Equations (8.19) and (8.20) because $L = f(h, k, F)$ with $F = f(h, k, \lambda)$. Once we derive the differential equations for the adjoints, we will not worry about the equinoctial frame because the optimal thrust direction will be directly computed in the polar coordinate frame with the thrust pitch and yaw angles θ_t and θ_h obtained directly from

$$\theta_t = \tan^{-1}(u_r/u_\theta) \tag{8.24}$$

$$\theta_h = \tan^{-1}(u_w/u_\theta) \tag{8.25}$$

with $\hat{\mathbf{u}} = (u_r, u_\theta, u_w)$ such that the thrust is given by $\mathbf{f} = f\hat{\mathbf{u}}$, with f its magnitude and $\hat{\mathbf{u}}$ its direction, and the acceleration given by $\boldsymbol{\Gamma} = \frac{\mathbf{f}}{m} = f_t\hat{\mathbf{u}} = (f_r, f_\theta, f_h)$, with m the vehicle mass and $f_t = f/m$. Considering continuous constant acceleration for the purpose of the discussion and without any loss of generality, let $\mathbf{z} = (a\ h\ k\ p\ q\ \lambda)^T$ be the state vector and let $\boldsymbol{\lambda}_z = \begin{pmatrix} \lambda_a & \lambda_h & \lambda_k & \lambda_p & \lambda_q & \lambda_\lambda \end{pmatrix}^T$ be the corresponding adjoint vector. Note that when λ appears with one of the elements as a subscript, it is the multiplier adjoint to that particular element, whereas without the subscript, it is the mean longitude. Let the 6×3 B matrix be such that

$$\begin{pmatrix} \dot{a} \\ \dot{h} \\ \dot{k} \\ \dot{p} \\ \dot{q} \\ \dot{\lambda} \end{pmatrix} = \begin{pmatrix} B_{11} & B_{12} & B_{13} \\ B_{21} & B_{22} & B_{23} \\ B_{31} & B_{32} & B_{33} \\ B_{41} & B_{42} & B_{43} \\ B_{51} & B_{52} & B_{53} \\ B_{61} & B_{62} & B_{63} \end{pmatrix} \begin{pmatrix} u_r \\ u_\theta \\ u_h \end{pmatrix} f_t + \begin{pmatrix} 0 \\ 0 \\ 0 \\ 0 \\ 0 \\ n \end{pmatrix} \tag{8.26}$$

The six rows of B consist of the partials of the six equinoctial elements with respect to the velocity vector, with the components along the polar $\hat{\mathbf{r}}$, $\hat{\boldsymbol{\theta}}$, $\hat{\mathbf{h}}$ axes. The Hamiltonian and the Euler–Lagrange equations can now be written as

$$H = \boldsymbol{\lambda}_z^T B(\mathbf{z}, L) f_t \hat{\mathbf{u}} + \lambda_\lambda n = \boldsymbol{\lambda}_z^T \dot{\mathbf{z}} \tag{8.27}$$

$$\dot{\boldsymbol{\lambda}}_z = -\frac{\partial H}{\partial \mathbf{z}} = -\boldsymbol{\lambda}_z^T \frac{\partial B}{\partial \mathbf{z}} f_t \hat{\mathbf{u}} - \lambda_\lambda \frac{\partial n}{\partial \mathbf{z}} \tag{8.28}$$

When taking the partial derivatives in Equation (8.28), we must allow for the variation of L with respect to h, k and λ, and because L is given in terms of F, we must also allow for the variation of F with respect to h, k and λ:

$$\frac{\partial L}{\partial h} = \left(\frac{\partial L}{\partial h}\right)_F + \frac{\partial L}{\partial F}\frac{\partial F}{\partial h} \tag{8.29}$$

$$\frac{\partial L}{\partial k} = \left(\frac{\partial L}{\partial k}\right)_F + \frac{\partial L}{\partial F}\frac{\partial F}{\partial k} \tag{8.30}$$

$$\frac{\partial L}{\partial \lambda} = \frac{\partial L}{\partial F}\frac{\partial F}{\partial \lambda} \tag{8.31}$$

with the subscript F indicating that F is held constant there. We also need $\partial r/\partial a$, $\partial r/\partial h$, $\partial r/\partial k$ and $\partial r/\partial \lambda$, and it will be shown that $\partial L/\partial a = 0$. These partial derivatives must be expressed in terms of L such that both the dynamic and adjoint equations are written consistently in terms of L. Because we are integrating λ, it effectively becomes independent of a, h and k such that, from Kepler's equation $\lambda = F - ks_F + hc_F$, we have from $\partial\lambda/\partial a = \frac{r}{a}\frac{\partial F}{\partial a} = 0$ that $\partial F/\partial a = 0$. In a similar way, from $\partial\lambda/\partial h = 0$ we obtain $\frac{\partial F}{\partial h} = -\frac{a}{r}c_F$, from $\partial\lambda/\partial k = 0$ we have $\partial F/\partial k = \frac{a}{r}s_F$, and finally from $\partial\lambda/\partial\lambda = 1$ we have $\partial F/\partial\lambda = \frac{a}{r}$. This in turn yields the partial derivatives of the radial distance r with respect to a, h, k and λ such that from $r = a\,(1 - kc_F - hs_F)$ we get

$$\frac{\partial r}{\partial a} = \frac{r}{a} \tag{8.32}$$

$$\frac{\partial r}{\partial h} = \frac{a^2}{r}\,(h - s_F) \tag{8.33}$$

$$\frac{\partial r}{\partial k} = \frac{a^2}{r}\,(k - c_F) \tag{8.34}$$

$$\frac{\partial r}{\partial F} = a\,(ks_F - hc_F) \tag{8.35}$$

All these partials are identical to those derived in [6] and Chapter 3 of this book, which used the elements a, h, k, p, q, λ and the equinoctial orbit frame $\hat{\mathbf{f}}$, $\hat{\mathbf{g}}$, $\hat{\mathbf{w}}$ and with all the differential equations written in terms of the eccentric longitude F. In particular, if we substitute the quantities rc_L and rs_L for X_1 and Y_1, then the expressions for s_F and c_F in [1–3, 5] and Chapter 2 of this book can be written as

$$s_F = h + \frac{r}{a}\frac{s_L\left(1 - h^2\beta\right) - hk\beta c_L}{\left(1 - h^2 - k^2\right)^{1/2}} \tag{8.36}$$

$$c_F = k + \frac{r}{a}\frac{c_L\left(1 - k^2\beta\right) - hk\beta s_L}{\left(1 - h^2 - k^2\right)^{1/2}} \tag{8.37}$$

yielding the following identity:

$$ks_F - hc_F = \frac{r}{a}\frac{(ks_L - hc_L)}{\left(1 - h^2 - k^2\right)^{1/2}} \tag{8.38}$$

These expressions now allow us to write $\partial r/\partial h$, $\partial r/\partial k$ and $\partial r/\partial F$ directly in terms of L, namely

$$\frac{\partial r}{\partial h} = -\frac{a\left[\left(1 - h^2\beta\right)s_L - hk\beta c_L\right]}{\left(1 - h^2 - k^2\right)^{1/2}} \tag{8.39}$$

$$\frac{\partial r}{\partial k} = -\frac{a\left[\left(1 - k^2\beta\right)c_L - hk\beta s_L\right]}{\left(1 - h^2 - k^2\right)^{1/2}} \tag{8.40}$$

$$\frac{\partial r}{\partial F} = \frac{a\,(ks_L - hc_L)}{\left(1 - h^2 - k^2\right)^{1/2}} \tag{8.41}$$

Now the partial derivative $\partial L/\partial F$ needed in Equations (8.29), (8.30) and (8.31) can be obtained from Equation (8.19) as follows. From $c_L = X_1/r$,

$$-s_L \frac{\partial L}{\partial F} = \frac{1}{r}\frac{\partial X_1}{\partial F} - \frac{X_1}{r^2}\frac{\partial r}{\partial F}$$

with $\partial r/\partial F$ given in Equation (8.41) and

$$\frac{\partial X_1}{\partial F} = a\left[hk\beta c_F - \left(1 - h^2\beta\right) s_F\right]$$

Using Equations (8.36) and (8.37) in $\partial X_1/\partial F$, we get

$$\frac{\partial L}{\partial F} = \frac{c_L\left(ks_L - hc_L - hk\right)}{s_L\left(1-h^2-k^2\right)^{1/2}} + \frac{ah}{rs_L}\left(1-h^2-k^2\right)^{1/2} + \frac{\left(1-h^2\right)}{\left(1-h^2-k^2\right)^{1/2}} \tag{8.42}$$

In a similar way, from Equation (8.20) and using

$$\frac{\partial Y_1}{\partial F} = a\left[\left(1 - k^2\beta\right) c_F - hk\beta s_F\right]$$

we have

$$\frac{\partial L}{\partial F} = \frac{ak}{rc_L}\left(1-h^2-k^2\right)^{1/2} + \frac{\left(1-k^2\right)}{\left(1-h^2-k^2\right)^{1/2}} - \frac{s_L\left(ks_L - hc_L + hk\right)}{c_L\left(1-h^2-k^2\right)^{1/2}} \tag{8.43}$$

The partials in Equations (8.42) and (8.43) can be shown to be identical, and they can further be reduced to the simpler form

$$\frac{\partial L}{\partial F} = \frac{a}{r}\left(1-h^2-k^2\right)^{1/2} \tag{8.44}$$

or

$$\frac{\partial L}{\partial F} = \frac{1 + hs_L + kc_L}{\left(1-h^2-k^2\right)^{1/2}} \tag{8.45}$$

We can also obtain $\partial L/\partial F$ more simply from

$$\tan L = Y_1/X_1$$

$$\frac{1}{c_L^2}\frac{\partial L}{\partial F} = \frac{X_1\left(\partial Y_1/\partial F\right) - Y_1\left(\partial X_1/\partial F\right)}{X_1^2}$$

and leaving X_1, Y_1, $\partial X_1/\partial F$ and $\partial Y_1/\partial F$ in terms of F while using $X_1^2 = r^2 c_L^2$ to get

$$\frac{r^2}{a^2}\frac{\partial L}{\partial F} = \left(1 - h^2\beta - k^2\beta\right)\left(1 - hs_F - kc_F\right)$$

We do not need to replace s_F and c_F by their expressions in terms of s_L and c_L because the last parenthesis is conveniently replaced by r/a such that, finally and more readily than before,

$$\begin{aligned}\frac{\partial L}{\partial F} &= \frac{a}{r}\left(1 - h^2\beta - k^2\beta\right)\\ &= \frac{a}{r}\frac{(1-\beta)}{\beta} = \frac{a}{r}\left(1-h^2-k^2\right)^{1/2}\end{aligned}$$

Now as far as the $\partial L/\partial a$ partial is concerned, we can use c_L in Equation (8.19) to get

$$-s_L\frac{\partial L}{\partial a}=\frac{\left(r-a\frac{\partial r}{\partial a}\right)}{r^2}\left(\frac{r}{a}c_L\right)+\frac{a}{r}\left[hk\beta c_F-\left(1-h^2\beta\right)s_F\right]\frac{\partial F}{\partial a} \tag{8.46}$$

with

$$\frac{\partial r}{\partial a}=\frac{r}{a}+a\left(ks_F-hc_F\right)\frac{\partial F}{\partial a} \tag{8.47}$$

However, $\partial F/\partial a=0$ and, therefore, $\partial r/\partial a=r/a$, and because s_L is not always equal to zero, we have

$$\frac{\partial L}{\partial a}=0 \tag{8.48}$$

If we delay taking $\partial F/\partial a=0$ in Equations (8.47) and (8.46) and make use of Equation (8.36) for s_F and Equation (8.37) for c_F as well as the identity in Equation (8.38), then the expression in Equation (8.46) will be reduced to

$$\frac{\partial L}{\partial a}=\frac{\left(1+hs_L+kc_L\right)}{\left(1-h^2-k^2\right)^{1/2}}\frac{\partial F}{\partial a}=\frac{\partial L}{\partial F}\frac{\partial F}{\partial a} \tag{8.49}$$

where the partial in Equation (8.45) has been recognized. Now because $\partial F/\partial a=0$, then $\partial L/\partial a$ must also be identically equal to zero because $\partial L/\partial F$ itself is always nonzero. There are several ways of generating the $\partial L/\partial h$ and $\partial L/\partial k$ partials. We can use Equation (8.19) or (8.20) and write, for example,

$$\frac{\partial L}{\partial h}=-\frac{1}{s_L}\frac{\partial}{\partial h}\left\{\frac{a}{r}\left[\left(1-h^2\beta\right)c_F+hk\beta s_F-k\right]\right\}$$

and allow for the variation of F with respect to h by using $\partial F/\partial h=-\frac{a}{r}c_F$, as well as using Equation (8.39) for $\partial r/\partial h$. Then s_F and c_F are replaced by their expressions in terms of L according to Equations (8.36) and (8.37), such that the preceding partial is given strictly in terms of L. We can also use $\tan L=Y_1(F)/X_1(F)$ and write

$$\frac{\partial L}{\partial h}=\frac{X_1\left(\partial Y_1/\partial h\right)-Y_1\left(\partial X_1/\partial h\right)}{r^2}$$

and express the right-hand side in terms of L. It is, however, much easier to use $r=a\left(1-h^2-k^2\right)\left(1+hs_L+kc_L\right)^{-1}$ and write

$$\frac{\partial r}{\partial h}=\frac{-2ah}{\left(1+hs_L+kc_L\right)}-\frac{a\left(1-h^2-k^2\right)}{\left(1+hs_L+kc_L\right)^2}\left[s_L+\left(hc_L-ks_L\right)\frac{\partial L}{\partial h}\right]$$

and make use of the $\partial r/\partial h$ partial in Equation (8.39) written directly in terms of L. Equating these two partials will yield the desired quantity

$$\frac{\partial L}{\partial h}=\frac{\frac{a^2}{r^2}\left(1-h^2-k^2\right)^{1/2}\left[\left(1-h^2\beta\right)s_L-hk\beta c_L\right]-s_L-\frac{2ah}{r}}{\left(hc_L-ks_L\right)} \tag{8.50}$$

Let us derive this particular partial in still another way. From $L=\omega+\Omega+\theta^*=\tan^{-1}(h/k)+\theta^*$, we have, with $e^2=h^2+k^2$,

$$\frac{\partial L}{\partial h}=\frac{k}{e^2}+\frac{\partial\theta^*}{\partial h} \tag{8.51}$$

Now from $r = a\left(1 - e^2\right)/(1 + ec_{\theta^*})$, we also have

$$\frac{\partial r}{\partial h} = \frac{-2ae\frac{\partial e}{\partial h}}{(1 + ec_{\theta^*})} - \frac{a(1-e^2)}{(1 + ec_{\theta^*})^2}\left[\frac{\partial e}{\partial h}c_{\theta^*} - es_{\theta^*}\frac{\partial \theta^*}{\partial h}\right] \tag{8.52}$$

However, $\partial e/\partial h = h/e$, and from orbital mechanics we have $rc_{\theta^*} = a(c_E - e)$ and $rs_{\theta^*} = a(1-e^2)^{1/2}s_E$, where E, the eccentric anomaly, is given by $E = F - (\omega + \Omega)$. Therefore,

$$c_E = \frac{k}{e}c_F + \frac{h}{e}s_F$$
$$s_E = \frac{k}{e}s_F - \frac{h}{e}c_F$$

such that with $e = \left(h^2 + k^2\right)^{1/2}$,

$$c_{\theta^*} = \frac{a}{er}\left[kc_F + hs_F - e^2\right]$$
$$s_{\theta^*} = \frac{a}{er}\left(1 - e^2\right)^{1/2}(ks_F - hc_F)$$

Once again using $\partial r/\partial h$ of Equation (8.39), we can solve for $\partial\theta^*/\partial h$ from Equation (8.52) so that finally Equation (8.51) is given directly in terms of L:

$$\frac{\partial L}{\partial h} = \frac{k}{\left(h^2 + k^2\right)} - \frac{a^2}{r^2}\frac{\left(1 - h^2 - k^2\right)^{1/2}}{(ks_L - hc_L)}\left[\left(1 - h^2\beta\right)s_L - hk\beta c_L\right]$$
$$+ \frac{a}{r}\frac{h}{(ks_L - hc_L)}\left[2 + \frac{\left(1 - h^2 - k^2\right)}{h^2 + k^2}\right] - \frac{h}{\left(h^2 + k^2\right)(ks_L - hc_L)}$$

After further regrouping of terms, it can be shown that this partial is identical to Equation (8.50). This partial can also be given in terms of F as

$$\frac{\partial L}{\partial h} = \frac{\frac{-a^2}{r^2}(s_F - h)\left(1 - h^2 - k^2\right) + Y_1/a + 2h}{\left(1 - h^2 - k^2\right)^{1/2}(ks_F - hc_F)} \tag{8.53}$$

In a similar manner, we can derive $\partial L/\partial k$ and get

$$\frac{\partial L}{\partial k} = \frac{\frac{a^2}{r^2}\left(1 - h^2 - k^2\right)^{1/2}\left[\left(1 - k^2\beta\right)c_L - hk\beta s_L\right] - c_L - 2\frac{a}{r}k}{(hc_L - ks_L)} \tag{8.54}$$

in terms of L, or as in Equation (8.55) in the next equation, in terms of F:

$$\frac{\partial L}{\partial k} = \frac{\frac{-a^2}{r^2}\left(1 - h^2 - k^2\right)(c_F - k) + X_1/a + 2k}{(ks_F - hc_F)\left(1 - h^2 - k^2\right)^{1/2}} \tag{8.55}$$

We now complete the derivation of the partials of L with respect to the integrated state variables by generating $\partial L/\partial\lambda$. From $L = \omega + \Omega + \theta^* = \tan^{-1}(h/k) + \theta^*$, we have $\partial L/\partial\lambda = \partial\theta^*/\partial\lambda$. We also have $\frac{\partial r}{\partial\lambda} = \frac{\partial r}{\partial F}\frac{\partial F}{\partial\lambda} = a\,(ks_F - hc_F)\,\partial F/\partial\lambda = \frac{a^2}{r}(ks_F - hc_F)$ because $\partial F/\partial\lambda = a/r$. From $r = a\left(1 - e^2\right)/(1 + ec_{\theta^*})$, we have

$$\frac{\partial r}{\partial\lambda} = \frac{ae\left(1 - e^2\right)s_{\theta^*}\partial\theta^*/\partial\lambda}{(1 + ec_{\theta^*})^2} = \frac{a^2}{r}(ks_F - hc_F)$$

which yields

$$\frac{\partial\theta^*}{\partial\lambda} = \frac{a^2}{r^2}\left(1 - e^2\right)^{1/2}$$

and therefore

$$\frac{\partial L}{\partial\lambda} = \frac{a^2}{r^2}\left(1 - h^2 - k^2\right)^{1/2} \tag{8.56}$$

This also shows that $\partial L/\partial\lambda = (\partial L/\partial F)(\partial F/\partial\lambda)$. The partial $\partial r/\partial\lambda$ can be written in terms of L as

$$\frac{\partial r}{\partial\lambda} = \frac{a\,(ks_L - hc_L)}{\left(1 - h^2 - k^2\right)^{1/2}} \tag{8.57}$$

We have thus generated all the partial derivatives of L, namely, $\partial L/\partial a$, $\partial L/\partial h$, $\partial L/\partial k$ and $\partial L/\partial\lambda$ given by Equations (8.48), (8.50), (8.54) and (8.56), respectively, as well as the partials of r, namely, $\partial r/\partial a$, $\partial r/\partial h$, $\partial r/\partial k$ and $\partial r/\partial\lambda$ in Equations (8.32), (8.39), (8.40) and (8.57), respectively. We can now generate all the $\partial B/\partial\mathbf{z}$ partials of interest using

$$\frac{\partial n}{\partial a} = -\frac{3n}{2a} = -\frac{3}{2}\mu^{1/2}a^{-5/2} \tag{8.58}$$

The nonzero elements of matrix B and their nonzero partials, with $G = \left(1 - h^2 - k^2\right)^{1/2}$ and $K = \left(1 + p^2 + q^2\right)$, are shown in Appendix 8.1.

8.3 Numerical Example

Let us now duplicate an example of a minimum-time low-Earth-orbit (LEO) to geostationary-Earth-orbit (GEO) transfer that appeared in [7] and Chapters 2, 5–7 of this book. The minimization of t_f or maximization of $-t_f$ requires the satisfaction of the transversality condition at the guessed final time, $H_f = 1$ for the Hamiltonian used in Equation (8.27). The orientation of the constant acceleration vector is optimized by selecting $\hat{\mathbf{u}} = \left[\boldsymbol{\lambda}_z^T B(\mathbf{z}, L)\right]^T / \left|\boldsymbol{\lambda}_z^T B(\mathbf{z}, L)\right|$. Given $a_0, e_0, i_0, \Omega_0, \omega_0$ or in an equivalent way $(a)_0$, $(h)_0$, $(k)_0$, $(p)_0$ and $(q)_0$, the corresponding $(\lambda_a)_0$, $(\lambda_h)_0$, $(\lambda_k)_0$, $\left(\lambda_p\right)_0$ and $\left(\lambda_q\right)_0$ as well as the transfer time t_f and the initial mean longitude $(\lambda)_0$, are guessed, and the dynamic Equations (8.13) through (8.18) and the adjoint Equation (8.28) are integrated forward with $(\lambda_\lambda)_0 = 0$, until t_f. The first five multipliers, $(\lambda)_0$, and t_f are then adjusted until $a_f, h_f, k_f, p_f, q_f, (\lambda_\lambda)_f = 0$, and $H_f = 1$ are satisfied for an optimal free–free transfer. By a free–free transfer, we mean a transfer in which the first five elements defining the orbit size, shape and orientation are fixed and the location or longitude of the spacecraft is free on both the initial and final orbits. An unconstrained minimization algorithm is used to this effect and the following objective function minimized:

$$F' = \omega_1\left(a - a_f\right)^2 + \omega_2\left(h - h_f\right)^2 + \omega_3\left(k - k_f\right)^2 + \omega_4\left(p - p_f\right)^2 + \omega_5\left(q - q_f\right)^2 + \omega_6\left(\lambda_\lambda - 0\right)^2 + \omega_7(H - 1)^2$$

The ω_i weights had to be adjusted manually between consecutive runs in order to control the convergence process. The minimization subroutine uses a quasi-Newton algorithm and is based on a general descent method. No attempt was made to apply the multiple shooting method, which could, in theory, increase the robustness of the iterative convergence process. However, as was demonstrated in [7] and Chapter 2 of this book, the transfer can be solved first by using the averaging technique in a robust way, and the initial values of the Lagrange multipliers thus generated are used as a starting guess to solve the exact unaveraged transfer much more efficiently because these starting values are very close to the exact solution. Let $a_0 = 7000\,\text{km}$, $e_0 = 0$, $i_0 = 28.5$ deg, $\Omega_0 = 0$ deg and $\omega_0 = 0$ deg, and let $a_f = 42{,}000\,\text{km}$, $e_f = 10^{-3}$, $i_f = 1$ deg, $\Omega_f = 0$ deg, $\omega_f = 0$ deg and $f_t = 9.8 \times 10^{-5}\,\text{km/s}^2$. The solution is given by $(\lambda_a)_0 = 4.675229762\,\text{s/km}$, $(\lambda_h)_0 = 5.413413947 \times 10^2$ s, $(\lambda_k)_0 = -9.202702084 \times 10^3$ s, $(\lambda_p)_0 = 1.778011878 \times 10^1$ s, $(\lambda_q)_0 = -2.258455855 \times 10^4$ s, $(\lambda)_0 = -2.274742851$ rad, which corresponds to $M_0 = -130.333164$ deg, and $t_f = 58{,}089.90058$ s. The final achieved parameters are $a_f = 42{,}000.003\,\text{km}$, $e_f = 9.986352 \times 10^{-4}$, $i_f = 0.999809$ deg, $\Omega_f = 1.099680$ deg $\times 10^{-4}$, $\omega_f = 1.827518$ deg $\times 10^{-2}$, $(\lambda_\lambda)_f = -1.823951 \times 10^{-4}$s/rad and $H_f = 1.003704$, indicating convergence. Also, $\lambda_f = 19.655283$ rad, corresponding to $M_f = 46.146408$ deg. The initial values of the multipliers are essentially identical, as they should be because of their definition as influence functions, to the solution obtained in [7] and Chapters 2, 5–7 of this book, thereby validating the mathematical derivations produced in this chapter.

8.4 Conclusion

The mathematical derivations leading to the generation of the complete set of the dynamic and adjoint differential equations needed in trajectory optimization work are derived for a set of nonsingular orbit elements and for the resolution of the perturbation accelerations in the Euler–Hill rotating orbital frame. The element set adopted here is that originally used by Broucke and Cefola [1], for which the perturbation equations for the geopotential, drag and third-body effects were developed. These analytic methods derived from celestial mechanics can be put to effective use in constructing efficient optimization computer programs, whether based on direct or indirect methods. Finally, an example of a free–free orbit transfer is duplicated with the present formulation in order to validate this simpler version, which also replaces the eccentric longitude by the true longitude as the accessory variable appearing in the right-hand side of all the variational and adjoint equations.

Appendix 8.1 The *B* Matrix and its Partial Derivatives

$$B_{11} = 2n^{-1}G^{-1}\left(ks_L - hc_L\right)$$

$$\frac{\partial B_{11}}{\partial a} = -2n^{-2}\frac{\partial n}{\partial a}G^{-1}\left(ks_L - hc_L\right) \tag{A8.1}$$

$$\frac{\partial B_{11}}{\partial h} = 2n^{-1}hG^{-3}\left(ks_L - hc_L\right) + 2n^{-1}G^{-1}\left[-c_L + \left(hs_L + kc_L\right)\frac{\partial L}{\partial h}\right] \tag{A8.2}$$

$$\frac{\partial B_{11}}{\partial k} = 2n^{-1}kG^{-3}(ks_L - hc_L) + 2n^{-1}G^{-1}\left[s_L + (kc_L + hs_L)\frac{\partial L}{\partial k}\right] \tag{A8.3}$$

$$\frac{\partial B_{11}}{\partial \lambda} = 2n^{-1}G^{-1}(kc_L + hs_L)\frac{\partial L}{\partial \lambda} \tag{A8.4}$$

$$B_{12} = 2n^{-1}ar^{-1}G$$

$$\frac{\partial B_{12}}{\partial a} = -2n^{-2}ar^{-1}\frac{\partial n}{\partial a}G \tag{A8.5}$$

$$\frac{\partial B_{12}}{\partial h} = -2n^{-1}ar^{-2}\frac{\partial r}{\partial h}G - 2n^{-1}ar^{-1}hG^{-1} \tag{A8.6}$$

$$\frac{\partial B_{12}}{\partial k} = -2n^{-1}ar^{-2}\frac{\partial r}{\partial k}G - 2n^{-1}ar^{-1}kG^{-1} \tag{A8.7}$$

$$\frac{\partial B_{12}}{\partial \lambda} = -2n^{-1}ar^{-2}\frac{\partial r}{\partial \lambda}G \tag{A8.8}$$

$$B_{21} = -n^{-1}a^{-1}Gc_L$$

$$\frac{\partial B_{21}}{\partial a} = -\frac{1}{2}n^{-1}a^{-2}Gc_L \tag{A8.9}$$

$$\frac{\partial B_{21}}{\partial h} = n^{-1}a^{-1}hG^{-1}c_L + n^{-1}a^{-1}Gs_L\frac{\partial L}{\partial h} \tag{A8.10}$$

$$\frac{\partial B_{21}}{\partial k} = n^{-1}a^{-1}kG^{-1}c_L + n^{-1}a^{-1}Gs_L\frac{\partial L}{\partial k} \tag{A8.11}$$

$$\frac{\partial B_{21}}{\partial \lambda} = n^{-1}a^{-1}Gs_L\frac{\partial L}{\partial \lambda} \tag{A8.12}$$

$$B_{22} = rn^{-1}a^{-2}G^{-1}(h + s_L) + n^{-1}a^{-1}Gs_L$$

$$\frac{\partial B_{22}}{\partial a} = \frac{1}{2}n^{-1}a^{-3}rG^{-1}(h + s_L) + \frac{1}{2}n^{-1}a^{-2}Gs_L \tag{A8.13}$$

$$\begin{aligned}\frac{\partial B_{22}}{\partial h} &= n^{-1}a^{-2}G^{-1}(h + s_L)\left[\frac{\partial r}{\partial h} + rhG^{-2}\right] + n^{-1}a^{-2}rG^{-1}\left(1 + c_L\frac{\partial L}{\partial h}\right)\\ &\quad - n^{-1}a^{-1}hs_LG^{-1} + n^{-1}a^{-1}Gc_L\frac{\partial L}{\partial h}\end{aligned} \tag{A8.14}$$

$$\begin{aligned}\frac{\partial B_{22}}{\partial k} &= n^{-1}a^{-2}G^{-1}(h + s_L)\left[\frac{\partial r}{\partial k} + rkG^{-2}\right] + n^{-1}a^{-2}rG^{-1}c_L\frac{\partial L}{\partial k}\\ &\quad - n^{-1}a^{-1}ks_LG^{-1} + n^{-1}a^{-1}Gc_L\frac{\partial L}{\partial k}\end{aligned} \tag{A8.15}$$

$$\frac{\partial B_{22}}{\partial \lambda} = \frac{\partial r}{\partial \lambda}n^{-1}a^{-2}G^{-1}(h + s_L) + rn^{-1}a^{-2}G^{-1}c_L\frac{\partial L}{\partial \lambda} + n^{-1}a^{-1}Gc_L\frac{\partial L}{\partial \lambda} \tag{A8.16}$$

$$B_{23} = -n^{-1}a^{-2}rG^{-1}k(pc_L - qs_L)$$

$$\frac{\partial B_{23}}{\partial a} = -\frac{1}{2}n^{-1}a^{-3}rG^{-1}k(pc_L - qs_L) \tag{A8.17}$$

$$\begin{aligned}\frac{\partial B_{23}}{\partial h} &= -n^{-1}a^{-2}G^{-1}k(pc_L - qs_L)\left[\frac{\partial r}{\partial h} + hrG^{-2}\right]\\ &\quad + n^{-1}a^{-2}rkG^{-1}(ps_L + qc_L)\frac{\partial L}{\partial h}\end{aligned} \tag{A8.18}$$

$$\frac{\partial B_{23}}{\partial k} = -n^{-1}a^{-2}G^{-1}k(pc_L - qs_L)\left[\frac{\partial r}{\partial k} + krG^{-2}\right] + n^{-1}a^{-2}rkG^{-1}(ps_L + qc_L)\frac{\partial L}{\partial k} - n^{-1}a^{-2}rG^{-1}(pc_L - qs_L) \quad \text{(A8.19)}$$

$$\frac{\partial B_{23}}{\partial p} = -n^{-1}a^{-2}rG^{-1}kc_L \quad \text{(A8.20)}$$

$$\frac{\partial B_{23}}{\partial q} = n^{-1}a^{-2}rG^{-1}ks_L \quad \text{(A8.21)}$$

$$\frac{\partial B_{23}}{\partial \lambda} = -n^{-1}a^{-2}\frac{\partial r}{\partial \lambda}G^{-1}k(pc_L - qs_L) + n^{-1}a^{-2}rG^{-1}k(ps_L + qc_L)\frac{\partial L}{\partial \lambda} \quad \text{(A8.22)}$$

$$B_{31} = n^{-1}a^{-1}Gs_L$$

$$\frac{\partial B_{31}}{\partial a} = \frac{1}{2}n^{-1}a^{-2}Gs_L \quad \text{(A8.23)}$$

$$\frac{\partial B_{31}}{\partial h} = -n^{-1}a^{-1}G^{-1}hs_L + n^{-1}a^{-1}Gc_L\frac{\partial L}{\partial h} \quad \text{(A8.24)}$$

$$\frac{\partial B_{31}}{\partial k} = -n^{-1}a^{-1}G^{-1}ks_L + n^{-1}a^{-1}Gc_L\frac{\partial L}{\partial k} \quad \text{(A8.25)}$$

$$\frac{\partial B_{31}}{\partial \lambda} = n^{-1}a^{-1}Gc_L\frac{\partial L}{\partial \lambda} \quad \text{(A8.26)}$$

$$B_{32} = n^{-1}a^{-2}rG^{-1}(k + c_L) + n^{-1}a^{-1}Gc_L$$

$$\frac{\partial B_{32}}{\partial a} = \frac{1}{2}n^{-1}a^{-3}rG^{-1}(k + c_L) + \frac{1}{2}n^{-1}a^{-2}Gc_L \quad \text{(A8.27)}$$

$$\frac{\partial B_{32}}{\partial h} = n^{-1}a^{-2}G^{-1}(k + c_L)\left[\frac{\partial r}{\partial h} + hrG^{-2}\right] - n^{-1}a^{-2}rG^{-1}s_L\frac{\partial L}{\partial h} - n^{-1}a^{-1}hG^{-1}c_L - n^{-1}a^{-1}Gs_L\frac{\partial L}{\partial h} \quad \text{(A8.28)}$$

$$\frac{\partial B_{32}}{\partial k} = n^{-1}a^{-2}G^{-1}(k + c_L)\left[\frac{\partial r}{\partial k} + krG^{-2}\right] - n^{-1}a^{-2}rG^{-1}s_L\frac{\partial L}{\partial k} - n^{-1}a^{-1}kG^{-1}c_L - n^{-1}a^{-1}Gs_L\frac{\partial L}{\partial k} + n^{-1}a^{-2}rG^{-1} \quad \text{(A8.29)}$$

$$\frac{\partial B_{32}}{\partial \lambda} = n^{-1}a^{-2}\frac{\partial r}{\partial \lambda}G^{-1}(k + c_L) - n^{-1}a^{-2}rG^{-1}s_L\frac{\partial L}{\partial \lambda} - n^{-1}a^{-1}Gs_L\frac{\partial L}{\partial \lambda} \quad \text{(A8.30)}$$

$$B_{33} = n^{-1}a^{-2}rG^{-1}h(pc_L - qs_L)$$

$$\frac{\partial B_{33}}{\partial a} = \frac{1}{2}n^{-1}a^{-3}rG^{-1}h(pc_L - qs_L) \quad \text{(A8.31)}$$

$$\frac{\partial B_{33}}{\partial h} = n^{-1}a^{-2}G^{-1}h(pc_L - qs_L)\left[\frac{\partial r}{\partial h} + hrG^{-2}\right] + n^{-1}a^{-2}rG^{-1}\left[(pc_L - qs_L) - h\,(ps_L + qc_L)\,\frac{\partial L}{\partial h}\right] \quad \text{(A8.32)}$$

$$\frac{\partial B_{33}}{\partial k} = n^{-1}a^{-2}G^{-1}h(pc_L - qs_L)\left[\frac{\partial r}{\partial k} + krG^{-2}\right] - n^{-1}a^{-2}rG^{-1}h(ps_L + qc_L)\frac{\partial L}{\partial k} \quad \text{(A8.33)}$$

$$\frac{\partial B_{33}}{\partial p} = n^{-1}a^{-2}rG^{-1}hc_L \quad \text{(A8.34)}$$

$$\frac{\partial B_{33}}{\partial q} = -n^{-1}a^{-2}rG^{-1}hs_L \quad \text{(A8.35)}$$

$$\frac{\partial B_{33}}{\partial \lambda} = n^{-1}a^{-2}\frac{\partial r}{\partial \lambda}G^{-1}h(pc_L - qs_L) - n^{-1}a^{-2}rG^{-1}h(ps_L + qc_L)\frac{\partial L}{\partial \lambda} \quad \text{(A8.36)}$$

$$B_{43} = 2^{-1}n^{-1}a^{-2}rG^{-1}Ks_L$$

$$\frac{\partial B_{43}}{\partial a} = \frac{1}{4}n^{-1}a^{-3}rG^{-1}Ks_L \quad \text{(A8.37)}$$

$$\frac{\partial B_{43}}{\partial h} = \frac{1}{2}n^{-1}a^{-2}G^{-1}Ks_L\left[\frac{\partial r}{\partial h} + hrG^{-2}\right] + \frac{1}{2}n^{-1}a^{-2}rG^{-1}Kc_L\frac{\partial L}{\partial h} \quad \text{(A8.38)}$$

$$\frac{\partial B_{43}}{\partial k} = \frac{1}{2}n^{-1}a^{-2}G^{-1}Ks_L\left[\frac{\partial r}{\partial k} + krG^{-2}\right] + \frac{1}{2}n^{-1}a^{-2}rG^{-1}Kc_L\frac{\partial L}{\partial k} \quad \text{(A8.39)}$$

$$\frac{\partial B_{43}}{\partial p} = n^{-1}a^{-2}rG^{-1}ps_L \quad \text{(A8.40)}$$

$$\frac{\partial B_{43}}{\partial q} = n^{-1}a^{-2}rG^{-1}qs_L \quad \text{(A8.41)}$$

$$\frac{\partial B_{43}}{\partial \lambda} = \frac{1}{2}n^{-1}a^{-2}\frac{\partial r}{\partial \lambda}G^{-1}Ks_L + \frac{1}{2}n^{-1}a^{-2}rG^{-1}Kc_L\frac{\partial L}{\partial \lambda} \quad \text{(A8.42)}$$

$$B_{53} = \frac{1}{2}n^{-1}a^{-2}rG^{-1}Kc_L$$

$$\frac{\partial B_{53}}{\partial a} = \frac{1}{4}n^{-1}a^{-3}rG^{-1}Kc_L \quad \text{(A8.43)}$$

$$\frac{\partial B_{53}}{\partial h} = \frac{1}{2}n^{-1}a^{-2}G^{-1}Kc_L\left[\frac{\partial r}{\partial h} + hrG^{-2}\right] - \frac{1}{2}n^{-1}a^{-2}rG^{-1}Ks_L\frac{\partial L}{\partial h} \quad \text{(A8.44)}$$

$$\frac{\partial B_{53}}{\partial k} = \frac{1}{2}n^{-1}a^{-2}G^{-1}Kc_L\left[\frac{\partial r}{\partial k} + krG^{-2}\right] - \frac{1}{2}n^{-1}a^{-2}rG^{-1}Ks_L\frac{\partial L}{\partial k} \quad \text{(A8.45)}$$

$$\frac{\partial B_{53}}{\partial p} = n^{-1}a^{-2}rG^{-1}pc_L \quad \text{(A8.46)}$$

$$\frac{\partial B_{53}}{\partial q} = n^{-1}a^{-2}rG^{-1}qc_L \quad \text{(A8.47)}$$

$$\frac{\partial B_{53}}{\partial \lambda} = \frac{1}{2}n^{-1}a^{-2}\frac{\partial r}{\partial \lambda}G^{-1}Kc_L - \frac{1}{2}n^{-1}a^{-2}rG^{-1}Ks_L\frac{\partial L}{\partial \lambda} \quad \text{(A8.48)}$$

$$B_{61} = -n^{-1}a^{-1}(1-\beta)(hs_L + kc_L) - 2n^{-1}a^{-2}r$$

$$\frac{\partial B_{61}}{\partial a} = -\frac{1}{2}n^{-1}a^{-2}(1-\beta)(hs_L + kc_L) - n^{-1}a^{-3}r \quad \text{(A8.49)}$$

$$\frac{\partial B_{61}}{\partial h} = n^{-1}a^{-1}\frac{h\beta^3}{(1-\beta)}(hs_L + kc_L) - n^{-1}a^{-1}(1-\beta) \times \left[s_L + (hc_L - ks_L)\frac{\partial L}{\partial h}\right] - 2n^{-1}a^{-2}\frac{\partial r}{\partial h} \quad \text{(A8.50)}$$

$$\frac{\partial B_{61}}{\partial k} = n^{-1}a^{-2}\frac{k\beta^3}{(1-\beta)}(hs_L + kc_L) - 2n^{-1}a^{-2}\frac{\partial r}{\partial k} - n^{-1}a^{-1}(1-\beta)\left[c_L + (hc_L - ks_L)\frac{\partial L}{\partial k}\right] \quad \text{(A8.51)}$$

$$\frac{\partial B_{61}}{\partial \lambda} = -n^{-1}a^{-1}(1-\beta)(hc_L - ks_L)\frac{\partial L}{\partial \lambda} - 2n^{-1}a^{-2}\frac{\partial r}{\partial \lambda} \quad \text{(A8.52)}$$

$$B_{62} = -n^{-1}a^{-1}(1-\beta)(hc_L - ks_L)\left[1 + ra^{-1}G^{-2}\right]$$

$$\frac{\partial B_{62}}{\partial a} = -\frac{1}{2}n^{-1}a^{-2}(1-\beta)(hc_L - ks_L)\left[1 + ra^{-1}G^{-2}\right] \quad \text{(A8.53)}$$

$$\frac{\partial B_{62}}{\partial h} = n^{-1}a^{-1}\frac{h\beta^3}{(1-\beta)}(hc_L - ks_L)\left[1 + ra^{-1}G^{-2}\right] - n^{-1}a^{-1}(1-\beta)\left[c_L - (hs_L + kc_L)\frac{\partial L}{\partial h}\right]\left[1 + ra^{-1}G^{-2}\right] - n^{-1}a^{-2}(1-\beta)(hc_L - ks_L)G^{-2}\left[\frac{\partial r}{\partial h} + 2hrG^{-2}\right] \quad \text{(A8.54)}$$

$$\frac{\partial B_{62}}{\partial k} = n^{-1}a^{-1}\frac{k\beta^3}{(1-\beta)}(hc_L - ks_L)\left[1 + ra^{-1}G^{-2}\right] + n^{-1}a^{-1}(1-\beta)\left[s_L + (hs_L + kc_L)\frac{\partial L}{\partial k}\right]\left[1 + ra^{-1}G^{-2}\right] - n^{-1}a^{-2}(1-\beta)(hc_L - ks_L)G^{-2}\left[\frac{\partial r}{\partial k} + 2krG^{-2}\right] \quad \text{(A8.55)}$$

$$\frac{\partial B_{62}}{\partial \lambda} = n^{-1}a^{-1}(1-\beta)(hs_L + kc_L)\frac{\partial L}{\partial \lambda}\left[1 + ra^{-1}G^{-2}\right] - n^{-1}a^{-2}(1-\beta)(hc_L - ks_L)\frac{\partial r}{\partial \lambda}G^{-2} \quad \text{(A8.56)}$$

$$B_{63} = -n^{-1}a^{-2}rG^{-1}(pc_L - qs_L)$$

$$\frac{\partial B_{63}}{\partial a} = -\frac{1}{2}n^{-1}a^{-3}rG^{-1}(pc_L - qs_L) \quad \text{(A8.57)}$$

$$\frac{\partial B_{63}}{\partial h} = -n^{-1}a^{-2}G^{-1}(pc_L - qs_L)\left[\frac{\partial r}{\partial h} + hrG^{-2}\right] + n^{-1}a^{-2}rG^{-1}(ps_L + qc_L)\frac{\partial L}{\partial h} \quad \text{(A8.58)}$$

$$\frac{\partial B_{63}}{\partial k} = -n^{-1}a^{-2}G^{-1}(pc_L - qs_L)\left[\frac{\partial r}{\partial k} + krG^{-2}\right] + n^{-1}a^{-2}rG^{-1}(ps_L + qc_L)\frac{\partial L}{\partial k} \tag{A8.59}$$

$$\frac{\partial B_{63}}{\partial p} = -n^{-1}a^{-2}rG^{-1}c_L \tag{A8.60}$$

$$\frac{\partial B_{63}}{\partial q} = n^{-1}a^{-2}rG^{-1}s_L \tag{A8.61}$$

$$\frac{\partial B_{63}}{\partial \lambda} = -n^{-1}a^{-2}\frac{\partial r}{\partial \lambda}G^{-1}(pc_L - qs_L) + n^{-1}a^{-2}rG^{-1}(ps_L + qc_L)\frac{\partial L}{\partial \lambda} \tag{A8.62}$$

References

[1] Broucke, R. A. and Cefola, P. J. (1972) On the Equinoctial Orbit Elements, *Celestial Mechanics*, Vol. 5, 303–310.

[2] Cefola, P. J. (1972) *Equinoctial Orbit Elements: Application to Artificial Satellite Orbits*, AIAA Paper 72–937, AIAA/AAS Astrodynamics Conference, Palo Alto, CA.

[3] Edelbaum, T. N., Sackett, L. L. and Malchow, H. L. (1973) *Optimal Low Thrust Geocentric Transfer*, AIAA Paper 73–1074, AIAA 10th Electric Propulsion Conference, Lake Tahoe, NV.

[4] Cefola, P. J., Long, A. C. and Holloway, G., Jr. (1974) *The Long-Term Prediction of Artificial Satellite Orbits*, AIAA Paper 74–170, AIAA 12th Aerospace Sciences Meeting, Washington, DC.

[5] Kechichian, J. A. (1990) *Equinoctial Orbit Elements: Application to Optimal Transfer Problems*, AIAA Paper 90–2976, AIAA/AAS Astrodynamics Conference, Portland, OR.

[6] Kechichian, J. A. (1992) *Optimal Low-Thrust Rendezvous Using Equinoctial Orbit Elements*, IAF Paper 92–0014, 43rd Congress of the International Astronautical Federation, Washington, DC.

[7] Kechichian, J. A. (1994) *Optimal LEO–GEO Intermediate Acceleration Orbit Transfer*, AAS Paper 94–125, AAS/AIAA Spaceflight Mechanics Meeting, Cocoa Beach, FL.

[8] Enright, P. J. and Conway, B. A. (1991) Optimal Finite-Thrust Spacecraft Trajectories Using Collocation and Nonlinear Programming, *Journal of Guidance, Control, and Dynamics*, Vol. 1, No. 5, 981–985.

[9] Betts, J. T. (1993) Using Sparse Nonlinear Programming to Compute Low Thrust Orbit Transfers, *Journal of the Astronautical Sciences*, Vol. 41, No. 3, 349–371.

[10] Betts, J. T. (1994) Optimal Interplanetary Orbit Transfers by Direct Transcription, *Journal of the Astronautical Sciences*, Vol. 42, No. 3, 247–268.

[11] Kluever, C. A. and Pierson, B. L. (1994) Optimal Low-Thrust Earth–Moon Transfers with a Switching Function Structure, *Journal of the Astronautical Sciences*, Vol. 42, No. 3, 269–283.

[12] Coverstone-Carroll, V. and Williams, S. N. (1994) Optimal Low-Thrust Trajectories Using Differential Inclusion Concepts, *Journal of the Astronautical Sciences*, Vol. 42, No. 4, 379–393.

[13] Battin, R. H. (1987) *An Introduction to the Mathematics and Methods of Astrodynamics*, AIAA, New York, NY, 490–494.

9 Trajectory Optimization Using Nonsingular Orbital Elements and True Longitude

9.1 Introduction

The system and adjoint differential equations needed to solve low-thrust transfer and rendezvous trajectories are derived in this chapter, in terms of nonsingular orbit elements, with the true longitude selected as the sixth variable that also defines the radial distance. The perturbation accelerations are resolved in the rotating polar frame such that the equations of motion are written in polar coordinates. This formulation is particularly convenient for the treatment of the J_2 perturbation acceleration whose components are easily expressed in terms of the true longitude. The iterations needed for the solution of Kepler's equation are also eliminated and the analytic form of the adjoint equations provided in their simplest aspect.

The use of orbital elements and in particular nonsingular equinoctial elements has been of great benefit in developing simulation software for trajectory propagation and optimization, especially for low-thrust applications [1–5]. Rapid numerical integration of the equations for the state and adjoint variables is made possible by the averaging technique, which is applicable only if these variables vary slowly in time, precluding thereby the adoption of the Cartesian coordinates preferred by flight mechanics specialists.

There are many ways of writing the equations of motion and their associated adjoint equations, because the sixth state variable can be chosen to be the mean longitude, the eccentric longitude or the true longitude at either the current time or epoch time. This is identical to the adoption of the mean anomaly, the eccentric anomaly, or the true anomaly when the more familiar classical orbit elements are used. The epoch formulations are fundamental in the sense that all the state variables are truly constant in the absence of any perturbative acceleration, whether it is conservative such as in the case of the geopotential or third-body effects, or nonconservative, such as in the case of thrust or atmospheric drag. The component resolution of these accelerations can be made in various rotating orbital frames, such as the Euler–Hill or polar frame, the equinoctial frame, or even the tangential frame. Furthermore, the right-hand sides of all these system and adjoint differential equations are written in terms of either the eccentric longitude or the true longitude, but not the mean longitude (the radial distance cannot be written in terms of the mean longitude), the equations of motion being defined by the partials of the radial distance with respect to the orbital elements of choice. Various sets of these elements were adopted, coupled with the selection of the equinoctial frame and with the radial distance expressed in terms of the eccentric longitude, to solve the low-thrust transfer problem with only the thrust perturbation acceleration active [6–8, and Chapters 2, 3, and 6].

In [9] and Chapter 8 of this book, it has been recognized that the treatment of the J_2 oblateness effect in the context of exact precision-integrated trajectories is made easier by the use of the true longitude in the right-hand side of all of our differential equations because it is more complicated to generate the analytic form of the adjoint equations if the J_2 accelerations are written in terms of the eccentric longitude. This chapter is an extension of the analysis presented in [9] and Chapter 8 of this book and makes possible a further simplification by adopting the true longitude as the sixth state variable instead of the mean longitude of [9]. The need to solve Kepler's transcendental equation is thus eliminated, and the adjoint equations are obtained in their simplest form. Much of the original theory of optimal low-thrust transfer using equinoctial elements developed by Edelbaum et al., as well as numerous solved transfers, is found in [10] and Chapters 2 and 3 of this book.

9.2 Equations of Motion with the True Longitude as the Sixth State Variable

The variational equations for the classical orbit elements and in the Gaussian form, with the disturbing acceleration resolved in the rotating Euler–Hill orbital frame (also known as the polar coordinate frame), are given by

$$\dot{a} = \frac{2a^2}{h'}\left(es_{\theta^*}f_r + \frac{p'}{r}f_\theta\right) \tag{9.1}$$

$$\dot{e} = \frac{1}{h'}\left[p's_{\theta^*}f_r + \left\{(p'+r)c_{\theta^*} + re\right\}f_\theta\right] \tag{9.2}$$

$$\dot{i} = \frac{rc_\theta}{h'}f_h \tag{9.3}$$

$$\dot{\Omega} = \frac{rs_\theta}{h's_i}f_h \tag{9.4}$$

$$\dot{\omega} = \frac{1}{h'e}\left[-p'c_{\theta^*}f_r + (p'+r)s_{\theta^*}f_\theta\right] - \frac{rs_\theta c_i}{h's_i}f_h \tag{9.5}$$

$$\dot{M} = n + \frac{\left(1-e^2\right)^{1/2}}{h'e}\left[\left(p'c_{\theta^*} - 2re\right)f_r - (p'+r)s_{\theta^*}f_\theta\right] \tag{9.6}$$

Here $\theta = \omega + \theta^*$ is the orbital angular position, where θ^* is the true anomaly; n is the orbit mean motion; $p' = a(1-e^2)$ is the orbit parameter; $h' = \left[\mu a\left(1-e^2\right)\right]^{1/2}$ is the orbit angular momentum with μ the Earth gravity constant; and f_r, f_θ, f_h are the components of the disturbing acceleration vector $\mathbf{\Gamma} = \mathbf{f}/m = f_t\hat{\mathbf{u}} = (f_r,f_\theta,f_h)$ with magnitude f_t such that m is the spacecraft mass and f is the magnitude of the disturbing force, or in our case the thrust vector $\mathbf{f}$. The direction of the acceleration vector is given by the unit vector $\hat{\mathbf{u}}$ such that $f_r = f_tu_r$, $f_\theta = f_tu_\theta$ and $f_h = f_tu_h$. The radial distance $r = p'/(1+ec_{\theta^*})$, with c_{θ^*} standing for $\cos\theta^*$ and s_{θ^*} standing for $\sin\theta^*$, etc. Equation (9.6) for the mean anomaly can be replaced by Equation (9.7) for the true anomaly:

$$\dot{\theta}^* = \frac{h'}{r^2} + \frac{1}{eh'}\left[p'c_{\theta^*}f_r - (p'+r)s_{\theta^*}f_\theta\right] \tag{9.7}$$

These equations can also be written for the nonsingular equinoctial elements a, h, k, p, q, λ, where $a = a$, $h = e\sin(\omega+\Omega)$, $k = e\cos(\omega+\Omega)$, $p = \tan(i/2)\sin\Omega$, $q = \tan(i/2)\cos\Omega$ and $\lambda = M+\omega+\Omega$, which is also called the mean longitude:

$$\dot{a} = \frac{2}{n\left(1-h^2-k^2\right)^{1/2}}\left[(ks_L - hc_L)f_r + (1+hs_L+kc_L)f_\theta\right] \tag{9.8}$$

$$\dot{h} = \frac{\left(1-h^2-k^2\right)^{1/2}}{na\left(1+hs_L+kc_L\right)}\left[-\left(1+hs_L+kc_L\right)c_L f_r \right. \\ \left. + \left\{h + \left(2+hs_L+kc_L\right)s_L\right\}f_\theta - k\left(pc_L - qs_L\right)f_h\right] \tag{9.9}$$

$$\dot{k} = \frac{\left(1-h^2-k^2\right)^{1/2}}{na\left(1+hs_L+kc_L\right)}\left[\left(1+hs_L+kc_L\right)s_L f_r \right. \\ \left. + \left\{k + \left(2+hs_L+kc_L\right)c_L\right\}f_\theta + h\left(pc_L - qs_L\right)f_h\right] \tag{9.10}$$

$$\dot{p} = \frac{\left(1-h^2-k^2\right)^{1/2}}{2na\left(1+hs_L+kc_L\right)}\left(1+p^2+q^2\right)s_L f_h \tag{9.11}$$

$$\dot{q} = \frac{\left(1-h^2-k^2\right)^{1/2}}{2na\left(1+hs_L+kc_L\right)}\left(1+p^2+q^2\right)c_L f_h \tag{9.12}$$

$$\dot{\lambda} = n - \frac{\left(1-h^2-k^2\right)^{1/2}}{na\left(1+hs_L+kc_L\right)}\left[\left\{\beta\left(1+hs_L+kc_L\right)\left(hs_L+kc_L\right) + 2\left(1-h^2-k^2\right)^{1/2}\right\}f_r \right. \\ \left. + \beta\left(2+hs_L+kc_L\right)\left(hc_L - ks_L\right)f_\theta + \left(pc_L - qs_L\right)f_h\right] \tag{9.13}$$

The true longitude L appearing in the right-hand side of the preceding differential equations is defined as $L = \theta^* + \omega + \Omega$ or $L = \theta + \Omega$, and the factor β is a function only of h and k with

$$\beta = \frac{1}{1+\left(1-h^2-k^2\right)^{1/2}} \tag{9.14}$$

The preceding equations have been transformed to their present form by making use of the identities $r = a\left(1-h^2-k^2\right)/(1+hs_L+kc_L)$, $h' = na^2\left(1-h^2-k^2\right)^{1/2}$ and

$$s_{\theta^*} = \frac{a}{er}\left[kc_F + hs_F - e^2\right] = -\frac{1}{e} + \frac{a}{er}\left(1-e^2\right) = \frac{1}{e}\left[\frac{p'}{r} - 1\right]$$

$$c_{\theta^*} = \frac{a}{er}\left(1-e^2\right)^{1/2}\left(ks_F - hc_F\right)$$

which are then written in terms of L with the use of the following identity:

$$\left(ks_F - hc_F\right) = \frac{r\left(ks_L - hc_L\right)}{a\left(1-h^2-k^2\right)^{1/2}}$$

We thus have $s_{\theta^*} = \left(ks_L - hc_L\right)/\left(h^2+k^2\right)^{1/2}$ and $c_{\theta^*} = \left(hs_L + kc_L\right)/\left(h^2+k^2\right)^{1/2}$, and from $\theta = L - \Omega$, we have $s_\theta = s_L c_\Omega - c_L s_\Omega$, $c_\theta = c_L c_\Omega + s_L s_\Omega$, which are written in their final form by replacing $s_\Omega = p/\left(p^2+q^2\right)^{1/2}$ and $c_\Omega = q/\left(p^2+q^2\right)^{1/2}$:

$$s_\theta = \frac{qs_L}{\left(p^2+q^2\right)^{1/2}} - \frac{pc_L}{\left(p^2+q^2\right)^{1/2}}$$
$$c_\theta = \frac{qc_L}{\left(p^2+q^2\right)^{1/2}} + \frac{ps_L}{\left(p^2+q^2\right)^{1/2}}$$

We also have $p'/r = (1 + hs_L + kc_L)$ and $r/h' = \left(1-h^2-k^2\right)^{1/2}/[na\,(1+hs_L+kc_L)]$ Because λ is being integrated, the current value of L must be computed from the identities [9, and Chapter 8]

$$c_L = \frac{a}{r}\left[\left(1-h^2\beta\right)c_F + hk\beta s_F - k\right] \tag{9.15}$$
$$s_L = \frac{a}{r}\left[hk\beta c_F + \left(1-k^2\beta\right)s_F - h\right] \tag{9.16}$$

after evaluating the eccentric longitude F from Kepler's equation $\lambda = F - ks_F + hc_F$. By definition, $F = E + \omega + \Omega$, where E is the classical eccentric anomaly. We can avoid these intermediate calculations concerning F if we replace Equation (9.13) for $\dot\lambda$ by the corresponding equation for $\dot L$ because we have $\dot L = \dot\theta^* + (\dot\omega + \dot\Omega) = \dot\theta^* + \dot{\bar\omega}$, with $\bar\omega = \omega + \Omega$ representing the longitude of pericenter. Making use of Equation (9.7) and with $h'/r^2 = n\,(1+hs_L+kc_L)^2/\left(1-h^2-k^2\right)^{3/2}$,

$$\dot{\bar\omega} = \frac{1}{h'e}\left[-p'c_{\theta^*}f_r + (p'+r)s_{\theta^*}f_\theta\right] + \frac{r}{h'}s_\theta \tan\frac{i}{2}f_h \tag{9.17}$$

we get

$$\dot L = \dot\theta^* + \dot{\bar\omega} = \frac{h'}{r^2} + \frac{r}{h'}s_\theta \tan\frac{i}{2}f_h \tag{9.18}$$

The term $s_\theta \tan i/2$ is evaluated from the definitions $s_{i/2} = \left(p^2+q^2\right)^{1/2}/\left(1+p^2+q^2\right)^{1/2}$ and $c_{i/2} = 1/\left(1+p^2+q^2\right)^{1/2}$, such that $\tan i/2 = \left(p^2+q^2\right)^{1/2}$ and finally $s_\theta \tan i/2 = (qs_L - pc_L)$. The definitions of p and q also provide directly the preceding expression for $\tan i/2$. We can now write the differential equation for L as

$$\dot L = \frac{na^2\left(1-h^2-k^2\right)^{1/2}}{r^2} + \frac{r\,(qs_L - pc_L)}{na^2\left(1-h^2-k^2\right)^{1/2}}f_h \tag{9.19}$$

or

$$\dot L = \frac{n\,(1+hs_L+kc_L)^2}{\left(1-h^2-k^2\right)^{3/2}} + \frac{\left(1-h^2-k^2\right)^{1/2}}{na\,(1+hs_L+kc_L)}(qs_L - pc_L)f_h \tag{9.20}$$

The full set of the equations of motion is now given by Equations (9.8) through (9.12), and (9.19) or (9.20), with the true longitude L obtained directly through integration. This set will provide us with the simplest form of the adjoint equations because now we can set $\partial L/\partial a = \partial L/\partial h = \partial L/\partial k = 0$ when generating these adjoint differential equations and eliminate the need for solving the transcendental Kepler equation for F through iteration. Let $\mathbf{z} = (a\,h\,k\,p\,q\,L)^T$ stand for the state vector with corresponding adjoint $\boldsymbol{\lambda}_z = \left(\lambda_a\,\lambda_h\,\lambda_k\,\lambda_p\,\lambda_q\,\lambda_L\right)^T$. The dynamic equations are given in compact form by

$$\begin{pmatrix} \dot{a} \\ \dot{h} \\ \dot{k} \\ \dot{p} \\ \dot{q} \\ \dot{L} \end{pmatrix} = \begin{pmatrix} B^L_{11} & B^L_{12} & B^L_{13} \\ B^L_{21} & B^L_{22} & B^L_{23} \\ B^L_{31} & B^L_{32} & B^L_{33} \\ B^L_{41} & B^L_{42} & B^L_{43} \\ B^L_{51} & B^L_{52} & B^L_{53} \\ B^L_{61} & B^L_{62} & B^L_{63} \end{pmatrix} \begin{pmatrix} u_r \\ u_\theta \\ u_h \end{pmatrix} f_t + \begin{pmatrix} 0 \\ 0 \\ 0 \\ 0 \\ 0 \\ \frac{na^2(1-h^2-k^2)^{1/2}}{r^2} \end{pmatrix} \tag{9.21}$$

The Hamiltonian and the differential equations for the adjoints are obtained from

$$H = \boldsymbol{\lambda}_z^T B^L(\mathbf{z}) f_t \hat{\mathbf{u}} + \lambda_L \frac{na^2\left(1-h^2-k^2\right)^{1/2}}{r^2} \tag{9.22}$$

$$\dot{\boldsymbol{\lambda}}_z = -\frac{\partial H}{\partial \mathbf{z}} = -\boldsymbol{\lambda}_z^T \frac{\partial B^L}{\partial \mathbf{z}} f_t \hat{\mathbf{u}} - \lambda_L \frac{\partial}{\partial \mathbf{z}} \left[\frac{na^2\left(1-h^2-k^2\right)^{1/2}}{r^2} \right] \tag{9.23}$$

where B^L is the 6×3 matrix shown in Equation (9.21). Because L is now independent of a, h, k, p, and q, we have in Equation (9.23) $\partial L/\partial a = \partial L/\partial h = \partial L/\partial k = \partial L/\partial p = \partial L/\partial q = 0$. The radial distance r appears in the matrix B^L such that we need the partials $\partial r/\partial a$, $\partial r/\partial h$, $\partial r/\partial k$ and $\partial r/\partial L$ to generate Equation (9.23). The partials $\partial r/\partial p = \partial r/\partial q = 0$ because r is an in-plane variable, as is also L. From the definition of r,

$$r = \frac{a\left(1-h^2-k^2\right)}{\left(1+hs_L+kc_L\right)} \tag{9.24}$$

it is straightforward to get

$$\frac{\partial r}{\partial a} = \frac{r}{a} \tag{9.25}$$

We also have

$$\frac{\partial r}{\partial h} = \frac{-2ah}{\left(1+hs_L+kc_L\right)} - \frac{a\left(1-h^2-k^2\right)s_L}{\left(1+hs_L+kc_L\right)^2} \tag{9.26}$$

which can also be written as

$$\frac{\partial r}{\partial h} = -\frac{r}{a\left(1-h^2-k^2\right)}\left(2ah+rs_L\right) \tag{9.27}$$

In a similar way, we have the following two forms for $\partial r/\partial k$:

$$\frac{\partial r}{\partial k} = \frac{-2ak}{\left(1+hs_L+kc_L\right)} - \frac{a\left(1-h^2-k^2\right)c_L}{\left(1+hs_L+kc_L\right)^2} \tag{9.28}$$

$$\frac{\partial r}{\partial k} = -\frac{r}{a\left(1-h^2-k^2\right)}\left(2ak+rc_L\right) \tag{9.29}$$

Finally,

$$\frac{\partial r}{\partial L} = -\frac{r^2}{a}\frac{\left(hc_L-ks_L\right)}{\left(1-h^2-k^2\right)} \tag{9.30}$$

This partial can also be written in terms of the eccentric longitude F because we have the identity [9, and Chapter 8]

$$ks_F - hc_F = \frac{r}{a}\frac{ks_L - hc_L}{\left(1 - h^2 - k^2\right)^{1/2}} \tag{9.31}$$

such that

$$\frac{\partial r}{\partial L} = \frac{r\left(ks_F - hc_F\right)}{\left(1 - h^2 - k^2\right)^{1/2}} \tag{9.32}$$

It is also shown in [9] and Chapter 8 of this book that

$$\frac{\partial L}{\partial F} = \frac{a}{r}\left(1 - h^2 - k^2\right)^{1/2} = \frac{\left(1 + hs_L + kc_L\right)}{\left(1 - h^2 - k^2\right)^{1/2}} \tag{9.33}$$

which is obtained by direct differentiation of the identities $c_L = \frac{a}{r}\left[\left(1 - h^2\beta\right)c_F + hk\beta s_F - k\right]$ and $s_L = \frac{a}{r}\left[hk\beta c_F + \left(1 - k^2\beta\right)s_F - h\right]$.

We can now verify that

$$\frac{\partial r}{\partial F} = \frac{\partial r}{\partial L}\frac{\partial L}{\partial F} = \frac{r(ks_F - hc_F)}{\left(1 - h^2 - k^2\right)^{1/2}}\frac{a}{r}\left(1 - h^2 - k^2\right)^{1/2} = a(ks_F - hc_F) \tag{9.34}$$

which is the same expression obtained directly from the definition of r in terms of F, namely, $r = a(1 - kc_F - hs_F)$ and therefore $\partial r/\partial F = a(ks_F - hc_F)$. It is also shown in [9] and Chapter 8 of this book that $\partial r/\partial\lambda = \frac{a^2}{r^2}\left(1 - h^2 - k^2\right)^{1/2}$ such that the chain rule

$$\frac{\partial r}{\partial\lambda} = \frac{\partial r}{\partial L}\frac{\partial L}{\partial\lambda} = \frac{a\left(ks_L - hc_L\right)}{\left(1 - h^2 - k^2\right)^{1/2}} \tag{9.35}$$

recovers the expression for the partial $\partial r/\partial\lambda$ obtained in [9] and Chapter 8 of this book.

Let us now digress momentarily from the task of deriving Equation (9.23) and discuss the similarities between the formulation developed in this chapter and the formulation based on the set (a, h, k, p, q, F) developed in [8] and Chapter 6 of this book. Neither of these two formulations requires solution of Kepler's transcendental equation iteratively at each integration step because the dynamic equations for the set (a, h, k, p, q, L) are written in terms of the sixth variable L, and the dynamic equations for the set (a, h, k, p, q, F) are written in terms of the sixth variable F. The fact that the polar frame is used in the present case and the equinoctial frame in the second case, for the component resolution of the disturbing acceleration, is irrelevant to our discussion. In both of these formulations, $\partial r/\partial a = r/a$. However, if we compare $\partial r/\partial h = -as_F$ of [8] and Chapter 6 of this book with $\partial r/\partial h$ of Equation (9.26) or (9.27), we will find, after expressing $-as_F$ in terms of L, that these two partials, as expected, are not identical. The same holds true for the $\partial r/\partial k$ partial of [8] and Chapter 6 of this book, which is equal to $-ac_F$ and the present partial in Equation (9.28) or (9.29). The expressions for s_F and c_F in terms of L are

$$s_F = h + \frac{r}{a}\frac{\left(1 - h^2\beta\right)s_L - hk\beta c_L}{\left(1 - h^2 - k^2\right)^{1/2}} \tag{9.36}$$

$$c_F = k + \frac{r}{a}\frac{\left(1 - k^2\beta\right)c_L - hk\beta s_L}{\left(1 - h^2 - k^2\right)^{1/2}} \tag{9.37}$$

Let us now go back to the original formulations of [7] and [9] and Chapters 3 and 8 of this book, which use the set (a, h, k, p, q, λ). In [9] and Chapter 8 of this book, L appears as the accessory variable in the right-hand side of the equations of motion. Therefore, $L = f(h, k, F)$ with $F = f(h, k, \lambda)$ such that

$$\left(\frac{\partial r}{\partial h}\right)_{tot} = \left(\frac{\partial r}{\partial h}\right)_L + \frac{\partial r}{\partial L}\left(\frac{\partial L}{\partial h}\right)_{tot} = \left(\frac{\partial r}{\partial h}\right)_L + \frac{\partial r}{\partial L}\left[\left(\frac{\partial L}{\partial h}\right)_F + \frac{\partial L}{\partial F}\frac{\partial F}{\partial h}\right] \tag{9.38}$$

On the other hand, from [7] and Chapter 3 of this book, we have with $r = a(1 - kc_F - hs_F)$,

$$\left(\frac{\partial r}{\partial h}\right)_{tot} = \left(\frac{\partial r}{\partial h}\right)_F + \frac{\partial r}{\partial F}\frac{\partial F}{\partial h} \tag{9.39}$$

However, $\partial r/\partial F$ can be written as

$$\frac{\partial r}{\partial F} = \frac{\partial r}{\partial L}\frac{\partial L}{\partial F} \tag{9.40}$$

such that Equation (9.39) takes the following form:

$$\left(\frac{\partial r}{\partial h}\right)_{tot} = \left(\frac{\partial r}{\partial h}\right)_F + \frac{\partial r}{\partial L}\frac{\partial L}{\partial F}\frac{\partial F}{\partial h} \tag{9.41}$$

Equations (9.38) and (9.41) must be identical, which in turn shows that

$$\left(\frac{\partial r}{\partial h}\right)_L + \left(\frac{\partial r}{\partial L}\right)\left(\frac{\partial L}{\partial h}\right)_F = \left(\frac{\partial r}{\partial h}\right)_F \tag{9.42}$$

Because $(\partial r/\partial h)_F = -as_F$, $\partial r/\partial L = -\frac{r^2}{a}\frac{hc_L - ks_L}{(1-h^2-k^2)} = \frac{r(ks_F - hc_F)}{(1-h^2-k^2)^{1/2}}$ and because from [9] and Chapter 8 of this book, $\partial L/\partial F = \frac{a}{r}\left(1 - h^2 - k^2\right)^{1/2}$ with $\partial F/\partial h = -\frac{a}{r}c_F$, we can now evaluate $(\partial L/\partial h)_F$ needed in Equation (9.42) by making use of the $(\partial L/\partial h)_{tot}$ partial developed in [9] and Chapter 8 of this book such that, in terms of F,

$$\left(\frac{\partial L}{\partial h}\right)_{tot} = \frac{-\frac{a^2}{r^2}(s_F - h)\left(1 - h^2 - k^2\right) + \frac{Y_1}{a} + 2h}{\left(1 - h^2 - k^2\right)^{1/2}(ks_F - hc_F)} = \left(\frac{\partial L}{\partial h}\right)_F + \frac{\partial L}{\partial F}\frac{\partial F}{\partial h} \tag{9.43}$$

Therefore, with $(\partial L/\partial h)_{tot}$ given by Equation (9.43),

$$\left(\frac{\partial L}{\partial h}\right)_F = \left(\frac{\partial L}{\partial h}\right)_{tot} - \frac{\partial L}{\partial F}\frac{\partial F}{\partial h} = \left(\frac{\partial L}{\partial h}\right)_{tot} + \frac{a^2}{r^2}\left(1 - h^2 - k^2\right)^{1/2} c_F$$

which is now used in Equation (9.42) with $(\partial r/\partial h)_L$ given by Equation (9.27). Because $Y_1 = rs_L$, the left-hand side of Equation (9.42) can be written as

$$\begin{aligned}
&-\frac{2hr}{\left(1 - h^2 - k^2\right)} - \frac{rY_1}{a\left(1 - h^2 - k^2\right)} - \frac{a^2}{r}(s_F - h) + \frac{rY_1}{a\left(1 - h^2 - k^2\right)} \\
&+ \frac{2hr}{\left(1 - h^2 - k^2\right)} + \frac{a^2}{r}(ks_F - hc_F)\, c_F \\
&= -\frac{a^2}{r}s_F + \frac{a^2}{r}h\left(1 - c_F^2\right) + \frac{a^2}{r}ks_F c_F \\
&= -\frac{a^2}{r}(1 - hs_F - kc_F)\, s_F \\
&= -as_F
\end{aligned}$$

which is the expression of $(\partial r/\partial h)_F$. Therefore, Equation (9.42) holds, showing how $(\partial r/\partial h)_L$ differs from $(\partial r/\partial h)_F$.

Identical arguments also show how $(\partial r/\partial k)_L$ differs from $(\partial r/\partial k)_F$. These manipulations show that when either L or F is selected as the sixth state variable, the variation of r due to the variations in h and k must not account for the variation of r due to the variations of the respective anomalies. Furthermore, holding L fixed or F fixed will yield different partials, as was just shown. When the orbit is circular, $e = 0$ or $h = k = 0$ and $r = a$, and the expression for $\partial r/\partial h$ in Equation (9.27) reduces to $\partial r/\partial h = -rs_L$. However, from Equation (9.36) we also get $s_F = s_L$ such that $\partial r/\partial h = -as_F$ and therefore $(\partial r/\partial h)_L = (\partial r/\partial h)_F$. When $e \neq 0$, these two partials are therefore not identical. The same observations in Equation (9.29) and Equation (9.37) show that $(\partial r/\partial k)_L = -rc_L$ with $c_L = c_F$ when $e = 0$, and in that case $(\partial r/\partial k)_L = (\partial r/\partial k)_F$, too. We can now derive the $\partial B^L/\partial \mathbf{z}$ partials. We have with $\partial r/\partial h$, $\partial r/\partial k$ and $\partial r/\partial L$ given by Equations (9.27), (9.29) and (9.30), respectively, and with $\partial n/\partial a = -3n/(2a)$, and letting $G = \left(1 - h^2 - k^2\right)^{1/2}$ and $K = \left(1 + p^2 + q^2\right)$, the elements of the B^L matrix and their partial derivatives with respect to all six orbital elements, are given in the appendix.

9.3 Numerical Results

Let us now establish the mathematical relationships between the Lagrange multipliers of the (a, h, k, p, q, λ) formulation of [9] and Chapter 8 of this book and the present (a, h, k, p, q, L) formulation. Using the λ superscript and the L superscript for the λ and L formulations, respectively, we can write

$$\begin{aligned} &\lambda_a^\lambda da + \lambda_h^\lambda dh + \lambda_k^\lambda dk + \lambda_p^\lambda dp + \lambda_q^\lambda dq + \lambda_\lambda^\lambda d\lambda + H^\lambda dt \\ &\quad = \lambda_a^L da + \lambda_h^L dh + \lambda_k^L dk + \lambda_p^L dp + \lambda_q^L dq + \lambda_L^L dL + H^L dt \end{aligned} \tag{9.44}$$

The Hamiltonians are both constant because both systems are autonomous and equal to the unit value to satisfy the transversality condition for a minimum-time transfer. We now need to express dL in terms of dh, dk and $d\lambda$. From $L = f(h, k, F)$, we have

$$dL = \left(\frac{\partial L}{\partial h}\right)_F dh + \left(\frac{\partial L}{\partial k}\right)_F dk + \frac{\partial L}{\partial F} dF \tag{9.45}$$

From $F = f(h, k, \lambda)$ we have

$$dF = \frac{\partial F}{\partial h} dh + \frac{\partial F}{\partial k} dk + \frac{\partial F}{\partial \lambda} d\lambda \tag{9.46}$$

which can also be written as

$$dF = \frac{a}{r}\left(d\lambda - c_F dh + s_F dk\right) \tag{9.47}$$

because we have $\partial F/\partial h = -\frac{a}{r} c_F$, $\partial F/\partial k = \frac{a}{r} s_F$ and $\partial F/\partial \lambda = a/r$. Replacing dF in Equation (9.45) we get

$$\begin{aligned} dL &= \left[\left(\frac{\partial L}{\partial h}\right)_F - \frac{a}{r} c_F \frac{\partial L}{\partial F}\right] dh + \left[\left(\frac{\partial L}{\partial k}\right)_F + \frac{a}{r} s_F \frac{\partial L}{\partial F}\right] dk + \frac{a}{r}\frac{\partial L}{\partial F} d\lambda \\ &= \left(\frac{\partial L}{\partial h}\right)_{tot} dh + \left(\frac{\partial L}{\partial k}\right)_{tot} dk + \frac{a}{r}\frac{\partial L}{\partial F} d\lambda \end{aligned} \tag{9.48}$$

where the partials $(\partial L/\partial h)_{tot}$ and $(\partial L/\partial k)_{tot}$ have been obtained in [9] and Chapter 8 of this book as

$$\frac{\partial L}{\partial h} = \frac{\frac{a^2}{r^2}\left(1-h^2-k^2\right)^{1/2}\left[\left(1-h^2\beta\right)s_L - hk\beta c_L\right] - s_L - 2\frac{a}{r}h}{(hc_L - ks_L)} \tag{9.49}$$

$$\frac{\partial L}{\partial k} = \frac{\frac{a^2}{r^2}\left(1-h^2-k^2\right)^{1/2}\left[\left(1-k^2\beta\right)c_L - hk\beta s_L\right] - c_L - 2\frac{a}{r}k}{(hc_L - ks_L)} \tag{9.50}$$

Replacing dL in Equation (9.48) in the identity given in Equation (9.44) and equating the coefficients of all the differentials, we get

$$\lambda_a^\lambda da + \lambda_h^\lambda dh + \lambda_k^\lambda dk + \lambda_p^\lambda dp + \lambda_q^\lambda dq + \lambda_\lambda^\lambda d\lambda$$
$$= \lambda_a^L da + \lambda_h^L dh + \lambda_k^L dk + \lambda_p^L dp + \lambda_q^L dq + \lambda_L^L\left[\left(\frac{\partial L}{\partial h}\right)_{tot} dh + \left(\frac{\partial L}{\partial k}\right)_{tot} dk + \frac{a}{r}\frac{\partial L}{\partial F}d\lambda\right]$$

and therefore

$$\lambda_a^\lambda = \lambda_a^L \tag{9.51}$$

$$\lambda_h^\lambda = \lambda_h^L + \lambda_L^L\left(\frac{\partial L}{\partial h}\right)_{tot} \tag{9.52}$$

$$\lambda_k^\lambda = \lambda_k^L + \lambda_L^L\left(\frac{\partial L}{\partial k}\right)_{tot} \tag{9.53}$$

$$\lambda_p^\lambda = \lambda_p^L \tag{9.54}$$

$$\lambda_q^\lambda = \lambda_q^L \tag{9.55}$$

$$\lambda_\lambda^\lambda = \lambda_L^L\frac{a}{r}\frac{\partial L}{\partial F} = \lambda_L^L\frac{a^2}{r^2}\left(1-h^2-k^2\right)^{1/2} \tag{9.56}$$

because $\partial L/\partial F = \frac{a}{r}\left(1-h^2-k^2\right)^{1/2}$ [4]. If the initial orbit is circular, then $(r)_0 = (a)_0$, $(h)_0 = (k)_0 = 0$, and therefore $\left(\lambda_L^L\right)_0 = \left(\lambda_\lambda^\lambda\right)_0$. Furthermore, if we optimize the initial and final orbital positions, we must require $\left(\lambda_L^L\right)_0 = 0$ and $\left(\lambda_L^L\right)_f = 0$. The first condition will in turn result in $\left(\lambda_h^\lambda\right)_0 = \left(\lambda_h^L\right)_0$ and $\left(\lambda_k^\lambda\right)_0 = \left(\lambda_k^L\right)_0$ such that the Lagrange multipliers of both formulations have identical initial values, namely, $\left(\lambda_a^L\right)_0 = \left(\lambda_a^\lambda\right)_0$, $\left(\lambda_h^L\right)_0 = \left(\lambda_h^\lambda\right)_0$, $\left(\lambda_k^L\right)_0 = \left(\lambda_k^\lambda\right)_0$, $\left(\lambda_p^L\right)_0 = \left(\lambda_p^\lambda\right)_0$, $\left(\lambda_q^L\right)_0 = \left(\lambda_q^\lambda\right)_0$ and $\left(\lambda_L^L\right)_0 = \left(\lambda_\lambda^\lambda\right)_0$.

We can verify that the mathematics presented in this chapter are indeed correct by duplicating the minimum-time transfer example presented in [9] and Chapter 8 of this book. Given the initial and target orbits shown in Table 9.1, with both M_0 and M_f free, and using the acceleration value of $f_t = 9.8 \times 10^{-5}$ km/s^2, the converged values of the multipliers at time zero, as well as the optimized value of the initial mean longitude $(\lambda)_0 = -2.2747428$ rad and the minimum time $t_f = 58089.9005$ s of the solution generated in [9] and Chapter 8 of this book are used to run a trajectory with the present formulation to verify that the desired transfer is achieved. Starting from

Table 9.1 Transfer parameters

Orbit	a, km	e	i, deg	Ω, deg	ω, deg	M, deg
Initial	7000	0	28.5	0	0	−130.333164 (optimized)
Target	42,000	10^{-3}	1	0	0	Free
Achieved	41,999.9929	9.983×10^{-4}	0.999797	0.000326	359.995148	46.169264 (optimized)

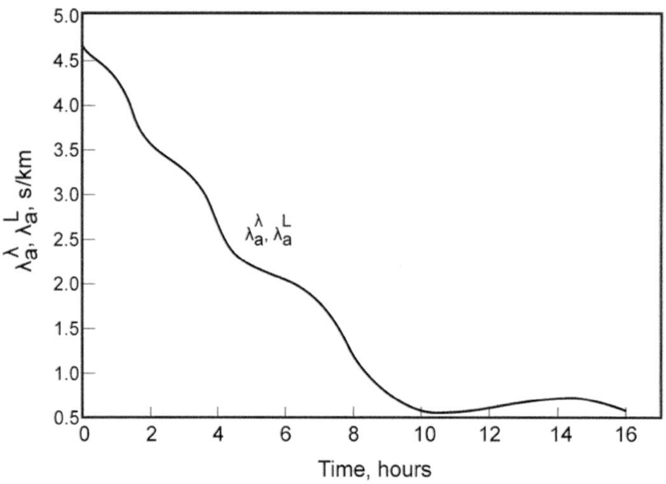

Figure 9.1 Evolution of λ_a multipliers during LEO to near-GEO minimum-time transfer.

$$\left.\begin{aligned}
\left(\lambda_a^L\right)_0 &= 4.675229762 \text{ s/km} \\
\left(\lambda_h^L\right)_0 &= 5.413413947 \times 10^2 \text{ s} \\
\left(\lambda_k^L\right)_0 &= -9.202702084 \times 10^3 \text{ s} \\
\left(\lambda_p^L\right)_0 &= 1.778011878 \times 10^1 \text{ s} \\
\left(\lambda_q^L\right)_0 &= -2.258455855 \times 10^4 \text{ s} \\
(L)_0 &= -2.274742851 \text{ rad}
\end{aligned}\right\} \tag{9.57}$$

and integrating both the system and adjoint equations until t_f, using the optimal control $\hat{\mathbf{u}} = \left[\boldsymbol{\lambda}_z^T B^L(\mathbf{z})\right]^T / \left|\boldsymbol{\lambda}_z^T B^L(\mathbf{z})\right|$, we get the achieved orbit listed in Table 9.1, with $\left(\lambda_L^L\right)_f = 8.3054694 \times 10^{-3}$ s/rad and optimized $(L)_f = 19.6567242$ rad corresponding to $M_f = 46.169264$ deg. Note that $(L)_0$ in Equation (9.57) corresponds to $M_0 = -130.333164$ deg, which is our initial optimized orbit location obtained in [9] and Chapter 8 of this book. The value of M_f achieved here is very close to the optimized solution of [9] and Chapter 8 of this book of $M_f = 46.146408$ deg, duplicating thereby the optimal transfer example of [9] and Chapter 8 of this book because $\lambda_f = M_f + \omega_f + \Omega_f = 19.655282$ rad in both cases for the final location.

We have also obtained an identical solution by iteration starting from guessed values for the various multipliers, $(\lambda)_0$, and flight time t_f. We now compare the multipliers' time histories of both formulations, namely the (a, h, k, p, q, λ) formulation of [9]

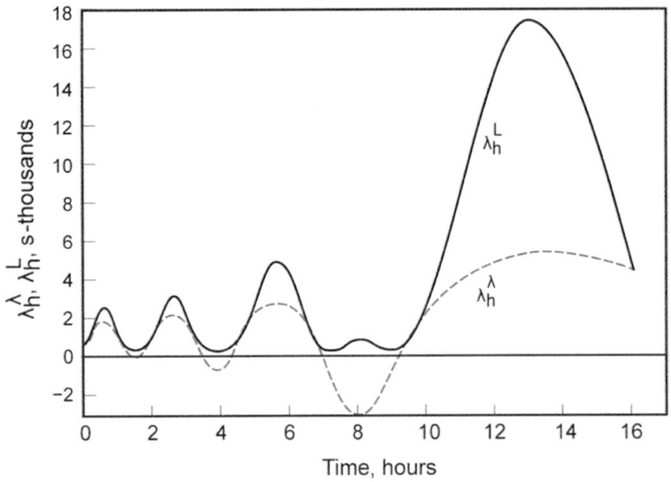

Figure 9.2 Evolution of λ_h multipliers during optimal transfer.

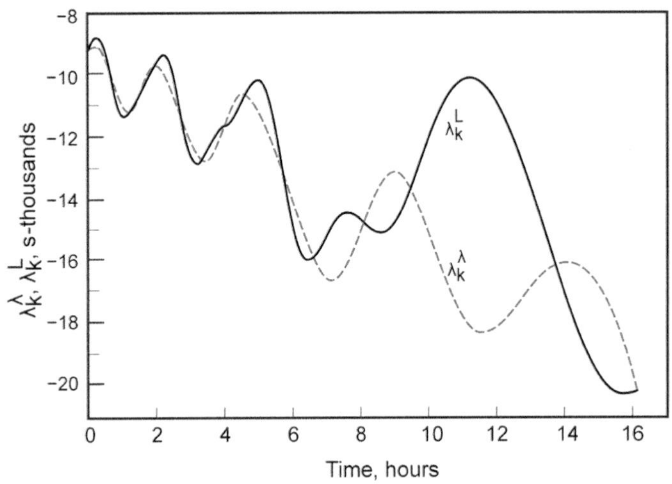

Figure 9.3 Evolution of λ_k multipliers during optimal transfer.

and Chapter 8 of this book and the present (a, h, k, p, q, L) formulation. Figure 9.1 shows the λ_a evolution with identical curves as shown in Equation (9.51). Figures 9.2 and 9.3 are for the λ_h and λ_k multipliers with $\left(\lambda_h^\lambda\right)_0 = \left(\lambda_h^L\right)_0$, $\left(\lambda_k^\lambda\right)_0 = \left(\lambda_k^L\right)_0$ and $\left(\lambda_h^\lambda\right)_f = \left(\lambda_h^L\right)_f$, $\left(\lambda_k^\lambda\right)_f = \left(\lambda_k^L\right)_f$, as predicted by Equations (9.52) and (9.53) because $\left(\lambda_L^L\right)_0 = \left(\lambda_L^L\right)_f = 0$ to satisfy the boundary conditions at the initial and final times, which have been optimized for a free–free minimum-time transfer. Figures 9.4 and 9.5 show that $\left(\lambda_p^\lambda\right) = \left(\lambda_p^L\right)$ and $\left(\lambda_q^\lambda\right) = \left(\lambda_q^L\right)$ at all times, whereas Figure 9.6 shows how $\left(\lambda_\lambda^\lambda\right)$ and $\left(\lambda_L^L\right)$ evolve during the transfer, starting and ending at zero with essentially identical behavior during the first spiraling stage of the transfer where eccentricity stays small,

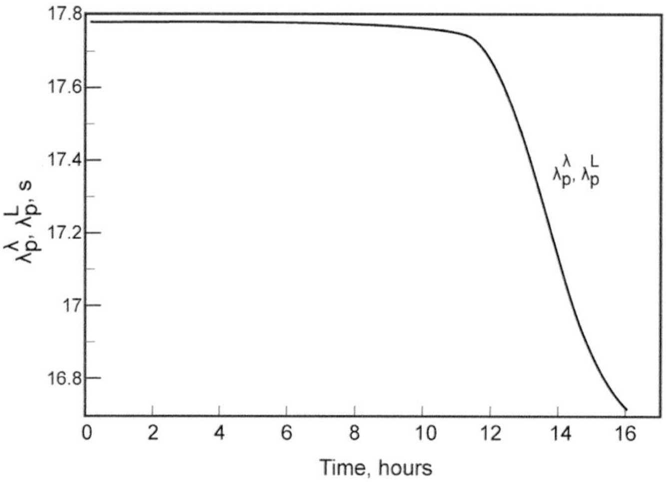

Figure 9.4 Evolution of λ_p multipliers during optimal transfer.

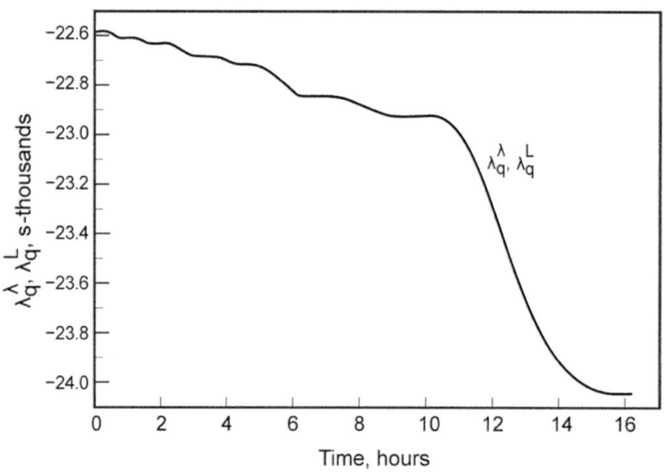

Figure 9.5 Evolution of λ_q multipliers during optimal transfer.

and differing in magnitude as the eccentricity builds up for the direct transfer orbit to geostationary Earth orbit (GEO). This is expected because the true and mean longitudes are identical for zero eccentricity, as is the case with the true and mean anomalies.

9.4 Conclusion

The derivations leading to the generation of the full set of the equations of motion and of their corresponding adjoints are presented in terms of nonsingular orbit elements, with the true longitude representing the sixth state variable. Polar coordinates are used for the resolution of the perturbing accelerations, with all 12 differential equations written directly in terms of the true longitude itself. This formulation is simple both in terms

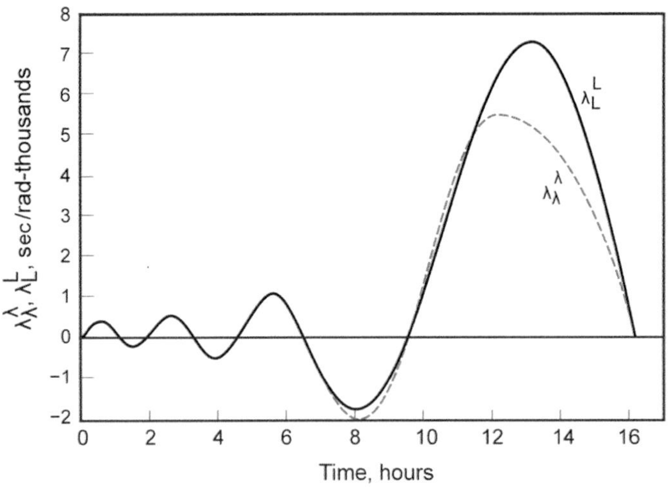

Figure 9.6 Evolution of λ_L and λ_λ multipliers during optimal transfer.

of the boundary conditions and the complexity of the analytic form of the equations for the adjoints. This is amenable to the straightforward treatment of the Earth oblateness perturbation, with the various acceleration components written simply in terms of the true longitude. An example of a minimum-time, free–free, low-thrust orbit transfer generated earlier is duplicated by the present analysis to validate the mathematical derivations shown in this chapter.

Appendix 9.1 The B^L Matrix and its Partial Derivatives

$$B_{11}^L = 2n^{-1}G^{-1}\left(ks_L - hc_L\right) \tag{A9.1}$$

$$\frac{\partial B_{11}^L}{\partial a} = -2n^{-2}\frac{\partial n}{\partial a}G^{-1}\left(ks_L - hc_L\right) \tag{A9.2}$$

$$\frac{\partial B_{11}^L}{\partial h} = 2n^{-1}hG^{-3}\left(ks_L - hc_L\right) - 2n^{-1}G^{-1}c_L \tag{A9.3}$$

$$\frac{\partial B_{11}^L}{\partial k} = 2n^{-1}kG^{-3}\left(ks_L - hc_L\right) + 2n^{-1}G^{-1}s_L \tag{A9.4}$$

$$\frac{\partial B_{11}^L}{\partial p} = \frac{\partial B_{11}^L}{\partial q} = 0 \tag{A9.5}$$

$$\frac{\partial B_{11}^L}{\partial L} = 2n^{-1}G^{-1}\left(kc_L + hs_L\right) \tag{A9.6}$$

$$B_{12}^L = 2n^{-1}ar^{-1}G \tag{A9.7}$$

$$\frac{\partial B_{12}^L}{\partial a} = -2n^{-2}ar^{-1}\frac{\partial n}{\partial a}G \tag{A9.8}$$

$$\frac{\partial B_{12}^L}{\partial h} = -2n^{-1}ar^{-2}\frac{\partial r}{\partial h}G - 2n^{-1}ar^{-1}h\,G^{-1} \tag{A9.9}$$

$$\frac{\partial B_{12}^L}{\partial k} = -2n^{-1}ar^{-2}G\frac{\partial r}{\partial k} - 2n^{-1}ar^{-1}k\,G^{-1} \tag{A9.10}$$

$$\frac{\partial B_{12}^L}{\partial p} = \frac{\partial B_{12}^L}{\partial q} = 0 \tag{A9.11}$$

$$\frac{\partial B_{12}^L}{\partial L} = -2n^{-1}ar^{-2}G\frac{\partial r}{\partial L} \tag{A9.12}$$

where $B_{13}^L = 0$ and therefore all its partials are equal to zero;

$$B_{21}^L = -n^{-1}a^{-1}Gc_L \tag{A9.13}$$

$$\frac{\partial B_{21}^L}{\partial a} = -2^{-1}n^{-1}a^{-2}Gc_L \tag{A9.14}$$

$$\frac{\partial B_{21}^L}{\partial h} = n^{-1}a^{-1}hG^{-1}c_L \tag{A9.15}$$

$$\frac{\partial B_{21}^L}{\partial k} = n^{-1}a^{-1}kG^{-1}c_L \tag{A9.16}$$

$$\frac{\partial B_{21}^L}{\partial p} = \frac{\partial B_{21}^L}{\partial q} = 0 \tag{A9.17}$$

$$\frac{\partial B_{21}^L}{\partial L} = n^{-1}a^{-1}Gs_L \tag{A9.18}$$

$$B_{22}^L = n^{-1}a^{-2}rG^{-1}\,(h+s_L) + n^{-1}a^{-1}Gs_L \tag{A9.19}$$

$$\frac{\partial B_{22}^L}{\partial a} = 2^{-1}n^{-1}a^{-3}rG^{-1}\,(h+s_L) + 2^{-1}n^{-1}a^{-2}Gs_L \tag{A9.20}$$

$$\frac{\partial B_{22}^L}{\partial h} = n^{-1}a^{-2}G^{-1}\,(h+s_L)\left[\frac{\partial r}{\partial h} + rhG^{-2}\right] + n^{-1}a^{-2}rG^{-1} - n^{-1}a^{-1}hs_LG^{-1} \tag{A9.21}$$

$$\frac{\partial B_{22}^L}{\partial k} = n^{-1}a^{-2}G^{-1}\,(h+s_L)\left[\frac{\partial r}{\partial k} + rkG^{-2}\right] - n^{-1}a^{-1}ks_LG^{-1} \tag{A9.22}$$

$$\frac{\partial B_{22}^L}{\partial p} = \frac{\partial B_{22}^L}{\partial q} = 0 \tag{A9.23}$$

$$\frac{\partial B_{22}^L}{\partial L} = n^{-1}a^{-2}\,(h+s_L)\,G^{-1}\frac{\partial r}{\partial L} + n^{-1}a^{-2}rc_LG^{-1} + n^{-1}a^{-1}c_LG \tag{A9.24}$$

$$B_{23}^L = -n^{-1}a^{-2}rG^{-1}k(pc_L - qs_L) \tag{A9.25}$$

$$\frac{\partial B_{23}^L}{\partial a} = -2^{-1}n^{-1}a^{-3}rG^{-1}k\,(pc_L - qs_L) \tag{A9.26}$$

$$\frac{\partial B_{23}^L}{\partial h} = -n^{-1}a^{-2}G^{-1}k\,(pc_L - qs_L)\left[\frac{\partial r}{\partial h} + hrG^{-2}\right] \tag{A9.27}$$

$$\frac{\partial B_{23}^L}{\partial k} = -n^{-1}a^{-2}G^{-1}k(pc_L - qs_L)\left[\frac{\partial r}{\partial k} + krG^{-2}\right] - n^{-1}a^{-2}rG^{-1}\,(pc_L - qs_L) \tag{A9.28}$$

$$\frac{\partial B_{23}^L}{\partial p} = -n^{-1}a^{-2}rG^{-1}kc_L \tag{A9.29}$$

$$\frac{\partial B_{23}^L}{\partial q} = n^{-1}a^{-2}rG^{-1}ks_L \tag{A9.30}$$

$$\frac{\partial B_{23}^L}{\partial L} = -n^{-1}a^{-2}G^{-1}k\left(pc_L - qs_L\right)\frac{\partial r}{\partial L} + n^{-1}a^{-2}rG^{-1}k\left(ps_L + qc_L\right) \tag{A9.31}$$

$$B_{31}^L = n^{-1}a^{-1}Gs_L \tag{A9.32}$$

$$\frac{\partial B_{31}^L}{\partial a} = 2^{-1}n^{-1}a^{-2}Gs_L \tag{A9.33}$$

$$\frac{\partial B_{31}^L}{\partial h} = -n^{-1}a^{-1}G^{-1}hs_L \tag{A9.34}$$

$$\frac{\partial B_{31}^L}{\partial k} = -n^{-1}a^{-1}G^{-1}ks_L \tag{A9.35}$$

$$\frac{\partial B_{31}^L}{\partial p} = \frac{\partial B_{31}^L}{\partial q} = 0 \tag{A9.36}$$

$$\frac{\partial B_{31}^L}{\partial L} = n^{-1}a^{-1}Gc_L \tag{A9.37}$$

$$B_{32}^L = n^{-1}a^{-2}rG^{-1}\left(k + c_L\right) + n^{-1}a^{-1}Gc_L \tag{A9.38}$$

$$\frac{\partial B_{32}^L}{\partial a} = 2^{-1}n^{-1}a^{-3}rG^{-1}\left(k + c_L\right) + 2^{-1}n^{-1}a^{-2}Gc_L \tag{A9.39}$$

$$\frac{\partial B_{32}^L}{\partial h} = n^{-1}a^{-2}G^{-1}\left(k + c_L\right)\left[\frac{\partial r}{\partial h} + hrG^{-2}\right] - n^{-1}a^{-1}hG^{-1}c_L \tag{A9.40}$$

$$\frac{\partial B_{32}^L}{\partial k} = n^{-1}a^{-2}G^{-1}\left(k + c_L\right)\left[\frac{\partial r}{\partial k} + krG^{-2}\right] - n^{-1}a^{-1}kG^{-1}c_L + n^{-1}a^{-2}rG^{-1} \tag{A9.41}$$

$$\frac{\partial B_{32}^L}{\partial p} = \frac{\partial B_{32}^L}{\partial q} = 0 \tag{A9.42}$$

$$\frac{\partial B_{32}^L}{\partial L} = n^{-1}a^{-2}G^{-1}\left(k + c_L\right)\frac{\partial r}{\partial L} - n^{-1}a^{-2}rG^{-1}s_L - n^{-1}a^{-1}Gs_L \tag{A9.43}$$

$$B_{33}^L = n^{-1}a^{-2}rG^{-1}h\left(pc_L - qs_L\right) \tag{A9.44}$$

$$\frac{\partial B_{33}^L}{\partial a} = 2^{-1}n^{-1}a^{-3}rG^{-1}h\left(pc_L - qs_L\right) \tag{A9.45}$$

$$\frac{\partial B_{33}^L}{\partial h} = n^{-1}a^{-2}G^{-1}h\left(pc_L - qs_L\right)\left(\frac{\partial r}{\partial h} + hrG^{-2}\right) + n^{-1}a^{-2}rG^{-1}\left(pc_L - qs_L\right) \tag{A9.46}$$

$$\frac{\partial B_{33}^L}{\partial k} = n^{-1}a^{-2}G^{-1}h\left(pc_L - qs_L\right)\left(\frac{\partial r}{\partial k} + krG^{-2}\right) \tag{A9.47}$$

$$\frac{\partial B_{33}^L}{\partial p} = n^{-1}a^{-2}rG^{-1}hc_L \tag{A9.48}$$

$$\frac{\partial B_{33}^L}{\partial q} = -n^{-1}a^{-2}rG^{-1}hs_L \tag{A9.49}$$

$$\frac{\partial B_{33}^L}{\partial L} = n^{-1}a^{-2}G^{-1}h\left(pc_L - qs_L\right)\frac{\partial r}{\partial L} - n^{-1}a^{-2}rG^{-1}h\left(ps_L + qc_L\right) \tag{A9.50}$$

$$B^L_{43} = 2^{-1}n^{-1}a^{-2}rG^{-1}Ks_L \tag{A9.51}$$

$$\frac{\partial B^L_{43}}{\partial a} = 4^{-1}n^{-1}a^{-3}rG^{-1}Ks_L \tag{A9.52}$$

$$\frac{\partial B^L_{43}}{\partial h} = 2^{-1}n^{-1}a^{-2}G^{-1}Ks_L\left[\frac{\partial r}{\partial h} + hrG^{-2}\right] \tag{A9.53}$$

$$\frac{\partial B^L_{43}}{\partial k} = 2^{-1}n^{-1}a^{-2}G^{-1}Ks_L\left[\frac{\partial r}{\partial k} + krG^{-2}\right] \tag{A9.54}$$

$$\frac{\partial B^L_{43}}{\partial p} = n^{-1}a^{-2}rG^{-1}ps_L \tag{A9.55}$$

$$\frac{\partial B^L_{43}}{\partial q} = n^{-1}a^{-2}rG^{-1}qs_L \tag{A9.56}$$

$$\frac{\partial B^L_{43}}{\partial L} = 2^{-1}n^{-1}a^{-2}G^{-1}Ks_L\frac{\partial r}{\partial L} + 2^{-1}n^{-1}a^{-2}rG^{-1}Kc_L \tag{A9.57}$$

$$B^L_{53} = 2^{-1}n^{-1}a^{-2}rG^{-1}Kc_L \tag{A9.58}$$

$$\frac{\partial B^L_{53}}{\partial a} = 4^{-1}n^{-1}a^{-3}rG^{-1}Kc_L \tag{A9.59}$$

$$\frac{\partial B^L_{53}}{\partial h} = 2^{-1}n^{-1}a^{-2}G^{-1}Kc_L\left[\frac{\partial r}{\partial h} + hrG^{-2}\right] \tag{A9.60}$$

$$\frac{\partial B^L_{53}}{\partial k} = 2^{-1}n^{-1}a^{-2}G^{-1}Kc_L\left[\frac{\partial r}{\partial k} + krG^{-2}\right] \tag{A9.61}$$

$$\frac{\partial B^L_{53}}{\partial p} = n^{-1}a^{-2}rG^{-1}pc_L \tag{A9.62}$$

$$\frac{\partial B^L_{53}}{\partial q} = n^{-1}a^{-2}rG^{-1}qc_L \tag{A9.63}$$

$$\frac{\partial B^L_{53}}{\partial L} = 2^{-1}n^{-1}a^{-2}G^{-1}Kc_L\frac{\partial r}{\partial L} - 2^{-1}n^{-1}a^{-2}rG^{-1}Ks_L \tag{A9.64}$$

$$B^L_{63} = n^{-1}a^{-2}rG^{-1}\left(qs_L - pc_L\right) \tag{A9.65}$$

$$\frac{\partial B^L_{63}}{\partial a} = 2^{-1}n^{-1}a^{-3}rG^{-1}\left(qs_L - pc_L\right) \tag{A9.66}$$

$$\frac{\partial B^L_{63}}{\partial h} = n^{-1}a^{-2}\left(qs_L - pc_L\right)G^{-1}\left[\frac{\partial r}{\partial h} + rhG^{-2}\right] \tag{A9.67}$$

$$\frac{\partial B^L_{63}}{\partial k} = n^{-1}a^{-2}\left(qs_L - pc_L\right)G^{-1}\left[\frac{\partial r}{\partial k} + rkG^{-2}\right] \tag{A9.68}$$

$$\frac{\partial B^L_{63}}{\partial p} = -n^{-1}a^{-2}rG^{-1}c_L \tag{A9.69}$$

$$\frac{\partial B^L_{63}}{\partial q} = n^{-1}a^{-2}rG^{-1}s_L \tag{A9.70}$$

$$\frac{\partial B^L_{63}}{\partial L} = n^{-1}a^{-2}\left(qs_L - pc_L\right)G^{-1}\frac{\partial r}{\partial L} + n^{-1}a^{-2}r\left(qc_L + ps_L\right)G^{-1} \tag{A9.71}$$

Finally,

$$\frac{\partial}{\partial a}\left[\frac{na^2\left(1-h^2-k^2\right)^{1/2}}{r^2}\right]=-\frac{3}{2}nar^{-2}G \tag{A9.72}$$

$$\frac{\partial}{\partial h}\left[\frac{na^2\left(1-h^2-k^2\right)^{1/2}}{r^2}\right]=-2na^2r^{-3}G\frac{\partial r}{\partial h}-na^2r^{-2}hG^{-1} \tag{A9.73}$$

$$\frac{\partial}{\partial k}\left[\frac{na^2\left(1-h^2-k^2\right)^{1/2}}{r^2}\right]=-2na^2r^{-3}G\frac{\partial r}{\partial k}-na^2r^{-2}kG^{-1} \tag{A9.74}$$

$$\frac{\partial}{\partial L}\left[\frac{na^2\left(1-h^2-k^2\right)^{1/2}}{r^2}\right]=-2na^2r^{-3}G\frac{\partial r}{\partial L} \tag{A9.75}$$

References

[1] Broucke, R. A. and Cefola, P. J. (1972) On the Equinoctial Orbit Elements, *Celestial Mechanics*, Vol. 5, 303–310.

[2] Cefola, P. J. (1972) *Equinoctial Orbit Elements: Application to Artificial Satellite Orbits*, AIAA Paper 72–937, AIAA/AAS Astrodynamics Conference, Palo Alto, CA.

[3] Edelbaum, T. N., Sackett, L. L. and Malchow, H. L. (1973) *Optimal Low Thrust Geocentric Transfer*, AIAA Paper 73–1074, AIAA 10th Electric Propulsion Conference, Lake Tahoe, NV.

[4] Cefola, P. J., Long, A. C. and Holloway, G., Jr. (1974) *The Long-Term Prediction of Artificial Satellite Orbits*, AIAA Paper 74–170, AIAA 12th Aerospace Sciences Meeting, Washington, DC.

[5] Sackett, L. L. and Edelbaum, T. N. (1975) *Effect of Attitude Constraints on Solar-Electric Geocentric Transfers*, AIAA Paper 75–350, AIAA 11th Electric Propulsion Conference, New Orleans, LA.

[6] Kechichian, J. A. (1990) *Equinoctial Orbit Elements: Application to Optimal Transfer Problems*, AIAA Paper 90–2976, AIAA/AAS Astrodynamics Conference, Portland, OR.

[7] Kechichian, J. A. (1992) *Optimal Low-Thrust Rendezvous Using Equinoctial Orbit Elements*, IAF Paper 92-0014, 43rd Congress of the International Astronautical Federation, Washington, DC.

[8] Kechichian, J. A. (1993) *Trajectory Optimization Using Eccentric Longitude Formulation*, AAS Paper 93–664, AAS/AIAA Astrodynamics Specialist Conference, Victoria, BC.

[9] Kechichian, J. A. (1995) *The Mechanics of Trajectory Optimization in Terms of Nonsingular Variational Equations in Polar Coordinates*, AAS Paper 95–121, AAS/AIAA Spaceflight Mechanics Meeting, Albuquerque, NM.

[10] Kechichian, J. A. (1996) Optimal Low-Thrust Orbit Transfer, *Orbital Mechanics*, edited by Chobotov, V. A., 2nd ed., AIAA, Reston, VA.

10 The Treatment of the Earth Oblateness Effect in Trajectory Optimization in Equinoctial Coordinates

10.1 Introduction

The effect of the first-order secular perturbation due to the zonal harmonic J_2 in optimal transfer problems is analyzed in this chapter. The J_2-induced acceleration is transformed from the polar frame to the direct equinoctial frame and the dynamic system for the thrust and oblateness perturbations formed. The differential equations for the adjoints are then derived in analytic form such that the full six-state optimal transfer system is integrated numerically. Furthermore, the complete averaged system is also presented with the averaged state and adjoint rates due to the thrust perturbation obtained by numerical quadrature, and the ones due to the oblateness effect derived analytically.

The mechanics of trajectory optimization in terms of nonsingular variational equations in equinoctial coordinates were treated initially in [1] and later extended and completed in [2] and [3] and Chapters 2 and 3 of this book. The first-order secular perturbation due to the second zonal harmonic J_2 is combined with the thrust perturbation to generate exact precision-integrated transfer and rendezvous trajectories, using continuous constant thrust acceleration. The exact six-state thrust-perturbed transfer problem has been analyzed earlier in [2] and [3] and Chapters 2 and 3 of this book by using a set of nonsingular equinoctial orbit elements, where the mean longitude represented the sixth state variable. The variational equations were written in terms of the eccentric longitude and the thrust acceleration vector resolved in the rotating orbital direct equinoctial frame. The J_2-induced acceleration components, given in polar coordinate form, are first transformed to the equinoctial frame, and written in terms of the elements described above. The variational Hamiltonian is then formed by adding the J_2-dependent part to the main thrust-dependent term, and the Euler–Lagrange differential equations for the adjoints derived analytically. This formulation is made consistent by using the same orbital elements to describe the thrust and J_2 perturbations, and by expressing the variational equations in the common equinoctial frame in terms of the eccentric longitude. This chapter shows the mathematical derivations that lead to the generation of the state equations of motion, as well as the equations of the corresponding adjoint variables, such that the full six-state transfer system is presented for numerically integrating the J_2-perturbed, thrust direction-optimized minimum-time trajectory. The J_2 effect was used previously in the context of averaging for rapid numerical integration of long-duration low-thrust orbit transfers in [1]. This in effect eliminated the orbital position dependency of the oblateness, as well as the thrust accelerations, such that the averaged rates of the five slowly varying elements were independent of orbital position. This chapter keeps this dependency intact and is therefore better suited to analyze transfers, which are simulated in more exacting fashion using precision integration.

10.2 Resolution of the J_2 Acceleration in Terms of the Equinoctial Elements

The potential U of the symmetrical model of the Earth is given in terms of the zonal harmonics J_n as

$$U = \frac{\mu}{r}\left[1 - \sum_{n=2}^{\infty} J_n \left(\frac{R}{r}\right)^n P_n(s_\delta)\right] \tag{10.1}$$

with r standing for the magnitude of the spacecraft position vector respective to the center of the Earth, R the equatorial radius of the Earth, δ the spacecraft equatorial declination, $P_n(s_\delta)$ the Legendre polynomials of order n in $\sin\delta$ and μ the Earth gravity constant. Here s_δ stands for $\sin\delta$, and likewise, c_δ for $\cos\delta$, etc. Taking only the J_2 zonal into consideration and not confusing F with the eccentric longitude, the disturbing potential is given by

$$F = U - U_0 = U - \frac{\mu}{r} = -\frac{\mu}{r}\sum_{n=2}^{\infty} J_n \left(\frac{R}{r}\right)^n P_n(s_\delta)$$

$$F = -\frac{\mu}{r} J_2 \left(\frac{R}{r}\right)^2 \left(\frac{3}{2}s_\delta^2 - \frac{1}{2}\right) \tag{10.2}$$

The perturbing acceleration $\mathbf{f}$, being the gradient of F, we have with $s_\delta = \frac{z}{r}$

$$\mathbf{f} = \nabla F = -\mu J_2 R^2 \nabla \left(\frac{3}{2}\frac{z^2}{r^5} - \frac{1}{2r^3}\right)$$

and because $\nabla r = \hat{\mathbf{r}}$, and $\nabla z = \hat{\mathbf{z}}$, this can be expressed in the rotating Euler–Hill frame $(\hat{\mathbf{r}}, \hat{\boldsymbol{\theta}}, \hat{\mathbf{h}})$ as

$$\mathbf{f} = -\mu J_2 R^2 \left[\frac{3z}{r^5}\hat{\mathbf{z}} + \left(\frac{3}{2r^4} - \frac{15}{2}\frac{z^2}{r^6}\right)\hat{\mathbf{r}}\right] \tag{10.3}$$

The unit vectors $\hat{\mathbf{r}}$, $\hat{\boldsymbol{\theta}}$ and $\hat{\mathbf{h}}$, are along the radial, normal and out-of-plane directions, and $\hat{\mathbf{x}}$, $\hat{\mathbf{y}}$ and $\hat{\mathbf{z}}$ are unit vectors defining the Earth-centered inertial equatorial frame. They are related through a rotation matrix involving the Eulerian angles Ω, i and $\theta = \omega + \theta^*$, which are the right ascension of the ascending node, inclination and angular position, with ω being the argument of perigee, and θ^* the true anomaly:

$$\begin{pmatrix}\hat{\mathbf{x}}\\ \hat{\mathbf{y}}\\ \hat{\mathbf{z}}\end{pmatrix} = \begin{pmatrix} c_\Omega c_\theta - s_\Omega c_i s_\theta & -c_\Omega s_\theta - s_\Omega c_i c_\theta & s_\Omega s_i \\ s_\Omega c_\theta + c_\Omega c_i s_\theta & -s_\Omega s_\theta + c_\Omega c_i c_\theta & -c_\Omega s_i \\ s_i s_\theta & s_i c_\theta & c_i \end{pmatrix} \begin{pmatrix}\hat{\mathbf{r}}\\ \hat{\boldsymbol{\theta}}\\ \hat{\mathbf{h}}\end{pmatrix} \tag{10.4}$$

If we use $\hat{\mathbf{z}} = s_i s_\theta \hat{\mathbf{r}} + s_i c_\theta \hat{\boldsymbol{\theta}} + c_i \hat{\mathbf{h}}$, $\mathbf{r} = r\hat{\mathbf{r}}$, such that $z = \hat{\mathbf{z}} \bullet \mathbf{r} = r s_i s_\theta$, then

$$\mathbf{f} = -\frac{3\mu J_2 R^2}{r^4}\left[\left(\frac{1}{2} - \frac{3}{2}s_i^2 s_\theta^2\right)\hat{\mathbf{r}} + s_i^2 s_\theta c_\theta \hat{\boldsymbol{\theta}} + s_i c_i s_\theta \hat{\mathbf{h}}\right] \tag{10.5}$$

Therefore, the $\hat{\mathbf{r}}$, $\hat{\boldsymbol{\theta}}$ and $\hat{\mathbf{h}}$ components of the acceleration $\mathbf{f}$ due to J_2 are simply given as

$$f_r = -\frac{3\mu J_2 R^2}{2r^4}\left(1 - 3s_i^2 s_\theta^2\right) \tag{10.6}$$

$$f_\theta = -\frac{3\mu J_2 R^2}{r^4} s_i^2 s_\theta c_\theta \tag{10.7}$$

$$f_h = -\frac{3\mu J_2 R^2}{r^4} s_i c_i s_\theta \tag{10.8}$$

Let us use the equinoctial elements defined as $a = a$, $h = e\sin(\omega + \Omega)$, $k = e\cos(\omega + \Omega)$, $p = \tan(i/2)\sin\Omega$, $q = \tan(i/2)\cos\Omega$ and $\lambda = M + \omega + \Omega$. The eccentric and true longitudes are defined in terms of the eccentric and true anomalies by $F = E + \omega + \Omega$ and $L = \theta^* + \omega + \Omega$. The mean anomaly M defines the mean longitude λ. From $L = \Omega + \theta$, we have

$$s_\theta = s_L c_\Omega - c_L s_\Omega \tag{10.9}$$

$$c_\theta = c_L c_\Omega + s_L s_\Omega \tag{10.10}$$

We can now make use of the inverse definitions, namely, $e = (h^2 + k^2)^{1/2}$, $i = 2\tan^{-1}(p^2 + q^2)^{1/2}$, $\Omega = \tan^{-1}(p/q)$, $\omega = \tan^{-1}(h/k) - \tan^{-1}(p/q)$, $M = \lambda - \tan^{-1}(h/k)$ and $E = F - \tan^{-1}(h/k)$, $\theta^* = L - \tan^{-1}(h/k)$ to obtain

$$s_\Omega = p/(p^2 + q^2)^{1/2} \tag{10.11}$$

$$c_\Omega = q/(p^2 + q^2)^{1/2} \tag{10.12}$$

$$s_i = 2(p^2 + q^2)^{1/2}/(1 + p^2 + q^2) \tag{10.13}$$

$$c_i = (1 - p^2 - q^2)/(1 + p^2 + q^2) \tag{10.14}$$

We can now express s_θ, c_θ, $s_i^2 s_\theta^2$, $s_i^2 s_\theta c_\theta$ and $s_i c_i s_\theta$ as

$$\begin{aligned}
s_\theta &= (q s_L - p c_L)/(p^2 + q^2)^{1/2}\\
c_\theta &= (q c_L + p s_L)/(p^2 + q^2)^{1/2}\\
s_i^2 s_\theta^2 &= 4(q s_L - p c_L)^2/(1 + p^2 + q^2)^2\\
s_i^2 s_\theta c_\theta &= 4(q s_L - p c_L)(q c_L + p s_L)/(1 + p^2 + q^2)^2\\
s_i c_i s_\theta &= 2(1 - p^2 - q^2)(q s_L - p c_L)/(1 + p^2 + q^2)^2
\end{aligned}$$

and write, f_r, f_θ and f_h directly in terms of a, h, k, p, q and L:

$$f_r = -\frac{3\mu J_2 R^2}{2r^4}\left[1 - 12(q s_L - p c_L)^2/(1 + p^2 + q^2)^2\right] \tag{10.15}$$

$$f_\theta = -\frac{12\mu J_2 R^2}{r^4}(q s_L - p c_L)(q c_L + p s_L)/(1 + p^2 + q^2)^2 \tag{10.16}$$

$$f_h = -\frac{6\mu J_2 R^2}{r^4}(q s_L - p c_L)(1 - p^2 - q^2)/(1 + p^2 + q^2)^2 \tag{10.17}$$

The radial distance r is given by

$$r = a(1 - h^2 - k^2)/(1 + h s_L + k c_L) \tag{10.18}$$

and directly in terms of F, as

$$r = a(1 - kc_F - hs_F) \tag{10.19}$$

In [2] and [3] and Chapters 2 and 3 of this book, the thrust acceleration, which was the only perturbation considered, was resolved along the direct equinoctial frame $\hat{\mathbf{f}}$, $\hat{\mathbf{g}}$, $\hat{\mathbf{w}}$, with $\hat{\mathbf{f}}$ and $\hat{\mathbf{g}}$ unit vectors defining the instantaneous orbit plane, and with the direction of $\hat{\mathbf{f}}$ obtained through a clockwise rotation of an angle Ω from the direction of the ascending node. The unit vector $\hat{\mathbf{w}}$ is along the direction of the instantaneous angular momentum vector. The position and velocity vectors are simply written as

$$\mathbf{r} = X_1\hat{\mathbf{f}} + Y_1\hat{\mathbf{g}}$$
$$\dot{\mathbf{r}} = \dot{X}_1\hat{\mathbf{f}} + \dot{Y}_1\hat{\mathbf{g}}$$

with

$$X_1 = a\left[(1 - h^2\beta)c_F + hk\beta s_F - k\right] \tag{10.20}$$

$$Y_1 = a\left[hk\beta c_F + (1 - k^2\beta)s_F - h\right] \tag{10.21}$$

$$\dot{X}_1 = a^2nr^{-1}\left[hk\beta c_F - (1 - h^2\beta)s_F\right] \tag{10.22}$$

$$\dot{Y}_1 = a^2nr^{-1}\left[(1 - k^2\beta)c_F - hk\beta s_F\right] \tag{10.23}$$

n is the orbit mean motion, and β is defined as

$$\beta = \frac{1}{1 + (1 - h^2 - k^2)^{1/2}} \tag{10.24}$$

Therefore, $(1 - h^2 - k^2)^{1/2} = (1 - \beta)/\beta$ and $(h^2 + k^2) = (2\beta - 1)/\beta^2$.

Because X_1 and Y_1 are the components of $\mathbf{r}$ along $\hat{\mathbf{f}}$ and $\hat{\mathbf{g}}$, respectively, they can also be written as $X_1 = rc_L$, and $Y_1 = rs_L$, so that $\mathbf{r} = rc_L\hat{\mathbf{f}} + rs_L\hat{\mathbf{g}}$, and therefore

$$\hat{\mathbf{r}} = \frac{X_1}{r}\hat{\mathbf{f}} + \frac{Y_1}{r}\hat{\mathbf{g}}$$
$$\hat{\boldsymbol{\theta}} = -\frac{Y_1}{r}\hat{\mathbf{f}} + \frac{X_1}{r}\hat{\mathbf{g}}$$
$$\hat{\mathbf{h}} = \hat{\mathbf{w}}$$

We have the transformation between the polar frame and the equinoctial frame as

$$\begin{pmatrix} \hat{\mathbf{f}} \\ \hat{\mathbf{g}} \\ \hat{\mathbf{w}} \end{pmatrix} = \begin{pmatrix} c_L & -s_L & 0 \\ s_L & c_L & 0 \\ 0 & 0 & 1 \end{pmatrix} \begin{pmatrix} \hat{\mathbf{r}} \\ \hat{\boldsymbol{\theta}} \\ \hat{\mathbf{h}} \end{pmatrix} \tag{10.25}$$

which allows us to express the J_2 acceleration in the $\hat{\mathbf{f}}$, $\hat{\mathbf{g}}$, $\hat{\mathbf{w}}$ frame of interest, simply as

$$f_f = c_Lf_r - s_Lf_\theta$$
$$f_g = s_Lf_r + c_Lf_\theta$$
$$f_w = f_h$$

or in terms of the eccentric longitude F

$$f_f = ar^{-1}\Big[(1-h^2\beta)c_F + hk\beta s_F - k\Big]f_r - ar^{-1}\Big[hk\beta c_F + (1-k^2\beta)s_F - h\Big]f_\theta \tag{10.26}$$

$$f_g = ar^{-1}\Big[hk\beta c_F + (1-k^2\beta)s_F - h\Big]f_r + ar^{-1}\Big[(1-h^2\beta)c_F + hk\beta s_F - k\Big]f_\theta \tag{10.27}$$

$$f_w = f_h \tag{10.28}$$

We need, however, to write f_r, f_θ and f_h in Equations (10.15), (10.16) and (10.17), in terms of Falso, and this is done by using the identities

$$(qs_L - pc_L) = \frac{1}{r}(qY_1 - pX_1)$$
$$(qc_L + ps_L) = \frac{1}{r}(qX_1 + pY_1)$$

with X_1 and Y_1 given by Equations (10.20) and (10.21) in terms of F. In summary, $\mathbf{f} = f_f\hat{\mathbf{f}} + f_g\hat{\mathbf{g}} + f_w\hat{\mathbf{w}}$, with

$$f_f = \frac{X_1}{r}f_r - \frac{Y_1}{r}f_\theta \tag{10.29}$$

$$f_g = \frac{Y_1}{r}f_r + \frac{X_1}{r}f_\theta \tag{10.30}$$

$$f_w = f_h \tag{10.31}$$

$$f_r = -\frac{3}{2}\mu J_2 R^2 r^{-4} + 18\mu J_2 R^2 r^{-6}(1+p^2+q^2)^{-2}(qY_1 - pX_1)^2 \tag{10.32}$$

$$f_\theta = -12\mu J_2 R^2 r^{-6}(1+p^2+q^2)^{-2}(qY_1 - pX_1)(qX_1 + pY_1) \tag{10.33}$$

$$f_h = -6\mu J_2 R^2 r^{-5}(1+p^2+q^2)^{-2}(1-p^2-q^2)(qY_1 - pX_1) \tag{10.34}$$

let $\mathbf{z} = (a\,h\,k\,p\,q\,\lambda)^T$ stand now for our six-state vector, with the mean longitude λ as the sixth orbit element. We now combine the effects of the thrust acceleration $f_t = f/m$ which we will assume to be constant, with $\mathbf{f}$ the acceleration due to J_2, and write the equations of motion as

$$\dot{a} = (\partial a/\partial\dot{\mathbf{r}})^T \bullet (\hat{\mathbf{u}}f_t + \mathbf{f}) \tag{10.35}$$

$$\dot{h} = (\partial h/\partial\dot{\mathbf{r}})^T \bullet (\hat{\mathbf{u}}f_t + \mathbf{f}) \tag{10.36}$$

$$\dot{k} = (\partial k/\partial\dot{\mathbf{r}})^T \bullet (\hat{\mathbf{u}}f_t + \mathbf{f}) \tag{10.37}$$

$$\dot{p} = (\partial p/\partial\dot{\mathbf{r}})^T \bullet (\hat{\mathbf{u}}f_t + \mathbf{f}) \tag{10.38}$$

$$\dot{q} = (\partial q/\partial\dot{\mathbf{r}})^T \bullet (\hat{\mathbf{u}}f_t + \mathbf{f}) \tag{10.39}$$

$$\dot{\lambda} = n + (\partial\lambda/\partial\dot{\mathbf{r}})^T \bullet (\hat{\mathbf{u}}f_t + \mathbf{f}) \tag{10.40}$$

Here f and m are the thrust magnitude and spacecraft mass, and $\hat{\mathbf{u}}$ is a unit vector along the thrust direction, with components u_f, u_g and u_w. The velocity partials are given by

$$\partial a/\partial \dot{\mathbf{r}} = 2a^{-1}n^{-2}\left(\dot{X}_1\hat{\mathbf{f}} + \dot{Y}_1\hat{\mathbf{g}}\right) = M_{11}\hat{\mathbf{f}} + M_{12}\hat{\mathbf{g}} + M_{13}\hat{\mathbf{w}} \tag{10.41}$$

$$\begin{aligned}\partial h/\partial \dot{\mathbf{r}} &= Gn^{-1}a^{-2}\left[\left(\frac{\partial X_1}{\partial k} - h\beta\frac{\dot{X}_1}{n}\right)\hat{\mathbf{f}} + \left(\frac{\partial Y_1}{\partial k} - h\beta\frac{\dot{Y}_1}{n}\right)\hat{\mathbf{g}}\right] \\ &\quad + k\left(qY_1 - pX_1\right)n^{-1}a^{-2}G^{-1}\hat{\mathbf{w}} = M_{21}\hat{\mathbf{f}} + M_{22}\hat{\mathbf{g}} + M_{23}\hat{\mathbf{w}}\end{aligned} \tag{10.42}$$

$$\begin{aligned}\partial k/\partial \dot{\mathbf{r}} &= -Gn^{-1}a^{-2}\left[\left(\frac{\partial X_1}{\partial h} + k\beta\frac{\dot{X}_1}{n}\right)\hat{\mathbf{f}} + \left(\frac{\partial Y_1}{\partial h} + k\beta\frac{\dot{Y}_1}{n}\right)\hat{\mathbf{g}}\right] \\ &\quad - h\left(qY_1 - pX_1\right)n^{-1}a^{-2}G^{-1}\hat{\mathbf{w}} = M_{31}\hat{\mathbf{f}} + M_{32}\hat{\mathbf{g}} + M_{33}\hat{\mathbf{w}}\end{aligned} \tag{10.43}$$

$$\partial p/\partial \dot{\mathbf{r}} = KY_1\frac{n^{-1}a^{-2}G^{-1}}{2}\hat{\mathbf{w}} = M_{41}\hat{\mathbf{f}} + M_{42}\hat{\mathbf{g}} + M_{43}\hat{\mathbf{w}} \tag{10.44}$$

$$\partial q/\partial \dot{\mathbf{r}} = KX_1\frac{n^{-1}a^{-2}G^{-1}}{2}\hat{\mathbf{w}} = M_{51}\hat{\mathbf{f}} + M_{52}\hat{\mathbf{g}} + M_{53}\hat{\mathbf{w}} \tag{10.45}$$

$$\begin{aligned}\partial \lambda/\partial \dot{\mathbf{r}} &= n^{-1}a^{-2}\left[-2X_1 + G\left(h\beta\frac{\partial X_1}{\partial h} + k\beta\frac{\partial X_1}{\partial k}\right)\right]\hat{\mathbf{f}} \\ &\quad + n^{-1}a^{-2}\left[-2Y_1 + G\left(h\beta\frac{\partial Y_1}{\partial h} + k\beta\frac{\partial Y_1}{\partial k}\right)\right]\hat{\mathbf{g}} + n^{-1}a^{-2}G^{-1}\left(qY_1 - pX_1\right)\hat{\mathbf{w}} \\ &= M_{61}\hat{\mathbf{f}} + M_{62}\hat{\mathbf{g}} + M_{63}\hat{\mathbf{w}}\end{aligned} \tag{10.46}$$

where, $G = (1 - h^2 - k^2)^{1/2}$, and $K = 1 + p^2 + q^2$, and where

$$\frac{\partial X_1}{\partial h} = -a\left[(hc_F - ks_F)\left(\beta + \frac{h^2\beta^3}{(1-\beta)}\right) + \frac{a}{r}c_F(h\beta - s_F)\right] \tag{10.47}$$

$$\frac{\partial X_1}{\partial k} = -a\left[(hc_F - ks_F)\frac{hk\beta^3}{(1-\beta)} + 1 + \frac{a}{r}s_F(s_F - h\beta)\right] \tag{10.48}$$

$$\frac{\partial Y_1}{\partial h} = a\left[(hc_F - ks_F)\frac{hk\beta^3}{(1-\beta)} - 1 + \frac{a}{r}c_F(k\beta - c_F)\right] \tag{10.49}$$

$$\frac{\partial Y_1}{\partial k} = a\left[(hc_F - ks_F)\left(\beta + \frac{k^2\beta^3}{(1-\beta)}\right) + \frac{a}{r}s_F(c_F - k\beta)\right] \tag{10.50}$$

The 6×3 M matrix is dependent on the elements $\mathbf{z}$, as well as F, which must be solved for by iteration from Kepler's equation $\lambda = F - ks_F + hc_F$ as λ is being integrated numerically. The partials in Equations (10.47) through (10.50) are derived by allowing for the variation of F with respect to h and k, because F is only an accessory variable and not an independent orbital element.

10.3 Minimum-Time Transfer Around the Oblate Earth

We define a Hamiltonian H for the full differential system in Equations (10.35) through (10.40) as

$$H = \boldsymbol{\lambda}_z^T M(\mathbf{z}, F) f_t \hat{\mathbf{u}} + \lambda_\lambda n + \boldsymbol{\lambda}_z^T M(\mathbf{z}, F)\mathbf{f} \tag{10.51}$$

where $\boldsymbol{\lambda}_z^T = \left(\lambda_a\,\lambda_h\,\lambda_k\,\lambda_p\,\lambda_q\,\lambda_\lambda\right)$ represents the Lagrange multipliers adjoint to the state variables $\mathbf{z}$, such that the Euler–Lagrange differential equations are given by

$$\dot{\boldsymbol{\lambda}}_z = -\frac{\partial H}{\partial \mathbf{z}} = -\boldsymbol{\lambda}_z^T \frac{\partial M}{\partial \mathbf{z}} f_t \hat{\mathbf{u}} - \lambda_\lambda \frac{\partial n}{\partial \mathbf{z}} - \boldsymbol{\lambda}_z^T \frac{\partial M}{\partial \mathbf{z}} \mathbf{f} - \boldsymbol{\lambda}_z^T M \frac{\partial \mathbf{f}}{\partial \mathbf{z}} \tag{10.52}$$

The partials $\partial M/\partial \mathbf{z}$ have been derived in [2] and Chapter 3 of this book under the same assumptions used here, namely with $F = f(h, k, \lambda)$, and $\partial F/\partial h = -(a/r)c_F$, $\partial F/\partial k = (a/r)s_F$ and $\partial F/\partial \lambda = a/r$. Furthermore, $\partial F/\partial a = 0$ and the partials of r are according to $\partial r/\partial a = r/a$, $\partial r/\partial h = a^2 r^{-1}(h - s_F)$, $\partial r/\partial k = a^2 r^{-1}(k - c_F)$, $\partial r/\partial \lambda = (\partial r/\partial F)(\partial F/\partial \lambda)$ with $\partial F/\partial \lambda = a/r$, and $\partial r/\partial F = a(k s_F - h c_F)$. Now the thrust direction is optimized by maximizing the Hamiltonian and selecting the direction $\hat{\mathbf{u}}$ along $\boldsymbol{\lambda}_z^T M(\mathbf{z}, F)$ during the transfer. Minimum-time solutions are obtained by maximizing the performance index $J = -\int_{t_0}^{t_f} dt = t_0 - t_f$, with fixed t_0. The transversality condition at the end time t_f is simply given by $H_f = 1$ [4–6]). We now need to generate the partials $\partial \mathbf{f}/\partial \mathbf{z}$ in Equation (10.52) to complete this analysis. For f_f, we have

$$\frac{\partial f_f}{\partial a} = \frac{\partial (X_1/r)}{\partial a} f_r + \frac{X_1}{r}\frac{\partial f_r}{\partial a} - \frac{\partial (Y_1/r)}{\partial a} f_\theta - \frac{Y_1}{r}\frac{\partial f_\theta}{\partial a} \tag{10.53}$$

$$\frac{\partial f_f}{\partial h} = \frac{\partial (X_1/r)}{\partial h} f_r + \frac{X_1}{r}\frac{\partial f_r}{\partial h} - \frac{\partial (Y_1/r)}{\partial h} f_\theta - \frac{Y_1}{r}\frac{\partial f_\theta}{\partial h} \tag{10.54}$$

$$\frac{\partial f_f}{\partial k} = \frac{\partial (X_1/r)}{\partial k} f_r + \frac{X_1}{r}\frac{\partial f_r}{\partial k} - \frac{\partial (Y_1/r)}{\partial k} f_\theta - \frac{Y_1}{r}\frac{\partial f_\theta}{\partial k} \tag{10.55}$$

$$\frac{\partial f_f}{\partial p} = \frac{X_1}{r}\frac{\partial f_r}{\partial p} - \frac{Y_1}{r}\frac{\partial f_\theta}{\partial p} \tag{10.56}$$

$$\frac{\partial f_f}{\partial q} = \frac{X_1}{r}\frac{\partial f_r}{\partial q} - \frac{Y_1}{r}\frac{\partial f_\theta}{\partial q} \tag{10.57}$$

$$\frac{\partial f_f}{\partial \lambda} = \frac{\partial (X_1/r)}{\partial \lambda} f_r + \frac{X_1}{r}\frac{\partial f_r}{\partial \lambda} - \frac{\partial (Y_1/r)}{\partial \lambda} f_\theta - \frac{Y_1}{r}\frac{\partial f_\theta}{\partial \lambda} \tag{10.58}$$

The partials with respect to λ are obtained by the application of the chain rule, such as $\partial(\bullet)/\partial \lambda = (\partial(\bullet)/\partial F)(\partial F/\partial \lambda)$, with $\partial F/\partial \lambda = a/r$. Thus,

$$\partial(X_1/r)/\partial a = r^{-1}\partial X_1/\partial a - r^{-2} X_1 \partial r/\partial a$$

$$\partial(X_1/r)/\partial h = r^{-1}\partial X_1/\partial h - r^{-2} X_1 \partial r/\partial h$$

$$\partial(X_1/r)/\partial k = r^{-1}\partial X_1/\partial k - r^{-2} X_1 \partial r/\partial k$$

$$\partial(X_1/r)/\partial F = r^{-1}\partial X_1/\partial F - r^{-2} X_1 \partial r/\partial F$$

and similarly,

$$\partial(Y_1/r)/\partial a = r^{-1}\partial Y_1/\partial a - r^{-2} Y_1 \partial r/\partial a$$

$$\partial(Y_1/r)/\partial h = r^{-1}\partial Y_1/\partial h - r^{-2} Y_1 \partial r/\partial h$$

$$\partial(Y_1/r)/\partial k = r^{-1}\partial Y_1/\partial k - r^{-2} Y_1 \partial r/\partial k$$

$$\partial(Y_1/r)/\partial F = r^{-1}\partial Y_1/\partial F - r^{-2} Y_1 \partial r/\partial F$$

with $\partial X_1/\partial h$, $\partial X_1/\partial k$, $\partial Y_1/\partial h$ and $\partial Y_1/\partial k$, given by Equations (10.47) through (10.50), with

$$\partial X_1/\partial F = a[hk\beta c_F - (1 - h^2\beta)s_F]$$
$$\partial Y_1/\partial F = a[-hk\beta s_F + (1 - k^2\beta)c_F]$$
$$\partial X_1/\partial a = X_1/a$$
$$\partial Y_1/\partial a = Y_1/a$$

The partials of r have already been derived above. In an identical manner, the partials of f_g with respect to the state variables are obtained as

$$\frac{\partial f_g}{\partial a} = \frac{\partial (Y_1/r)}{\partial a} f_r + \frac{Y_1}{r}\frac{\partial f_r}{\partial a} + \frac{\partial (X_1/r)}{\partial a} f_\theta + \frac{X_1}{r}\frac{\partial f_\theta}{\partial a} \tag{10.59}$$
$$\frac{\partial f_g}{\partial h} = \frac{\partial (Y_1/r)}{\partial h} f_r + \frac{Y_1}{r}\frac{\partial f_r}{\partial h} + \frac{\partial (X_1/r)}{\partial h} f_\theta + \frac{X_1}{r}\frac{\partial f_\theta}{\partial h} \tag{10.60}$$
$$\frac{\partial f_g}{\partial k} = \frac{\partial (Y_1/r)}{\partial k} f_r + \frac{Y_1}{r}\frac{\partial f_r}{\partial k} + \frac{\partial (X_1/r)}{\partial k} f_\theta + \frac{X_1}{r}\frac{\partial f_\theta}{\partial k} \tag{10.61}$$
$$\frac{\partial f_g}{\partial p} = \frac{Y_1}{r}\frac{\partial f_r}{\partial p} + \frac{X_1}{r}\frac{\partial f_\theta}{\partial p} \tag{10.62}$$
$$\frac{\partial f_g}{\partial q} = \frac{Y_1}{r}\frac{\partial f_r}{\partial q} + \frac{X_1}{r}\frac{\partial f_\theta}{\partial q} \tag{10.63}$$
$$\frac{\partial f_g}{\partial \lambda} = \frac{\partial (Y_1/r)}{\partial \lambda} f_r + \frac{Y_1}{r}\frac{\partial f_r}{\partial \lambda} + \frac{\partial (X_1/r)}{\partial \lambda} f_\theta + \frac{X_1}{r}\frac{\partial f_\theta}{\partial \lambda} \tag{10.64}$$

and finally, the partials of f_w are simply given as

$$\frac{\partial f_w}{\partial a} = \frac{\partial f_h}{\partial a} \tag{10.65}$$
$$\frac{\partial f_w}{\partial h} = \frac{\partial f_h}{\partial h} \tag{10.66}$$
$$\frac{\partial f_w}{\partial k} = \frac{\partial f_h}{\partial k} \tag{10.67}$$
$$\frac{\partial f_w}{\partial p} = \frac{\partial f_h}{\partial p} \tag{10.68}$$
$$\frac{\partial f_w}{\partial q} = \frac{\partial f_h}{\partial q} \tag{10.69}$$
$$\frac{\partial f_w}{\partial \lambda} = \frac{\partial f_h}{\partial \lambda} \tag{10.70}$$

All we need now are the partials of f_r, f_θ and f_h with respect to a, h, k, p, q and λ, and these are derived directly from Equations (10.32) through (10.34). We have:

$$\begin{aligned}\frac{\partial f_r}{\partial a} &= 6\mu J_2 R^2 r^{-5}\frac{\partial r}{\partial a} - 108\mu J_2 R^2 r^{-7} K^{-2}\left(qY_1 - pX_1\right)^2 \frac{\partial r}{\partial a} \\ &\quad + 36\mu J_2 R^2 r^{-6} K^{-2}\left(qY_1 - pX_1\right)\left(q\frac{\partial Y_1}{\partial a} - p\frac{\partial X_1}{\partial a}\right)\end{aligned} \tag{10.71}$$

$$\frac{\partial f_r}{\partial h} = 6\mu J_2 R^2 r^{-5}\frac{\partial r}{\partial h} - 108\mu J_2 R^2 r^{-7} K^{-2}(qY_1 - pX_1)^2 \frac{\partial r}{\partial h} + 36\mu J_2 R^2 r^{-6} K^{-2}(qY_1 - pX_1)\left(q\frac{\partial Y_1}{\partial h} - p\frac{\partial X_1}{\partial h}\right) \tag{10.72}$$

$$\frac{\partial f_r}{\partial k} = 6\mu J_2 R^2 r^{-5}\frac{\partial r}{\partial k} - 108\mu J_2 R^2 r^{-7} K^{-2}(qY_1 - pX_1)^2 \frac{\partial r}{\partial k} + 36\mu J_2 R^2 r^{-6} K^{-2}(qY_1 - pX_1)\left(q\frac{\partial Y_1}{\partial k} - p\frac{\partial X_1}{\partial k}\right) \tag{10.73}$$

$$\frac{\partial f_r}{\partial p} = 18\mu J_2 R^2 r^{-6}\left[-4pK^{-3}(qY_1 - pX_1)^2 - 2X_1 K^{-2}(qY_1 - pX_1)\right] \tag{10.74}$$

$$\frac{\partial f_r}{\partial q} = 18\mu J_2 R^2 r^{-6}\left[-4qK^{-3}(qY_1 - pX_1)^2 + 2Y_1 K^{-2}(qY_1 - pX_1)\right] \tag{10.75}$$

$$\frac{\partial f_r}{\partial F} = 6\mu J_2 R^2 r^{-5}\frac{\partial r}{\partial F} - 108\mu J_2 R^2 r^{-7} K^{-2}(qY_1 - pX_1)^2 \frac{\partial r}{\partial F} + 36\mu J_2 R^2 r^{-6} K^{-2}(qY_1 - pX_1)\left(q\frac{\partial Y_1}{\partial F} - p\frac{\partial X_1}{\partial F}\right) \tag{10.76}$$

The partial $\frac{\partial f_r}{\partial \lambda}$ is obtained from $\frac{\partial f_r}{\partial F}\frac{\partial F}{\partial \lambda}$ with $\frac{\partial F}{\partial \lambda} = \frac{a}{r}$. The same chain rule will apply for the partials of f_θ and f_h with respect to λ given below. For the f_θ component, we have:

$$\frac{\partial f_\theta}{\partial a} = 72\mu J_2 R^2 r^{-7} K^{-2}(qY_1 - pX_1)(qX_1 + pY_1)\frac{\partial r}{\partial a} - 12\mu J_2 R^2 r^{-6} K^{-2} \times \left[\left(q\frac{\partial Y_1}{\partial a} - p\frac{\partial X_1}{\partial a}\right)(qX_1 + pY_1) + (qY_1 - pX_1)\left(q\frac{\partial X_1}{\partial a} + p\frac{\partial Y_1}{\partial a}\right)\right] \tag{10.77}$$

$$\frac{\partial f_\theta}{\partial h} = 72\mu J_2 R^2 r^{-7} K^{-2}(qY_1 - pX_1)(qX_1 + pY_1)\frac{\partial r}{\partial h} - 12\mu J_2 R^2 r^{-6} K^{-2} \times \left[\left(q\frac{\partial Y_1}{\partial h} - p\frac{\partial X_1}{\partial h}\right)(qX_1 + pY_1) + (qY_1 - pX_1)\left(q\frac{\partial X_1}{\partial h} + p\frac{\partial Y_1}{\partial h}\right)\right] \tag{10.78}$$

$$\frac{\partial f_\theta}{\partial k} = 72\mu J_2 R^2 r^{-7} K^{-2}(qY_1 - pX_1)(qX_1 + pY_1)\frac{\partial r}{\partial k} - 12\mu J_2 R^2 r^{-6} K^{-2} \times \left[\left(q\frac{\partial Y_1}{\partial k} - p\frac{\partial X_1}{\partial k}\right)(qX_1 + pY_1) + (qY_1 - pX_1)\left(q\frac{\partial X_1}{\partial k} + p\frac{\partial Y_1}{\partial k}\right)\right] \tag{10.79}$$

$$\frac{\partial f_\theta}{\partial p} = -12\mu J_2 R^2 r^{-6}\left[-4pK^{-3}(qY_1 - pX_1)(qX_1 + pY_1) + K^{-2}\{-X_1(qX_1 + pY_1) + Y_1(qY_1 - pX_1)\}\right] \tag{10.80}$$

$$\frac{\partial f_\theta}{\partial q} = -12\mu J_2 R^2 r^{-6}\left[-4qK^{-3}(qY_1 - pX_1)(qX_1 + pY_1) + K^{-2}\{Y_1(qX_1 + pY_1) + X_1(qY_1 - pX_1)\}\right] \tag{10.81}$$

$$\frac{\partial f_\theta}{\partial F} = 72\mu J_2 R^2 r^{-7} K^{-2}(qY_1 - pX_1)(qX_1 + pY_1)\frac{\partial r}{\partial F} - 12\mu J_2 R^2 r^{-6} K^{-2} \times \left[\left(q\frac{\partial Y_1}{\partial F} - p\frac{\partial X_1}{\partial F}\right)(qX_1 + pY_1) + (qY_1 - pX_1)\left(q\frac{\partial X_1}{\partial F} + p\frac{\partial Y_1}{\partial F}\right)\right] \tag{10.82}$$

Finally, for the f_h component, we get:

$$\frac{\partial f_h}{\partial a} = 30\mu J_2 R^2 r^{-6} K^{-2}\left(1 - p^2 - q^2\right)(qY_1 - pX_1)\frac{\partial r}{\partial a} - 6\mu J_2 R^2 r^{-5} K^{-2}\left(1 - p^2 - q^2\right)\left(q\frac{\partial Y_1}{\partial a} - p\frac{\partial X_1}{\partial a}\right) \tag{10.83}$$

$$\frac{\partial f_h}{\partial h} = 30\mu J_2 R^2 r^{-6} K^{-2}\left(1 - p^2 - q^2\right)(qY_1 - pX_1)\frac{\partial r}{\partial h} - 6\mu J_2 R^2 r^{-5} K^{-2}\left(1 - p^2 - q^2\right)\left(q\frac{\partial Y_1}{\partial h} - p\frac{\partial X_1}{\partial h}\right) \tag{10.84}$$

$$\frac{\partial f_h}{\partial k} = 30\mu J_2 R^2 r^{-6} K^{-2}\left(1 - p^2 - q^2\right)(qY_1 - pX_1)\frac{\partial r}{\partial k} - 6\mu J_2 R^2 r^{-5} K^{-2}\left(1 - p^2 - q^2\right)\left(q\frac{\partial Y_1}{\partial k} - p\frac{\partial X_1}{\partial k}\right) \tag{10.85}$$

$$\frac{\partial f_h}{\partial p} = -6\mu J_2 R^2 r^{-5}\left[-2pK^{-3}(qY_1 - pX_1)\left(3 - p^2 - q^2\right) - X_1 K^{-2}\left(1 - p^2 - q^2\right)\right] \tag{10.86}$$

$$\frac{\partial f_h}{\partial q} = -6\mu J_2 R^2 r^{-5}\left[-2qK^{-3}(qY_1 - pX_1)\left(3 - p^2 - q^2\right) + Y_1 K^{-2}\left(1 - p^2 - q^2\right)\right] \tag{10.87}$$

$$\frac{\partial f_h}{\partial F} = 30\mu J_2 R^2 r^{-6} K^{-2}\left(1 - p^2 - q^2\right)(qY_1 - pX_1)\frac{\partial r}{\partial F} - 6\mu J_2 R^2 r^{-5} K^{-2}\left(1 - p^2 - q^2\right)\left(q\frac{\partial Y_1}{\partial F} - p\frac{\partial X_1}{\partial F}\right) \tag{10.88}$$

10.4 The Thrust and J_2-Perturbed Averaged Dynamic and Adjoint System of Equations

The unaveraged Hamiltonian in Equation (10.51) can be written as $H = H_T + H_{J_2}$, with $H_T = \boldsymbol{\lambda}_z^T M(\mathbf{z}, F) f_t \hat{\mathbf{u}} + \lambda_\lambda n$, and $H_{J_2} = \boldsymbol{\lambda}_z^T M(\mathbf{z}, F)\,\mathbf{f}$, representing the thrust and J_2 contributions. The averaged Hamiltonian for the thrust perturbation is obtained by averaging H_T over the orbital period T_0:

$$\tilde{H}_T = \frac{1}{T_0}\int_0^{T_0} H_T dt = \frac{1}{T_0}\int_{-\pi}^{\pi} \frac{H_T dF}{\dot{F}(\tilde{\mathbf{z}}, F)} \tag{10.89}$$

with $T_0 = 2\pi/\tilde{n}$, $\tilde{n} = \mu^{1/2}\tilde{a}^{-3/2}$, and with $\tilde{a}$ indicating the averaged value of the semimajor axis at time t, etc. Using Kepler's equation $\tilde{\lambda} = F - \tilde{k}s_F + \tilde{h}c_F = \tilde{n}t$, we have $\dot{F} = 2\pi/\left[T_0\left(1 - \tilde{k}c_F - \tilde{h}s_F\right)\right]$, and with $s = (1/T_0)/\dot{F} = \left(1 - \tilde{k}c_F - \tilde{h}s_F\right)/2\pi$, the differential equations for the approximation for the state and costate variables are given by [1]:

$$\dot{\tilde{\mathbf{z}}}_T = \left(\frac{\partial \tilde{H}_T}{\partial \tilde{\boldsymbol{\lambda}}_z}\right)^T = \int_{-\pi}^{\pi} \left(\frac{\partial H_T}{\partial \tilde{\boldsymbol{\lambda}}_z}\right)^T s\left(\tilde{\mathbf{z}}, F\right) dF \tag{10.90}$$

$$\dot{\tilde{\boldsymbol{\lambda}}}_z = -\left(\frac{\partial \tilde{H}_T}{\partial \tilde{\mathbf{z}}}\right)^T = -\int_{-\pi}^{\pi} \left[\left(\frac{\partial H_T}{\partial \tilde{\mathbf{z}}}\right)^T s\left(\tilde{\mathbf{z}}, F\right) + H_T\left(\frac{\partial s}{\partial \tilde{\mathbf{z}}}\right)^T\right] dF \tag{10.91}$$

We have $\partial s/\partial \tilde{a} = 0$, $\partial s/\partial \tilde{h} = -s_F/2\pi$, $\partial s/\partial \tilde{k} = -c_F/2\pi$, $\partial s/\partial \tilde{p} = \partial s/\partial \tilde{q} = 0$, and $\partial s/\partial \tilde{\lambda} = \left(\tilde{k}s_F - \tilde{h}c_F\right)/\left[\left(1 - \tilde{k}c_F - \tilde{h}s_F\right)2\pi\right]$, because $\partial s/\partial \tilde{\lambda} = (\partial s/\partial F)\left(\partial F/\partial \tilde{\lambda}\right)$. For continuous thrusting, the $\partial s/\partial \tilde{\mathbf{z}}$ terms will give a zero contribution, such that $H_T\left(\partial s/\partial \tilde{\mathbf{z}}\right)^T$ will also give a zero net contribution, with H_T being constant on an optimal trajectory. The averaged equations for the thrust perturbation are therefore obtained from Equations (10.92) through (10.103) by numerical quadrature:

$$\dot{\tilde{a}}_T = \frac{1}{2\pi} f_t \int_{-\pi}^{\pi} \left(\frac{\partial a}{\partial \dot{\mathbf{r}}}\right)^T \bullet \hat{\mathbf{u}}\left(1 - \tilde{k}c_F - \tilde{h}s_F\right) dF \tag{10.92}$$

$$\dot{\tilde{h}}_T = \frac{1}{2\pi} f_t \int_{-\pi}^{\pi} \left(\frac{\partial h}{\partial \dot{\mathbf{r}}}\right)^T \bullet \hat{\mathbf{u}}\left(1 - \tilde{k}c_F - \tilde{h}s_F\right) dF \tag{10.93}$$

$$\dot{\tilde{k}}_T = \frac{1}{2\pi} f_t \int_{-\pi}^{\pi} \left(\frac{\partial k}{\partial \dot{\mathbf{r}}}\right)^T \bullet \hat{\mathbf{u}}\left(1 - \tilde{k}c_F - \tilde{h}s_F\right) dF \tag{10.94}$$

$$\dot{\tilde{p}}_T = \frac{1}{2\pi} f_t \int_{-\pi}^{\pi} \left(\frac{\partial p}{\partial \dot{\mathbf{r}}}\right)^T \bullet \hat{\mathbf{u}}\left(1 - \tilde{k}c_F - \tilde{h}s_F\right) dF \tag{10.95}$$

$$\dot{\tilde{q}}_T = \frac{1}{2\pi} f_t \int_{-\pi}^{\pi} \left(\frac{\partial q}{\partial \dot{\mathbf{r}}}\right)^T \bullet \hat{\mathbf{u}}\left(1 - \tilde{k}c_F - \tilde{h}s_F\right) dF \tag{10.96}$$

$$\dot{\tilde{\lambda}}_T = \frac{1}{2\pi} f_t \int_{-\pi}^{\pi} \left(\frac{\partial \lambda}{\partial \dot{\mathbf{r}}}\right)^T \bullet \hat{\mathbf{u}}\left(1 - \tilde{k}c_F - \tilde{h}s_F\right) dF + \frac{1}{2\pi}\int_{-\pi}^{\pi} \tilde{n}\left(1 - \tilde{k}c_F - \tilde{h}s_F\right) dF \tag{10.97}$$

$$\dot{\tilde{\lambda}}_a = \frac{1}{2\pi} f_t \int_{-\pi}^{\pi} -\tilde{\boldsymbol{\lambda}}_z^T \frac{\partial M}{\partial \tilde{a}} \bullet \hat{\mathbf{u}}\left(1 - \tilde{k}c_F - \tilde{h}s_F\right) dF + \frac{1}{2\pi}\int_{-\pi}^{\pi} -\tilde{\lambda}_\lambda \frac{\partial \tilde{n}}{\partial \tilde{a}}\left(1 - \tilde{k}c_F - \tilde{h}s_F\right) dF \tag{10.98}$$

$$\dot{\tilde{\lambda}}_h = \frac{1}{2\pi} f_t \int_{-\pi}^{\pi} -\tilde{\boldsymbol{\lambda}}_z^T \frac{\partial M}{\partial \tilde{h}} \bullet \hat{\mathbf{u}}\left(1 - \tilde{k}c_F - \tilde{h}s_F\right) dF \tag{10.99}$$

$$\dot{\tilde{\lambda}}_k = \frac{1}{2\pi} f_t \int_{-\pi}^{\pi} -\tilde{\boldsymbol{\lambda}}_z^T \frac{\partial M}{\partial \tilde{k}} \bullet \hat{\mathbf{u}}\left(1 - \tilde{k}c_F - \tilde{h}s_F\right) dF \tag{10.100}$$

$$\dot{\tilde{\lambda}}_p = \frac{1}{2\pi} f_t \int_{-\pi}^{\pi} -\tilde{\boldsymbol{\lambda}}_z^T \frac{\partial M}{\partial \tilde{p}} \bullet \hat{\mathbf{u}}\left(1 - \tilde{k}c_F - \tilde{h}s_F\right) dF \tag{10.101}$$

$$\dot{\tilde{\lambda}}_q = \frac{1}{2\pi} f_t \int_{-\pi}^{\pi} -\tilde{\boldsymbol{\lambda}}_z^T \frac{\partial M}{\partial \tilde{q}} \bullet \hat{\mathbf{u}} \left(1 - \tilde{k} c_F - \tilde{h} s_F\right) dF \tag{10.102}$$

$$\dot{\tilde{\lambda}}_\lambda = \frac{1}{2\pi} f_t \int_{-\pi}^{\pi} -\tilde{\boldsymbol{\lambda}}_z^T \frac{\partial M}{\partial \tilde{\lambda}} \bullet \hat{\mathbf{u}} \left(1 - \tilde{k} c_F - \tilde{h} s_F\right) dF \tag{10.103}$$

The thrust direction is optimized by using $\hat{\mathbf{u}} = \tilde{\boldsymbol{\lambda}}_z^T M\left(\tilde{\mathbf{z}}, F\right) / \left|\tilde{\boldsymbol{\lambda}}_z^T M\left(\tilde{\mathbf{z}}, F\right)\right|$, and is therefore a function of $\tilde{\mathbf{z}}$, F and $\tilde{\boldsymbol{\lambda}}_z$. Furthermore, the second term in Equation (10.97) will give a zero net contribution. If we now add the J_2 contribution, then the complete averaged Hamiltonian $\tilde{H}$ will be given by

$$\tilde{H} = \tilde{H}_T + \tilde{H}_{J_2} \tag{10.104}$$

with

$$\tilde{H}_{J_2} = \tilde{\boldsymbol{\lambda}}_z^T \dot{\tilde{\mathbf{z}}}_{J_2} \tag{10.105}$$

and, therefore, the augmented Euler–Lagrange differential equations,

$$\dot{\tilde{\boldsymbol{\lambda}}}_z = -\tilde{\boldsymbol{\lambda}}_z^T \frac{\partial \dot{\tilde{\mathbf{z}}}_{J_2}}{\partial \tilde{\mathbf{z}}} - \int_{-\pi}^{\pi} \left[\left(\frac{\partial H_T}{\partial \tilde{\mathbf{z}}}\right)^T s\left(\tilde{\mathbf{z}}, F\right) + H_T \left(\frac{\partial s}{\partial \tilde{\mathbf{z}}}\right)^T\right] dF \tag{10.106}$$

and $\tilde{H} = \tilde{\boldsymbol{\lambda}}_z^T \dot{\tilde{\mathbf{z}}}_T + \tilde{\boldsymbol{\lambda}}_z^T \dot{\tilde{\mathbf{z}}}_{J_2}$. We thus need both $\dot{\tilde{\mathbf{z}}}_{J_2}$ and $\partial \dot{\tilde{\mathbf{z}}}_{J_2} / \partial \tilde{\mathbf{z}}$ for all six elements, and this can be derived analytically by making use of the analytic form of the secular perturbations of the first order in terms of the classical elements, namely $\dot{\tilde{a}}_{J_2} = \dot{\tilde{e}}_{J_2} = \dot{\tilde{i}}_{J_2} = 0$ and $\dot{\tilde{\omega}}_{J_2} = \frac{3}{2} J_2 \frac{R^2}{\tilde{p}'^2} \tilde{n} \left(2 - \frac{5}{2} s_{\tilde{i}}^2\right)$, $\dot{\tilde{\Omega}}_{J_2} = -\frac{3}{2} J_2 \frac{R^2}{\tilde{p}'^2} \tilde{n} c_{\tilde{i}}$. Here, $\tilde{p}'$ is the mean orbit parameter given by $\tilde{p}' = \tilde{a}\left(1 - \tilde{e}^2\right)$. From the definition of $h = e \sin(\omega + \Omega)$, and using the identities for s_Ω, c_Ω, s_i and c_i given in Equations (10.11) through (10.14), we have:

$$\begin{aligned}
\dot{\tilde{h}}_{J_2} &= \dot{\tilde{e}}_{J_2} \frac{\tilde{h}}{\tilde{e}} + \tilde{k}\left(\dot{\tilde{\omega}}_{J_2} + \dot{\tilde{\Omega}}_{J_2}\right) \\
\dot{\tilde{h}}_{J_2} &= \frac{3}{4} \tilde{k} J_2 R^2 \tilde{p}'^{-2} \tilde{n} \left(5 c_{\tilde{i}}^2 - 1 - 2 c_{\tilde{i}}\right) \\
\dot{\tilde{h}}_{J_2} &= \frac{3}{2} \frac{\tilde{k} J_2 R^2 \mu}{\tilde{n} \tilde{a}^5 \left(1 - \tilde{h}^2 - \tilde{k}^2\right)^2} \left[\frac{1 + 3\left(\tilde{p}^2 + \tilde{q}^2\right)^2 - 6\left(\tilde{p}^2 + \tilde{q}^2\right)}{\left(1 + \tilde{p}^2 + \tilde{q}^2\right)^2}\right]
\end{aligned} \tag{10.107}$$

In a similar way, from $k = e \cos(\omega + \Omega)$, we have:

$$\begin{aligned}
\dot{\tilde{k}}_{J_2} &= -\tilde{h}\left(\dot{\tilde{\omega}}_{J_2} + \dot{\tilde{\Omega}}_{J_2}\right) \\
\dot{\tilde{k}}_{J_2} &= -\frac{3}{2} \frac{\tilde{h} J_2 R^2 \mu}{\tilde{n} \tilde{a}^5 \left(1 - \tilde{h}^2 - \tilde{k}^2\right)^2} \left[\frac{1 + 3\left(\tilde{p}^2 + \tilde{q}^2\right)^2 - 6\left(\tilde{p}^2 + \tilde{q}^2\right)}{\left(1 + \tilde{p}^2 + \tilde{q}^2\right)^2}\right]
\end{aligned} \tag{10.108}$$

From the definition of $p = \tan(i/2) s_\Omega$, we get, with $\dot{\tilde{i}}_{J_2} = 0$,

$$\begin{aligned}
\dot{\tilde{p}}_{J_2} &= \tan\left(\tilde{i}/2\right) c_{\tilde{\Omega}} \dot{\tilde{\Omega}}_{J_2} \\
\dot{\tilde{p}}_{J_2} &= \tilde{q} \dot{\tilde{\Omega}}_{J_2}
\end{aligned}$$

$$\dot{\tilde{p}}_{J_2} = -\frac{3}{2}\frac{\tilde{q}J_2R^2\mu\left(1-\tilde{p}^2-\tilde{q}^2\right)}{\tilde{n}\tilde{a}^5\left(1-\tilde{h}^2-\tilde{k}^2\right)^2\left(1+\tilde{p}^2+\tilde{q}^2\right)} \tag{10.109}$$

From $q = \tan(i/2)\,c_\Omega$, we get

$$\dot{\tilde{q}}_{J_2} = -\tan\left(\tilde{i}/2\right)s_{\tilde{\Omega}}\dot{\tilde{\Omega}}_{J_2}$$

$$\dot{\tilde{q}}_{J_2} = -\tilde{p}\dot{\tilde{\Omega}}_{J_2}$$

$$\dot{\tilde{q}}_{J_2} = \frac{3}{2}\frac{\tilde{p}J_2R^2\mu\left(1-\tilde{p}^2-\tilde{q}^2\right)}{\tilde{n}\tilde{a}^5\left(1-\tilde{h}^2-\tilde{k}^2\right)^2\left(1+\tilde{p}^2+\tilde{q}^2\right)} \tag{10.110}$$

Finally, from $\lambda = M + \omega + \Omega$, we have $\dot{\tilde{\lambda}}_{J_2} = \dot{\tilde{M}}_{J_2} + \dot{\tilde{\omega}}_{J_2} + \dot{\tilde{\Omega}}_{J_2}$, with $\dot{\tilde{M}}_{J_2}$ given by

$$\dot{\tilde{M}}_{J_2} = \frac{3}{2}J_2R^2\tilde{p}'^{-2}\tilde{n}\left(1-\frac{3}{2}s_{\tilde{i}}^2\right)\left(1-\tilde{e}^2\right)^{1/2}$$

which can be written in terms of the equinoctial elements as

$$\dot{\tilde{M}}_{J_2} = \frac{3}{2}\frac{J_2R^2\mu\left(1-\tilde{h}^2-\tilde{k}^2\right)^{-3/2}}{\tilde{n}\tilde{a}^5\left(1+\tilde{p}^2+\tilde{q}^2\right)^2}\left[1+\left(\tilde{p}^2+\tilde{q}^2\right)^2-4\left(\tilde{p}^2+\tilde{q}^2\right)\right] \tag{10.111}$$

Therefore, we have

$$\dot{\tilde{\lambda}}_{J_2} = \dot{\tilde{M}}_{J_2} + \frac{\dot{\tilde{h}}_{J_2}}{\tilde{k}} = \dot{\tilde{M}}_{J_2} - \frac{\dot{\tilde{k}}_{J_2}}{\tilde{h}} \tag{10.112}$$

with $\dot{\tilde{M}}_{J_2}$ given by Equation (10.111). We can now derive the partials $\partial\dot{\tilde{\mathbf{z}}}_{J_2}/\partial\tilde{\mathbf{z}}$ needed in Equation (10.106) analytically. They are, for $\dot{\tilde{h}}_{J_2}$:

$$\partial\dot{\tilde{h}}_{J_2}/\partial\tilde{a} = -7\dot{\tilde{h}}_{J_2}/(2\tilde{a}) \tag{10.113}$$

$$\partial\dot{\tilde{h}}_{J_2}/\partial\tilde{h} = 4\tilde{h}\dot{\tilde{h}}_{J_2}\left(1-\tilde{h}^2-\tilde{k}^2\right)^{-1} \tag{10.114}$$

$$\partial\dot{\tilde{h}}_{J_2}/\partial\tilde{k} = \dot{\tilde{h}}_{J_2}/\tilde{k} + 4\tilde{k}\dot{\tilde{h}}_{J_2}\left(1-\tilde{h}^2-\tilde{k}^2\right)^{-1} \tag{10.115}$$

$$\partial\dot{\tilde{h}}_{J_2}/\partial\tilde{p} = \frac{12\mu J_2R^2\tilde{k}\tilde{p}\left[3\left(\tilde{p}^2+\tilde{q}^2\right)-2\right]}{\tilde{n}\tilde{a}^5\left(1-\tilde{h}^2-\tilde{k}^2\right)^2\left(1+\tilde{p}^2+\tilde{q}^2\right)^3} \tag{10.116}$$

$$\partial\dot{\tilde{h}}_{J_2}/\partial\tilde{q} = \frac{12\mu J_2R^2\tilde{k}\tilde{q}\left[3\left(\tilde{p}^2+\tilde{q}^2\right)-2\right]}{\tilde{n}\tilde{a}^5\left(1-\tilde{h}^2-\tilde{k}^2\right)^2\left(1+\tilde{p}^2+\tilde{q}^2\right)^3} \tag{10.117}$$

For the $\dot{\tilde{k}}_{J_2}$ rate, we have:

$$\partial\dot{\tilde{k}}_{J_2}/\partial\tilde{a} = -7\dot{\tilde{k}}_{J_2}/(2\tilde{a}) \tag{10.118}$$

$$\partial\dot{\tilde{k}}_{J_2}/\partial\tilde{h} = \dot{\tilde{k}}_{J_2}/\tilde{h} + 4\tilde{h}\dot{\tilde{k}}_{J_2}\left(1-\tilde{h}^2-\tilde{k}^2\right)^{-1} \tag{10.119}$$

$$\partial\dot{\tilde{k}}_{J_2}/\partial\tilde{k} = 4\tilde{k}\dot{\tilde{k}}_{J_2}\left(1-\tilde{h}^2-\tilde{k}^2\right)^{-1} \tag{10.120}$$

$$\partial\dot{\tilde{k}}_{J_2}/\partial\tilde{p} = -\frac{12\mu J_2 R^2\tilde{h}\tilde{p}\left[3\left(\tilde{p}^2+\tilde{q}^2\right)-2\right]}{\tilde{n}\tilde{a}^5\left(1-\tilde{h}^2-\tilde{k}^2\right)^2\left(1+\tilde{p}^2+\tilde{q}^2\right)^3} \tag{10.121}$$

$$\partial\dot{\tilde{k}}_{J_2}/\partial\tilde{q} = -\frac{12\mu J_2 R^2\tilde{h}\tilde{q}\left[3\left(\tilde{p}^2+\tilde{q}^2\right)-2\right]}{\tilde{n}\tilde{a}^5\left(1-\tilde{h}^2-\tilde{k}^2\right)^2\left(1+\tilde{p}^2+\tilde{q}^2\right)^3} \tag{10.122}$$

For the $\dot{\tilde{p}}_{J_2}$ rate, we have:

$$\partial\dot{\tilde{p}}_{J_2}/\partial\tilde{a} = -7\dot{\tilde{p}}_{J_2}/(2\tilde{a}) \tag{10.123}$$

$$\partial\dot{\tilde{p}}_{J_2}/\partial\tilde{h} = 4\tilde{h}\dot{\tilde{p}}_{J_2}\left(1-\tilde{h}^2-\tilde{k}^2\right)^{-1} \tag{10.124}$$

$$\partial\dot{\tilde{p}}_{J_2}/\partial\tilde{k} = 4\tilde{k}\dot{\tilde{p}}_{J_2}\left(1-\tilde{h}^2-\tilde{k}^2\right)^{-1} \tag{10.125}$$

$$\partial\dot{\tilde{p}}_{J_2}/\partial\tilde{p} = -\frac{4\tilde{p}\dot{\tilde{p}}_{J_2}}{\left[1-\left(\tilde{p}^2+\tilde{q}^2\right)^2\right]} \tag{10.126}$$

$$\partial\dot{\tilde{p}}_{J_2}/\partial\tilde{q} = -\frac{\dot{\tilde{p}}_{J_2}\left[\left(\tilde{p}^2+\tilde{q}^2\right)^2+4\tilde{q}^2-1\right]}{\tilde{q}\left[1-\left(\tilde{p}^2+\tilde{q}^2\right)^2\right]} \tag{10.127}$$

For the $\dot{\tilde{q}}_{J_2}$ rate, we have:

$$\partial\dot{\tilde{q}}_{J_2}/\partial\tilde{a} = -7\dot{\tilde{q}}_{J_2}/(2\tilde{a}) \tag{10.128}$$

$$\partial\dot{\tilde{q}}_{J_2}/\partial\tilde{h} = 4\tilde{h}\dot{\tilde{q}}_{J_2}\left(1-\tilde{h}^2-\tilde{k}^2\right)^{-1} \tag{10.129}$$

$$\partial\dot{\tilde{q}}_{J_2}/\partial\tilde{k} = 4\tilde{k}\dot{\tilde{q}}_{J_2}\left(1-\tilde{h}^2-\tilde{k}^2\right)^{-1} \tag{10.130}$$

$$\partial\dot{\tilde{q}}_{J_2}/\partial\tilde{p} = \frac{\dot{\tilde{q}}_{J_2}\left[1-\left(\tilde{p}^2+\tilde{q}^2\right)^2-4\tilde{p}^2\right]}{\tilde{p}\left[1-\left(\tilde{p}^2+\tilde{q}^2\right)^2\right]} \tag{10.131}$$

$$\partial\dot{\tilde{q}}_{J_2}/\partial\tilde{q} = -\frac{4\tilde{q}\dot{\tilde{q}}_{J_2}}{\left[1-\left(\tilde{p}^2+\tilde{q}^2\right)^2\right]} \tag{10.132}$$

and for the $\dot{\tilde{\lambda}}_{J_2}$ rate,

$$\partial\dot{\tilde{\lambda}}_{J_2}/\partial\tilde{a} = -\frac{7}{2\tilde{a}}\dot{\tilde{M}}_{J_2} - \frac{7}{2\tilde{a}}\frac{\dot{\tilde{h}}_{J_2}}{\tilde{k}} = -\frac{7}{2\tilde{a}}\dot{\tilde{M}}_{J_2} + \frac{7}{2\tilde{a}}\frac{\dot{\tilde{k}}_{J_2}}{\tilde{h}} \tag{10.133}$$

$$\begin{aligned}\partial\dot{\tilde{\lambda}}_{J_2}/\partial\tilde{h} &= \frac{3\tilde{h}\dot{\tilde{M}}_{J_2}}{\left(1-\tilde{h}^2-\tilde{k}^2\right)} + \frac{4\tilde{h}}{\left(1-\tilde{h}^2-\tilde{k}^2\right)}\frac{\dot{\tilde{h}}_{J_2}}{\tilde{k}} \\ &= \frac{3\tilde{h}\dot{\tilde{M}}_{J_2}}{\left(1-\tilde{h}^2-\tilde{k}^2\right)} - \frac{4\tilde{h}}{\left(1-\tilde{h}^2-\tilde{k}^2\right)}\frac{\dot{\tilde{k}}_{J_2}}{\tilde{h}}\end{aligned} \tag{10.134}$$

$$\partial\dot{\tilde{\lambda}}_{J_2}/\partial\tilde{k} = \frac{3\tilde{k}\dot{\tilde{M}}_{J_2}}{\left(1-\tilde{h}^2-\tilde{k}^2\right)} + \frac{4\tilde{k}}{\left(1-\tilde{h}^2-\tilde{k}^2\right)}\frac{\dot{\tilde{h}}_{J_2}}{\tilde{k}}$$
$$= \frac{3\tilde{k}\dot{\tilde{M}}_{J_2}}{\left(1-\tilde{h}^2-\tilde{k}^2\right)} - \frac{4\tilde{k}}{\left(1-\tilde{h}^2-\tilde{k}^2\right)}\frac{\dot{\tilde{k}}_{J_2}}{\tilde{h}} \tag{10.135}$$

$$\partial\dot{\tilde{\lambda}}_{J_2}/\partial\tilde{p} = \frac{12J_2R^2\mu\tilde{p}\left[3\left(\tilde{p}^2+\tilde{q}^2\right)-2\right]}{\tilde{n}\tilde{a}^5\left(1-\tilde{h}^2-\tilde{k}^2\right)^2\left(1+\tilde{p}^2+\tilde{q}^2\right)^3}$$
$$+\frac{6J_2R^2\mu\tilde{p}\left(1-\tilde{h}^2-\tilde{k}^2\right)^{-3/2}\left[3\left(\tilde{p}^2+\tilde{q}^2\right)-3\right]}{\tilde{n}\tilde{a}^5\left(1+\tilde{p}^2+\tilde{q}^2\right)^3} \tag{10.136}$$

$$\partial\dot{\tilde{\lambda}}_{J_2}/\partial\tilde{q} = \frac{12J_2R^2\mu\tilde{q}\left[3\left(\tilde{p}^2+\tilde{q}^2\right)-2\right]}{\tilde{n}\tilde{a}^5\left(1-\tilde{h}^2-\tilde{k}^2\right)^2\left(1+\tilde{p}^2+\tilde{q}^2\right)^3}$$
$$+\frac{6J_2R^2\mu\tilde{q}\left(1-\tilde{h}^2-\tilde{k}^2\right)^{-3/2}\left[3\left(\tilde{p}^2+\tilde{q}^2\right)-3\right]}{\tilde{n}\tilde{a}^5\left(1+\tilde{p}^2+\tilde{q}^2\right)^3} \tag{10.137}$$

The remaining partials of $\dot{\tilde{h}}_{J_2}$, $\dot{\tilde{k}}_{J_2}$, $\dot{\tilde{p}}_{J_2}$, $\dot{\tilde{q}}_{J_2}$ and $\dot{\tilde{\lambda}}_{J_2}$ with respect to $\tilde{\lambda}$ are all equal to zero, because $\tilde{\lambda}$ does not appear explicitly in these rate equations. Furthermore, $\dot{\tilde{a}}_{J_2}$ is identically zero and so are all its partials.

10.5 Numerical Example

Let us revisit the low-Earth orbit–geostationary orbit minimum-time transfer example used in [3] and the previous chapters. The initial orbit is specified by $a_0 = 7000\,\text{km}$, $e_0 = 0$, $i_0 = 28.5\,\text{deg}$, $\Omega_0 = 0\,\text{deg}$, $\omega_0 = 0\,\text{deg}$ and M_0 free. The final orbit is specified by $a_f = 42{,}000\text{km}$, $e_f = 10^{-3}$, $i_f = 1\,\text{deg}$, $\Omega_f = 0\,\text{deg}$, $\omega_f = 0\,\text{deg}$ and M_f free. Using only the thrust perturbation, and neglecting the **f** acceleration due to J_2 in both the dynamic and adjoint Equations (10.35) through (10.40), and Equation (10.52), this optimal transfer of [3] was solved by iterating on the guessed values of the Lagrange multipliers at time zero, namely $(\lambda_a)_0$, $(\lambda_h)_0$, $(\lambda_k)_0$, $(\lambda_p)_0$, $(\lambda_q)_0$, as well as the initial longitude $(\lambda)_0$ and transfer time t_f, and using $(\lambda_\lambda)_0 = 0$, until the boundary conditions at the final time are satisfied. These end conditions are given by the specified values of a_f, h_f, k_f, p_f and q_f, as well as $(\lambda_\lambda)_f = 0$, and $H_f = 1$. The dynamic and adjoint systems are precision-integrated, with the thrust direction obtained from $\hat{\mathbf{u}} = \boldsymbol{\lambda}_z^T M(\mathbf{z}, F)/|\boldsymbol{\lambda}_z^T M(\mathbf{z}, F)|$, and the objective function to minimize chosen as

$$F' = w_1(a-a_f)^2 + w_2(h-h_f)^2 + w_3(k-k_f)^2 + w_4(p-p_f)^2 + w_5(q-q_f)^2$$
$$+ w_6(\lambda_\lambda - 0)^2 + w_7(H-1)^2$$

The w_i are weights that can be adjusted for scaling purposes. The software of [7], which was adopted to solve this unconstrained minimization problem, is based on a general descent method using a quasi-Newton algorithm. The solution was obtained with $\mu = 398{,}601.3\,\text{km}^3/\text{s}^2$ and $f_t = 9.8 \times 10^{-5}\,\text{km/s}^2$, as

$$\begin{aligned}(\lambda_a)_0 &= 4.675229762\,\text{s/km}\\(\lambda_h)_0 &= 5.413413947 \times 10^2\,\text{s}\\(\lambda_k)_0 &= -9.202702084 \times 10^3\,\text{s}\\(\lambda_p)_0 &= 1.778011878 \times 10^1\,\text{s}\\(\lambda_q)_0 &= -2.258455855 \times 10^4\,\text{s}\end{aligned} \tag{10.138}$$

with $(\lambda)_0 = -2.274742851$ rad, corresponding to the optimal initial mean anomaly $M_0 = -130.333164$ deg and the minimum time t_f of 58,089.90058 s. If we now carry out an integration of the canonical system in Equations (10.35) through (10.40), and Equation (10.52), using $J_2 = 1.08263 \times 10^{-3}$, and starting with the initial parameters in Equation (10.138), and $(\lambda)_0 = -2.274742851$ rad, terminating at $t_f = 58{,}089.90058$ s, and with the thrust direction still optimized by $\hat{\mathbf{u}} = \boldsymbol{\lambda}_z^T M(\mathbf{z}, F)/|\boldsymbol{\lambda}_z^T M(\mathbf{z}, F)|$, then we can establish a measure of the effect of ignoring the J_2 acceleration on the achieved transfer, even though this is a rather rapid transfer that is not overly affected by the J_2 precession. Figure 10.1 shows the evolution of the eccentricity versus the semimajor axis, with $a_f = 40{,}427.5184$ km, and $e_f = 4.024759 \times 10^{-2}$, well off our desired target.

We now use the solution in Equation (10.138) with its corresponding optimized initial longitude of $(\lambda)_0 = -2.274742851$ rad, as the initial guess to start the iterations for the solution of the optimal free–free transfer, for the canonical system in Equations (10.35) through (10.40), and Equation (10.52). The solution is now given by

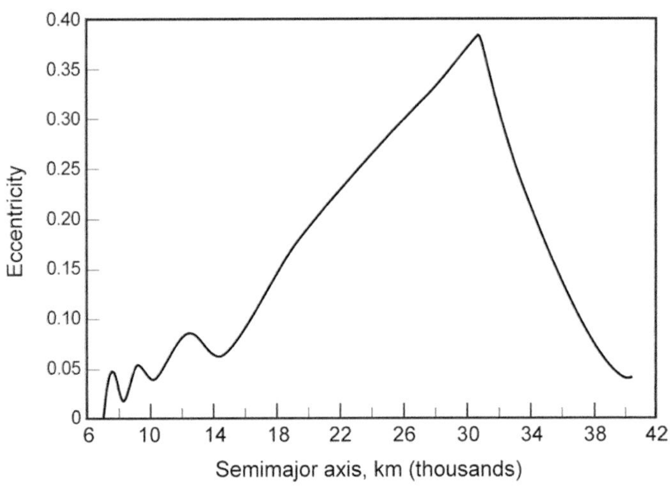

Figure 10.1 Evolution of eccentricity versus semimajor axis for transfer with J_2-unoptimized initial multipliers.

$$
\begin{aligned}
(\lambda_a)_0 &= 4.800100306\,\text{s/km} \\
(\lambda_h)_0 &= 8.060772261 \times 10^2\,\text{s} \\
(\lambda_k)_0 &= -9.150040837 \times 10^3\,\text{s} \\
(\lambda_p)_0 &= 3.281827358 \times 10^1\,\text{s} \\
(\lambda_q)_0 &= -2.254928992 \times 10^4\,\text{s}
\end{aligned}
\tag{10.139}
$$

with $(\lambda)_0 = -2.299291130$ rad, corresponding to $M_0 = -131.7396776$ deg, and the overall minimum time $t_f = 58{,}104.83438$ s, which requires a ΔV of 5.694273769 km/s for $f_t = 9.8 \times 10^{-5}\,\text{km/s}^2$, or roughly 10^{-2} g. The final achieved parameters are obtained as $a_f = 41{,}999.99992$ km, $e_f = 1.000022 \times 10^{-3}$, $i_f = 1.000001$ deg, $\Omega_f = 359.999569$ deg, $\omega_f = 359.999132$ deg and $M_f = 45.411543$ deg, with imposed $(\lambda_\lambda)_0 = 0$ s/rad and iterated $(\lambda_\lambda)_f = 4.07864 \times 10^{-5}$ s/rad and $H_f = 1.000000$, indicating a converged transfer solution. The effect of adding the J_2 acceleration is to increase the duration of the minimum-time transfer by 14.9338 s, when compared to the solution given in [3] and the preceding chapters for the thrust perturbation alone. This precision integration transfer is obtained by numerically integrating Equations (10.35) through (10.40) for the state variables, and Equation (10.52) for the adjoint variables, and therefore corresponds to the exact J_2-perturbed transfer solution. The optimized departure location on the initial orbit has now moved by some 1.406 deg when compared to the solution without the oblateness effect. The evolution of all the multipliers for this exact J_2-perturbed optimal transfer from $t_0 = 0$ to $t_f = 58{,}104.83438$ s is compared with the multipliers histories using the same canonical system of Equations (10.35) through (10.40), and Equation (10.52), starting from the initial values in Equation (10.138) and the corresponding $(\lambda)_0 = -2.274742851$ rad, and terminating at $t_f = 58{,}089.90058$ s, in Figures 10.2 through 10.7. Figure 10.8 shows the variation of the semimajor axis in both cases, with only the optimal solution reaching the desired target of 42,000 km. In

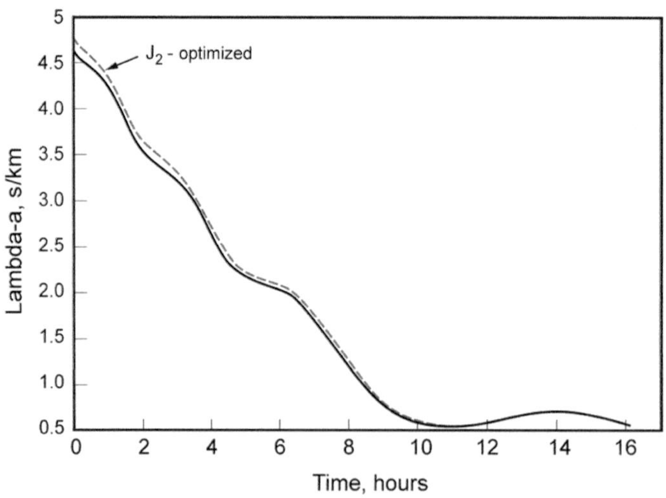

Figure 10.2 Evolution of λ_a multipliers with J_2-optimized and J_2-unoptimized initial values.

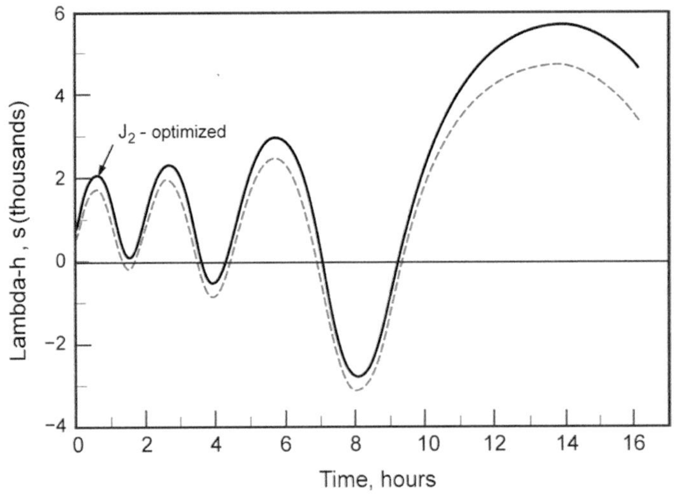

Figure 10.3 Evolution of λ_h multipliers with J_2-optimized and J_2-unoptimized initial values.

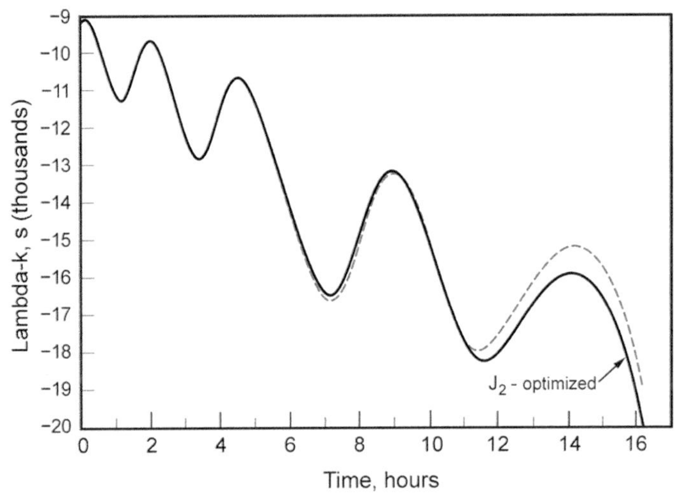

Figure 10.4 Evolution of λ_k multipliers with J_2-optimized and J_2-unoptimized initial values.

practice, it is better to first solve the averaged transfer described in Section 10.4 because the averaged dynamics exhibit much more robust convergence characteristics than the exact system of Sections 10.2 and 10.3, and use the solution thus obtained as initial guess to start the final iterations, leading to the solution of the exact system.

This technique of first solving the averaged transfer is already of great benefit in the thrust-only problem, as illustrated by several examples in [8] and Chapter 2 of this book, and is even more important when additional perturbations such as due to J_2 are considered, even though we have not attempted to show this by an example in this present chapter.

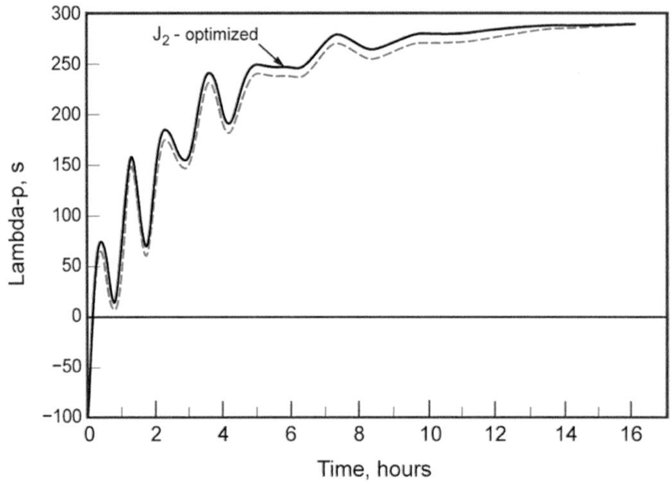

Figure 10.5 Evolution of λ_p multipliers with J_2-optimized and J_2-unoptimized initial values.

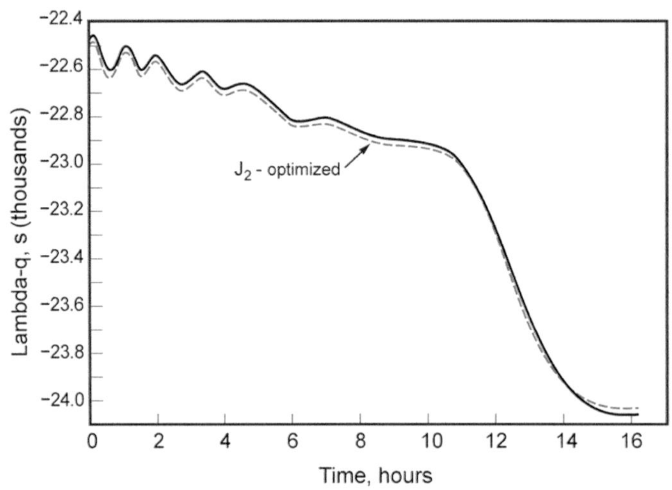

Figure 10.6 Evolution of λ_q multipliers with J_2-optimized and J_2-unoptimized initial values.

10.6 Conclusion

The analysis of the problem of optimal low-thrust transfer and rendezvous with continuous constant acceleration, and based on the use of the nonsingular equinoctial orbit elements, has been extended to include the effect of the first-order secular perturbation due to the second zonal harmonic J_2. The mean longitude has been adopted as the sixth state variable, and both dynamic and adjoint systems of differential equations written in terms of the accessory eccentric longitude variable. The thrust, as well as

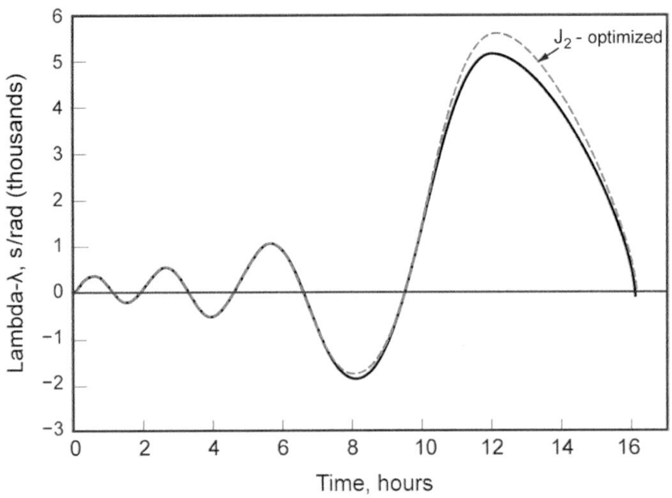

Figure 10.7 Evolution of λ_λ multipliers with J_2-optimized and J_2-unoptimized initial values.

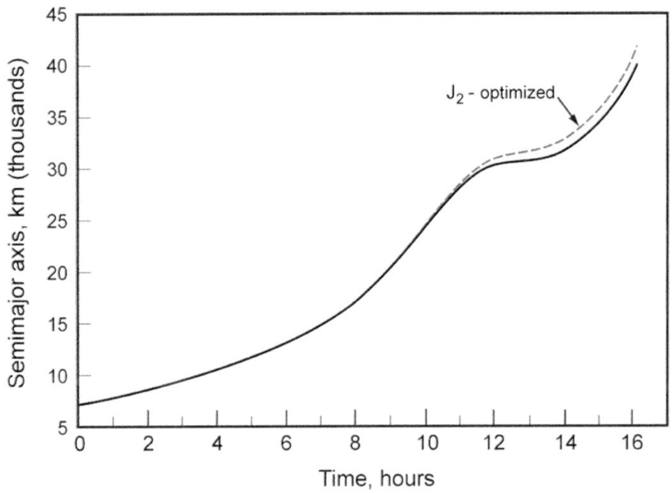

Figure 10.8 Evolution of semimajor axis for transfer with J_2-optimized and J_2-unoptimized initial multipliers.

the oblateness perturbation accelerations, are resolved in the orbital direct equinoctial system, in a consistent way, such that the complete differential system can be numerically integrated to generate optimal transfer and rendezvous trajectories. Furthermore, the full system of averaged equations is also derived for rapid integration. Some simplifications in these derivations are possible by selecting the true longitude as the fast orbital element, and expressing the differential system in terms of this unique variable.

References

[1] Edelbaum, T. N., Sackett, L. L. and Malchow, H. L. (1973) *Optimal Low Thrust Geocentric Transfer*, AIAA Paper 73–1074, AIAA 10th Electric Propulsion Conference, Lake Tahoe, NV.

[2] Kechichian, J. A. (1992) *Optimal Low-Thrust Rendezvous Using Equinoctial Orbit Elements*, IAF Paper 92–0014 43rd Congress of the International Astronautical Federation, Washington, DC.

[3] Kechichian, J. A. (1994) *Optimal LEO–GEO Intermediate Acceleration Orbit Transfer*, AAS Paper 94–125, AAS/AIAA Spaceflight Mechanics Meeting, Cocoa Beach, FL.

[4] Bryson, A. E. and Ho, Y.-C. (1969) *Applied Optimal Control*, Ginn, Waltham, MA.

[5] Marec, J.-P. (1979) *Optimal Space Trajectories*, Elsevier, Amsterdam.

[6] Vinh, N. X. (1981) *Optimal Trajectories in Atmospheric Flight*, Elsevier, Amsterdam.

[7] Kahaner, D., Moler, C. and Nash, S. (1989) *Numerical Methods and Software*, Prentice-Hall, Englewood Cliffs, NJ.

[8] Kechichian, J. A. (1996) Optimal Low-Thrust Orbit Transfer, *Orbital Mechanics*, edited by Chobotov, V. A., 2nd ed., AIAA, Reston, VA, 327–402.

11 Minimum-Time Constant Acceleration Orbit Transfer with First-Order Oblateness Effect

11.1 Introduction

The analysis of the minimum-time thrust and J_2-perturbed orbit transfer between elliptic orbits of arbitrary size, shape and orientation, and using continuous constant acceleration whose direction is optimized, is presented for two different formulations using nonsingular equinoctial orbit elements. A previous formulation that used the equinoctial orbital rotating frame for the component resolution of the perturbation vector is extended to the formulation that uses the true longitude as the sixth state variable, and the polar frame for the component resolution of the thrust and J_2-induced accelerations. This analysis is much simpler, because it provides the simplest form of the differential equations for the adjoints, and removes the need for solving Kepler's transcendental equation during the numerical integration of the dynamic and adjoint system of equations. Because of these simpler features, the corresponding software is also more robust, resulting in improved convergence to the optimal solution of interest. The constant power and constant I_{sp} case, which results in constant thrust, is a trivial extension of the constant acceleration case studied here, because the acceleration is easily updated by updating the mass of the vehicle during the numerical integration.

The problem of minimum-time orbit transfer using continuous constant acceleration whose direction is optimized, and the full set of the nonsingular equinoctial orbit elements as coordinates, has been presented in [1] and Chapter 2 of this book. The differential equations for the coordinates a, h, k, p, q and λ, which were precision integrated, considered only the thrust perturbation which was resolved along the rotating direct equinoctial orbital frame, and they were generated and written in terms of F, which had to be determined through iteration at each integration step from Kepler's equation. The analysis was later made simpler by adopting the Euler–Hill or polar frame as the orbital frame for the thrust perturbation vector resolution, and the differential equations for a, h, k, p, q and λ were written directly in terms of L, which resulted in a simpler form for the equations of the adjoint variables [2, and Chapter 8]. A further simplification was achieved [3, and Chapter 9] by adopting the Euler–Hill frame [2], but also by replacing the sixth element λ by L and by expressing the dynamic equations in terms of L. This change of variable has provided the simplest form of the differential equations for the adjoint variables and eliminated the need to solve for F through iteration from Kepler's equation at each integration step. This optimization analysis was later extended to include the effect of the oblateness of the Earth and the J_2-induced acceleration was added to the thrust acceleration within the original mathematical framework [2]. The analysis of this augmented problem was presented [4, and Chapter 10] with the J_2 acceleration first resolved along the directions of the equinoctial frame, before generating the corresponding equations for the adjoints. This chapter extends the analysis of this problem by casting the mathematics within the new framework [2, 3] for further verification of the results, for simplifying the equations for the adjoints, for again eliminating the need for solving Kepler's equation, and in general for producing a more robust software with better convergence characteristics. The true longitude has also been adopted as the fast element by Geffroy and Epenoy [5]. The next section shows how the full set of the dynamic and adjoint differential equations for the a, h, k, p, q, λ and a, h, k, p, q, L sets are derived, extending the original analysis [2, 3] for the thrust-only case, and thereby providing two additional formulations for comparison with the previous results [4]. After establishing the exact correspondence between the three various versions, and duplicating a previous transfer example [4], it is concluded that the a, h, k, p, q, L formulation, besides being the simpler one, also exhibits the more robust convergence, as the transfer is generated by a single run, as opposed to several runs needed with the previous software [4, and Chapter 10] where the iterated values of the Lagrange multipliers from one run are used as the initial guess to start the next run, until final convergence is achieved. Finally, for the purpose of generating averaged transfers, the averaged rate of L, due to J_2, is derived for the formulation using the a, h, k, p, q, L set, after which the analytic partial derivatives of that particular $\dot{L}_{J_2}$ rate, with respect to the first five slowly varying elements, are also derived.

11.2 The Analysis of the Second Zonal Perturbation Effect in Minimum-Time Low-Thrust Transfers

The perturbing acceleration $\mathbf{f}$ due to the second zonal harmonic J_2 is resolved along the Euler–Hill axes $\hat{\mathbf{r}}$, $\hat{\boldsymbol{\theta}}$ and $\hat{\mathbf{h}}$ and given by

$$f_r = -\frac{3\mu J_2 R^2}{2r^4}\left(1 - 3s_i^2 s_\theta^2\right) \tag{11.1}$$

$$f_\theta = -\frac{3\mu J_2 R^2}{r^4} s_i^2 s_\theta c_\theta \tag{11.2}$$

$$f_h = -\frac{3\mu J_2 R^2}{r^4} s_i c_i s_\theta \tag{11.3}$$

These acceleration components given in terms of the classical elements can also be written in terms of the equinoctial elements a, h, k, p, q and L as

$$f_r = -\frac{3\mu J_2 R^2}{2r^4}\left[1 - 12\frac{(qs_L - pc_L)^2}{\left(1+p^2+q^2\right)^2}\right] \tag{11.4}$$

$$f_\theta = -\frac{12\mu J_2 R^2}{r^4}\frac{(qs_L - pc_L)(qc_L + ps_L)}{\left(1+p^2+q^2\right)^2} \tag{11.5}$$

$$f_h = -\frac{6\mu J_2 R^2}{r^4} \frac{(q s_L - p c_L)\left(1 - p^2 - q^2\right)}{\left(1 + p^2 + q^2\right)^2} \tag{11.6}$$

where the radial distance is given by

$$r = \frac{a\left(1 - h^2 - k^2\right)}{(1 + h s_L + k c_L)} \tag{11.7}$$

The mean longitude λ was selected as the sixth state variable [4, and Chapter 10] such that the state vector was given by $\mathbf{z} = (a\ h\ k\ p\ q\ \lambda)^T$, and the equations of motion by

$$\dot{a} = (\partial a/\partial \dot{\mathbf{r}})\left(\hat{\mathbf{u}} f_t + \mathbf{f}\right) \tag{11.8}$$

$$\dot{h} = (\partial h/\partial \dot{\mathbf{r}})\left(\hat{\mathbf{u}} f_t + \mathbf{f}\right) \tag{11.9}$$

$$\dot{k} = (\partial k/\partial \dot{\mathbf{r}})\left(\hat{\mathbf{u}} f_t + \mathbf{f}\right) \tag{11.10}$$

$$\dot{p} = (\partial p/\partial \dot{\mathbf{r}})\left(\hat{\mathbf{u}} f_t + \mathbf{f}\right) \tag{11.11}$$

$$\dot{q} = (\partial q/\partial \dot{\mathbf{r}})\left(\hat{\mathbf{u}} f_t + \mathbf{f}\right) \tag{11.12}$$

$$\dot{\lambda} = n + (\partial \lambda/\partial \dot{\mathbf{r}})\left(\hat{\mathbf{u}} f_t + \mathbf{f}\right) \tag{11.13}$$

The partial derivatives $\partial a/\partial \dot{\mathbf{r}}$, $\partial h/\partial \dot{\mathbf{r}}$, $\partial k/\partial \dot{\mathbf{r}}$, $\partial p/\partial \dot{\mathbf{r}}$, $\partial q/\partial \dot{\mathbf{r}}$ and $\partial \lambda/\partial \dot{\mathbf{r}}$ are shown in Equations (7) through (12) in [1] and Chapter 2 of this book, and Equations (41) through (46) in [4] and Chapter 10 of this book. They are given in terms of the orbit elements, as well as X_1, Y_1, $\dot{X}_1$, $\dot{Y}_1$ and the partial derivatives $\partial X_1/\partial h$, $\partial X_1/\partial k$, $\partial Y_1/\partial h$, $\partial Y_1/\partial k$. The expressions of the last six quantities are shown in Equations (20) through (23) and (47) through (50) in [4] and Chapter 10 of this book, and they are also listed in [1] on p. 804, as well as in Chapter 2 of this book. X_1, Y_1, $\dot{X}_1$ and $\dot{Y}_1$ are the components of the position and velocity vectors $\mathbf{r} = X_1\hat{\mathbf{f}} + Y_1\hat{\mathbf{g}}$, $\dot{\mathbf{r}} = \dot{X}_1\hat{\mathbf{f}} + \dot{Y}_1\hat{\mathbf{g}}$, respectively, and they are given in terms of the elements a, h, k, as well as the eccentric longitude F. The unit vectors $\hat{\mathbf{f}}$ and $\hat{\mathbf{g}}$ are complemented by $\hat{\mathbf{w}}$ to form the direct equinoctial frame that is obtained by rotating the polar or Euler–Hill frame $(\hat{\mathbf{r}}, \hat{\boldsymbol{\theta}}, \hat{\mathbf{h}})$ through the angle L

$$\begin{pmatrix} \hat{\mathbf{f}} \\ \hat{\mathbf{g}} \\ \hat{\mathbf{w}} \end{pmatrix} = \begin{pmatrix} c_L & -s_L & 0 \\ s_L & c_L & 0 \\ 0 & 0 & 1 \end{pmatrix} \begin{pmatrix} \hat{\mathbf{r}} \\ \hat{\boldsymbol{\theta}} \\ \hat{\mathbf{h}} \end{pmatrix} \tag{11.14}$$

Because the position vector $\mathbf{r}$ can also be written as $\mathbf{r} = r c_L \hat{\mathbf{f}} + r s_L \hat{\mathbf{g}}$, we have $\hat{\mathbf{r}} = \frac{X_1}{r}\hat{\mathbf{f}} + \frac{Y_1}{r}\hat{\mathbf{g}}$, $\hat{\boldsymbol{\theta}} = -\frac{Y_1}{r}\hat{\mathbf{f}} + \frac{X_1}{r}\hat{\mathbf{g}}$ and $\hat{\mathbf{h}} = \hat{\mathbf{w}}$, where $r = a(1 - k c_F - h s_F)$ is given in terms of F. Because of the equivalencies $(q s_L - p c_L) = \frac{1}{r}(q Y_1 - p X_1)$ and $(q c_L + p s_L) = \frac{1}{r}(q X_1 + p Y_1)$, the components of $\mathbf{f}$ in Equations (11.4) through (11.6) can be written directly [4] and Chapter 10 of this book in terms of quantities that are expressed in terms of F:

$$f_r = -\frac{3}{2}\mu J_2 R^2 r^{-4} + 18\mu J_2 R^2 r^{-6}\left(1 + p^2 + q^2\right)^{-2}(q Y_1 - p X_1)^2 \tag{11.15}$$

$$f_\theta = -12\mu J_2 R^2 r^{-6}\left(1 + p^2 + q^2\right)^{-2}(q Y_1 - p X_1)(q X_1 + p Y_1) \tag{11.16}$$

$$f_h = -6\mu J_2 R^2 r^{-5}\left(1 + p^2 + q^2\right)^{-2}\left(1 - p^2 - q^2\right)(q Y_1 - p X_1) \tag{11.17}$$

Thus, the $\hat{\mathbf{f}}$, $\hat{\mathbf{g}}$ and $\hat{\mathbf{w}}$ components of $\mathbf{f}$ are readily obtained from

$$f_f = \frac{X_1}{r} f_r - \frac{Y_1}{r} f_\theta \tag{11.18}$$

$$f_g = \frac{Y_1}{r} f_r + \frac{X_1}{r} f_\theta \tag{11.19}$$

$$f_w = f_h \tag{11.20}$$

with f_r, f_θ and f_h given by Equations (11.15), (11.16) and (11.17). The equations of motion, Equations (11.8) through (11.13), are now determined with $\mathbf{f} = f_f \hat{\mathbf{f}} + f_g \hat{\mathbf{g}} + f_w \hat{\mathbf{w}}$ resolved along the axes of the equinoctial frame, as in the case for the thrust perturbation [4, and Chapter 10]. The Hamiltonian of this differential system is given [4, and Chapter 10] by

$$H = \boldsymbol{\lambda}_z^T M(\mathbf{z}, F) f_t \hat{\mathbf{u}} + \lambda_\lambda n + \boldsymbol{\lambda}_z^T M(\mathbf{z}, F) \mathbf{f} \tag{11.21}$$

where $\boldsymbol{\lambda}_z^T = (\lambda_a \lambda_h \lambda_k \lambda_p \lambda_q \lambda_\lambda)$ is the adjoint variables vector that obeys the following Euler–Lagrange vector differential equation [4, and Chapter 10]:

$$\dot{\boldsymbol{\lambda}}_z = -\frac{\partial H}{\partial \mathbf{z}} = -\boldsymbol{\lambda}_z^T \frac{\partial M}{\partial \mathbf{z}} f_t \hat{\mathbf{u}} - \lambda_\lambda \frac{\partial n}{\partial \mathbf{z}} - \boldsymbol{\lambda}_z^T \frac{\partial M}{\partial \mathbf{z}} \mathbf{f} - \boldsymbol{\lambda}_z^T M \frac{\partial \mathbf{f}}{\partial \mathbf{z}} \tag{11.22}$$

Unlike the mean longitude λ, the adjoint variables appear with an element subscript as both use the same Greek letter in the classical celestial mechanics and optimal control literatures, respectively. The differential Equations (11.8) through (11.13) are given in compact form by

$$\begin{pmatrix} \dot{a} \\ \dot{h} \\ \dot{k} \\ \dot{p} \\ \dot{q} \\ \dot{\lambda} \end{pmatrix} = \begin{pmatrix} M_{11} & M_{12} & M_{13} \\ M_{21} & M_{22} & M_{23} \\ M_{31} & M_{32} & M_{33} \\ M_{41} & M_{42} & M_{43} \\ M_{51} & M_{52} & M_{53} \\ M_{61} & M_{62} & M_{63} \end{pmatrix} \left[\begin{pmatrix} u_f \\ u_g \\ u_w \end{pmatrix} f_t + \begin{pmatrix} f_f \\ f_g \\ f_w \end{pmatrix} \right] + \begin{pmatrix} 0 \\ 0 \\ 0 \\ 0 \\ 0 \\ n \end{pmatrix} \tag{11.23}$$

The $\partial M/\partial \mathbf{z}$ partials are shown in the appendix of [1], except that Equation (A96) has an error in the form of an h instead of k in [1] but correctly in Chapter 3 of this book, and should read as

$$\frac{\partial^2 X_1}{\partial F \partial k} = -a \left[-(h s_F + k c_F) \frac{h k \beta^3}{1-\beta} + \frac{a^2}{r^2} (s_F - h\beta)(c_F - k) + \frac{a}{r} s_F c_F \right] \tag{11.24}$$

The $\partial M/\partial \mathbf{z}$ partials in Equation (11.22) were derived by using the assumption that $F = f(h, k, \lambda)$, such that

$$\begin{aligned} \frac{\partial F}{\partial a} &= 0 & \frac{\partial F}{\partial h} &= -\frac{a}{r} c_F \\ \frac{\partial F}{\partial k} &= \frac{a}{r} s_F & \frac{\partial F}{\partial \lambda} &= \frac{a}{r} \end{aligned} \tag{11.25}$$

Because r is also a function of a, h, k and F, its partials are derived using the same assumption as earlier, such that

$$\begin{aligned} \frac{\partial r}{\partial a} &= \frac{a}{r} & \frac{\partial r}{\partial h} &= \frac{a^2}{r} (h - s_F) \\ \frac{\partial r}{\partial k} &= \frac{a^2}{r} (k - c_F) & \frac{\partial r}{\partial \lambda} &= \frac{\partial r}{\partial F} \frac{\partial F}{\partial \lambda} = \frac{a^2}{r} (k s_F - h c_F) \end{aligned} \tag{11.26}$$

because

$$\frac{\partial r}{\partial F} = a\left(ks_F - hc_F\right) \qquad \frac{\partial F}{\partial \lambda} = \frac{a}{r} \tag{11.27}$$

The $\partial \mathbf{f}/\partial \mathbf{z}$ partials needed in Equation (11.22) involve the partial derivatives of f_r, f_θ and f_h with respect to all six elements a, h, k, p, q and λ, and they have been derived in [4], where they are given by Equations (53) through (58), (59) through (64) and (65) through (70). They also appear in Chapter 10.

The equations of motion, [2] and Chapter 8 of this book, were written directly in terms of L, and the perturbation acceleration resolved along the rotating Euler–Hill frame, and λ was selected as the sixth element. Thus Equations (13) through (18) of [2] were written as

$$\dot{a} = \frac{2f_t}{n\left(1-h^2-k^2\right)^{1/2}}\left[\left(ks_L - hc_L\right)u_r + \left(1 + hs_L + kc_L\right)u_\theta\right] \tag{11.28}$$

$$\begin{aligned}\dot{h} = {} & \frac{\left(1-h^2-k^2\right)^{1/2} f_t}{na\left(1+hs_L+kc_L\right)}\{-(1+hs_L+kc_L)c_L u_r \\ & + \left[h + \left(2 + hs_L + kc_L\right)s_L\right]u_\theta - k\left(pc_L - qs_L\right)u_h\}\end{aligned} \tag{11.29}$$

$$\begin{aligned}\dot{k} = {} & \frac{\left(1-h^2-k^2\right)^{1/2} f_t}{na\left(1+hs_L+kc_L\right)}\{\left(1+hs_L+kc_L\right)s_L u_r \\ & + \left[k + \left(2 + hs_L + kc_L\right)c_L\right]u_\theta + h\left(pc_L - qs_L\right)u_h\}\end{aligned} \tag{11.30}$$

$$\dot{p} = \frac{(1-h^2-k^2)^{1/2} f_t}{2na(1+hs_L+kc_L)}(1+p^2+q^2)s_L u_h \tag{11.31}$$

$$\dot{q} = \frac{(1-h^2-k^2)^{1/2} f_t}{2na(1+hs_L+kc_L)}(1+p^2+q^2)c_L u_h \tag{11.32}$$

$$\begin{aligned}\dot{\boldsymbol{\lambda}} = n - {} & \frac{\left(1-h^2-k^2\right)^{1/2} f_t}{na\left(1+hs_L+kc_L\right)}\left\{\left[\beta\left(1+hs_L+kc_L\right)\left(hs_L+kc_L\right)\right.\right. \\ & \left.\left. +2\left(1-h^2-k^2\right)^{1/2}\right]u_r + \beta\left(2+hs_L+kc_L\right)\left(hc_L-ks_L\right)u_\theta + \left(pc_L-qs_L\right)u_h\right\}\end{aligned} \tag{11.33}$$

Because λ is being integrated, F must be calculated from Kepler's equation, after which L is evaluated from

$$c_L = \left(\frac{a}{r}\right)\left[(1-h^2\beta)c_F + hk\beta s_F - k\right] \tag{11.34}$$

$$s_L = \left(\frac{a}{r}\right)\left[hk\beta c_F + (1-k^2\beta)s_F - h\right] \tag{11.35}$$

as shown in Equations (19) and (20) of [2] and also in Chapter 8 of this book. The equations of motion for this set of state variables $\mathbf{z} = (a\ h\ k\ p\ q\ \lambda)^T$ are written in a compact form as in Equation (26) of [2]:

$$\begin{pmatrix} \dot{a} \\ \dot{h} \\ \dot{k} \\ \dot{p} \\ \dot{q} \\ \dot{\lambda} \end{pmatrix} = \begin{pmatrix} B_{11} & B_{12} & B_{13} \\ B_{21} & B_{22} & B_{23} \\ B_{31} & B_{32} & B_{33} \\ B_{41} & B_{42} & B_{43} \\ B_{51} & B_{52} & B_{53} \\ B_{61} & B_{62} & B_{63} \end{pmatrix} \begin{pmatrix} u_r \\ u_\theta \\ u_h \end{pmatrix} f_t + \begin{pmatrix} 0 \\ 0 \\ 0 \\ 0 \\ 0 \\ n \end{pmatrix} \tag{11.36}$$

Its Hamiltonian is given by

$$H = \boldsymbol{\lambda}_z^T B\,(\mathbf{z}, L) f_t \hat{\mathbf{u}} + \lambda_\lambda n \tag{11.37}$$

and, when the J_2 perturbation is also considered, by

$$H = \boldsymbol{\lambda}_z^T B\,(\mathbf{z}, L) f_t \hat{\mathbf{u}} + \lambda_\lambda n + \boldsymbol{\lambda}_z^T B\,(\mathbf{z}, L)\,\mathbf{f} \tag{11.38}$$

with the components of $\mathbf{f}$ given by Equations (11.4)–(11.6). The unit vector $\hat{\mathbf{u}}$ is now resolved along the $\hat{\mathbf{r}}$, $\hat{\boldsymbol{\theta}}$ and $\hat{\mathbf{h}}$ directions, too. The Euler–Lagrange differential equations for the elements of the $\boldsymbol{\lambda}_z$ vector, namely, $\boldsymbol{\lambda}_z^T = \left(\lambda_a \lambda_h \lambda_k \lambda_p \lambda_q \lambda_\lambda\right)$, are given for the thrust and J_2 perturbations as

$$\dot{\boldsymbol{\lambda}}_z = -\frac{\partial H}{\partial \mathbf{z}} = -\boldsymbol{\lambda}_z^T \frac{\partial B}{\partial \mathbf{z}} f_t \hat{\mathbf{u}} - \lambda_\lambda \frac{\partial n}{\partial \mathbf{z}} - \boldsymbol{\lambda}_z^T \frac{\partial B}{\partial \mathbf{z}} \mathbf{f} - \boldsymbol{\lambda}_z^T B \frac{\partial \mathbf{f}}{\partial \mathbf{z}} \tag{11.39}$$

The partials $\frac{\partial B}{\partial \mathbf{z}}$ have a simpler form [2, and Chapter 8] than the partials $\frac{\partial M}{\partial \mathbf{z}}$ appearing in Equation (11.22), and shown explicitly in [1] and Chapter 3 of this book. They were derived by allowing for the variation of L with respect to h, k and λ, and because L is expressed in terms of F as shown in Equations (11.34) through (11.35), F is also allowed to vary with respect to h, k and λ, as in [2] and Chapter 8 of this book. Thus, $\frac{\partial B}{\partial \mathbf{z}}$ was derived using

$$\begin{aligned} &\frac{\partial F}{\partial a} = 0 && \frac{\partial F}{\partial h} = -\frac{a}{r} c_F \\ &\frac{\partial F}{\partial k} = \frac{a}{r} s_F && \frac{\partial F}{\partial \lambda} = \frac{a}{r} \end{aligned} \tag{11.40}$$

and also using the following partials of the radial distance:

$$\begin{aligned} &\frac{\partial r}{\partial a} = \frac{r}{a} && \frac{\partial r}{\partial h} = -\frac{a[(1-h^2\beta)s_L - hk\beta c_L]}{(1-h^2-k^2)^{1/2}} \\ &\frac{\partial r}{\partial k} = -\frac{a[(1-k^2\beta)c_L - hk\beta s_L]}{(1-h^2-k^2)^{1/2}} && \frac{\partial r}{\partial \lambda} = \frac{a(ks_L - hc_L)}{(1-h^2-k^2)^{1/2}} \end{aligned} \tag{11.41}$$

and finally, the partial derivatives of L, namely

$$\left.\begin{aligned} &\frac{\partial L}{\partial a} = 0 \\ &\frac{\partial L}{\partial h} = \frac{\frac{a^2}{r^2}(1-h^2-k^2)^{1/2}\left[\left(1-h^2\beta\right)s_L - hk\beta c_L\right] - s_L - \frac{2ah}{r}}{(hc_L - ks_L)} \\ &\frac{\partial L}{\partial k} = \frac{\frac{a^2}{r^2}(1-h^2-k^2)^{1/2}\left[\left(1-k^2\beta\right)c_L - hk\beta s_L\right] - c_L - 2\frac{a}{r}k}{(hc_L - ks_L)} \\ &\frac{\partial L}{\partial \lambda} = \frac{a^2}{r^2}(1-h^2-k^2)^{1/2} \end{aligned}\right\} \tag{11.42}$$

The partials in Equations (11.40) through (11.42) are fully derived in [2] and Chapter 8 of this book in great detail. We still need the partials $\frac{\partial \mathbf{f}}{\partial \mathbf{z}}$ in Equation (11.39) to complete the analysis of this particular choice of the dynamic equations. From Equations (11.4) through (11.6) the following are obtained:

$$\frac{\partial f_r}{\partial a} = 6\mu J_2 R^2 r^{-5} \frac{\partial r}{\partial a} - 72\mu J_2 R^2 r^{-5} \frac{\partial r}{\partial a} \left(q s_L - p c_L\right)^2 K^{-2} \tag{11.43}$$

$$\begin{aligned} \frac{\partial f_r}{\partial h} &= 6\mu J_2 R^2 r^{-5} \frac{\partial r}{\partial h} - 72\mu J_2 R^2 r^{-5} \frac{\partial r}{\partial h} \left(q c_L - p s_L\right)^2 K^{-2} \\ &+ 36\mu J_2 R^2 r^{-4} \left(q s_L - p c_L\right) \left(q c_L + p s_L\right) \frac{\partial L}{\partial h} K^{-2} \end{aligned} \tag{11.44}$$

$$\begin{aligned} \frac{\partial f_r}{\partial k} &= 6\mu J_2 R^2 r^{-5} \frac{\partial r}{\partial k} - 72\mu J_2 R^2 r^{-5} \frac{\partial r}{\partial k} \left(q s_L - p c_L\right)^2 K^{-2} \\ &+ 36\mu J_2 R^2 r^{-4} \left(q s_L - p c_L\right) \left(q c_L + p s_L\right) \frac{\partial L}{\partial k} K^{-2} \end{aligned} \tag{11.45}$$

$$\frac{\partial f_r}{\partial p} = -36\mu J_2 R^2 r^{-4} \left(q s_L - p c_L\right) c_L K^{-2} - 72\mu J_2 R^2 r^{-4} p \left(q s_L - p c_L\right)^2 K^{-3} \tag{11.46}$$

$$\frac{\partial f_r}{\partial q} = 36\mu J_2 R^2 r^{-4} \left(q s_L - p c_L\right) s_L K^{-2} - 72\mu J_2 R^2 r^{-4} q \left(q s_L - p c_L\right)^2 K^{-3} \tag{11.47}$$

$$\begin{aligned} \frac{\partial f_r}{\partial \lambda} &= 6\mu J_2 R^2 r^{-5} \frac{\partial r}{\partial \lambda} - 72\mu J_2 R^2 r^{-5} \frac{\partial r}{\partial \lambda} \left(q s_L - p c_L\right)^2 K^{-2} \\ &+ 36\mu J_2 R^2 r^{-4} \left(q s_L - p c_L\right) \left(q c_L + p s_L\right) \frac{\partial L}{\partial \lambda} K^{-2} \end{aligned} \tag{11.48}$$

The f_θ partials are given by

$$\frac{\partial f_\theta}{\partial a} = 48\mu J_2 R^2 r^{-5} \frac{\partial r}{\partial a} \left(q s_L - p c_L\right) \left(q c_L + p s_L\right) K^{-2} \tag{11.49}$$

$$\begin{aligned} \frac{\partial f_\theta}{\partial h} &= 48\mu J_2 R^2 r^{-5} \frac{\partial r}{\partial h} \left(q s_L - p c_L\right) \left(q c_L + p s_L\right) K^{-2} \\ &- 12\mu J_2 R^2 r^{-4} \left[\left(q c_L + p s_L\right)^2 - \left(q s_L - p c_L\right)^2\right] \frac{\partial L}{\partial h} K^{-2} \end{aligned} \tag{11.50}$$

$$\begin{aligned} \frac{\partial f_\theta}{\partial k} &= 48\mu J_2 R^2 r^{-5} \frac{\partial r}{\partial k} \left(q s_L - p c_L\right) \left(q c_L + p s_L\right) K^{-2} \\ &- 12\mu J_2 R^2 r^{-4} \left[\left(q c_L + p s_L\right)^2 - \left(q s_L - p c_L\right)^2\right] \frac{\partial L}{\partial k} K^{-2} \end{aligned} \tag{11.51}$$

$$\begin{aligned} \frac{\partial f_\theta}{\partial p} &= -12\mu J_2 R^2 r^{-4} \Big\{ \big[s_L \left(q s_L - p c_L\right) - c_L \left(q c_L + p s_L\right) \big] K^{-2} \\ &- 4p \left(q s_L - p c_L\right) \left(q c_L + p s_L\right) K^{-3} \Big\} \end{aligned} \tag{11.52}$$

$$\begin{aligned} \frac{\partial f_\theta}{\partial q} &= -12\mu J_2 R^2 r^{-4} \Big\{ \big[s_L \left(q c_L + p s_L\right) + c_L \left(q s_L - p c_L\right) \big] K^{-2} \\ &- 4q \left(q s_L - p c_L\right) \left(q c_L + p s_L\right) K^{-3} \Big\} \end{aligned} \tag{11.53}$$

$$\frac{\partial f_\theta}{\partial \lambda} = 48\mu J_2 R^2 r^{-5} \frac{\partial r}{\partial \lambda} (qs_L - pc_L)(qc_L + ps_L) K^{-2}$$
$$- 12\mu J_2 R^2 r^{-4} \left[(qc_L + ps_L)^2 - (qs_L - pc_L)^2 \right] \frac{\partial L}{\partial \lambda} K^{-2} \tag{11.54}$$

Finally, the f_h partials are given by

$$\frac{\partial f_h}{\partial a} = 24\mu J_2 R^2 r^{-5} \frac{\partial r}{\partial a} (qs_L - pc_L)(2 - K) K^{-2} \tag{11.55}$$

$$\frac{\partial f_h}{\partial h} = 24\mu J_2 R^2 r^{-5} \frac{\partial r}{\partial h} (qs_L - pc_L)(2 - K) K^{-2}$$
$$- 6\mu J_2 R^2 r^{-4} (qc_L + ps_L) \frac{\partial L}{\partial h} (2 - K) K^{-2} \tag{11.56}$$

$$\frac{\partial f_h}{\partial k} = 24\mu J_2 R^2 r^{-5} \frac{\partial r}{\partial k} (qs_L - pc_L)(2 - K) K^{-2}$$
$$- 6\mu J_2 R^2 r^{-4} (qc_L + ps_L) \frac{\partial L}{\partial k} (2 - K) K^{-2} \tag{11.57}$$

$$\frac{\partial f_h}{\partial p} = -6\mu J_2 R^2 r^{-4} \Big\{ \big[-c_L (2 - K) - 2p (qs_L - pc_L) \big] K^{-2}$$
$$- 4p (qs_L - pc_L)(2 - K) K^{-3} \Big\} \tag{11.58}$$

$$\frac{\partial f_h}{\partial q} = -6\mu J_2 R^2 r^{-4} \Big\{ \big[s_L (2 - K) - 2q (qs_L - pc_L) \big] K^{-2}$$
$$- 4q (qs_L - pc_L)(2 - K) K^{-3} \Big\} \tag{11.59}$$

$$\frac{\partial f_h}{\partial \lambda} = 24\mu J_2 R^2 r^{-5} \frac{\partial r}{\partial \lambda} (qs_L - pc_L)(2 - K) K^{-2}$$
$$- 6\mu J_2 R^2 r^{-4} \left[(qc_L + ps_L) \frac{\partial L}{\partial \lambda} (2 - K) K^{-2} \right] \tag{11.60}$$

The partials of r and L with respect to the elements, which appear in the partial derivatives of f_r, f_θ and f_h shown earlier, are given by Equations (11.41) through (11.42). Further simplifications to this analysis are possible if we use the equations of motion developed previously [3, and Chapter 9], and that use L as the sixth state variable. The first five equations are identical to Equations (11.28) through (11.32), and the sixth equation given by Equation (19) or (20) of [3] and also in Chapter 9 of this book, or

$$\dot{L} = \frac{na^2 \left(1 - h^2 - k^2\right)^{1/2}}{r^2} + \frac{r (qs_L - pc_L)}{na^2 \left(1 - h^2 - k^2\right)^{1/2}} f_h \tag{11.61}$$

The basic dynamic equations for the thrust perturbation are written in compact form as

$$\begin{pmatrix} \dot{a} \\ \dot{h} \\ \dot{k} \\ \dot{p} \\ \dot{q} \\ \dot{L} \end{pmatrix} = \begin{pmatrix} B^L_{11} & B^L_{12} & B^L_{13} \\ B^L_{21} & B^L_{22} & B^L_{23} \\ B^L_{31} & B^L_{32} & B^L_{33} \\ B^L_{41} & B^L_{42} & B^L_{43} \\ B^L_{51} & B^L_{52} & B^L_{53} \\ B^L_{61} & B^L_{62} & B^L_{63} \end{pmatrix} \begin{pmatrix} u_r \\ u_\theta \\ u_h \end{pmatrix} f_t + \begin{pmatrix} 0 \\ 0 \\ 0 \\ 0 \\ 0 \\ \frac{na^2\left(1-h^2-k^2\right)^{1/2}}{r^2} \end{pmatrix} \tag{11.62}$$

The Hamiltonian involving the thrust perturbation is given by $H = \boldsymbol{\lambda}_z^T B^L(\mathbf{z}) f_t \hat{\mathbf{u}} + \lambda_L \frac{na^2(1-h^2-k^2)^{1/2}}{r^2}$ and for the combined thrust and J_2 perturbations by

$$H = \boldsymbol{\lambda}_z^T B^L(\mathbf{z}) f_t \hat{\mathbf{u}} + \lambda_L \frac{na^2\left(1-h^2-k^2\right)^{1/2}}{r^2} + \boldsymbol{\lambda}_z^T B^L(\mathbf{z})\,\mathbf{f} \tag{11.63}$$

The adjoint equations take the form

$$\begin{aligned}\dot{\boldsymbol{\lambda}}_z = -\frac{\partial H}{\partial \mathbf{z}} &= -\boldsymbol{\lambda}_z^T \frac{\partial B^L}{\partial \mathbf{z}} f_t \hat{\mathbf{u}} - \lambda_L \frac{\partial}{\partial \mathbf{z}}\left[\frac{na^2\left(1-h^2-k^2\right)^{1/2}}{r^2}\right] \\ &- \boldsymbol{\lambda}_z^T \frac{\partial B^L}{\partial \mathbf{z}}\mathbf{f} - \boldsymbol{\lambda}_z^T B^L \frac{\partial \mathbf{f}}{\partial \mathbf{z}}\end{aligned} \tag{11.64}$$

As shown earlier [3, and Chapter 9], the partials of L with respect to a, h and k are identically equal to zero, with $\frac{\partial L}{\partial L} = 1$, and the partials of r with respect to a, h, k and L are given by

$$\begin{aligned}\frac{\partial r}{\partial a} &= \frac{r}{a} & \frac{\partial r}{\partial h} &= -\frac{r}{a\left(1-h^2-k^2\right)}(2ah + rs_L) \\ \frac{\partial r}{\partial k} &= -\frac{r}{a\left(1-h^2-k^2\right)}(2ak + rc_L) & \frac{\partial r}{\partial L} &= -\frac{r^2}{a}\frac{(hc_L - ks_L)}{\left(1-h^2-k^2\right)}\end{aligned} \tag{11.65}$$

The 6×3 matrix B^L, as well as its $\frac{\partial B^L}{\partial \mathbf{z}}$ partials, are listed in [3] and Chapter 9 of this book, and this formulation using the state vector $\mathbf{z} = (a\ h\ k\ p\ q\ L)^T$ provides the simplest form of the $\frac{\partial B^L}{\partial \mathbf{z}}$ partials when compared to either the $\frac{\partial B}{\partial \mathbf{z}}$ or the $\frac{\partial M}{\partial \mathbf{z}}$ partials because, unlike $M(\mathbf{z}, F)$ and $B(\mathbf{z}, L)$, the present matrix B^L is only a direct function of $\mathbf{z}$, namely $B^L(\mathbf{z})$, which allows us to consider $\frac{\partial L}{\partial a} = \frac{\partial L}{\partial h} = \frac{\partial L}{\partial k} = 0$, thus simplifying to a considerable extent the form taken by the differential equations for the adjoints. This simplification is still effective when the $\frac{\partial \mathbf{f}}{\partial \mathbf{z}}$ partials are generated in Equation (11.64) because there, too, we are allowed to ignore the variation of L with respect to a, h and k. These partial derivatives are now written explicitly from Equations (11.4) through (11.6) as

$$\frac{\partial f_r}{\partial a} = 6\mu J_2 R^2 r^{-5}\frac{\partial r}{\partial a} - 72\mu J_2 R^2 r^{-5}\frac{\partial r}{\partial a}(qs_L - pc_L)^2 K^{-2} \tag{11.66}$$

$$\frac{\partial f_r}{\partial h} = 6\mu J_2 R^2 r^{-5}\frac{\partial r}{\partial h} - 72\mu J_2 R^2 r^{-5}\frac{\partial r}{\partial h}(qs_L - pc_L)^2 K^{-2} \tag{11.67}$$

$$\frac{\partial f_r}{\partial k} = 6\mu J_2 R^2 r^{-5}\frac{\partial r}{\partial k} - 72\mu J_2 R^2 r^{-5}\frac{\partial r}{\partial k}(qs_L - pc_L)^2 K^{-2} \tag{11.68}$$

$$\frac{\partial f_r}{\partial p} = -36\mu J_2 R^2 r^{-4}(qs_L - pc_L)\,c_L K^{-2} - 72\mu J_2 R^2 r^{-4} p\,(qs_L - pc_L)^2 K^{-3} \tag{11.69}$$

$$\frac{\partial f_r}{\partial q} = 36\mu J_2 R^2 r^{-4}(qs_L - pc_L)\,s_L K^{-2} - 72\mu J_2 R^2 r^{-4} q\,(qs_L - pc_L)^2 K^{-3} \tag{11.70}$$

$$\begin{aligned}\frac{\partial f_r}{\partial L} &= 6\mu J_2 R^2 r^{-5}\frac{\partial r}{\partial L} - 72\mu J_2 R^2 r^{-5}\frac{\partial r}{\partial L}(qs_L - pc_L)^2 K^{-2} \\ &+ 36\mu J_2 R^2 r^{-4}(qs_L - pc_L)(qc_L + ps_L)K^{-2}\end{aligned} \tag{11.71}$$

The f_θ partials are given by

$$\frac{\partial f_\theta}{\partial a} = 48\mu J_2 R^2 r^{-5} \frac{\partial r}{\partial a} (q s_L - p c_L)(q c_L + p s_L) K^{-2} \tag{11.72}$$

$$\frac{\partial f_\theta}{\partial h} = 48\mu J_2 R^2 r^{-5} \frac{\partial r}{\partial h} (q s_L - p c_L)(q c_L + p s_L) K^{-2} \tag{11.73}$$

$$\frac{\partial f_\theta}{\partial k} = 48\mu J_2 R^2 r^{-5} \frac{\partial r}{\partial k} (q s_L - p c_L)(q c_L + p s_L) K^{-2} \tag{11.74}$$

$$\begin{aligned}\frac{\partial f_\theta}{\partial p} &= -12\mu J_2 R^2 r^{-4} \Big\{ \big[s_L (q s_L - p c_L) - c_L (q c_L + p s_L) \big] K^{-2} \\ &\quad - 4p (q s_L - p c_L)(q c_L + p s_L) K^{-3} \Big\}\end{aligned} \tag{11.75}$$

$$\begin{aligned}\frac{\partial f_\theta}{\partial q} &= -12\mu J_2 R^2 r^{-4} \Big\{ \big[s_L (q c_L + p s_L) + c_L (q s_L - p c_L) \big] K^{-2} \\ &\quad - 4q (q s_L - p c_L)(q c_L + p s_L) K^{-3} \Big\}\end{aligned} \tag{11.76}$$

$$\begin{aligned}\frac{\partial f_\theta}{\partial L} &= 48\mu J_2 R^2 r^{-5} \frac{\partial r}{\partial L} (q s_L - p c_L)(q c_L + p s_L) K^{-2} \\ &\quad - 12\mu J_2 R^2 r^{-4} \left[(q c_L + p s_L)^2 - (q s_L - p c_L)^2 \right] K^{-2}\end{aligned} \tag{11.77}$$

Finally, the f_h partials are given by

$$\frac{\partial f_h}{\partial a} = 24\mu J_2 R^2 r^{-5} \frac{\partial r}{\partial a} (q s_L - p c_L)(2 - K) K^{-2} \tag{11.78}$$

$$\frac{\partial f_h}{\partial h} = 24\mu J_2 R^2 r^{-5} \frac{\partial r}{\partial h} (q s_L - p c_L)(2 - K) K^{-2} \tag{11.79}$$

$$\frac{\partial f_h}{\partial k} = 24\mu J_2 R^2 r^{-5} \frac{\partial r}{\partial k} (q s_L - p c_L)(2 - K) K^{-2} \tag{11.80}$$

$$\begin{aligned}\frac{\partial f_h}{\partial p} &= -6\mu J_2 R^2 r^{-4} \Big\{ \big[-c_L (2 - K) - 2p (q s_L - p c_L) \big] K^{-2} \\ &\quad - 4p (q s_L - p c_L)(2 - K) K^{-3} \Big\}\end{aligned} \tag{11.81}$$

$$\begin{aligned}\frac{\partial f_h}{\partial q} &= -6\mu J_2 R^2 r^{-4} \Big\{ \big[s_L (2 - K) - 2q (q s_L - p c_L) \big] K^{-2} \\ &\quad - 4q (q s_L - p c_L)(2 - K) K^{-3} \Big\}\end{aligned} \tag{11.82}$$

$$\begin{aligned}\frac{\partial f_h}{\partial L} &= 24\mu J_2 R^2 r^{-5} \frac{\partial r}{\partial L} (q s_L - p c_L)(2 - K) K^{-2} \\ &\quad - 6\mu J_2 R^2 r^{-4} (q c_L + p s_L)(2 - K) K^{-2}\end{aligned} \tag{11.83}$$

The partial derivatives of r, namely, $\frac{\partial r}{\partial a}$, $\frac{\partial r}{\partial h}$, $\frac{\partial r}{\partial k}$ and $\frac{\partial r}{\partial L}$ that appear in Equations (11.66) through (11.83) are according to Equation (11.65). Thus, the combined use of the polar or Euler–Hill frame instead of the more cumbersome equinoctial $\hat{\mathbf{f}}$, $\hat{\mathbf{g}}$, $\hat{\mathbf{w}}$ frame, and the use of L as the sixth state variable instead of λ, has eliminated the need to calculate F from Kepler's equation by iteration at each integration step and has

Table 11.1 Orbit parameters

Orbit	a, km	e	i, deg	Ω, deg	ω, deg	M, deg
Initial	7000	0	28.5	0	0	−131.7396776 (optimized)
Target	42,000	10^{-3}	1	0	0	Free
Achieved	41,999.99992	1.000022×10^{-3}	1.000001	359.999569	359.999132	45.411543 (optimized)

resulted in the simplest form of the equations for the adjoint differential equations that are integrated numerically and simultaneously with the dynamic equations. The optimal orientation of the thrust acceleration vector is provided by the optimal control $\hat{\mathbf{u}} = \left[\boldsymbol{\lambda}_z^T B^L(\mathbf{z})\right]^T / \left|\boldsymbol{\lambda}_z^T B^L(\mathbf{z})\right|$ in this case. Because of the error in the $\partial^2 X_1/\partial F \partial k$ partial, the example shown in [4] and Chapter 10 of this book is generated to provide the comparison needed to validate the versions that have been derived earlier for the case in which L effectively replaces F in the right-hand sides of the dynamic and adjoint equations. The initial and target orbit parameters are shown in Table 11.1.

This free–free transfer is solved by guessing the initial values of the first five Lagrange multipliers, namely, $(\lambda_a)_0$, $(\lambda_h)_0$, $(\lambda_k)_0$, $(\lambda_p)_0$, $(\lambda_q)_0$, as well as the initial longitude $(\lambda)_0$ and the transfer time t_f along with imposed $(\lambda_\lambda)_0 = 0$, and by iterating on the guessed quantities until the final parameters a_f, h_f, k_f, p_f, q_f, which correspond to the target orbit shown in Table 11.1, and $(\lambda_\lambda)_f = 0$ and $H_f = 1$ are met to within a small tolerance. The orientation of the continuous constant acceleration vector is obtained from Pontryagin's maximum principle by selecting $\hat{\mathbf{u}}$ parallel to $\boldsymbol{\lambda}_z^T M(\mathbf{z}, F)$. In practice, an unconstrained minimization algorithm is used to minimize the quadratic objective function F':

$$F' = w_1(a - a_f)^2 + w_2(h - h_f)^2 + w_3(k - k_f)^2 + w_4(p - p_f)^2 + w_5(q - q_f)^2 + w_6(\lambda_\lambda - 0)^2 + w_7(H - 1)^2 \tag{11.84}$$

where the w_i are adjustable weights. Let $f_t = 9.8 \times 10^{-5}\ \mathrm{km/s^2}$, and use the solution corresponding to the minimum-time transfer without the J_2 perturbation of [2] and [3] and Chapters 8 and 9 of this book, namely

$$\left.\begin{array}{ll} (\lambda_a)_0 = 4.675229762\ \mathrm{s/km} & (\lambda_p)_0 = 1.778011878 \times 10^1\ \mathrm{s} \\ (\lambda_h)_0 = 5.413413947 \times 10^2\ \mathrm{s} & (\lambda_q)_0 = -2.258455855 \times 10^4\ \mathrm{s} \\ (\lambda_k)_0 = -9.202702084 \times 10^3\ \mathrm{s} & (\lambda)_0 = -2.274742851\ \mathrm{rad} \\ t_f = 58{,}089.90058\ \mathrm{s} & \end{array}\right\} \tag{11.85}$$

with $H_f = 1.003704$. The value of $(\lambda)_0$ corresponds to $M_0 = -130.333164$ deg, and the multipliers in Equation (11.85) are scaled to provide $H = 1$, resulting in

$$\left.\begin{array}{ll} (\lambda_a)_0 = 4.657973438\ \mathrm{s/km} & (\lambda_p)_0 = 1.771449217 \times 10^1\ \mathrm{s} \\ (\lambda_h)_0 = 5.393432977 \times 10^2\ \mathrm{s} & (\lambda_q)_0 = -2.250119870 \times 10^4\ \mathrm{s} \\ (\lambda_k)_0 = -9.168734810 \times 10^3\ \mathrm{s} & (\lambda)_0 = -2.274742851\ \mathrm{rad} \\ t_f = 58{,}089.90058\ \mathrm{s} & \end{array}\right\} \tag{11.86}$$

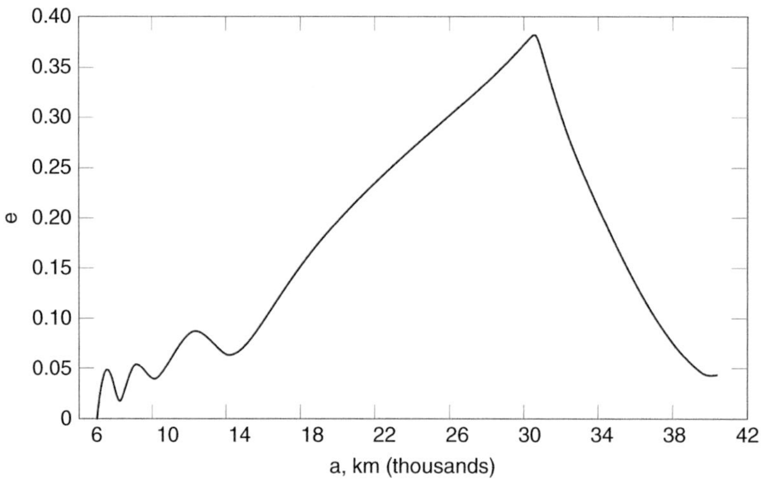

Figure 11.1 Effects of J_2 perturbation on minimum-time transfer solution generated without oblateness consideration.

A propagation is now carried out using Equations (11.8) through (11.13) and Equation (11.22), with the optimal control law obtained from $\hat{\mathbf{u}} = \boldsymbol{\lambda}_z^T M(\mathbf{z}, F)/|\boldsymbol{\lambda}_z^T M(\mathbf{z}, F)|$, and starting at time zero with the initial values in Equation (11.86), with $(\lambda_\lambda)_0 = 0$, and terminating the numerical integration at $t_f = 58{,}089.90058$ s, with $J_2 = 1.08263 \times 10^{-3}$. The initial conditions corresponding to the minimum-time solution without the J_2 effect will fail to integrate the orbit to the desired target orbit if we activate the J_2 perturbation as we just did. This failure is shown in Figure 11.1, which depicts the evolution of the eccentricity versus the semimajor axis with $a_f = 40{,}427.5184$ km and $e_f = 4.024759 \times 10^{-2}$, well short of the desired parameters $a_f = 42{,}000$ km and $e_f = 10^{-3}$ at time t_f.

The minimum-time solution with J_2 active is now obtained by iteration starting from the guess provided in Equation (11.86) and still using the nonlinear system of equations that correspond to Equations (11.22) and (11.23). The result is the solution given by

$$\left.\begin{array}{ll} (\lambda_a)_0 = 4.800100306\,\text{s/km} & (\lambda_p)_0 = 3.281827358 \times 10^1\,\text{s} \\ (\lambda_h)_0 = 8.060772261 \times 10^2\,\text{s} & (\lambda_q)_0 = -2.254928992 \times 10^4\,\text{s} \\ (\lambda_k)_0 = -9.150040837 \times 10^3\,\text{s} & (\lambda)_0 = -2.299291130\,\text{rad} \\ t_f = 58{,}104.83438\,\text{s} & \end{array}\right\} \quad (11.87)$$

The value of $(\lambda)_0$ corresponds to $M_0 = -131.7396776$ deg, and the final achieved parameters are shown in Table 11.1, with $M_f = 45.411543$ deg, with imposed $(\lambda_\lambda)_0 = 0$ s/rad and iterated $(\lambda_\lambda)_f = 4.07864 \times 10^{-5}$ s/rad, which is effectively near the desired value of zero for an optimized arrival point on the target orbit. The Hamiltonian $H = 1.000000$ is constant throughout, and the total $\Delta V = f_t t_f$ is equal to 5.694273769 km/s. This solution is slightly different from the one shown in [4] because there, the error in $\partial^2 X_1/\partial F \partial k$ has affected and contaminated the exact solution here but it is identical to the one in Chapter 10. The net effect in considering the J_2 perturbation has resulted in a transfer duration increased by some 14.9338 s when compared to the thrust-only case, and the location of the departure point has now moved by some 1.406 deg with respect to the location corresponding to the solution without the oblateness effect. We now verify the soundness of the last two versions of this optimization problem by carrying out integrations of the system of equations corresponding to the Hamiltonian in Equation (11.38), namely

$$\begin{pmatrix} \dot{a} \\ \dot{h} \\ \dot{k} \\ \dot{p} \\ \dot{q} \\ \dot{\lambda} \end{pmatrix} = \begin{pmatrix} B_{11} & B_{12} & B_{13} \\ B_{21} & B_{22} & B_{23} \\ B_{31} & B_{32} & B_{33} \\ B_{41} & B_{42} & B_{43} \\ B_{51} & B_{52} & B_{53} \\ B_{61} & B_{62} & B_{63} \end{pmatrix} \left[\begin{pmatrix} u_r \\ u_\theta \\ u_h \end{pmatrix} f_t + \begin{pmatrix} f_r \\ f_\theta \\ f_h \end{pmatrix} \right] + \begin{pmatrix} 0 \\ 0 \\ 0 \\ 0 \\ 0 \\ n \end{pmatrix} \tag{11.88}$$

and Equation (11.39), and starting from the initial conditions in Equation (11.87), until the final time $t_f = 58{,}104.83438$ s. The trajectory is integrated to $a_f = 41{,}999.99992$ km, $e_f = 1.000022 \times 10^{-3}$, $i_f = 1.000001$ deg, $\Omega_f = 359.999569$ deg, $\omega_f = 359.999137$ deg, $M_f = 45.411538$ deg matching the achieved parameters in Table 11.1 very closely. In a similar way, the dynamic equations corresponding to the Hamiltonian in Equation (11.63), namely

$$\begin{pmatrix} \dot{a} \\ \dot{h} \\ \dot{k} \\ \dot{p} \\ \dot{q} \\ \dot{L} \end{pmatrix} = \begin{pmatrix} B^L_{11} & B^L_{12} & B^L_{13} \\ B^L_{21} & B^L_{22} & B^L_{23} \\ B^L_{31} & B^L_{32} & B^L_{33} \\ B^L_{41} & B^L_{42} & B^L_{43} \\ B^L_{51} & B^L_{52} & B^L_{53} \\ B^L_{61} & B^L_{62} & B^L_{63} \end{pmatrix} \left[\begin{pmatrix} u_r \\ u_\theta \\ u_h \end{pmatrix} f_t + \begin{pmatrix} f_r \\ f_\theta \\ f_h \end{pmatrix} \right] + \begin{pmatrix} 0 \\ 0 \\ 0 \\ 0 \\ 0 \\ \frac{na^2\left(1-h^2-k^2\right)^{1/2}}{r^2} \end{pmatrix} \tag{11.89}$$

as well as the adjoint system in Equation (11.64), are also integrated forward in time starting with the same initial conditions listed in Equation (11.87), where $L_0 = -2.299291130$ rad replaces the identical value of $(\lambda)_0$ shown there, with $(\lambda_L)_0 = 0$ instead of $(\lambda_\lambda)_0$, until $t_f = 58{,}104.83438$ s. The trajectory is integrated to $a_f = 41{,}999.99992$ km, $e_f = 1.000022 \times 10^{-3}$, $i_f = 1.000001$ deg, $\Omega_f = 359.999569$ deg, $\omega_f = 359.999137$ deg, $M_f = 45.411538$ deg, effectively showing that all three formulations are identical and provide the same trajectory. All the pertinent integrations are carried out by setting the relative and absolute error controls at the 10^{-9} level. The solution generated in Equation (11.87) required several runs with different w_i weights for each run for a successful convergence. However, the converged solution was obtained with a single run using the L formulation developed in Equations (11.63), (11.89) and (11.64) with $w_1 = 1$, $w_2 = 10^7$, $w_3 = 10^8$, $w_4 = 10^7$, $w_5 = 10^8$, $w_6 = 1$ and $w_7 = 10^2$, and with the function value F', where the sixth term in Equation (11.84) is now replaced by $w_6(\lambda_L - 1)^2$, minimized to the level of 10^{-3}. The minimization algorithm uses an internal tolerance value for the convergence tests that is equal to the square root of the error in the function values, which are essentially equal to the machine roundoff level ε_{mach}. Unlike the algorithms that compute the zeros of a function to nearly full machine precision, the minimization algorithms can find the local minima to

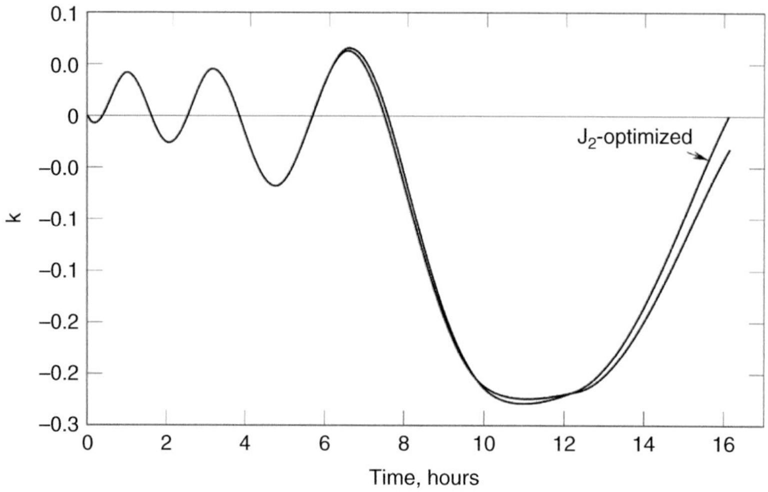

Figure 11.2 Evolution of k versus time for J_2-optimized and J_2-nonoptimized perturbed minimum-time transfers.

only about half-precision because now the zeros of the derivatives of F' are being computed. Another advantage of the L formulation for minimum-time problems lies in the increased compactness of the resulting equations, which would be further beneficial if other perturbations, such as those due to third-body gravity are also considered. Finally, it is possible to modulate the value of J_2 from zero to its actual value and obtain a series of converged runs starting from the known thrust-only solution, if the nominal J_2-perturbed case proved difficult to generate. We finally compare this J_2-optimized trajectory given in Equation (11.87) with the trajectory that is generated by using the initial conditions shown in Equation (11.85) that correspond to the minimum-time transfer solution for the thrust perturbation only. However, we integrate the same system of Equations (11.22) and (11.23) with the J_2 perturbation on until $t_f = 58{,}089.90058$ s to compare the time histories of the various transfer parameters even for this short-duration transfer, where J_2 introduces only a small integrated perturbation effect as compared to the roughly 10^{-2} g thrust acceleration used here. Figures 11.2 and 11.3 compare the evolutions of the equinoctial elements k and p, whereas Figures 11.4 and 11.5 compare the evolution of the classical elements e and M, showing clearly that the fully J_2-optimized solution achieves the required target conditions. Figures 11.6, 11.7 and 11.8 show the evolutions of the Lagrange multipliers λ_k, λ_p and λ_λ, effectively showing that the optimal $(\lambda_\lambda)_f$ alone terminates at the zero value as it should for a free–free transfer in which both the initial and final orbital locations are indeed optimal, providing the absolute minimum-time transfer solution. In Figure 11.9, only the optimal transfer satisfies the $H = 1$ value throughout the transfer, whereas the other transfer results in a slightly lower value of the Hamiltonian, indicating that the transfer time is not optimized and in this case is optimistic.

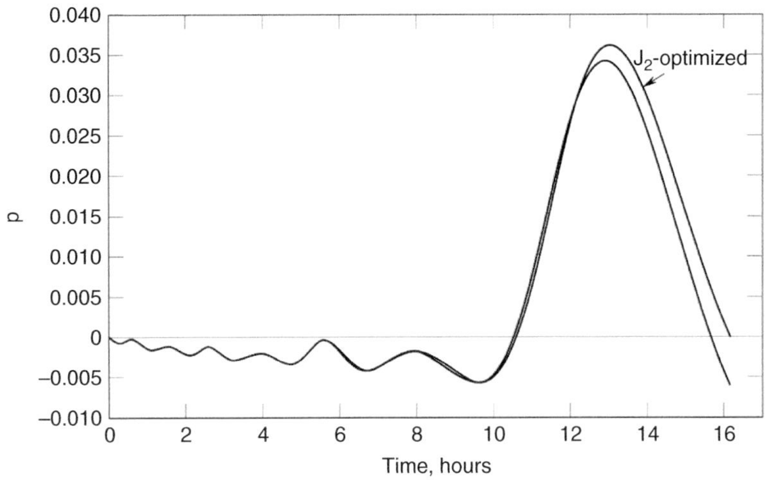

Figure 11.3 Evolution of p versus time for J_2-optimized and J_2-nonoptimized perturbed minimum-time transfers.

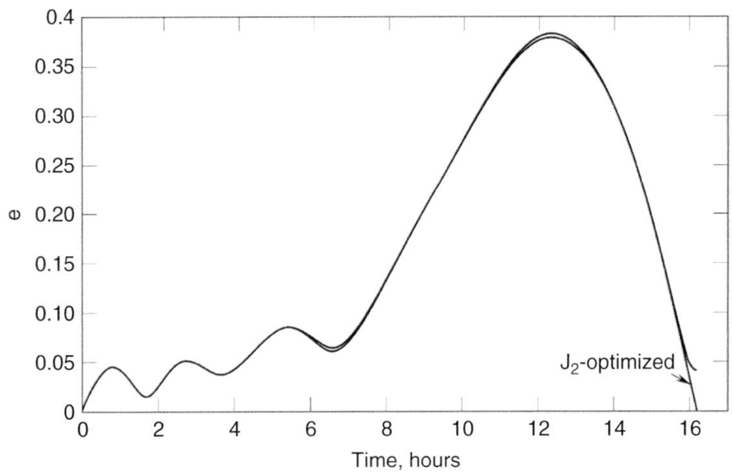

Figure 11.4 Evolution of e versus time for J_2-optimized and J_2-nonoptimized perturbed minimum-time transfers.

11.3 The Averaged Rates of the Elements Due to J_2 and Their Partial Derivatives

The averaged Hamiltonian for the thrust and J_2 perturbations can be written as

$$\tilde{H} = \tilde{H}_T + \tilde{H}_{J_2} \tag{11.90}$$

where

$$\tilde{H}_{J_2} = \tilde{\boldsymbol{\lambda}}_z^T \dot{\tilde{\mathbf{z}}}_{J_2} \tag{11.91}$$

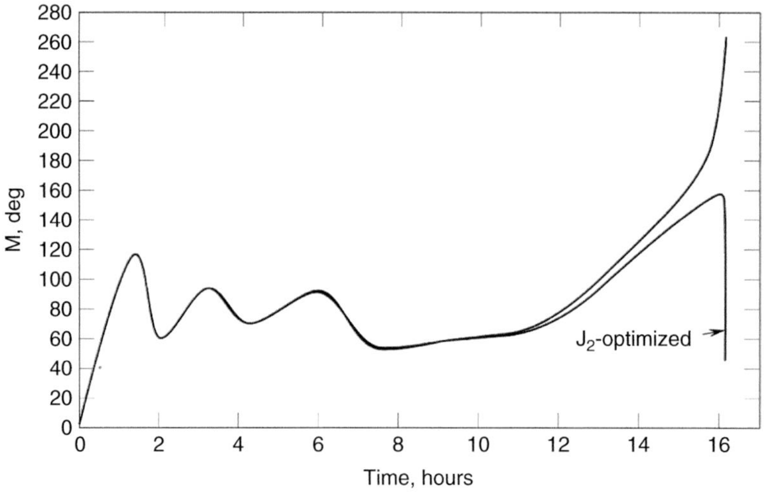

Figure 11.5 Evolution of M versus time for J_2-optimized and J_2-nonoptimized perturbed minimum-time transfers.

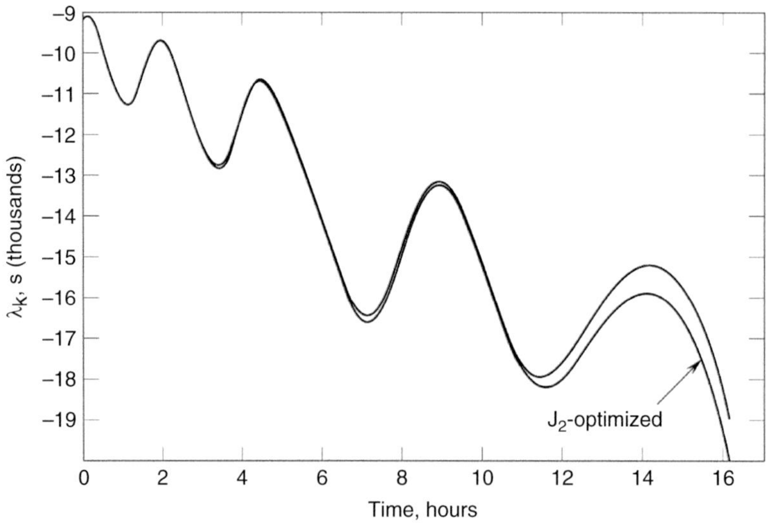

Figure 11.6 Evolution of λ_k versus time for J_2-optimized and J_2-nonoptimized perturbed minimum-time transfers.

The differential equations for the adjoints involve the $\partial\dot{\tilde{\mathbf{z}}}_{J_2}/\partial\tilde{\mathbf{z}}$ partial derivatives, and both $\dot{\tilde{\mathbf{z}}}_{J_2}$ and $\partial\dot{\tilde{\mathbf{z}}}_{J_2}/\partial\tilde{\mathbf{z}}$ have been derived for the $\tilde{a}$, $\tilde{h}$, $\tilde{k}$, $\tilde{p}$, $\tilde{q}$ and $\tilde{\lambda}$ elements [4, and Chapter 10]. Because we are now replacing $\tilde{\lambda}$ by $\tilde{L}$, an expression for $\dot{\tilde{L}}_{J_2}$ must be developed first before its partial derivatives can be generated. By the use of the averaged rates of ω, Ω and M in terms of the classical elements [4, and Chapter 10],

$$\dot{\tilde{\omega}}_{J_2} = \frac{3}{2} J_2 \frac{R^2}{\tilde{p}'^2} \tilde{n} \left(2 - \frac{5}{2} s_{\tilde{i}}^2 \right) \tag{11.92}$$

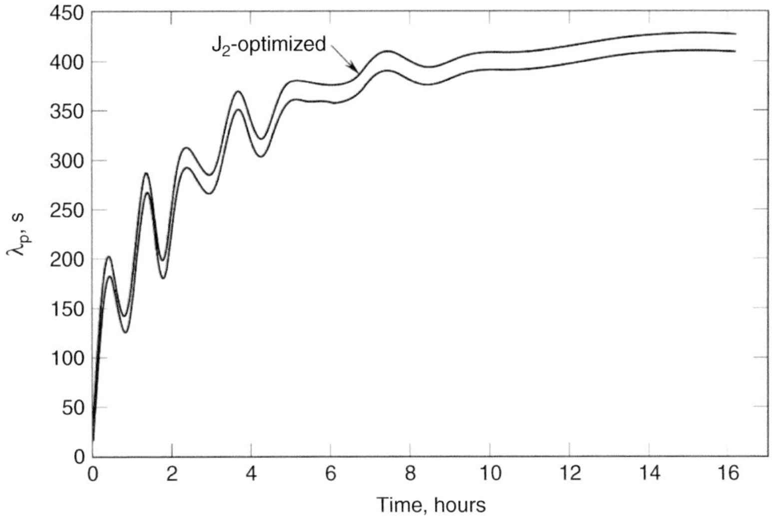

Figure 11.7 Evolution of λ_p versus time for J_2-optimized and J_2-nonoptimized perturbed minimum-time transfers.

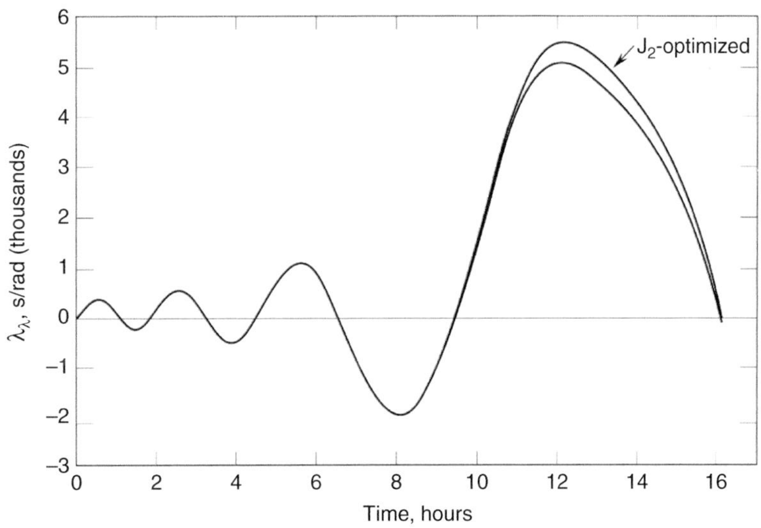

Figure 11.8 Evolution of λ_λ versus time for J_2-optimized and J_2-nonoptimized perturbed minimum-time transfers.

$$\dot{\tilde{\Omega}}_{J_2} = -\frac{3}{2} J_2 \frac{R^2}{\tilde{p}'^2} \tilde{n}\, c_{\tilde{i}} \tag{11.93}$$

$$\dot{\tilde{M}}_{J_2} = \frac{3}{2} J_2 \frac{R^2}{\tilde{p}'^2} \tilde{n} \left(1 - \frac{3}{2} s_{\tilde{i}}^2\right)\left(1 - \tilde{e}^2\right)^{1/2} \tag{11.94}$$

The following expressions for $\dot{\tilde{h}}_{J_2}$, $\dot{\tilde{k}}_{J_2}$, $\dot{\tilde{p}}_{J_2}$, $\dot{\tilde{q}}_{J_2}$ and $\dot{\tilde{\lambda}}_{J_2}$ were developed by also making use of $\dot{\tilde{a}}_{J_2} = \dot{\tilde{e}}_{J_2} = \dot{\tilde{i}}_{J_2} = 0$, and by noting that $\tilde{p}'$ is the mean orbit parameter given by $\tilde{p}' = \tilde{a}\left(1 - \tilde{e}^2\right)$:

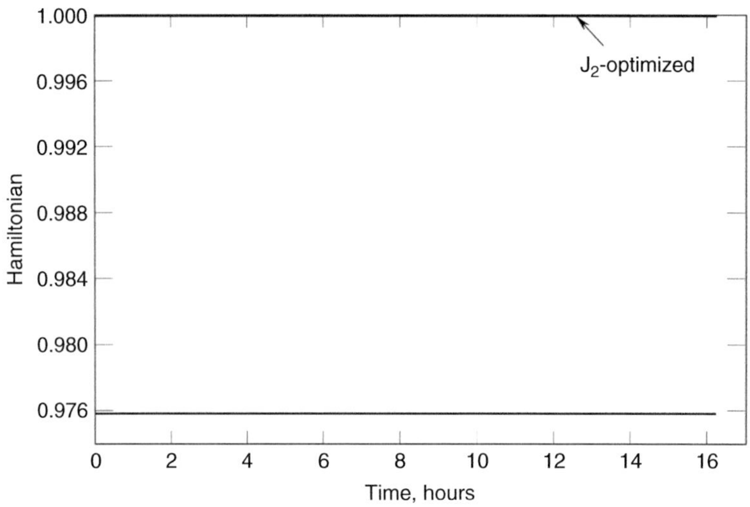

Figure 11.9 Evolution of Hamiltonian versus time for J_2-optimized and J_2-nonoptimized perturbed minimum-time transfers.

$$\dot{\tilde{h}}_{J_2} = \frac{3}{2}\frac{\tilde{k}J_2R^2\mu}{\tilde{n}\tilde{a}^5\left(1-\tilde{h}^2-\tilde{k}^2\right)^2}\left[\frac{1+3\left(\tilde{p}^2+\tilde{q}^2\right)^2-6\left(\tilde{p}^2+\tilde{q}^2\right)}{\left(1+\tilde{p}^2+\tilde{q}^2\right)^2}\right] \tag{11.95}$$

$$\dot{\tilde{k}}_{J_2} = -\frac{3}{2}\frac{\tilde{h}J_2R^2\mu}{\tilde{n}\tilde{a}^5\left(1-\tilde{h}^2-\tilde{k}^2\right)^2}\left[\frac{1+3\left(\tilde{p}^2+\tilde{q}^2\right)^2-6\left(\tilde{p}^2+\tilde{q}^2\right)}{\left(1+\tilde{p}^2+\tilde{q}^2\right)^2}\right] \tag{11.96}$$

$$\dot{\tilde{p}}_{J_2} = -\frac{3}{2}\frac{\tilde{q}J_2R^2\mu\left(1-\tilde{p}^2-\tilde{q}^2\right)}{\tilde{n}\tilde{a}^5\left(1-\tilde{h}^2-\tilde{k}^2\right)^2\left(1+\tilde{p}^2+\tilde{q}^2\right)} \tag{11.97}$$

$$\dot{\tilde{q}}_{J_2} = \frac{3}{2}\frac{\tilde{p}J_2R^2\mu\left(1-\tilde{p}^2-\tilde{q}^2\right)}{\tilde{n}\tilde{a}^5\left(1-\tilde{h}^2-\tilde{k}^2\right)^2\left(1+\tilde{p}^2+\tilde{q}^2\right)} \tag{11.98}$$

$$\dot{\tilde{\lambda}}_{J_2} = \dot{\tilde{M}}_{J_2} + \frac{\dot{\tilde{h}}_{J_2}}{\tilde{k}} = \dot{\tilde{M}}_{J_2} - \frac{\dot{\tilde{k}}_{J_2}}{\tilde{h}} \tag{11.99}$$

where

$$\dot{\tilde{M}}_{J_2} = \frac{3}{2}\frac{J_2R^2\mu\left(1-\tilde{h}^2-\tilde{k}^2\right)^{-3/2}}{\tilde{n}\tilde{a}^5\left(1+\tilde{p}^2+\tilde{q}^2\right)^2}\left[1+\left(\tilde{p}^2+\tilde{q}^2\right)^2-4\left(\tilde{p}^2+\tilde{q}^2\right)\right] \tag{11.100}$$

There is a typographical error in Equation (108) of [4], which should have $\left(1+\tilde{p}^2+\tilde{q}^2\right)^2$ instead of $\left(1+\tilde{p}^2+\tilde{q}^2\right)$ as the term in the denominator, and this is corrected in Equation (11.96) here. Equations (11.95) through (11.100) are derived as Equations (107) through (110), and (112) and (111), respectively in [4], where two typographical errors in Equation (111) of [4] have been corrected in Equation (11.100) here.

The partial derivatives of $\dot{\tilde{h}}_{J_2}$, $\dot{\tilde{k}}_{J_2}$, $\dot{\tilde{p}}_{J_2}$, $\dot{\tilde{q}}_{J_2}$ and $\dot{\tilde{\lambda}}_{J_2}$ with respect to $\tilde{a}$, $\tilde{h}$, $\tilde{k}$, $\tilde{p}$ and $\tilde{q}$ have also been derived in [4] and given there by Equations (113) through (137). Note that Equation (135) of [4] also has a typographical error and should read

$$\frac{\partial \dot{\tilde{\lambda}}_{J_2}}{\partial \tilde{k}} = \frac{3\,\tilde{k}\,\dot{\tilde{M}}_{J_2}}{\left(1-\tilde{h}^2-\tilde{k}^2\right)} - \frac{4\tilde{k}}{\left(1-\tilde{h}^2-\tilde{k}^2\right)}\frac{\dot{\tilde{k}}_{J_2}}{\tilde{h}} \tag{11.101}$$

Equation (121) of [4] should read as Equation (10.121), whereas Equations (136) and (137) should read

$$\begin{aligned}\frac{\partial \dot{\tilde{\lambda}}_{J_2}}{\partial \tilde{p}} &= \frac{12J_2R^2\mu\tilde{p}\left[3\left(\tilde{p}^2+\tilde{q}^2\right)-2\right]}{\tilde{n}\tilde{a}^5\left(1-\tilde{h}^2-\tilde{k}^2\right)^2\left(1+\tilde{p}^2+\tilde{q}^2\right)^3} \\ &\quad + \frac{6J_2R^2\mu\tilde{p}\left(1-\tilde{h}^2-\tilde{k}^2\right)^{-3/2}\left[3\left(\tilde{p}^2+\tilde{q}^2\right)-3\right]}{\tilde{n}\tilde{a}^5\left(1+\tilde{p}^2+\tilde{q}^2\right)^3} \end{aligned} \tag{11.102}$$

$$\begin{aligned}\frac{\partial \dot{\tilde{\lambda}}_{J_2}}{\partial \tilde{q}} &= \frac{12J_2R^2\mu\tilde{q}\left[3\left(\tilde{p}^2+\tilde{q}^2\right)-2\right]}{\tilde{n}\tilde{a}^5\left(1-\tilde{h}^2-\tilde{k}^2\right)^2\left(1+\tilde{p}^2+\tilde{q}^2\right)^3} \\ &\quad + \frac{6J_2R^2\mu\tilde{q}\left(1-\tilde{h}^2-\tilde{k}^2\right)^{-3/2}\left[3\left(\tilde{p}^2+\tilde{q}^2\right)-3\right]}{\tilde{n}\tilde{a}^5\left(1+\tilde{p}^2+\tilde{q}^2\right)^3} \end{aligned} \tag{11.103}$$

It can also be shown that $\dot{\tilde{M}}_{0_{J_2}} = \dot{\tilde{M}}_{J_2}$ and $\dot{\tilde{\lambda}}_{0_{J_2}} = \dot{\tilde{\lambda}}_{J_2}$. From $\dot{M}_0$ written in terms of the partial derivatives of the disturbing function R, namely $\dot{M}_0 = -\frac{2}{na}\frac{\partial R}{\partial a} - \frac{(1-e^2)}{na^2e}\frac{\partial R}{\partial e}$, we have for the J_2 perturbation

$$\tilde{R} = \frac{\tilde{n}^2J_2R^2}{4}(1-\tilde{e}^2)^{-3/2}(2-3s^2_{\tilde{i}}) \tag{11.104}$$

Therefore,

$$\frac{\partial \tilde{R}}{\partial \tilde{a}} = -\frac{3}{4}\frac{\tilde{n}^2}{\tilde{a}}J_2R^2\left(1-\tilde{e}^2\right)^{-3/2}\left(2-3s^2_{\tilde{i}}\right) \tag{11.105}$$

$$\frac{\partial \tilde{R}}{\partial \tilde{e}} = \frac{3}{4}\tilde{n}^2J_2R^2\left(1-\tilde{e}^2\right)^{-5/2}\tilde{e}\left(2-3s^2_{\tilde{i}}\right) \tag{11.106}$$

which yields after some manipulations the expression for $\dot{\tilde{M}}_{0_{J_2}}$ as

$$\dot{\tilde{M}}_{0_{J_2}} = \frac{3}{2}J_2R^2\tilde{p}'^{-2}\tilde{n}\left(1-\frac{3}{2}s^2_{\tilde{i}}\right)(1-\tilde{e}^2)^{1/2} \tag{11.107}$$

which is identical to Equation (11.94). Similarly, from

$$\dot{\tilde{\lambda}}_{0_{J_2}} = -\frac{2}{\tilde{n}\tilde{a}}\frac{\partial \tilde{R}}{\partial \tilde{a}} + \frac{\left(1-\tilde{e}^2\right)^{1/2}\left[1-\left(1-\tilde{e}^2\right)^{1/2}\right]}{\tilde{n}\tilde{a}^2\tilde{e}}\frac{\partial \tilde{R}}{\partial \tilde{e}} + \frac{\tan\left(\tilde{i}/2\right)}{\tilde{n}\tilde{a}^2\left(1-\tilde{e}^2\right)^{1/2}}\frac{\partial \tilde{R}}{\partial \tilde{i}} \tag{11.108}$$

and using

$$\frac{\partial\tilde{R}}{\partial\tilde{i}} = -\frac{3\tilde{n}^2 J_2 R^2}{2}(1-\tilde{e}^2)^{-3/2} s_{\tilde{i}} c_{\tilde{i}} \tag{11.109}$$

the expression for $\dot{\tilde{\lambda}}_{0_{J_2}}$ is obtained after some manipulations:

$$\begin{aligned}\dot{\tilde{\lambda}}_{0_{J_2}} &= \frac{3}{4}\frac{\tilde{n}}{\tilde{a}^2} J_2 R^2 (1-\tilde{e}^2)^{-3/2}\left(2-3s_{\tilde{i}}^2\right)\\ &+ \frac{3}{2}\frac{\tilde{n}}{\tilde{a}^2} J_2 R^2 (1-\tilde{e}^2)^{-2}\left(2-c_{\tilde{i}}-\frac{5}{2}s_{\tilde{i}}^2\right)\end{aligned} \tag{11.110}$$

This expression is equivalent to $\dot{\tilde{\lambda}}_{J_2} = \dot{\tilde{M}}_{J_2} + \dot{\tilde{\omega}}_{J_2} + \dot{\tilde{\Omega}}_{J_2}$, which is obtained by adding the quantities in Equations (11.92) through (11.94), thereby showing that $\dot{\tilde{\lambda}}_{0_{J_2}} = \dot{\tilde{\lambda}}_{J_2}$ too. We now use $\dot{L}$ in Equation (18) of [3], which is also given by Equation (11.61) in terms of equinoctial elements in the present chapter. With f_h given by Equation (11.3), we have with $h' = \mu^{1/2} a^{1/2}(1-e^2)^{1/2}$, $r = a(1-e^2)(1+ec_{\theta^*})^{-1}$, and $\theta = \omega + \theta^*$,

$$\dot{L} = \frac{h'}{r^2} + \frac{r}{h'} s_\theta \tan\left(\frac{i}{2}\right) f_h \tag{11.111}$$

$$\dot{L} = n\left(1-e^2\right)^{-3/2}(1+ec_{\theta^*})^2 - \frac{3nJ_2R^2}{a^2}\left(1-e^2\right)^{-7/2}\tan\left(\frac{i}{2}\right) s_i c_i s_{\omega+\theta^*}^2 (1+ec_{\theta^*})^3$$

such that

$$\begin{aligned}\dot{\tilde{L}} &= n\left(1-e^2\right)^{-3/2}\frac{1}{2\pi}\int_0^{2\pi}(1+ec_{\theta^*})^2\, d\theta^*\\ &- \frac{3nJ_2R^2}{a^2}\left(1-e^2\right)^{-7/2}\tan\left(\frac{i}{2}\right) s_i c_i \frac{1}{2\pi}\int_0^{2\pi} s_{\omega+\theta^*}^2 (1+ec_{\theta^*})^3\, d\theta^*\end{aligned}$$

The integrations yield

$$\frac{1}{2\pi}\int_0^{2\pi}(1+ec_{\theta^*})^2\, d\theta^* = \frac{1}{2}\left(2+e^2\right)$$

$$\frac{1}{2\pi}\int_0^{2\pi} s_{\omega+\theta^*}^2 (1+ec_{\theta^*})^3\, d\theta^* = \frac{1}{2} + \frac{3}{8}e^2 + \frac{3}{4}e^2 s_\omega^2$$

which are replaced in $\dot{\tilde{L}}$ to yield

$$\dot{\tilde{L}} = n\left(1-e^2\right)^{-3/2}\left(1+\frac{e^2}{2}\right) - \frac{3nJ_2R^2}{a^2}\left(1-e^2\right)^{-7/2}(1-c_i)\, c_i\left(\frac{1}{2}+\frac{3}{8}e^2+\frac{3}{4}e^2 s_\omega^2\right) \tag{11.112}$$

This expression can also be converted to a form that involves the equinoctial elements themselves instead of the classical elements. However, we need an expression for s_ω in terms of h, k, p and q. From the definitions of h and k, we have

$$s_{\omega+\Omega} = \frac{h}{\left(h^2+k^2\right)^{1/2}} = \frac{h}{e} \qquad c_{\omega+\Omega} = \frac{k}{\left(h^2+k^2\right)^{1/2}} = \frac{k}{e}$$

which yield

$$s_\omega = \frac{h}{e}c_\Omega - \frac{k}{e}s_\Omega \qquad c_\omega = \frac{h}{e}s_\Omega + \frac{k}{e}c_\Omega$$

and because

$$s_\Omega = \frac{p}{\left(p^2+q^2\right)^{1/2}} \qquad c_\Omega = \frac{q}{\left(p^2+q^2\right)^{1/2}}$$

they can be written as

$$s_\omega = \frac{hq - kp}{\left(h^2+k^2\right)^{1/2}\left(p^2+q^2\right)^{1/2}} \tag{11.113}$$

$$c_\omega = \frac{hp + kq}{\left(h^2+k^2\right)^{1/2}\left(p^2+q^2\right)^{1/2}} \tag{11.114}$$

If we use $c_i = \left(1-p^2-q^2\right)\left(1+p^2+q^2\right)^{-1}$ and the s_ω expression in Equation (11.113), then the $\dot{\bar{L}}$ Equation (11.112) will be converted to the form

$$\begin{aligned}\dot{\bar{L}} = n\left(1-e^2\right)^{-3/2}\left(1+\frac{e^2}{2}\right) - \frac{3n\left(1-e^2\right)^{-7/2}}{a^2}\frac{J_2R^2\left(1-p^2-q^2\right)}{\left(1+p^2+q^2\right)} \\ \times\left[p^2+q^2+\frac{9}{4}\left(h^2q^2+k^2p^2\right)+\frac{3}{4}\left(k^2q^2+h^2p^2\right)-3hkpq\right]\end{aligned} \tag{11.115}$$

As a further verification, we can start from $\dot{L}$ in Equation (11.61) and f_h in Equation (11.6), which are directly written in terms of h, k, p, q and L, and using $r = a\left(1-h^2-k^2\right)\left(1+hs_L+kc_L\right)^{-1}$ write

$$\begin{aligned}\dot{\bar{L}} = &\frac{n\left(1+hs_L+kc_L\right)^2}{\left(1-e^2\right)^{3/2}} \\ &- \frac{6n\left(1-e^2\right)^{-7/2}J_2R^2\left(1-p^2-q^2\right)}{a^2\left(1+p^2+q^2\right)^2}\left(qs_L-pc_L\right)^2\left(1+hs_L+kc_L\right)^3\end{aligned}$$

The averaging procedure requires the following integrations:

$$\frac{1}{2\pi}\int_0^{2\pi}\left(1+hs_L+kc_L\right)^2 dL = 1+\left(\frac{h^2+k^2}{2}\right) = 1+\frac{e^2}{2}$$

$$\begin{aligned}\frac{1}{2\pi}\int_0^{2\pi}\left(qs_L-pc_L\right)^2\left(1+hs_L+kc_L\right)^3 dL = &\left(\frac{p^2+q^2}{2}\right)+\frac{9}{8}\left(h^2q^2+k^2p^2\right) \\ &+\frac{3}{8}\left(k^2q^2+h^2p^2\right)-\frac{3}{2}hkpq\end{aligned}$$

providing the averaged rate for L as in Equation (11.115).

Therefore,

$$\dot{\bar{L}} = n\left(1-e^2\right)^{-3/2}\left(1+\frac{e^2}{2}\right)+\dot{\bar{L}}_{J_2} \tag{11.116}$$

with

$$\dot{\tilde{L}}_{J_2} = -\frac{3\tilde{n}\tilde{a}^{-2}J_2R^2\left(1-\tilde{p}^2-\tilde{q}^2\right)}{\left(1-\tilde{h}^2-\tilde{k}^2\right)^{7/2}\left(1+\tilde{p}^2+\tilde{q}^2\right)^2}\Big[\tilde{p}^2+\tilde{q}^2 + \frac{9}{4}\left(\tilde{h}^2\tilde{q}^2+\tilde{k}^2\tilde{p}^2\right)+\frac{3}{4}\left(\tilde{k}^2\tilde{q}^2+\tilde{h}^2\tilde{p}^2\right)-3\tilde{h}\tilde{k}\tilde{p}\tilde{q}\Big] \tag{11.117}$$

The partial derivatives of $\dot{\tilde{L}}_{J_2}$ are now readily obtained as

$$\frac{\partial\dot{\tilde{L}}_{J_2}}{\partial\tilde{a}} = -\frac{7}{2\tilde{a}}\dot{\tilde{L}}_{J_2} \tag{11.118}$$

$$\frac{\partial\dot{\tilde{L}}_{J_2}}{\partial\tilde{h}} = \frac{7\tilde{h}}{\left(1-\tilde{h}^2-\tilde{k}^2\right)}\dot{\tilde{L}}_{J_2} - \frac{3\tilde{n}\tilde{a}^{-2}J_2R^2\left(1-\tilde{p}^2-\tilde{q}^2\right)}{\left(1-\tilde{h}^2-\tilde{k}^2\right)^{7/2}\left(1+\tilde{p}^2+\tilde{q}^2\right)^2}\left[\frac{9}{2}\tilde{h}\tilde{q}^2+\frac{3}{2}\tilde{h}\tilde{p}^2-3\tilde{k}\tilde{p}\tilde{q}\right] \tag{11.119}$$

$$\frac{\partial\dot{\tilde{L}}_{J_2}}{\partial\tilde{k}} = \frac{7\tilde{k}}{\left(1-\tilde{h}^2-\tilde{k}^2\right)}\dot{\tilde{L}}_{J_2} - \frac{3\tilde{n}\tilde{a}^{-2}J_2R^2\left(1-\tilde{p}^2-\tilde{q}^2\right)}{\left(1-\tilde{h}^2-\tilde{k}^2\right)^{7/2}\left(1+\tilde{p}^2+\tilde{q}^2\right)^2}\left[\frac{9}{2}\tilde{k}\tilde{p}^2+\frac{3}{2}\tilde{k}\tilde{q}^2-3\tilde{h}\tilde{p}\tilde{q}\right] \tag{11.120}$$

$$\frac{\partial\dot{\tilde{L}}_{J_2}}{\partial\tilde{p}} = -\frac{2\tilde{p}}{\left(1-\tilde{p}^2-\tilde{q}^2\right)}\dot{\tilde{L}}_{J_2} - \frac{4\tilde{p}}{\left(1+\tilde{p}^2+\tilde{q}^2\right)}\dot{\tilde{L}}_{J_2} - \frac{3\tilde{n}\tilde{a}^{-2}J_2R^2\left(1-\tilde{p}^2-\tilde{q}^2\right)}{\left(1-\tilde{h}^2-\tilde{k}^2\right)^{7/2}\left(1+\tilde{p}^2+\tilde{q}^2\right)^2}\left[2\tilde{p}+\frac{9}{2}\tilde{p}\tilde{k}^2+\frac{3}{2}\tilde{p}\tilde{h}^2-3\tilde{h}\tilde{k}\tilde{q}\right] \tag{11.121}$$

$$\frac{\partial\dot{\tilde{L}}_{J_2}}{\partial\tilde{q}} = -\frac{2\tilde{q}}{\left(1-\tilde{p}^2-\tilde{q}^2\right)}\dot{\tilde{L}}_{J_2} - \frac{4\tilde{q}}{\left(1+\tilde{p}^2+\tilde{q}^2\right)}\dot{\tilde{L}}_{J_2} - \frac{3\tilde{n}\tilde{a}^{-2}J_2R^2\left(1-\tilde{p}^2-\tilde{q}^2\right)}{\left(1-\tilde{h}^2-\tilde{k}^2\right)^{7/2}\left(1+\tilde{p}^2+\tilde{q}^2\right)^2}\left[2\tilde{q}+\frac{9}{2}\tilde{h}^2\tilde{q}+\frac{3}{2}\tilde{k}^2\tilde{q}-3\tilde{h}\tilde{k}\tilde{p}\right] \tag{11.122}$$

11.4 Conclusion

Precision integrated minimum-time orbit transfers between general elliptic orbits using continuous constant acceleration and subjected to the perturbing effects of the second zonal harmonic J_2 are generated with two different formulations using two different sets of equinoctial elements. The use of the true longitude as the sixth element, as well as the variable with respect to which all of the dynamic and adjoint differential equations are expressed, coupled with the use of the polar coordinate frame as the orbital frame for component resolution of the thrust and J_2-induced perturbations result in an analysis that is substantially simpler than previously developed versions and lead to simulation software that exhibits better convergence characteristics. The full six-state analysis is used to generate a free–free minimum-time transfer that is verified and compared with the results generated by a previously derived formulation. The partial derivatives of the averaged rate of the true longitude due to J_2, with respect to the five slowly varying elements, are also derived analytically for use in software that uses averaged dynamics for approximate and rapid generation of minimum-time transfers.

References

[1] Kechichian, J. A. (1997) Optimal Low-Earth-Orbit–Geostationary-Earth-Orbit Intermediate Acceleration Orbit Transfer, *Journal of Guidance, Control, and Dynamics*, Vol. 20, No. 4, 803–811.

[2] Kechichian, J. A. (1997) Mechanics of Trajectory Optimization Using Nonsingular Variational Equations in Polar Coordinates, *Journal of Guidance, Control, and Dynamics*, Vol. 20, No. 4, 812–818.

[3] Kechichian, J. A. (1997) Trajectory Optimization Using Nonsingular Orbital Elements and True Longitude, *Journal of Guidance, Control, and Dynamics*, Vol. 20, No. 5, 1003–1009.

[4] Kechichian, J. A. (1997) The Treatment of the Earth Oblateness Effect in Trajectory Optimization in Equinoctial Coordinates, *Acta Astronautica*, Vol. 40, No. 1, 69–82.

[5] Geffroy, S. and Epenoy, R. (1997) Optimal Low-Thrust Transfers with Constraints-Generalization of Averaging Techniques, *Acta Astronautica*, Vol. 41, No. 3, 133–149.

12 The Streamlined and Complete Set of the Nonsingular J_2-Perturbed Dynamic and Adjoint Equations for Trajectory Optimization in Terms of Eccentric Longitude

12.1 Introduction

The complete set of the dynamic and multipliers differential equations for a nonsingular set of equinoctial elements in which the eccentric longitude stands for the sixth or fast orbital element are derived in a more straightforward and streamlined manner as compared to the previous derivation presented in Chapter 6. The mathematics are validated through comparison with the original formulation as well as with two other formulations using the mean longitude and the true longitude elements as the respective sixth element of the equinoctial set. Furthermore, this new formulation is used to derive the costate or adjoint differential equations by fully accounting for the secular first-order perturbative effect of the second zonal harmonic J_2, and the complete set of the perturbed dynamic and adjoint system of equations are also validated by direct comparison with the two previously derived formulations using the mean and true longitudes respectively. The present formulation, as well as the one using the true longitude as the sixth orbital element, remove the need to solve Kepler's transcendental equation at each integration step, a need that is inevitable when the mean longitude formulation is used, because in the latter case the right-hand sides of the various differential equations cannot be written directly in terms of the mean longitude. The inclusion of the J_3 and J_4 terms can be similarly accounted for with both the eccentric and true longitude sets and mutually validated also. This particular formulation has been adopted by several aerospace contractors in the United States to build specialized flight guidance software to steer payloads released in high-energy orbits to their final destinations using highly efficient low-thrust propulsion.

The mathematics and mechanics of orbital dynamics using nonsingular elements has benefited from the contributions of Broucke, Cefola and Edelbaum et al. [1–3] who fully derived the necessary equations for trajectory propagation and orbit transfer applications. [4–7] are representative of the many excellent papers addressing the optimal transfer problem that recently appeared in the literature using the numerically efficient and robust direct method based on the techniques of collocation and nonlinear programming. The application of the theory of optimal control to trajectory optimization using the traditional indirect shooting method has also been thoroughly exposed in the classical books by Bryson and Ho [8] and Marec [9]. It is well known that the variational equations for the classical elements in the Gaussian form with the components of the disturbing acceleration due to the thrust and the Earth zonal harmonics resolved along the radial, transverse and out-of-plane directions representing the Euler–Hill rotating frame attached to the spacecraft are singular at zero eccentricity and zero inclination, making it impossible to effectively integrate such trajectories numerically. It is thanks to the adoption of the equinoctial elements instead of the classical elements in [1–3] that this difficulty has been circumvented, especially when applied to the important low Earth orbit (LEO) to geosynchronous Earth orbit (GEO) transfer problem.

This chapter is an enhanced and extended version of a previous contribution in Chapter 6 and [10] that showed the derivations that lead to the successful generation of the complete set of the differential equations for the thrust-perturbation-only state and adjoint variables used in the simulation of low-thrust minimum-time transfers between any given elliptic or circular orbits. The state variables used are the well-known equinoctial orbit elements constructed from the classical elements a, $h = e\sin(\Omega + \omega)$, $k = e\cos(\Omega + \omega)$, $p = \tan(i/2)\sin\Omega$, $q = \tan(i/2)\cos\Omega$ and the eccentric longitude $F = E + \Omega + \omega$. The use of equinoctial elements leads to differential equations for both the state variables and the corresponding multipliers that are free of singularities for the important zero eccentricity and zero inclination orbits, or in other words circular and equatorial orbits. This system of equations is thus ideal for integrating continuous constant acceleration as well as constant thrust magnitude optimal trajectories from low inclined circular orbits to equatorial geosynchronous circular orbits without singularity-induced loss of precision during the numerical integrations. An essential simplification of the analysis of [10] and Chapter 6 of this book is obtained here by using an explicit form of the $\partial F/\partial \dot{\mathbf{r}}$ partial derivative, also called velocity partial where $\dot{\mathbf{r}}$ is the velocity vector, instead of the more elaborate partial $\partial F/\partial \dot{\mathbf{r}}$ given in terms of the $\partial \lambda/\partial \dot{\mathbf{r}}$, $\partial h/\partial \dot{\mathbf{r}}$ and $\partial k/\partial \dot{\mathbf{r}}$ partials used in [10] that lead to more complicated forms for the adjoint differential equations. This was done then in order to take advantage of the existing and previously derived partial derivatives of the $\partial \lambda/\partial \dot{\mathbf{r}}$, $\partial h/\partial \dot{\mathbf{r}}$ and $\partial k/\partial \dot{\mathbf{r}}$ velocity partials with respect to the first five elements of the equinoctial set as well as the mean longitude $\lambda = M + \Omega + \omega$ adopted in [11] and Chapter 10 of this book. Minor modifications of some of these partials lead to obtaining the full set of the Lagrange multiplier equations for the corresponding set which uses F instead of λ as the sixth element. When the explicit form of the $\partial F/\partial \dot{\mathbf{r}}$ velocity partial that defines $\dot{F}$ (the sixth state equation) is used instead, the relevant derivations and the form of the various partial derivatives needed in the description of the adjoint equations are made simpler and in a more streamlined manner. The next section shows all the details of this more direct formulation. This new formulation is next used in conjunction with the additional terms needed to define the set of dynamic and costate equations for the case in which the perturbation due to the second zonal harmonic J_2 is accounted for both in the state and adjoint equations. The subsequent section shows all the necessary derivations that result in the full set of differential equations with F as the sixth or fast element. The optimal trajectory generated in [12] and Chapter 11 of this book is exactly duplicated here in order to validate all the mathematical derivations. Note that Equation (120) of [12] has a minor typographical error in its last terms in the form of $3\tilde{k}\tilde{p}\tilde{q}$ instead of $3\tilde{h}\tilde{p}\tilde{q}$ and is shown correctly in Chapter 11.

12.2 System and Adjoint Differential Equations in Terms of the Eccentric Longitude

Resolving the thrust acceleration vector $\mathbf{f}_t = f_t\hat{\mathbf{u}}$ in the direct equinoctial frame $(\hat{\mathbf{f}}, \hat{\mathbf{g}}, \hat{\mathbf{w}})$ with unit vectors $\hat{\mathbf{f}}$ and $\hat{\mathbf{g}}$ in the current orbit plane, and with $\hat{\mathbf{f}}$ rotated clockwise through an angle Ω from the direction of the ascending node, and letting $\mathbf{r} = X_1\hat{\mathbf{f}} + Y_1\hat{\mathbf{g}}$ and $\dot{\mathbf{r}} = \dot{X}_1\hat{\mathbf{f}} + \dot{Y}_1\hat{\mathbf{g}}$, the basic equations of motion for the thrust perturbation are as in [10] and Chapter 6 of this book, with $n = (\mu/a^3)^{1/2}$ standing for the orbit mean motion, r the radial distance and μ the Earth gravity constant:

$$\dot{a} = (\partial a/\partial \dot{\mathbf{r}})^T \cdot \hat{\mathbf{u}} f_t \tag{12.1}$$

$$\dot{h} = (\partial h/\partial \dot{\mathbf{r}})^T \cdot \hat{\mathbf{u}} f_t \tag{12.2}$$

$$\dot{k} = (\partial k/\partial \dot{\mathbf{r}})^T \cdot \hat{\mathbf{u}} f_t \tag{12.3}$$

$$\dot{p} = (\partial p/\partial \dot{\mathbf{r}})^T \cdot \hat{\mathbf{u}} f_t \tag{12.4}$$

$$\dot{q} = (\partial q/\partial \dot{\mathbf{r}})^T \cdot \hat{\mathbf{u}} f_t \tag{12.5}$$

$$\dot{F} = \frac{na}{r} + (\partial F/\partial \dot{\mathbf{r}})^T \cdot \hat{\mathbf{u}} f_t \tag{12.6}$$

The geometry of the direct equinoctial frame, as well as that of the more familiar Euler–Hill frame $(\hat{\mathbf{r}}, \hat{\boldsymbol{\theta}}, \hat{\mathbf{h}})$ attached to the spacecraft, are shown in Figure 12.1.

With the unit vector $\hat{\mathbf{u}}$ being along the instantaneous thrust vector $\mathbf{f}_t$, the position and velocity vector components in terms of the equinoctial elements a, h, k, p, q and F are, as in [10] and Chapter 6 of this book,

$$X_1 = a\left[\left(1 - h^2\beta\right)c_F + hk\beta s_F - k\right] \tag{12.7}$$

$$Y_1 = a\left[hk\beta c_F + \left(1 - k^2\beta\right)s_F - h\right] \tag{12.8}$$

$$\dot{X}_1 = a^2 n r^{-1}\left[hk\beta c_F - \left(1 - h^2\beta\right)s_F\right] \tag{12.9}$$

$$\dot{Y}_1 = a^2 n r^{-1}\left[\left(1 - k^2\beta\right)c_F - hk\beta s_F\right] \tag{12.10}$$

with s_F and c_F standing for $\sin F$ and $\cos F$ respectively, $\beta = 1/(1+G)$, $G = (1 - h^2 - k^2)^{\frac{1}{2}}$ and $r = a(1 - kc_F - hs_F)$.

The partials appearing in Equations (12.1) through (12.6) are derived as

$$\frac{\partial a}{\partial \dot{\mathbf{r}}} = 2a^{-1}n^{-2}\left(\dot{X}_1\hat{\mathbf{f}} + \dot{Y}_1\hat{\mathbf{g}}\right) = M_{11}\hat{\mathbf{f}} + M_{12}\hat{\mathbf{g}} + M_{13}\hat{\mathbf{w}} \tag{12.11}$$

$$\frac{\partial h}{\partial \dot{\mathbf{r}}} = Gn^{-1}a^{-2}\left[\left(\frac{\partial X_1}{\partial k} - h\beta\frac{\dot{X}_1}{n}\right)\hat{\mathbf{f}} + \left(\frac{\partial Y_1}{\partial k} - h\beta\frac{\dot{Y}_1}{n}\right)\hat{\mathbf{g}}\right] + k\left(qY_1 - pX_1\right)n^{-1}a^{-2}G^{-1}\hat{\mathbf{w}} = M_{21}\hat{\mathbf{f}} + M_{22}\hat{\mathbf{g}} + M_{23}\hat{\mathbf{w}} \tag{12.12}$$

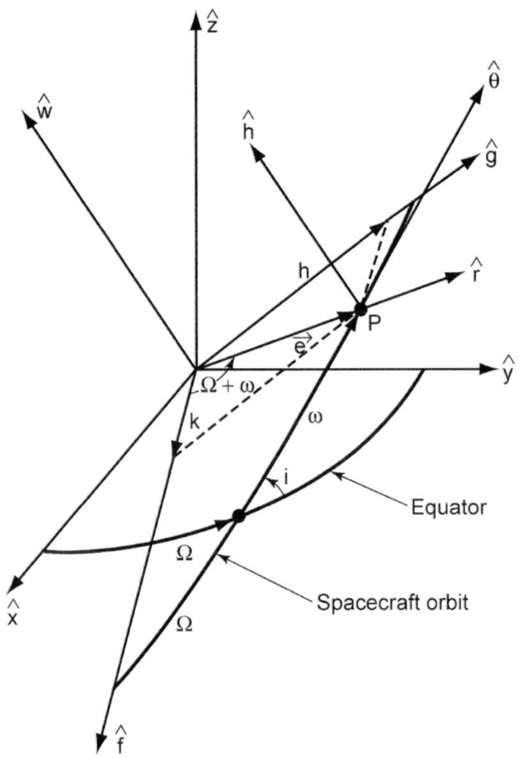

Figure 12.1 Direct equinoctial and Euler-Hill frames with spacecraft at perigee.

$$\frac{\partial k}{\partial \dot{\mathbf{r}}} = -Gn^{-1}a^{-2}\left[\left(\frac{\partial X_1}{\partial h} + k\beta\frac{\dot{X}_1}{n}\right)\hat{\mathbf{f}} + \left(\frac{\partial Y_1}{\partial h} + k\beta\frac{\dot{Y}_1}{n}\right)\hat{\mathbf{g}}\right]$$
$$- h\left(qY_1 - pX_1\right)n^{-1}a^{-2}G^{-1}\hat{\mathbf{w}} = M_{31}\hat{\mathbf{f}} + M_{32}\hat{\mathbf{g}} + M_{33}\hat{\mathbf{w}} \quad (12.13)$$

$$\frac{\partial p}{\partial \dot{\mathbf{r}}} = KY_1\frac{n^{-1}a^{-2}G^{-1}}{2}\hat{\mathbf{w}} = M_{41}\hat{\mathbf{f}} + M_{42}\hat{\mathbf{g}} + M_{43}\hat{\mathbf{w}} \quad (12.14)$$

$$\frac{\partial q}{\partial \dot{\mathbf{r}}} = KX_1\frac{n^{-1}a^{-2}G^{-1}}{2}\hat{\mathbf{w}} = M_{51}\hat{\mathbf{f}} + M_{52}\hat{\mathbf{g}} + M_{53}\hat{\mathbf{w}} \quad (12.15)$$

$$\frac{\partial F}{\partial \dot{\mathbf{r}}} = \frac{a}{r}\left[\left(\frac{\partial \lambda}{\partial \dot{\mathbf{r}}} + s_F\frac{\partial k}{\partial \dot{\mathbf{r}}} - c_F\frac{\partial h}{\partial \dot{\mathbf{r}}}\right)\right] = M_{61}^F\hat{\mathbf{f}} + M_{62}^F\hat{\mathbf{g}} + M_{63}^F\hat{\mathbf{w}} \quad (12.16)$$

with $K = 1 + p^2 + q^2$ and with

$$\frac{\partial \lambda}{\partial \dot{\mathbf{r}}} = n^{-1}a^{-2}\left[-2X_1 + G\left(h\beta\frac{\partial X_1}{\partial h} + k\beta\frac{\partial X_1}{\partial k}\right)\right]\hat{\mathbf{f}}$$
$$+ n^{-1}a^{-2}\left[-2Y_1 + G\left(h\beta\frac{\partial Y_1}{\partial h} + k\beta\frac{\partial Y_1}{\partial k}\right)\right]\hat{\mathbf{g}}$$
$$+ n^{-1}a^{-2}G^{-1}\left(qY_1 - pX_1\right)\hat{\mathbf{w}} = M_{61}\hat{\mathbf{f}} + M_{62}\hat{\mathbf{g}} + M_{63}\hat{\mathbf{w}} \quad (12.17)$$

The position vector component partials with respect to h and k are obtained as

$$\frac{\partial X_1}{\partial h} = a\left[-\left(hc_F - ks_F\right)\left(\beta + \frac{h^2\beta^3}{1-\beta}\right) - \frac{a}{r}c_F\left(h\beta - s_F\right)\right] \tag{12.18}$$

$$\frac{\partial X_1}{\partial k} = -a\left[\left(hc_F - ks_F\right)\frac{hk\beta^3}{1-\beta} + 1 + \frac{a}{r}s_F\left(s_F - h\beta\right)\right] \tag{12.19}$$

$$\frac{\partial Y_1}{\partial h} = a\left[\left(hc_F - ks_F\right)\frac{hk\beta^3}{1-\beta} - 1 + \frac{a}{r}c_F\left(k\beta - c_F\right)\right] \tag{12.20}$$

$$\frac{\partial Y_1}{\partial k} = a\left[\left(hc_F - ks_F\right)\left(\beta + \frac{k^2\beta^3}{1-\beta}\right) + \frac{a}{r}s_F\left(c_F - k\beta\right)\right] \tag{12.21}$$

These equations are fully derived in [13] and [14] with $\partial F/\partial \dot{\mathbf{r}}$ written directly in terms of $\partial\lambda/\partial\dot{\mathbf{r}}$, $\partial h/\partial\dot{\mathbf{r}}$ and $\partial k/\partial\dot{\mathbf{r}}$. This was done in order to take advantage of the partials of matrix M with respect to the elements using λ as the sixth or fast element which are needed to generate the adjoint differential equations and which appeared in [14] and Chapter 3 of this book. It was then an easy matter to use these partials with minor modifications to generate the partials of $\partial F/\partial\dot{\mathbf{r}}$ with respect to the elements needed in the Euler–Lagrange or adjoint equations. However, it is possible to streamline the presentation of the adjoint equations by using the explicit form of $\partial F/\partial\dot{\mathbf{r}}$ written as

$$\begin{aligned}\frac{\partial F}{\partial \dot{\mathbf{r}}} &= \frac{1}{nar}\left\{\left[-2X_1 + G\left(h\beta - s_F\right)\frac{\partial X_1}{\partial h}\right.\right.\\ &\left. + G\left(k\beta - c_F\right)\frac{\partial X_1}{\partial k} - \beta G\left(ks_F - hc_F\right)\frac{\dot{X}_1}{n}\right]\hat{\mathbf{f}} + \left[-2Y_1 + G\left(h\beta - s_F\right)\frac{\partial Y_1}{\partial h}\right.\\ &\left.\left. + G\left(k\beta - c_F\right)\frac{\partial Y_1}{\partial k} - \beta G\left(ks_F - hc_F\right)\frac{\dot{Y}_1}{n}\right]\hat{\mathbf{g}} + \frac{r}{aG}\left(qY_1 - pX_1\right)\hat{\mathbf{w}}\right\}\\ &= M^F_{61}\hat{\mathbf{f}} + M^F_{62}\hat{\mathbf{g}} + M^F_{63}\hat{\mathbf{w}}\end{aligned} \tag{12.22}$$

The equations of motion are left in their present compact form, meaning that X_1, Y_1, $\dot{X}_1$ and $\dot{Y}_1$, as well as $\frac{\partial X_1}{\partial h}$, $\frac{\partial X_1}{\partial k}$, $\frac{\partial Y_1}{\partial h}$ and $\frac{\partial Y_1}{\partial k}$ are not replaced by their explicit expressions given above in order to prevent the velocity partials $\partial a/\partial\dot{\mathbf{r}}$ through $\partial F/\partial\dot{\mathbf{r}}$ from adopting exceedingly long forms.

It is then straightforward to write the dynamic or state equations as

$$\dot{\mathbf{z}} = M\hat{\mathbf{u}}f_t + \begin{pmatrix}0 & 0 & 0 & 0 & 0 & \frac{na}{r}\end{pmatrix}^T \tag{12.23}$$

with $\mathbf{z} = (a\ h\ k\ p\ q\ F)^T$ representing the state vector and with the 6×3 matrix M given by

$$M = \begin{bmatrix}(\partial a/\partial\dot{\mathbf{r}})^T\\ (\partial h/\partial\dot{\mathbf{r}})^T\\ (\partial k/\partial\dot{\mathbf{r}})^T\\ (\partial p/\partial\dot{\mathbf{r}})^T\\ (\partial q/\partial\dot{\mathbf{r}})^T\\ (\partial F/\partial\dot{\mathbf{r}})^T\end{bmatrix} = \begin{bmatrix}M_{11} & M_{12} & M_{13}\\ M_{21} & M_{22} & M_{23}\\ M_{31} & M_{32} & M_{33}\\ M_{41} & M_{42} & M_{43}\\ M_{51} & M_{52} & M_{53}\\ M^F_{61} & M^F_{62} & M^F_{63}\end{bmatrix} \tag{12.24}$$

The maximization of the following Hamiltonian is carried out next by aligning $\hat{\mathbf{u}}$ with $\boldsymbol{\lambda}_z^T M(\mathbf{z}, F)$ and the Euler–Lagrange equations generated by making use of the partials shown below:

$$
\begin{aligned}
H &= \boldsymbol{\lambda}_z^T \dot{\mathbf{z}} = \boldsymbol{\lambda}_z^T M(\mathbf{z}, F) f_t \hat{\mathbf{u}} + \lambda_F \frac{na}{r} \\
\frac{\partial r}{\partial a} &= \frac{r}{a}, \frac{\partial r}{\partial h} = -a s_F, \frac{\partial r}{\partial k} = -a c_F \\
\frac{\partial r}{\partial F} &= a\left(k s_F - h c_F\right), \frac{\partial F}{\partial a} = 0, \frac{\partial F}{\partial h} = 0, \frac{\partial F}{\partial k} = 0
\end{aligned}
\tag{12.25}
$$

Because F is an orbital element, its partials with respect to the other elements are effectively zero.

As discussed in [10] and Chapter 6 of this book, when taking the partials of H with respect to the elements, the eccentric longitude F appearing in the right-hand side of all six differential equations, in X_1, Y_1, $\dot{X}_1$, $\dot{Y}_1$, $\partial X_1/\partial h$, $\partial X_1/\partial k$, $\partial Y_1/\partial h$ and $\partial Y_1/\partial k$ as well as r, s_F and c_F must be held fixed such that

$$
\dot{\boldsymbol{\lambda}}_z = -\frac{\partial H}{\partial \mathbf{z}} = -\boldsymbol{\lambda}_z^T \frac{\partial M}{\partial \mathbf{z}} f_t \hat{\mathbf{u}} - \lambda_F \frac{\partial (na/r)}{\partial \mathbf{z}} \tag{12.26}
$$

Letting $u_1 = u_f$, $u_2 = u_g$ and $u_3 = u_w$, the adjoint equations are, as in [10] and Chapter 6 of this book,

$$
\begin{aligned}
\dot{\lambda}_a = -\frac{\partial H}{\partial a} = \Bigg[&-\lambda_a \frac{\partial M_{1i}}{\partial a} u_i - \lambda_h \frac{\partial M_{2i}}{\partial a} u_i - \lambda_k \frac{\partial M_{3i}}{\partial a} u_i - \lambda_p \frac{\partial M_{4i}}{\partial a} u_i \\
&-\lambda_q \frac{\partial M_{5i}}{\partial a} u_i - \lambda_F \frac{\partial M_{6i}^F}{\partial a} u_i \Bigg] f_t - \lambda_F \frac{\partial}{\partial a}\left(\frac{na}{r}\right)
\end{aligned}
\tag{12.27}
$$

$$
\begin{aligned}
\dot{\lambda}_h = -\frac{\partial H}{\partial h} = \Bigg[&-\lambda_a \frac{\partial M_{1i}}{\partial h} u_i - \lambda_h \frac{\partial M_{2i}}{\partial h} u_i - \lambda_k \frac{\partial M_{3i}}{\partial h} u_i - \lambda_p \frac{\partial M_{4i}}{\partial h} u_i \\
&-\lambda_q \frac{\partial M_{5i}}{\partial h} u_i - \lambda_F \frac{\partial M_{6i}^F}{\partial h} u_i \Bigg] f_t - \lambda_F \frac{\partial}{\partial h}\left(\frac{na}{r}\right)
\end{aligned}
\tag{12.28}
$$

$$
\begin{aligned}
\dot{\lambda}_k = -\frac{\partial H}{\partial k} = \Bigg[&-\lambda_a \frac{\partial M_{1i}}{\partial k} u_i - \lambda_h \frac{\partial M_{2i}}{\partial k} u_i - \lambda_k \frac{\partial M_{3i}}{\partial k} u_i - \lambda_p \frac{\partial M_{4i}}{\partial k} u_i \\
&-\lambda_q \frac{\partial M_{5i}}{\partial k} u_i - \lambda_F \frac{\partial M_{6i}^F}{\partial k} u_i \Bigg] f_t - \lambda_F \frac{\partial}{\partial k}\left(\frac{na}{r}\right)
\end{aligned}
\tag{12.29}
$$

$$
\begin{aligned}
\dot{\lambda}_p = -\frac{\partial H}{\partial p} = \Bigg[&-\lambda_a \frac{\partial M_{1i}}{\partial p} u_i - \lambda_h \frac{\partial M_{2i}}{\partial p} u_i - \lambda_k \frac{\partial M_{3i}}{\partial p} u_i - \lambda_p \frac{\partial M_{4i}}{\partial p} u_i \\
&-\lambda_q \frac{\partial M_{5i}}{\partial p} u_i - \lambda_F \frac{\partial M_{6i}^F}{\partial p} u_i \Bigg] f_t
\end{aligned}
\tag{12.30}
$$

$$
\begin{aligned}
\dot{\lambda}_q = -\frac{\partial H}{\partial q} = \Bigg[&-\lambda_a \frac{\partial M_{1i}}{\partial q} u_i - \lambda_h \frac{\partial M_{2i}}{\partial q} u_i - \lambda_k \frac{\partial M_{3i}}{\partial q} u_i - \lambda_p \frac{\partial M_{4i}}{\partial q} u_i \\
&-\lambda_q \frac{\partial M_{5i}}{\partial q} u_i - \lambda_F \frac{\partial M_{6i}^F}{\partial q} u_i \Bigg] f_t
\end{aligned}
\tag{12.31}
$$

$$\dot{\lambda}_F = -\frac{\partial H}{\partial F} = \left[-\lambda_a \frac{\partial M_{1i}}{\partial F} u_i - \lambda_h \frac{\partial M_{2i}}{\partial F} u_i - \lambda_k \frac{\partial M_{3i}}{\partial F} u_i - \lambda_p \frac{\partial M_{4i}}{\partial F} u_i \right.$$
$$\left. -\lambda_q \frac{\partial M_{5i}}{\partial F} u_i - \lambda_F \frac{\partial M_{6i}^F}{\partial F} u_i \right] f_t - \lambda_F \frac{\partial}{\partial F}\left(\frac{na}{r}\right) \tag{12.32}$$

As stated above, the $\partial M/\partial \mathbf{z}$ partials are derived with $\partial F/\partial h = \partial F/\partial k = \partial F/\partial a = 0$, and whenever X_1 and Y_1 appear explicitly in the matrix M, their partials $(\partial X_1/\partial h)_F$, $(\partial X_1/\partial k)_F$, $(\partial Y_1/\partial h)_F$ and $(\partial Y_1/\partial k)_F$ must be generated by holding F constant. These partials, written below, must be recognized as being different from the partials $\partial X_1/\partial h$, $\partial X_1/\partial k$, $\partial Y_1/\partial h$ and $\partial Y_1/\partial k$ in Equations (12.18) through (12.21), used in the generation of the dynamic equations where the dependence of F on h and k was fully accounted for; [10] and Chapter 6 of this book discuss thoroughly the mathematics leading to these derivations.

$$\left(\frac{\partial X_1}{\partial h}\right)_F = a\left[-(hc_F - ks_F)\left(\beta + \frac{h^2\beta^3}{1-\beta}\right) - h\beta c_F\right]$$
$$\left(\frac{\partial X_1}{\partial k}\right)_F = -a\left[(hc_F - ks_F)\frac{hk\beta^3}{1-\beta} + 1 - h\beta s_F\right]$$
$$\left(\frac{\partial Y_1}{\partial h}\right)_F = a\left[(hc_F - ks_F)\frac{hk\beta^3}{1-\beta} - 1 + k\beta c_F\right]$$
$$\left(\frac{\partial Y_1}{\partial k}\right)_F = a\left[(hc_F - ks_F)\left(\beta + \frac{k^2\beta^3}{1-\beta}\right) - k\beta s_F\right]$$

The nonzero partials of the first five rows of M are, as in [10] and Chapter 6 of this book,

$$\frac{\partial M_{11}}{\partial h} = \frac{2}{n^2 a}\frac{\partial \dot{X}_1}{\partial h} \tag{12.33}$$
$$\frac{\partial M_{12}}{\partial h} = \frac{2}{n^2 a}\frac{\partial \dot{Y}_1}{\partial h} \tag{12.34}$$

$$\frac{\partial M_{21}}{\partial h} = -\frac{h}{na^2G}\left(\frac{\partial X_1}{\partial k} - \frac{h\beta}{n}\dot{X}_1\right) + \frac{G}{na^2}\left[\frac{\partial^2 X_1}{\partial h \partial k} - \frac{\dot{X}_1}{n}\left(\beta + \frac{h^2\beta^3}{1-\beta}\right) - \frac{h\beta}{n}\frac{\partial \dot{X}_1}{\partial h}\right] \tag{12.35}$$

Also,

$$\frac{\partial M_{22}}{\partial h} = -\frac{h}{na^2G}\left(\frac{\partial Y_1}{\partial k} - \frac{h\beta}{n}\dot{Y}_1\right) + \frac{G}{na^2}\left[\frac{\partial^2 Y_1}{\partial h \partial k} - \frac{\dot{Y}_1}{n}\left(\beta + \frac{h^2\beta^3}{1-\beta}\right) - \frac{h\beta}{n}\frac{\partial \dot{Y}_1}{\partial h}\right] \tag{12.36}$$

$$\frac{\partial M_{23}}{\partial h} = \frac{hkG^{-3}}{na^2}(qY_1 - pX_1) + \frac{k\left[q(\partial Y_1/\partial h)_F - p(\partial X_1/\partial h)_F\right]}{na^2G} \tag{12.37}$$

$$\frac{\partial M_{31}}{\partial h} = \frac{h}{na^2G}\left(\frac{\partial X_1}{\partial h} + k\beta\frac{\dot{X}_1}{n}\right) - \frac{G}{na^2}\left[\frac{\partial^2 X_1}{\partial h^2} + \frac{hk\beta^3}{1-\beta}\frac{\dot{X}_1}{n} + \frac{k\beta}{n}\frac{\partial \dot{X}_1}{\partial h}\right] \tag{12.38}$$

$$\frac{\partial M_{32}}{\partial h} = \frac{h}{na^2G}\left(\frac{\partial Y_1}{\partial h} + k\beta\frac{\dot{Y}_1}{n}\right) - \frac{G}{na^2}\left[\frac{\partial^2 Y_1}{\partial h^2} + \frac{hk\beta^3}{1-\beta}\frac{\dot{Y}_1}{n} + \frac{k\beta}{n}\frac{\partial \dot{Y}_1}{\partial h}\right] \tag{12.39}$$

$$\frac{\partial M_{33}}{\partial h} = -\frac{1}{na^2G}\left\{(qY_1 - pX_1) + h\left[q\left(\frac{\partial Y_1}{\partial h}\right)_F - p\left(\frac{\partial X_1}{\partial h}\right)_F\right]\right\} - \frac{h^2(qY_1 - pX_1)}{na^2G^3} \tag{12.40}$$

$$\frac{\partial M_{43}}{\partial h} = \frac{K}{2na^2G}\left[\left(\frac{\partial Y_1}{\partial h}\right)_F + \frac{hY_1}{G^2}\right] \tag{12.41}$$

$$\frac{\partial M_{53}}{\partial h} = \frac{K}{2na^2G}\left[\left(\frac{\partial X_1}{\partial h}\right)_F + \frac{hX_1}{G^2}\right] \tag{12.42}$$

Likewise, the nonzero partials of M with respect to k are

$$\frac{\partial M_{11}}{\partial k} = \frac{2}{n^2a}\frac{\partial \dot{X}_1}{\partial k} \tag{12.43}$$

$$\frac{\partial M_{12}}{\partial k} = \frac{2}{n^2a}\frac{\partial \dot{Y}_1}{\partial k} \tag{12.44}$$

$$\frac{\partial M_{21}}{\partial k} = -\frac{k}{na^2G}\left(\frac{\partial X_1}{\partial k} - \frac{h\beta}{n}\dot{X}_1\right) + \frac{G}{na^2}\left[\frac{\partial^2 X_1}{\partial k^2} - \frac{hk\beta^3}{n(1-\beta)}\dot{X}_1 - \frac{h\beta}{n}\frac{\partial \dot{X}_1}{\partial k}\right] \tag{12.45}$$

$$\frac{\partial M_{22}}{\partial k} = -\frac{k}{na^2G}\left(\frac{\partial Y_1}{\partial k} - \frac{h\beta}{n}\dot{Y}_1\right) + \frac{G}{na^2}\left[\frac{\partial^2 Y_1}{\partial k^2} - \frac{hk\beta^3}{n(1-\beta)}\dot{Y}_1 - \frac{h\beta}{n}\frac{\partial \dot{Y}_1}{\partial k}\right] \tag{12.46}$$

$$\frac{\partial M_{23}}{\partial k} = \frac{(qY_1 - pX_1)}{na^2G} + \frac{1}{na^2G}\left\{k\left[q\left(\frac{\partial Y_1}{\partial k}\right)_F - p\left(\frac{\partial X_1}{\partial k}\right)_F\right] + \frac{k^2(qY_1 - pX_1)}{G^2}\right\} \tag{12.47}$$

$$\frac{\partial M_{31}}{\partial k} = \frac{k}{na^2G}\left(\frac{\partial X_1}{\partial h} + k\beta\frac{\dot{X}_1}{n}\right) - \frac{G}{na^2}\left[\frac{\partial^2 X_1}{\partial k \partial h} + \left(\beta + \frac{k^2\beta^3}{1-\beta}\right)\frac{\dot{X}_1}{n} + \frac{k\beta}{n}\frac{\partial \dot{X}_1}{\partial k}\right] \tag{12.48}$$

$$\frac{\partial M_{32}}{\partial k} = \frac{k}{na^2G}\left(\frac{\partial Y_1}{\partial h} + k\beta\frac{\dot{Y}_1}{n}\right) - \frac{G}{na^2}\left[\frac{\partial^2 Y_1}{\partial k \partial h} + \left(\beta + \frac{k^2\beta^3}{1-\beta}\right)\frac{\dot{Y}_1}{n} + \frac{k\beta}{n}\frac{\partial \dot{Y}_1}{\partial k}\right] \tag{12.49}$$

$$\frac{\partial M_{33}}{\partial k} = -\frac{h}{na^2G}\left[q\left(\frac{\partial Y_1}{\partial k}\right)_F - p\left(\frac{\partial X_1}{\partial k}\right)_F\right] - \frac{hk}{na^2G^3}(qY_1 - pX_1) \tag{12.50}$$

$$\frac{\partial M_{43}}{\partial k} = \frac{K}{2na^2G}\left(\frac{\partial Y_1}{\partial k}\right)_F + \frac{kK}{2na^2G^3}Y_1 \tag{12.51}$$

$$\frac{\partial M_{53}}{\partial k} = \frac{K}{2na^2G}\left(\frac{\partial X_1}{\partial k}\right)_F + \frac{kK}{2na^2G^3}X_1 \tag{12.52}$$

Next, the nonzero partials with respect to p and q are obtained as

$$\partial M_{23}/\partial p = -kX_1/na^2G \tag{12.53}$$

$$\partial M_{33}/\partial p = hX_1/na^2G \tag{12.54}$$

$$\partial M_{43}/\partial p = pY_1/na^2G \tag{12.55}$$

$$\partial M_{53}/\partial p = pX_1/na^2G \tag{12.56}$$

$$\partial M_{23}/\partial q = kY_1/na^2G \tag{12.57}$$

$$\partial M_{33}/\partial q = -hY_1/na^2G \tag{12.58}$$

$$\partial M_{43}/\partial q = qY_1/na^2G \tag{12.59}$$

$$\partial M_{53}/\partial q = qX_1/na^2G \tag{12.60}$$

The accessory partials $\partial\dot{X}_1/\partial h$, $\partial\dot{X}_1/\partial k$, $\partial\dot{Y}_1/\partial h$ and $\partial\dot{Y}_1/\partial k$ appearing in some of the above derived derivatives are directly obtained from the expressions for $\dot{X}_1$ and $\dot{Y}_1$ in Equations (12.9) and (12.10) by once again letting $\partial F/\partial h = \partial F/\partial k = 0$, such that

$$\frac{\partial\dot{X}_1}{\partial h} = \frac{a}{r}\dot{X}_1 s_F + \frac{na^2}{r}\left[h\beta s_F + (kc_F + hs_F)\left(\beta + \frac{h^2\beta^3}{1-\beta}\right)\right] \tag{12.61}$$

$$\frac{\partial\dot{X}_1}{\partial k} = \frac{a}{r}\dot{X}_1 c_F + \frac{na^2}{r}\left[h\beta c_F + (kc_F + hs_F)\frac{hk\beta^3}{1-\beta}\right] \tag{12.62}$$

$$\frac{\partial\dot{Y}_1}{\partial h} = \frac{a}{r}\dot{Y}_1 s_F + \frac{na^2}{r}\left[-k\beta s_F - (kc_F + hs_F)\frac{hk\beta^3}{1-\beta}\right] \tag{12.63}$$

$$\frac{\partial\dot{Y}_1}{\partial k} = \frac{a}{r}\dot{Y}_1 c_F + \frac{na^2}{r}\left[-k\beta c_F - (kc_F + hs_F)\left(\beta + \frac{k^2\beta^3}{1-\beta}\right)\right] \tag{12.64}$$

As for the second-order partials also appearing in some of the above derived derivatives, they are obtained directly from Equations (12.18)–(12.21) by once again setting $\partial F/\partial h = \partial F/\partial k = \partial F/\partial a = 0$, such that

$$\frac{\partial^2 X_1}{\partial h^2} = a\left\{-\left(1+\frac{a}{r}\right)c_F\left(\beta + \frac{h^2\beta^3}{1-\beta}\right) - \frac{h\beta^3}{1-\beta}(hc_F - ks_F)\left[3 + \frac{h^2\beta^2(3-2\beta)}{(1-\beta)^2}\right]\right.$$
$$\left. - \frac{a^2}{r^2}s_F c_F(h\beta - s_F)\right\} \tag{12.65}$$

$$\frac{\partial^2 X_1}{\partial k^2} = -a\left\{-s_F\left(1+\frac{a}{r}\right)\frac{hk\beta^3}{1-\beta} + \frac{h\beta^3}{1-\beta}(hc_F - ks_F)\left[1 + \frac{k^2\beta^2(3-2\beta)}{(1-\beta)^2}\right]\right.$$
$$\left. - \frac{a^2}{r^2}s_F c_F(h\beta - s_F)\right\} \tag{12.66}$$

$$\frac{\partial^2 X_1}{\partial h\partial k} = -a\left\{c_F\frac{hk\beta^3}{1-\beta} + \frac{k\beta^3}{1-\beta}(hc_F - ks_F)\left[1 + \frac{h^2\beta^2(3-2\beta)}{(1-\beta)^2}\right]\right.$$
$$\left. + \frac{a^2}{r^2}s_F^2(s_F - h\beta) - \frac{a}{r}s_F\left(\beta + \frac{h^2\beta^3}{1-\beta}\right)\right\} \tag{12.67}$$

$$\frac{\partial^2 X_1}{\partial k \partial h} = a\left\{s_F\left(\beta + \frac{h^2\beta^3}{1-\beta}\right) - \frac{k\beta^3}{1-\beta}\left(hc_F - ks_F\right)\left[1 + \frac{h^2\beta^2\left(3-2\beta\right)}{\left(1-\beta\right)^2}\right]\right.$$
$$\left. -\frac{a^2}{r^2}c_F^2\left(h\beta - s_F\right) - \frac{a}{r}c_F\frac{hk\beta^3}{1-\beta}\right\} \tag{12.68}$$

$$\frac{\partial^2 Y_1}{\partial h^2} = a\left\{c_F\left(1+\frac{a}{r}\right)\frac{hk\beta^3}{1-\beta} + \frac{k\beta^3}{1-\beta}\left(hc_F - ks_F\right)\left[1 + \frac{h^2\beta^2\left(3-2\beta\right)}{\left(1-\beta\right)^2}\right]\right.$$
$$\left. +\frac{a^2}{r^2}s_F c_F\left(k\beta - c_F\right)\right\} \tag{12.69}$$

$$\frac{\partial^2 Y_1}{\partial k^2} = a\left\{-s_F\left(1+\frac{a}{r}\right)\left(\beta + \frac{k^2\beta^3}{1-\beta}\right) - \frac{k\beta^3}{1-\beta}\left(hc_F - ks_F\right)\left[3 + \frac{k^2\beta^2\left(3-2\beta\right)}{\left(1-\beta\right)^2}\right]\right.$$
$$\left. +\frac{a^2}{r^2}s_F c_F\left(c_F - k\beta\right)\right\} \tag{12.70}$$

$$\frac{\partial^2 Y_1}{\partial h \partial k} = a\left\{c_F\left(\beta + \frac{k^2\beta^3}{1-\beta}\right) + \frac{h\beta^3}{1-\beta}\left(hc_F - ks_F\right)\left[1 + \frac{k^2\beta^2\left(3-2\beta\right)}{\left(1-\beta\right)^2}\right]\right.$$
$$\left. +\frac{a^2}{r^2}s_F^2\left(c_F - k\beta\right) - \frac{as_F}{r}\frac{hk\beta^3}{1-\beta}\right\} \tag{12.71}$$

$$\frac{\partial^2 Y_1}{\partial k \partial h} = a\left\{-s_F\frac{hk\beta^3}{1-\beta} + \frac{h\beta^3}{1-\beta}\left(hc_F - ks_F\right)\left[1 + \frac{k^2\beta^2\left(3-2\beta\right)}{\left(1-\beta\right)^2}\right]\right.$$
$$\left. +\frac{a^2}{r^2}c_F^2\left(k\beta - c_F\right) + \frac{a}{r}c_F\left(\beta + \frac{k^2\beta^3}{1-\beta}\right)\right\} \tag{12.72}$$

$$\frac{\partial^2 X_1}{\partial a \partial h} = \frac{1}{a}\frac{\partial X_1}{\partial h} \tag{12.73}$$
$$\frac{\partial^2 X_1}{\partial a \partial k} = \frac{1}{a}\frac{\partial X_1}{\partial k} \tag{12.74}$$
$$\frac{\partial^2 Y_1}{\partial a \partial h} = \frac{1}{a}\frac{\partial Y_1}{\partial h} \tag{12.75}$$
$$\frac{\partial^2 Y_1}{\partial a \partial k} = \frac{1}{a}\frac{\partial Y_1}{\partial k} \tag{12.76}$$

The above four second-order partials are written in terms of $\partial X_1/\partial h$, $\partial X_1/\partial k$, $\partial Y_1/\partial h$ and $\partial Y_1/\partial k$, given in Equations (12.18)–(12.21). These four mixed second-order partials as well as the partials of X_1, Y_1, $\dot{X}_1$ and $\dot{Y}_1$ with respect to a, shown below, are needed to generate the $\partial M/\partial a$ partials with the use of $\partial F/\partial a = 0$, such that

$$\partial X_1/\partial a = X_1/a \tag{12.77}$$
$$\partial Y_1/\partial a = Y_1/a \tag{12.78}$$

$$\frac{\partial \dot{X}_1}{\partial a} = -\frac{na}{2r}\left[hk\beta c_F - \left(1 - h^2\beta\right)s_F\right] = -\frac{\dot{X}_1}{2a} \tag{12.79}$$

$$\frac{\partial \dot{Y}_1}{\partial a} = \frac{na}{2r}\left[hk\beta s_F - \left(1 - k^2\beta\right)c_F\right] = -\frac{\dot{Y}_1}{2a} \tag{12.80}$$

$$\frac{\partial M}{\partial a} = \frac{1}{2a}\begin{bmatrix} 3 & 0 & 0 & 0 & 0 \\ & 1 & 0 & 0 & 0 \\ & & 1 & 0 & 0 \\ & & & 1 & 0 \\ 0 & & & & 1 \end{bmatrix} M \tag{12.81}$$

In [10] and Chapter 6 of this book, the nonzero partials of the sixth row of matrix M were generated by making use of the $\partial\lambda/\partial\dot{\mathbf{r}}$, $\partial h/\partial\dot{\mathbf{r}}$ and $\partial k/\partial\dot{\mathbf{r}}$ partials because $\partial F/\partial\dot{\mathbf{r}}$ was written as $\frac{a}{r}\left[\frac{\partial\lambda}{\partial\dot{\mathbf{r}}} + s_F\frac{\partial k}{\partial\dot{\mathbf{r}}} - c_F\frac{\partial h}{\partial\dot{\mathbf{r}}}\right]$ involving the known velocity partials whose partial derivatives with respect to the six elements were developed earlier in [15], where λ was used as the sixth equinoctial element. This has led to a more complicated form of the partials of M_{61}^F, M_{62}^F and M_{63}^F with respect to all six elements used in this chapter, namely a, h, k, p, q and F, shown in Appendix B of [10] and Chapter 6 of this book. This earlier form is simplified here by using the explicit form of $\partial F/\partial\dot{\mathbf{r}}$ given in Equation (12.22) and by taking the partials of its three components directly with respect to a, h, k, p, q and F. Because β and $G = (1-\beta)/\beta$ appear in the $\partial F/\partial\dot{\mathbf{r}}$ equation, their partials with respect to h and k are needed. These partials are obtained as

$$\frac{\partial\beta}{\partial h} = \frac{h\beta^3}{1-\beta} \tag{12.82}$$

$$\frac{\partial\beta}{\partial k} = \frac{k\beta^3}{1-\beta} \tag{12.83}$$

$$\frac{\partial G}{\partial h} = \frac{\partial G}{\partial\beta}\frac{\partial\beta}{\partial h} = -\frac{1}{\beta^2}\frac{h\beta^3}{1-\beta} = -\frac{h\beta}{1-\beta} \tag{12.84}$$

$$\frac{\partial G}{\partial k} = \frac{\partial G}{\partial\beta}\frac{\partial\beta}{\partial k} = -\frac{1}{\beta^2}\frac{k\beta^3}{1-\beta} = -\frac{k\beta}{1-\beta} \tag{12.85}$$

Also $\frac{\partial r}{\partial a} = \frac{r}{a}$, $\frac{\partial n}{\partial a} = -\frac{3}{2}\frac{n}{a}$, and

$$\frac{\partial}{\partial a}\left(\frac{1}{nar}\right) = -\frac{1}{2na^2r} \tag{12.86}$$

$$\frac{\partial}{\partial h}\left(\frac{1}{nar}\right) = \frac{s_F}{nr^2} \tag{12.87}$$

$$\frac{\partial}{\partial k}\left(\frac{1}{nar}\right) = \frac{c_F}{nr^2} \tag{12.88}$$

where use is made of $\partial r/\partial h = -as_F$ and $\partial r/\partial k = -ac_F$. Therefore,

$$\frac{\partial M_{61}^F}{\partial a} = -\frac{1}{2na^2r}\left[-2X_1 + G(h\beta - s_F)\frac{\partial X_1}{\partial h} + G(k\beta - c_F)\frac{\partial X_1}{\partial k} - \beta G(ks_F - hc_F)\frac{\dot{X}_1}{n}\right] + \frac{1}{nar} \times \left[-2\frac{\partial X_1}{\partial a} + G(h\beta - s_F)\frac{\partial^2 X_1}{\partial a\partial h} + G(k\beta - c_F)\frac{\partial^2 X_1}{\partial a\partial k} - \beta G(ks_F - hc_F)\frac{1}{n}\frac{\partial \dot{X}_1}{\partial a} - \beta G(ks_F - hc_F)\dot{X}_1\left(\frac{3}{2na}\right)\right] \tag{12.89}$$

$$\frac{\partial M_{62}^F}{\partial a} = -\frac{1}{2na^2r}\left[-2Y_1 + G(h\beta - s_F)\frac{\partial Y_1}{\partial h} + G(k\beta - c_F)\frac{\partial Y_1}{\partial k} - \beta G(ks_F - hc_F)\frac{\dot{Y}_1}{n}\right] + \frac{1}{nar}\left[-2\frac{\partial Y_1}{\partial a} + G(h\beta - s_F) \times \frac{\partial^2 Y_1}{\partial a\partial h} + G(k\beta - c_F)\frac{\partial^2 Y_1}{\partial a\partial k} - \beta G(ks_F - hc_F)\frac{1}{n}\frac{\partial \dot{Y}_1}{\partial a} - \beta G(ks_F - hc_F)\dot{Y}_1\left(\frac{3}{2na}\right)\right] \tag{12.90}$$

$$\frac{\partial M_{63}^F}{\partial a} = \frac{1}{nar}\left[\frac{r}{aG}\left(q\frac{\partial Y_1}{\partial a} - p\frac{\partial X_1}{\partial a}\right)\right] - \frac{1}{2na^3G}(qY_1 - pX_1) \tag{12.91}$$

The partials ofM_{61}^F, M_{62}^F and M_{63}^F with respect to h are reduced to the form

$$\frac{\partial M_{61}^F}{\partial h} = \frac{s_F}{nr^2}\left[-2X_1 + G(h\beta - s_F)\frac{\partial X_1}{\partial h} + G(k\beta - c_F)\frac{\partial X_1}{\partial k} - \beta G(ks_F - hc_F)\frac{\dot{X}_1}{n}\right] + \frac{1}{nar}\left[-2\left(\frac{\partial X_1}{\partial h}\right)_F + \left\{\frac{h\beta}{1-\beta}\left(s_F - h\beta^2\right) + (1-\beta)\right\}\frac{\partial X_1}{\partial h} + G(h\beta - s_F)\frac{\partial^2 X_1}{\partial h^2} + \frac{h\beta\left(c_F - k\beta^2\right)}{1-\beta}\frac{\partial X_1}{\partial k} + G(k\beta - c_F)\frac{\partial^2 X_1}{\partial h\partial k} + \left\{(ks_F - hc_F)\frac{h\beta^3}{1-\beta} + c_F(1-\beta)\right\}\frac{\dot{X}_1}{n} - (1-\beta)(ks_F - hc_F)\frac{1}{n}\frac{\partial \dot{X}_1}{\partial h}\right] \tag{12.92}$$

$$\frac{\partial M_{62}^F}{\partial h} = \frac{s_F}{nr^2}\left[-2Y_1 + G(h\beta - s_F)\frac{\partial Y_1}{\partial h} + G(k\beta - c_F)\frac{\partial Y_1}{\partial k} - \beta G(ks_F - hc_F)\frac{\dot{Y}_1}{n}\right] + \frac{1}{nar}\left[-2\left(\frac{\partial Y_1}{\partial h}\right)_F + \left\{\frac{h\beta}{1-\beta}\left(s_F - h\beta^2\right) + (1-\beta)\right\}\frac{\partial Y_1}{\partial h} + G(h\beta - s_F)\frac{\partial^2 Y_1}{\partial h^2} + \frac{h\beta\left(c_F - k\beta^2\right)}{1-\beta}\frac{\partial Y_1}{\partial k} + G(k\beta - c_F)\frac{\partial^2 Y_1}{\partial h\partial k} + \left\{(ks_F - hc_F)\frac{h\beta^3}{1-\beta} + c_F(1-\beta)\right\}\frac{\dot{Y}_1}{n} - (1-\beta)(ks_F - hc_F)\frac{1}{n}\frac{\partial \dot{Y}_1}{\partial h}\right] \tag{12.93}$$

$$\frac{\partial M_{63}^F}{\partial h} = \frac{1}{nar}\left\{\left[-\frac{s_F}{G} + \frac{r}{a}\frac{h\beta}{G^2(1-\beta)}\right](qY_1 - pX_1) + \frac{r}{aG}\left[q\left(\frac{\partial Y_1}{\partial h}\right)_F - p\left(\frac{\partial X_1}{\partial h}\right)_F\right]\right\} + \frac{s_F}{nraG}(qY_1 - pX_1) \quad (12.94)$$

Likewise, the partials of M_{61}^F, M_{62}^F and M_{63}^F with respect to k are derived to yield

$$\begin{aligned}\frac{\partial M_{61}^F}{\partial k} = {} & \frac{c_F}{nr^2}\left[-2X_1 + G(h\beta - s_F)\frac{\partial X_1}{\partial h} + G(k\beta - c_F)\frac{\partial X_1}{\partial k}\right. \\ & \left. -\beta G(ks_F - hc_F)\frac{\dot{X}_1}{n}\right] + \frac{1}{nar}\left[-2\left(\frac{\partial X_1}{\partial k}\right)_F + \frac{k\beta}{1-\beta}\left(s_F - h\beta^2\right)\frac{\partial X_1}{\partial h}\right. \\ & + G(h\beta - s_F)\frac{\partial^2 X_1}{\partial k \partial h} + \left\{\frac{k\beta}{1-\beta}\left(c_F - k\beta^2\right) + (1-\beta)\right\}\frac{\partial X_1}{\partial k} \\ & + G(k\beta - c_F)\frac{\partial^2 X_1}{\partial k^2} + \left\{(ks_F - hc_F)\frac{k\beta^3}{1-\beta} - (1-\beta)s_F\right\}\frac{\dot{X}_1}{n} \\ & \left. -(1-\beta)(ks_F - hc_F)\frac{1}{n}\frac{\partial \dot{X}_1}{\partial k}\right]\end{aligned} \quad (12.95)$$

$$\begin{aligned}\frac{\partial M_{62}^F}{\partial k} = {} & \frac{c_F}{nr^2}\left[-2Y_1 + G(h\beta - s_F)\frac{\partial Y_1}{\partial h} + G(k\beta - c_F)\frac{\partial Y_1}{\partial k}\right. \\ & \left. -\beta G(ks_F - hc_F)\frac{\dot{Y}_1}{n}\right] + \frac{1}{nar}\left[-2\left(\frac{\partial Y_1}{\partial k}\right)_F + \frac{k\beta}{1-\beta}\left(s_F - h\beta^2\right)\frac{\partial Y_1}{\partial h}\right. \\ & + G(h\beta - s_F)\frac{\partial^2 Y_1}{\partial k \partial h} + \left\{\frac{k\beta}{1-\beta}\left(c_F - k\beta^2\right) + (1-\beta)\right\}\frac{\partial Y_1}{\partial k} \\ & + G(k\beta - c_F)\frac{\partial^2 Y_1}{\partial k^2} + \left\{(ks_F - hc_F)\frac{k\beta^3}{1-\beta} - (1-\beta)s_F\right\}\frac{\dot{Y}_1}{n} \\ & \left. -(1-\beta)(ks_F - hc_F)\frac{1}{n}\frac{\partial \dot{Y}_1}{\partial k}\right]\end{aligned} \quad (12.96)$$

$$\frac{\partial M_{63}^F}{\partial k} = \frac{1}{nar}\left\{\left[-\frac{c_F}{G} + \frac{r}{a}\frac{k\beta}{G^2(1-\beta)}\right](qY_1 - pX_1) + \frac{r}{aG}\left[q\left(\frac{\partial Y_1}{\partial k}\right)_F - p\left(\frac{\partial X_1}{\partial k}\right)_F\right]\right\} + \frac{c_F}{nraG}(qY_1 - pX_1) \quad (12.97)$$

The nonzero partials of M_{61}^F, M_{62}^F and M_{63}^F with respect to p and q are

$$\partial M_{63}^F/\partial p = -X_1/na^2G \quad (12.98)$$

$$\partial M_{63}^F/\partial q = Y_1/na^2G \quad (12.99)$$

As in [10] and Chapter 6 of this book, additional partials are also needed in the Euler–Lagrange equations as

$$\frac{\partial}{\partial a}\left(\frac{na}{r}\right) = -\frac{3}{2}\frac{\mu^{1/2}a^{-3/2}}{r} \quad (12.100)$$

$$\frac{\partial}{\partial h}\left(\frac{na}{r}\right) = \frac{na^2}{r^2} s_F \tag{12.101}$$

$$\frac{\partial}{\partial k}\left(\frac{na}{r}\right) = \frac{na^2}{r^2} c_F \tag{12.102}$$

This long list of the relevant partials is now completed by generating the partials of M with respect to F, the fast variable. The nonzero partials are

$$\frac{\partial M_{11}}{\partial F} = \frac{2}{n^2 a}\frac{\partial \dot{X}_1}{\partial F} \tag{12.103}$$

$$\frac{\partial M_{12}}{\partial F} = \frac{2}{n^2 a}\frac{\partial \dot{Y}_1}{\partial F} \tag{12.104}$$

$$\frac{\partial M_{21}}{\partial F} = \frac{G}{na^2}\left(\frac{\partial^2 X_1}{\partial F \partial k} - \frac{h\beta}{n}\frac{\partial \dot{X}_1}{\partial F}\right) \tag{12.105}$$

$$\frac{\partial M_{22}}{\partial F} = \frac{G}{na^2}\left(\frac{\partial^2 Y_1}{\partial F \partial k} - \frac{h\beta}{n}\frac{\partial \dot{Y}_1}{\partial F}\right) \tag{12.106}$$

$$\frac{\partial M_{23}}{\partial F} = \frac{k\left[q\left(\partial Y_1/\partial F\right) - p\left(\partial X_1/\partial F\right)\right]}{na^2 G} \tag{12.107}$$

$$\frac{\partial M_{31}}{\partial F} = -\frac{G}{na^2}\left(\frac{\partial^2 X_1}{\partial F \partial h} + \frac{k\beta}{n}\frac{\partial \dot{X}_1}{\partial F}\right) \tag{12.108}$$

$$\frac{\partial M_{32}}{\partial F} = -\frac{G}{na^2}\left(\frac{\partial^2 Y_1}{\partial F \partial h} + \frac{k\beta}{n}\frac{\partial \dot{Y}_1}{\partial F}\right) \tag{12.109}$$

$$\frac{\partial M_{33}}{\partial F} = \frac{-h\left[q\left(\partial Y_1/\partial F\right) - p\left(\partial X_1/\partial F\right)\right]}{na^2 G} \tag{12.110}$$

$$\frac{\partial M_{43}}{\partial F} = \frac{K}{2na^2 G}\frac{\partial Y_1}{\partial F} \tag{12.111}$$

$$\frac{\partial M_{53}}{\partial F} = \frac{K}{2na^2 G}\frac{\partial X_1}{\partial F} \tag{12.112}$$

and with the use of $\partial r/\partial F = a\left(k s_F - h c_F\right)$ and

$$\frac{\partial}{\partial F}\left(\frac{1}{nar}\right) = -\frac{\left(k s_F - h c_F\right)}{nr^2} \tag{12.113}$$

the partials of M_{61}^F, M_{62}^F and M_{63}^F with respect to F are obtained as

$$\begin{aligned}\frac{\partial M_{61}^F}{\partial F} &= \frac{\left(h c_F - k s_F\right)}{nr^2}\left[-2X_1 + G\left(h\beta - s_F\right)\frac{\partial X_1}{\partial h}\right.\\ &\left. + G\left(k\beta - c_F\right)\frac{\partial X_1}{\partial k} - \beta G\left(k s_F - h c_F\right)\frac{\dot{X}_1}{n}\right]\\ &+ \frac{1}{nar}\left[-2\frac{\partial X_1}{\partial F} - G c_F \frac{\partial X_1}{\partial h} + G\left(h\beta - s_F\right)\frac{\partial^2 X_1}{\partial F \partial h}\right.\\ &+ G s_F \frac{\partial X_1}{\partial k} + G\left(k\beta - c_F\right)\frac{\partial^2 X_1}{\partial F \partial k} - \beta G\left(k c_F + h s_F\right)\frac{\dot{X}_1}{n}\\ &\left. - \beta G\left(k s_F - h c_F\right)\frac{1}{n}\frac{\partial \dot{X}_1}{\partial F}\right]\end{aligned} \tag{12.114}$$

$$\frac{\partial M_{62}^F}{\partial F} = \frac{(hc_F - ks_F)}{nr^2}\left[-2Y_1 + G(h\beta - s_F)\frac{\partial Y_1}{\partial h}\right.$$
$$\left. + G(k\beta - c_F)\frac{\partial Y_1}{\partial k} - \beta G(ks_F - hc_F)\frac{\dot{Y}_1}{n}\right]$$
$$+ \frac{1}{nar}\left[-2\frac{\partial Y_1}{\partial F} - Gc_F\frac{\partial Y_1}{\partial h} + G(h\beta - s_F)\frac{\partial^2 Y_1}{\partial F \partial h}\right.$$
$$+ Gs_F\frac{\partial Y_1}{\partial k} + G(k\beta - c_F)\frac{\partial^2 Y_1}{\partial F \partial k} - \beta G(kc_F + hs_F)\frac{\dot{Y}_1}{n}$$
$$\left. -\beta G(ks_F - hc_F)\frac{1}{n}\frac{\partial \dot{Y}_1}{\partial F}\right] \tag{12.115}$$

$$\frac{\partial M_{63}^F}{\partial F} = \frac{1}{nar}\left[\frac{(ks_F - hc_F)}{G}(qY_1 - pX_1) + \frac{r}{aG}\left(q\frac{\partial Y_1}{\partial F} - p\frac{\partial X_1}{\partial F}\right)\right]$$
$$+ \frac{(hc_F - ks_F)}{nraG}(qY_1 - pX_1) \tag{12.116}$$

Embedded within the partial derivatives of M with respect to F are the partials of X_1, Y_1, $\dot{X}_1$ and $\dot{Y}_1$ with respect to F, as well as the second-order mixed partials $\partial^2 X_1/\partial F \partial h$, $\partial^2 X_1/\partial F \partial k$, $\partial^2 Y_1/\partial F \partial h$ and $\partial^2 Y_1/\partial F \partial k$. They are derived as

$$\frac{\partial X_1}{\partial F} = a\left[hk\beta c_F - \left(1 - h^2\beta\right)s_F\right] \tag{12.117}$$

$$\frac{\partial Y_1}{\partial F} = a\left[-hk\beta s_F + \left(1 - k^2\beta\right)c_F\right] \tag{12.118}$$

$$\frac{\partial \dot{X}_1}{\partial F} = -\frac{a}{r}(ks_F - hc_F)\dot{X}_1 + \frac{a^2 n}{r}\left[-hk\beta s_F - \left(1 - h^2\beta\right)c_F\right] \tag{12.119}$$

$$\frac{\partial \dot{Y}_1}{\partial F} = -\frac{a}{r}(ks_F - hc_F)\dot{Y}_1 + \frac{a^2 n}{r}\left[-hk\beta c_F - \left(1 - k^2\beta\right)s_F\right] \tag{12.120}$$

$$\frac{\partial^2 X_1}{\partial F \partial h} = a\left[(hs_F + kc_F)\left(\beta + \frac{h^2\beta^3}{1-\beta}\right) + \frac{a^2}{r^2}(h\beta - s_F)(s_F - h) + \frac{a}{r}c_F^2\right] \tag{12.121}$$

$$\frac{\partial^2 X_1}{\partial F \partial k} = -a\left[-(hs_F + kc_F)\frac{hk\beta^3}{1-\beta} + \frac{a^2}{r^2}(s_F - h\beta)(c_F - k) + \frac{a}{r}s_F c_F\right] \tag{12.122}$$

$$\frac{\partial^2 Y_1}{\partial F \partial h} = a\left[-(hs_F + kc_F)\frac{hk\beta^3}{1-\beta} - \frac{a^2}{r^2}(k\beta - c_F)(s_F - h) + \frac{a}{r}s_F c_F\right] \tag{12.123}$$

$$\frac{\partial^2 Y_1}{\partial F \partial k} = a\left[-(hs_F + kc_F)\left(\beta + \frac{k^2\beta^3}{1-\beta}\right) + \frac{a^2}{r^2}(c_F - k\beta)(c_F - k) - \frac{a}{r}s_F^2\right] \tag{12.124}$$

and finally

$$\frac{\partial}{\partial F}\left(\frac{na}{r}\right) = -\frac{na^2}{r^2}(ks_F - hc_F) \tag{12.125}$$

12.3 Accounting of the J_2 Perturbation

It has been shown in [11] and Chapter 10 of this book that the components of the acceleration $\mathbf{f}_{J_2}$ due to J_2 along the directions $\hat{\mathbf{r}}$, $\hat{\boldsymbol{\theta}}$, and $\hat{\mathbf{h}}$ of the polar frame can be written as

$$f_r = -\frac{3}{2}\mu J_2 R^2 r^{-4} + 18\mu J_2 R^2 r^{-6}\left(1+p^2+q^2\right)^{-2}(qY_1 - pX_1)^2 \tag{12.126}$$

$$f_\theta = -12\mu J_2 R^2 r^{-6}\left(1+p^2+q^2\right)^{-2}(qY_1 - pX_1)(qX_1 + pY_1) \tag{12.127}$$

$$f_h = -6\mu J_2 R^2 r^{-5}\left(1+p^2+q^2\right)^{-2}\left(1-p^2-q^2\right)(qY_1 - pX_1) \tag{12.128}$$

where R stands for the equatorial radius of the Earth. In the $\hat{\mathbf{f}}$, $\hat{\mathbf{g}}$, $\hat{\mathbf{w}}$ system, the acceleration components are given by

$$f_f = \frac{X_1}{r}f_r - \frac{Y_1}{r}f_\theta \tag{12.129}$$

$$f_g = \frac{Y_1}{r}f_r + \frac{X_1}{r}f_\theta \tag{12.130}$$

$$f_w = f_h \tag{12.131}$$

Then the dynamic equations are readily written as

$$\dot{a} = (\partial a/\partial \dot{\mathbf{r}})^T\left(\hat{\mathbf{u}}f_t + \mathbf{f}_{J_2}\right) \tag{12.132}$$

$$\dot{h} = (\partial h/\partial \dot{\mathbf{r}})^T\left(\hat{\mathbf{u}}f_t + \mathbf{f}_{J_2}\right) \tag{12.133}$$

$$\dot{k} = (\partial k/\partial \dot{\mathbf{r}})^T\left(\hat{\mathbf{u}}f_t + \mathbf{f}_{J_2}\right) \tag{12.134}$$

$$\dot{p} = (\partial p/\partial \dot{\mathbf{r}})^T\left(\hat{\mathbf{u}}f_t + \mathbf{f}_{J_2}\right) \tag{12.135}$$

$$\dot{q} = (\partial q/\partial \dot{\mathbf{r}})^T\left(\hat{\mathbf{u}}f_t + \mathbf{f}_{J_2}\right) \tag{12.136}$$

$$\dot{F} = na/r + (\partial F/\partial \dot{\mathbf{r}})^T\left(\hat{\mathbf{u}}f_t + \mathbf{f}_{J_2}\right) \tag{12.137}$$

The Euler–Lagrange equations are thus given by

$$\begin{aligned}\dot{\boldsymbol{\lambda}}_z = -\frac{\partial H}{\partial \mathbf{z}} &= -\boldsymbol{\lambda}_z^T\frac{\partial M}{\partial \mathbf{z}}f_t\hat{\mathbf{u}} - \lambda_F\frac{\partial}{\partial \mathbf{z}}\left(\frac{na}{r}\right)\\ &- \boldsymbol{\lambda}_z^T\frac{\partial M(\mathbf{z},F)}{\partial \mathbf{z}}\mathbf{f}_{J_2} - \boldsymbol{\lambda}_z^T M(\mathbf{z},F)\frac{\partial \mathbf{f}_{J_2}}{\partial \mathbf{z}}\end{aligned} \tag{12.138}$$

where the Hamiltonian is given by

$$H = \boldsymbol{\lambda}_z^T M(\mathbf{z},F) f_t\hat{\mathbf{u}} + \lambda_F\frac{na}{r} + \boldsymbol{\lambda}_z^T M(\mathbf{z},F)\,\mathbf{f}_{J_2} \tag{12.139}$$

Having already derived the $\partial M(\mathbf{z},F)/\partial \mathbf{z}$ and $\partial(na/r)/\partial \mathbf{z}$ partials, all we need are the $\partial \mathbf{f}_{J_2}/\partial \mathbf{z}$ partials by once again taking $\partial F/\partial a = \partial F/\partial h = \partial F/\partial k = 0$ and as before $\partial r/\partial a = r/a$, $\partial r/\partial h = -as_F$, $\partial r/\partial k = -ac_F$ and $\partial r/\partial F = a(ks_F - hc_F)$ consistent with F as an independent variable representing the sixth element of $\mathbf{z}$.

Because $\mathbf{f}_{J_2} = f_f\hat{\mathbf{f}} + f_g\hat{\mathbf{g}} + f_w\hat{\mathbf{w}}$, Equations (12.129)–(12.131) yield the partial derivatives of the f_f component as

$$\frac{\partial f_f}{\partial a} = \frac{\partial (X_1/r)}{\partial a} f_r + \frac{X_1}{r} \frac{\partial f_r}{\partial a} - \frac{\partial (Y_1/r)}{\partial a} f_\theta - \frac{Y_1}{r} \frac{\partial f_\theta}{\partial a} \tag{12.140}$$

$$\frac{\partial f_f}{\partial h} = \frac{\partial (X_1/r)}{\partial h} f_r + \frac{X_1}{r} \frac{\partial f_r}{\partial h} - \frac{\partial (Y_1/r)}{\partial h} f_\theta - \frac{Y_1}{r} \frac{\partial f_\theta}{\partial h} \tag{12.141}$$

$$\frac{\partial f_f}{\partial k} = \frac{\partial (X_1/r)}{\partial k} f_r + \frac{X_1}{r} \frac{\partial f_r}{\partial k} - \frac{\partial (Y_1/r)}{\partial k} f_\theta - \frac{Y_1}{r} \frac{\partial f_\theta}{\partial k} \tag{12.142}$$

$$\frac{\partial f_f}{\partial p} = \frac{X_1}{r} \frac{\partial f_r}{\partial p} - \frac{Y_1}{r} \frac{\partial f_\theta}{\partial p} \tag{12.143}$$

$$\frac{\partial f_f}{\partial q} = \frac{X_1}{r} \frac{\partial f_r}{\partial q} - \frac{Y_1}{r} \frac{\partial f_\theta}{\partial q} \tag{12.144}$$

$$\frac{\partial f_f}{\partial F} = \frac{\partial (X_1/r)}{\partial F} f_r + \frac{X_1}{r} \frac{\partial f_r}{\partial F} - \frac{\partial (Y_1/r)}{\partial F} f_\theta - \frac{Y_1}{r} \frac{\partial f_\theta}{\partial F} \tag{12.145}$$

The nonzero partials of X_1/r and Y_1/r appearing above with respect to the equinoctial elements a, h, k and F are given by

$$\frac{\partial (X_1/r)}{\partial a} = \frac{1}{r} \frac{\partial X_1}{\partial a} - \frac{1}{r^2} X_1 \frac{\partial r}{\partial a} \tag{12.146}$$

$$\frac{\partial (X_1/r)}{\partial h} = \frac{1}{r} \left(\frac{\partial X_1}{\partial h} \right)_F - \frac{1}{r^2} X_1 \frac{\partial r}{\partial h} \tag{12.147}$$

$$\frac{\partial (X_1/r)}{\partial k} = \frac{1}{r} \left(\frac{\partial X_1}{\partial k} \right)_F - \frac{1}{r^2} X_1 \frac{\partial r}{\partial k} \tag{12.148}$$

$$\frac{\partial (X_1/r)}{\partial F} = \frac{1}{r} \frac{\partial X_1}{\partial F} - \frac{1}{r^2} X_1 \frac{\partial r}{\partial F} \tag{12.149}$$

$$\frac{\partial (Y_1/r)}{\partial a} = \frac{1}{r} \frac{\partial Y_1}{\partial a} - \frac{1}{r^2} Y_1 \frac{\partial r}{\partial a} \tag{12.150}$$

$$\frac{\partial (Y_1/r)}{\partial h} = \frac{1}{r} \left(\frac{\partial Y_1}{\partial h} \right)_F - \frac{1}{r^2} Y_1 \frac{\partial r}{\partial h} \tag{12.151}$$

$$\frac{\partial (Y_1/r)}{\partial k} = \frac{1}{r} \left(\frac{\partial Y_1}{\partial k} \right)_F - \frac{1}{r^2} Y_1 \frac{\partial r}{\partial k} \tag{12.152}$$

$$\frac{\partial (Y_1/r)}{\partial F} = \frac{1}{r} \frac{\partial Y_1}{\partial F} - \frac{1}{r^2} Y_1 \frac{\partial r}{\partial F} \tag{12.153}$$

with $\partial X_1/\partial a = X_1/a$ and $\partial Y_1/\partial a = Y_1/a$ given in Equations (12.77) and (12.78) respectively, and with $\partial X_1/\partial F$ and $\partial Y_1/\partial F$ given in Equations (12.117) and (12.118). As before, $\partial r/\partial a = r/a$, $\partial r/\partial h = -as_F$, $\partial r/\partial k = -ac_F$ and $\partial r/\partial F = a\,(ks_F - hc_F)$ are derived from the orbit equation $r = a\,(1 - kc_F - hs_F)$ written in terms of the fast element F. The remaining partials of f_g and f_w with respect to the elements are listed below and have the same form as in [11] and Chapter 10 of this book:

$$\frac{\partial f_g}{\partial a} = \frac{\partial (Y_1/r)}{\partial a} f_r + \frac{Y_1}{r} \frac{\partial f_r}{\partial a} + \frac{\partial (X_1/r)}{\partial a} f_\theta + \frac{X_1}{r} \frac{\partial f_\theta}{\partial a} \tag{12.154}$$

$$\frac{\partial f_g}{\partial h} = \frac{\partial (Y_1/r)}{\partial h} f_r + \frac{Y_1}{r} \frac{\partial f_r}{\partial h} + \frac{\partial (X_1/r)}{\partial h} f_\theta + \frac{X_1}{r} \frac{\partial f_\theta}{\partial h} \tag{12.155}$$

$$\frac{\partial f_g}{\partial k} = \frac{\partial (Y_1/r)}{\partial k} f_r + \frac{Y_1}{r} \frac{\partial f_r}{\partial k} + \frac{\partial (X_1/r)}{\partial k} f_\theta + \frac{X_1}{r} \frac{\partial f_\theta}{\partial k} \tag{12.156}$$

$$\frac{\partial f_g}{\partial p} = \frac{Y_1}{r}\frac{\partial f_r}{\partial p} + \frac{X_1}{r}\frac{\partial f_\theta}{\partial p} \tag{12.157}$$

$$\frac{\partial f_g}{\partial q} = \frac{Y_1}{r}\frac{\partial f_r}{\partial q} + \frac{X_1}{r}\frac{\partial f_\theta}{\partial q} \tag{12.158}$$

$$\frac{\partial f_g}{\partial F} = \frac{\partial (Y_1/r)}{\partial F}f_r + \frac{Y_1}{r}\frac{\partial f_r}{\partial F} + \frac{\partial (X_1/r)}{\partial F}f_\theta + \frac{X_1}{r}\frac{\partial f_\theta}{\partial F} \tag{12.159}$$

$$\partial f_w/\partial a = \partial f_h/\partial a \tag{12.160}$$

$$\partial f_w/\partial h = \partial f_h/\partial h \tag{12.161}$$

$$\partial f_w/\partial k = \partial f_h/\partial k \tag{12.162}$$

$$\partial f_w/\partial p = \partial f_h/\partial p \tag{12.163}$$

$$\partial f_w/\partial q = \partial f_h/\partial q \tag{12.164}$$

$$\partial f_w/\partial F = \partial f_h/\partial F \tag{12.165}$$

It is worthwhile noting that the $\partial X_1/\partial h$, $\partial X_1/\partial k$, $\partial Y_1/\partial h$ and $\partial Y_1/\partial k$ used in [11] and Chapter 10 of this book in the description of the partials of X_1/r and Y_1/r with respect to h and k have a different form than the $(\partial X_1/\partial h)_F$, $(\partial X_1/\partial k)_F$, $(\partial Y_1/\partial h)_F$ and $(\partial Y_1/\partial k)_F$ partials used in this chapter to describe the X_1/r and Y_1/r partials shown in Equations (12.147), (12.148), (12.151) and (12.152) respectively, because F is an independent variable here, as opposed to the mean longitude λ in [11] and Chapter 10 of this book. Also the $\partial r/\partial h$, $\partial r/\partial k$ partials appearing in the above-cited equations have a different form in [11] and Chapter 10 of this book for the same reason. The derivations shown in this chapter are now concluded by generating the partial derivatives of f_r, f_θ and f_h with respect to all six elements because they appear in the partials of f_f, f_g and f_w. Going back to Equations (12.126)–(12.128) for the f_r, f_θ and f_h expressions, we have

$$\begin{aligned}\frac{\partial f_r}{\partial a} &= 6\mu J_2 R^2 r^{-5}\frac{\partial r}{\partial a} - 108\mu J_2 R^2 r^{-7} K^{-2}\left(qY_1 - pX_1\right)^2\frac{\partial r}{\partial a}\\ &+ 36\mu J_2 R^2 r^{-6} K^{-2}\left(qY_1 - pX_1\right)\left(q\frac{\partial Y_1}{\partial a} - p\frac{\partial X_1}{\partial a}\right)\end{aligned} \tag{12.166}$$

$$\begin{aligned}\frac{\partial f_r}{\partial h} &= 6\mu J_2 R^2 r^{-5}\frac{\partial r}{\partial h} - 108\mu J_2 R^2 r^{-7} K^{-2}\left(qY_1 - pX_1\right)^2\frac{\partial r}{\partial h}\\ &+ 36\mu J_2 R^2 r^{-6} K^{-2}\left(qY_1 - pX_1\right)\left[q\left(\frac{\partial Y_1}{\partial h}\right)_F - p\left(\frac{\partial X_1}{\partial h}\right)_F\right]\end{aligned} \tag{12.167}$$

$$\begin{aligned}\frac{\partial f_r}{\partial k} &= 6\mu J_2 R^2 r^{-5}\frac{\partial r}{\partial k} - 108\mu J_2 R^2 r^{-7} K^{-2}\left(qY_1 - pX_1\right)^2\frac{\partial r}{\partial k}\\ &+ 36\mu J_2 R^2 r^{-6} K^{-2}\left(qY_1 - pX_1\right)\left[q\left(\frac{\partial Y_1}{\partial k}\right)_F - p\left(\frac{\partial X_1}{\partial k}\right)_F\right]\end{aligned} \tag{12.168}$$

$$\frac{\partial f_r}{\partial p} = 18\mu J_2 R^2 r^{-6}\left[-4pK^{-3}\left(qY_1 - pX_1\right)^2 - 2X_1 K^{-2}\left(qY_1 - pX_1\right)\right] \tag{12.169}$$

$$\frac{\partial f_r}{\partial q} = 18\mu J_2 R^2 r^{-6}\left[-4qK^{-3}\left(qY_1 - pX_1\right)^2 + 2Y_1 K^{-2}\left(qY_1 - pX_1\right)\right] \tag{12.170}$$

$$\begin{aligned}\frac{\partial f_r}{\partial F} &= 6\mu J_2 R^2 r^{-5}\frac{\partial r}{\partial F} - 108\mu J_2 R^2 r^{-7} K^{-2}\left(qY_1 - pX_1\right)^2\frac{\partial r}{\partial F}\\ &+ 36\mu J_2 R^2 r^{-6} K^{-2}\left(qY_1 - pX_1\right)\left[q\frac{\partial Y_1}{\partial F} - p\frac{\partial X_1}{\partial F}\right]\end{aligned} \tag{12.171}$$

Similarly,

$$\frac{\partial f_\theta}{\partial a} = 72\mu J_2 R^2 r^{-7} K^{-2} (qY_1 - pX_1)(qX_1 + pY_1)\frac{\partial r}{\partial a} - 12\mu J_2 R^2 r^{-6} K^{-2} \times \left[\left(q\frac{\partial Y_1}{\partial a} - p\frac{\partial X_1}{\partial a}\right)(qX_1 + pY_1) + (qY_1 - pX_1)\left(q\frac{\partial X_1}{\partial a} + p\frac{\partial Y_1}{\partial a}\right)\right] \tag{12.172}$$

$$\frac{\partial f_\theta}{\partial h} = 72\mu J_2 R^2 r^{-7} K^{-2} (qY_1 - pX_1)(qX_1 + pY_1)\frac{\partial r}{\partial h} - 12\mu J_2 R^2 r^{-6} K^{-2} \left\{\left[q\left(\frac{\partial Y_1}{\partial h}\right)_F - p\left(\frac{\partial X_1}{\partial h}\right)_F\right](qX_1 + pY_1) + (qY_1 - pX_1)\left[q\left(\frac{\partial X_1}{\partial h}\right)_F + p\left(\frac{\partial Y_1}{\partial h}\right)_F\right]\right\} \tag{12.173}$$

$$\frac{\partial f_\theta}{\partial k} = 72\mu J_2 R^2 r^{-7} K^{-2} (qY_1 - pX_1)(qX_1 + pY_1)\frac{\partial r}{\partial k} - 12\mu J_2 R^2 r^{-6} K^{-2} \left\{\left[q\left(\frac{\partial Y_1}{\partial k}\right)_F - p\left(\frac{\partial X_1}{\partial k}\right)_F\right](qX_1 + pY_1) + (qY_1 - pX_1)\left[q\left(\frac{\partial X_1}{\partial k}\right)_F + p\left(\frac{\partial Y_1}{\partial k}\right)_F\right]\right\} \tag{12.174}$$

$$\frac{\partial f_\theta}{\partial p} = -12\mu J_2 R^2 r^{-6}\left[-4pK^{-3}(qY_1 - pX_1)(qX_1 + pY_1) + K^{-2}\{-X_1(qX_1 + pY_1) + Y_1(qY_1 - pX_1)\}\right] \tag{12.175}$$

$$\frac{\partial f_\theta}{\partial q} = -12\mu J_2 R^2 r^{-6}\left[-4qK^{-3}(qY_1 - pX_1)(qX_1 + pY_1) + K^{-2}\{Y_1(qX_1 + pY_1) + X_1(qY_1 - pX_1)\}\right] \tag{12.176}$$

$$\frac{\partial f_\theta}{\partial F} = 72\mu J_2 R^2 r^{-7} K^{-2} (qY_1 - pX_1)(qX_1 + pY_1)\frac{\partial r}{\partial F} - 12\mu J_2 R^2 r^{-6} K^{-2} \times \left\{\left(q\frac{\partial Y_1}{\partial F} - p\frac{\partial X_1}{\partial F}\right)(qX_1 + pY_1) + (qY_1 - pX_1)\left(q\frac{\partial X_1}{\partial F} + p\frac{\partial Y_1}{\partial F}\right)\right\} \tag{12.177}$$

For the f_h component, the partials are given by

$$\frac{\partial f_h}{\partial a} = 30\mu J_2 R^2 r^{-6} K^{-2}\left(1 - p^2 - q^2\right)(qY_1 - pX_1)\frac{\partial r}{\partial a} - 6\mu J_2 R^2 r^{-5} K^{-2}\left(1 - p^2 - q^2\right)\left(q\frac{\partial Y_1}{\partial a} - p\frac{\partial X_1}{\partial a}\right) \tag{12.178}$$

$$\frac{\partial f_h}{\partial h} = 30\mu J_2 R^2 r^{-6} K^{-2}\left(1 - p^2 - q^2\right)(qY_1 - pX_1)\frac{\partial r}{\partial h} - 6\mu J_2 R^2 r^{-5} K^{-2}\left(1 - p^2 - q^2\right)\left[q\left(\frac{\partial Y_1}{\partial h}\right)_F - p\left(\frac{\partial X_1}{\partial h}\right)_F\right] \tag{12.179}$$

$$\frac{\partial f_h}{\partial k} = 30\mu J_2 R^2 r^{-6} K^{-2}\left(1 - p^2 - q^2\right)(qY_1 - pX_1)\frac{\partial r}{\partial k} - 6\mu J_2 R^2 r^{-5} K^{-2}\left(1 - p^2 - q^2\right)\left[q\left(\frac{\partial Y_1}{\partial k}\right)_F - p\left(\frac{\partial X_1}{\partial k}\right)_F\right] \quad (12.180)$$

$$\frac{\partial f_h}{\partial p} = -6\mu J_2 R^2 r^{-5}\left[-2pK^{-3}(qY_1 - pX_1)\left(3 - p^2 - q^2\right) - X_1 K^{-2}\left(1 - p^2 - q^2\right)\right] \quad (12.181)$$

$$\frac{\partial f_h}{\partial q} = -6\mu J_2 R^2 r^{-5}\left[-2qK^{-3}(qY_1 - pX_1)\left(3 - p^2 - q^2\right) + Y_1 K^{-2}\left(1 - p^2 - q^2\right)\right] \quad (12.182)$$

$$\frac{\partial f_h}{\partial F} = 30\mu J_2 R^2 r^{-6} K^{-2}\left(1 - p^2 - q^2\right)(qY_1 - pX_1)\frac{\partial r}{\partial F} - 6\mu J_2 R^2 r^{-5} K^{-2}\left(1 - p^2 - q^2\right)\left(q\frac{\partial Y_1}{\partial F} - p\frac{\partial X_1}{\partial F}\right) \quad (12.183)$$

These partials of f_r, f_θ and f_h also involve the partials $(\partial X_1/\partial h)_F$, $(\partial X_1/\partial k)_F$, $(\partial Y_1/\partial h)_F$ and $(\partial Y_1/\partial k)_F$ instead of the partials $\partial X_1/\partial h$, $\partial X_1/\partial k$, $\partial Y_1/\partial h$ and $\partial Y_1/\partial k$ used in [11] and Chapter 10 of this book, as well as the $\partial r/\partial h$, $\partial r/\partial k$ partials used in this chapter, which have a different form from the $\partial r/\partial h$ and $\partial r/\partial k$ partials used in [11] and Chapter 10 of this book because once again F is the fast variable here as opposed to λ in [11] and Chapter 10 of this book. Otherwise the f_r, f_θ and f_h partials here have the same functional form as in [11] and Chapter 10 of this book. Going back to the example transfer shown in [12] and Chapter 11 of this book and in Table 12.1, J_2 is first set to zero, and the converged values of the various multipliers at time zero generated in [12] and Chapter 11 of this book are used to integrate the optimal trajectory forward in time until $t_f = 58{,}089.90058$ s, using the optimal firing orientation given by $\hat{\mathbf{u}} = \boldsymbol{\lambda}_z^T M(\mathbf{z}, F)/|\boldsymbol{\lambda}_z^T M(\mathbf{z}, F)|$ which effectively maximizes the Hamiltonian in Equation (12.25). The initial and final orbital positions are also optimized in this minimum-time transfer example, such that

$$(\lambda_a)_0 = 4.657973438\ \text{s/km} \qquad (\lambda_p)_0 = 1.771449217 \times 10^1\ \text{s}$$
$$(\lambda_h)_0 = 5.393432977 \times 10^2\ \text{s} \qquad (\lambda_q)_0 = -2.250119870 \times 10^4\ \text{s}$$
$$(\lambda_k)_0 = -9.168734810 \times 10^3\ \text{s} \qquad (\lambda)_0 = -2.274742851\ \text{rad}$$

with $(\lambda)_0$ above corresponding to $M_0 = -130.3331648$ deg. An acceleration $f_t = 9.8 \times 10^{-5}$ km/s^2 is used to generate the state vector at t_f shown in Table 12.1. The variable-step integrator relative and absolute error controls are set at the 10^{-9} level.

The "Achieved [10]" parameters are generated using the state and multipliers differential equations developed in [10] and Chapter 6 of this book, while the "Achieved [Present]" parameters use the more direct and streamlined corresponding differential equations developed in this chapter. The present derivations are thus validated due to the perfect matching of the final state obtained both ways. The validation of the state and multipliers differential equations used in this chapter and shown in Equations (12.132)–(12.138) with the J_2 terms accounted for is carried out next by running the trajectory

Table 12.1 Initial, target and achieved orbit parameters without J_2 perturbation

Orbit	a, km	e	i, deg	Ω, deg	ω, deg	M, F deg
Initial	7000	0	28.5	0	0	$M_0 = -130.333164$ (opt.)
Target	42000	10^{-3}	1	0	0	Free
Achieved [10]	41,999.99289	9.9834741×10^{-4}	9.9979725×10^{-1}	$3.32097378 \times 10^{-4}$	359.994535	$F = 46.2060306$
Achieved [Present]	41,999.99289	9.9834746×10^{-4}	9.9979725×10^{-1}	$3.32097784 \times 10^{-4}$	359.994534	$F = 46.2060306$

Table 12.2 Initial, target and achieved orbit parameters with J_2 perturbation

Orbit	a, km	e	i, deg	Ω, deg	ω, deg	M, F, L deg
Initial	7000	0	28.5	0	0	-131.7396776 (opt.)
Target	42000	10^{-3}	1	0	0	Free
Achieved [λ]	41,999.99992	1.000022×10^{-3}	1.0000013	359.999569	359.999137	$\lambda = 45.4102445$ (opt.)
Achieved [L]	41,999.99992	1.000022×10^{-3}	1.0000013	359.999569	359.999136	$L = 45.4919264$ (opt.)
Achieved [F]	41,999.99992	1.000022×10^{-3}	1.0000013	359.999569	359.999130	$F = 45.45107830$ (opt.)

forward in time until t_f = 58,104.83438 s, the minimum-time solution obtained in [12] and Chapter 11 of this book using a true longitude derived formulation. The true longitude is defined in terms of the true anomaly θ^* and $L = \theta^* + \Omega + \omega$. In [12], some minor errors that were present in an earlier derivation [11] that used the mean longitude formulation were corrected and both formulations agreed perfectly once these errors were corrected in this book. We now use the converged values of the various multipliers generated in [12] and Chapter 11 of this book:

$$(\lambda_a)_0 = 4.800100306\,\text{s/km} \qquad (\lambda_p)_0 = 3.281827358 \times 10^1\,\text{s}$$
$$(\lambda_h)_0 = 8.060772261 \times 10^2\,\text{s} \qquad (\lambda_q)_0 = -2.254928992 \times 10^4\,\text{s}$$
$$(\lambda_k)_0 = -9.150040837 \times 10^3\,\text{s} \qquad (\lambda)_0 = -2.299291130\,\text{rad}$$

where $(\lambda)_0$ corresponds to $M_0 = -131.7396776$ deg, and propagate forward the state and costate equations as stated above, using the optimal thrust orientation given by $\hat{\mathbf{u}} = \boldsymbol{\lambda}_z^T M(\mathbf{z}, F)/|\boldsymbol{\lambda}_z^T M(\mathbf{z}, F)|$. Table 12.2 shows the achieved state at t_f from all three formulations with once again a perfect match validating the analysis shown in the present chapter.

The value of $\lambda = 45.4102445$ deg corresponds to $L = 45.4919264$ deg, which is identical to the value of L obtained directly from the "Achieved [L]" formulation. The same λ also corresponds to $M = 45.4115380$ deg, which compares also very well with the value of $M = 45.411543$ deg obtained in [12] and Chapter 11 of this book after extraction from the integrated L value. The initial values of λ_λ, λ_F and λ_L of the three formulations are identically equal to zero because the initial location on the departing orbit is optimized, resulting in $M_0 = -131.7396776$ deg. The first five multipliers, namely λ_a, λ_h, λ_k, λ_p and λ_q have the same initial values at time zero regardless of the formulation in use because the initial orbit in our example is circular and from [16] and Chapter 5 of this book, $\lambda_\lambda = \frac{a^2(1-h^2-k^2)^{1/2}}{r^2}\lambda_L$, $\lambda_\lambda = \lambda_M$, $\lambda_F = \frac{r}{a}\lambda_M$, and because $r = a$ and $h = k = 0$ on the initial circular orbit, it follows that $(\lambda_\lambda)_0 = (\lambda_L)_0 = (\lambda_F)_0 = (\lambda_M)_0$. Figure 12.2 shows the evolution of λ, F and L during this short transfer to the GEO orbit with marked differences during the final ascent to the GEO altitude when the powered spacecraft's orbit becomes elliptical before the final circularization and entry into the GEO.

12.4 Conclusion

A previously derived system of dynamic and adjoint equations based on the adoption of the convenient eccentric longitude variable as the sixth element of a nonsingular equinoctial set of elements that led to singularity-free differential equations for the important zero eccentricity and zero inclination orbits, and valid for the thrust-perturbation-only case, has been further streamlined in this chapter in order to simplify both the state and costate equations. The analysis is further extended by including the secular J_2 effect within the framework of this streamlined theory, and the mathematics fully validated by direct comparison with a similar formulation based on the use of the

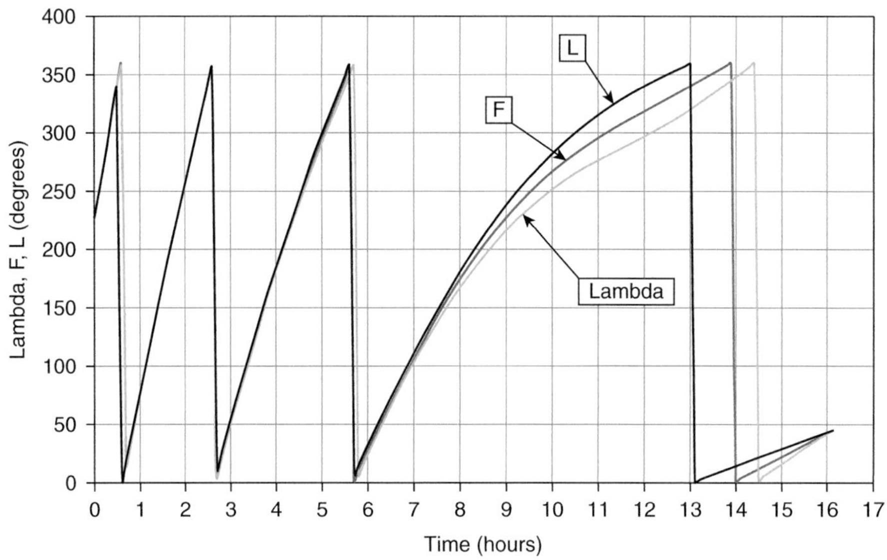

Figure 12.2 Evolution of λ, F and L during optimal transfer.

equally convenient true longitude as the sixth element instead. The present streamlined eccentric longitude theory, as well as the true longitude theory, are readily amenable to the inclusion of the higher-order harmonic terms providing a mutual validation as in the present chapter, which only considers the effect of the second zonal harmonic.

References

[1] Broucke, R. A. and Cefola, P. J. (1972) On the Equinoctial Orbit Elements, *Celestial Mechanics*, Vol. 5, 303–310.

[2] Cefola, P. J. (1972) *Equinoctial Orbit Elements: Application to Artificial Satellite Orbits*, AIAA Paper 72–937, AIAA/AAS Astrodynamics Conference, Palo Alto, CA.

[3] Edelbaum, T. N., Sackett, L. L. and Malchow, H. L. (1973) *Optimal Low Thrust Geocentric Transfer*, AIAA Paper 73–1074, AIAA 10th Electric Propulsion Conference, Lake Tahoe, NV.

[4] Enright, P. J. and Conway, B. A. (1991) Optimal Finite-Thrust Spacecraft Trajectories Using Collocation and Nonlinear Programming, *Journal of Guidance, Control, and Dynamics*, Vol. 14, No. 5, 981–985.

[5] Betts, J. T. (1994) Optimal Interplanetary Orbit Transfers by Direct Transcription, *Journal of the Astronautical Sciences*, Vol. 42, No. 3, 247–268.

[6] Kluever, C. A. (1997) Optimal Low-Thrust Interplanetary Trajectories by Direct Method Techniques, *The Journal of the Astronautical Sciences*, Vol. 45, No. 3, 247–262.

[7] Coverstone-Carroll, V. and Williams, S. N. (1994) Optimal Low-Thrust Trajectories Using Differential Inclusion Concepts, *Journal of the Astronautical Sciences*, Vol. 42, No. 4, 379–393.

[8] Bryson, A. E., Jr. and Ho, Y.-C. (1969) *Applied Optimal Control*, Ginn and Company, Waltham, MA.

[9] Marec, J.-P. (1979) *Optimal Space Trajectories*, Elsevier, Amsterdam.

[10] Kéchichian, J. A. (1998) Trajectory Optimization Using Eccentric Longitude Formulation, *Journal of Spacecraft and Rockets*, Vol. 35, No. 3, 317–326.

[11] Kéchichian, J. A. (1997) The Treatment of the Earth Oblateness Effect in Trajectory Optimization in Equinoctial Coordinates, *Acta Astronautica*, Vol. 40, No. 1, 69–82.

[12] Kéchichian, J. A. (2000) Minimum-Time Constant Acceleration Orbit Transfer with First-Order Oblateness Effect, *Journal of Guidance, Control, and Dynamics*, Vol. 23, No. 4, 595–603.

[13] Kéchichian, J. A. (1990) *Equinoctial Orbit Elements: Application to Optimal Transfer Problems*, AIAA Paper 90–2976, AIAA/AAS Astrodynamics Conference, Portland, OR.

[14] Kéchichian, J. A. (1991) *Trajectory Optimization with a Modified Set of Equinoctial Orbit Elements*, AAS Paper 91–524, AIAA/AAS Astrodynamics Specialist Conference, Durango, CO.

[15] Kéchichian, J. A. (1992) *Optimal Low-Thrust Rendezvous Using Equinoctial Orbit Elements*, IAF Paper 92–0014, 43rd Congress of the International Astronautical Federation, Washington, DC.

[16] Kéchichian, J. A. (1999) Minimum-Time Low-Thrust Rendezvous and Transfer Using Epoch Mean Longitude Formulation, *Journal of Guidance, Control, and Dynamics*, Vol. 22, No. 3, 421–432.

13 The Inclusion of the Higher-Order Harmonics in the Modeling of Optimal Low-Thrust Orbit Transfer

13.1 Introduction

The higher-fidelity modeling of minimum-time transfers using continuous constant acceleration low-thrust is depicted by including the higher zonal harmonics J_3 and J_4 for the Earth gravity model. The inclusion of these higher-order harmonics is of great benefit in carrying out accurate transfer simulations, especially for long-duration flights dwelling in low altitudes, where the effects of these zonals are greatest. The analysis presented here can also be coded in the flight guidance computer of spacecraft for autonomous operations and on ground computers for solution uploads and resetting during low-thrust transfers. Equinoctial elements are used to avoid singularities when orbits are circular or equational and the applicability of the theory is of a general nature regardless of the size, shape and spatial orientation of the orbits, provided they are not of the parabolic or hyperbolic types. To this end, two sets of dynamic and adjoint differential equations in terms of nonsingular orbit elements are derived by further considering a more accurate perturbation model in the form of the higher-order Earth zonal harmonics J_3 and J_4. Previous analyses involved only the first-order J_2 term in order to model optimal low-thrust transfers between any two given circular or elliptic orbits. The first formulation uses the eccentric longitude as the sixth element of an equinoctial set of elements while describing the thrust as well as the zonal accelerations in the so-called direct equinoctial frame. The second formulation makes use of the true longitude as the sixth element instead, while resolving the thrust and the zonal accelerations in the rotating Euler–Hill frame, simplifying considerably the algebraic derivations leading to the generation of the nonsingular differential equations that are also free of any singularity for the important zero-eccentricity and zero-inclination cases often encountered in Earth orbit transfer problems. The derivations of both nonsingular formulations are mutually validated by generating an optimal transfer example that achieves the same target conditions regardless of which formulation is used.

This chapter is a direct extension of several previous contributions made by this author in the field of optimal orbit transfer using unaveraged precision-integrated dynamics. The theory of orbital dynamics and optimal transfer has greatly benefited from the pioneering contributions of Broucke, Cefola and Edelbaum [1–3], who introduced the use of nonsingular equinoctial orbital elements that are immune to the nasty singularities inherent in circular and/or equational orbits. These elements were also adopted by Betts [4] and Walker [5] with a slight modification validating their applicability for parabolic and hyperbolic orbits as well. Battin [6], in his classical masterpiece, discusses and defines equinoctial-based differential equations to integrate spacecraft trajectories. This author provided a series of contributions [7–13], also shown in the preceding chapters, directly applicable for optimal low-thrust orbit transfer work using a variety of equinoctial elements sets while also considering the perturbative effects of the oblateness of the Earth, such as J_2. More recently, Feistel [14] made use of a form for the various zonal accelerations resolved in the inertial system and given in terms of the radius vector components along the inertial directions [15] and derived the partial derivatives of these acceleration vectors with respect to the radius vector itself, which in turn allowed him to generate the partial derivatives of these accelerations with respect to the equinoctial elements in order to finally produce the adjoint differential equations needed to solve the two-point boundary value problem orbit transfer. Besides J_2, Feistel also considered the J_3, J_4, J_5 and J_6 terms while obtaining a quasi-perfect numerical agreement for J_2 with this author's results, the latter based on the Gaussian formulation.

In this chapter, two fundamental sets of equinoctial elements have been used to validate the mathematical derivations that lead to the generation of the corresponding nonsingular state and adjoint differential equations for direct use in operational guidance applications. The elements used are a, $h = e\sin(\Omega + \omega)$, $k = e\cos(\Omega + \omega)$, $p = \tan(i/2)\sin\Omega$, $q = \tan(i/2)\cos\Omega$ and either $F = E + \Omega + \omega$, the eccentric longitude, or $L = \theta^* + \Omega + \omega$, the true longitude, as the sixth element of the six-element sets. Furthermore, the direct equinoctial frame $(\hat{\mathbf{f}}, \hat{\mathbf{g}}, \hat{\mathbf{w}})$, which has been historically used first such as in [1–3] is adopted for the a, h, k, p, q, F formulation to resolve the various accelerations. Because the right-hand sides of the differential equations can also be cast in terms of F, there is no need to solve Kepler's transcendental equation at each integration step to extract F itself because it is being directly integrated. The $\hat{\mathbf{f}}$, $\hat{\mathbf{g}}$, $\hat{\mathbf{w}}$ unit vectors are such that $\hat{\mathbf{f}}$ and $\hat{\mathbf{g}}$ are in the current orbit plane with $\hat{\mathbf{f}}$ rotated clockwise through the angle Ω from the direction of the ascending node, $\hat{\mathbf{g}}$ is 90 deg ahead of $\hat{\mathbf{f}}$ in the direction of motion, while $\hat{\mathbf{w}}$ is along the instantaneous normal to the orbit plane.

The second, more compact formulation adopts the set a, h, k, p, q, L and resolves the various accelerations along the rotating $\mathbf{r}$, $\hat{\boldsymbol{\theta}}$, $\hat{\mathbf{h}}$ Euler–Hill orbital frame, with $\hat{\mathbf{r}}$ along the radius vector $\mathbf{r}$, $\hat{\boldsymbol{\theta}}$ in the instantaneous orbit plane and 90 deg ahead of $\hat{\mathbf{r}}$ along the direction of motion, and $\hat{\mathbf{h}}$ along the out-of-plane direction. Contrary to the F formulation, there is no need to rotate the accelerations due to J_2, J_3 and J_4 from the $\mathbf{r}$, $\hat{\boldsymbol{\theta}}$, $\hat{\mathbf{h}}$ frame to the $\hat{\mathbf{f}}$, $\hat{\mathbf{g}}$, $\hat{\mathbf{w}}$ frame, thus simplifying to a large extent the algebraic derivations required to produce the adjoint differential equations used in the steering of the thrust vector, when the L formulation is used instead. The subsequent sections deal with the development of the acceleration components due to the zonals both in terms of F and L, before deriving the Euler–Lagrange equations for the various multipliers. A numerical comparison is shown at the end to show that both formulations lead to the exact same transfers.

A short-duration transfer example that was used before [7–13, and the preceding chapters] is adopted here for illustration purposes and in order to avoid very long integrations. The results presented at the end of this chapter show only small differences between the solutions using J_2 only, and those more elaborate ones using additionally the higher harmonics. However, these differences become considerable for long-duration transfers that spend a long time at lower altitudes before spiraling out to the higher orbits where the perturbations' effects decrease very quickly. Additional discussions are provided in Section 13.5 to make the case for the adoption of the present higher-fidelity modeling.

13.2 Zonal Harmonics Perturbation Acceleration Components in the Euler–Hill Frame

Assuming that Earth is symmetrical about its polar axis, its potential U can then be written in terms of the zonal harmonics J_n as

$$U = \frac{\mu}{r}\left[1 - \sum_{n=2}^{\infty} J_n \left(\frac{R}{r}\right)^n P_n(s_\delta)\right] \tag{13.1}$$

where r is the magnitude of the spacecraft position vector with respect to the center of the Earth, R the radius of the Earth at the equator, δ the spacecraft declination with respect to the equator, and $P_n(s_\delta)$ the Legendre polynomials of order n in $\sin\delta$, and μ is the Earth gravity constant. The disturbing potential is then given by

$$F' = U - U_0 = U - \frac{\mu}{r} = -\frac{\mu}{r}\sum_{n=2}^{\infty} J_n \left(\frac{R}{r}\right)^n P_n\left(s_\delta\right) \tag{13.2}$$

Neglecting the J_5 and higher zonals, F' takes the form

$$F' = -\frac{\mu}{r}\left[J_2\left(\frac{R}{r}\right)^2\left(\frac{3}{2}s_\delta^2 - \frac{1}{2}\right) + J_3\left(\frac{R}{r}\right)^3\left(\frac{5}{2}s_\delta^3 - \frac{3}{2}s_\delta\right)\right.$$
$$\left. + J_4\left(\frac{R}{r}\right)^4\left(\frac{35}{8}s_\delta^4 - \frac{30}{8}s_\delta^2 + \frac{3}{8}\right)\right] \tag{13.3}$$

Replacing s_δ by $\frac{z}{r}$, the perturbing acceleration $\mathbf{f}_{J_i}$ is then given by the gradient of the disturbing potential

$$\mathbf{f}_{J_i} = \nabla F' = -\mu J_2 R^2 \nabla\left(\frac{3}{2}\frac{z^2}{r^5} - \frac{1}{2r^3}\right) - \mu J_3 R^3 \nabla\left(\frac{5}{2}\frac{z^3}{r^7} - \frac{3}{2}\frac{z}{r^5}\right)$$
$$- \mu J_4 R^4 \nabla\left(\frac{35}{8}\frac{z^4}{r^9} - \frac{30}{8}\frac{z^2}{r^7} + \frac{3}{8r^5}\right) \tag{13.4}$$

Because $\nabla r = \hat{\mathbf{r}}$ and $\nabla z = \hat{\mathbf{z}}$, where $\hat{\mathbf{r}}$ and $\hat{\mathbf{z}}$ are unit vectors along $\mathbf{r}$ and $\mathbf{z}$, $\mathbf{f}_{J_i}$ can then be expressed in the Euler–Hill frame $(\hat{\mathbf{r}}, \hat{\boldsymbol{\theta}}, \hat{\mathbf{h}})$ as

$$\mathbf{f}_{J_i} = -\mu J_2 R^2\left[\frac{3z}{r^5}\hat{\mathbf{z}} + \left(\frac{3}{2r^4} - \frac{15}{2}\frac{z^2}{r^6}\right)\hat{\mathbf{r}}\right]$$
$$- \mu J_3 R^3\left[\left(\frac{15}{2}\frac{z^2}{r^7} - \frac{3}{2r^5}\right)\hat{\mathbf{z}} + \left(\frac{15}{2}\frac{z}{r^6} - \frac{35}{2}\frac{z^3}{r^8}\right)\hat{\mathbf{r}}\right]$$
$$- \mu J_4 R^4\left[\left(\frac{35}{2}\frac{z^3}{r^9} - \frac{15}{2}\frac{z}{r^7}\right)\hat{\mathbf{z}} + \left(\frac{105}{4}\frac{z^2}{r^8} - \frac{315}{8}\frac{z^4}{r^{10}} - \frac{15}{8r^6}\right)\hat{\mathbf{r}}\right] \tag{13.5}$$

The transformation from the Euler–Hill frame to the inertial equational frame $(\hat{\mathbf{x}}, \hat{\mathbf{y}}, \hat{\mathbf{z}})$ is obtained through the three rotations involving the Eulerian angles Ω, i and $\theta = \omega + \theta^*$, where θ^* stands for the true anomaly

$$\begin{pmatrix} \hat{\mathbf{x}} \\ \hat{\mathbf{y}} \\ \hat{\mathbf{z}} \end{pmatrix} = \begin{pmatrix} c_\Omega c_\theta - s_\Omega c_i s_\theta & -c_\Omega s_\theta - s_\Omega c_i c_\theta & s_\Omega s_i \\ s_\Omega c_\theta + c_\Omega c_i s_\theta & -s_\Omega s_\theta + c_\Omega c_i c_\theta & -c_\Omega s_i \\ s_i s_\theta & s_i c_\theta & c_i \end{pmatrix} \begin{pmatrix} \hat{\mathbf{r}} \\ \hat{\boldsymbol{\theta}} \\ \hat{\mathbf{h}} \end{pmatrix} \tag{13.6}$$

Because $\hat{\mathbf{z}} = s_i s_\theta \hat{\mathbf{r}} + s_i c_\theta \hat{\boldsymbol{\theta}} + c_i \hat{\mathbf{h}}$ and $z = \hat{\mathbf{z}} \cdot \mathbf{r} = r s_i s_\theta$ where $\mathbf{r} = r\hat{\mathbf{r}}$, the acceleration $\mathbf{f}_{J_i}$ can then be cast into the following final form:

$$\begin{aligned} \mathbf{f}_{J_i} = &-\frac{3\mu J_2 R^2}{r^4}\left[\left(\frac{1}{2} - \frac{3}{2}s_i^2 s_\theta^2\right)\hat{\mathbf{r}} + s_i^2 s_\theta c_\theta \hat{\boldsymbol{\theta}} + s_i c_i s_\theta \hat{\mathbf{h}}\right] \\ &- \frac{\mu J_3 R^3}{r^5}\left[\left(6 s_i s_\theta - 10 s_i^3 s_\theta^3\right)\hat{\mathbf{r}} + \frac{1}{2}\left(15 s_i^3 s_\theta^2 c_\theta - 3 s_i c_\theta\right)\hat{\boldsymbol{\theta}}\right. \\ &\left.+ \frac{1}{2}\left(15 s_i^2 c_i s_\theta^2 - 3c_i\right)\hat{\mathbf{h}}\right] - \frac{5\mu J_4 R^4}{2r^6}\left[\left(-\frac{35}{4}s_i^4 s_\theta^4 + \frac{15}{2}s_i^2 s_\theta^2 - \frac{3}{4}\right)\hat{\mathbf{r}}\right. \\ &\left.+ \left(7 s_i^4 s_\theta^3 c_\theta - 3 s_i^2 s_\theta c_\theta\right)\hat{\boldsymbol{\theta}} + \left(7 s_i^3 c_i s_\theta^3 - 3 s_i c_i s_\theta\right)\hat{\mathbf{h}}\right] \end{aligned} \tag{13.7}$$

Using the trigonometric power relations to express s_θ^2, s_θ^3, s_θ^4 in terms of the multiple-angle expressions involving only first powers, and rearranging terms, the r, θ, h components of $\mathbf{f}_{J_i}$ due to J_2 can be written as

$$(f_r)_{J_2} = -\frac{3\mu J_2 R^2}{2r^4}\left(1 - 3 s_i^2 s_\theta^2\right) \tag{13.8}$$

$$(f_\theta)_{J_2} = -\frac{3\mu J_2 R^2}{r^4} s_i^2 s_\theta c_\theta \tag{13.9}$$

$$(f_h)_{J_2} = -\frac{3\mu J_2 R^2}{r^4} s_i c_i s_\theta \tag{13.10}$$

For the J_3 zonal,

$$(f_r)_{J_3} = -\frac{\mu J_3 R^3}{r^5}\left[6 s_i s_\theta - \frac{5}{2}s_i^3\left(3 s_\theta - s_{3\theta}\right)\right] \tag{13.11}$$

$$(f_\theta)_{J_3} = -\frac{\mu J_3 R^3}{2r^5}\left[\frac{15}{4}s_i^3\left(c_\theta - c_{3\theta}\right) - 3 s_i c_\theta\right] \tag{13.12}$$

$$(f_h)_{J_3} = -\frac{\mu J_3 R^3}{2r^5}\left[\frac{15}{2}s_i^2 c_i\left(1 - c_{2\theta}\right) - 3c_i\right] \tag{13.13}$$

And finally for the J_4 zonal,

$$(f_r)_{J_4} = -\frac{5\mu J_4 R^4}{8r^6}\left[15 s_i^2\left(1 - c_{2\theta}\right) - \frac{35}{8}s_i^4\left(3 - 4c_{2\theta} + c_{4\theta}\right) - 3\right] \tag{13.14}$$

$$(f_\theta)_{J_4} = -\frac{5\mu J_4 R^4}{2r^6}\left[\frac{7}{4}s_i^4 c_\theta\left(3 s_\theta - s_{3\theta}\right) - 3 s_i^2 s_\theta c_\theta\right] \tag{13.15}$$

$$(f_h)_{J_4} = -\frac{5\mu J_4 R^4}{2r^6}\left[\frac{7}{4}s_i^3 c_i\left(3 s_\theta - s_{3\theta}\right) - 3 s_i c_i s_\theta\right] \tag{13.16}$$

These acceleration components can also be obtained directly from their inertial representations as shown in Appendix 13.1. The above derivations are thus validated.

13.3 The Treatment of the J_3, J_4 Perturbations within the Eccentric Longitude Formulation

Following the analysis presented in [12] and [13] and Chapter 12 of this book, the dynamic equations for the set a, h, k, p, q, F are readily augmented to include the higher zonals' contributions as

$$\dot{a} = (\partial a/\partial \dot{\mathbf{r}})^T \left(\hat{\mathbf{u}} f_t + \mathbf{f}_{J_2} + \mathbf{f}_{J_3} + \mathbf{f}_{J_4}\right) \tag{13.17}$$

$$\dot{h} = (\partial h/\partial \dot{\mathbf{r}})^T \left(\hat{\mathbf{u}} f_t + \mathbf{f}_{J_2} + \mathbf{f}_{J_3} + \mathbf{f}_{J_4}\right) \tag{13.18}$$

$$\dot{k} = (\partial k/\partial \dot{\mathbf{r}})^T \left(\hat{\mathbf{u}} f_t + \mathbf{f}_{J_2} + \mathbf{f}_{J_3} + \mathbf{f}_{J_4}\right) \tag{13.19}$$

$$\dot{p} = (\partial p/\partial \dot{\mathbf{r}})^T \left(\hat{\mathbf{u}} f_t + \mathbf{f}_{J_2} + \mathbf{f}_{J_3} + \mathbf{f}_{J_4}\right) \tag{13.20}$$

$$\dot{q} = (\partial q/\partial \dot{\mathbf{r}})^T \left(\hat{\mathbf{u}} f_t + \mathbf{f}_{J_2} + \mathbf{f}_{J_3} + \mathbf{f}_{J_4}\right) \tag{13.21}$$

$$\dot{F} = na/r + (\partial F/\partial \dot{\mathbf{r}})^T \left(\hat{\mathbf{u}} f_t + \mathbf{f}_{J_2} + \mathbf{f}_{J_3} + \mathbf{f}_{J_4}\right) \tag{13.22}$$

The Hamiltonian of this system of equations can be written as

$$\begin{aligned} H = {} & \boldsymbol{\lambda}_z^T M(\mathbf{z}, F) f_t \hat{\mathbf{u}} + \lambda_F \frac{na}{r} + \boldsymbol{\lambda}_z^T M(\mathbf{z}, F)\, \mathbf{f}_{J_2} \\ & + \boldsymbol{\lambda}_z^T M(\mathbf{z}, F)\, \mathbf{f}_{J_3} + \boldsymbol{\lambda}_z^T M(\mathbf{z}, F)\, \mathbf{f}_{J_4} \end{aligned} \tag{13.23}$$

and the Euler–Lagrange or adjoint equations are thus given by

$$\begin{aligned} \dot{\boldsymbol{\lambda}}_z = -\partial H/\partial \mathbf{z} = {} & -\boldsymbol{\lambda}_z^T \frac{\partial M}{\partial \mathbf{z}} f_t \hat{\mathbf{u}} - \lambda_F \frac{\partial}{\partial \mathbf{z}}\left(\frac{na}{r}\right) - \boldsymbol{\lambda}_z^T \frac{\partial M(\mathbf{z}, F)}{\partial \mathbf{z}} \mathbf{f}_{J_2} \\ & - \boldsymbol{\lambda}_z^T M(\mathbf{z}, F) \frac{\partial \mathbf{f}_{J_2}}{\partial \mathbf{z}} - \boldsymbol{\lambda}_z^T \frac{\partial M(\mathbf{z}, F)}{\partial \mathbf{z}} \mathbf{f}_{J_3} - \boldsymbol{\lambda}_z^T M(\mathbf{z}, F) \frac{\partial \mathbf{f}_{J_3}}{\partial \mathbf{z}} \\ & - \boldsymbol{\lambda}_z^T \frac{\partial M(\mathbf{z}, F)}{\partial \mathbf{z}} \mathbf{f}_{J_4} - \boldsymbol{\lambda}_z^T M(\mathbf{z}, F) \frac{\partial \mathbf{f}_{J_4}}{\partial \mathbf{z}} \end{aligned} \tag{13.24}$$

The treatment of the J_2 perturbation is given in full detail for the particular set $\mathbf{z} = (a, h, k, p, q, F)$ in [12] and [13], and in the preceding chapter, with all the necessary details of the derivations leading to the complete description of the dynamic and adjoint equations including the elements of the M matrix and its various partials with respect to the equinoctial elements of interest. In [12] there are some typographical errors concerning the following equations, which should read correctly as

$$\frac{\partial \dot{Y}_1}{\partial a} = \frac{na}{2r}\left[hk\beta s_F - \left(1 - k^2\beta\right) c_F\right] = -\frac{\dot{Y}_1}{2a} \qquad \text{(80) of [12]}$$

$$\begin{aligned} \frac{\partial M_{62}^F}{\partial k} = {} & \frac{c_F}{nr^2}\left[-2Y_1 + G(h\beta - s_F)\frac{\partial Y_1}{\partial h} + G(k\beta - c_F)\frac{\partial Y_1}{\partial k} \right. \\ & \left. -\beta G(k s_F - h c_F)\frac{\dot{Y}_1}{n}\right] + \frac{1}{nar}\left[-2\left(\frac{\partial Y_1}{\partial k}\right)_F + \frac{k\beta}{1-\beta}\left(s_F - h\beta^2\right)\frac{\partial Y_1}{\partial h}\right. \end{aligned}$$

$$+ G\left(h\beta - s_F\right)\frac{\partial^2 Y_1}{\partial k \partial h} + \left\{\frac{k\beta}{1-\beta}\left(c_F - k\beta^2\right) + (1-\beta)\right\}\frac{\partial Y_1}{\partial k}$$
$$+ G\left(k\beta - c_F\right)\frac{\partial^2 Y_1}{\partial k^2} + \left\{\left(ks_F - hc_F\right)\frac{k\beta^3}{1-\beta} - (1-\beta)\, s_F\right\}\frac{\dot{Y}_1}{n}$$
$$- (1-\beta)\left(ks_F - hc_F\right)\frac{1}{n}\frac{\partial \dot{Y}_1}{\partial k}\Bigg] \qquad \text{(96) of [12]}$$

$$f_\theta = -12\mu J_2 R^2 r^{-6}\left(1 + p^2 + q^2\right)^{-2}\left(qY_1 - pX_1\right)\left(qX_1 + pY_1\right) \qquad \text{(127) of [12]}$$

and finally

$$\frac{\partial f_\theta}{\partial q} = -12\mu J_2 R^2 r^{-6}\Big[-4qK^{-3}\left(qY_1 - pX_1\right)\left(qX_1 + pY_1\right)$$
$$+ K^{-2}\left\{Y_1\left(qX_1 + pY_1\right) + X_1\left(qY_1 - pX_1\right)\right\}\Big] \qquad \text{(176) of [12]}$$

The journal version [13] of the original work [12] is, however, free of these typographic errors. This is also true for the preceding chapter. These preceding four equations are also free of typographical errors in Chapter 12 and preceding chapters.

It has been shown in [8] and Chapter 10 of this book that the classical elements i and θ are related to p, q, L by the expressions

$$s_i = \frac{2\left(p^2 + q^2\right)^{1/2}}{1 + p^2 + q^2} \tag{13.25}$$

$$c_i = \frac{1 - p^2 - q^2}{1 + p^2 + q^2} \tag{13.26}$$

$$s_\theta = \frac{qs_L - pc_L}{\left(p^2 + q^2\right)^{1/2}} \tag{13.27}$$

$$c_\theta = \frac{qc_L + ps_L}{\left(p^2 + q^2\right)^{1/2}} \tag{13.28}$$

We also have $3s_\theta - s_{3\theta} = 4s_\theta^3$, $c_\theta - c_{3\theta} = 4c_\theta s_\theta^2$ such that

$$s_i s_\theta = 2\left(qs_L - pc_L\right)/\left(1 + p^2 + q^2\right) \tag{13.29}$$

$$s_i c_\theta = 2\left(qc_L + ps_L\right)/\left(1 + p^2 + q^2\right) \tag{13.30}$$

$$s_i^3\left(3s_\theta - s_{3\theta}\right) = \frac{32\left(qs_L - pc_L\right)^3}{\left(1 + p^2 + q^2\right)^3} \tag{13.31}$$

$$s_i^3\left(c_\theta - c_{3\theta}\right) = \frac{32\left(qc_L + ps_L\right)\left(qs_L - pc_L\right)^2}{\left(1 + p^2 + q^2\right)^3} \tag{13.32}$$

$$s_i^2 c_i\left(1 - c_{2\theta}\right) = \frac{8\left(qs_L - pc_L\right)^2\left(1 - p^2 - q^2\right)}{\left(1 + p^2 + q^2\right)^3} \tag{13.33}$$

Expressions (13.11), (13.12) and (13.13) can now be written directly in terms of the nonsingular elements as

$$(f_r)_{J_3} = -\frac{\mu J_3 R^3}{r^5}\frac{(qs_L - pc_L)}{(1+p^2+q^2)}\left[12 - \frac{80\,(qs_L - pc_L)^2}{\left(1+p^2+q^2\right)^2}\right] \tag{13.34}$$

$$(f_\theta)_{J_3} = -\frac{\mu J_3 R^3}{2r^5}\frac{(qc_L + ps_L)}{(1+p^2+q^2)}\left[120\frac{(qs_L - pc_L)^2}{\left(1+p^2+q^2\right)^2} - 6\right] \tag{13.35}$$

$$(f_h)_{J_3} = -\frac{\mu J_3 R^3}{2r^5}\frac{(1-p^2-q^2)}{(1+p^2+q^2)}\left[60\frac{(qs_L - pc_L)^2}{\left(1+p^2+q^2\right)^2} - 3\right] \tag{13.36}$$

The above three acceleration components are readily converted to a form involving F instead of L through the use of the relations [8], also shown in Chapter 10 as

$$qs_L - pc_L = \frac{1}{r}(qY_1 - pX_1) \tag{13.37}$$

$$qc_L + ps_L = \frac{1}{r}(qX_1 + pY_1) \tag{13.38}$$

These read as

$$\begin{aligned}(f_r)_{J_3} &= -12\mu J_3 R^3 r^{-6}(qY_1 - pX_1)\left(1+p^2+q^2\right)^{-1}\\ &\quad + 80\mu J_3 R^3 r^{-8}(qY_1 - pX_1)^3\left(1+p^2+q^2\right)^{-3}\end{aligned} \tag{13.39}$$

$$\begin{aligned}(f_\theta)_{J_3} &= -60\mu J_3 R^3 r^{-8}(qX_1 + pY_1)(qY_1 - pX_1)^2\left(1+p^2+q^2\right)^{-3}\\ &\quad + 3\mu J_3 R^3 r^{-6}(qX_1 + pY_1)\left(1+p^2+q^2\right)^{-1}\end{aligned} \tag{13.40}$$

$$\begin{aligned}(f_h)_{J_3} &= -30\mu J_3 R^3 r^{-7}\left(1-p^2-q^2\right)\left(1+p^2+q^2\right)^{-3}(qY_1 - pX_1)^2\\ &\quad + \frac{3}{2}\mu J_3 R^3 r^{-5}\left(1-p^2-q^2\right)\left(1+p^2+q^2\right)^{-1}\end{aligned} \tag{13.41}$$

We have $f_f = c_L f_r - s_L f_\theta, f_g = s_L f_r + c_L f_\theta, f_w = f_h$ or in terms of F, $f_f = (X_1/r)f_r - (Y_1/r)f_\theta, f_g = (Y_1/r)f_r + (X_1/r)f_\theta, f_w = f_h$.

The partial derivatives $\partial \mathbf{f}_{J_3}/\partial \mathbf{z}$ have the same structure as in [8], also shown in Chapter 10, namely with $\mathbf{f}_{J_3} = (f_f)_{J_3}\hat{\mathbf{f}} + (f_g)_{J_3}\hat{\mathbf{g}} + (f_w)_{J_3}\hat{\mathbf{w}}$. The partials of f_f are given by

$$\frac{\partial (f_f)_{J_3}}{\partial a} = \frac{\partial (X_1/r)}{\partial a}(f_r)_{J_3} + \frac{X_1}{r}\frac{\partial (f_r)_{J_3}}{\partial a} - \frac{\partial (Y_1/r)}{\partial a}(f_\theta)_{J_3} - \frac{Y_1}{r}\frac{\partial (f_\theta)_{J_3}}{\partial a} \tag{13.42}$$

$$\frac{\partial (f_f)_{J_3}}{\partial h} = \frac{\partial (X_1/r)}{\partial h}(f_r)_{J_3} + \frac{X_1}{r}\frac{\partial (f_r)_{J_3}}{\partial h} - \frac{\partial (Y_1/r)}{\partial h}(f_\theta)_{J_3} - \frac{Y_1}{r}\frac{\partial (f_\theta)_{J_3}}{\partial h} \tag{13.43}$$

$$\frac{\partial (f_f)_{J_3}}{\partial k} = \frac{\partial (X_1/r)}{\partial k}(f_r)_{J_3} + \frac{X_1}{r}\frac{\partial (f_r)_{J_3}}{\partial k} - \frac{\partial (Y_1/r)}{\partial k}(f_\theta)_{J_3} - \frac{Y_1}{r}\frac{\partial (f_\theta)_{J_3}}{\partial k} \tag{13.44}$$

$$\frac{\partial (f_f)_{J_3}}{\partial p} = \frac{X_1}{r}\frac{\partial (f_r)_{J_3}}{\partial p} - \frac{Y_1}{r}\frac{\partial (f_\theta)_{J_3}}{\partial p} \tag{13.45}$$

$$\frac{\partial \left(f_f\right)_{J_3}}{\partial q} = \frac{X_1}{r}\frac{\partial \left(f_r\right)_{J_3}}{\partial q} - \frac{Y_1}{r}\frac{\partial \left(f_\theta\right)_{J_3}}{\partial q} \tag{13.46}$$

$$\frac{\partial \left(f_f\right)_{J_3}}{\partial F} = \frac{\partial \left(X_1/r\right)}{\partial F}\left(f_r\right)_{J_3} + \frac{X_1}{r}\frac{\partial \left(f_r\right)_{J_3}}{\partial F} - \frac{\partial \left(Y_1/r\right)}{\partial F}\left(f_\theta\right)_{J_3} - \frac{Y_1}{r}\frac{\partial \left(f_\theta\right)_{J_3}}{\partial F} \tag{13.47}$$

The partial derivatives of X_1/r and Y_1/r with respect to a, h, k and F appearing above are given in Equations (146)–(153) of [12] and [13] and also in Chapter 12. They involve the partials $\partial X_1/\partial a = X_1/a$ and $\partial Y_1/\partial a = Y_1/a$ given also in [13] as Equations (77) and (78), and the partials $\partial X_1/\partial F$ and $\partial Y_1/\partial F$ given in Equations (117) and (118) in [13]. These partials also appear in Chapter 12. As in [13] and Chapter 12 of this book, $\partial r/\partial a = r/a$, $\partial r/\partial h = -as_F$, $\partial r/\partial k = -ac_F$ and $\partial r/\partial F = a(ks_F - hc_F)$ directly derived from the orbit equation $r = a(1 - kc_F - hs_F)$. $\partial r/\partial h$, $\partial r/\partial k$ appear as $(\partial r/\partial h)_F$ and $(\partial r/\partial k)_F$ in the following.

In a similar way, the partials of f_g are given by

$$\frac{\partial \left(f_g\right)_{J_3}}{\partial a} = \frac{\partial \left(Y_1/r\right)}{\partial a}\left(f_r\right)_{J_3} + \frac{Y_1}{r}\frac{\partial \left(f_r\right)_{J_3}}{\partial a} + \frac{\partial \left(X_1/r\right)}{\partial a}\left(f_\theta\right)_{J_3} + \frac{X_1}{r}\frac{\partial \left(f_\theta\right)_{J_3}}{\partial a} \tag{13.48}$$

$$\frac{\partial \left(f_g\right)_{J_3}}{\partial h} = \frac{\partial \left(Y_1/r\right)}{\partial h}\left(f_r\right)_{J_3} + \frac{Y_1}{r}\frac{\partial \left(f_r\right)_{J_3}}{\partial h} + \frac{\partial \left(X_1/r\right)}{\partial h}\left(f_\theta\right)_{J_3} + \frac{X_1}{r}\frac{\partial \left(f_\theta\right)_{J_3}}{\partial h} \tag{13.49}$$

$$\frac{\partial \left(f_g\right)_{J_3}}{\partial k} = \frac{\partial \left(Y_1/r\right)}{\partial k}\left(f_r\right)_{J_3} + \frac{Y_1}{r}\frac{\partial \left(f_r\right)_{J_3}}{\partial k} + \frac{\partial \left(X_1/r\right)}{\partial k}\left(f_\theta\right)_{J_3} + \frac{X_1}{r}\frac{\partial \left(f_\theta\right)_{J_3}}{\partial k} \tag{13.50}$$

$$\frac{\partial \left(f_g\right)_{J_3}}{\partial p} = \frac{Y_1}{r}\frac{\partial \left(f_r\right)_{J_3}}{\partial p} + \frac{X_1}{r}\frac{\partial \left(f_\theta\right)_{J_3}}{\partial p} \tag{13.51}$$

$$\frac{\partial \left(f_g\right)_{J_3}}{\partial q} = \frac{Y_1}{r}\frac{\partial \left(f_r\right)_{J_3}}{\partial q} + \frac{X_1}{r}\frac{\partial \left(f_\theta\right)_{J_3}}{\partial q} \tag{13.52}$$

$$\frac{\partial \left(f_g\right)_{J_3}}{\partial F} = \frac{\partial \left(Y_1/r\right)}{\partial F}\left(f_r\right)_{J_3} + \frac{Y_1}{r}\frac{\partial \left(f_r\right)_{J_3}}{\partial F} + \frac{\partial \left(X_1/r\right)}{\partial F}\left(f_\theta\right)_{J_3} + \frac{X_1}{r}\frac{\partial \left(f_\theta\right)_{J_3}}{\partial F} \tag{13.53}$$

And finally for f_w,

$$\partial \left(f_w\right)_{J_3}/\partial a = \partial \left(f_h\right)_{J_3}/\partial a \tag{13.54}$$

$$\partial \left(f_w\right)_{J_3}/\partial h = \partial \left(f_h\right)_{J_3}/\partial h \tag{13.55}$$

$$\partial \left(f_w\right)_{J_3}/\partial k = \partial \left(f_h\right)_{J_3}/\partial k \tag{13.56}$$

$$\partial \left(f_w\right)_{J_3}/\partial p = \partial \left(f_h\right)_{J_3}/\partial p \tag{13.57}$$

$$\partial \left(f_w\right)_{J_3}/\partial q = \partial \left(f_h\right)_{J_3}/\partial q \tag{13.58}$$

$$\partial \left(f_w\right)_{J_3}/\partial F = \partial \left(f_h\right)_{J_3}/\partial F \tag{13.59}$$

Note that the partials of X_1/r and Y_1/r with respect to h and k involve the following partials $(\partial X_1/\partial h)_F$, $(\partial X_1/\partial k)_F$, $(\partial Y_1/\partial h)_F$, $(\partial Y_1/\partial k)_F$, given in [13] and also in Chapter 12. We can now generate the partials of $(f_r)_{J_3}$, $(f_\theta)_{J_3}$ and $(f_h)_{J_3}$ with respect to a, h, k, p, q and F directly from Equations (13.39)–(13.41) of this chapter.

For $(f_r)_{J_3}$, we have the following partials:

$$\begin{aligned}\frac{\partial (f_r)_{J_3}}{\partial a} &= 72\mu J_3 R^3 r^{-7} \frac{\partial r}{\partial a} (qY_1 - pX_1) \left(1 + p^2 + q^2\right)^{-1} \\ &- 12\mu J_3 R^3 r^{-6} \left(q\frac{\partial Y_1}{\partial a} - p\frac{\partial X_1}{\partial a}\right) \left(1 + p^2 + q^2\right)^{-1} \\ &- 640\mu J_3 R^3 r^{-9} \frac{\partial r}{\partial a} (qY_1 - pX_1)^3 \left(1 + p^2 + q^2\right)^{-3} \\ &+ 240\mu J_3 R^3 r^{-8} (qY_1 - pX_1)^2 \left(q\frac{\partial Y_1}{\partial a} - p\frac{\partial X_1}{\partial a}\right) \left(1 + p^2 + q^2\right)^{-3} \end{aligned} \tag{13.60}$$

$$\begin{aligned}\frac{\partial (f_r)_{J_3}}{\partial h} &= 72\mu J_3 R^3 r^{-7} \left(\frac{\partial r}{\partial h}\right)_F (qY_1 - pX_1) \left(1 + p^2 + q^2\right)^{-1} \\ &- 12\mu J_3 R^3 r^{-6} \left[q\left(\frac{\partial Y_1}{\partial h}\right)_F - p\left(\frac{\partial X_1}{\partial h}\right)_F\right] \left(1 + p^2 + q^2\right)^{-1} \\ &- 640\mu J_3 R^3 r^{-9} \left(\frac{\partial r}{\partial h}\right)_F (qY_1 - pX_1)^3 \left(1 + p^2 + q^2\right)^{-3} \\ &+ 240\mu J_3 R^3 r^{-8} (qY_1 - pX_1)^2 \left[q\left(\frac{\partial Y_1}{\partial h}\right)_F - p\left(\frac{\partial X_1}{\partial h}\right)_F\right] \left(1 + p^2 + q^2\right)^{-3} \end{aligned} \tag{13.61}$$

$$\begin{aligned}\frac{\partial (f_r)_{J_3}}{\partial k} &= 72\mu J_3 R^3 r^{-7} \left(\frac{\partial r}{\partial k}\right)_F (qY_1 - pX_1) \left(1 + p^2 + q^2\right)^{-1} \\ &- 12\mu J_3 R^3 r^{-6} \left[q\left(\frac{\partial Y_1}{\partial k}\right)_F - p\left(\frac{\partial X_1}{\partial k}\right)_F\right] \left(1 + p^2 + q^2\right)^{-1} \\ &- 640\mu J_3 R^3 r^{-9} \left(\frac{\partial r}{\partial k}\right)_F (qY_1 - pX_1)^3 \left(1 + p^2 + q^2\right)^{-3} \\ &+ 240\mu J_3 R^3 r^{-8} (qY_1 - pX_1)^2 \left[q\left(\frac{\partial Y_1}{\partial k}\right)_F - p\left(\frac{\partial X_1}{\partial k}\right)_F\right] \left(1 + p^2 + q^2\right)^{-3} \end{aligned} \tag{13.62}$$

$$\begin{aligned}\frac{\partial (f_r)_{J_3}}{\partial p} &= 12\mu J_3 R^3 r^{-6} X_1 \left(1 + p^2 + q^2\right)^{-1} \\ &+ 12\mu J_3 R^3 r^{-6} (qY_1 - pX_1) \left(1 + p^2 + q^2\right)^{-2} 2p \\ &- 240\mu J_3 R^3 r^{-8} (qY_1 - pX_1)^2 X_1 \left(1 + p^2 + q^2\right)^{-3} \\ &- 240\mu J_3 R^3 r^{-8} (qY_1 - pX_1)^3 \left(1 + p^2 + q^2\right)^{-4} 2p \end{aligned} \tag{13.63}$$

$$\begin{aligned}\frac{\partial (f_r)_{J_3}}{\partial q} &= -12\mu J_3 R^3 r^{-6} Y_1 \left(1 + p^2 + q^2\right)^{-1} \\ &+ 12\mu J_3 R^3 r^{-6} (qY_1 - pX_1) \left(1 + p^2 + q^2\right)^{-2} 2q \\ &+ 240\mu J_3 R^3 r^{-8} (qY_1 - pX_1)^2 Y_1 \left(1 + p^2 + q^2\right)^{-3} \\ &- 240\mu J_3 R^3 r^{-8} (qY_1 - pX_1)^3 \left(1 + p^2 + q^2\right)^{-4} 2q \end{aligned} \tag{13.64}$$

$$\begin{aligned}\frac{\partial (f_r)_{J_3}}{\partial F} &= 72\mu J_3 R^3 r^{-7} \frac{\partial r}{\partial F} (qY_1 - pX_1) \left(1+p^2+q^2\right)^{-1} \\ &- 12\mu J_3 R^3 r^{-6} \left(q\frac{\partial Y_1}{\partial F} - p\frac{\partial X_1}{\partial F}\right) \left(1+p^2+q^2\right)^{-1} \\ &- 640\mu J_3 R^3 r^{-9} \frac{\partial r}{\partial F} (qY_1 - pX_1)^3 \left(1+p^2+q^2\right)^{-3} \\ &+ 240\mu J_3 R^3 r^{-8} (qY_1 - pX_1)^2 \left(q\frac{\partial Y_1}{\partial F} - p\frac{\partial X_1}{\partial F}\right) \left(1+p^2+q^2\right)^{-3} \end{aligned} \quad (13.65)$$

For $(f_\theta)_{J_3}$, we have

$$\begin{aligned}\frac{\partial (f_\theta)_{J_3}}{\partial a} &= 480\mu J_3 R^3 r^{-9} \frac{\partial r}{\partial a} (qX_1 + pY_1)(qY_1 - pX_1)^2 \left(1+p^2+q^2\right)^{-3} \\ &- 60\mu J_3 R^3 r^{-8} \left(q\frac{\partial X_1}{\partial a} + p\frac{\partial Y_1}{\partial a}\right) (qY_1 - pX_1)^2 \left(1+p^2+q^2\right)^{-3} \\ &- 120\mu J_3 R^3 r^{-8} (qX_1 + pY_1)(qY_1 - pX_1) \left(q\frac{\partial Y_1}{\partial a} - p\frac{\partial X_1}{\partial a}\right) \left(1+p^2+q^2\right)^{-3} \\ &- 18\mu J_3 R^3 r^{-7} \frac{\partial r}{\partial a} (qX_1 + pY_1) \left(1+p^2+q^2\right)^{-1} \\ &+ 3\mu J_3 R^3 r^{-6} \left(q\frac{\partial X_1}{\partial a} + p\frac{\partial Y_1}{\partial a}\right) \left(1+p^2+q^2\right)^{-1} \end{aligned} \quad (13.66)$$

$$\begin{aligned}\frac{\partial (f_\theta)_{J_3}}{\partial h} &= 480\mu J_3 R^3 r^{-9} \left(\frac{\partial r}{\partial h}\right)_F (qX_1 + pY_1)(qY_1 - pX_1)^2 \left(1+p^2+q^2\right)^{-3} \\ &- 60\mu J_3 R^3 r^{-8} \left[q\left(\frac{\partial X_1}{\partial h}\right)_F + p\left(\frac{\partial Y_1}{\partial h}\right)_F\right] (qY_1 - pX_1)^2 \left(1+p^2+q^2\right)^{-3} \\ &- 120\mu J_3 R^3 r^{-8} (qX_1 + pY_1)(qY_1 - pX_1) \left[q\left(\frac{\partial Y_1}{\partial h}\right)_F - p\left(\frac{\partial X_1}{\partial h}\right)_F\right] \\ &\times \left(1+p^2+q^2\right)^{-3} - 18\mu J_3 R^3 r^{-7} \left(\frac{\partial r}{\partial h}\right)_F (qX_1 + pY_1) \left(1+p^2+q^2\right)^{-1} \\ &+ 3\mu J_3 R^3 r^{-6} \left[q\left(\frac{\partial X_1}{\partial h}\right)_F + p\left(\frac{\partial Y_1}{\partial h}\right)_F\right] \left(1+p^2+q^2\right)^{-1} \end{aligned} \quad (13.67)$$

$$\begin{aligned}\frac{\partial (f_\theta)_{J_3}}{\partial k} &= 480\mu J_3 R^3 r^{-9} \left(\frac{\partial r}{\partial k}\right)_F (qX_1 + pY_1)(qY_1 - pX_1)^2 \left(1+p^2+q^2\right)^{-3} \\ &- 60\mu J_3 R^3 r^{-8} \left[q\left(\frac{\partial X_1}{\partial k}\right)_F + p\left(\frac{\partial Y_1}{\partial k}\right)_F\right] (qY_1 - pX_1)^2 \left(1+p^2+q^2\right)^{-3} \\ &- 120\mu J_3 R^3 r^{-8} (qX_1 + pY_1)(qY_1 - pX_1) \left[q\left(\frac{\partial Y_1}{\partial k}\right)_F - p\left(\frac{\partial X_1}{\partial k}\right)_F\right] \\ &\times \left(1+p^2+q^2\right)^{-3} - 18\mu J_3 R^3 r^{-7} \left(\frac{\partial r}{\partial k}\right)_F (qX_1 + pY_1) \left(1+p^2+q^2\right)^{-1} \\ &+ 3\mu J_3 R^3 r^{-6} \left[q\left(\frac{\partial X_1}{\partial k}\right)_F + p\left(\frac{\partial Y_1}{\partial k}\right)_F\right] \left(1+p^2+q^2\right)^{-1} \end{aligned} \quad (13.68)$$

$$\begin{aligned}\frac{\partial (f_\theta)_{J_3}}{\partial p} &= -60\mu J_3 R^3 r^{-8} Y_1 (qY_1 - pX_1)^2 \left(1+p^2+q^2\right)^{-3} \\ &+ 120\mu J_3 R^3 r^{-8} (qX_1 + pY_1)(qY_1 - pX_1) X_1 \left(1+p^2+q^2\right)^{-3} \\ &+ 180\mu J_3 R^3 r^{-8} (qX_1 + pY_1)(qY_1 - pX_1)^2 \left(1+p^2+q^2\right)^{-4} 2p \\ &+ 3\mu J_3 R^3 r^{-6} Y_1 \left(1+p^2+q^2\right)^{-1} - 3\mu J_3 R^3 r^{-6} (qX_1 + pY_1)\left(1+p^2+q^2\right)^{-2} 2p\end{aligned} \tag{13.69}$$

$$\begin{aligned}\frac{\partial (f_\theta)_{J_3}}{\partial q} &= -60\mu J_3 R^3 r^{-8} X_1 (qY_1 - pX_1)^2 \left(1+p^2+q^2\right)^{-3} \\ &- 120\mu J_3 R^3 r^{-8} (qX_1 + pY_1)(qY_1 - pX_1) Y_1 \left(1+p^2+q^2\right)^{-3} \\ &+ 180\mu J_3 R^3 r^{-8} (qX_1 + pY_1)(qY_1 - pX_1)^2 \left(1+p^2+q^2\right)^{-4} 2q \\ &+ 3\mu J_3 R^3 r^{-6} X_1 \left(1+p^2+q^2\right)^{-1} - 3\mu J_3 R^3 r^{-6} (qX_1 + pY_1)\left(1+p^2+q^2\right)^{-2} 2q\end{aligned} \tag{13.70}$$

$$\begin{aligned}\frac{\partial (f_\theta)_{J_3}}{\partial F} &= 480\mu J_3 R^3 r^{-9} \frac{\partial r}{\partial F} (qX_1 + pY_1)(qY_1 - pX_1)^2 \left(1+p^2+q^2\right)^{-3} \\ &- 60\mu J_3 R^3 r^{-8} \left[q\left(\frac{\partial X_1}{\partial F}\right) + p\left(\frac{\partial Y_1}{\partial F}\right)\right](qY_1 - pX_1)^2 \left(1+p^2+q^2\right)^{-3} \\ &- 120\mu J_3 R^3 r^{-8} (qX_1 + pY_1)(qY_1 - pX_1)\left[q\left(\frac{\partial Y_1}{\partial F}\right) - p\left(\frac{\partial X_1}{\partial F}\right)\right]\left(1+p^2+q^2\right)^{-3} \\ &- 18\mu J_3 R^3 r^{-7} \frac{\partial r}{\partial F} (qX_1 + pY_1)\left(1+p^2+q^2\right)^{-1} \\ &+ 3\mu J_3 R^3 r^{-6} \left[q\left(\frac{\partial X_1}{\partial F}\right) + p\left(\frac{\partial Y_1}{\partial F}\right)\right]\left(1+p^2+q^2\right)^{-1}\end{aligned} \tag{13.71}$$

For $(f_h)_{J_3}$, we have the partials

$$\begin{aligned}\frac{\partial (f_h)_{J_3}}{\partial a} &= 210\mu J_3 R^3 r^{-8} \frac{\partial r}{\partial a}\left(1-p^2-q^2\right)\left(1+p^2+q^2\right)^{-3}(qY_1 - pX_1)^2 \\ &- 60\mu J_3 R^3 r^{-7}\left(1-p^2-q^2\right)\left(1+p^2+q^2\right)^{-3}(qY_1 - pX_1)\left[q\frac{\partial Y_1}{\partial a} - p\frac{\partial X_1}{\partial a}\right] \\ &- \frac{15}{2}\mu J_3 R^3 r^{-6} \frac{\partial r}{\partial a}\left(1-p^2-q^2\right)\left(1+p^2+q^2\right)^{-1}\end{aligned} \tag{13.72}$$

$$\begin{aligned}\frac{\partial (f_h)_{J_3}}{\partial h} &= 210\mu J_3 R^3 r^{-8} \left(\frac{\partial r}{\partial h}\right)_F \left(1-p^2-q^2\right)\left(1+p^2+q^2\right)^{-3}(qY_1 - pX_1)^2 \\ &\quad - 60\mu J_3 R^3 r^{-7}\left(1-p^2-q^2\right)\left(1+p^2+q^2\right)^{-3}(qY_1 - pX_1)\left[q\left(\frac{\partial Y_1}{\partial h}\right)_F\right. \\ &\quad \left. -p\left(\frac{\partial X_1}{\partial h}\right)_F\right] - \frac{15}{2}\mu J_3 R^3 r^{-6}\left(\frac{\partial r}{\partial h}\right)_F \left(1-p^2-q^2\right)\left(1+p^2+q^2\right)^{-1}\end{aligned} \tag{13.73}$$

$$\frac{\partial (f_h)_{J_3}}{\partial k} = 210\mu J_3 R^3 r^{-8}\left(\frac{\partial r}{\partial k}\right)_F \left(1-p^2-q^2\right)\left(1+p^2+q^2\right)^{-3}(qY_1-pX_1)^2$$
$$- 60\mu J_3 R^3 r^{-7}\left(1-p^2-q^2\right)\left(1+p^2+q^2\right)^{-3}(qY_1-pX_1)\left[q\left(\frac{\partial Y_1}{\partial k}\right)_F\right.$$
$$\left.-p\left(\frac{\partial X_1}{\partial k}\right)_F\right] - \frac{15}{2}\mu J_3 R^3 r^{-6}\left(\frac{\partial r}{\partial k}\right)_F\left(1-p^2-q^2\right)\left(1+p^2+q^2\right)^{-1} \tag{13.74}$$

$$\frac{\partial (f_h)_{J_3}}{\partial p} = 60\mu J_3 R^3 r^{-7} p\left(1+p^2+q^2\right)^{-3}(qY_1-pX_1)^2$$
$$+ 180\mu J_3 R^3 r^{-7}\left(1-p^2-q^2\right)\left(1+p^2+q^2\right)^{-4} p\,(qY_1-pX_1)^2$$
$$+ 60\mu J_3 R^3 r^{-7}\left(1-p^2-q^2\right)\left(1+p^2+q^2\right)^{-3}(qY_1-pX_1)\,X_1$$
$$- 3\mu J_3 R^3 r^{-5} p\left(1+p^2+q^2\right)^{-1} - 3\mu J_3 R^3 r^{-5}\left(1-p^2-q^2\right)\left(1+p^2+q^2\right)^{-2} p \tag{13.75}$$

$$\frac{\partial (f_h)_{J_3}}{\partial q} = 60\mu J_3 R^3 r^{-7} q\left(1+p^2+q^2\right)^{-3}(qY_1-pX_1)^2$$
$$+ 180\mu J_3 R^3 r^{-7}\left(1-p^2-q^2\right)\left(1+p^2+q^2\right)^{-4} q\,(qY_1-pX_1)^2$$
$$- 60\mu J_3 R^3 r^{-7}\left(1-p^2-q^2\right)\left(1+p^2+q^2\right)^{-3}(qY_1-pX_1)\,Y_1$$
$$- 3\mu J_3 R^3 r^{-5} q\left(1+p^2+q^2\right)^{-1} - 3\mu J_3 R^3 r^{-5}\left(1-p^2-q^2\right)\left(1+p^2+q^2\right)^{-2} q \tag{13.76}$$

$$\frac{\partial (f_h)_{J_3}}{\partial F} = 210\mu J_3 R^3 r^{-8}\frac{\partial r}{\partial F}\left(1-p^2-q^2\right)\left(1+p^2+q^2\right)^{-3}(qY_1-pX_1)^2$$
$$- 60\mu J_3 R^3 r^{-7}\left(1-p^2-q^2\right)\left(1+p^2+q^2\right)^{-3}(qY_1-pX_1)\left[q\left(\frac{\partial Y_1}{\partial F}\right)-p\left(\frac{\partial X_1}{\partial F}\right)\right]$$
$$- \frac{15}{2}\mu J_3 R^3 r^{-6}\frac{\partial r}{\partial F}\left(1-p^2-q^2\right)\left(1+p^2+q^2\right)^{-1} \tag{13.77}$$

The J_4 accelerations given in Equations (13.14)–(13.16) in terms of the angles i and θ are converted to a form involving the equinoctial elements by using the following equivalencies:

$$s_i^2\left(1-c_{2\theta}\right) = 8\left(qs_L-pc_L\right)^2\Big/\left(1+p^2+q^2\right)^2 \tag{13.78}$$

$$s_i^2 s_\theta c_\theta = 4\left(qs_L-pc_L\right)\left(qc_L+ps_L\right)\Big/\left(1+p^2+q^2\right)^2 \tag{13.79}$$

$$s_i c_i s_\theta = 2\left(1-p^2-q^2\right)\left(qs_L-pc_L\right)\Big/\left(1+p^2+q^2\right)^2 \tag{13.80}$$

$$s_i^4 c_\theta\left(3s_\theta-s_{3\theta}\right) = 64\left(qs_L-pc_L\right)^3\left(qc_L+ps_L\right)\Big/\left(1+p^2+q^2\right)^4 \tag{13.81}$$

$$s_i^3 c_i \left(3s_\theta - s_{3\theta}\right) = 32\left(qs_L - pc_L\right)^3 \left(1 - p^2 - q^2\right) \Big/ \left(1 + p^2 + q^2\right)^4 \tag{13.82}$$

$$s_i^4 \left(3 - 4c_{2\theta} + c_{4\theta}\right) = 128\left(qs_L - pc_L\right)^4 \Big/ \left(1 + p^2 + q^2\right)^4 \tag{13.83}$$

Therefore, the J_4 acceleration components can be written as

$$\begin{aligned}(f_r)_{J_4} &= -75\mu J_4 R^4 r^{-6}\left(qs_L - pc_L\right)^2\left(1 + p^2 + q^2\right)^{-2} \\ &\quad + 350\mu J_4 R^4 r^{-6}\left(qs_L - pc_L\right)^4\left(1 + p^2 + q^2\right)^{-4} + \frac{15}{8}\mu J_4 R^4 r^{-6}\end{aligned} \tag{13.84}$$

$$\begin{aligned}(f_\theta)_{J_4} &= -280\mu J_4 R^4 r^{-6}\left(qs_L - pc_L\right)^3\left(qc_L + ps_L\right)\left(1 + p^2 + q^2\right)^{-4} \\ &\quad + 30\mu J_4 R^4 r^{-6}\left(qs_L - pc_L\right)\left(qc_L + ps_L\right)\left(1 + p^2 + q^2\right)^{-2}\end{aligned} \tag{13.85}$$

$$\begin{aligned}(f_h)_{J_4} &= -140\mu J_4 R^4 r^{-6}\left(qs_L - pc_L\right)^3\left(1 - p^2 - q^2\right)\left(1 + p^2 + q^2\right)^{-4} \\ &\quad + 15\mu J_4 R^4 r^{-6}\left(qs_L - pc_L\right)\left(1 - p^2 - q^2\right)\left(1 + p^2 + q^2\right)^{-2}\end{aligned} \tag{13.86}$$

And in terms of the fast variable F, the J_4 acceleration components take the form

$$\begin{aligned}(f_r)_{J_4} &= -75\mu J_4 R^4 r^{-8}\left(qY_1 - pX_1\right)^2\left(1 + p^2 + q^2\right)^{-2} \\ &\quad + 350\mu J_4 R^4 r^{-10}\left(qY_1 - pX_1\right)^4\left(1 + p^2 + q^2\right)^{-4} + \frac{15}{8}\mu J_4 R^4 r^{-6}\end{aligned} \tag{13.87}$$

$$\begin{aligned}(f_\theta)_{J_4} &= -280\mu J_4 R^4 r^{-10}\left(qY_1 - pX_1\right)^3\left(qX_1 + pY_1\right)\left(1 + p^2 + q^2\right)^{-4} \\ &\quad + 30\mu J_4 R^4 r^{-8}\left(qY_1 - pX_1\right)\left(qX_1 + pY_1\right)\left(1 + p^2 + q^2\right)^{-2}\end{aligned} \tag{13.88}$$

$$\begin{aligned}(f_h)_{J_4} &= -140\mu J_4 R^4 r^{-9}\left(qY_1 - pX_1\right)^3\left(1 - p^2 - q^2\right)\left(1 + p^2 + q^2\right)^{-4} \\ &\quad + 15\mu J_4 R^4 r^{-7}\left(qY_1 - pX_1\right)\left(1 - p^2 - q^2\right)\left(1 + p^2 + q^2\right)^{-2}\end{aligned} \tag{13.89}$$

The partial derivatives $\partial \mathbf{f}_{J_4}/\partial \mathbf{z}$ with $\mathbf{f}_{J_4} = \left(f_f\right)_{J_4}\hat{\mathbf{f}} + \left(f_g\right)_{J_4}\hat{\mathbf{g}} + \left(f_h\right)_{J_4}\hat{\mathbf{h}}$ are identical to Equations (13.43)–(13.59), except that the J_3 accelerations are replaced by the J_4 counterparts shown in Equations (13.87)–(13.89). These last three equations are used to generate the partials of $(f_r)_{J_4}$, $(f_\theta)_{J_4}$ and $(f_h)_{J_4}$ with respect to a, h, k, p, q and F. For $(f_r)_{J_4}$, the six partial derivatives can be written as

$$\begin{aligned}\frac{\partial\left(f_r\right)_{J_4}}{\partial a} &= 600\mu J_4 R^4 r^{-9}\frac{\partial r}{\partial a}\left(qY_1 - pX_1\right)^2\left(1 + p^2 + q^2\right)^{-2} \\ &\quad - 150\mu J_4 R^4 r^{-8}\left(qY_1 - pX_1\right)\left(q\frac{\partial Y_1}{\partial a} - p\frac{\partial X_1}{\partial a}\right)\left(1 + p^2 + q^2\right)^{-2} \\ &\quad - 3500\mu J_4 R^4 r^{-11}\frac{\partial r}{\partial a}\left(qY_1 - pX_1\right)^4\left(1 + p^2 + q^2\right)^{-4} \\ &\quad + 1400\mu J_4 R^4 r^{-10}\left(qY_1 - pX_1\right)^3\left(q\frac{\partial Y_1}{\partial a} - p\frac{\partial X_1}{\partial a}\right)\left(1 + p^2 + q^2\right)^{-4} \\ &\quad - \frac{45}{4}\mu J_4 R^4 r^{-7}\frac{\partial r}{\partial a}\end{aligned} \tag{13.90}$$

$$\begin{aligned}\frac{\partial (f_r)_{J_4}}{\partial h} &= 600\mu J_4 R^4 r^{-9}\left(\frac{\partial r}{\partial h}\right)_F (qY_1 - pX_1)^2\left(1+p^2+q^2\right)^{-2}\\
&- 150\mu J_4 R^4 r^{-8}(qY_1 - pX_1)\left[q\left(\frac{\partial Y_1}{\partial h}\right)_F - p\left(\frac{\partial X_1}{\partial h}\right)_F\right]\left(1+p^2+q^2\right)^{-2}\\
&- 3500\mu J_4 R^4 r^{-11}\left(\frac{\partial r}{\partial h}\right)_F (qY_1 - pX_1)^4\left(1+p^2+q^2\right)^{-4}\\
&+ 1400\mu J_4 R^4 r^{-10}(qY_1 - pX_1)^3\left[q\left(\frac{\partial Y_1}{\partial h}\right)_F - p\left(\frac{\partial X_1}{\partial h}\right)_F\right]\left(1+p^2+q^2\right)^{-4}\\
&- \frac{45}{4}\mu J_4 R^4 r^{-7}\left(\frac{\partial r}{\partial h}\right)_F \qquad (13.91)\end{aligned}$$

$$\begin{aligned}\frac{\partial (f_r)_{J_4}}{\partial k} &= 600\mu J_4 R^4 r^{-9}\left(\frac{\partial r}{\partial k}\right)_F (qY_1 - pX_1)^2\left(1+p^2+q^2\right)^{-2}\\
&- 150\mu J_4 R^4 r^{-8}(qY_1 - pX_1)\left[q\left(\frac{\partial Y_1}{\partial k}\right)_F - p\left(\frac{\partial X_1}{\partial k}\right)_F\right]\left(1+p^2+q^2\right)^{-2}\\
&- 3500\mu J_4 R^4 r^{-11}\left(\frac{\partial r}{\partial k}\right)_F (qY_1 - pX_1)^4\left(1+p^2+q^2\right)^{-4}\\
&+ 1400\mu J_4 R^4 r^{-10}(qY_1 - pX_1)^3\left[q\left(\frac{\partial Y_1}{\partial k}\right)_F - p\left(\frac{\partial X_1}{\partial k}\right)_F\right]\left(1+p^2+q^2\right)^{-4}\\
&- \frac{45}{4}\mu J_4 R^4 r^{-7}\left(\frac{\partial r}{\partial k}\right)_F \qquad (13.92)\end{aligned}$$

$$\begin{aligned}\frac{\partial (f_r)_{J_4}}{\partial p} &= 150\mu J_4 R^4 r^{-8}(qY_1 - pX_1)X_1\left(1+p^2+q^2\right)^{-2}\\
&+ 300\mu J_4 R^4 r^{-8}(qY_1 - pX_1)^2\left(1+p^2+q^2\right)^{-3} p\\
&- 1400\mu J_4 R^4 r^{-10}(qY_1 - pX_1)^3 X_1\left(1+p^2+q^2\right)^{-4}\\
&- 2800\mu J_4 R^4 r^{-10}(qY_1 - pX_1)^4\left(1+p^2+q^2\right)^{-5} p \qquad (13.93)\end{aligned}$$

$$\begin{aligned}\frac{\partial (f_r)_{J_4}}{\partial q} &= -150\mu J_4 R^4 r^{-8}(qY_1 - pX_1)Y_1\left(1+p^2+q^2\right)^{-2}\\
&+ 300\mu J_4 R^4 r^{-8}(qY_1 - pX_1)^2\left(1+p^2+q^2\right)^{-3} q\\
&+ 1400\mu J_4 R^4 r^{-10}(qY_1 - pX_1)^3 Y_1\left(1+p^2+q^2\right)^{-4}\\
&- 2800\mu J_4 R^4 r^{-10}(qY_1 - pX_1)^4\left(1+p^2+q^2\right)^{-5} q \qquad (13.94)\end{aligned}$$

$$\begin{aligned}\frac{\partial (f_r)_{J_4}}{\partial F} &= 600\mu J_4 R^4 r^{-9}\frac{\partial r}{\partial F}(qY_1 - pX_1)^2\left(1+p^2+q^2\right)^{-2}\\
&- 150\mu J_4 R^4 r^{-8}(qY_1 - pX_1)\left[q\frac{\partial Y_1}{\partial F} - p\frac{\partial X_1}{\partial F}\right]\left(1+p^2+q^2\right)^{-2}\\
&- 3500\mu J_4 R^4 r^{-11}\frac{\partial r}{\partial F}(qY_1 - pX_1)^4\left(1+p^2+q^2\right)^{-4}\end{aligned}$$

$$+ 1400\mu J_4 R^4 r^{-10} (qY_1 - pX_1)^3 \left[q\frac{\partial Y_1}{\partial F} - p\frac{\partial X_1}{\partial F} \right] \left(1 + p^2 + q^2\right)^{-4}$$
$$- \frac{45}{4}\mu J_4 R^4 r^{-7} \frac{\partial r}{\partial F} \tag{13.95}$$

For $(f_\theta)_{J_4}$, in a similar way, the following partial derivatives are obtained as

$$\frac{\partial (f_\theta)_{J_4}}{\partial a} = 2800\mu J_4 R^4 r^{-11} \frac{\partial r}{\partial a} (qY_1 - pX_1)^3 (qX_1 + pY_1) \left(1 + p^2 + q^2\right)^{-4}$$
$$- 840\mu J_4 R^4 r^{-10} (qY_1 - pX_1)^2 \left(q\frac{\partial Y_1}{\partial a} - p\frac{\partial X_1}{\partial a} \right) (qX_1 + pY_1) \left(1 + p^2 + q^2\right)^{-4}$$
$$- 280\mu J_4 R^4 r^{-10} (qY_1 - pX_1)^3 \left[q\frac{\partial X_1}{\partial a} + p\frac{\partial Y_1}{\partial a} \right] \left(1 + p^2 + q^2\right)^{-4}$$
$$- 240\mu J_4 R^4 r^{-9} \frac{\partial r}{\partial a} (qY_1 - pX_1)(qX_1 + pY_1) \left(1 + p^2 + q^2\right)^{-2}$$
$$+ 30\mu J_4 R^4 r^{-8} \left(q\frac{\partial Y_1}{\partial a} - p\frac{\partial X_1}{\partial a} \right) (qX_1 + pY_1) \left(1 + p^2 + q^2\right)^{-2}$$
$$+ 30\mu J_4 R^4 r^{-8} (qY_1 - pX_1) \left(q\frac{\partial X_1}{\partial a} + p\frac{\partial Y_1}{\partial a} \right) \left(1 + p^2 + q^2\right)^{-2} \tag{13.96}$$

$$\frac{\partial (f_\theta)_{J_4}}{\partial h} = 2800\mu J_4 R^4 r^{-11} \left(\frac{\partial r}{\partial h}\right)_F (qY_1 - pX_1)^3 (qX_1 + pY_1) \left(1 + p^2 + q^2\right)^{-4}$$
$$- 840\mu J_4 R^4 r^{-10} (qY_1 - pX_1)^2 \left[q\left(\frac{\partial Y_1}{\partial h}\right)_F - p\left(\frac{\partial X_1}{\partial h}\right)_F \right]$$
$$\times (qX_1 + pY_1) \left(1 + p^2 + q^2\right)^{-4} - 280\mu J_4 R^4 r^{-10} (qY_1 - pX_1)^3 \left[q\left(\frac{\partial X_1}{\partial h}\right)_F \right.$$
$$\left. + p\left(\frac{\partial Y_1}{\partial h}\right)_F \right] \left(1 + p^2 + q^2\right)^{-4} - 240\mu J_4 R^4 r^{-9} \left(\frac{\partial r}{\partial h}\right)_F (qY_1 - pX_1)$$
$$\times (qX_1 + pY_1) \left(1 + p^2 + q^2\right)^{-2} + 30\mu J_4 R^4 r^{-8} \left[q\left(\frac{\partial Y_1}{\partial h}\right)_F - p\left(\frac{\partial X_1}{\partial h}\right)_F \right]$$
$$\times (qX_1 + pY_1) \left(1 + p^2 + q^2\right)^{-2} + 30\mu J_4 R^4 r^{-8} (qY_1 - pX_1)$$
$$\times \left[q\left(\frac{\partial X_1}{\partial h}\right)_F + p\left(\frac{\partial Y_1}{\partial h}\right)_F \right] \left(1 + p^2 + q^2\right)^{-2} \tag{13.97}$$

$$\frac{\partial (f_\theta)_{J_4}}{\partial k} = 2800\mu J_4 R^4 r^{-11} \left(\frac{\partial r}{\partial k}\right)_F (qY_1 - pX_1)^3 (qX_1 + pY_1) \left(1 + p^2 + q^2\right)^{-4}$$
$$- 840\mu J_4 R^4 r^{-10} (qY_1 - pX_1)^2 \left[q\left(\frac{\partial Y_1}{\partial k}\right)_F - p\left(\frac{\partial X_1}{\partial k}\right)_F \right]$$
$$\times (qX_1 + pY_1) \left(1 + p^2 + q^2\right)^{-4} - 280\mu J_4 R^4 r^{-10} (qY_1 - pX_1)^3$$
$$\times \left[q\left(\frac{\partial X_1}{\partial k}\right)_F + p\left(\frac{\partial Y_1}{\partial k}\right)_F \right] \left(1 + p^2 + q^2\right)^{-4}$$

$$
\begin{aligned}
&- 240\mu J_4 R^4 r^{-9}\left(\frac{\partial r}{\partial k}\right)_F (qY_1 - pX_1)(qX_1 + pY_1)\left(1 + p^2 + q^2\right)^{-2} \\
&+ 30\mu J_4 R^4 r^{-8}\left[q\left(\frac{\partial Y_1}{\partial k}\right)_F - p\left(\frac{\partial X_1}{\partial k}\right)_F\right](qX_1 + pY_1)\left(1 + p^2 + q^2\right)^{-2} \\
&+ 30\mu J_4 R^4 r^{-8}(qY_1 - pX_1)\left[q\left(\frac{\partial X_1}{\partial k}\right)_F + p\left(\frac{\partial Y_1}{\partial k}\right)_F\right]\left(1 + p^2 + q^2\right)^{-2}
\end{aligned}
\tag{13.98}
$$

$$
\begin{aligned}
\frac{\partial (f_\theta)_{J_4}}{\partial p} &= 840\mu J_4 R^4 r^{-10}(qY_1 - pX_1)^2 X_1 (qX_1 + pY_1)\left(1 + p^2 + q^2\right)^{-4} \\
&- 280\mu J_4 R^4 r^{-10}(qY_1 - pX_1)^3 Y_1\left(1 + p^2 + q^2\right)^{-4} \\
&+ 2240\mu J_4 R^4 r^{-10}(qY_1 - pX_1)^3 (qX_1 + pY_1)\left(1 + p^2 + q^2\right)^{-5} p \\
&- 30\mu J_4 R^4 r^{-8} X_1 (qX_1 + pY_1)\left(1 + p^2 + q^2\right)^{-2} \\
&+ 30\mu J_4 R^4 r^{-8}(qY_1 - pX_1) Y_1\left(1 + p^2 + q^2\right)^{-2} \\
&- 120\mu J_4 R^4 r^{-8}(qY_1 - pX_1)(qX_1 + pY_1)\left(1 + p^2 + q^2\right)^{-3} p
\end{aligned}
\tag{13.99}
$$

$$
\begin{aligned}
\frac{\partial (f_\theta)_{J_4}}{\partial q} &= -840\mu J_4 R^4 r^{-10}(qY_1 - pX_1)^2 Y_1 (qX_1 + pY_1)\left(1 + p^2 + q^2\right)^{-4} \\
&- 280\mu J_4 R^4 r^{-10}(qY_1 - pX_1)^3 X_1\left(1 + p^2 + q^2\right)^{-4} \\
&+ 2240\mu J_4 R^4 r^{-10}(qY_1 - pX_1)^3 (qX_1 + pY_1)\left(1 + p^2 + q^2\right)^{-5} q \\
&+ 30\mu J_4 R^4 r^{-8} Y_1 (qX_1 + pY_1)\left(1 + p^2 + q^2\right)^{-2} \\
&+ 30\mu J_4 R^4 r^{-8}(qY_1 - pX_1) X_1\left(1 + p^2 + q^2\right)^{-2} \\
&- 120\mu J_4 R^4 r^{-8}(qY_1 - pX_1)(qX_1 + pY_1)\left(1 + p^2 + q^2\right)^{-3} q
\end{aligned}
\tag{13.100}
$$

$$
\begin{aligned}
\frac{\partial (f_\theta)_{J_4}}{\partial F} &= 2800\mu J_4 R^4 r^{-11}\frac{\partial r}{\partial F}(qY_1 - pX_1)^3 (qX_1 + pY_1)\left(1 + p^2 + q^2\right)^{-4} \\
&- 840\mu J_4 R^4 r^{-10}(qY_1 - pX_1)^2\left(q\frac{\partial Y_1}{\partial F} - p\frac{\partial X_1}{\partial F}\right)(qX_1 + pY_1)\left(1 + p^2 + q^2\right)^{-4} \\
&- 280\mu J_4 R^4 r^{-10}(qY_1 - pX_1)^3\left(q\frac{\partial X_1}{\partial F} + p\frac{\partial Y_1}{\partial F}\right)\left(1 + p^2 + q^2\right)^{-4} \\
&- 240\mu J_4 R^4 r^{-9}\frac{\partial r}{\partial F}(qY_1 - pX_1)(qX_1 + pY_1)\left(1 + p^2 + q^2\right)^{-2} \\
&+ 30\mu J_4 R^4 r^{-8}\left(q\frac{\partial Y_1}{\partial F} - p\frac{\partial X_1}{\partial F}\right)(qX_1 + pY_1)\left(1 + p^2 + q^2\right)^{-2} \\
&+ 30\mu J_4 R^4 r^{-8}(qY_1 - pX_1)\left(q\frac{\partial X_1}{\partial F} + p\frac{\partial Y_1}{\partial F}\right)\left(1 + p^2 + q^2\right)^{-2}
\end{aligned}
\tag{13.101}
$$

And finally for $(f_h)_{J_4}$, the relevant partials can be cast in the following form:

$$\begin{aligned}\frac{\partial (f_h)_{J_4}}{\partial a} &= 1260\mu J_4 R^4 r^{-10}\frac{\partial r}{\partial a}(qY_1 - pX_1)^3\left(1-p^2-q^2\right)\left(1+p^2+q^2\right)^{-4}\\ &- 420\mu J_4 R^4 r^{-9}(qY_1 - pX_1)^2\left(q\frac{\partial Y_1}{\partial a} - p\frac{\partial X_1}{\partial a}\right)\left(1-p^2-q^2\right)\left(1+p^2+q^2\right)^{-4}\\ &- 105\mu J_4 R^4 r^{-8}\frac{\partial r}{\partial a}(qY_1 - pX_1)\left(1-p^2-q^2\right)\left(1+p^2+q^2\right)^{-2}\\ &+ 15\mu J_4 R^4 r^{-7}\left(q\frac{\partial Y_1}{\partial a} - p\frac{\partial X_1}{\partial a}\right)\left(1-p^2-q^2\right)\left(1+p^2+q^2\right)^{-2} \end{aligned} \tag{13.102}$$

$$\begin{aligned}\frac{\partial (f_h)_{J_4}}{\partial h} &= 1260\mu J_4 R^4 r^{-10}\left(\frac{\partial r}{\partial h}\right)_F(qY_1 - pX_1)^3\left(1-p^2-q^2\right)\left(1+p^2+q^2\right)^{-4}\\ &- 420\mu J_4 R^4 r^{-9}(qY_1 - pX_1)^2\left[q\left(\frac{\partial Y_1}{\partial h}\right)_F - p\left(\frac{\partial X_1}{\partial h}\right)_F\right]\\ &\times\left(1-p^2-q^2\right)\left(1+p^2+q^2\right)^{-4} - 105\mu J_4 R^4 r^{-8}\left(\frac{\partial r}{\partial h}\right)_F(qY_1 - pX_1)\\ &\times\left(1-p^2-q^2\right)\left(1+p^2+q^2\right)^{-2} + 15\mu J_4 R^4 r^{-7}\left[q\left(\frac{\partial Y_1}{\partial h}\right)_F - p\left(\frac{\partial X_1}{\partial h}\right)_F\right]\\ &\times\left(1-p^2-q^2\right)\left(1+p^2+q^2\right)^{-2} \end{aligned} \tag{13.103}$$

$$\begin{aligned}\frac{\partial (f_h)_{J_4}}{\partial k} &= 1260\mu J_4 R^4 r^{-10}\left(\frac{\partial r}{\partial k}\right)_F(qY_1 - pX_1)^3\left(1-p^2-q^2\right)\left(1+p^2+q^2\right)^{-4}\\ &- 420\mu J_4 R^4 r^{-9}(qY_1 - pX_1)^2\left[q\left(\frac{\partial Y_1}{\partial k}\right)_F - p\left(\frac{\partial X_1}{\partial k}\right)_F\right]\left(1-p^2-q^2\right)\\ &\times\left(1+p^2+q^2\right)^{-4} - 105\mu J_4 R^4 r^{-8}\left(\frac{\partial r}{\partial k}\right)_F(qY_1 - pX_1)\left(1-p^2-q^2\right)\\ &\times\left(1+p^2+q^2\right)^{-2} + 15\mu J_4 R^4 r^{-7}\left[q\left(\frac{\partial Y_1}{\partial k}\right)_F - p\left(\frac{\partial X_1}{\partial k}\right)_F\right]\\ &\times\left(1-p^2-q^2\right)\left(1+p^2+q^2\right)^{-2} \end{aligned} \tag{13.104}$$

$$\begin{aligned}\frac{\partial (f_h)_{J_4}}{\partial p} &= 420\mu J_4 R^4 r^{-9}(qY_1 - pX_1)^2 X_1\left(1-p^2-q^2\right)\left(1+p^2+q^2\right)^{-4}\\ &+ 280\mu J_4 R^4 r^{-9}(qY_1 - pX_1)^3 p\left(1+p^2+q^2\right)^{-4}\\ &+ 1120\mu J_4 R^4 r^{-9}(qY_1 - pX_1)^3\left(1-p^2-q^2\right)\left(1+p^2+q^2\right)^{-5}p\\ &- 15\mu J_4 R^4 r^{-7}X_1\left(1-p^2-q^2\right)\left(1+p^2+q^2\right)^{-2}\\ &- 30\mu J_4 R^4 r^{-7}(qY_1 - pX_1)p\left(1+p^2+q^2\right)^{-2}\\ &- 60\mu J_4 R^4 r^{-7}(qY_1 - pX_1)\left(1-p^2-q^2\right)\left(1+p^2+q^2\right)^{-3}p \end{aligned} \tag{13.105}$$

$$\frac{\partial (f_h)_{J_4}}{\partial q} = -420\mu J_4 R^4 r^{-9} (qY_1 - pX_1)^2 Y_1 \left(1 - p^2 - q^2\right)\left(1 + p^2 + q^2\right)^{-4}$$
$$+ 280\mu J_4 R^4 r^{-9} (qY_1 - pX_1)^3 q \left(1 + p^2 + q^2\right)^{-4}$$
$$+ 1120\mu J_4 R^4 r^{-9} (qY_1 - pX_1)^3 \left(1 - p^2 - q^2\right)\left(1 + p^2 + q^2\right)^{-5} q$$
$$+ 15\mu J_4 R^4 r^{-7} Y_1 \left(1 - p^2 - q^2\right)\left(1 + p^2 + q^2\right)^{-2}$$
$$- 30\mu J_4 R^4 r^{-7} (qY_1 - pX_1) q \left(1 + p^2 + q^2\right)^{-2}$$
$$- 60\mu J_4 R^4 r^{-7} (qY_1 - pX_1) \left(1 - p^2 - q^2\right)\left(1 + p^2 + q^2\right)^{-3} q \quad (13.106)$$

$$\frac{\partial (f_h)_{J_4}}{\partial F} = 1260\mu J_4 R^4 r^{-10} \frac{\partial r}{\partial F} (qY_1 - pX_1)^3 \left(1 - p^2 - q^2\right)\left(1 + p^2 + q^2\right)^{-4}$$
$$- 420\mu J_4 R^4 r^{-9} (qY_1 - pX_1)^2 \left(q\frac{\partial Y_1}{\partial F} - p\frac{\partial X_1}{\partial F}\right)\left(1 - p^2 - q^2\right)\left(1 + p^2 + q^2\right)^{-4}$$
$$- 105\mu J_4 R^4 r^{-8} \frac{\partial r}{\partial F} (qY_1 - pX_1) \left(1 - p^2 - q^2\right)\left(1 + p^2 + q^2\right)^{-2}$$
$$+ 15\mu J_4 R^4 r^{-7} \left(q\frac{\partial Y_1}{\partial F} - p\frac{\partial X_1}{\partial F}\right)\left(1 - p^2 - q^2\right)\left(1 + p^2 + q^2\right)^{-2} \quad (13.107)$$

13.4 The Treatment of the J_3, J_4 Perturbations within the True Longitude Formulation

The dynamic and adjoint equations for the true longitude formulation were given in [7] and [11] and also in Chapters 9 and 11. The dynamic equations are given by

$$\begin{pmatrix} \dot{a} \\ \dot{h} \\ \dot{k} \\ \dot{p} \\ \dot{q} \\ \dot{L} \end{pmatrix} = \begin{pmatrix} B^L_{11} & B^L_{12} & B^L_{13} \\ B^L_{21} & B^L_{22} & B^L_{23} \\ B^L_{31} & B^L_{32} & B^L_{33} \\ B^L_{41} & B^L_{42} & B^L_{43} \\ B^L_{51} & B^L_{52} & B^L_{53} \\ B^L_{61} & B^L_{62} & B^L_{63} \end{pmatrix} \left[\begin{pmatrix} u_r \\ u_\theta \\ u_h \end{pmatrix} (f_t) + \mathbf{f}_{J_2} + \mathbf{f}_{J_3} + \mathbf{f}_{J_4}\right]$$
$$+ \begin{pmatrix} 0 \\ 0 \\ 0 \\ 0 \\ 0 \\ \frac{na^2\left(1-h^2-k^2\right)^{1/2}}{r^2} \end{pmatrix} \quad (13.108)$$

Unlike the eccentric longitude formulation of the previous section, where the thrust acceleration, the components of the rows of matrix M, and the J_2, J_3 and J_4 accelerations were resolved along the equinoctial frame $\hat{\mathbf{f}}$, $\hat{\mathbf{g}}$, $\hat{\mathbf{w}}$, this true longitude formulation uses the Euler–Hill $\hat{\mathbf{r}}$, $\hat{\boldsymbol{\theta}}$, $\hat{\mathbf{h}}$ frame such that the rows of matrix B^L above as well as the thrust and zonal accelerations are resolved in this Hill frame. The Hamiltonian and the adjoint differential equations for the combined thrust and J_2 perturbations given in [11] and Chapter 11 are completed below by the addition of the higher-order J_3 and J_4 terms such that with the unit vector $\hat{\mathbf{u}} = (u_r, u_\theta, u_h)$,

$$H = \boldsymbol{\lambda}_z^T B^L(\mathbf{z}) f_t \hat{\mathbf{u}} + \lambda_L \left[na^2 \left(1 - h^2 - k^2\right)^{1/2} / r^2 \right] + \boldsymbol{\lambda}_z^T B^L(\mathbf{z}) \left(\mathbf{f}_{J_2} + \mathbf{f}_{J_3} + \mathbf{f}_{J_4}\right) \tag{13.109}$$

$$\dot{\boldsymbol{\lambda}}_z = -\frac{\partial H}{\partial \mathbf{z}} = -\boldsymbol{\lambda}_z^T \frac{\partial B^L}{\partial \mathbf{z}} f_t \hat{\mathbf{u}} - \lambda_L \frac{\partial}{\partial \mathbf{z}} \left[\frac{na^2 \left(1 - h^2 - k^2\right)^{1/2}}{r^2} \right] - \boldsymbol{\lambda}_z^T \frac{\partial B^L}{\partial \mathbf{z}} \left(\mathbf{f}_{J_2} + \mathbf{f}_{J_3} + \mathbf{f}_{J_4}\right) - \boldsymbol{\lambda}_z^T B^L \left(\frac{\partial \mathbf{f}_{J_2}}{\partial \mathbf{z}} + \frac{\partial \mathbf{f}_{J_3}}{\partial \mathbf{z}} + \frac{\partial \mathbf{f}_{J_4}}{\partial \mathbf{z}} \right) \tag{13.110}$$

In [11] and Chapter 11, J_3 and J_4 were not included in the analysis, and there $\mathbf{f}_{J_2}$ was simply written as $\mathbf{f}$. The $\partial \mathbf{f}_{J_3}/\partial \mathbf{z}$ and $\partial \mathbf{f}_{J_4}/\partial \mathbf{z}$ partials are needed to complete this analysis. Note that the partial derivatives of matrix B^L with respect to $\mathbf{z} = (a, h, k, p, q, L)$ as given in the appendix of [7] have two typographic errors in Equations (A65) and (A74) that should read as

$$B^L_{63} = n^{-1} a^{-2} r G^{-1} (q s_L - p c_L) \qquad \text{(A65) of [2]}$$

$$\frac{\partial}{\partial k} \left[\frac{na^2 \left(1 - h^2 - k^2\right)^{1/2}}{r^2} \right] = -2na^2 r^{-3} G \frac{\partial r}{\partial k} - na^2 r^{-2} k G^{-1} \qquad \text{(A74) of [2]}$$

Chapter 9 is, however, free of these typographical errors.

Now using the expressions for $(f_r)_{J_3}$, $(f_\theta)_{J_3}$ and $(f_h)_{J_3}$ in Equations (13.34)–(13.36), directly in terms of a, h, k, p, q and L, we have for $(f_r)_{J_3}$, the following partials with respect to the elements a, h, k, p, q and L as

$$\frac{\partial (f_r)_{J_3}}{\partial a} = 60\mu J_3 R^3 r^{-6} \left(\frac{\partial r}{\partial a}\right)_L (q s_L - p c_L) \left(1 + p^2 + q^2\right)^{-1} - 400\mu J_3 R^3 r^{-6} \left(\frac{\partial r}{\partial a}\right)_L (q s_L - p c_L)^3 \left(1 + p^2 + q^2\right)^{-3} \tag{13.111}$$

$$\frac{\partial (f_r)_{J_3}}{\partial h} = 60\mu J_3 R^3 r^{-6} \left(\frac{\partial r}{\partial h}\right)_L (q s_L - p c_L) \left(1 + p^2 + q^2\right)^{-1} - 400\mu J_3 R^3 r^{-6} \left(\frac{\partial r}{\partial h}\right)_L (q s_L - p c_L)^3 \left(1 + p^2 + q^2\right)^{-3} \tag{13.112}$$

$$\frac{\partial (f_r)_{J_3}}{\partial k} = 60\mu J_3 R^3 r^{-6} \left(\frac{\partial r}{\partial k}\right)_L (q s_L - p c_L) \left(1 + p^2 + q^2\right)^{-1} - 400\mu J_3 R^3 r^{-6} \left(\frac{\partial r}{\partial k}\right)_L (q s_L - p c_L)^3 \left(1 + p^2 + q^2\right)^{-3} \tag{13.113}$$

$$\frac{\partial (f_r)_{J_3}}{\partial p} = 12\mu J_3 R^3 r^{-5} c_L \left(1 + p^2 + q^2\right)^{-1}$$
$$+ 24\mu J_3 R^3 r^{-5} (q s_L - p c_L) \left(1 + p^2 + q^2\right)^{-2} p$$
$$- 240\mu J_3 R^3 r^{-5} (q s_L - p c_L)^2 c_L \left(1 + p^2 + q^2\right)^{-3}$$
$$- 480\mu J_3 R^3 r^{-5} (q s_L - p c_L)^3 \left(1 + p^2 + q^2\right)^{-4} p \qquad (13.114)$$

$$\frac{\partial (f_r)_{J_3}}{\partial q} = -12\mu J_3 R^3 r^{-5} s_L \left(1 + p^2 + q^2\right)^{-1}$$
$$+ 24\mu J_3 R^3 r^{-5} (q s_L - p c_L) \left(1 + p^2 + q^2\right)^{-2} q$$
$$+ 240\mu J_3 R^3 r^{-5} (q s_L - p c_L)^2 s_L \left(1 + p^2 + q^2\right)^{-3}$$
$$- 480\mu J_3 R^3 r^{-5} (q s_L - p c_L)^3 \left(1 + p^2 + q^2\right)^{-4} q \qquad (13.115)$$

$$\frac{\partial (f_r)_{J_3}}{\partial L} = 60\mu J_3 R^3 r^{-6} \left(\frac{\partial r}{\partial L}\right)_L (q s_L - p c_L) \left(1 + p^2 + q^2\right)^{-1}$$
$$- 12\mu J_3 R^3 r^{-5} (q c_L + p s_L) \left(1 + p^2 + q^2\right)^{-1}$$
$$- 400\mu J_3 R^3 r^{-6} \left(\frac{\partial r}{\partial L}\right)_L (q s_L - p c_L)^3 \left(1 + p^2 + q^2\right)^{-3}$$
$$+ 240\mu J_3 R^3 r^{-5} (q s_L - p c_L)^2 (q c_L + p s_L) \left(1 + p^2 + q^2\right)^{-3} \qquad (13.116)$$

For the $(f_\theta)_{J_3}$ partials, we have in a similar way

$$\frac{\partial (f_\theta)_{J_3}}{\partial a} = 300\mu J_3 R^3 r^{-6} \left(\frac{\partial r}{\partial a}\right)_L (q c_L + p s_L)(q s_L - p c_L)^2 \left(1 + p^2 + q^2\right)^{-3}$$
$$- 15\mu J_3 R^3 r^{-6} \left(\frac{\partial r}{\partial a}\right)_L (q c_L + p s_L) \left(1 + p^2 + q^2\right)^{-1} \qquad (13.117)$$

$$\frac{\partial (f_\theta)_{J_3}}{\partial h} = 300\mu J_3 R^3 r^{-6} \left(\frac{\partial r}{\partial h}\right)_L (q c_L + p s_L)(q s_L - p c_L)^2 \left(1 + p^2 + q^2\right)^{-3}$$
$$- 15\mu J_3 R^3 r^{-6} \left(\frac{\partial r}{\partial h}\right)_L (q c_L + p s_L) \left(1 + p^2 + q^2\right)^{-1} \qquad (13.118)$$

$$\frac{\partial (f_\theta)_{J_3}}{\partial k} = 300\mu J_3 R^3 r^{-6} \left(\frac{\partial r}{\partial k}\right)_L (q c_L + p s_L)(q s_L - p c_L)^2 \left(1 + p^2 + q^2\right)^{-3}$$
$$- 15\mu J_3 R^3 r^{-6} \left(\frac{\partial r}{\partial k}\right)_L (q c_L + p s_L) \left(1 + p^2 + q^2\right)^{-1} \qquad (13.119)$$

$$\frac{\partial (f_\theta)_{J_3}}{\partial p} = -60\mu J_3 R^3 r^{-5} s_L (q s_L - p c_L)^2 \left(1 + p^2 + q^2\right)^{-3}$$
$$+ 120\mu J_3 R^3 r^{-5} (q c_L + p s_L)(q s_L - p c_L) c_L \left(1 + p^2 + q^2\right)^{-3}$$

$$+ 360\mu J_3 R^3 r^{-5} (qc_L + ps_L)(qs_L - pc_L)^2 \left(1 + p^2 + q^2\right)^{-4} p$$
$$+ 3\mu J_3 R^3 r^{-5} s_L \left(1 + p^2 + q^2\right)^{-1} - 6\mu J_3 R^3 r^{-5} (qc_L + ps_L) \left(1 + p^2 + q^2\right)^{-2} p \tag{13.120}$$

$$\frac{\partial (f_\theta)_{J_3}}{\partial q} = -60\mu J_3 R^3 r^{-5} c_L (qs_L - pc_L)^2 \left(1 + p^2 + q^2\right)^{-3}$$
$$- 120\mu J_3 R^3 r^{-5} (qc_L + ps_L)(qs_L - pc_L) s_L \left(1 + p^2 + q^2\right)^{-3}$$
$$+ 360\mu J_3 R^3 r^{-5} (qc_L + ps_L)(qs_L - pc_L)^2 \left(1 + p^2 + q^2\right)^{-4} q$$
$$+ 3\mu J_3 R^3 r^{-5} c_L \left(1 + p^2 + q^2\right)^{-1} - 6\mu J_3 R^3 r^{-5} (qc_L + ps_L) \left(1 + p^2 + q^2\right)^{-2} q \tag{13.121}$$

$$\frac{\partial (f_\theta)_{J_3}}{\partial L} = 300\mu J_3 R^3 r^{-6} \left(\frac{\partial r}{\partial L}\right)_L (qc_L + ps_L)(qs_L - pc_L)^2 \left(1 + p^2 + q^2\right)^{-3}$$
$$+ 60\mu J_3 R^3 r^{-5} (qs_L - pc_L)^3 \left(1 + p^2 + q^2\right)^{-3}$$
$$- 120\mu J_3 R^3 r^{-5} (qs_L - pc_L)(qc_L + ps_L)^2 \left(1 + p^2 + q^2\right)^{-3}$$
$$- 15\mu J_3 R^3 r^{-6} \left(\frac{\partial r}{\partial L}\right)_L (qc_L + ps_L) \left(1 + p^2 + q^2\right)^{-1}$$
$$- 3\mu J_3 R^3 r^{-5} (qs_L - pc_L) \left(1 + p^2 + q^2\right)^{-1} \tag{13.122}$$

For $(f_h)_{J_3}$, the relevant partial derivatives take the form

$$\frac{\partial (f_h)_{J_3}}{\partial a} = 150\mu J_3 R^3 r^{-6} \left(\frac{\partial r}{\partial a}\right)_L (qs_L - pc_L)^2 \left(1 - p^2 - q^2\right) \left(1 + p^2 + q^2\right)^{-3}$$
$$- \frac{15}{2}\mu J_3 R^3 r^{-6} \left(\frac{\partial r}{\partial a}\right)_L \left(1 - p^2 - q^2\right) \left(1 + p^2 + q^2\right)^{-1} \tag{13.123}$$

$$\frac{\partial (f_h)_{J_3}}{\partial h} = 150\mu J_3 R^3 r^{-6} \left(\frac{\partial r}{\partial h}\right)_L (qs_L - pc_L)^2 \left(1 - p^2 - q^2\right) \left(1 + p^2 + q^2\right)^{-3}$$
$$- \frac{15}{2}\mu J_3 R^3 r^{-6} \left(\frac{\partial r}{\partial h}\right)_L \left(1 - p^2 - q^2\right) \left(1 + p^2 + q^2\right)^{-1} \tag{13.124}$$

$$\frac{\partial (f_h)_{J_3}}{\partial k} = 150\mu J_3 R^3 r^{-6} \left(\frac{\partial r}{\partial k}\right)_L (qs_L - pc_L)^2 \left(1 - p^2 - q^2\right) \left(1 + p^2 + q^2\right)^{-3}$$
$$- \frac{15}{2}\mu J_3 R^3 r^{-6} \left(\frac{\partial r}{\partial k}\right)_L \left(1 - p^2 - q^2\right) \left(1 + p^2 + q^2\right)^{-1} \tag{13.125}$$

$$\frac{\partial (f_h)_{J_3}}{\partial p} = 60\mu J_3 R^3 r^{-5} (qs_L - pc_L) c_L \left(1 - p^2 - q^2\right) \left(1 + p^2 + q^2\right)^{-3}$$
$$+ 60\mu J_3 R^3 r^{-5} (qs_L - pc_L)^2 p \left(1 + p^2 + q^2\right)^{-3}$$

$$+ 180\mu J_3 R^3 r^{-5} (qs_L - pc_L)^2 \left(1 - p^2 - q^2\right)\left(1 + p^2 + q^2\right)^{-4} p$$
$$- 3\mu J_3 R^3 r^{-5} p \left(1 + p^2 + q^2\right)^{-1} - 3\mu J_3 R^3 r^{-5} \left(1 - p^2 - q^2\right)\left(1 + p^2 + q^2\right)^{-2} p \tag{13.126}$$

$$\frac{\partial (f_h)_{J_3}}{\partial q} = -60\mu J_3 R^3 r^{-5} (qs_L - pc_L)\, s_L \left(1 - p^2 - q^2\right)\left(1 + p^2 + q^2\right)^{-3}$$
$$+ 60\mu J_3 R^3 r^{-5} (qs_L - pc_L)^2 q \left(1 + p^2 + q^2\right)^{-3}$$
$$+ 180\mu J_3 R^3 r^{-5} (qs_L - pc_L)^2 \left(1 - p^2 - q^2\right)\left(1 + p^2 + q^2\right)^{-4} q$$
$$- 3\mu J_3 R^3 r^{-5} q \left(1 + p^2 + q^2\right)^{-1} - 3\mu J_3 R^3 r^{-5} \left(1 - p^2 - q^2\right)\left(1 + p^2 + q^2\right)^{-2} q \tag{13.127}$$

$$\frac{\partial (f_h)_{J_3}}{\partial L} = 150\mu J_3 R^3 r^{-6} \left(\frac{\partial r}{\partial L}\right)_L (qs_L - pc_L)^2 \left(1 - p^2 - q^2\right)\left(1 + p^2 + q^2\right)^{-3}$$
$$- 60\mu J_3 R^3 r^{-5} (qs_L - pc_L)(qc_L + ps_L) \left(1 - p^2 - q^2\right)\left(1 + p^2 + q^2\right)^{-3}$$
$$- \frac{15}{2}\mu J_3 R^3 r^{-6} \left(\frac{\partial r}{\partial L}\right)_L \left(1 - p^2 - q^2\right)\left(1 + p^2 + q^2\right)^{-1} \tag{13.128}$$

In a similar way, using the $\mathbf{f}_{J_4}$ acceleration components given in Equations (13.84)–(13.86) in terms of a, h, k, p, q and L, we have for the $(f_r)_{J_4}$ component the partials

$$\frac{\partial (f_r)_{J_4}}{\partial a} = 450\mu J_4 R^4 r^{-7} \left(\frac{\partial r}{\partial a}\right)_L (qs_L - pc_L)^2 \left(1 + p^2 + q^2\right)^{-2} - 2100\mu J_4 R^4 r^{-7}$$
$$\times \left(\frac{\partial r}{\partial a}\right)_L (qs_L - pc_L)^4 \left(1 + p^2 + q^2\right)^{-4} - \frac{45}{4}\mu J_4 R^4 r^{-7} \left(\frac{\partial r}{\partial a}\right)_L \tag{13.129}$$

$$\frac{\partial (f_r)_{J_4}}{\partial h} = 450\mu J_4 R^4 r^{-7} \left(\frac{\partial r}{\partial h}\right)_L (qs_L - pc_L)^2 \left(1 + p^2 + q^2\right)^{-2} - 2100\mu J_4 R^4 r^{-7}$$
$$\times \left(\frac{\partial r}{\partial h}\right)_L (qs_L - pc_L)^4 \left(1 + p^2 + q^2\right)^{-4} - \frac{45}{4}\mu J_4 R^4 r^{-7} \left(\frac{\partial r}{\partial h}\right)_L \tag{13.130}$$

$$\frac{\partial (f_r)_{J_4}}{\partial k} = 450\mu J_4 R^4 r^{-7} \left(\frac{\partial r}{\partial k}\right)_L (qs_L - pc_L)^2 \left(1 + p^2 + q^2\right)^{-2} - 2100\mu J_4 R^4 r^{-7}$$
$$\times \left(\frac{\partial r}{\partial k}\right)_L (qs_L - pc_L)^4 \left(1 + p^2 + q^2\right)^{-4} - \frac{45}{4}\mu J_4 R^4 r^{-7} \left(\frac{\partial r}{\partial k}\right)_L \tag{13.131}$$

$$\frac{\partial (f_r)_{J_4}}{\partial p} = 150\mu J_4 R^4 r^{-6} (qs_L - pc_L)\, c_L \left(1 + p^2 + q^2\right)^{-2}$$
$$+ 300\mu J_4 R^4 r^{-6} (qs_L - pc_L)^2 \left(1 + p^2 + q^2\right)^{-3} p$$

$$- 1400\mu J_4 R^4 r^{-6} \left(qs_L - pc_L\right)^3 c_L \left(1 + p^2 + q^2\right)^{-4}$$

$$- 2800\mu J_4 R^4 r^{-6} \left(qs_L - pc_L\right)^4 \left(1 + p^2 + q^2\right)^{-5} p \qquad (13.132)$$

$$\frac{\partial \left(f_r\right)_{J_4}}{\partial q} = -150\mu J_4 R^4 r^{-6} \left(qs_L - pc_L\right) s_L \left(1 + p^2 + q^2\right)^{-2}$$

$$+ 300\mu J_4 R^4 r^{-6} \left(qs_L - pc_L\right)^2 \left(1 + p^2 + q^2\right)^{-3} q$$

$$+ 1400\mu J_4 R^4 r^{-6} \left(qs_L - pc_L\right)^3 s_L \left(1 + p^2 + q^2\right)^{-4}$$

$$- 2800\mu J_4 R^4 r^{-6} \left(qs_L - pc_L\right)^4 \left(1 + p^2 + q^2\right)^{-5} q \qquad (13.133)$$

$$\frac{\partial \left(f_r\right)_{J_4}}{\partial L} = 450\mu J_4 R^4 r^{-7} \left(\frac{\partial r}{\partial L}\right)_L \left(qs_L - pc_L\right)^2 \left(1 + p^2 + q^2\right)^{-2}$$

$$- 150\mu J_4 R^4 r^{-6} \left(qs_L - pc_L\right) \left(qc_L + ps_L\right) \left(1 + p^2 + q^2\right)^{-2}$$

$$- 2100\mu J_4 R^4 r^{-7} \left(\frac{\partial r}{\partial L}\right)_L \left(qs_L - pc_L\right)^4 \left(1 + p^2 + q^2\right)^{-4}$$

$$+ 1400\mu J_4 R^4 r^{-6} \left(qs_L - pc_L\right)^3 \left(qc_L + ps_L\right) \left(1 + p^2 + q^2\right)^{-4}$$

$$- \frac{45}{4}\mu J_4 R^4 r^{-7} \left(\frac{\partial r}{\partial L}\right)_L \qquad (13.134)$$

For the $(f_\theta)_{J_4}$ partials, we have

$$\frac{\partial \left(f_\theta\right)_{J_4}}{\partial a} = 1680\mu J_4 R^4 r^{-7} \left(\frac{\partial r}{\partial a}\right)_L \left(qs_L - pc_L\right)^3 \left(qc_L + ps_L\right) \left(1 + p^2 + q^2\right)^{-4}$$

$$- 180\mu J_4 R^4 r^{-7} \left(\frac{\partial r}{\partial a}\right)_L \left(qs_L - pc_L\right) \left(qc_L + ps_L\right) \left(1 + p^2 + q^2\right)^{-2} \qquad (13.135)$$

$$\frac{\partial \left(f_\theta\right)_{J_4}}{\partial h} = 1680\mu J_4 R^4 r^{-7} \left(\frac{\partial r}{\partial h}\right)_L \left(qs_L - pc_L\right)^3 \left(qc_L + ps_L\right) \left(1 + p^2 + q^2\right)^{-4}$$

$$- 180\mu J_4 R^4 r^{-7} \left(\frac{\partial r}{\partial h}\right)_L \left(qs_L - pc_L\right) \left(qc_L + ps_L\right) \left(1 + p^2 + q^2\right)^{-2} \qquad (13.136)$$

$$\frac{\partial \left(f_\theta\right)_{J_4}}{\partial k} = 1680\mu J_4 R^4 r^{-7} \left(\frac{\partial r}{\partial k}\right)_L \left(qs_L - pc_L\right)^3 \left(qc_L + ps_L\right) \left(1 + p^2 + q^2\right)^{-4}$$

$$- 180\mu J_4 R^4 r^{-7} \left(\frac{\partial r}{\partial k}\right)_L \left(qs_L - pc_L\right) \left(qc_L + ps_L\right) \left(1 + p^2 + q^2\right)^{-2} \qquad (13.137)$$

$$\frac{\partial \left(f_\theta\right)_{J_4}}{\partial p} = 840\mu J_4 R^4 r^{-6} \left(qs_L - pc_L\right)^2 c_L \left(qc_L + ps_L\right) \left(1 + p^2 + q^2\right)^{-4}$$

$$- 280\mu J_4 R^4 r^{-6} \left(qs_L - pc_L\right)^3 s_L \left(1 + p^2 + q^2\right)^{-4}$$

$$+ 2240\mu J_4 R^4 r^{-6} (qs_L - pc_L)^3 (qc_L + ps_L) \left(1 + p^2 + q^2\right)^{-5} p$$
$$- 30\mu J_4 R^4 r^{-6} c_L (qc_L + ps_L) \left(1 + p^2 + q^2\right)^{-2}$$
$$+ 30\mu J_4 R^4 r^{-6} (qs_L - pc_L) s_L \left(1 + p^2 + q^2\right)^{-2}$$
$$- 120\mu J_4 R^4 r^{-6} (qs_L - pc_L)(qc_L + ps_L) \left(1 + p^2 + q^2\right)^{-3} p \quad (13.138)$$

$$\frac{\partial (f_\theta)_{J_4}}{\partial q} = -840\mu J_4 R^4 r^{-6} (qs_L - pc_L)^2 s_L (qc_L + ps_L) \left(1 + p^2 + q^2\right)^{-4}$$
$$- 280\mu J_4 R^4 r^{-6} (qs_L - pc_L)^3 c_L \left(1 + p^2 + q^2\right)^{-4}$$
$$+ 2240\mu J_4 R^4 r^{-6} (qs_L - pc_L)^3 (qc_L + ps_L) \left(1 + p^2 + q^2\right)^{-5} q$$
$$+ 30\mu J_4 R^4 r^{-6} s_L (qc_L + ps_L) \left(1 + p^2 + q^2\right)^{-2}$$
$$+ 30\mu J_4 R^4 r^{-6} (qs_L - pc_L) c_L \left(1 + p^2 + q^2\right)^{-2}$$
$$- 120\mu J_4 R^4 r^{-6} (qs_L - pc_L)(qc_L + ps_L) \left(1 + p^2 + q^2\right)^{-3} q \quad (13.139)$$

$$\frac{\partial (f_\theta)_{J_4}}{\partial L} = 1680\mu J_4 R^4 r^{-7} \left(\frac{\partial r}{\partial L}\right)_L (qs_L - pc_L)^3 (qc_L + ps_L) \left(1 + p^2 + q^2\right)^{-4}$$
$$- 840\mu J_4 R^4 r^{-6} (qs_L - pc_L)^2 (qc_L + ps_L)^2 \left(1 + p^2 + q^2\right)^{-4}$$
$$+ 280\mu J_4 R^4 r^{-6} (qs_L - pc_L)^4 \left(1 + p^2 + q^2\right)^{-4}$$
$$- 180\mu J_4 R^4 r^{-7} \left(\frac{\partial r}{\partial L}\right)_L (qs_L - pc_L)(qc_L + ps_L) \left(1 + p^2 + q^2\right)^{-2}$$
$$+ 30\mu J_4 R^4 r^{-6} (qc_L + ps_L)^2 \left(1 + p^2 + q^2\right)^{-2}$$
$$- 30\mu J_4 R^4 r^{-6} (qs_L - pc_L)^2 \left(1 + p^2 + q^2\right)^{-2} \quad (13.140)$$

And finally for $(f_h)_{J_4}$, the six partial derivatives are obtained as

$$\frac{\partial (f_h)_{J_4}}{\partial a} = 840\mu J_4 R^4 r^{-7} \left(\frac{\partial r}{\partial a}\right)_L (qs_L - pc_L)^3 \left(1 - p^2 - q^2\right) \left(1 + p^2 + q^2\right)^{-4}$$
$$- 90\mu J_4 R^4 r^{-7} \left(\frac{\partial r}{\partial a}\right)_L (qs_L - pc_L) \left(1 - p^2 - q^2\right) \left(1 + p^2 + q^2\right)^{-2} \quad (13.141)$$

$$\frac{\partial (f_h)_{J_4}}{\partial h} = 840\mu J_4 R^4 r^{-7} \left(\frac{\partial r}{\partial h}\right)_L (qs_L - pc_L)^3 \left(1 - p^2 - q^2\right) \left(1 + p^2 + q^2\right)^{-4}$$
$$- 90\mu J_4 R^4 r^{-7} \left(\frac{\partial r}{\partial h}\right)_L (qs_L - pc_L) \left(1 - p^2 - q^2\right) \left(1 + p^2 + q^2\right)^{-2} \quad (13.142)$$

$$\begin{aligned}\frac{\partial (f_h)_{J_4}}{\partial k} &= 840\mu J_4 R^4 r^{-7}\left(\frac{\partial r}{\partial k}\right)_L (qs_L - pc_L)^3\left(1-p^2-q^2\right)\left(1+p^2+q^2\right)^{-4}\\ &\quad - 90\mu J_4 R^4 r^{-7}\left(\frac{\partial r}{\partial k}\right)_L (qs_L - pc_L)\left(1-p^2-q^2\right)\left(1+p^2+q^2\right)^{-2}\end{aligned} \tag{13.143}$$

$$\begin{aligned}\frac{\partial (f_h)_{J_4}}{\partial p} &= 420\mu J_4 R^4 r^{-6}(qs_L - pc_L)^2 c_L\left(1-p^2-q^2\right)\left(1+p^2+q^2\right)^{-4}\\ &\quad + 280\mu J_4 R^4 r^{-6}(qs_L - pc_L)^3 p\left(1+p^2+q^2\right)^{-4}\\ &\quad + 1120\mu J_4 R^4 r^{-6}(qs_L - pc_L)^3\left(1-p^2-q^2\right)\left(1+p^2+q^2\right)^{-5} p\\ &\quad - 15\mu J_4 R^4 r^{-6} c_L\left(1-p^2-q^2\right)\left(1+p^2+q^2\right)^{-2}\\ &\quad - 30\mu J_4 R^4 r^{-6}(qs_L - pc_L) p\left(1+p^2+q^2\right)^{-2}\\ &\quad - 60\mu J_4 R^4 r^{-6}(qs_L - pc_L)\left(1-p^2-q^2\right)\left(1+p^2+q^2\right)^{-3} p\end{aligned} \tag{13.144}$$

$$\begin{aligned}\frac{\partial (f_h)_{J_4}}{\partial q} &= -420\mu J_4 R^4 r^{-6}(qs_L - pc_L)^2 s_L\left(1-p^2-q^2\right)\left(1+p^2+q^2\right)^{-4}\\ &\quad + 280\mu J_4 R^4 r^{-6}(qs_L - pc_L)^3 q\left(1+p^2+q^2\right)^{-4}\\ &\quad + 1120\mu J_4 R^4 r^{-6}(qs_L - pc_L)^3\left(1-p^2-q^2\right)\left(1+p^2+q^2\right)^{-5} q\\ &\quad + 15\mu J_4 R^4 r^{-6} s_L\left(1-p^2-q^2\right)\left(1+p^2+q^2\right)^{-2}\\ &\quad - 30\mu J_4 R^4 r^{-6}(qs_L - pc_L) q\left(1+p^2+q^2\right)^{-2}\\ &\quad - 60\mu J_4 R^4 r^{-6}(qs_L - pc_L)\left(1-p^2-q^2\right)\left(1+p^2+q^2\right)^{-3} q\end{aligned} \tag{13.145}$$

$$\begin{aligned}\frac{\partial (f_h)_{J_4}}{\partial L} &= 840\mu J_4 R^4 r^{-7}\left(\frac{\partial r}{\partial L}\right)_L (qs_L - pc_L)^3\left(1-p^2-q^2\right)\left(1+p^2+q^2\right)^{-4}\\ &\quad - 420\mu J_4 R^4 r^{-6}(qs_L - pc_L)^2(qc_L + ps_L)\left(1-p^2-q^2\right)\left(1+p^2+q^2\right)^{-4}\\ &\quad - 90\mu J_4 R^4 r^{-7}\left(\frac{\partial r}{\partial L}\right)_L (qs_L - pc_L)\left(1-p^2-q^2\right)\left(1+p^2+q^2\right)^{-2}\\ &\quad + 15\mu J_4 R^4 r^{-6}(qc_L + ps_L)\left(1-p^2-q^2\right)\left(1+p^2+q^2\right)^{-2}\end{aligned} \tag{13.146}$$

Note that the $\left(\frac{\partial r}{\partial a}\right)_L$, $\left(\frac{\partial r}{\partial h}\right)_L$, $\left(\frac{\partial r}{\partial k}\right)_L$, $\left(\frac{\partial r}{\partial L}\right)_L$ partials appearing in the above equations are as in Equation (65) of [11] and also in Chapter 11, namely:

$$\left(\frac{\partial r}{\partial a}\right)_L = \frac{r}{a}, \left(\frac{\partial r}{\partial h}\right)_L = -\frac{r}{a\left(1-h^2-k^2\right)}(2ah + rs_L)$$

$$\left(\frac{\partial r}{\partial k}\right)_L = -\frac{r}{a\left(1-h^2-k^2\right)}(2ak + rc_L), \left(\frac{\partial r}{\partial L}\right)_L = -\frac{r^2}{a}\frac{hc_L - ks_L}{\left(1-h^2-k^2\right)}$$

13.5 Numerical Results

We first show the need to include the J_3 and J_4 accelerations by running an optimal trajectory generated in [12] and [13] and also in Chapter 12, with only J_2 taken into account. The initial and final orbits for the example transfer are shown in Table 13.1. The overall minimum-time solution is obtained by also searching for the optimal departure point on the initial orbit at time $t = 0$, as well as the optimal arrival location on the final orbit at time t_f, which is itself one of the search parameters. The shooting method is used by starting from guessed values for the five multipliers at time $t = 0$, namely $(\lambda_a)_0$, $(\lambda_h)_0$, $(\lambda_k)_0$, $(\lambda_p)_0$, $(\lambda_q)_0$ as well as the initial mean longitude $(\lambda)_0$ or rather eccentric longitude $(F)_0$ corresponding to the initial angular position $(\theta)_0$. The value of t_f is also guessed and the dynamic and adjoint differential equations with J_2 only are integrated forward in time simultaneously starting from $(\lambda_F)_0 = 0$ until t_f while thrusting along the optimal $\hat{\mathbf{u}}$ orientation given by $\hat{\mathbf{u}} = \boldsymbol{\lambda}_z^T M(\mathbf{z}, F) / \left|\boldsymbol{\lambda}_z^T M(\mathbf{z}, F)\right|$ that effectively maximizes the Hamiltonian H in Equation (13.23) with f_{J_3} and f_{J_4} turned off. The quantities $(\lambda_a)_0$, $(\lambda_h)_0$, $(\lambda_k)_0$, $(\lambda_p)_0$, $(\lambda_q)_0$ as well as $(F)_0$ and t_f are slowly adjusted until the final orbit parameters $(a)_f$, $(h)_f$, $(k)_f$, $(p)_f$, $(q)_f$ that correspond to the target a, e, i, Ω, ω given in Table 13.1 are met to within a reasonably small tolerance, with $(F)_f = 0$ and $H_f = 1$ also satisfied.

The initial and final orbit parameters are given as classical elements in Table 13.1, but are converted at time $t = 0$ to obtain the equinoctial equivalents to start the integrations, and converted from the equinoctials at t_f to the equivalent classicals for tabulation purposes.

The initial values of the converged multipliers are given by $(\lambda_a)_0 = 4.800100306$ s/km, $(\lambda_h)_0 = 8.060772261 \times 10^2$ s, $(\lambda_k)_0 = -9.150040837 \times 10^3$ s, $(\lambda_p)_0 = 3.281827358 \times 10^1$ s, $(\lambda_q)_0 = -2.254928992 \times 10^4$ s, the optimal initial location was obtained as $(\lambda)_0 = -2.299291130$ rad, and the minimum time t_f was obtained as 58,104.83438 s.

With $(\lambda)_0$ corresponding to the mean anomaly $M_0 = -131.7396776$ deg, the state and costate equations are propagated forward in time starting from the converged multipliers above and using Equations (13.108) and (13.110) by firing along the optimal direction $\hat{\mathbf{u}} = \boldsymbol{\lambda}_z^T B^L(\mathbf{z}) / \left|\boldsymbol{\lambda}_z^T B^L(\mathbf{z})\right|$ until $t_f = 58{,}104.83438$ s. Both t_f and the converged multipliers as well as the optimal $(\lambda)_0$ correspond to the optimal solution obtained with J_2 only [12, 13] (see also Chapter 12). Thus three runs are made using these converged values, the first two using the original J_2 formulation of [11], [12] and [13] respectively and the third run using Equations (13.108) and (13.110) of the present chapter with J_3 and J_4 turned on in order to fly the trajectory using a more accurate Earth gravity model. The first two runs from time $t = 0$ to t_f will achieve the desired end conditions because the initial converged multiplier values are the optimal ones corresponding to the simplified gravity model that uses J_2 only. Obviously using these converged values to fly with a more accurate gravity model that uses J_2, J_3 and J_4 and terminating the flight at the J_2-optimal flight time t_f will fail to meet the end conditions.

Table 13.1 Initial, target and achieved orbit parameters using J_2-compliant solution, with J_2 only, and with J_2, J_3, J_4 perturbations

Orbit	a, km	e	i, deg	Ω, deg	ω, deg	M, F, L deg
Initial	7000	0	28.5	0	0	$M_o = -131.7396776$ (opt.)
Target	42,000	10^{-3}	1	0	0	Free
Achieved [L] J_2	41,999.99992	1.000022×10^{-3}	1.0000013	359.999569	359.999136	$L = 45.4919264$ (opt.)
Achieved [F] J_2	41,999.99992	1.000022×10^{-3}	1.0000013	359.999569	359.999130	$F = 45.45107830$ (opt.)
Achieved [L] J_2, J_3, J_4	41,996.09349	$9.222074532 \times 10^{-4}$	$9.995354342 \times 10^{-1}$	359.9363898	4.3177075	$M = 41.17416016$

Table 13.1 shows the achieved final parameters for this low Earth orbit (LEO) to geosynchronous Earth orbit (GEO) transfer using the Achieved [L] of [11] and Chapter 11, and Achieved [F] of [12] and [13] and Chapter 12 formulations using J_2 only, and the present formulation Achieved [L] using J_2, J_3 and J_4, that is, Equations (13.108)–(13.110). As in all runs, we use $f_t = 9.8 \times 10^{-5}$ km/s^2 and $J_2 = 1.08263 \times 10^{-3}$, $J_3 = -2.56 \times 10^{-6}$, $J_4 = -1.58 \times 10^{-6}$ and a Runge–Kutta 78 integration method as given by the Fehlberg coefficients with relative and absolute error controls set at the 10^{-9} level.

As expected, the Achieved [L]-J_2 and Achieved [F]-J_2 forward runs meet the target conditions effectively as shown in Table 13.1, while the Achieved [L]-J_2 J_3 J_4 run misses the final semimajor axis by 4 km and the argument of perigee by 4 deg or equivalently the final position by as much. These results show that restricting ourselves to a reduced model with J_2 only and using the corresponding optimal solution to fly in the more realistic and accurate model with J_2, J_3 and J_4 will miss the target even for this very short-duration transfer that experiences very little perturbation from the zonal harmonics. These differences will be dramatically more enhanced for long-duration transfers that dwell for long periods of time in the gravity well at LEO where the zonal perturbations affect the trajectories considerably.

Note that the integrated value of L at the final time t_f, namely $L = 45.4919264$ deg from Achieved [L], corresponds to $M = 45.411543$ deg, which corresponds to the optimal insertion point on the target orbit with J_2 only. As discussed above, the value of M as well as the values of the other parameters from Achieved [L]-J_2 J_3 J_4 are clearly significantly different due to the addition of the J_3 and J_4 terms.

We now generate an optimal minimum-time trajectory using Equations (13.108)–(13.110) and searching on the initial values of $(\lambda_a)_0$, $(\lambda_h)_0$, $(\lambda_k)_0$, $(\lambda_p)_0$, $(\lambda_q)_0$ and L_0 and starting with $(\lambda_L)_0 = 0$ until the final a, h, k, p, q, $(\lambda_L) = 0$, and $H = 1$ are satisfied. The result is $(\lambda_a)_0 = 4.800362521$ s/km, $(\lambda_h)_0 = 8.064085075 \times 10^2$ s, $(\lambda_k)_0 = -9.149918686 \times 10^3$ s, $(\lambda_p)_0 = 3.284285354 \times 10^1$ s, $(\lambda_q)_0 = -2.254952264 \times 10^4$ s, $(L)_0 = -2.299361237$ rad and $t_f = 58{,}104.96895$ s.

Because $e_0 = 0$, $\omega_0 = 0$ and $\Omega_0 = 0$, $(L)_0$ is identical to $(M)_0 = -131.7436944$ deg. Using these initial values, the Achieved [L]-J_2 J_3 J_4 (Equations [13.108]–[13.110]), and Achieved [F]-J_2 J_3 J_4 (Equations [13.17]–[13.24]) parameters at time t_f are in perfect agreement as seen in Table 13.2, thus validating both the L and F formulations of the present chapter. Note that $(\lambda_L)_0 = (\lambda_F)_0 = 0$.

As stated above, this fast transfer trajectory experiences very little zonal perturbation effects and that is why the J_2-optimized and the present J_2-J_3-J_4-optimized solutions differ very little in total flight time t_f as well as in the final insertion location. These differences can become quite substantial and cannot be neglected both during preliminary designs and actual operations. Even the small differences seen in this present example can be misleading because if the J_3 and J_4 perturbations are not accounted for, and the solutions rely only on the J_2-only model, they cannot be flown accurately, as shown in Table 13.1, due to the presence of the higher-order harmonics. In fact, further harmonics such as J_5 and J_6 must also be included, extending the present analysis especially for

Table 13.2 Initial, target and achieved orbit parameters using J_2, J_3, J_4 perturbations-compliant solution

Orbit	a, km	e	i, deg	Ω, deg	ω, deg	M, F, L deg
Initial	7000	0	28.5	0	0	$M_0 = -131.7436944$
Target	42,000	10^{-3}	1	0	0	Free
Achieved [L] $J_2J_3J_4$	41,999.99995	$1.000001445 \times 10^{-3}$	1.000000093	359.9999474	$1.107722739 \times 10^{-4}$	$L = 45.49094391$
Achieved [F] $J_2J_3J_4$	41,999.99995	$1.000001445 \times 10^{-3}$	1.000000093	359.9999474	$1.107673521 \times 10^{-4}$	$F = 45.45009831$

long-duration transfers spanning several months with a considerable amount spent in LEO.

However, it is possible to use a cruder J_2-only model and continuously update the transfer solution during an actual flight in order to eventually correct for the effects of the unmodeled higher-order perturbations as the spacecraft approaches its higher destination orbit such as GEO. It is, however, better to use a more refined model such as the one exposed here even if continuous updates are planned for actual flights because more accurate and economical trajectories can then be flown. If low thrust is used instead to transfer to lower orbits or if the transfers are done entirely in low orbit and of long duration, then higher-fidelity models are even more beneficial to implement.

The value of $F = 45.45009839$ deg corresponds to $M = 45.40920885$ deg and the value of $L = 45.49094391$ deg corresponds to $M = 45.40920884$ deg, indicating perfect agreement for the optimized insertion location on the target orbit.

13.6 Conclusion

A higher-fidelity modeling of optimal low-thrust transfers between any given circular or elliptic orbits using a more complete Earth gravity model in the form of higher-order harmonics has thus been derived based on the Gaussian form of the dynamic system equations in terms of the nonsingular equinoctial elements.

Two previously developed formulations based on the use of the eccentric and true longitude as the sixth or fast orbit element respectively, and fully accounting for the J_2 perturbation both in the dynamic and adjoint differential equations, have been extended in this present chapter by the inclusion of the higher zonal harmonics terms J_3 and J_4 in order to generate even higher-accuracy optimal low-thrust trajectories than with J_2 alone. A fast LEO to GEO minimum-time transfer using a relatively high constant thrust acceleration is generated for illustration purposes and the achieved trajectories from both formulations are compared, leading to a perfect match, thus validating both formulations at once. The consideration of two different formulations is necessary in order to make sure that the nonsingular state and adjoint differential equations are both free of algebraic and coding errors regardless of which formulation is adopted for operational use. The true longitude formulation, which also resolves the various accelerations in the Euler–Hill frame, is more concise and easier to derive than the eccentric longitude formulation which resolves the accelerations along the direct equinoctial frame. However, the latter formulation was the one that was historically developed first. All the derivations were done by hand without the use of any symbolic manipulation software. The numerical results and their discussion show the benefits of adopting this more refined theory for both simulations and flight guidance applications. The consideration of additional harmonics such as J_5 and J_6 within the present context of the Gaussian formulation can lead to even higher-accuracy modeling of optimal low-thrust transfers.

Appendix 13.1 Transformation of the J_3, J_4 Inertial Accelerations to the Rotating Frame

The perturbation accelerations due to J_3 and J_4 in the Euler–Hill frame can also be obtained directly from the inertial system $\hat{\mathbf{x}}$, $\hat{\mathbf{y}}$, $\hat{\mathbf{z}}$. Going back to Equation (13.3) for the disturbing potential F', we have with $r = (x^2 + y^2 + z^2)^{1/2}$, $\partial r/\partial x = x/r$, $\partial r/\partial y = y/r$, $\partial r/\partial \mathbf{z} = z/r$, such that for the J_3 terms, and in view of $s_\theta = z/r$,

$$\ddot{x} = \partial F'/\partial x = -\frac{\mu x}{r^3} J_3 \frac{5}{2}\left(\frac{R}{r}\right)^3 \left(3\frac{z}{r} - 7\frac{z^3}{r^3}\right) \tag{A13.1}$$

$$\ddot{y} = \partial F'/\partial y = -\frac{\mu y}{r^3} J_3 \frac{5}{2}\left(\frac{R}{r}\right)^3 \left(3\frac{z}{r} - 7\frac{z^3}{r^3}\right) \tag{A13.2}$$

$$\ddot{z} = \partial F'/\partial z = -\frac{\mu z}{r^3} J_3 \frac{3}{2}\left(\frac{R}{r}\right)^3 \left(10\frac{z}{r} - \frac{35}{3}\frac{z^3}{r^3} - \frac{r}{z}\right) \tag{A13.3}$$

Replacing x, y and z above by their respective expressions in terms of the Eulerian angles with $x = r(c_\Omega c_\theta - s_\Omega c_i s_\theta)$, $y = r(s_\Omega c_\theta + c_\Omega c_i s_\theta)$ and $z = r s_i s_\theta$, and in view of the rotation matrix in Equation (13.6), and after the same algebra,

$$\begin{aligned}(f_r)_{J_3} &= (c_\Omega c_\theta - s_\Omega c_i s_\theta)\,\ddot{x} + (s_\Omega c_\theta + c_\Omega c_i s_\theta)\,\ddot{y} + s_i s_\theta \ddot{z} \\ &= -\frac{\mu J_3 R^3}{r^5}\left(6 s_i s_\theta - 10 s_i^3 s_\theta^3\right)\end{aligned} \tag{A13.4}$$

In a similar way,

$$\begin{aligned}(f_\theta)_{J_3} &= (-c_\Omega s_\theta - s_\Omega c_i c_\theta)\,\ddot{x} + (-s_\Omega s_\theta + c_\Omega c_i c_\theta)\,\ddot{y} + s_i c_\theta \ddot{z} \\ &= -\frac{\mu J_3 R^3}{2r^5}\left[\frac{15}{4} s_i^3 \left(c_\theta - c_{3\theta}\right) - 3 s_i c_\theta\right]\end{aligned} \tag{A13.5}$$

$$\begin{aligned}(f_h)_{J_3} &= s_\Omega s_i \ddot{x} - c_\Omega s_i \ddot{y} + c_i \ddot{z} \\ &= -\frac{\mu J_3 R^3}{2r^5}\left(15 s_i^2 c_i s_\theta^2 - 3 c_i\right)\end{aligned} \tag{A13.6}$$

For the J_4 accelerations

$$\ddot{x} = \partial F'/\partial x = -\frac{\mu x}{r^3} J_4 \left(\frac{R}{r}\right)^4 \frac{5}{8}\left(-63\frac{z^4}{r^4} + 42\frac{z^2}{r^2} - 3\right) \tag{A13.7}$$

$$\ddot{y} = \partial F'/\partial y = -\frac{\mu y}{r^3} J_4 \left(\frac{R}{r}\right)^4 \frac{5}{8}\left(-63\frac{z^4}{r^4} + 42\frac{z^2}{r^2} - 3\right) \tag{A13.8}$$

$$\ddot{z} = \partial F'/\partial z = -\frac{\mu z}{r^3} J_4 \left(\frac{R}{r}\right)^4 \frac{5}{8}\left(-63\frac{z^4}{r^4} + 70\frac{z^2}{r^2} - 15\right) \tag{A13.9}$$

which leads to the expressions

$$\begin{aligned}(f_r)_{J_4} &= (c_\Omega c_\theta - s_\Omega c_i s_\theta)\,\ddot{x} + (s_\Omega c_\theta + c_\Omega c_i s_\theta)\,\ddot{y} + s_i s_\theta \ddot{z} \\ &= -\frac{5\mu J_4 R^4}{8r^6}\left(-35 s_i^4 s_\theta^4 + 30 s_i^2 s_\theta^2 - 3\right)\end{aligned} \tag{A13.10}$$

where s_θ^4 can be written as $\frac{1}{8}(3 - 4c_{2\theta} + c_{4\theta})$, and

$$\begin{aligned}(f_\theta)_{J_4} &= (-c_\Omega s_\theta - s_\Omega c_i c_\theta)\ddot{x} + (-s_\Omega s_\theta + c_\Omega c_i c_\theta)\ddot{y} + s_i c_\theta \ddot{z} \\ &= -\frac{5\mu J_4 R^4}{2r^6}\left(7s_i^4 s_\theta^3 c_\theta - 3s_i^2 s_\theta c_\theta\right)\end{aligned} \tag{A13.11}$$

with $s_\theta^3 = \frac{1}{4}(3s_\theta - s_{3\theta})$

$$\begin{aligned}(f_h)_{J_4} &= s_\Omega s_i \ddot{x} - c_\Omega s_i \ddot{y} + c_i \ddot{z} \\ &= -\frac{5\mu J_4 R^4}{2r^6}\left(7s_i^3 c_i s_\theta^3 - 3s_i c_i s_\theta\right)\end{aligned} \tag{A13.12}$$

These expressions are equivalent to those in Equations (13.11)–(13.16). The inertial accelerations namely $\ddot{x}$, $\ddot{y}$, $\ddot{z}$, for both the J_3 and J_4 terms can be found in [15]. However, the J_2 terms in Equations (9.7-4) and (9.7-6) of that reference have a typographical error as they should read $J_2 \frac{3}{2}\left(\frac{r_e}{r}\right)^2\left(5\frac{z^2}{r^2} - 1\right)$ and $J_2 \frac{3}{2}\left(\frac{r_e}{r}\right)^2\left(3 - 5\frac{z^2}{r^2}\right)$ respectively. The symbol r_e is used as R in this chapter.

References

[1] Broucke, R. A. and Cefola, P. J. (1972) On the Equinoctial Orbit Elements, *Celestial Mechanics*, Vol. 5, 303–310.

[2] Cefola, P. J. (1972) *Equinoctial Orbit Elements: Application to Artificial Satellite Orbits*, AIAA Paper 72–937, AIAA/AAS Astrodynamics Conference, Palo Alto, CA.

[3] Edelbaum, T. N., Sackett, L. L. and Malchow, H. L. (1973) *Optimal Low Thrust Geocentric Transfer*, AIAA Paper 73–1074, AIAA 10th Electric Propulsion Conference, Lake Tahoe, NV.

[4] Betts, J. T. (1994) Optimal Interplanetary Orbit Transfers by Direct Transcription, *The Journal of the Astronautical Sciences*, Vol. 42, No. 3, 247–268.

[5] Walker, M. J. H., Ireland, B. and Owens, J. (1985) A Set of Modified Equinoctial Orbital Elements, *Celestial Mechanics*, Vol. 36, 409–419.

[6] Battin, R. H. (1987) *An Introduction to the Mathematics and Methods of Astrodynamics*, AIAA, New York, NY, 490–494.

[7] Kéchichian, J. A. (1997) Trajectory Optimization Using Nonsingular Orbital Elements and True Longitude, *Journal of Guidance, Control, and Dynamics*, Vol. 20, No. 5, 1003–1009.

[8] Kéchichian, J. A. (1997) The Treatment of the Earth Oblateness Effect in Trajectory Optimization in Equinoctial Coordinates, *Acta Astronautica*, Vol. 40, No. 1, 69–82.

[9] Kéchichian, J. A. (1998) Trajectory Optimization Using Eccentric Longitude Formulation, *Journal of Spacecraft and Rockets*, Vol. 35, No. 3, 317–326.

[10] Kéchichian, J. A. (1999) Minimum-Time Low-Thrust Rendezvous and Transfer Using Epoch Mean Longitude Formulation, *Journal of Guidance, Control, and Dynamics*, Vol. 22, No. 3, 421–432.

[11] Kéchichian, J. A. (2000) Minimum-Time Constant Acceleration Orbit Transfer with First-Order Oblateness Effect, *Journal of Guidance, Control, and Dynamics*, Vol. 23, No. 4, 595–603.

[12] Kéchichian, J. A. (2007) *The Streamlined and Complete Set of the Nonsingular J_2-Perturbed Dynamic and Adjoint Equations for Trajectory Optimization in Terms of Eccentric Longitude*, AAS Paper 07-120, AAS/AIAA Space Flight Mechanics Meeting, Sedona, AZ.

[13] Kéchichian, J. A. (2007) The Streamlined and Complete Set of the Nonsingular J_2-Perturbed Dynamic and Adjoint Equations for Trajectory Optimization in Terms of Eccentric Longitude, *The Journal of the Astronautical Sciences*, Vol. 55, No. 3, 325–348.

[14] Feistel, B. S. (2007) *Perturbing Accelerations in Low Thrust Trajectory Optimization*, M.S. Thesis, University of Texas at Austin.

[15] Vallado, D. A. (2004) *Fundamentals of Astrodynamics and Applications*, 2nd ed., Microcosm Press, El Segundo.

14 Analytic Expansions of Luni-Solar Gravity Perturbations Along Rotating Axes for Trajectory Optimization: The Dynamic System

14.1 Introduction

An analytic form of the accelerations due to the luni-solar perturbations resolved along the rotating Euler–Hill frame is devised by using the expansion method. The addition of higher-order terms to the main gravity gradient term linear in the spacecraft radial distance, carried out to the third order, provides a very high level of accuracy in accounting for the gravity perturbations experienced by a vehicle in orbit due to the Sun and the Moon. The nodal precession as well as the perigee advance of the lunar orbit are taken into account analytically by using the analytic lunar theory of de Pontécoulant. The analytic description of the apparent solar orbit and the motion of the Moon remove the need to call an ephermeris generator at each integration step during the numerical integration of the spacecraft trajectory, leading to the self-contained software for rapid and efficient optimal trajectory generation through iterations. Equinoctial elements are used to describe the spacecraft state and the luni-solar accelerations are given in terms of the apparent solar and lunar longitudes, as well as the Eulerian angles of the spacecraft orbit with respect to the inertial ecliptic system. The analysis is useful in optimal low-thrust orbit transfers, complementing previous analyses carried out in the preceding chapters in which thrust and Earth zonal perturbations such as J_2, J_3 and J_4 in terms of the nonsingular equinoctial elements are included.

Following the resolution of both the thrust and zonal harmonics perturbation accelerations along rotating axes for use in high fidelity optimal trajectory generation within the context of precision integration, this effort considers the perturbative effects due to luni-solar gravity by resolving also these perturbation accelerations along the same rotating axes for further improvements in accuracy. Recent works [1, 2], also shown in Chapters 9 and 13, and [3] show how to factor in some or all of the above-mentioned perturbations, with [4] providing a method that also encompasses the luni-solar gravity, based on a method of explicit calculation of the disturbing acceleration. The second-order vector differential equation for the motion of the spacecraft involves a forcing term given by the partial derivative of a scalar quantity named the disturbing function with respect to the spacecraft position vector. This perturbation associated with each disturbing body is expressed explicitly in [4] without expressing the disturbing function as a power series in the ratio of the radial distance to the Earth–third-body distance.

However, a third-body ephemeris-generating software is needed to call the Earth–third-body position vector in an inertial space at each integration step. This work uses an expansion version of the disturbing function instead in order to express the perturbation acceleration due to a third body in terms of its longitude along the ecliptic and its radial distance to the Earth center, as well as the spacecraft orbit Eulerian angles in the inertial ecliptic system for the Sun, and the longitude along the lunar orbit, its radial distance to the Earth center and the spacecraft Eulerian angles in the Moon orbit system for the Moon. Thus the longitudes of the Sun and Moon can be described analytically using the de Pontécoulant lunar theory, in which the moon's orbit node is allowed to precess and its perigee to advance due to solar gravity itself. The expansions are carried out to the third order beyond the well-known linear gravity gradient terms in the spacecraft radial distance in order to obtain a very high level of accuracy due to the rapid convergence characteristics of the power series. A cruder attempt at resolving the perturbation accelerations keeping only the linear gravity gradient terms was made in [5] based on a method shown by Breakwell in [6], adequate for short-term propagations in which the sun's and moon's orbits are held fixed without accounting for the variations in the lunar orbit itself. It was then further possible to solve analytically for the perturbed relative motion in the particular but important near-circular case [5]. However, in the more general elliptic orbit case of this work, the equations of motion are integrated numerically using a set of nonsingular orbit elements for more general applications. The next sections show how the luni-solar gravity accelerations are derived and resolved along the rotating axes. The last section dwells on the details of de Pontécoulant's lunar theory and its use in computing certain parameters such as longitude, distance and Eulerian angles needed in the evaluation of the luni-solar accelerations.

14.2 Solar Gravity Perturbation Acceleration Components in the Euler–Hill Frame

The inertial equatorial frame centered at the Earth is shown in Figure 14.1 as $(\hat{\mathbf{x}}\,\hat{\mathbf{y}}\,\hat{\mathbf{z}})$ with $\hat{\mathbf{x}}$ and $\hat{\mathbf{y}}$ in the equatorial frame and with $\hat{\mathbf{x}}$ pointing toward the vernal equinox γ, and $\hat{\mathbf{z}}$ along the north pole. Also depicted in the figure is the spacecraft orbit and the rotating frame $(\hat{\mathbf{r}}, \hat{\boldsymbol{\theta}}, \hat{\mathbf{h}})$ attached to the spacecraft at O', with Eulerian angles Ω, i, ω, true anomaly θ^* and declination δ. The perigee is at P and the unit vectors $\hat{\mathbf{r}}$ and $\hat{\boldsymbol{\theta}}$ are such that $\hat{\mathbf{r}}$ is along the radius $\mathbf{r}$, and $\hat{\boldsymbol{\theta}}$ in the instantaneous orbit plane and in the direction of motion. The sun's apparent orbit is contained in the ecliptic plane defined by $(\hat{\mathbf{x}}_e\,\hat{\mathbf{y}}_e)$ with $\hat{\mathbf{x}}_e$ along $\hat{\mathbf{x}}$, and this plane is inclined with respect to the equator by the angle $\varepsilon = i_s$, also called the obliquity of the ecliptic. Letting $\mathbf{r}_\odot$ represent the radius vector of the Sun measured from the Earth center O and letting ϕ_o designate the sun's latitude with respect to the equatorial plane, the sun's position is defined by the angle $\theta_s = \omega_s + \theta_s^*$, with P_s representing the perigee of the sun's apparent orbit, as shown in Figure 14.1.

In Figure 14.2 the spacecraft orbit is defined in the inertial ecliptic system by the angles Ω', i', ω' and θ^*, and the sun's longitude measured from $\hat{\mathbf{x}}_e$ is given by $\lambda_\odot$, while

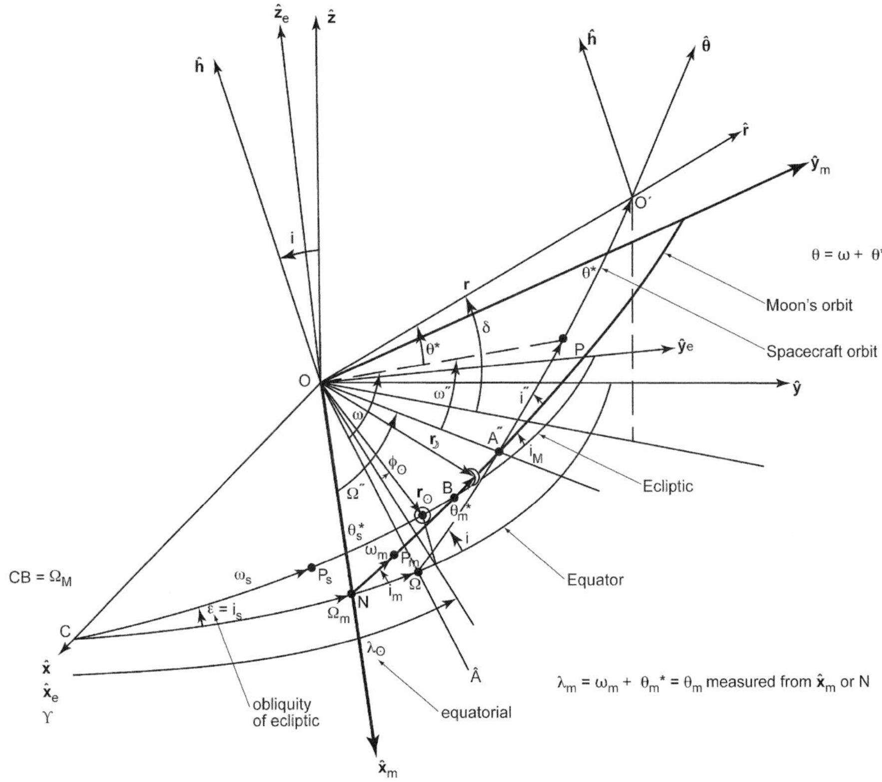

Figure 14.1 Rotating and inertial coordinate frames.

$\mathbf{r}_\odot - \mathbf{r}$ is the vector from the spacecraft to the Sun. The total acceleration due to the solar gravity is given by

$$\mathbf{F} = \nabla \left(\frac{\mu_\odot}{|\mathbf{r}_\odot - \mathbf{r}|} \right) \tag{14.1}$$

with $\mu_\odot$ the gravity constant of the Sun. The denominator given below is then expanded in terms of the Legendre polynomials in $c_\phi = \cos\phi$ for short, with $c_\phi = \hat{\mathbf{r}} \cdot \hat{\mathbf{r}}_\odot = \frac{\mathbf{r}}{r} \cdot \hat{\mathbf{r}}_\odot$ such that with $x' = r/r_\odot$, a small parameter,

$$|\mathbf{r}_\odot - \mathbf{r}| = \left(r_\odot^2 - 2\mathbf{r}_\odot \mathbf{r} + r^2 \right)^{1/2} = r_\odot \left(1 - 2\frac{\hat{\mathbf{r}}_\odot \mathbf{r}}{r_\odot} + \frac{r^2}{r_\odot^2} \right)^{1/2} \tag{14.2}$$

$$\begin{aligned} |\mathbf{r}_\odot - \mathbf{r}|^{-1} &= r_\odot^{-1} \left(1 - 2\frac{r}{r_\odot} c_\phi + \frac{r^{\,2}}{r_\odot^2} \right)^{-1/2} = r_\odot^{-1} \left(1 - 2x' c_\phi + x'^2 \right)^{-1/2} \\ &= r_\odot^{-1} \left[1 + x' c_\phi + x'^2 \left(\frac{3}{2} c_\phi^2 - \frac{1}{2} \right) + x'^3 \left(\frac{5}{2} c_\phi^3 - \frac{3}{2} c_\phi \right) \right. \\ &\quad \left. + x'^4 \left(\frac{35}{8} c_\phi^4 - \frac{30}{8} c_\phi^2 + \frac{3}{8} \right) + \ldots \right] \end{aligned} \tag{14.3}$$

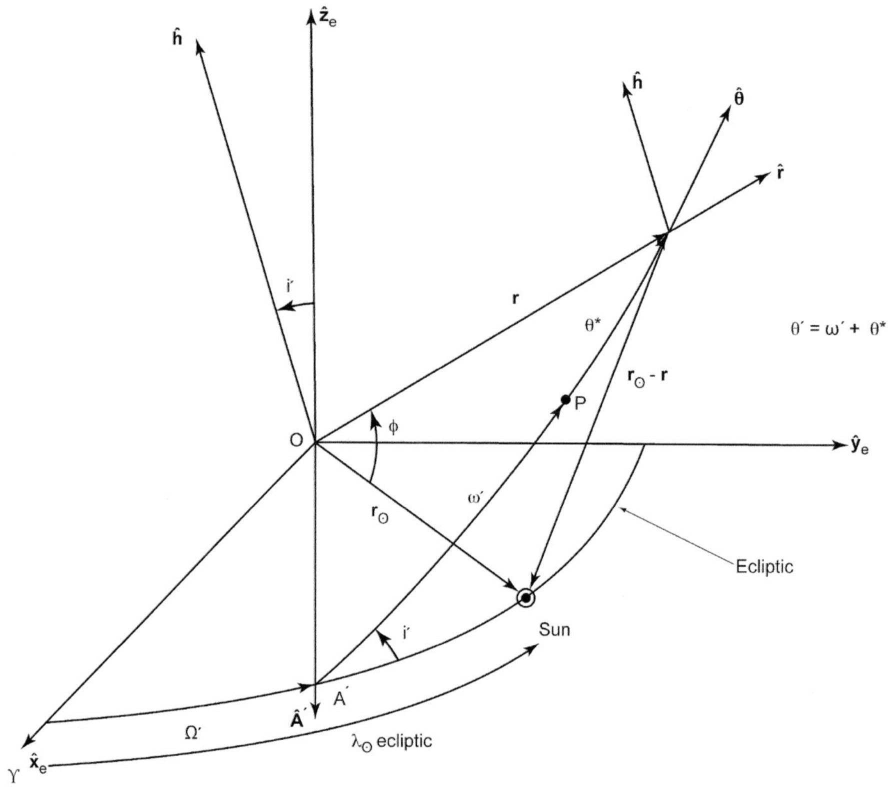

Figure 14.2 Relative geometry of rotating and ecliptic frames.

The vector function $\mathbf{F}$ is then reduced to

$$
\begin{aligned}
\mathbf{F} = \nabla \Bigg\{ & \frac{\mu_\odot}{r_\odot} + \frac{\mu_\odot}{r_\odot}\frac{(\mathbf{r}\,\hat{\mathbf{r}}_\odot)}{r_\odot} + \frac{\mu_\odot}{r_\odot}\left[\frac{3}{2}\frac{(\mathbf{r}\,\hat{\mathbf{r}}_\odot)^2}{r_\odot^2} - \frac{1}{2}\frac{r^2}{r_\odot^2}\right] \\
& + \frac{\mu_\odot}{r_\odot}\left[\frac{5}{2}\frac{(\mathbf{r}\,\hat{\mathbf{r}}_\odot)^3}{r_\odot^3} - \frac{3}{2}\frac{(\mathbf{r}\,\hat{\mathbf{r}}_\odot)}{r_\odot^3}r^2\right] \\
& + \frac{\mu_\odot}{r_\odot}\left[\frac{35}{8}\frac{(\mathbf{r}\,\hat{\mathbf{r}}_\odot)^4}{r_\odot^4} - \frac{30}{8}\frac{(\mathbf{r}\,\hat{\mathbf{r}}_\odot)^2}{r_\odot^4}r^2 + \frac{3}{8}\frac{r^4}{r_\odot^4}\right] + \cdots \Bigg\}
\end{aligned} \tag{14.4}
$$

The third-body gravity perturbation acceleration due to the Sun is then obtained from

$$
\begin{aligned}
\mathbf{f}_\odot = \mathbf{F} - \frac{\mu_\odot}{r_\odot^2}\hat{\mathbf{r}}_\odot = & \frac{\mu_\odot}{r_\odot^3}\left[3(\mathbf{r}\,\hat{\mathbf{r}}_\odot)\hat{\mathbf{r}}_\odot - \frac{1}{2}(\mathbf{r} + \mathbf{r})\right] \\
& + \frac{\mu_\odot}{r_\odot^4}\left\{\frac{5}{2}3(\mathbf{r}\,\hat{\mathbf{r}}_\odot)^2\hat{\mathbf{r}}_\odot - \frac{3}{2}\left[\hat{\mathbf{r}}_\odot r^2 + (\mathbf{r}\,\hat{\mathbf{r}}_\odot)2\mathbf{r}\right]\right\} \\
& + \frac{\mu_\odot}{r_\odot^5}\left[\frac{35}{8}4(\mathbf{r}\,\hat{\mathbf{r}}_\odot)^3\hat{\mathbf{r}}_\odot - \frac{30}{8}2(\mathbf{r}\,\hat{\mathbf{r}}_\odot)\hat{\mathbf{r}}_\odot r^2\right. \\
& \left. - \frac{30}{8}(\mathbf{r}\,\hat{\mathbf{r}}_\odot)^2(2\mathbf{r}) + \frac{3}{8}2(\mathbf{r}\,\mathbf{r})(2\mathbf{r})\right] + \cdots
\end{aligned}
$$

which can be further reduced to the form

$$\begin{aligned}\mathbf{f}_\odot = \frac{\mu_\odot r}{r_\odot^3}\left[3\left(\hat{\mathbf{r}}\,\hat{\mathbf{r}}_\odot\right)\hat{\mathbf{r}}_\odot - \hat{\mathbf{r}}\right] + \frac{\mu_\odot r^2}{r_\odot^4}\left\{\frac{15}{2}\left(\hat{\mathbf{r}}\,\hat{\mathbf{r}}_\odot\right)^2\hat{\mathbf{r}}_\odot\right.\\ \left.-\frac{3}{2}\left[\hat{\mathbf{r}}_\odot + 2\left(\hat{\mathbf{r}}\,\hat{\mathbf{r}}_\odot\right)\hat{\mathbf{r}}\right]\right\} + \frac{\mu_\odot}{2r_\odot^5}r^3\left[35\left(\hat{\mathbf{r}}\,\hat{\mathbf{r}}_\odot\right)^3\hat{\mathbf{r}}_\odot - 15\left(\hat{\mathbf{r}}\,\hat{\mathbf{r}}_\odot\right)\hat{\mathbf{r}}_\odot\right.\\ \left.-15\left(\hat{\mathbf{r}}\,\hat{\mathbf{r}}_\odot\right)^2\hat{\mathbf{r}} + 3\hat{\mathbf{r}}\right] + \cdots\end{aligned} \tag{14.5}$$

The first term in Equation (14.5) represents the well-known gravity gradient term linear in the spacecraft radial distance $\mathbf{r}$, while the second and third terms stand, respectively, for the quadratic and cubic contributions.

The acceleration components along the rotating axes $\hat{\mathbf{r}}$, $\hat{\boldsymbol{\theta}}$ and $\hat{\mathbf{h}}$ are now obtained by forming the respective dot products such that $f_r = \mathbf{f}_\odot \cdot \hat{\mathbf{r}}$, $f_\theta = \mathbf{f}_\odot \cdot \hat{\boldsymbol{\theta}}$ and $f_h = \mathbf{f}_\odot \cdot \hat{\mathbf{h}}$ leading to the following expressions with $n_\odot^2 = \mu_\odot / r_\odot^3$:

$$\begin{aligned}f_r = n_\odot^2 r\left[3\left(\hat{\mathbf{r}}\,\hat{\mathbf{r}}_\odot\right)\left(\hat{\mathbf{r}}_\odot\,\hat{\mathbf{r}}\right) - \left(\hat{\mathbf{r}}\,\hat{\mathbf{r}}\right)\right] + \frac{n_\odot^2}{r_\odot}r^2\left\{\frac{15}{2}\left(\hat{\mathbf{r}}\hat{\mathbf{r}}_\odot\right)^2\left(\hat{\mathbf{r}}_\odot\hat{\mathbf{r}}\right)\right.\\ \left.-\frac{3}{2}\left[\left(\hat{\mathbf{r}}_\odot\,\hat{\mathbf{r}}\right) + 2\left(\hat{\mathbf{r}}\,\hat{\mathbf{r}}_\odot\right)\left(\hat{\mathbf{r}}\,\hat{\mathbf{r}}\right)\right]\right\} + \frac{n_\odot^2}{2r_\odot^2}r^3\left[35\left(\hat{\mathbf{r}}\,\hat{\mathbf{r}}_\odot\right)^3\left(\hat{\mathbf{r}}_\odot\hat{\mathbf{r}}\right)\right.\\ \left.-15\left(\hat{\mathbf{r}}\,\hat{\mathbf{r}}_\odot\right)\left(\hat{\mathbf{r}}_\odot\,\hat{\mathbf{r}}\right) - 15\left(\hat{\mathbf{r}}\,\hat{\mathbf{r}}_\odot\right)^2\left(\hat{\mathbf{r}}\,\hat{\mathbf{r}}\right) + 3\left(\hat{\mathbf{r}}\,\hat{\mathbf{r}}\right)\right]\end{aligned} \tag{14.6}$$

$$\begin{aligned}f_\theta = n_\odot^2 r\left[3\left(\hat{\mathbf{r}}\,\hat{\mathbf{r}}_\odot\right)\left(\hat{\mathbf{r}}_\odot\,\hat{\boldsymbol{\theta}}\right) - \left(\hat{\mathbf{r}}\,\hat{\boldsymbol{\theta}}\right)\right] + \frac{n_\odot^2}{r_\odot}r^2\left\{\frac{15}{2}\left(\hat{\mathbf{r}}\,\hat{\mathbf{r}}_\odot\right)^2\left(\hat{\mathbf{r}}_\odot\hat{\boldsymbol{\theta}}\right)\right.\\ \left.-\frac{3}{2}\left[\left(\hat{\mathbf{r}}_\odot\,\hat{\boldsymbol{\theta}}\right) + 2\left(\hat{\mathbf{r}}\,\hat{\mathbf{r}}_\odot\right)\left(\hat{\mathbf{r}}\,\hat{\boldsymbol{\theta}}\right)\right]\right\} + \frac{n_\odot^2}{2r_\odot^2}r^3\left[35\left(\hat{\mathbf{r}}\,\hat{\mathbf{r}}_\odot\right)^3\left(\hat{\mathbf{r}}_\odot\hat{\boldsymbol{\theta}}\right)\right.\\ \left.-15\left(\hat{\mathbf{r}}\,\hat{\mathbf{r}}_\odot\right)\left(\hat{\mathbf{r}}_\odot\,\hat{\boldsymbol{\theta}}\right) - 15\left(\hat{\mathbf{r}}\,\hat{\mathbf{r}}_\odot\right)^2\left(\hat{\mathbf{r}}\,\hat{\boldsymbol{\theta}}\right) + 3\left(\hat{\mathbf{r}}\,\hat{\boldsymbol{\theta}}\right)\right]\end{aligned} \tag{14.7}$$

$$\begin{aligned}f_h = n_\odot^2 r\left[3\left(\hat{\mathbf{r}}\,\hat{\mathbf{r}}_\odot\right)\left(\hat{\mathbf{r}}_\odot\,\hat{\mathbf{h}}\right) - \left(\hat{\mathbf{r}}\,\hat{\mathbf{h}}\right)\right] + \frac{n_\odot^2}{r_\odot}r^2\left\{\frac{15}{2}\left(\hat{\mathbf{r}}\,\hat{\mathbf{r}}_\odot\right)^2\left(\hat{\mathbf{r}}_\odot\hat{\mathbf{h}}\right)\right.\\ \left.-\frac{3}{2}\left[2\left(\hat{\mathbf{r}}\,\hat{\mathbf{r}}_\odot\right)\left(\hat{\mathbf{r}}\,\hat{\mathbf{h}}\right) + \left(\hat{\mathbf{r}}_\odot\,\hat{\mathbf{h}}\right)\right]\right\} + \frac{n_\odot^2}{2r_\odot^2}r^3\left[35\left(\hat{\mathbf{r}}\,\hat{\mathbf{r}}_\odot\right)^3\left(\hat{\mathbf{r}}_\odot\hat{\mathbf{h}}\right)\right.\\ \left.-15\left(\hat{\mathbf{r}}\,\hat{\mathbf{r}}_\odot\right)\left(\hat{\mathbf{r}}_\odot\,\hat{\mathbf{h}}\right) - 15\left(\hat{\mathbf{r}}\,\hat{\mathbf{r}}_\odot\right)^2\left(\hat{\mathbf{r}}\,\hat{\mathbf{h}}\right) + 3\left(\hat{\mathbf{r}}\,\hat{\mathbf{h}}\right)\right]\end{aligned} \tag{14.8}$$

Of course, $(\hat{\mathbf{r}}\,\hat{\boldsymbol{\theta}}) = 0$ and $(\hat{\mathbf{r}}\,\hat{\mathbf{h}}) = 0$ leading to some simplifying of the above three expressions. Now $\hat{\mathbf{r}}_\odot$ can be written as $\hat{\mathbf{r}}_\odot = c_{\lambda_\odot}\hat{\mathbf{x}}_e + s_{\lambda_\odot}\hat{\mathbf{y}}_e$ in terms of the ecliptic longitude of the Sun, and in view of three Eulerian rotations involving Ω', i' and $\theta' = \omega' + \theta^*$, that is, which carry out the transformation from the rotating system to the inertial system, namely

$$\hat{\mathbf{x}}_e = \left(c_{\Omega'}c_{\theta'} - s_{\Omega'}c_{i'}s_{\theta'}\right)\hat{\mathbf{r}} + \left(-c_{\Omega'}s_{\theta'} - s_{\Omega'}c_{i'}c_{\theta'}\right)\hat{\boldsymbol{\theta}} + s_{\Omega'}s_{i'}\hat{\mathbf{h}} \tag{14.9}$$

$$\hat{\mathbf{y}}_e = \left(s_{\Omega'}c_{\theta'} + c_{\Omega'}c_{i'}s_{\theta'}\right)\hat{\mathbf{r}} + \left(-s_{\Omega'}s_{\theta'} + c_{\Omega'}c_{i'}c_{\theta'}\right)\hat{\boldsymbol{\theta}} - c_{\Omega'}s_{i'}\hat{\mathbf{h}} \tag{14.10}$$

$$\hat{\mathbf{z}}_e = s_{i'}s_{\theta'}\hat{\mathbf{r}} + s_{i'}c_{\theta'}\hat{\boldsymbol{\theta}} + c_{i'}\hat{\mathbf{h}} \tag{14.11}$$

The unit vector $\hat{\mathbf{r}}_\odot$ can be cast into the following form involving $\lambda_\odot$ and Eulerian angles:

$$\hat{\mathbf{r}}_\odot = \left(c_{\lambda_\odot-\Omega'}c_{\theta'} + c_{i'}s_{\theta'}s_{\lambda_\odot-\Omega'}\right)\hat{\mathbf{r}} + \left(-s_{\theta'}c_{\lambda_\odot-\Omega'} + c_{i'}c_{\theta'}s_{\lambda_\odot-\Omega'}\right)\hat{\boldsymbol{\theta}} - s_{i'}s_{\lambda_\odot-\Omega'}\hat{\mathbf{h}} \tag{14.12}$$

Therefore, the nonzero dot products in f_r, f_θ and f_h can be reduced to the following forms:

$$\hat{\mathbf{r}}_\odot\hat{\mathbf{r}} = c_{\lambda_\odot-\Omega'}c_{\theta'} + c_{i'}s_{\theta'}s_{\lambda_\odot-\Omega'} \tag{14.13}$$

$$\hat{\mathbf{r}}_\odot\hat{\boldsymbol{\theta}} = -s_{\theta'}c_{\lambda_\odot-\Omega'} + c_{i'}c_{\theta'}s_{\lambda_\odot-\Omega'} \tag{14.14}$$

$$\hat{\mathbf{r}}_\odot\hat{\mathbf{h}} = -s_{i'}s_{\lambda_\odot-\Omega'} \tag{14.15}$$

which in turn reduce the acceleration components in Equations (14.6)–(14.8) to the following expressions, which include terms of order up to three, as before. Thus $f_r = f_{r_1} + f_{r_2} + f_{r_3}, f_\theta = f_{\theta_1} + f_{\theta_2} + f_{\theta_3}$ and $f_h = f_{h_1} + f_{h_2} + f_{h_3}$.

$$\begin{aligned}
f_r = {} & rn_\odot^2\left[3\left(c^2_{\lambda_\odot-\Omega'}c^2_{\theta'} + c^2_{i'}s^2_{\theta'}s^2_{\lambda_\odot-\Omega'} + 2c_{i'}s_{\theta'}c_{\theta'}s_{\lambda_\odot-\Omega'}c_{\lambda_\odot-\Omega'}\right) - 1\right] \\
& + \frac{3}{2}r^2\frac{n_\odot^2}{r_\odot}\left[5\left(c^3_{\lambda_\odot-\Omega'}c^3_{\theta'} + c^3_{i'}s^3_{\theta'}s^3_{\lambda_\odot-\Omega'} + 3c^2_{\lambda_\odot-\Omega'}c^2_{\theta'}c_{i'}s_{\theta'}s_{\lambda_\odot-\Omega'}\right.\right. \\
& \left.\left. + 3c^2_{i'}s^2_{\theta'}c_{\theta'}s^2_{\lambda_\odot-\Omega'}c_{\lambda_\odot-\Omega'}\right) - 3\left(c_{\lambda_\odot-\Omega'}c_{\theta'} + c_{i'}s_{\theta'}s_{\lambda_\odot-\Omega'}\right)\right] \\
& + \frac{r^3}{2}\frac{n_\odot^2}{r_\odot^2}\left[35\left(c^4_{\theta'}c^4_{\lambda_\odot-\Omega'} + 4c_{i'}c^3_{\theta'}s_{\theta'}s_{\lambda_\odot-\Omega'}c^3_{\lambda_\odot-\Omega'}\right.\right. \\
& \left. + 4c^2_{i'}s^2_{\theta'}c^2_{\theta'}s^2_{\lambda_\odot-\Omega'}c^2_{\lambda_\odot-\Omega'} + c^4_{i'}s^4_{\theta'}s^4_{\lambda_\odot-\Omega'} + 4c^3_{i'}s^3_{\theta'}c_{\theta'}s^3_{\lambda_\odot-\Omega'}c_{\lambda_\odot-\Omega'}\right) \\
& \left. -30\left(c^2_{\lambda_\odot-\Omega'}c^2_{\theta'} + c^2_{i'}s^2_{\theta'}s^2_{\lambda_\odot-\Omega'} + 2c_{i'}s_{\theta'}c_{\theta'}s_{\lambda_\odot-\Omega'}c_{\lambda_\odot-\Omega'}\right) + 3\right]
\end{aligned} \tag{14.16}$$

$$\begin{aligned}
f_\theta = {} & 3rn_\odot^2\left[-s_{\theta'}c_{\theta'}c^2_{\lambda_\odot-\Omega'} - c_{i'}s^2_{\theta'}s_{\lambda_\odot-\Omega'}c_{\lambda_\odot-\Omega'} + c_{i'}c^2_{\theta'}s_{\lambda_\odot-\Omega'}c_{\lambda_\odot-\Omega'}\right. \\
& \left. + c^2_{i'}s_{\theta'}c_{\theta'}s^2_{\lambda_\odot-\Omega'}\right] + \frac{3}{2}r^2\frac{n_\odot^2}{r_\odot}\left[5\left(-s_{\theta'}c^2_{\theta'}c^3_{\lambda_\odot-\Omega'} + c_{i'}c^3_{\theta'}s_{\lambda_\odot-\Omega'}\,c^2_{\lambda_\odot-\Omega'}\right.\right. \\
& - c^2_{i'}s^3_{\theta'}c_{\lambda_\odot-\Omega'}s^2_{\lambda_\odot-\Omega'} + c^3_{i'}c_{\theta'}s^2_{\theta'}s^3_{\lambda_\odot-\Omega'} - 2c_{i'}s^2_{\theta'}c_{\theta'}s_{\lambda_\odot-\Omega'}c^2_{\lambda_\odot-\Omega'} \\
& \left.\left. + 2c^2_{i'}s_{\theta'}c^2_{\theta'}s^2_{\lambda_\odot-\Omega'}c_{\lambda_\odot-\Omega'}\right) - \left(-s_{\theta'}c_{\lambda_\odot-\Omega'} + c_{i'}c_{\theta'}s_{\lambda_\odot-\Omega'}\right)\right] + \frac{r^3}{2}\frac{n_\odot^2}{r_\odot^2} \\
& \times\left[35\left(-s_{\theta'}c^3_{\theta'}c^4_{\lambda_\odot-\Omega'} - s^4_{\theta'}c^3_{i'}c_{\lambda_\odot-\Omega'}s^3_{\lambda_\odot-\Omega'} - 3c_{i'}s^2_{\theta'}c^2_{\theta'}c^3_{\lambda_\odot-\Omega'}s_{\lambda_\odot-\Omega'}\right.\right. \\
& - 3c^2_{i'}s^3_{\theta'}c_{\theta'}s^2_{\lambda_\odot-\Omega'}c^2_{\lambda_\odot-\Omega'} + c_{i'}c^4_{\theta'}s_{\lambda_\odot-\Omega'}c^3_{\lambda_\odot-\Omega'} + c^4_{i'}c_{\theta'}s^3_{\theta'}s^4_{\lambda_\odot-\Omega'} \\
& \left. + 3c^2_{i'}c^3_{\theta'}s_{\theta'}c^2_{\lambda_\odot-\Omega'}s^2_{\lambda_\odot-\Omega'} + 3c^3_{i'}s^2_{\theta'}c^2_{\theta'}s^3_{\lambda_\odot-\Omega'}c_{\lambda_\odot-\Omega'}\right) - 15\left(-s_{\theta'}c_{\theta'}c^2_{\lambda_\odot-\Omega'}\right. \\
& \left.\left. - c_{i'}s^2_{\theta'}s_{\lambda_\odot-\Omega'}c_{\lambda_\odot-\Omega'} + c_{i'}c^2_{\theta'}s_{\lambda_\odot-\Omega'}c_{\lambda_\odot-\Omega'} + c^2_{i'}s_{\theta'}c_{\theta'}s^2_{\lambda_\odot-\Omega'}\right)\right]
\end{aligned} \tag{14.17}$$

$$\begin{aligned}
f_h = {} & -3rn_\odot^2\left[s_{i'}c_{\theta'}s_{\lambda_\odot-\Omega'}c_{\lambda_\odot-\Omega'} + s_{i'}c_{i'}s_{\theta'}s^2_{\lambda_\odot-\Omega'}\right] \\
& - \frac{3}{2}r^2\frac{n_\odot^2}{r_\odot}\left[5s_{i'}c^2_{\theta'}s_{\lambda_\odot-\Omega'}c^2_{\lambda_\odot-\Omega'} + 5s_{i'}c^2_{i'}s^2_{\theta'}s^3_{\lambda_\odot-\Omega'}\right.
\end{aligned}$$

$$
\begin{aligned}
&+10 s_{i'} c_{i'} s_{\theta'} c_{\theta'} s^2_{\lambda_\odot - \Omega'} c_{\lambda_\odot - \Omega'} - s_{i'} s_{\lambda_\odot - \Omega'} \Big] \\
&+ \frac{5}{2} \frac{r^3 n_\odot^2}{r_\odot^2} \Big[-7 s_{i'} c^3_{\theta'} s_{\lambda_\odot - \Omega'} c^3_{\lambda_\odot - \Omega'} - 7 s_{i'} c^3_{i'} s^3_{\theta'} s^4_{\lambda_\odot - \Omega'} \\
&- 21 s_{i'} c_{i'} c^2_{\theta'} s_{\theta'} s^2_{\lambda_\odot - \Omega'} c^2_{\lambda_\odot - \Omega'} - 21 s_{i'} c^2_{i'} s^2_{\theta'} c_{\theta'} s^3_{\lambda_\odot - \Omega'} c_{\lambda_\odot - \Omega'} \\
&+ 3 s_{i'} c_{\theta'} s_{\lambda_\odot - \Omega'} c_{\lambda_\odot - \Omega'} + 3 s_{i'} c_{i'} s_{\theta'} s^2_{\lambda_\odot - \Omega'} \Big]
\end{aligned} \tag{14.18}
$$

The angles Ω', i' and ω' are of course related to the equinoctial elements a, h, k, p, q and L through the angles Ω, i and ω of the spacecraft orbit. The numerical integration of the differential equations for the equinoctial elements will provide the angles Ω, i and ω referred to the equatorial system, and then Ω', i' and ω' are obtained from the nine vectorial orbital constants written in the ecliptic and equatorial systems. Because $\hat{\mathbf{x}}$, $\hat{\mathbf{y}}$, $\hat{\mathbf{z}}$ may be obtained by a rotation of the $\hat{\mathbf{x}}_e$, $\hat{\mathbf{y}}_e$, $\hat{\mathbf{z}}_e$ system about the $\hat{\mathbf{x}}_e$ axis, clockwise by the angle ε through

$$\hat{\mathbf{x}} = \hat{\mathbf{x}}_e \tag{14.19}$$

$$\hat{\mathbf{y}} = \hat{\mathbf{y}}_e c_\varepsilon - \hat{\mathbf{z}}_e s_\varepsilon \tag{14.20}$$

$$\hat{\mathbf{z}} = \hat{\mathbf{y}}_e s_\varepsilon + \hat{\mathbf{z}}_e c_\varepsilon \tag{14.21}$$

the nine direction cosines of the axes $\hat{\mathbf{x}}$, $\hat{\mathbf{y}}$, $\hat{\mathbf{z}}$ relative to $\hat{\mathbf{x}}_e$, $\hat{\mathbf{y}}_e$, $\hat{\mathbf{z}}_e$ will be given by [7]

$$P_x^{(e)} = c_{\omega'} c_{\Omega'} - s_{\omega'} s_{\Omega'} c_{i'} \tag{14.22}$$

$$Q_x^{(e)} = -s_{\omega'} c_{\Omega'} - c_{\omega'} s_{\Omega'} c_{i'} \tag{14.23}$$

$$R_x^{(e)} = s_{\Omega'} s_{i'} \tag{14.24}$$

$$P_y^{(e)} = c_{\omega'} s_{\Omega'} + s_{\omega'} c_{\Omega'} c_{i'} \tag{14.25}$$

$$Q_y^{(e)} = -s_{\omega'} s_{\Omega'} + c_{\omega'} c_{\Omega'} c_{i'} \tag{14.26}$$

$$R_y^{(e)} = -c_{\Omega'} s_{i'} \tag{14.27}$$

$$P_z^{(e)} = s_{\omega'} s_{i'} \tag{14.28}$$

$$Q_z^{(e)} = c_{\omega'} s_{i'} \tag{14.29}$$

$$R_z^{(e)} = c_{i'} \tag{14.30}$$

The vectorial orbital constants for the equator may then be obtained from the $P_x^{(e)}$, $Q_x^{(e)}$, $R_x^{(e)}$, ... expressions [7]:

$$P_x = P_x^{(e)} \tag{14.31}$$

$$Q_x = Q_x^{(e)} \tag{14.32}$$

$$R_x = R_x^{(e)} \tag{14.33}$$

$$P_y = P_y^{(e)} c_\varepsilon - P_z^{(e)} s_\varepsilon \tag{14.34}$$

$$Q_y = Q_y^{(e)} c_\varepsilon - Q_z^{(e)} s_\varepsilon \tag{14.35}$$

$$R_y = R_y^{(e)} c_\varepsilon - R_z^{(e)} s_\varepsilon \tag{14.36}$$

$$P_z = P_y^{(e)} s_\varepsilon + P_z^{(e)} c_\varepsilon \tag{14.37}$$

$$Q_z = Q_y^{(e)} s_\varepsilon + Q_z^{(e)} c_\varepsilon \tag{14.38}$$

$$R_z = R_y^{(e)} s_\varepsilon + R_z^{(e)} c_\varepsilon \tag{14.39}$$

Finally, the following expressions are needed to extract the Ω', i' and ω' parameters:

$$c_\omega c_\Omega - s_\omega s_\Omega c_i = c_{\omega'} c_{\Omega'} - s_{\omega'} s_{\Omega'} c_{i'} \tag{14.40}$$

$$-s_\omega c_\Omega - c_\omega s_\Omega c_i = -s_{\omega'} c_{\Omega'} - c_{\omega'} s_{\Omega'} c_{i'} \tag{14.41}$$

$$s_\Omega s_i = s_{\Omega'} s_{i'} \tag{14.42}$$

$$c_\omega s_\Omega + s_\omega c_\Omega c_i = (c_{\omega'} s_{\Omega'} + s_{\omega'} c_{\Omega'} c_{i'})\, c_\varepsilon - s_{\omega'} s_{i'} s_\varepsilon \tag{14.43}$$

$$-s_\omega s_\Omega + c_\omega c_\Omega c_i = (-s_{\omega'} s_{\Omega'} + c_{\omega'} c_{\Omega'} c_{i'})\, c_\varepsilon - c_{\omega'} s_{i'} s_\varepsilon \tag{14.44}$$

$$-c_\Omega s_i = -c_{\Omega'} s_{i'} c_\varepsilon - c_{i'} s_\varepsilon \tag{14.45}$$

$$s_\omega s_i = (c_{\omega'} s_{\Omega'} + s_{\omega'} c_{\Omega'} c_{i'})\, s_\varepsilon + s_{\omega'} s_{i'} c_\varepsilon \tag{14.46}$$

$$c_\omega s_i = (-s_{\omega'} s_{\Omega'} + c_{\omega'} c_{\Omega'} c_{i'})\, s_\varepsilon + c_{\omega'} s_{i'} c_\varepsilon \tag{14.47}$$

$$c_i = -c_{\Omega'} s_{i'} s_\varepsilon + c_{i'} c_\varepsilon \tag{14.48}$$

Equations (14.45) and (14.48) can be used to solve for $c_{i'}$:

$$c_{i'} = c_\Omega s_i s_\varepsilon + c_i c_\varepsilon \tag{14.49}$$

The same equations can also be used in conjunction with Equation (14.42) to eliminate $s_{i'}$ and arrive at

$$\tan \Omega' = s_\Omega s_i / (c_\Omega s_i c_\varepsilon - c_i s_\varepsilon) = s_{\Omega'} / c_{\Omega'} \tag{14.50}$$

Finally, Equations (14.44) and (14.47) yield an expression for $c_{\omega'} s_{i'}$ in terms of Ω, i, ω and ε, and in a similar way, Equations (14.43) and (14.46) are manipulated to yield an expression for $s_{\omega'} s_{i'}$ also in terms of Ω, i, ω and ε, after which the following expression for ω' is obtained:

$$\tan \omega' = [-c_\omega s_\Omega s_\varepsilon - s_\omega c_\Omega c_i s_\varepsilon + s_\omega s_i c_\varepsilon] \,/\, [s_\omega s_\Omega s_\varepsilon - c_\omega c_\Omega c_i s_\varepsilon + c_\omega s_i c_\varepsilon] = s_{\omega'} / c_{\omega'} \tag{14.51}$$

Because i' lies in the $(0, \pi)$ interval, $c_{i'}$ in Equation (14.49) will return i' without ambiguity. The more elaborate expressions for Ω' and ω' in Equations (14.50) and (14.51) must be used with a quadrant resolution feature, because their range is the $(0, 2\pi)$ interval instead. The trigonometric functions in the accelerations f_r, f_θ, f_h involve $s_{\Omega'}$, $c_{\Omega'}$, $s_{i'}$, $c_{i'}$, $s_{\omega'}$, $c_{\omega'}$ which can in turn be expressed in terms of Ω, i and ω. Equations (14.45) and (14.48) will yield

$$c_{\Omega'} s_{i'} = c_\Omega s_i c_\varepsilon - c_i s_\varepsilon \tag{14.52}$$

which can be used with Equation (14.42) to solve for $s_{i'}$, which is then itself used to solve for $s_{\Omega'}$ and $c_{\Omega'}$:

$$s_{i'} = \left[s_\Omega^2 s_i^2 + (c_\Omega s_i c_\varepsilon - c_i s_\varepsilon)^2 \right]^{1/2} \tag{14.53}$$

$$s_{\Omega'} = s_\Omega s_i / s_{i'} \tag{14.54}$$

$$c_{\Omega'} = (c_\Omega s_i c_\varepsilon - c_i s_\varepsilon) / s_{i'} \tag{14.55}$$

Because i' is in the $(0, \pi)$ interval, the plus sign in front of the radical in Equation (14.53) is adequate. In a similar way, the expressions that lead to $\tan\omega'$ in Equation (14.51) will yield

$$c_{\omega'} = [s_\omega s_\Omega s_\varepsilon - c_\omega c_\Omega c_i s_\varepsilon + c_\omega s_i c_\varepsilon]/s_{i'} \tag{14.56}$$

$$s_{\omega'} = [-c_\omega s_\Omega s_\varepsilon - s_\omega c_\Omega c_i s_\varepsilon + s_\omega s_i c_\varepsilon]/s_{i'} \tag{14.57}$$

We also need $s_{\theta*}$ and $c_{\theta*}$ in terms of the elements h, k and L because $s_{\theta'} = s_{\omega'}c_{\theta*} + s_{\theta*}c_{\omega'}$ and $c_{\theta'} = c_{\omega'}c_{\theta*} - s_{\omega'}s_{\theta*}$ and their powers appear in the accelerations. They are given by the well-known expressions

$$s_{\theta*} = (ks_L - hc_L)/(h^2 + k^2)^{1/2} \tag{14.58}$$

$$c_{\theta*} = (hs_L + kc_L)/(h^2 + k^2)^{1/2} \tag{14.59}$$

complementing the other well-known relationships, namely

$$s_\omega = (hq - kp) \Big/ \left[\left(h^2 + k^2\right)^{1/2} \left(p^2 + q^2\right)^{1/2}\right] \tag{14.60}$$

$$c_\omega = (hp + kq) \Big/ \left[\left(h^2 + k^2\right)^{1/2} \left(p^2 + q^2\right)^{1/2}\right] \tag{14.61}$$

$$s_\Omega = p \Big/ \left(p^2 + q^2\right)^{1/2} \tag{14.62}$$

$$c_\Omega = q \Big/ \left(p^2 + q^2\right)^{1/2} \tag{14.63}$$

$$s_i = 2\left(p^2 + q^2\right)^{1/2} \Big/ \left(1 + p^2 + q^2\right) \tag{14.64}$$

$$c_i = \left(1 - p^2 - q^2\right) \Big/ \left(1 + p^2 + q^2\right) \tag{14.65}$$

We can now express the primed functions directly in terms of the equinoctial elements themselves, arriving at the following expressions after some algebra:

$$s_{i'} = \left\{4p^2 + \left[2qc_\varepsilon - \left(1 - p^2 - q^2\right)s_\varepsilon\right]^2\right\}^{1/2} \Big/ \left(1 + p^2 + q^2\right) \tag{14.66}$$

$$s_{\Omega'} = 2p \Big/ \left\{4p^2 + \left[2qc_\varepsilon - \left(1 - p^2 - q^2\right)s_\varepsilon\right]^2\right\}^{1/2} \tag{14.67}$$

$$c_{\Omega'} = \left[2qc_\varepsilon - \left(1 - p^2 - q^2\right)s_\varepsilon\right] \Big/ \left\{4p^2 + \left[2qc_\varepsilon - \left(1 - p^2 - q^2\right)s_\varepsilon\right]^2\right\}^{1/2} \tag{14.68}$$

as well as

$$c_{\omega'} = \left[\frac{2hpq - k\left(1 + p^2 - q^2\right)}{\left(h^2 + k^2\right)^{1/2}}s_\varepsilon + \frac{2(hp + kq)}{(h^2 + k^2)^{1/2}}c_\varepsilon\right] \Big/ \left\{4p^2 + \left[2qc_\varepsilon - (1 - p^2 - q^2)s_\varepsilon\right]^2\right\}^{1/2} \tag{14.69}$$

$$s_{\omega'} = \left[\frac{-2kpq - h(1+p^2-q^2)}{(h^2+k^2)^{1/2}} s_\varepsilon + \frac{2(hq-kp)}{(h^2+k^2)^{1/2}} c_\varepsilon \right] \Big/ \left\{ 4p^2 + \left[2qc_\varepsilon - (1-p^2-q^2)s_\varepsilon \right]^2 \right\}^{1/2} \quad (14.70)$$

A simple verification exercise shows that $s^2_{\omega'} + c^2_{\omega'} = 1$. In theory, it is then possible to transform the expressions for the accelerations f_r, f_θ, f_h directly in terms of the equinoctial elements, because all the basic trigonometric functions involving the primed quantities Ω', i', ω' are expressible directly in terms of h, k, p, q and L.

Given the classical elements of the apparent orbit of the Sun at the initial time $t_o = 0$ that corresponds to a given calendar date, that is, a_s, e_s, i_s, Ω_s, ω_s and M_s where $i_s = \varepsilon$, $\Omega_s = 0$, the location of the Sun at time t is defined by $M_s(t) = M_s(0) + n_s t$ with n_s the mean motion of the apparent solar orbit. Kepler's equation provides E_s from which θ^*_s is derived. Then $\lambda_\odot = \Omega_s + \theta^*_s = \theta^*_s$ is the value of the longitude of the Sun along the ecliptic needed to complete the evaluation of the f_r, f_θ and f_h accelerations.

14.3 Lunar Gravity Perturbation Acceleration Components in the Euler–Hill Frame

The geometry of the lunar orbit is depicted in Figure 14.3 with once again $\hat{\mathbf{x}}$, $\hat{\mathbf{y}}$, $\hat{\mathbf{z}}$ representing the Earth-centered inertial Earth-centered inertial (ECI) equatorial frame. The moon's ascending node is at N and Eulerian angles are Ω_m, i_m, ω_m with P_m standing for perigee, θ^*_m for true anomaly and $\mathbf{r}_{☽} = \mathbf{r}_m$ for radial distance. Letting A'' define the ascending node of the spacecraft orbit with respect to the Moon orbit, the angles Ω'' defined by NA$''$, i'' and ω'' depict the orientation of the spacecraft orbit in the $\hat{\mathbf{x}}_m$, $\hat{\mathbf{y}}_m$, $\hat{\mathbf{z}}_m$ lunar orbital frame. It is clear that $\hat{\mathbf{x}}_m$, $\hat{\mathbf{y}}_m$, $\hat{\mathbf{z}}_m$ play the same role as the $\hat{\mathbf{x}}_e$, $\hat{\mathbf{y}}_e$, $\hat{\mathbf{z}}_e$ frame of the solar gravity perturbation of the previous section. Figure 14.4 depicts the spacecraft orbit geometry with respect to the ($\hat{\mathbf{x}}_m$, $\hat{\mathbf{y}}_m$, $\hat{\mathbf{z}}_m$) frame, with the Moon longitude λ_m measured from $\hat{\mathbf{x}}_m$. Both the lunar and apparent solar orbits are shown simultaneously in Figure 14.1 with all the relevant angles, including $\theta'' = \omega'' + \theta^*$.

The transformation from the rotating frame ($\hat{\mathbf{r}}$, $\hat{\boldsymbol{\theta}}$, $\hat{\mathbf{h}}$) to $\left(\hat{\mathbf{x}}_m, \hat{\mathbf{y}}_m, \hat{\mathbf{z}}_m\right)$ is given by

$$\hat{\mathbf{x}}_m = \left(c_{\Omega''}c_{\theta''} - s_{\Omega''}c_{i''}s_{\theta''}\right)\hat{\mathbf{r}} + \left(-c_{\Omega''}s_{\theta''} - s_{\Omega''}c_{i''}c_{\theta''}\right)\hat{\boldsymbol{\theta}} + s_{\Omega''}s_{i''}\hat{\mathbf{h}} \quad (14.71)$$

$$\hat{\mathbf{y}}_m = \left(s_{\Omega''}c_{\theta''} + c_{\Omega''}c_{i''}s_{\theta''}\right)\hat{\mathbf{r}} + \left(-s_{\Omega''}s_{\theta''} + c_{\Omega''}c_{i''}c_{\theta''}\right)\hat{\boldsymbol{\theta}} - c_{\Omega''}s_{i''}\hat{\mathbf{h}} \quad (14.72)$$

$$\hat{\mathbf{z}}_m = s_{i''}s_{\theta''}\hat{\mathbf{r}} + s_{i''}c_{\theta''}\hat{\boldsymbol{\theta}} + c_{i''}\hat{\mathbf{h}} \quad (14.73)$$

The rotation from the relative coordinates to the $\hat{\mathbf{x}}$, $\hat{\mathbf{y}}$, $\hat{\mathbf{z}}$ coordinates therefore involves the angle i_m, which replaces ε of the solar case, and a final rotation of Ω_m about $\hat{\mathbf{z}}$ because $\hat{\mathbf{x}}_m$ and $\hat{\mathbf{x}}$ do not coincide as in the Sun case, where $\hat{\mathbf{x}}_e$ and $\hat{\mathbf{x}}$ are aligned due to $\Omega_s = 0$. The final detail still unresolved in the lunar case is the transformation from Ω, i, ω to Ω'', i'', ω''. To this end, Equations (14.40)–(14.48) are still applicable here if we replace Ω', i', ω' and ε by Ω'', i'', ω'' and i_m. However, Ω must also be replaced by $\Omega - \Omega_m$ to make these equivalencies still valid and help extract the Ω'', i'', ω'' angles from

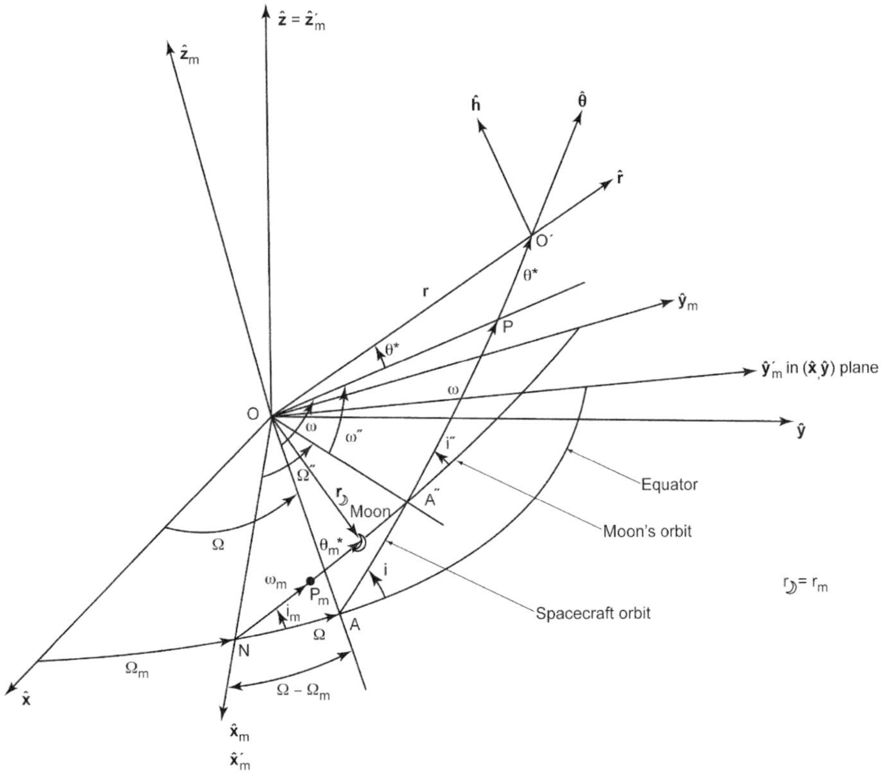

Figure 14.3 Relative geometry of the spacecraft and Moon orbits.

the knowledge of the Ω, i, ω values directly extracted from the equinoctial elements h, k, p, q and L. The reason why we need to replace Ω by $\Omega - -\Omega_m$ in these equivalencies is visible in Figure 14.3, where the angle $\widehat{NA} = \Omega - \Omega_m$ is such that $\Omega - \Omega_m$, i, ω describe the orientation of the spacecraft orbit in the $(\hat{\mathbf{x}}_m, \hat{\mathbf{y}}'_m, \hat{\mathbf{z}}_m)$ frame where $\hat{\mathbf{y}}'_m$ is now in the equatorial plane, and Ω'', i'', ω'' describe the same orbit in the $(\hat{\mathbf{x}}_m, \hat{\mathbf{y}}_m, \hat{\mathbf{z}}_m)$ frame. These two frames having the same relative geometry as the pair $(\hat{\mathbf{x}}, \hat{\mathbf{y}}, \hat{\mathbf{z}})$ and $(\hat{\mathbf{x}}_e, \hat{\mathbf{y}}_e, \hat{\mathbf{z}}_e)$, the pertinent relations valid for the solar case can then be used in the lunar case following the substitutions just mentioned. They are:

$$c_{i''} = c_{\Omega-\Omega_m} s_i s_{i_m} + c_i c_{i_m} \tag{14.74}$$

$$\tan\Omega'' = \left(s_{\Omega-\Omega_m} s_i\right) \Big/ \left[c_{\Omega-\Omega_m} s_i c_{i_m} - c_i s_i\right] \tag{14.75}$$

$$\begin{aligned}\tan\omega'' = {} & \left[-c_\omega s_{\Omega-\Omega_m} s_{i_m} - s_\omega c_{\Omega-\Omega_m} c_i s_{i_m} + s_\omega s_i c_{i_m}\right] \Big/ \\ & \left[s_\omega s_{\Omega-\Omega_m} s_{i_m} - c_\omega c_{\Omega-\Omega_m} c_i s_{i_m} + c_\omega s_i c_{i_m}\right]\end{aligned} \tag{14.76}$$

For example, from

$$s_{\Omega''} s_{i''} = s_i s_\Omega \tag{14.77}$$

$$c_{\Omega''} s_{i''} = c_{\Omega-\Omega_m} s_i c_{i_m} - c_i s_{i_m} \tag{14.78}$$

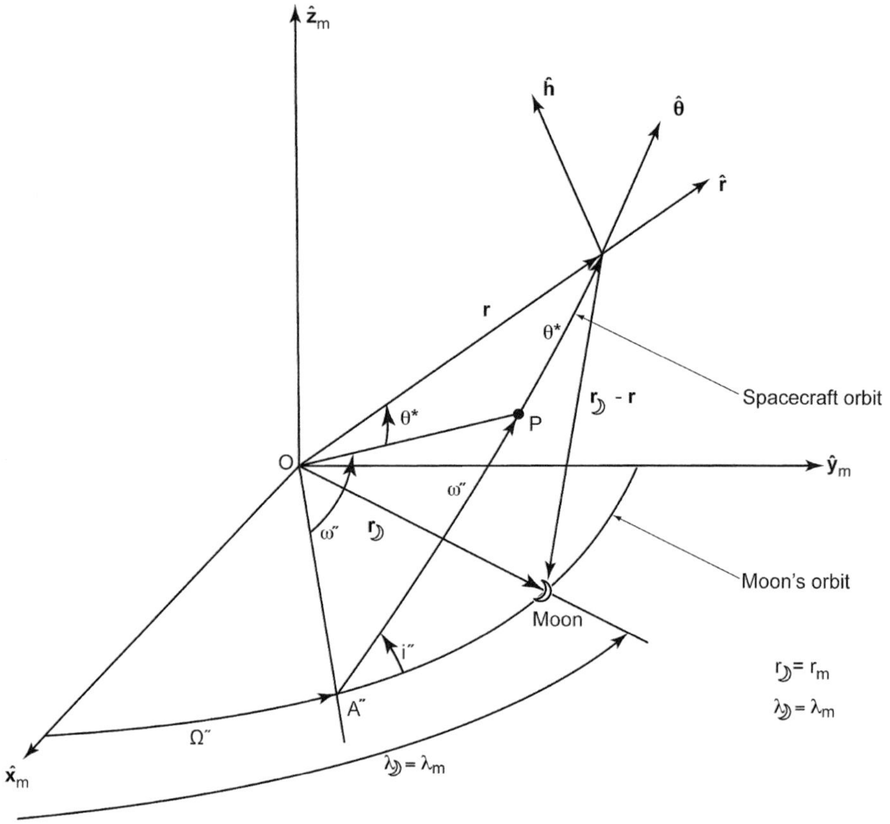

Figure 14.4 Relative geometry of rotating and Moon orbital frames.

we can obtain the expression for $s_{i''}$ as

$$s_{i''} = \left[s^2_{\Omega-\Omega_m} s^2_i + \left(c_{\Omega-\Omega_m} s_i c_{i_m} - c_i s_{i_m} \right)^2 \right]^{1/2} \quad (14.79)$$

which will be needed in the generation of the differential equations for the multipliers. Let n'_m stand for the mean motion of the lunar orbit to be defined according to the lunar theory of de Pontécoulant [8]. By analogy to the expressions for the solar gravity acceleration components along the rotating axes $\hat{\mathbf{r}}$, $\hat{\boldsymbol{\theta}}$, $\hat{\mathbf{h}}$, the lunar accelerations expanded to the third order in the spacecraft radial distance r can be written as follows, with $n^2_m = \mu_☽ / r^3_m$. As for the solar case, we have $f_r = f_{r_1} + f_{r_2} + f_{r_3}$, $f_\theta = f_{\theta_1} + f_{\theta_2} + f_{\theta_3}$, and $f_h = f_{h_1} + f_{h_2} + f_{h_3}$:

$$\begin{aligned} f_r = {} & r n^2_m \left[3 \left(c^2_{\lambda_m-\Omega''} c^2_{\theta''} + c^2_{i''} s^2_{\theta''} s^2_{\lambda_m-\Omega''} + 2 c_{i''} s_{\theta''} c_{\theta''} s_{\lambda_m-\Omega''} c_{\lambda_m-\Omega''} \right) - 1 \right] \\ & + \frac{3}{2} r^2 \frac{n^2_m}{r_m} \left[5 \left(c^3_{\lambda_m-\Omega''} c^3_{\theta''} + c^3_{i''} s^3_{\theta''} s^3_{\lambda_m-\Omega''} + 3 c^2_{\lambda_m-\Omega''} c^2_{\theta''} c_{i''} s_{\theta''} s_{\lambda_m-\Omega''} \right.\right. \\ & \left.\left. + 3 c^2_{i''} s^2_{\theta''} c_{\theta''} s^2_{\lambda_m-\Omega'} c_{\lambda_m-\Omega''} \right) - 3 \left(c_{\lambda_m-\Omega''} c_{\theta''} + c_{i''} s_{\theta''} s_{\lambda_m-\Omega''} \right) \right] \end{aligned}$$

$$
\begin{aligned}
&+\frac{r^3}{2}\frac{n_m^2}{r_m^2}\Big[35\Big(c_{\theta''}^4 c_{\lambda_m-\Omega''}^4 + 4c_{i''}c_{\theta''}^3 s_{\theta''} s_{\lambda_m-\Omega''} c_{\lambda_m-\Omega''}^3 \\
&+4c_{i''}^2 s_{\theta''}^2 c_{\theta''}^2 s_{\lambda_m-\Omega''}^2 c_{\lambda_m-\Omega''}^2 + c_{i''}^4 s_{\theta''}^4 s_{\lambda_m-\Omega''}^4 + 4c_{i''}^3 s_{\theta''}^3 c_{\theta''} s_{\lambda_m-\Omega''}^3 c_{\lambda_m-\Omega''}\Big) \\
&-30\Big(c_{\lambda_m-\Omega''}^2 c_{\theta''}^2 + c_{i''}^2 s_{\theta''}^2 s_{\lambda_m-\Omega''}^2 + 2c_{i''} s_{\theta''} c_{\theta''} s_{\lambda_m-\Omega''} c_{\lambda_m-\Omega''}\Big) + 3\Big]
\end{aligned}
\tag{14.80}
$$

$$
\begin{aligned}
f_\theta = 3rn_m^2&\Big[-s_{\theta''}c_{\theta''}c_{\lambda_m-\Omega''}^2 - c_{i''}s_{\theta''}^2 s_{\lambda_m-\Omega''} c_{\lambda_m-\Omega''} \\
&+ c_{i''}c_{\theta''}^2 s_{\lambda_m-\Omega''} c_{\lambda_m-\Omega''} + c_{i''}^2 s_{\theta''} c_{\theta''} s_{\lambda_m-\Omega''}^2\Big] \\
&+\frac{3}{2}r^2\frac{n_m^2}{r_m}\Big[5\Big(-s_{\theta''}c_{\theta''}^2 c_{\lambda_m-\Omega''}^3 + c_{i''}c_{\theta''}^3 s_{\lambda_m-\Omega''} c_{\lambda_m-\Omega''}^2 - c_{i''}^2 s_{\theta''}^3 c_{\lambda_m-\Omega''} s_{\lambda_m-\Omega''}^2 \\
&+ c_{i''}^3 c_{\theta''} s_{\theta''}^2 s_{\lambda_m-\Omega''}^3 - 2c_{i''} s_{\theta''}^2 c_{\theta''} s_{\lambda_m-\Omega''} c_{\lambda_m-\Omega''}^2 + 2c_{i''}^2 s_{\theta''} c_{\theta''}^2 s_{\lambda_m-\Omega''}^2 c_{\lambda_m-\Omega''}\Big) \\
&-\left(-s_{\theta''}c_{\lambda_m-\Omega''} + c_{i''}c_{\theta''}s_{\lambda_m-\Omega''}\right)\Big] + \frac{r^3}{2}\frac{n_m^2}{r_m^2}\Big[35\Big(-s_{\theta''}c_{\theta''}^3 c_{\lambda_m-\Omega''}^4 \\
&- s_{\theta''}^4 c_{i''}^3 c_{\lambda_m-\Omega''} s_{\lambda_m-\Omega''}^3 - 3c_{i''}s_{\theta''}^2 c_{\theta''}^2 c_{\lambda_m-\Omega''}^3 s_{\lambda_m-\Omega''} - 3c_{i''}^2 s_{\theta''}^3 c_{\theta''} s_{\lambda_m-\Omega''}^2 c_{\lambda_m-\Omega''}^2 \\
&+ c_{i''}c_{\theta''}^4 s_{\lambda_m-\Omega''} c_{\lambda_m-\Omega''}^3 + c_{i''}^4 c_{\theta''} s_{\theta''}^3 s_{\lambda_m-\Omega''}^4 + 3c_{i''}^2 c_{\theta''}^3 s_{\theta''} c_{\lambda_m-\Omega''}^2 s_{\lambda_m-\Omega''}^2 \\
&+ 3c_{i''}^3 s_{\theta''}^2 c_{\theta''}^2 s_{\lambda_m-\Omega''}^3 c_{\lambda_m-\Omega''}\Big) - 15\Big(- s_{\theta''}c_{\theta''}c_{\lambda_m-\Omega''}^2 \\
&- c_{i''}s_{\theta''}^2 s_{\lambda_m-\Omega''} c_{\lambda_m-\Omega''} + c_{i''}c_{\theta''}^2 s_{\lambda_m-\Omega''} c_{\lambda_m-\Omega''} + c_{i''}^2 s_{\theta''} c_{\theta''} s_{\lambda_m-\Omega''}^2\Big)\Big]
\end{aligned}
\tag{14.81}
$$

$$
\begin{aligned}
f_h = -3rn_m^2&\Big[s_{i''}c_{\theta''}s_{\lambda_m-\Omega''}c_{\lambda_m-\Omega''} + s_{i''}c_{i''}s_{\theta''}s_{\lambda_m-\Omega''}^2\Big] - \frac{3}{2}r^2\frac{n_m^2}{r_m} \\
&\times\Big[5s_{i''}c_{\theta''}^2 s_{\lambda_m-\Omega''} c_{\lambda_m-\Omega''}^2 + 5s_{i''}c_{i''}^2 s_{\theta''}^2 s_{\lambda_m-\Omega''}^3 + 10s_{i''}c_{i''}s_{\theta''}c_{\theta''}s_{\lambda_m-\Omega''}^2 c_{\lambda_m-\Omega''} \\
&- s_{i''}s_{\lambda_m-\Omega''}\Big] + \frac{5}{2}\frac{r^3 n_m^2}{r_m^2}\Big[-7s_{i''}c_{\theta''}^3 s_{\lambda_m-\Omega''} c_{\lambda_m-\Omega''}^3 - 7s_{i''}c_{i''}^3 s_{\theta''}^3 s_{\lambda_m-\Omega''}^4 \\
&- 21s_{i''}c_{i''}c_{\theta''}^2 s_{\theta''} s_{\lambda_m-\Omega''}^2 c_{\lambda_m-\Omega''}^2 - 21s_{i''}c_{i''}^2 s_{\theta''}^2 c_{\theta''} s_{\lambda_m-\Omega''}^3 c_{\lambda_m-\Omega''} \\
&+ 3s_{i''}c_{\theta''}s_{\lambda_m-\Omega''}c_{\lambda_m-\Omega''} + 3s_{i''}c_{i''}s_{\theta''}s_{\lambda_m-\Omega''}^2\Big]
\end{aligned}
\tag{14.82}
$$

Here, n_m is not the lunar orbit mean motion. The lunar mean motion n'_m is defined in the next section.

If we evaluate the ratios of the coefficient of the first-order term with the second- and third-order terms in f_r, f_θ and f_h for the solar gravity with $r_\odot \simeq 1{,}500{,}000$ km, and for $r \simeq 42{,}000$ km corresponding to the GEO altitude, we have for f_r and f_θ, ratios of 1.4 percent and 0.013 percent, respectively, while for f_h we get $\simeq$1.4 percent and 0.065 percent. It is clear that the third-order terms are $<$ 1 percent of the leading gravity gradient terms. If r is increased to $\simeq$100,000 km, these ratios increase to 3.33 percent, 0.074 percent, and 3.33 percent, 0.370 percent, still within $<$ 1 percent for the third-order contribution. For the lunar gravity, with $r_m \simeq 400{,}000$ km, the values corresponding to $r \simeq 42{,}000$ km are for f_r and f_θ, 5.25 percent and 0.183 percent, and for f_h, 5.25 percent and 0.918 percent. However, for $r \simeq 100{,}000$ km, these values increase to 12.5 percent,

1.041 percent and 12.5 percent, 5.208 percent Thus, only f_h exceeds the 1 percent level for the third-order term when r is large, that is, 5 percent for $r \simeq$ 100,000 km. This value can be brought down by adding a fourth term in the expansion for $\mathbf{f}_\odot$ or rather $\mathbf{f}_{☽}$.

Going back to Equation (14.4), we have

$$\mathbf{F} = \nabla \left\{ \cdots + \frac{\mu_\odot}{r_\odot} \left[\frac{63}{8} \frac{(\mathbf{r}\,\hat{\mathbf{r}}_\odot)^5}{r_\odot^5} - \frac{70}{8} \frac{(\mathbf{r}\,\hat{\mathbf{r}}_\odot)^3}{r_\odot^5} r^2 + \frac{15}{8} \frac{(\mathbf{r}\,\hat{\mathbf{r}}_\odot)}{r_\odot^5} r^4 \right] \right\} \tag{14.83}$$

where only the fourth term in the expansion is shown; thus

$$\begin{aligned} \mathbf{f}_\odot = \cdots + \frac{\mu_\odot}{8 r_\odot^6} r^4 \Big[& 315 (\hat{\mathbf{r}}\,\hat{\mathbf{r}}_\odot)^4 \hat{\mathbf{r}}_\odot - 210 (\hat{\mathbf{r}}\,\hat{\mathbf{r}}_\odot)^2 \hat{\mathbf{r}}_\odot \\ & -140 (\hat{\mathbf{r}}\,\hat{\mathbf{r}}_\odot)^3 \hat{\mathbf{r}} + 15 \hat{\mathbf{r}}_\odot + 60 (\hat{\mathbf{r}}\,\hat{\mathbf{r}}_\odot) \hat{\mathbf{r}} \Big] \end{aligned} \tag{14.84}$$

For the f_h component, that is, $f_h = \mathbf{f}_\odot \cdot \hat{\mathbf{h}}$, we get for the fourth-order term

$$\begin{aligned} (f_h)_4 = \left(\mathbf{f}_\odot \hat{\mathbf{h}}\right)_4 = \frac{\mu_\odot}{8 r_\odot^6} r^4 \Big[& 315 (\hat{\mathbf{r}}\,\hat{\mathbf{r}}_\odot)^4 \left(\hat{\mathbf{r}}_\odot \hat{\mathbf{h}}\right) \\ & -210 (\hat{\mathbf{r}}\,\hat{\mathbf{r}}_\odot)^2 \left(\hat{\mathbf{r}}_\odot \hat{\mathbf{h}}\right) + 15 \left(\hat{\mathbf{r}}_\odot \hat{\mathbf{h}}\right) \Big] \end{aligned} \tag{14.85}$$

and with the use of Equations (14.13) and (14.15),

$$\begin{aligned} (f_h)_4 &= \frac{15 \mu_\odot}{8 r_\odot^6} r^4 \left[21 (\hat{\mathbf{r}}\,\hat{\mathbf{r}}_\odot)^4 - 14 (\hat{\mathbf{r}}\,\hat{\mathbf{r}}_\odot)^2 + 1 \right] \left(\hat{\mathbf{r}}_\odot \hat{\mathbf{h}}\right) \\ &= \frac{15 \mu_\odot}{8 r_\odot^6} r^4 \Big[21 \left(c_{\lambda_\odot - \Omega'} c_{\theta'} + c_{i'} s_{\theta'} s_{\lambda_\odot - \Omega'} \right)^4 \\ &\quad -14 \left(c_{\lambda_\odot - \Omega'} c_{\theta'} + c_{i'} s_{\theta'} s_{\lambda_\odot - \Omega'} \right)^2 + 1 \Big] \left(-s_{i'} s_{\lambda_\odot - \Omega'} \right) \end{aligned} \tag{14.86}$$

We need to add the $(f_h)_4$ term in the lunar case only, in which case we write it as

$$\begin{aligned} (f_h)_4 = \frac{15 n_m^2}{8 r_m^3} r^4 \Big[& 21 \left(c_{\lambda_m - \Omega''} c_{\theta''} + c_{i''} s_{\theta''} s_{\lambda_m - \Omega''} \right)^4 \\ & -14 \left(c_{\lambda_m - \Omega''} c_{\theta''} + c_{i''} s_{\theta''} s_{\lambda_m - \Omega''} \right)^2 + 1 \Big] \left(-s_{i''} s_{\lambda_m - \Omega''} \right) \end{aligned} \tag{14.87}$$

The ratio of the coefficient of this term with the leading gravity gradient term for the large value of r $\simeq$ 100,000 km is now down to 0.976 percent within $<$ 1 percent as for the other acceleration components f_r, f_θ, where only third-order terms are sufficient for good accuracy.

14.4 The Use of de Pontécoulant's Lunar Theory

We still need to compute r_m, λ_m as a function of time by making use of de Pontécoulant's analytic lunar theory. Unlike the apparent solar orbit, the lunar orbit experiences important perturbations due to solar gravity such that its node Ω_M measured along the ecliptic regresses by a complete revolution of 360 deg in about 18.6 years, and its perigee advances by 360 deg in about 8.85 years. There exist small oscillations in longitude, called evection, variation and parallactic inequality, as well as oscillations in anomaly called the annual equation, but their amplitudes are small such that they are neglected here. Letting $n_\odot = 2\pi/T_\odot$, where $T_\odot = 365.2564$ days or $365.2564 \times 86{,}400$ s, and using $T_☽ = 27.321661$ mean solar days for the lunar period, and computing $m = n_\odot/n'_m$, with $n'_m = 2\pi/T_☽$ standing for the mean motion of the Moon, the pertinent quantities describing the motion of the lunar orbit can now be evaluated, in particular the parameters

$$\bar{\omega}_M = (\bar{\omega}_M)_0 + (1-c)n'_m t \tag{14.88}$$

$$\Omega_M = (\Omega_M)_0 + (1-g)n'_m t \tag{14.89}$$

where $\bar{\omega} = \Omega + \omega$ stands for longitude of perigee, and where c and g are constants such that $(1-c) > 0$ and $(1-g) < 0$. In [8], de Pontécoulant computed the values of c and g to the order $O(m^7)$ in the small parameter m with values of $c = 1 - 0.008450821$ and $g = 1 + 0.004021678$, and a bit less accurately to $O(m^2)$, $c = 1 - \frac{3}{4}m^2$, $g = 1 + \frac{3}{4}m^2$. The geometry of the instantaneous lunar orbit is depicted in Figure 14.5 at both the initial time $t = 0$ and at time t. Given the elements a_m, e_m, i_m, $(\Omega_m)_0$, $(\omega_m)_0$ and $(M_m)_0$ at $t = 0$ in the ECI equatorial system, the angle $\widehat{ED}$ is first evaluated from $c_{\widehat{ED}} = c_{(\Omega_m)_0} c_{(\Omega_M)_0} + s_{(\Omega_m)_0} s_{(\Omega_M)_0} c_{i_s}$ after computing $(\Omega_M)_0$ from

$$c_{(\Omega_M)_0} = \left[c_{i_M} c_{i_s} - c_{(i_m)_0}\right] \Big/ \left(s_{i_M} s_{i_s}\right) \tag{14.90}$$

$$s_{(\Omega_M)_0} = s_{(\Omega_m)_0} s_{(i_m)_0} \Big/ s_{i_M} \tag{14.91}$$

Both i_M and i_s are constant, and $(\Omega_M)_0$ is obtained by way of a quadrant resolution feature. Note that $i_s - i_M \le i_m \le i_s + i_M$ such that with $i_M = 5.145$ deg, $i_s = 23.445$ deg, $18.3 \le i_m \le 28.59$ deg and $-13.0 \le \Omega_m \le 13.0$ deg. The ecliptic $(\omega_M)_0 = (\omega_m)_0 - \widehat{ED}$ is obtained next, and through Kepler's equation $M_m = E_m - e_m s_{E_m}$, the eccentric anomaly $(E_m)_0$ is extracted from the given $(M_m)_0$, leading to $(\theta^*_m)_0$ and finally to $(\theta_M)_0 = (\omega_M)_0 + (\theta^*_m)_0$ and $(r_m)_0 = a_m(1 - e_m c_{(E_m)_0})$.

Next, $(\bar{\omega}_M)_0 = (\Omega_M)_0 + (\omega_M)_0$ is evaluated and using $M_m = n'_m t + \varepsilon' - \bar{\omega}_M$ to describe the mean anomaly of the Moon measured from the moving perigee, at time t, where $\varepsilon' = \bar{\omega} - n\tau$ stands for the moon's longitude at epoch or $t = 0$ with τ obtained from $(M_m)_0 = n'_m \tau$ or $\tau = (M_m)_0/n'_m$, the value of the constant ε' is more directly obtained from $\varepsilon' = (M_m)_0 + (\Omega_M)_0 + (\omega_M)_0$ because M_m can also be written as $M_m = n'_m t + \varepsilon' - (\bar{\omega}_M)_0 - (1-c)\, n'_m t = c n'_m t + \varepsilon' - (\bar{\omega}_M)_0$. Now $\bar{\omega}_M = (\bar{\omega}_M)_0 + (1-c)\, n'_m t$ is evaluated at future times t, such that $M_m(t) = n'_m t + \varepsilon' - \bar{\omega}_M$ leads to $E_m(t)$ via Kepler's equation and thereby $\theta^*_m(t)$ and $r_m(t)$ from the orbit equation. Also $\theta_M(t) = \omega_M(t) + \theta^*_m(t)$ as well as $\omega_M(t) = (\omega_M)_0 + (-c+g)\, n'_m t$ are evaluated next. The latter equation being true because from $\bar{\omega}_M = \Omega_M + \omega_M = (\bar{\omega}_M)_0 + (1-c)n'_m t$, we get $\omega_M = \bar{\omega}_M - \Omega_M = (\bar{\omega}_M)_0 + (1-c)n'_m t - \left[(\Omega_M)_0 + (1-g)\, n'_m t\right] = (\bar{\omega}_M)_0 - (\Omega_M)_0 + (-c+g)\, n'_m t$. This expression simplifies to $\omega_M = (\bar{\omega}_M)_0 - (\Omega_M)_0 + \frac{3}{2}m^2 n'_m t = (\omega_M)_0 + \frac{3}{2}m^2 n'_m t$ if c and g are evaluated to $O\left(m^2\right)$ only. We still need to evaluate $\lambda_m(t) = \omega_m(t) + \theta^*_m(t)$, and to this end $\omega_m(t)$ measured from the equator must be computed first. Thus, using de Pontécoulant's expression in Equation (14.89) for $\Omega_M(t) = (\Omega_M)_0 + (1-g)n'_m t$, this value is used to get the equatorial i_m from

$$c_{i_m} = c_{i_M} c_{i_s} - s_{i_s} s_{i_M} c_{\Omega_M} \tag{14.92}$$

via an inverse cosine routine without ambiguity, because $i_m > 0$ by definition, after which $\Omega_m(t)$ measured along the equator is computed from

$$s_{\Omega_m} = s_{i_M} s_{\Omega_M} / s_{i_m} \tag{14.93}$$

using an inverse sine routine, also without ambiguity because the Ω_m range spans the interval $(-13.0, 13.0 \text{ deg})$. From the triangle ABC in Figure 14.5, the arc $\widehat{CB}$ is evaluated from $c_{\widehat{CB}} = c_{\Omega_m} c_{\Omega_M} + s_{\Omega_m} s_{\Omega_M} c_{i_s}$ such that $\omega_m(t) = \widehat{CB} + \omega_M(t)$, where $\omega_M(t)$ is measured from the ecliptic. All the pertinent angles at time t are now known in order to evaluate λ_m, as well as Eulerian angles Ω'', ω'', i'' and $\theta'' = \omega'' + \theta_m^*$ in order to compute the acceleration components f_r, f_θ, f_h due to the lunar gravity. Following [1, 2], and also shown in Chapters 9 and 13, the dynamic system of equations can be cast as follows:

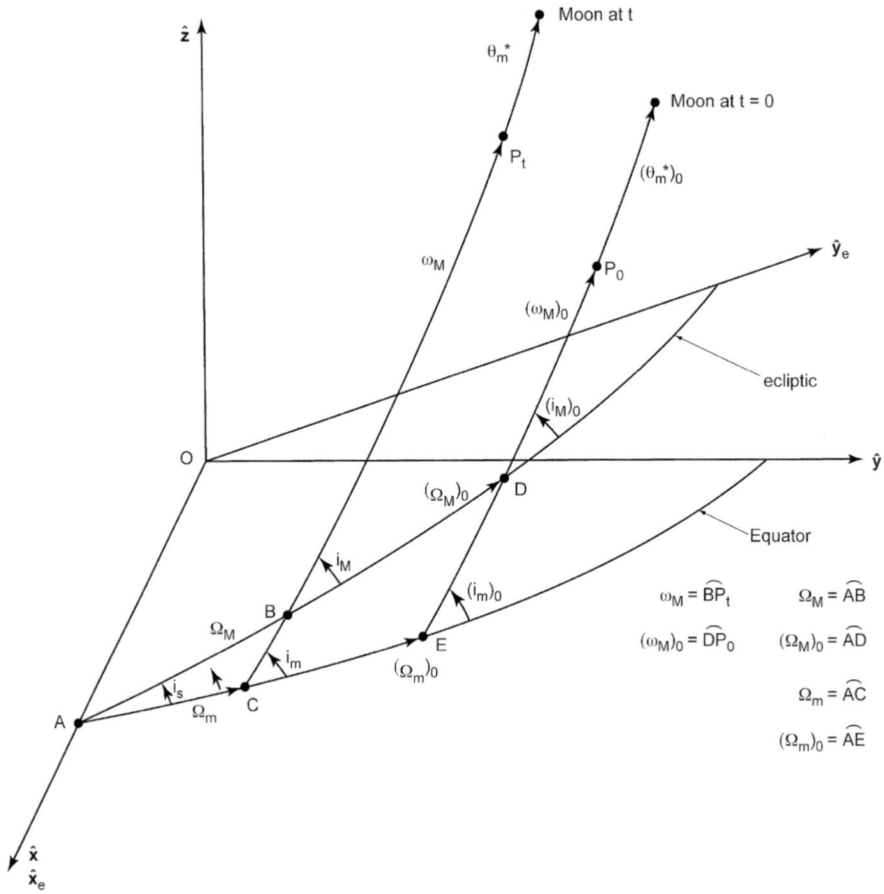

Figure 14.5 Moon orbit geometry.

$$\begin{pmatrix} \dot{a} \\ \dot{h} \\ \dot{k} \\ \dot{p} \\ \dot{q} \\ \dot{L} \end{pmatrix} = \begin{pmatrix} B^L_{11} & B^L_{12} & B^L_{13} \\ B^L_{21} & B^L_{22} & B^L_{23} \\ B^L_{31} & B^L_{32} & B^L_{33} \\ B^L_{41} & B^L_{42} & B^L_{43} \\ B^L_{51} & B^L_{52} & B^L_{53} \\ B^L_{61} & B^L_{62} & B^L_{63} \end{pmatrix} \left[\begin{pmatrix} u_r \\ u_\theta \\ u_h \end{pmatrix} (f_t) + \mathbf{f}_{J_2} + \mathbf{f}_{J_3} + \mathbf{f}_{J_4} + \mathbf{f}_\odot + \mathbf{f}_{\rightmoon} \right] + \begin{bmatrix} 0 \\ 0 \\ 0 \\ 0 \\ 0 \\ \frac{na^2(1-h^2-k^2)^{1/2}}{r^2} \end{bmatrix} \tag{14.94}$$

The equinoctial elements, as well as all the other relevant symbols, are defined in [1, 2] as well as Chapters 9 and 13, with all the various acceleration vectors resolved along the rotating $(\hat{\mathbf{r}}, \hat{\boldsymbol{\theta}}, \hat{\mathbf{h}})$ axes.

14.5 Numerical Results

An optimal transfer example using low thrust is generated next and later compared to a solution obtained using the method in [3], where the various accelerations due to the Earth zonal harmonics as well as those due to the Sun and the Moon are first computed in an inertial reference frame before transformation to the Euler–Hill rotating frame for subsequent use by the numerical integrator at each integration step to generate the trajectories needed by the optimizer.

Let the initial and final orbits be given by $a_0 = 7000\,\text{km}$, $e_0 = 0$, $i_0 = 28.5$ deg, $\Omega_0 = 0$ deg, $\omega_0 = 0$ deg, $M_0 = 164$ deg, $a_f = 42{,}000\,\text{km}$, $e_f = 10^{-3}$, $i_f = 1$ deg, $\Omega_f = 0$ deg, $\omega_f = 0$ deg and $L_f = 247$ radians. A continuous constant thrust acceleration $f_t = 9.8 \times 10^{-6}\,\text{km/s}^2$ is applied optimally to reach the desired final orbit under the influence of $J_2 = 1.08263 \times 10^{-3}$, $J_3 = -2.56 \times 10^{-6}$, $J_4 = -1.58 \times 10^{-6}$ and the luni-solar gravity. We further use $\mu_\odot = 0.13271244 \times 10^{12}\,\text{km}^3/\text{s}^2$, $\mu_{\rightmoon} = 0.490279 \times 10^4\,\text{km}^3/\text{s}^2$ and the following classical elements for the Sun and the Moon referred to the ECI equatorial system, namely, $a_s = 1.4960000 \times 10^8\,\text{km}$, $e_s = 1.67052259 \times 10^{-2}$, $i_s = 23.438077$ deg, $\Omega_s = 0$ deg, $\omega_s = 283.099535$ deg, $M_s = 86.3903095$ deg, $a_m = 385{,}220.2364\,\text{km}$, $e_m = 3.88151582 \times 10^{-2}$, $i_m = 26.79150504$ deg, $\Omega_m = 350.9778897$ deg, $\omega_m = 96.21592566$ deg, $M_m = 352.7031062$ deg, corresponding to the calendar date of April 1, 2009 at 0 h, 0 min, 0 s, or the Julian date of 2,454,922.5. These luni-solar elements were generated by W. Hallman of the Aerospace Corporation, and as such they represent osculating values. However, for the analytic models of luni-solar motions used here, we shall assume that these initial values are mean values instead. The system of the multiplier equations are derived in part 2 of this analysis in [9], also shown in Chapter 15, and the minimum-time transfer is thus

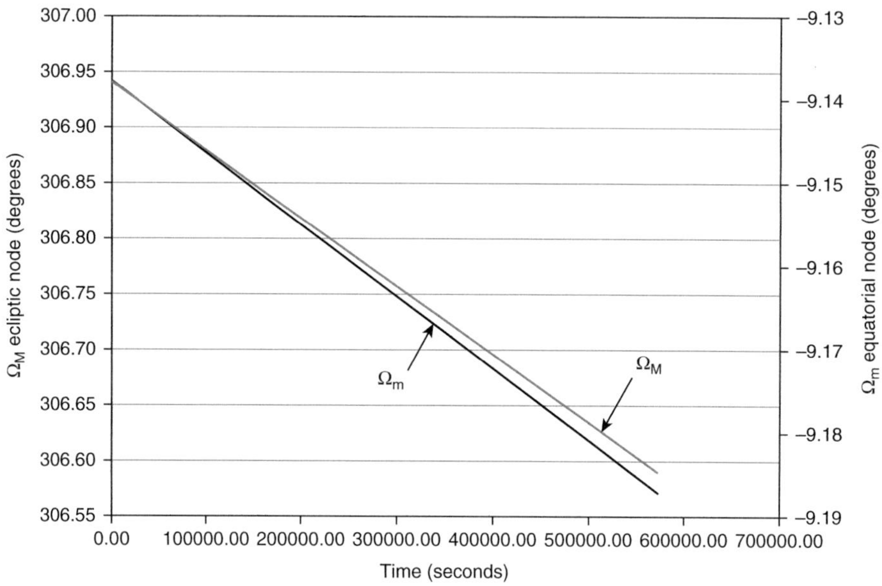

Figure 14.6 Evolution of the moon's ecliptic and equatorial nodes during transfer.

generated for this transfer with fixed initial and final states searching on the initial values of the six multipliers at time zero, λ_a, λ_h, λ_k, λ_p, λ_q, λ_L as well as the transfer time t_f in order to satisfy the six end constraints a_f, h_f, k_f, p_f, q_f and L_f, as well as $H_f = 1$ for the system Hamiltonian [9], also shown in Chapter 15. The evolutions of the moon's ecliptic and equatorial nodes Ω_M and Ω_m are depicted in Figure 14.6 with Ω_M precessing at the required rate of 360 deg per 18.6 years, while the moon's equatorial i_m is shown to change in Figure 14.7 due to the precession of the lunar orbit. The mean anomaly of the Moon is advancing in time as depicted in Figure 14.8, while its perigee motion is according to Figure 14.9 with ω_M the ecliptic value advancing at the rate of 360 deg per 8.85 years, as required. Figure 14.10 shows the progression of the longitudes $\lambda_\odot$ and λ_m, while Figures 14.11 and 14.12 depict the variations experienced by the spacecraft orbit Eulerian angles Ω', Ω'' and i', i'', respectively, needed in the evaluation of the luni-solar acceleration components along the rotating axes. The first-, second- and third-order radial component accelerations due to the Sun gravity are shown in Figure 14.13 during the transfer to the GEO orbit. Because of the large distance to the Sun, only the first-order component is of consequence as it is essentially equal to the sum $f_r = (f_r)_1 + (f_r)_2 + (f_r)_3$ with $(f_r)_2$ and $(f_r)_3$ very small. The same observations are also true for the normal and out-of-plane components, as shown in Figures 14.14 and 14.15. However, the benefit of expanding these accelerations to third order is clearly visible in Figures 14.17–14.27 for the case of the much closer Moon, with the third-order contributions already quite small but rightly accounted for, for an accurate evaluation of these lunar accelerations. As is clear from these figures, the total components are a bit different from their first-order counterparts, necessitating thereby the inclusion of the higher-order contributions. Figures 14.16 and 14.28 show the total values of the acceleration components due to the Sun and the Moon,

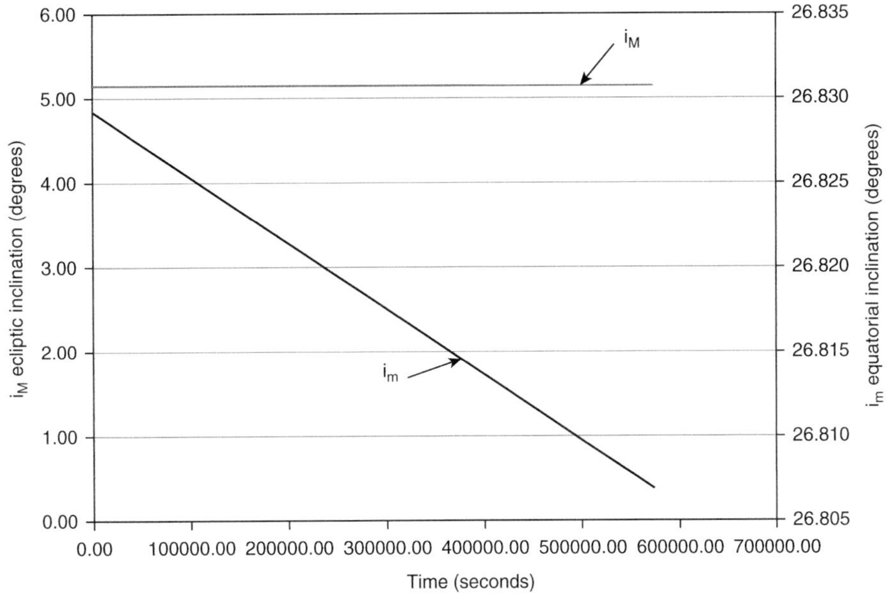

Figure 14.7 Evolution of the moon's ecliptic and equatorial inclinations during transfer.

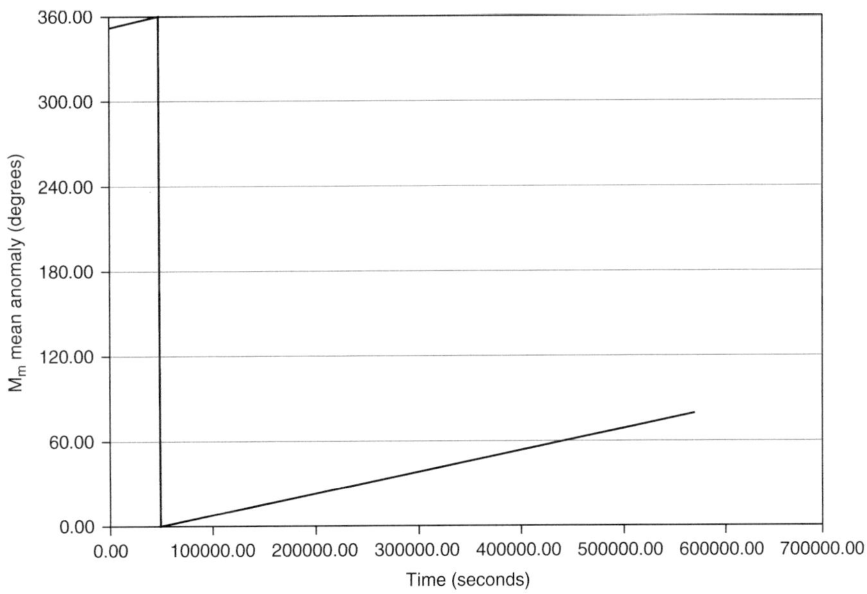

Figure 14.8 Variation of the moon's moving mean anomaly measured from its perigee during transfer.

respectively, with the radial components being the largest and the out-of-plane components the smallest, at least for the luni-solar geometry of this particular transfer example.

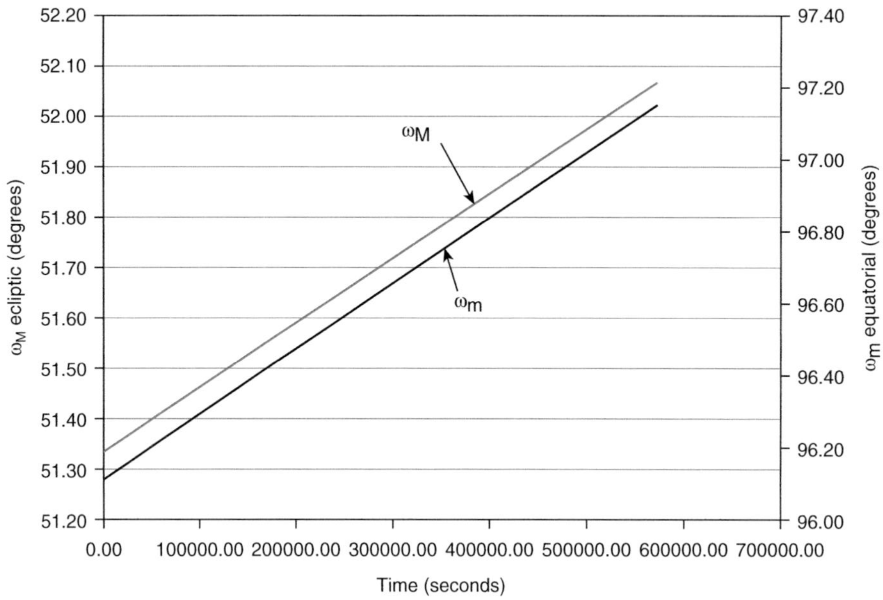

Figure 14.9 Advancement of the moon's perigee measured by its orbit ecliptic argument of perigee during transfer.

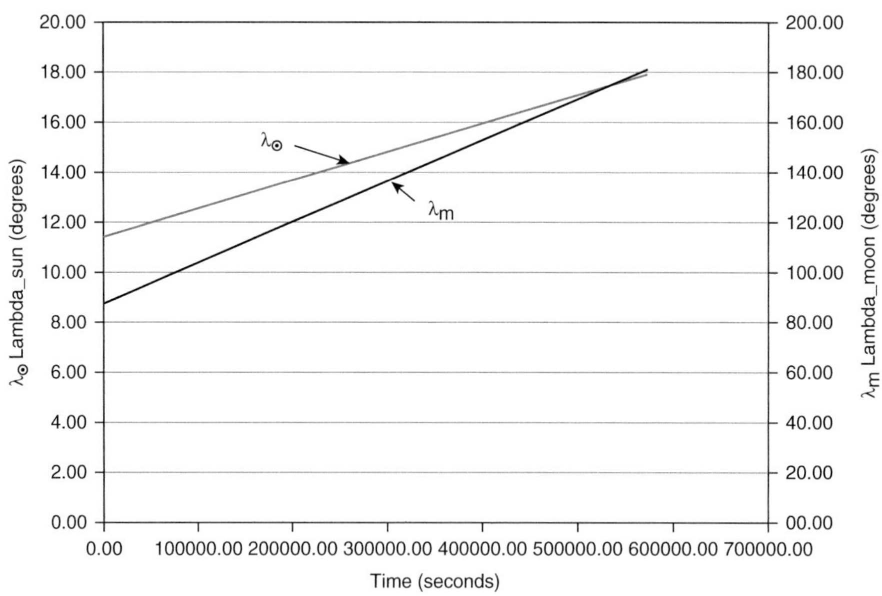

Figure 14.10 The sun's and moon's longitudes along ecliptic and lunar orbits respectively during transfer.

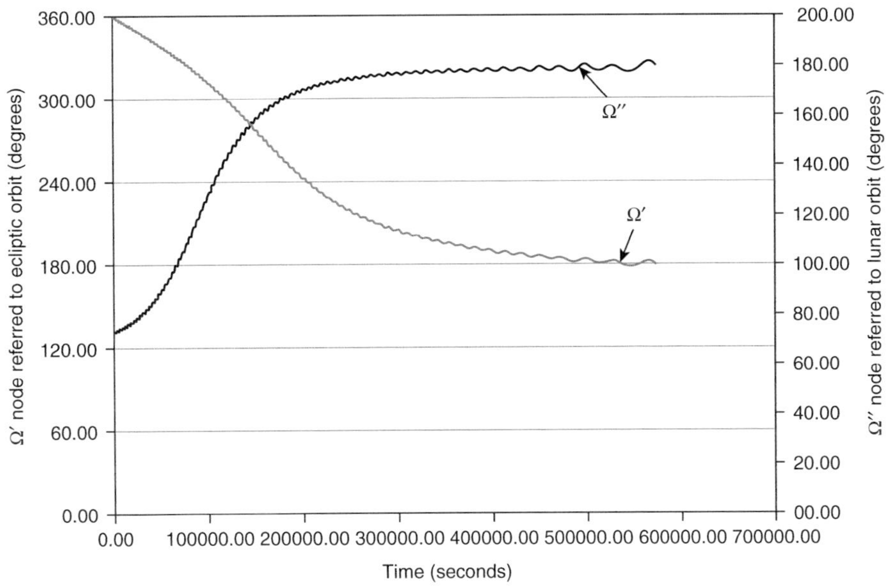

Figure 14.11 Variations of spacecraft orbit node along ecliptic and lunar orbits during transfer.

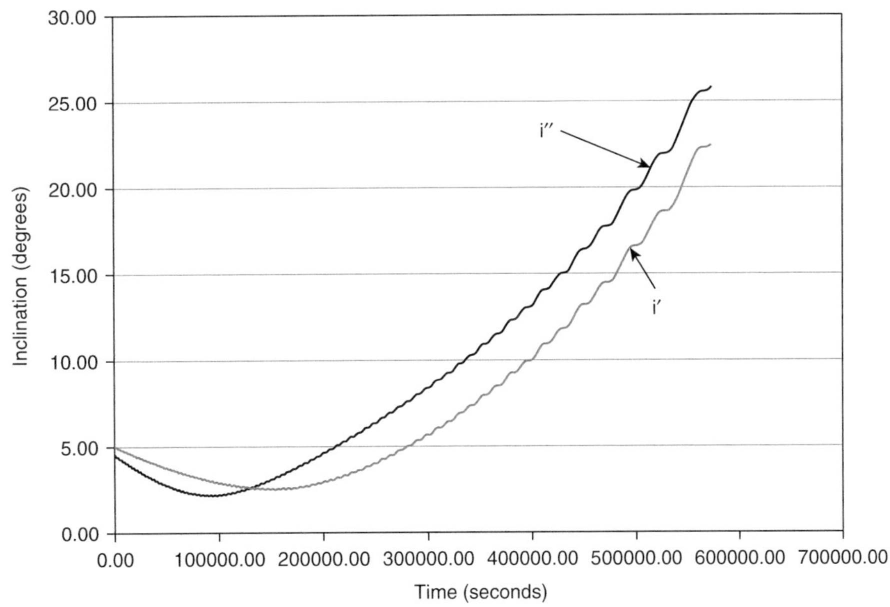

Figure 14.12 Variations of spacecraft orbit inclination referred to ecliptic and lunar orbits during transfer.

Figures 14.17 and 14.18 show the first- and second-order radial lunar accelerations, as well as the third-order and total ones, two curves at a time for ease of reading. Thus fmr_1, etc., stand for the first-order lunar acceleration and so forth, with Figure 14.19

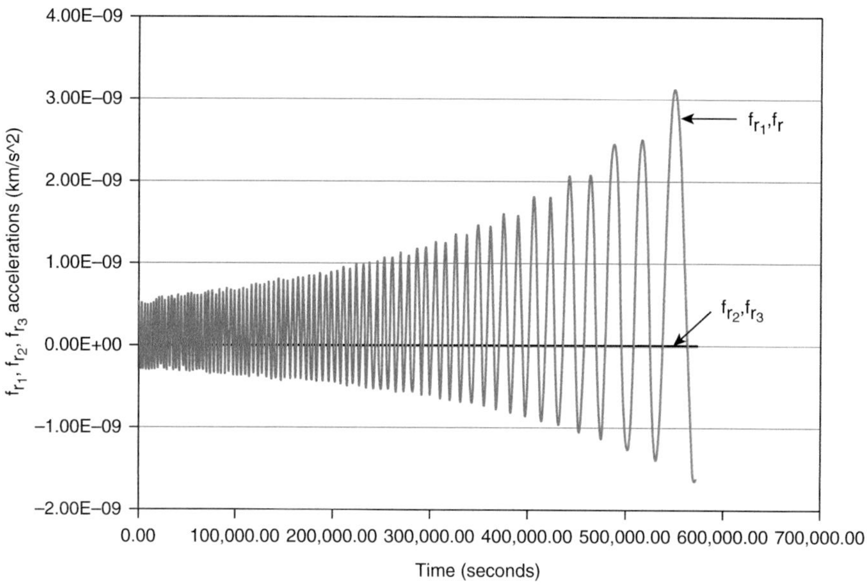

Figure 14.13 First-, second-, third-order and total radial solar accelerations during transfer.

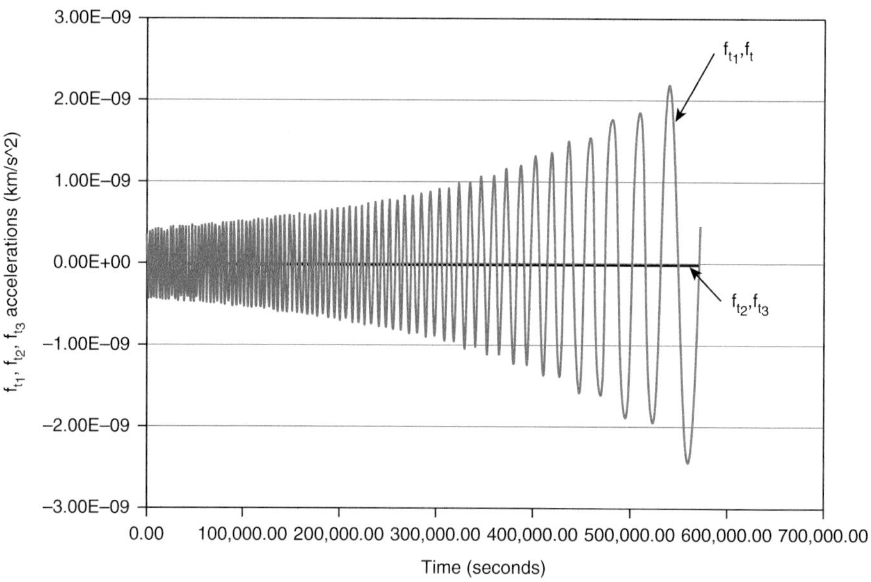

Figure 14.14 First-, second-, third-order and total normal solar accelerations during transfer.

depicting all four curves. Figures 14.20–14.25 show similar depictions for the normal and out-of-plane accelerations, namely *fmt* and *fmh*, with once again first and second order, as well as third and total, two curves at a time, with Figures 14.22 and 14.25 showing all four curves simultaneously. Figures 14.26 and 14.27 show the total radial

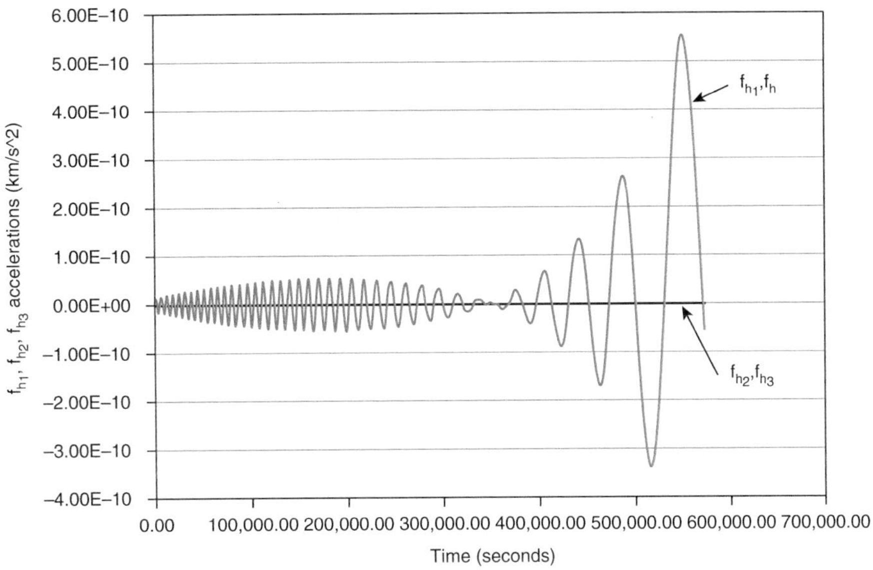

Figure 14.15 First-, second-, third-order and total out-of-plane solar accelerations during transfer.

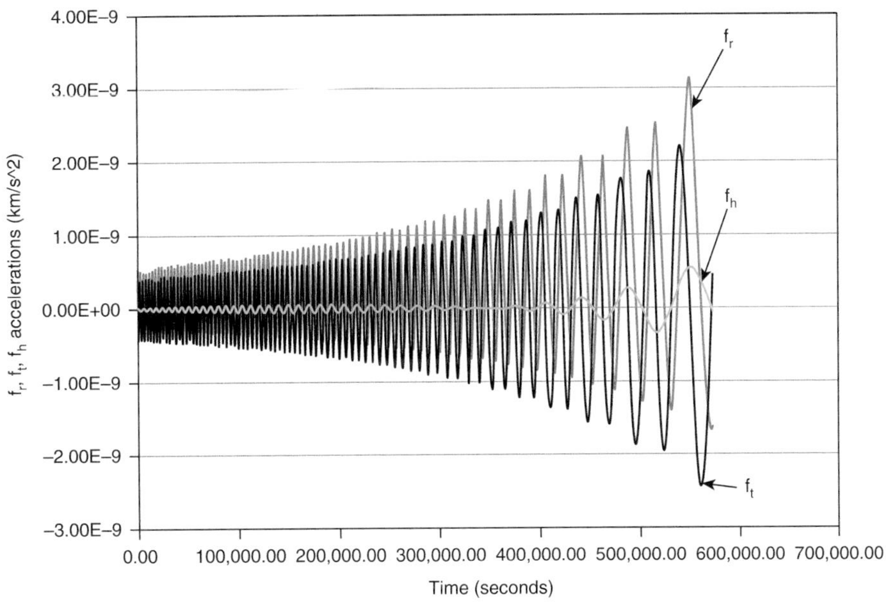

Figure 14.16 Total radial, normal and out-of-plane solar accelerations during transfer.

lunar acceleration, and the combined normal and out-of-plane counterparts respectively for ease of reading and visualization.

14.6 Conclusion

A high-order analytic expansion of the luni-solar disturbing function is carried out and the resulting accelerations are resolved along a set of rotating axes in order to arrive at

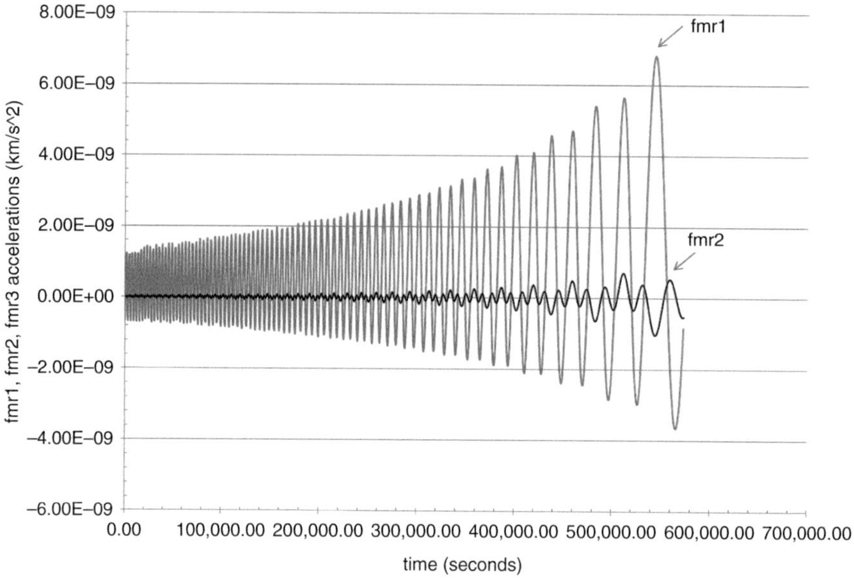

Figure 14.17 First- and second-order radial lunar accelerations during transfer.

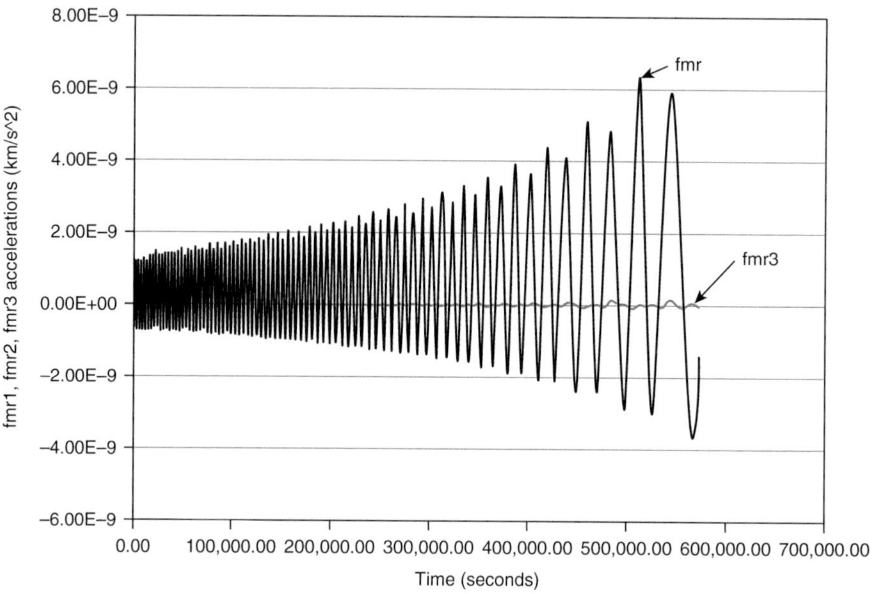

Figure 14.18 Third and total radial lunar accelerations during transfer.

a set of differential equations for precision integration. The apparent solar orbit is modeled as a conic, but the more complex motion of the moon's orbit whose ecliptic node regresses and perilune advances due to solar gravity is modeled analytically following de Pontécoulant's lunar theory. The resulting dynamic system is entirely self-contained

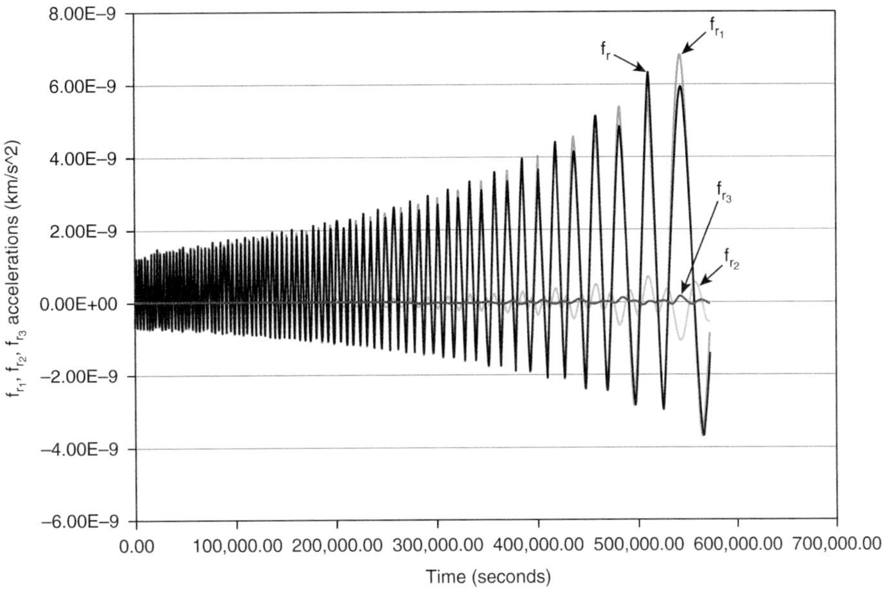

Figure 14.19 First-, second-, third-order and total radial lunar accelerations during transfer.

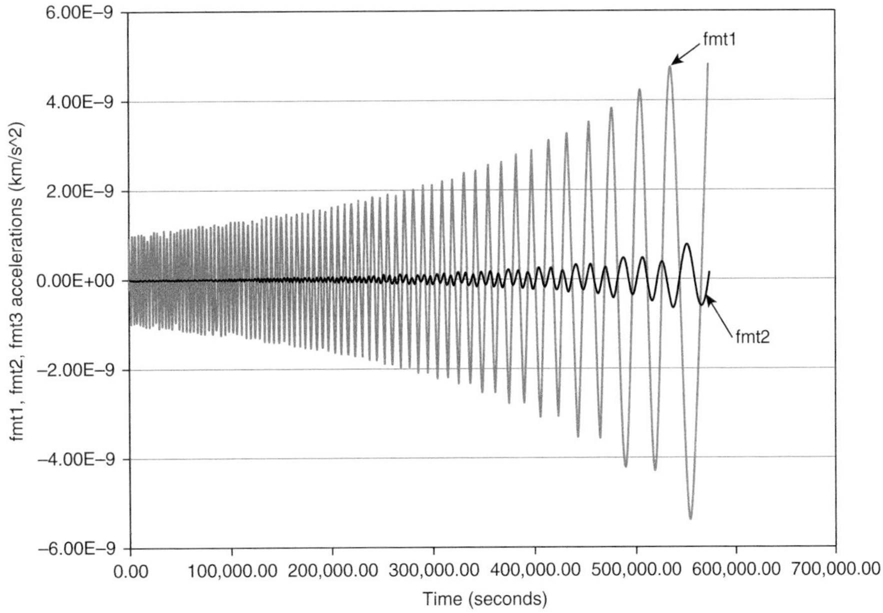

Figure 14.20 First- and second-order normal lunar accelerations during transfer.

and it does not require an outside luni-solar ephemeris generator, which must be called at each integration step to evaluate the third-body perturbation accelerations leading thereby to more efficient software for applications such as optimal low-thrust transfers in the general elliptic orbit. A set of nonsingular elements is also used to avoid singularities inherent in LEO–GEO transfers.

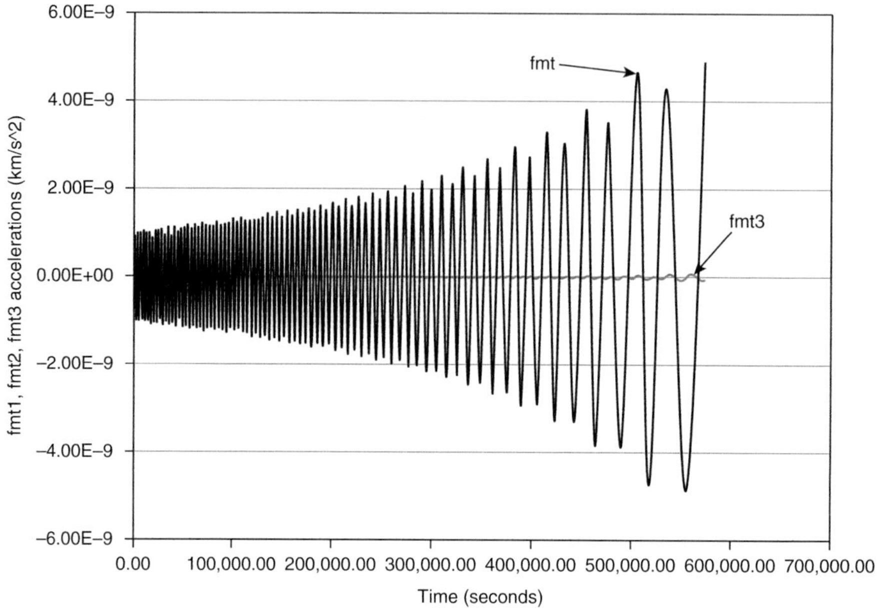

Figure 14.21 Third-order and total normal lunar accelerations during transfer.

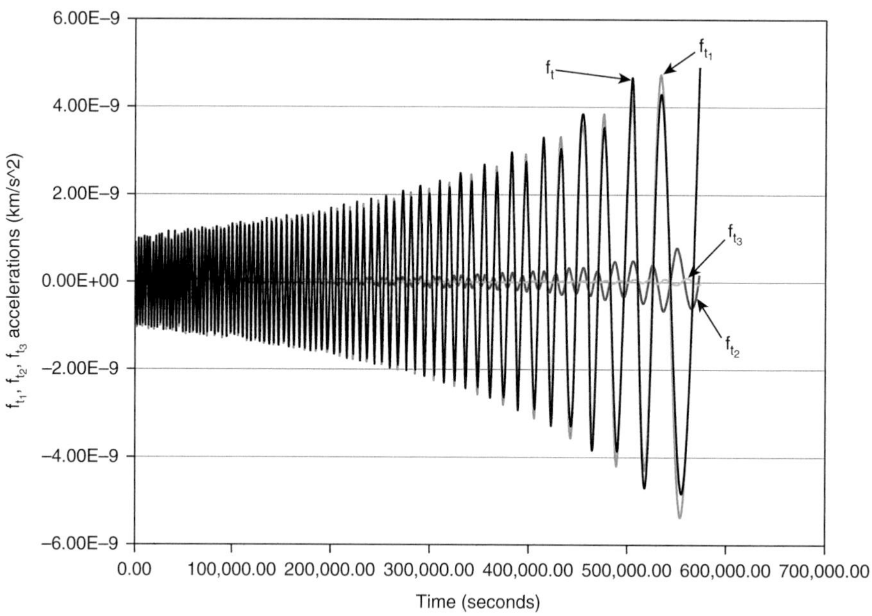

Figure 14.22 First-, second-, third-order and total normal lunar accelerations during transfer.

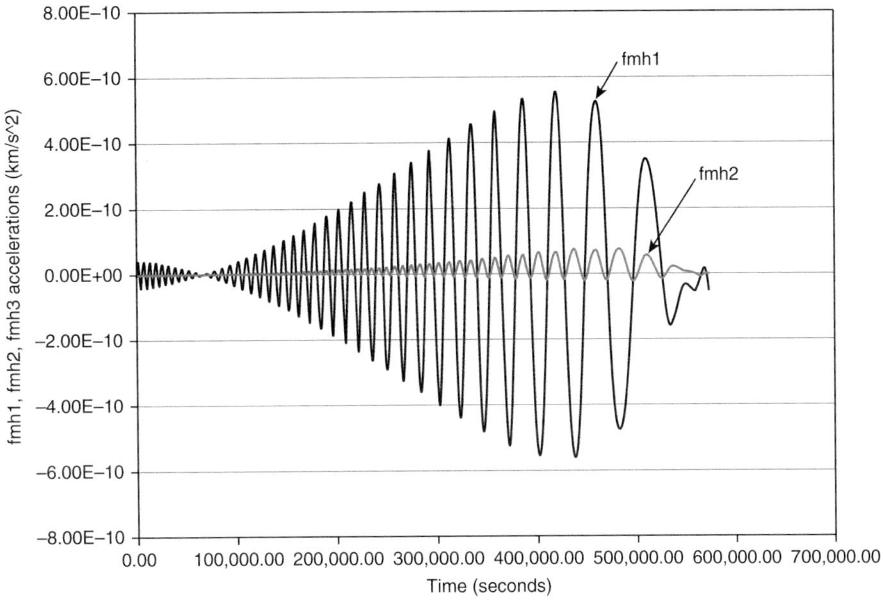

Figure 14.23 First- and second-order out-of-plane lunar accelerations during transfer.

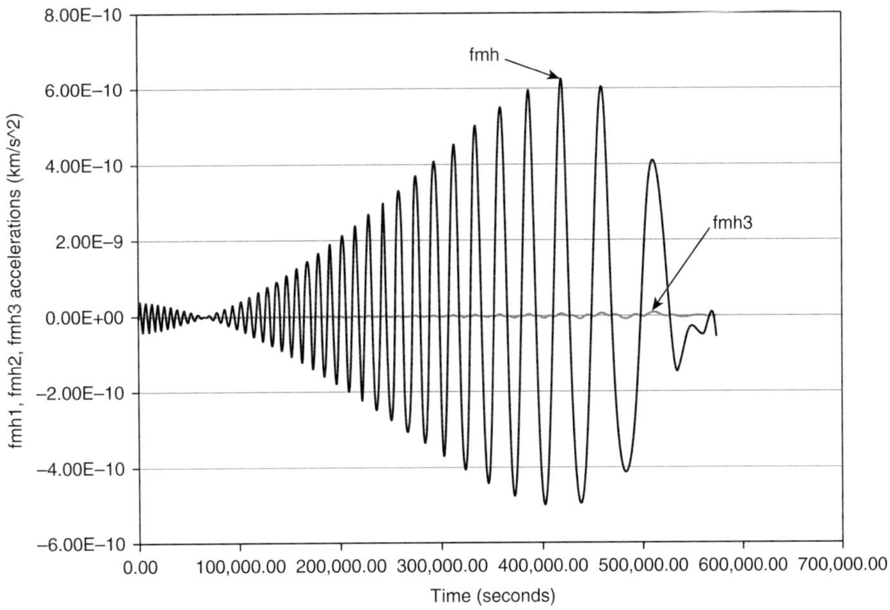

Figure 14.24 Third-order and total out-of-plane lunar accelerations during transfer.

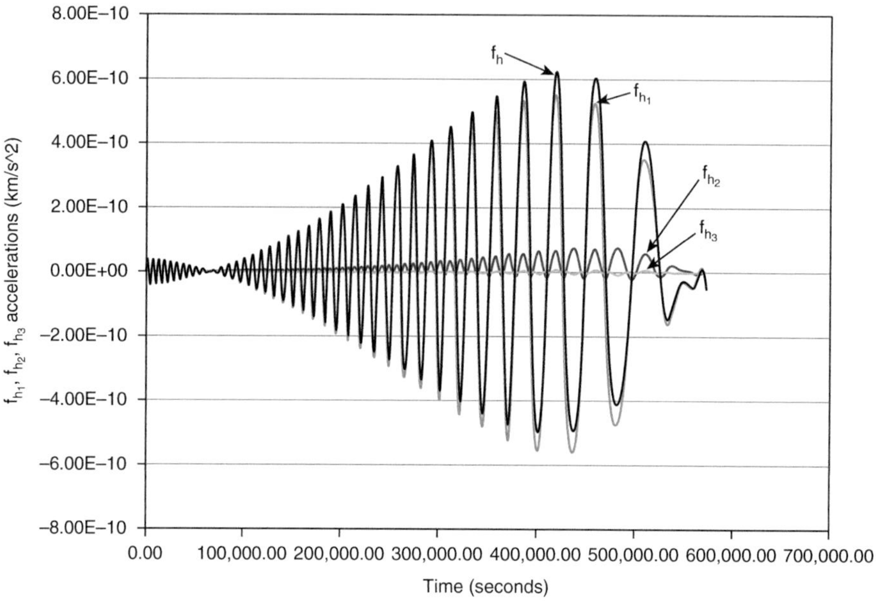

Figure 14.25 First-, second-, third-order and total out-of-plane lunar accelerations during transfer.

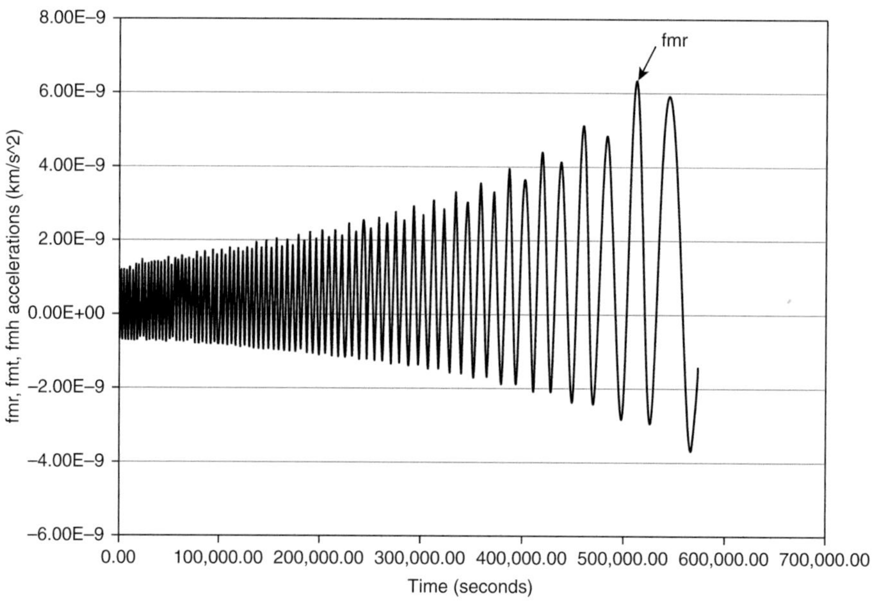

Figure 14.26 Radial lunar acceleration during transfer.

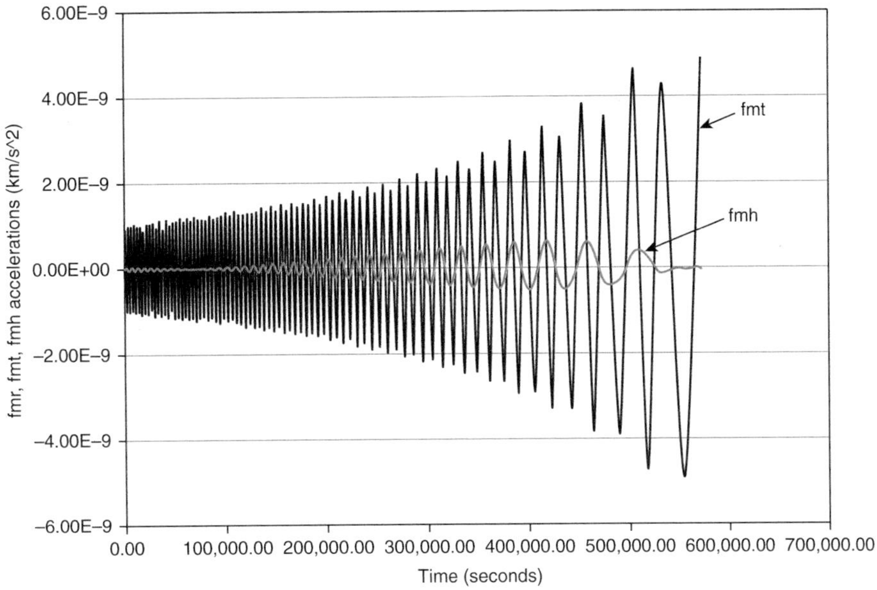

Figure 14.27 Normal and out-of-plane lunar acceleration during transfer.

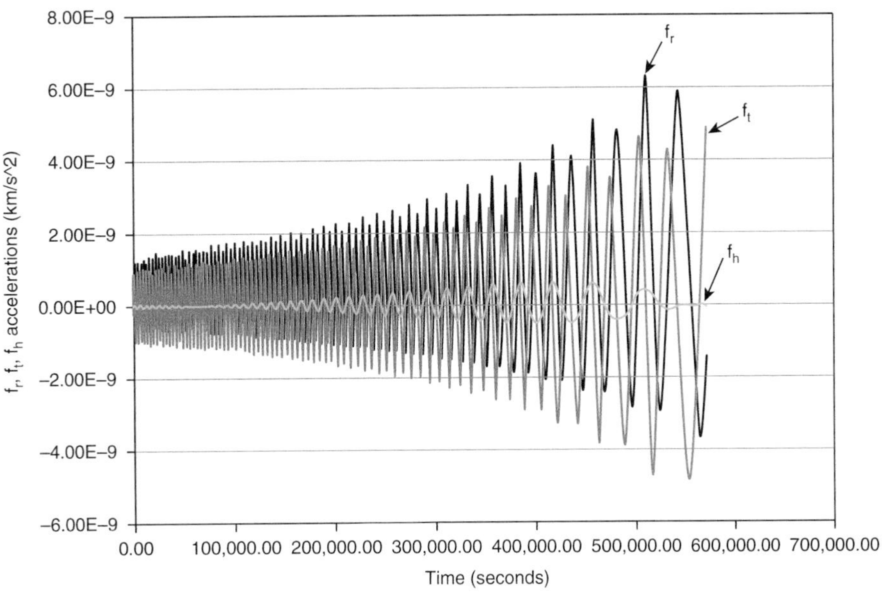

Figure 14.28 Total radial, normal and out-of-plane lunar accelerations during transfer.

References

[1] Kéchichian, J. A. (1997) Trajectory Optimization Using Nonsingular Orbital Elements and True Longitude, *Journal of Guidance, Control, and Dynamics*, Vol. 20, No. 5, 1003–1009.

[2] Kéchichian, J. A. (2008) Inclusion of Higher Order Harmonics in the Modeling of Optimal Low-Thrust Orbit Transfer, *The Journal of the Astronautical Sciences*, Vol. 56, No. 1, 41–70.

[3] Feistel, B. S. (2007) *Perturbing Accelerations in Low Thrust Trajectory Optimization*, M.S. Thesis, University of Texas at Austin.

[4] Battin, R. H. (1987) *An Introduction to the Mathematics and Methods of Astrodynamics*, AIAA, New York, NY, 387–391.

[5] Kéchichian, J. A. (1988) *Closed Form Solution of Perturbed Motion in Near-Circular Orbit Due to Luni-Solar Gravity in Rotating and Inertial Cartesian Frames*, AIAA Paper 88–4295-CP, AIAA/AAS Astrodynamics Conference, Minneapolis, MN.

[6] Breakwell, J. V. (1974) *Lecture Notes*: *Space Mechanics*, Department of Aeronautics and Astronautics, Stanford University, Palo Alto, CA.

[7] Brouwer, D. and Clemence, G. M. (1961) *Methods of Celestial Mechanics*, Academic Press, New York, NY, 31–35.

[8] Smart, W. M. (1961) *Celestial Mechanics*, John Wiley & Sons, Inc., New York, NY, 263–290.

[9] Kéchichian, J. A. (2010) *Analytic Expansions of Luni-Solar Gravity Perturbations along Rotating Axes for Trajectory Optimization: Part 2: The Multipliers System and Simulations*, IAF Paper IAC 10–C1.1.1, 61st International Astronautical Congress, Prague, Czech Republic.

15 Analytic Expansions of Luni-Solar Gravity Perturbations Along Rotating Axes for Trajectory Optimization: The Multipliers System and Simulations

15.1 Introduction

An efficient self-contained trajectory optimization software is generated by making use of de Pontécoulant's analytic lunar theory, removing the need for an outside third-body ephemeris program to compute the lunar and solar position vectors at each integration step. The accelerations being further resolved along the rotating Euler–Hill frame after expansion to third order in the spacecraft radial distance, the adjoint differential equations are derived in a direct manner complementing the generation of the dynamic system of equations for full compatibility. Because the variation of parameters equations are cast in terms of the nonsingular equinoctial elements with the perturbation accelerations resolved in their analytic form along the rotating axes, the adjoint equations are also derived in the same manner, providing a highly efficient and accurate system of equations for rapid computations in conjunction with Aerospace Corporation's NLP2 nonlinear programming codes to search for the initial values of the multipliers that steer the spacecraft toward its target orbit in minimum time. Numerical simulations show that the solutions obtained by the analysis developed in this chapter are essentially identical to the more indirect approach based on the use of inertial accelerations obtained from a separate ephemeris generator and subsequent conversions to the thrust frame and equinoctial system.

This chapter constitutes the second part of a two-part exposition of a method of efficient code development for trajectory optimization, being a companion chapter to the first part of [1], also shown in the Chapter 14 here. Part 1 focused on the derivations of the expressions for the dynamic equations under the influence of the third bodies, namely the Moon and the Sun, while part 2 dwells on the more difficult derivations that lead to the generation of the differential equations for the multipliers set such that both systems of equations can be efficiently coded and further associated with a parameter optimization software to search for the initial values of the multipliers that steer the spacecraft thrust orientation to reach the desired target orbit. The formulation employed matches the one used previously in [2] and [3], and also shown in Chapters 9 and 13, inasmuch as the rotating Euler–Hill frame is used to resolve the perturbing accelerations for direct integration of both the dynamic and adjoint equations in terms of the spacecraft nonsingular equinoctial elements. The luni-solar orbital motions are depicted according to the analytic theory of de Pontécoulant, as discussed in Chapter 14, such that the software is fully integrated without the need for an outside ephemeris generator that must be called at each integration step in order to extract the third-body position parameters before the evaluation of the resulting inertial accelerations which must then be converted to the rotating system as well as in terms of the spacecraft equinoctial elements as is done in [4]. In the subsequent sections, the partial derivatives of both the solar and lunar gravity perturbation vectors with respect to the spacecraft equinoctial elements are fully derived as they are needed in the description of the adjoint differential equations. The accuracy of this present method is established by comparing a fairly long-duration low-thrust transfer of about seven days from a low Earth orbit (LEO) to geosynchronous Earth orbit (GEO), to the solution generated using the method developed in [4]. Graphical representations of the various orbit parameters shown by their classical counterparts, as well as the evolutions of the associated multiplier variables, are depicted next. A longer-duration transfer of about three months from a geosynchronous transfer orbit (GTO) to GEO is also shown with a significant three-body perturbation effect and found to be within 2 min in total transfer time to the solution developed by W. Hallman of the Flight Mechanics Department at Aerospace, based on Feistel's method in [4]. Appendix 15.1 shows the long series of the analytic partials of the lunar and solar accelerations components along the rotating axes with respect to the spacecraft nonsingular elements.

15.2 The Hamiltonian and Euler–Lagrange Equations

The system of dynamic equations for the perturbed motion of a thrusting vehicle in general elliptic orbit given in terms of nonsingular equinoctial elements is shown in [1] and repeated below for convenience.

The elements a, $h = e\sin(\Omega + \omega)$, $k = e\cos(\Omega + \omega)$, $p = \tan(i/2)\sin\Omega$, $q = \tan(i/2)\cos\Omega$, $L = \theta^* + \Omega$ are used as before [1, 3] (see also Chapters 14, 9 and 13), defined in terms of the classical elements a, e, i, Ω, ω and θ^*, the true anomaly. The thrust vector as well as the perturbations $\mathbf{f}_{J_2}$, $\mathbf{f}_{J_3}$, $\mathbf{f}_{J_4}$, $\mathbf{f}_{\odot}$ and $\mathbf{f}_{☽}$ for the zonal harmonics J_2, J_3, J_4 and the solar and lunar gravity are resolved along the rotating Euler–Hill frame $(\hat{\mathbf{r}}, \hat{\boldsymbol{\theta}}, \hat{\mathbf{h}})$, with u_r, u_θ and u_h standing for the components of the unit vector $\hat{\mathbf{u}}$ along the direction of the thrust vector.

$$\begin{pmatrix} \dot{a} \\ \dot{h} \\ \dot{k} \\ \dot{p} \\ \dot{q} \\ \dot{L} \end{pmatrix} = \begin{pmatrix} B_{11}^L & B_{12}^L & B_{13}^L \\ B_{21}^L & B_{22}^L & B_{23}^L \\ B_{31}^L & B_{32}^L & B_{33}^L \\ B_{41}^L & B_{42}^L & B_{43}^L \\ B_{51}^L & B_{52}^L & B_{53}^L \\ B_{61}^L & B_{62}^L & B_{63}^L \end{pmatrix} \left[\begin{pmatrix} u_r \\ u_\theta \\ u_h \end{pmatrix} (f_t) + \mathbf{f}_{J_2} + \mathbf{f}_{J_3} + \mathbf{f}_{J_4} + \mathbf{f}_{\odot} + \mathbf{f}_{☽} \right] + \begin{bmatrix} 0 \\ 0 \\ 0 \\ 0 \\ 0 \\ \frac{na^2\left(1-h^2-k^2\right)^{1/2}}{r^2} \end{bmatrix} \tag{15.1}$$

The rows of matrix B^L are also resolved along the rotating axes and they are given in [2, 3] and Chapters 9 and 13. Letting $\mathbf{z} = (a, h, k, p, q, L)$ stand for the spacecraft state vector, the Hamiltonian of this perturbed system can be written as

$$H = \boldsymbol{\lambda}_z^T B^L(\mathbf{z}) f_t \hat{\mathbf{u}} + \lambda_L \left[\frac{na^2 \left(1 - h^2 - k^2\right)^{1/2}}{r^2} \right] + \boldsymbol{\lambda}_z^T B^L(\mathbf{z}) \left(\mathbf{f}_{J_2} + \mathbf{f}_{J_3} + \mathbf{f}_{J_4} + \mathbf{f}_{\odot} + \mathbf{f}_{\rightmoon} \right) \tag{15.2}$$

leading to the Euler–Lagrange differential equations for the multipliers

$$\begin{aligned} \dot{\boldsymbol{\lambda}}_z = -\frac{\partial H}{\partial \mathbf{z}} &= -\boldsymbol{\lambda}_z^T \frac{\partial B^L}{\partial \mathbf{z}} f_t \hat{\mathbf{u}} - \lambda_L \frac{\partial}{\partial \mathbf{z}} \left[\frac{na^2 \left(1 - h^2 - k^2\right)^{1/2}}{r^2} \right] \\ &- \boldsymbol{\lambda}_z^T \frac{\partial B^L}{\partial \mathbf{z}} \left(\mathbf{f}_{J_2} + \mathbf{f}_{J_3} + \mathbf{f}_{J_4} + \mathbf{f}_{\odot} + \mathbf{f}_{\rightmoon} \right) \\ &- \boldsymbol{\lambda}_z^T B^L \left(\frac{\partial \mathbf{f}_{J_2}}{\partial \mathbf{z}} + \frac{\partial \mathbf{f}_{J_3}}{\partial \mathbf{z}} + \frac{\partial \mathbf{f}_{J_4}}{\partial \mathbf{z}} + \frac{\partial \mathbf{f}_{\odot}}{\partial \mathbf{z}} + \frac{\partial \mathbf{f}_{\rightmoon}}{\partial \mathbf{z}} \right) \end{aligned} \tag{15.3}$$

In [1] and the preceding chapter, the vectors $\mathbf{f}_{\odot}$ and $\mathbf{f}_{\rightmoon}$ have been expanded to third order in the spacecraft radial distance r along the $\hat{\mathbf{r}}$, $\hat{\boldsymbol{\theta}}$ and $\hat{\mathbf{h}}$ directions, namely f_r, f_θ and f_h, and the partials $\frac{\partial B^L}{\partial \mathbf{z}}, \frac{\partial \mathbf{f}_{J_2}}{\partial \mathbf{z}}, \frac{\partial \mathbf{f}_{J_3}}{\partial \mathbf{z}}, \frac{\partial \mathbf{f}_{J_4}}{\partial \mathbf{z}}$ have already been derived in [2, 3] and Chapters 9 and 13. The description of $\dot{\boldsymbol{\lambda}}_z$ in Equation (15.3) will be completed once we derive the remaining $\frac{\partial \mathbf{f}_{\odot}}{\partial \mathbf{z}}$ and $\frac{\partial \mathbf{f}_{\rightmoon}}{\partial \mathbf{z}}$ partials.

Starting with solar gravity, we use an s superscript to differentiate it from the lunar gravity where the m superscript is used instead, that is, $\mathbf{f}_{\odot}^s$ and $\mathbf{f}_{\rightmoon}^m$, or f_r^s, f_θ^s, f_h^s, and f_r^m, f_θ^m, f_h^m for their respective components. Equations (16)–(18) of [1] and (14.16)–(14.18) of Chapter 14 for f_r^s, f_θ^s and f_h^s involve r, and the angles i', Ω' and θ'. It is clear that $r = f(a, h, k, L)$, $s_{i'}$ and $c_{i'}$ are $f(p, q)$, $s_{\Omega'}$ and $c_{\Omega'}$ also $f(p, q)$, while $s_{\theta'}$ and $c_{\theta'}$ are $f(h, k, p, q, L)$ because $\theta' = \omega' + \theta^*$ with $\omega' = f(h, k, p, q)$ and $\theta^* = f(h, k, L)$. It is thus necessary to compute the partial derivatives $\partial r/\partial a$, $\partial r/\partial h$, $\partial r/\partial k$ and $\partial r/\partial L$ for the radial distance r, the partials $\partial s_{\theta'}/\partial h$, $\partial s_{\theta'}/\partial k$, $\partial s_{\theta'}/\partial p$, $\partial s_{\theta'}/\partial q$ and $\partial s_{\theta'}/\partial L$ for $s_{\theta'}$ and similar partials for $c_{\theta'}$, and finally $\partial s_{i'}/\partial p$, $\partial s_{i'}/\partial q$ and similar partials for $c_{i'}$, as well as $\partial s_{\lambda_\odot - \Omega'}/\partial p$, $\partial s_{\lambda_\odot - \Omega'}/\partial q$ and similar partials for $c_{\lambda_\odot - \Omega'}$ with respect to p and q.

Starting with the expressions of $c_{i'}$ and $s_{i'}$ given in Equations (49) and (53) of [1] and (14.49) and (14.53) of Chapter 14, namely

$$c_{i'} = c_\Omega s_i s_\varepsilon + c_i c_\varepsilon = f(\Omega, i) \tag{15.4}$$

$$s_{i'} = \left[s_\Omega^2 s_i^2 + (c_\Omega s_i c_\varepsilon - c_i s_\varepsilon)^2 \right]^{1/2} = f(\Omega, i) \tag{15.5}$$

written in terms of the spacecraft orbit classical elements, the relevant partials can be cast as

$$\frac{\partial s_{i'}}{\partial p} = \frac{\partial s_{i'}}{\partial \Omega} \frac{\partial \Omega}{\partial p} + \frac{\partial s_{i'}}{\partial i} \frac{\partial i}{\partial p} \tag{15.6}$$

$$\frac{\partial s_{i'}}{\partial q} = \frac{\partial s_{i'}}{\partial \Omega} \frac{\partial \Omega}{\partial q} + \frac{\partial s_{i'}}{\partial i} \frac{\partial i}{\partial q} \tag{15.7}$$

and

$$\frac{\partial c_{i'}}{\partial p} = \frac{\partial c_{i'}}{\partial \Omega}\frac{\partial \Omega}{\partial p} + \frac{\partial c_{i'}}{\partial i}\frac{\partial i}{\partial p} \tag{15.8}$$

$$\frac{\partial c_{i'}}{\partial q} = \frac{\partial c_{i'}}{\partial \Omega}\frac{\partial \Omega}{\partial q} + \frac{\partial c_{i'}}{\partial i}\frac{\partial i}{\partial q} \tag{15.9}$$

From the inverse definition $\Omega = \tan^{-1}(p/q)$,

$$\frac{\partial \Omega}{\partial p} = \frac{q}{p^2+q^2} \tag{15.10}$$

$$\frac{\partial \Omega}{\partial q} = \frac{-p}{p^2+q^2} \tag{15.11}$$

which can also be obtained by taking the derivatives of $s_\Omega = p/\left(p^2+q^2\right)^{1/2}$ with respect to p and q and using $c_\Omega = q/\left(p^2+q^2\right)^{1/2}$. In a similar way, from the inverse expression $i = 2\tan^{-1}\left(p^2+q^2\right)^{1/2}$

$$\frac{\partial i}{\partial p} = \frac{2p}{\left(p^2+q^2\right)^{1/2}\left(1+p^2+q^2\right)} \tag{15.12}$$

$$\frac{\partial i}{\partial q} = \frac{2q}{\left(p^2+q^2\right)^{1/2}\left(1+p^2+q^2\right)} \tag{15.13}$$

which can also be obtained by taking the derivatives of the expression for $s_i = 2\left(p^2+q^2\right)^{1/2}/\left(1+p^2+q^2\right)$ with respect to p and q and using $c_i = \left(1-p^2-q^2\right)/\left(1+p^2+q^2\right)$.

From the expression for $s_{i'}$ in Equation (15.5)

$$\frac{\partial s_{i'}}{\partial \Omega} = \frac{s_\Omega c_\Omega s_i^2 - s_\Omega s_i c_\varepsilon\left(c_\Omega s_i c_\varepsilon - c_i s_\varepsilon\right)}{\left[s_\Omega^2 s_i^2 + \left(c_\Omega s_i c_\varepsilon - c_i s_\varepsilon\right)^2\right]^{1/2}} \tag{15.14}$$

$$\frac{\partial s_{i'}}{\partial i} = \frac{s_\Omega^2 s_i c_i + \left(c_\Omega s_i c_\varepsilon - c_i s_\varepsilon\right)\left(c_\Omega c_i c_\varepsilon + s_i s_\varepsilon\right)}{\left[s_\Omega^2 s_i^2 + \left(c_\Omega s_i c_\varepsilon - c_i s_\varepsilon\right)^2\right]^{1/2}} \tag{15.15}$$

where the denominator in the above two equations is $s_{i'}$ itself.

Similarly from Equation (15.4) for $c_{i'}$

$$\frac{\partial c_{i'}}{\partial \Omega} = -s_\Omega s_i s_\varepsilon \tag{15.16}$$

$$\frac{\partial c_{i'}}{\partial i} = c_\Omega c_i s_\varepsilon - s_i c_\varepsilon \tag{15.17}$$

Note that $c_{i'}$ can be written directly in terms of the equinoctial elements p and q as

$$c_{i'} = \frac{2q s_\varepsilon + \left(1-p^2-q^2\right)c_\varepsilon}{1+p^2+q^2} \tag{15.18}$$

so that its partials with respect to p and q can be generated directly instead of using the chain rule of Equations (15.8) and (15.9). However, in view of the complexity of the expressions for $s_{i'}$, $s_{\Omega'}$, $c_{\Omega'}$, $s_{\omega'}$ and $c_{\omega'}$ of Equations (66)–(70) of [1] and

(14.66)–(14.70) of Chapter 14 in terms of the equinoctials, it is clear that the adoption of the chain rule in deriving their partial derivatives is a much easier task. Next, we have for $s_{\Omega'}$ and $c_{\Omega'}$

$$\frac{\partial s_{\Omega'}}{\partial p} = \frac{\partial s_{\Omega'}}{\partial \Omega}\frac{\partial \Omega}{\partial p} + \frac{\partial s_{\Omega'}}{\partial i}\frac{\partial i}{\partial p} \tag{15.19}$$

$$\frac{\partial s_{\Omega'}}{\partial q} = \frac{\partial s_{\Omega'}}{\partial \Omega}\frac{\partial \Omega}{\partial q} + \frac{\partial s_{\Omega'}}{\partial i}\frac{\partial i}{\partial q} \tag{15.20}$$

and

$$\frac{\partial c_{\Omega'}}{\partial p} = \frac{\partial c_{\Omega'}}{\partial \Omega}\frac{\partial \Omega}{\partial p} + \frac{\partial c_{\Omega'}}{\partial i}\frac{\partial i}{\partial p} \tag{15.21}$$

$$\frac{\partial c_{\Omega'}}{\partial q} = \frac{\partial c_{\Omega'}}{\partial \Omega}\frac{\partial \Omega}{\partial q} + \frac{\partial c_{\Omega'}}{\partial i}\frac{\partial i}{\partial q} \tag{15.22}$$

Using Equations (54) and (55) of [1] and (14.54) and (14.55) of Chapter 14 for $s_{\Omega'}$ and $c_{\Omega'}$, namely

$$s_{\Omega'} = s_\Omega s_i / s_{i'} = f(\Omega, i) \tag{15.23}$$

$$c_{\Omega'} = (c_\Omega s_i c_\varepsilon - c_i s_\varepsilon)/s_{i'} = f(\Omega, i) \tag{15.24}$$

with $s_{i'}$ given itself in Equation (15.5) in terms of the classical elements Ω and i, we have

$$\frac{\partial s_{\Omega'}}{\partial \Omega} = \frac{c_\Omega s_i s_{i'} - s_\Omega s_i \frac{\partial s_{i'}}{\partial \Omega}}{s_{i'}^2} \tag{15.25}$$

$$\frac{\partial s_{\Omega'}}{\partial i} = \frac{s_\Omega c_i s_{i'} - s_\Omega s_i \frac{\partial s_{i'}}{\partial i}}{s_{i'}^2} \tag{15.26}$$

and

$$\frac{\partial c_{\Omega'}}{\partial \Omega} = \frac{-s_\Omega s_i c_\varepsilon s_{i'} - (c_\Omega s_i c_\varepsilon - c_i s_\varepsilon)\frac{\partial s_{i'}}{\partial \Omega}}{s_{i'}^2} \tag{15.27}$$

$$\frac{\partial c_{\Omega'}}{\partial i} = \frac{(c_\Omega c_i c_\varepsilon + s_i s_\varepsilon)\, s_{i'} - (c_\Omega s_i c_\varepsilon - c_i s_\varepsilon)\frac{\partial s_{i'}}{\partial i}}{s_{i'}^2} \tag{15.28}$$

For $s_{\omega'}$, $c_{\omega'}$, their expressions are shown in Equations (56) and (57) of [1] and (14.56) and (14.57) of Chapter 14 in terms of Ω, i, ω, namely

$$s_{\omega'} = \frac{-c_\omega s_\Omega s_\varepsilon - s_\omega c_\Omega c_i s_\varepsilon + s_\omega s_i c_\varepsilon}{s_{i'}} = f(\Omega, i, \omega) \tag{15.29}$$

$$c_{\omega'} = \frac{s_\omega s_\Omega s_\varepsilon - c_\omega c_\Omega c_i s_\varepsilon + c_\omega s_i c_\varepsilon}{s_{i'}} = f(\Omega, i, \omega) \tag{15.30}$$

However, s_ω and c_ω are functions of h, k, p and q because

$$s_\omega = \frac{hq - kp}{\left(h^2 + k^2\right)^{1/2}\left(p^2 + q^2\right)^{1/2}} = f(h, k, p, q) \tag{15.31}$$

$$c_\omega = \frac{hp + kq}{\left(h^2 + k^2\right)^{1/2}\left(p^2 + q^2\right)^{1/2}} = f(h, k, p, q) \tag{15.32}$$

and therefore $s_{\omega'}$ and $c_{\omega'}$ are $f(h,k,p,q)$, such that we need their partials with respect to these four elements. Thus

$$\frac{\partial s_{\omega'}}{\partial h} = \frac{\partial s_{\omega'}}{\partial \Omega}\frac{\partial \Omega}{\partial h} + \frac{\partial s_{\omega'}}{\partial i}\frac{\partial i}{\partial h} + \frac{\partial s_{\omega'}}{\partial \omega}\frac{\partial \omega}{\partial h} = \frac{\partial s_{\omega'}}{\partial \omega}\frac{\partial \omega}{\partial h} \tag{15.33}$$

because $\partial\Omega/\partial h$ and $\partial i/\partial h$ are also equal to zero

$$\frac{\partial s_{\omega'}}{\partial k} = \frac{\partial s_{\omega'}}{\partial \Omega}\frac{\partial \Omega}{\partial k} + \frac{\partial s_{\omega'}}{\partial i}\frac{\partial i}{\partial k} + \frac{\partial s_{\omega'}}{\partial \omega}\frac{\partial \omega}{\partial k} = \frac{\partial s_{\omega'}}{\partial \omega}\frac{\partial \omega}{\partial k} \tag{15.34}$$

because $\partial\Omega/\partial k$ and $\partial i/\partial k$ are also equal to zero

$$\frac{\partial s_{\omega'}}{\partial p} = \frac{\partial s_{\omega'}}{\partial \Omega}\frac{\partial \Omega}{\partial p} + \frac{\partial s_{\omega'}}{\partial i}\frac{\partial i}{\partial p} + \frac{\partial s_{\omega'}}{\partial \omega}\frac{\partial \omega}{\partial p} + \frac{\partial s_{\omega'}}{\partial s_{i'}}\frac{\partial s_{i'}}{\partial \Omega}\frac{\partial \Omega}{\partial p} + \frac{\partial s_{\omega'}}{\partial s_{i'}}\frac{\partial s_{i'}}{\partial i}\frac{\partial i}{\partial p} \tag{15.35}$$

with $\partial s_{\omega'}/\partial s_{i'}$ derived from Equation (15.29)

$$\frac{\partial s_{\omega'}}{\partial s_{i'}} = \frac{-s_{\omega'}}{s_{i'}} \tag{15.36}$$

and likewise:

$$\frac{\partial s_{\omega'}}{\partial \Omega} = \frac{(-c_\omega c_\Omega s_\varepsilon + s_\omega s_\Omega c_i s_\varepsilon) - s_{\omega'}\dfrac{\partial s_{i'}}{\partial \Omega}}{s_{i'}} \tag{15.37}$$

$$\frac{\partial s_{\omega'}}{\partial i} = \frac{(s_\omega c_\Omega s_i s_\varepsilon + s_\omega c_i c_\varepsilon) - s_{\omega'}\dfrac{\partial s_{i'}}{\partial i}}{s_{i'}} \tag{15.38}$$

$$\frac{\partial s_{\omega'}}{\partial \omega} = \frac{s_\omega s_\Omega s_\varepsilon - c_\omega c_\Omega c_i s_\varepsilon + c_\omega s_i c_\varepsilon}{s_{i'}} \tag{15.39}$$

Using the inverse definition $\omega = \tan^{-1}\left(\frac{h}{k}\right) - \tan^{-1}\left(\frac{p}{q}\right)$, we have

$$\frac{\partial \omega}{\partial h} = \frac{k}{h^2+k^2} \tag{15.40}$$

$$\frac{\partial \omega}{\partial k} = \frac{-h}{h^2+k^2} \tag{15.41}$$

$$\frac{\partial \omega}{\partial p} = \frac{-q}{(p^2+q^2)} \tag{15.42}$$

$$\frac{\partial \omega}{\partial q} = \frac{p}{(p^2+q^2)} \tag{15.43}$$

Because the luni-solar orbits are formulated in terms of classical elements, the partials $\partial\Omega/\partial p$, $\partial\Omega/\partial q$ in Equations (15.10) and (15.11), as well as the partials $\partial i/\partial p$, $\partial i/\partial q$, $\partial\omega/\partial p$ and $\partial\omega/\partial q$ in Equations (15.12), (15.13), (15.42) and (15.43) can tend toward 0/0 when both p and q approach 0. This condition is handled in the codes by setting p^2+q^2 to 1 in their denominator so that these partials are evaluated as 0 without ambiguity. Similarly, when both h and k tend toward 0, the partials $\partial\omega/\partial h$ and $\partial\omega/\partial k$ in Equations (15.40) and (15.41) are also evaluated as 0 by setting h^2+k^2 to 1 in their denominator.

Finally, if i' becomes less than 10^{-16}, it is set to 10^{-16} to avoid singularity conditions in the partials $\partial s_{i'}/\partial\Omega$, $\partial s_{i'}/\partial i$, $\partial s_{\Omega'}/\partial\Omega$, $\partial s_{\Omega'}/\partial i$, $\partial c_{\Omega'}/\partial\Omega$, $\partial c_{\Omega'}/\partial i$, $\partial s_{\omega'}/\partial\Omega$, $\partial s_{\omega'}/\partial i$

and $\partial s_{\omega'}/\partial\omega$ in Equations (15.14), (15.15), (15.25)–(15.28) and (15.37)–(15.39), with the understanding that the denominator in Equations (15.14) and (15.15) is $s_{i'}$ itself, as mentioned earlier.

Finally, from $s_{\omega'}$ and using the same manipulations

$$\frac{\partial s_{\omega'}}{\partial q} = \frac{\partial s_{\omega'}}{\partial \Omega}\frac{\partial \Omega}{\partial q} + \frac{\partial s_{\omega'}}{\partial i}\frac{\partial i}{\partial q} + \frac{\partial s_{\omega'}}{\partial \omega}\frac{\partial \omega}{\partial q} + \frac{\partial s_{\omega'}}{\partial s_{i'}}\frac{\partial s_{i'}}{\partial \Omega}\frac{\partial \Omega}{\partial q} + \frac{\partial s_{\omega'}}{\partial s_{i'}}\frac{\partial s_{i'}}{\partial i}\frac{\partial i}{\partial q} \tag{15.44}$$

with all the relevant partials clearly available. For $c_{\omega'}$ in Equation (15.30), an identical procedure leads to the following partial derivatives:

$$\frac{\partial c_{\omega'}}{\partial h} = \frac{\partial c_{\omega'}}{\partial \omega}\frac{\partial \omega}{\partial h} \tag{15.45}$$

$$\frac{\partial c_{\omega'}}{\partial k} = \frac{\partial c_{\omega'}}{\partial \omega}\frac{\partial \omega}{\partial k} \tag{15.46}$$

$$\frac{\partial c_{\omega'}}{\partial p} = \frac{\partial c_{\omega'}}{\partial \Omega}\frac{\partial \Omega}{\partial p} + \frac{\partial c_{\omega'}}{\partial i}\frac{\partial i}{\partial p} + \frac{\partial c_{\omega'}}{\partial \omega}\frac{\partial \omega}{\partial p} + \frac{\partial c_{\omega'}}{\partial s_{i'}}\frac{\partial s_{i'}}{\partial \Omega}\frac{\partial \Omega}{\partial p} + \frac{\partial c_{\omega'}}{\partial s_{i'}}\frac{\partial s_{i'}}{\partial i}\frac{\partial i}{\partial p} \tag{15.47}$$

$$\frac{\partial c_{\omega'}}{\partial q} = \frac{\partial c_{\omega'}}{\partial \Omega}\frac{\partial \Omega}{\partial q} + \frac{\partial c_{\omega'}}{\partial i}\frac{\partial i}{\partial q} + \frac{\partial c_{\omega'}}{\partial \omega}\frac{\partial \omega}{\partial q} + \frac{\partial c_{\omega'}}{\partial s_{i'}}\frac{\partial s_{i'}}{\partial \Omega}\frac{\partial \Omega}{\partial q} + \frac{\partial c_{\omega'}}{\partial s_{i'}}\frac{\partial s_{i'}}{\partial i}\frac{\partial i}{\partial q} \tag{15.48}$$

where

$$\frac{\partial c_{\omega'}}{\partial s_{i'}} = \frac{-c_{\omega'}}{s_{i'}} \tag{15.49}$$

$$\frac{\partial c_{\omega'}}{\partial \Omega} = \frac{(s_\omega c_\Omega s_\varepsilon + c_\omega s_\Omega c_i s_\varepsilon) - c_{\omega'}\frac{\partial s_{i'}}{\partial \Omega}}{s_{i'}} \tag{15.50}$$

$$\frac{\partial c_{\omega'}}{\partial i} = \frac{(c_\omega c_\Omega s_i s_\varepsilon + c_\omega c_i c_\varepsilon) - c_{\omega'}\frac{\partial s_{i'}}{\partial i}}{s_{i'}} \tag{15.51}$$

$$\frac{\partial c_{\omega'}}{\partial \omega} = \frac{c_\omega s_\Omega s_\varepsilon + s_\omega c_\Omega c_i s_\varepsilon - s_\omega s_i c_\varepsilon}{s_{i'}} \tag{15.52}$$

As far as $s_{\theta'}$ and $c_{\theta'}$ are concerned, the angle θ' is given by $\theta' = \omega' + \theta^*$, such that $s_{\theta'} = s_{\omega'+\theta^*}$ and $c_{\theta'} = c_{\omega'+\theta^*}$ are $f(h,k,p,q,L)$ because $\omega' = f(h,k,p,q)$ and $\theta^* = f(h,k,L)$. Therefore, we have

$$\frac{\partial s_{\theta'}}{\partial h} = \frac{\partial s_{\omega'}}{\partial h}c_{\theta^*} + s_{\omega'}\frac{\partial c_{\theta^*}}{\partial h} + \frac{\partial s_{\theta^*}}{\partial h}c_{\omega'} + s_{\theta^*}\frac{\partial c_{\omega'}}{\partial h} \tag{15.53}$$

$$\frac{\partial s_{\theta'}}{\partial k} = \frac{\partial s_{\omega'}}{\partial k}c_{\theta^*} + s_{\omega'}\frac{\partial c_{\theta^*}}{\partial k} + \frac{\partial s_{\theta^*}}{\partial k}c_{\omega'} + s_{\theta^*}\frac{\partial c_{\omega'}}{\partial k} \tag{15.54}$$

$$\frac{\partial s_{\theta'}}{\partial p} = \frac{\partial s_{\omega'}}{\partial p}c_{\theta^*} + \frac{\partial c_{\omega'}}{\partial p}s_{\theta^*} \tag{15.55}$$

$$\frac{\partial s_{\theta'}}{\partial q} = \frac{\partial s_{\omega'}}{\partial q}c_{\theta^*} + \frac{\partial c_{\omega'}}{\partial q}s_{\theta^*} \tag{15.56}$$

and for $c_{\theta'}$,

$$\frac{\partial c_{\theta'}}{\partial h} = \frac{\partial c_{\omega'}}{\partial h}c_{\theta^*} + c_{\omega'}\frac{\partial c_{\theta^*}}{\partial h} - \frac{\partial s_{\omega'}}{\partial h}s_{\theta^*} - s_{\omega'}\frac{\partial s_{\theta^*}}{\partial h} \tag{15.57}$$

$$\frac{\partial c_{\theta'}}{\partial k} = \frac{\partial c_{\omega'}}{\partial k}c_{\theta^*} + c_{\omega'}\frac{\partial c_{\theta^*}}{\partial k} - \frac{\partial s_{\omega'}}{\partial k}s_{\theta^*} - s_{\omega'}\frac{\partial s_{\theta^*}}{\partial k} \tag{15.58}$$

$$\frac{\partial c_{\theta'}}{\partial p} = \frac{\partial c_{\omega'}}{\partial p} c_{\theta^*} - \frac{\partial s_{\omega'}}{\partial p} s_{\theta^*} \tag{15.59}$$

$$\frac{\partial c_{\theta'}}{\partial q} = \frac{\partial c_{\omega'}}{\partial q} c_{\theta^*} - \frac{\partial s_{\omega'}}{\partial q} s_{\theta^*} \tag{15.60}$$

The partials of $s_{\theta'}$ and $c_{\theta'}$ with respect to L must also be derived. They are

$$\frac{\partial s_{\theta'}}{\partial L} = s_{\omega'} \frac{\partial c_{\theta^*}}{\partial L} + c_{\omega'} \frac{\partial s_{\theta^*}}{\partial L} \tag{15.61}$$

$$\frac{\partial c_{\theta'}}{\partial L} = c_{\omega'} \frac{\partial c_{\theta^*}}{\partial L} - s_{\omega'} \frac{\partial s_{\theta^*}}{\partial L} \tag{15.62}$$

To complete the derivations of the partials of the trigonometric functions in sine and cosine with respect to the equinoctials, we need from

$$s_{\theta^*} = \frac{k s_L - h c_L}{\left(h^2 + k^2\right)^{1/2}} \tag{15.63}$$

$$\frac{\partial s_{\theta^*}}{\partial h} = \frac{-k\left(k c_L + h s_L\right)}{\left(h^2 + k^2\right)^{3/2}}, \quad \frac{\partial s_{\theta^*}}{\partial k} = \frac{h\left(h s_L + k c_L\right)}{\left(h^2 + k^2\right)^{3/2}} \tag{15.64}$$

$$\frac{\partial s_{\theta^*}}{\partial L} = \frac{k c_L + h s_L}{\left(h^2 + k^2\right)^{1/2}} \tag{15.65}$$

and from

$$c_{\theta^*} = \frac{h s_L + k c_L}{\left(h^2 + k^2\right)^{1/2}} \tag{15.66}$$

$$\frac{\partial c_{\theta^*}}{\partial h} = \frac{k\left(k s_L - h c_L\right)}{\left(h^2 + k^2\right)^{3/2}}, \quad \frac{\partial c_{\theta^*}}{\partial k} = \frac{h\left(h c_L - k s_L\right)}{\left(h^2 + k^2\right)^{3/2}} \tag{15.67}$$

$$\frac{\partial c_{\theta^*}}{\partial L} = \frac{h c_L - k s_L}{\left(h^2 + k^2\right)^{1/2}} \tag{15.68}$$

The partials $\partial s_{\theta^*}/\partial h$, $\partial s_{\theta^*}/\partial k$, $\partial s_{\theta^*}/\partial L$, $\partial c_{\theta^*}/\partial h$, $\partial c_{\theta^*}/\partial k$ and $\partial c_{\theta^*}/\partial L$ are prevented from returning 0/0 in the codes by setting $h^2 + k^2 = 1$ in their denominator as soon as $h^2 + k^2$ becomes less than 10^{-16}, such that these partials are returned as 0 without ambiguity. Also for i' becoming less than 10^{-16}, it is as before set to 10^{-16} to avoid singularity conditions for $\partial c_{\omega'}/\partial \Omega$, $\partial c_{\omega'}/\partial i$ and $\partial c_{\omega'}/\partial \omega$ in Equations (15.50)–(15.52).

The spacecraft radial distance r also appears in the components of the gravity accelerations of the Sun and the Moon and its partials are also needed. From $r = a\left(1 - h^2 - k^2\right)/(1 + h s_L + k c_L)$,

$$\left(\frac{\partial r}{\partial a}\right)_L = \frac{r}{a}, \left(\frac{\partial r}{\partial h}\right)_L = \frac{-r\left(2ah + r s_L\right)}{a\left(1 - h^2 - k^2\right)},$$

$$(\partial r/\partial k)_L = \frac{-r\left(2ak + r c_L\right)}{a\left(1 - h^2 - k^2\right)}, (\partial r/\partial L)_L = \frac{-r^2}{a} \frac{\left(h c_L - k s_L\right)}{\left(1 - h^2 - k^2\right)} \tag{15.69}$$

as derived in earlier papers [2, 3] and Chapters 9 and 13. Because Ω' also appears in $s_{\lambda_\odot - \Omega'}$, $c_{\lambda_\odot - \Omega'}$, and their powers in the perturbation functions, we need the following partials with respect to p and q:

$$\frac{\partial\left(s_{\lambda_\odot - \Omega'}\right)}{\partial p} = s_{\lambda_\odot}\frac{\partial c_{\Omega'}}{\partial p} - c_{\lambda_\odot}\frac{\partial s_{\Omega'}}{\partial p} \tag{15.70}$$

$$\frac{\partial\left(s_{\lambda_\odot - \Omega'}\right)}{\partial q} = s_{\lambda_\odot}\frac{\partial c_{\Omega'}}{\partial q} - c_{\lambda_\odot}\frac{\partial s_{\Omega'}}{\partial q} \tag{15.71}$$

$$\frac{\partial\left(c_{\lambda_\odot - \Omega'}\right)}{\partial p} = c_{\lambda_\odot}\frac{\partial c_{\Omega'}}{\partial p} + s_{\lambda_\odot}\frac{\partial s_{\Omega'}}{\partial p} \tag{15.72}$$

$$\frac{\partial\left(c_{\lambda_\odot - \Omega'}\right)}{\partial q} = c_{\lambda_\odot}\frac{\partial c_{\Omega'}}{\partial q} + s_{\lambda_\odot}\frac{\partial s_{\Omega'}}{\partial q} \tag{15.73}$$

15.3 The $\partial \mathbf{f}_\odot / \partial \mathbf{z}$ Partial Derivatives for the Solar Gravity

Equations (16)–(18) of [1] and (14.16)–(14.18) of Chapter 14 show the expanded form of the solar gravity perturbation components f_r, f_θ and f_h, respectively, up to the third order in the spacecraft radial distance r. They involve the angles $\Omega' = f(p,q)$, $i' = f(p,q)$, as well as $\theta' = f(h,k,p,q,L)$, and the scalar $r = f(a,h,k,L)$. We rewrite f_r, f_θ and f_h as

$$f_r^s = rn_\odot^2 f_{r_1}^s + \frac{3}{2}r^2\frac{n_\odot^2}{r_\odot}f_{r_2}^s + \frac{r^3}{2}\frac{n_\odot^2}{r_\odot^2}f_{r_3}^s \tag{15.74}$$

$$f_\theta^s = 3rn_\odot^2 f_{\theta_1}^s + \frac{3}{2}r^2\frac{n_\odot^2}{r_\odot}f_{\theta_2}^s + \frac{r^3}{2}\frac{n_\odot^2}{r_\odot^2}f_{\theta_3}^s \tag{15.75}$$

$$f_h^s = -3rn_\odot^2 f_{h_1}^s - \frac{3}{2}r^2\frac{n_\odot^2}{r_\odot}f_{h_2}^s + \frac{5}{2}r^3\frac{n_\odot^2}{r_\odot^2}f_{h_3}^s \tag{15.76}$$

We can now generate the partials of $f_{r_1}^s, \ldots, f_{h_3}^s$ with respect to the equinoctials. They are for f_r^s:

$$\frac{\partial f_r^s}{\partial a} = \frac{\partial r}{\partial a}n_\odot^2 f_{r_1}^s + 3r\frac{\partial r}{\partial a}\frac{n_\odot^2}{r_\odot}f_{r_2}^s + \frac{3}{2}r^2\frac{\partial r}{\partial a}\frac{n_\odot^2}{r_\odot^2}f_{r_3}^s \tag{15.77}$$

$$\begin{aligned}\frac{\partial f_r^s}{\partial h} &= \frac{\partial r}{\partial h}n_\odot^2 f_{r_1}^s + 3r\frac{\partial r}{\partial h}\frac{n_\odot^2}{r_\odot}f_{r_2}^s + \frac{3}{2}r^2\frac{\partial r}{\partial h}\frac{n_\odot^2}{r_\odot^2}f_{r_3}^s \\ &\quad + rn_\odot^2\frac{\partial f_{r_1}^s}{\partial h} + \frac{3}{2}r^2\frac{n_\odot^2}{r_\odot}\frac{\partial f_{r_2}^s}{\partial h} + \frac{r^3}{2}\frac{n_\odot^2}{r_\odot^2}\frac{\partial f_{r_3}^s}{\partial h}\end{aligned} \tag{15.78}$$

$$\begin{aligned}\frac{\partial f_r^s}{\partial k} &= \frac{\partial r}{\partial k}n_\odot^2 f_{r_1}^s + 3r\frac{\partial r}{\partial k}\frac{n_\odot^2}{r_\odot}f_{r_2}^s + \frac{3}{2}r^2\frac{\partial r}{\partial k}\frac{n_\odot^2}{r_\odot^2}f_{r_3}^s \\ &\quad + rn_\odot^2\frac{\partial f_{r_1}^s}{\partial k} + \frac{3}{2}r^2\frac{n_\odot^2}{r_\odot}\frac{\partial f_{r_2}^s}{\partial k} + \frac{r^3}{2}\frac{n_\odot^2}{r_\odot^2}\frac{\partial f_{r_3}^s}{\partial k}\end{aligned} \tag{15.79}$$

$$\frac{\partial f_r^s}{\partial p} = rn_\odot^2\frac{\partial f_{r_1}^s}{\partial p} + \frac{3}{2}r^2\frac{n_\odot^2}{r_\odot}\frac{\partial f_{r_2}^s}{\partial p} + \frac{r^3}{2}\frac{n_\odot^2}{r_\odot^2}\frac{\partial f_{r_3}^s}{\partial p} \tag{15.80}$$

$$\frac{\partial f_r^s}{\partial q} = rn_\odot^2\frac{\partial f_{r_1}^s}{\partial q} + \frac{3}{2}r^2\frac{n_\odot^2}{r_\odot}\frac{\partial f_{r_2}^s}{\partial q} + \frac{r^3}{2}\frac{n_\odot^2}{r_\odot^2}\frac{\partial f_{r_3}^s}{\partial q} \tag{15.81}$$

$$\frac{\partial f_r^s}{\partial L} = \frac{\partial r}{\partial L} n_\odot^2 f_{r_1}^s + 3r \frac{\partial r}{\partial L} \frac{n_\odot^2}{r_\odot} f_{r_2}^s + \frac{3}{2} r^2 \frac{\partial r}{\partial L} \frac{n_\odot^2}{r_\odot^2} f_{r3}^s$$
$$+ r n_\odot^2 \frac{\partial f_{r_1}^s}{\partial L} + \frac{3}{2} r^2 \frac{n_\odot^2}{r_\odot} \frac{\partial f_{r_2}^s}{\partial L} + \frac{r^3}{2} \frac{n_\odot^2}{r_\odot^2} \frac{\partial f_{r_3}^s}{\partial L} \tag{15.82}$$

For f_θ^s, we have the following six partials:

$$\frac{\partial f_\theta^s}{\partial a} = \frac{\partial f_\theta^s}{\partial r} \frac{\partial r}{\partial a} = 3n_\odot^2 \frac{\partial r}{\partial a} f_{\theta_1}^s + 3r \frac{n_\odot^2}{r_\odot} \frac{\partial r}{\partial a} f_{\theta_2}^s + \frac{3}{2} r^2 \frac{n_\odot^2}{r_\odot^2} \frac{\partial r}{\partial a} f_{\theta_3}^s \tag{15.83}$$

$$\frac{\partial f_\theta^s}{\partial h} = 3 \frac{\partial r}{\partial h} n_\odot^2 f_{\theta_1}^s + 3r \frac{\partial r}{\partial h} \frac{n_\odot^2}{r_\odot} f_{\theta_2}^s + \frac{3}{2} r^2 \frac{\partial r}{\partial h} \frac{n_\odot^2}{r_\odot^2} f_{\theta_3}^s$$
$$+ 3 r n_\odot^2 \frac{\partial f_{\theta_1}^s}{\partial h} + \frac{3}{2} r^2 \frac{n_\odot^2}{r_\odot} \frac{\partial f_{\theta_2}^s}{\partial h} + \frac{r^3}{2} \frac{n_\odot^2}{r_\odot^2} \frac{\partial f_{\theta_3}^s}{\partial h} \tag{15.84}$$

$$\frac{\partial f_\theta^s}{\partial k} = 3 \frac{\partial r}{\partial k} n_\odot^2 f_{\theta_1}^s + 3r \frac{\partial r}{\partial k} \frac{n_\odot^2}{r_\odot} f_{\theta_2}^s + \frac{3}{2} r^2 \frac{\partial r}{\partial k} \frac{n_\odot^2}{r_\odot^2} f_{\theta_3}^s$$
$$+ 3 r n_\odot^2 \frac{\partial f_{\theta_1}^s}{\partial k} + \frac{3}{2} r^2 \frac{n_\odot^2}{r_\odot} \frac{\partial f_{\theta_2}^s}{\partial k} + \frac{r^3}{2} \frac{n_\odot^2}{r_\odot^2} \frac{\partial f_{\theta_3}^s}{\partial k} \tag{15.85}$$

$$\frac{\partial f_\theta^s}{\partial p} = 3 r n_\odot^2 \frac{\partial f_{\theta_1}^s}{\partial p} + \frac{3}{2} r^2 \frac{n_\odot^2}{r_\odot} \frac{\partial f_{\theta_2}^s}{\partial p} + \frac{r^3}{2} \frac{n_\odot^2}{r_\odot^2} \frac{\partial f_{\theta_3}^s}{\partial p} \tag{15.86}$$

$$\frac{\partial f_\theta^s}{\partial q} = 3 r n_\odot^2 \frac{\partial f_{\theta_1}^s}{\partial q} + \frac{3}{2} r^2 \frac{n_\odot^2}{r_\odot} \frac{\partial f_{\theta_2}^s}{\partial q} + \frac{r^3}{2} \frac{n_\odot^2}{r_\odot^2} \frac{\partial f_{\theta_3}^s}{\partial q} \tag{15.87}$$

$$\frac{\partial f_\theta^s}{\partial L} = 3 \frac{\partial r}{\partial L} n_\odot^2 f_{\theta_1}^s + 3r \frac{\partial r}{\partial L} \frac{n_\odot^2}{r_\odot} f_{\theta_2}^s + \frac{3}{2} r^2 \frac{\partial r}{\partial L} \frac{n_\odot^2}{r_\odot^2} f_{\theta_3}^s$$
$$+ 3 r n_\odot^2 \frac{\partial f_{\theta_1}^s}{\partial L} + \frac{3}{2} r^2 \frac{n_\odot^2}{r_\odot} \frac{\partial f_{\theta_2}^s}{\partial L} + \frac{r^3}{2} \frac{n_\odot^2}{r_\odot^2} \frac{\partial f_{\theta_3}^s}{\partial L} \tag{15.88}$$

and finally for f_h^s, we have

$$\frac{\partial f_h^s}{\partial a} = \frac{\partial f_h^s}{\partial r} \frac{\partial r}{\partial a} = -3n_\odot^2 \frac{\partial r}{\partial a} f_{h_1}^s - 3r \frac{n_\odot^2}{r_\odot} \frac{\partial r}{\partial a} f_{h_2}^s + \frac{15}{2} r^2 \frac{n_\odot^2}{r_\odot^2} \frac{\partial r}{\partial a} f_{h_3}^s \tag{15.89}$$

$$\frac{\partial f_h^s}{\partial h} = -3 \frac{\partial r}{\partial h} n_\odot^2 f_{h_1}^s - 3r \frac{\partial r}{\partial h} \frac{n_\odot^2}{r_\odot} f_{h_2}^s + \frac{15}{2} r^2 \frac{\partial r}{\partial h} \frac{n_\odot^2}{r_\odot^2} f_{h_3}^s$$
$$- 3 r n_\odot^2 \frac{\partial f_{h_1}^s}{\partial h} - \frac{3}{2} r^2 \frac{n_\odot^2}{r_\odot} \frac{\partial f_{h_2}^s}{\partial h} + \frac{5}{2} r^3 \frac{n_\odot^2}{r_\odot^2} \frac{\partial f_{h_3}^s}{\partial h} \tag{15.90}$$

$$\frac{\partial f_h^s}{\partial k} = -3 \frac{\partial r}{\partial k} n_\odot^2 f_{h_1}^s - 3r \frac{\partial r}{\partial k} \frac{n_\odot^2}{r_\odot} f_{h_2}^s + \frac{15}{2} r^2 \frac{\partial r}{\partial k} \frac{n_\odot^2}{r_\odot^2} f_{h_3}^s$$
$$- 3 r n_\odot^2 \frac{\partial f_{h_1}^s}{\partial k} - \frac{3}{2} r^2 \frac{n_\odot^2}{r_\odot} \frac{\partial f_{h_2}^s}{\partial k} + \frac{5}{2} r^3 \frac{n_\odot^2}{r_\odot^2} \frac{\partial f_{h_3}^s}{\partial k} \tag{15.91}$$

$$\frac{\partial f_h^s}{\partial p} = -3 r n_\odot^2 \frac{\partial f_{h_1}^s}{\partial p} - \frac{3}{2} r^2 \frac{n_\odot^2}{r_\odot} \frac{\partial f_{h_2}^s}{\partial p} + \frac{5}{2} r^3 \frac{n_\odot^2}{r_\odot^2} \frac{\partial f_{h_3}^s}{\partial p} \tag{15.92}$$

$$\frac{\partial f_h^s}{\partial q} = -3rn_\odot^2\frac{\partial f_{h_1}^s}{\partial q} - \frac{3}{2}r^2\frac{n_\odot^2}{r_\odot}\frac{\partial f_{h_2}^s}{\partial q} + \frac{5}{2}r^3\frac{n_\odot^2}{r_\odot^2}\frac{\partial f_{h_3}^s}{\partial q} \tag{15.93}$$

$$\frac{\partial f_h^s}{\partial L} = -3\frac{\partial r}{\partial L}n_\odot^2 f_{h_1}^s - 3r\frac{\partial r}{\partial L}\frac{n_\odot^2}{r_\odot}f_{h_2}^s + \frac{15}{2}r^2\frac{\partial r}{\partial L}\frac{n_\odot^2}{r_\odot^2}f_{h_3}^s$$
$$- 3rn_\odot^2\frac{\partial f_{h_1}^s}{\partial L} - \frac{3}{2}r^2\frac{n_\odot^2}{r_\odot}\frac{\partial f_{h_2}^s}{\partial L} + \frac{5}{2}r^3\frac{n_\odot^2}{r_\odot^2}\frac{\partial f_{h_3}^s}{\partial L} \tag{15.94}$$

The partial derivatives appearing in Equations (15.77)–(15.94) are shown in Appendix 15.1.

15.4 The $\partial f_{☽}/\partial z$ Partial Derivatives for the Lunar Gravity

The lunar gravity perturbations are shown in Equations (80)–(82) in [1], and also (14.80)–(14.82) in Chapter 14, having essentially the same form as the solar counterparts. The angles λ_m, Ω'', i'', $\theta'' = \omega'' + \theta^*$ and i_m replace $\lambda_\odot$, Ω', i', $\theta' = \omega' + \theta^*$ and ε of the solar case, and n_m is used instead of $n_\odot$, while r_m replaces $r_\odot$. These changes are also made in the vectorial orbital constants by replacing Ω with $\Omega - \Omega_m$ [1] and Chapter 14 of this book. It is clear that i_m and Ω_m vary in time due to the precession of the lunar orbit [1]. As discussed in [1] and Chapter 14, $n_\odot$ is defined as $\sqrt{\mu_\odot/r_\odot^3}$ as the apparent mean motion of the Sun, while $n_m = \sqrt{\mu_{☽}/r_m^3}$ is obviously not the mean motion of the Moon. The mean motion of the Moon is given by $n'_m = \frac{2\pi}{T_{☽}}$, with $T_{☽} = 27.321661$ mean solar days defining the lunar period [1] and Chapter 14 of this book.

First, we write the lunar perturbations in a compact form as

$$f_r^m = rn_m^2 f_{r_1}^m + \frac{3}{2}r^2\frac{n_m^2}{r_m}f_{r_2}^m + \frac{r^3}{2}\frac{n_m^2}{r_m^2}f_{r_3}^m \tag{15.95}$$

$$f_\theta^m = 3rn_m^2 f_{\theta_1}^m + \frac{3}{2}r^2\frac{n_m^2}{r_m}f_{\theta_2}^m + \frac{r^3}{2}\frac{n_m^2}{r_m^2}f_{\theta_3}^m \tag{15.96}$$

$$f_h^m = -3rn_m^2 f_{h_1}^m - \frac{3}{2}r^2\frac{n_m^2}{r_m}f_{h_2}^m + \frac{5}{2}r^3\frac{n_m^2}{r_m^2}f_{h_3}^m \tag{15.97}$$

The partials of these three components with respect to the six equinoctials have the same form as in the solar case, that is, Equations (15.77)–(15.94) including the partials listed in Appendix 15.1, provided that the angles Ω'', i'', ω'', i_m and $\Omega - \Omega_m$, instead of Ω', i', ω', ε and Ω are used. The angle $s_{i''}$ is given by

$$s_{i''} = \left[s_{\Omega-\Omega_m}^2 s_i^2 + \left(c_{\Omega-\Omega_m}s_i c_{i_m} - c_i s_{i_m}\right)^2\right]^{1/2} \tag{15.98}$$

leading to

$$\frac{\partial s_{i''}}{\partial \Omega} = \frac{s_{\Omega-\Omega_m}c_{\Omega-\Omega_m}s_i^2 - s_{\Omega-\Omega_m}s_i c_{i_m}\left(c_{\Omega-\Omega_m}s_i c_{i_m} - c_i s_{i_m}\right)}{s_{i''}} \tag{15.99}$$

$$\frac{\partial s_{i''}}{\partial i} = \frac{s_{\Omega-\Omega_m}^2 s_i c_i + \left(c_{\Omega-\Omega_m}s_i c_{i_m} - c_i s_{i_m}\right)\left(c_{\Omega-\Omega_m}c_i c_{i_m} + s_i s_{i_m}\right)}{s_{i''}} \tag{15.100}$$

and having the same form as the $\frac{\partial s_{i'}}{\partial \Omega}$ and $\frac{\partial s_{i'}}{\partial i}$ partials in Equations (15.14) and (15.15), except that $\Omega - \Omega_m$ is used instead of Ω, as discussed above and also in [1] and Chapter 14. In a similar way, from

$$c_{i''} = c_{\Omega-\Omega_m} s_i s_{i_m} + c_i c_{i_m} \tag{15.101}$$

we get

$$\frac{\partial c_{i''}}{\partial \Omega} = -s_{\Omega-\Omega_m} s_i s_{i_m} \tag{15.102}$$

$$\frac{\partial c_{i''}}{\partial i} = c_{\Omega-\Omega_m} c_i s_{i_m} - s_i c_{i_m} \tag{15.103}$$

From the $s_{\omega''}$ expression

$$s_{\omega''} = \frac{-c_\omega s_{\Omega-\Omega_m} s_{i_m} - s_\omega c_{\Omega-\Omega_m} c_i s_{i_m} + s_\omega s_i c_{i_m}}{s_{i''} = f(i, \Omega)} \tag{15.104}$$

we get

$$\frac{\partial s_{\omega''}}{\partial \omega} = \frac{s_\omega s_{\Omega-\Omega_m} s_{i_m} - c_\omega c_{\Omega-\Omega_m} c_i s_{i_m} + c_\omega s_i c_{i_m}}{s_{i''} = f(i, \Omega)} \tag{15.105}$$

$$\frac{\partial s_{\omega''}}{\partial \Omega} = \frac{\left(-c_\omega c_{\Omega-\Omega_m} s_{i_m} + s_\omega s_{\Omega-\Omega_m} c_i s_{i_m}\right) - s_{\omega''} \frac{\partial s_{i''}}{\partial \Omega}}{s_{i''}} \tag{15.106}$$

$$\frac{\partial s_{\omega''}}{\partial i} = \frac{\left(s_\omega c_{\Omega-\Omega_m} s_i s_{i_m} + s_\omega c_i c_{i_m}\right) - s_{\omega''} \frac{\partial s_{i''}}{\partial i}}{s_{i''}} \tag{15.107}$$

and from the $c_{\omega''}$ expression,

$$c_{\omega''} = \frac{s_\omega s_{\Omega-\Omega_m} s_{i_m} - c_\omega c_{\Omega-\Omega_m} c_i s_{i_m} + c_\omega s_i c_{i_m}}{s_{i''}} \tag{15.108}$$

we get

$$\frac{\partial c_{\omega''}}{\partial \omega} = \frac{c_\omega s_{\Omega-\Omega_m} s_{i_m} + s_\omega c_{\Omega-\Omega_m} c_i s_{i_m} - s_\omega s_i c_{i_m}}{s_{i''}} \tag{15.109}$$

$$\frac{\partial c_{\omega''}}{\partial \Omega} = \frac{\left(s_\omega c_{\Omega-\Omega_m} s_{i_m} + c_\omega s_{\Omega-\Omega_m} c_i s_{i_m}\right) - c_{\omega''} \frac{\partial s_{i''}}{\partial \Omega}}{s_{i''}} \tag{15.110}$$

$$\frac{\partial c_{\omega''}}{\partial i} = \frac{\left(c_\omega c_{\Omega-\Omega_m} s_i s_{i_m} + c_\omega c_i c_{i_m}\right) - c_{\omega''} \frac{\partial s_{i''}}{\partial i}}{s_{i''}} \tag{15.111}$$

From $\theta'' = \omega'' + \theta^* = f(h, k, p, q, L)$ where $\omega'' = f(h, k, p, q)$ and $\theta^* = f(h, k, L)$, we have

$$\frac{\partial c_{\theta''}}{\partial p} = \frac{\partial c_{\omega''}}{\partial p} c_{\theta^*} - \frac{\partial s_{\omega''}}{\partial p} s_{\theta^*} \tag{15.112}$$

$$\frac{\partial c_{\theta''}}{\partial q} = \frac{\partial c_{\omega''}}{\partial q} c_{\theta^*} - \frac{\partial s_{\omega''}}{\partial q} s_{\theta^*} \tag{15.113}$$

and

$$\frac{\partial s_{\theta''}}{\partial p} = \frac{\partial s_{\omega''}}{\partial p} c_{\theta^*} + \frac{\partial c_{\omega''}}{\partial p} s_{\theta^*} \tag{15.114}$$

$$\frac{\partial s_{\theta''}}{\partial q} = \frac{\partial s_{\omega''}}{\partial q} c_{\theta^*} + \frac{\partial c_{\omega''}}{\partial q} s_{\theta^*} \tag{15.115}$$

as well as

$$\frac{\partial c_{\theta''}}{\partial L} = c_{\omega''} \frac{\partial c_{\theta^*}}{\partial L} - s_{\omega''} \frac{\partial s_{\theta^*}}{\partial L} \tag{15.116}$$

$$\frac{\partial s_{\theta''}}{\partial L} = s_{\omega''} \frac{\partial c_{\theta^*}}{\partial L} + c_{\omega''} \frac{\partial s_{\theta^*}}{\partial L} \tag{15.117}$$

Therefore, it is now possible to evaluate the following partial derivatives, namely $\frac{\partial s_{i''}}{\partial p} = \frac{\partial s_{i''}}{\partial \Omega}\frac{\partial \Omega}{\partial p} + \frac{\partial s_{i''}}{\partial i}\frac{\partial i}{\partial p}$, $\frac{\partial s_{i''}}{\partial q} = \frac{\partial s_{i''}}{\partial \Omega}\frac{\partial \Omega}{\partial q} + \frac{\partial s_{i''}}{\partial i}\frac{\partial i}{\partial q}$, and $\frac{\partial c_{i''}}{\partial p} = \frac{\partial c_{i''}}{\partial \Omega}\frac{\partial \Omega}{\partial p} + \frac{\partial c_{i''}}{\partial i}\frac{\partial i}{\partial p}$, $\frac{\partial c_{i''}}{\partial q} = \frac{\partial c_{i''}}{\partial \Omega}\frac{\partial \Omega}{\partial q} + \frac{\partial c_{i''}}{\partial i}\frac{\partial i}{\partial q}$. As for the partials of $s_{\Omega''}$, $\frac{\partial s_{\Omega''}}{\partial p} = \frac{\partial s_{\Omega''}}{\partial \Omega}\frac{\partial \Omega}{\partial p} + \frac{\partial s_{\Omega''}}{\partial i}\frac{\partial i}{\partial p}$, $\frac{\partial s_{\Omega''}}{\partial q} = \frac{\partial s_{\Omega''}}{\partial \Omega}\frac{\partial \Omega}{\partial q} + \frac{\partial s_{\Omega''}}{\partial i}\frac{\partial i}{\partial q}$, we still need $\frac{\partial s_{\Omega''}}{\partial \Omega}$, and $\frac{\partial s_{\Omega''}}{\partial i}$.

From $s_{\Omega''} = \frac{s_{\Omega-\Omega_m} s_i}{s_{i''}}$

$$\frac{\partial s_{\Omega''}}{\partial \Omega} = \frac{c_{\Omega-\Omega_m} s_i s_{i''} - s_{\Omega-\Omega_m} s_i \frac{\partial s_{i''}}{\partial \Omega}}{s_{i''}^2} \tag{15.118}$$

$$\frac{\partial s_{\Omega''}}{\partial i} = \frac{s_{\Omega-\Omega_m} c_i s_{i''} - s_{\Omega-\Omega_m} s_i \frac{\partial s_{i''}}{\partial i}}{s_{i''}^2} \tag{15.119}$$

For the evaluation of $\frac{\partial c_{\Omega''}}{\partial p} = \frac{\partial c_{\Omega''}}{\partial \Omega}\frac{\partial \Omega}{\partial p} + \frac{\partial c_{\Omega''}}{\partial i}\frac{\partial i}{\partial p}$ and $\frac{\partial c_{\Omega''}}{\partial q} = \frac{\partial c_{\Omega''}}{\partial \Omega}\frac{\partial \Omega}{\partial q} + \frac{\partial c_{\Omega''}}{\partial i}\frac{\partial i}{\partial q}$, we have from $c_{\Omega''} = \frac{c_{\Omega-\Omega_m} s_i c_{i_m} - c_i s_{i_m}}{s_{i''}}$

$$\frac{\partial c_{\Omega''}}{\partial \Omega} = \frac{-s_{\Omega-\Omega_m} s_i c_{i_m} s_{i''} - \left(c_{\Omega-\Omega_m} s_i c_{i_m} - c_i s_{i_m}\right) \frac{\partial s_{i''}}{\partial \Omega}}{s_{i''}^2} \tag{15.120}$$

$$\frac{\partial c_{\Omega''}}{\partial i} = \frac{\left(c_{\Omega-\Omega_m} c_i c_{i_m} + s_i s_{i_m}\right) s_{i''} - \left(c_{\Omega-\Omega_m} s_i c_{i_m} - c_i s_{i_m}\right) \frac{\partial s_{i''}}{\partial i}}{s_{i''}^2} \tag{15.121}$$

Also, $\frac{\partial s_{\omega''}}{\partial h} = \frac{\partial s_{\omega''}}{\partial \Omega}\frac{\partial \Omega}{\partial h} + \frac{\partial s_{\omega''}}{\partial i}\frac{\partial i}{\partial h} + \frac{\partial s_{\omega''}}{\partial \omega}\frac{\partial \omega}{\partial h}$ reduces to

$$\frac{\partial s_{\omega''}}{\partial h} = \frac{\partial s_{\omega''}}{\partial \omega}\frac{\partial \omega}{\partial h} \tag{15.122}$$

and $\frac{\partial s_{\omega''}}{\partial k} = \frac{\partial s_{\omega''}}{\partial \Omega}\frac{\partial \Omega}{\partial k} + \frac{\partial s_{\omega''}}{\partial i}\frac{\partial i}{\partial k} + \frac{\partial s_{\omega''}}{\partial \omega}\frac{\partial \omega}{\partial k}$ reduces to

$$\frac{\partial s_{\omega''}}{\partial k} = \frac{\partial s_{\omega''}}{\partial \omega}\frac{\partial \omega}{\partial k} \tag{15.123}$$

because $\partial\Omega/\partial h = 0$, $\partial i/\partial h = 0$, $\partial\Omega/\partial k = 0$, and $\partial i/\partial k = 0$. We can now evaluate

$$\frac{\partial s_{\omega''}}{\partial p} = \frac{\partial s_{\omega''}}{\partial \Omega}\frac{\partial \Omega}{\partial p} + \frac{\partial s_{\omega''}}{\partial i}\frac{\partial i}{\partial p} + \frac{\partial s_{\omega''}}{\partial \omega}\frac{\partial \omega}{\partial p} + \frac{\partial s_{\omega''}}{\partial s_{i''}}\frac{\partial s_{i''}}{\partial \Omega}\frac{\partial \Omega}{\partial p} + \frac{\partial s_{\omega''}}{\partial s_{i''}}\frac{\partial s_{i''}}{\partial i}\frac{\partial i}{\partial p} \tag{15.124}$$

where $\frac{\partial s_{\omega''}}{\partial s_{i''}} = -\frac{-c_{\Omega-\Omega_m} s_{\Omega-\Omega_m} s_{i_m} - s_\omega c_{\Omega-\Omega_m} c_i s_{i_m} + s_\omega s_i c_{i_m}}{s_{i''}^2}$ is used such that

$$\frac{\partial s_{\omega''}}{\partial s_{i''}} = -\frac{s_{\omega''}}{s_{i''}} \tag{15.125}$$

also leads to

$$\frac{\partial s_{\omega''}}{\partial q} = \frac{\partial s_{\omega''}}{\partial \Omega}\frac{\partial \Omega}{\partial q} + \frac{\partial s_{\omega''}}{\partial i}\frac{\partial i}{\partial q} + \frac{\partial s_{\omega''}}{\partial \omega}\frac{\partial \omega}{\partial q} + \frac{\partial s_{\omega''}}{\partial s_{i''}}\frac{\partial s_{i''}}{\partial \Omega}\frac{\partial \Omega}{\partial q} + \frac{\partial s_{\omega''}}{\partial s_{i''}}\frac{\partial s_{i''}}{\partial i}\frac{\partial i}{\partial q} \tag{15.126}$$

From Equation (15.108) for $c_{\omega''}$, we have

$$\frac{\partial c_{\omega''}}{\partial h} = \frac{\partial c_{\omega''}}{\partial \omega}\frac{\partial \omega}{\partial h} \tag{15.127}$$

$$\frac{\partial c_{\omega''}}{\partial k} = \frac{\partial c_{\omega''}}{\partial \omega}\frac{\partial \omega}{\partial k} \tag{15.128}$$

Finally,

$$\frac{\partial c_{\omega''}}{\partial p} = \frac{\partial c_{\omega''}}{\partial \Omega}\frac{\partial \Omega}{\partial p} + \frac{\partial c_{\omega''}}{\partial i}\frac{\partial i}{\partial p} + \frac{\partial c_{\omega''}}{\partial \omega}\frac{\partial \omega}{\partial p} + \frac{\partial c_{\omega''}}{\partial s_{i''}}\frac{\partial s_{i''}}{\partial \Omega}\frac{\partial \Omega}{\partial p} + \frac{\partial c_{\omega''}}{\partial s_{i''}}\frac{\partial s_{i''}}{\partial i}\frac{\partial i}{\partial p} \tag{15.129}$$

where $\frac{\partial c_{\omega''}}{\partial s_{i''}} = -\frac{s_\omega s_{\Omega-\Omega_m} s_{i_m} - c_\omega c_{\Omega-\Omega_m} c_i s_{i_m} + c_\omega s_i c_{i_m}}{s_{i''}^2}$ is used, that is,

$$\frac{\partial c_{\omega''}}{\partial s_{i''}} = -\frac{c_{\omega''}}{s_{i''}} \tag{15.130}$$

$$\frac{\partial c_{\omega''}}{\partial q} = \frac{\partial c_{\omega''}}{\partial \Omega}\frac{\partial \Omega}{\partial q} + \frac{\partial c_{\omega''}}{\partial i}\frac{\partial i}{\partial q} + \frac{\partial c_{\omega''}}{\partial \omega}\frac{\partial \omega}{\partial q} + \frac{\partial c_{\omega''}}{\partial s_{i''}}\frac{\partial s_{i''}}{\partial \Omega}\frac{\partial \Omega}{\partial q} + \frac{\partial c_{\omega''}}{\partial s_{i''}}\frac{\partial s_{i''}}{\partial i}\frac{\partial i}{\partial q} \tag{15.131}$$

Note that the partials of $s_{\lambda_m-\Omega''}$ and $c_{\lambda_m-\Omega''}$ are identical to the partials of $s_{\lambda_\odot-\Omega'}$ and $c_{\lambda_\odot-\Omega'}$ of Equations (15.70)–(15.73) except that $\lambda_\odot$ and Ω' are replaced by λ_m and Ω''. As in the solar case of the previous section involving the i' angle, the angle i'' is set to 10^{-16} when it becomes less than 10^{-16} to avoid singularities in the corresponding partial derivatives involving $s_{i''}$ or $s_{i''}^2$.

15.5 Transfer Simulations

The transfer example is the one shown in [1] and in Chapter 14, and repeated here for convenience and completeness. The minimum-time solution using continuous constant thrust acceleration $f_t = 9.8 \times 10^{-6}\,\text{km/s}^2$ between the initial and final orbits, namely $a_0 = 7000\,\text{km}$, $e_0 = 0$, $i_0 = 28.5$ deg, $\Omega_0 = \omega_0 = 0$ deg, $M_0 = 164$ deg and $a_f = 42{,}000$ km, $e_f = 10^{-3}$, $i_f = 1$ deg, $\Omega_f = \omega_f = 0$ deg and $L_f = 247$ radians, the latter representing the accumulated final true longitude, is obtained by searching on the initial values of the Lagrange multipliers $(\lambda_a)_0$, $(\lambda_h)_0$, $(\lambda_k)_0$, $(\lambda_p)_0$, $(\lambda_q)_0$, $(\lambda_L)_0$ and the transfer time t_f, until the final constraints a_f, h_f, k_f, p_f, q_f, L_f and $H_f = 1$ are satisfied to within a desired tolerance. A Runge–Kutta 7/8 integrator is used alongside Aerospace Corporation's NLP2 nonlinear programming software with relative and absolute error controls of 10^{-16} on the integrator and 10^{-5} constraint and objective function tolerance on the optimizer. The numerical values of the second, third and fourth zonal harmonics, as well as the initial lunar and apparent solar orbit elements, are shown in [1] and Chapter 14.

An overview of the NLP2 algorithm as described in its Aerospace Internal Documentation is summarized as follows. NLP2 is a general-purpose method for solving nonlinear programming problems. It computes the vector $\mathbf{x}$ which minimizes the objective function $f(\mathbf{x})$ subject to both equality and inequality constraints $c_i(\mathbf{x}) = 0$ and $c_i(\mathbf{x}) \geq 0$. Two major operations are performed by the algorithm. The constraint satisfaction phase is performed first in order to determine a feasible point for the nonlinear constraint functions. Once such a point is located, a series of equality-constrained optimization problems is solved next using an orthogonal decomposition of the problem variables, with one subset of the variables used for eliminating the active constraints from the problem while the other subset is adjusted to minimize the objective function restricted to the surface of the active constraints. If a previously satisfied inequality constraint is encountered during the minimization process, the iterations are terminated and the offending constraint is added to the active set. The minimization proceeds otherwise to the optimum point at which deletions from the active set are considered. The algorithm terminates at the local solution when no deletions or additions are made from the active set.

The converged values of the multipliers at the initial time are obtained as $(\lambda_a)_0 = 0.261374574 \times 10^2$ s/km, $(\lambda_h)_0 = -0.688900980 \times 10^4$ s, $(\lambda_k)_0 = -0.175811457 \times 10^5$ s, $(\lambda_p)_0 = 0.174791943 \times 10^5$ s, $(\lambda_q)_0 = -0.710033665 \times 10^6$ s, $(\lambda_L)_0 = 0.313617552 \times 10^3$ s/rad , and a flight time of $t_f = 572{,}872.255$ s or 6.630465914 days. The final constraints $a_f = 42{,}000$ km, $h_f = 0$, $k_f = 0.1 \times 10^{-2}$, $p_f = 0$, $q_f = 0.872686779 \times 10^{-2}$, $L_f = 247$ rad and $H_f = 1$ are met to within $0.357977115 \times 10^{-8}$, $-0.361310329 \times 10^{-11}$, $0.685506990 \times 10^{-12}$, $-0.103925801 \times 10^{-11}$, $-0.392569657 \times 10^{-12}$, $-0.535749223 \times 10^{-10}$ and $-0.723310301 \times 10^{-12}$, showing a very strong convergence. An independent verification was performed by Hallman using Feistel's method of evaluating the perturbing accelerations, whether emanating from the zonals or third bodies in the inertial system following Battin's description [5], before conversion to the appropriate rotating system and equinoctial variables representation at each integration step. Hallman uses an external routine to evaluate the osculating orbital elements of the lunar and apparent solar orbits at each integration step in order to compute the inertial accelerations needed for the numerical integration of the state and adjoint differential equations. The software used for this chapter assumes that the semimajor axes and eccentricities of the lunar and solar orbits are mean values instead, and it propagates these orbits in the mean sense. Hallman's minimum-time solution for this example transfer yields a transfer time $t_f = 572{,}869.615$ s or some 2.64 s less than the solution given here. The figures here illustrate the geometry of our transfer, with Figure 15.1 showing the evolutions of a and e with increasing oscillation amplitude of e as the orbit ascends to the GEO altitude. Because the Hamiltonian of the system becomes explicitly dependent on time with the introduction of the third-body perturbations, it is no longer a constant of the motion and its variations are shown in Figure 15.2 together with the i-history of the spacecraft orbit. The variations of the node and the argument of perigee are depicted in Figure 15.3 with an important oscillation experienced by ω. The mean anomaly M cycles between 0 and 360 deg, while the accumulated value of the true longitude L increases steadily to attain the target value of 247 rad, as shown in Figure 15.4.

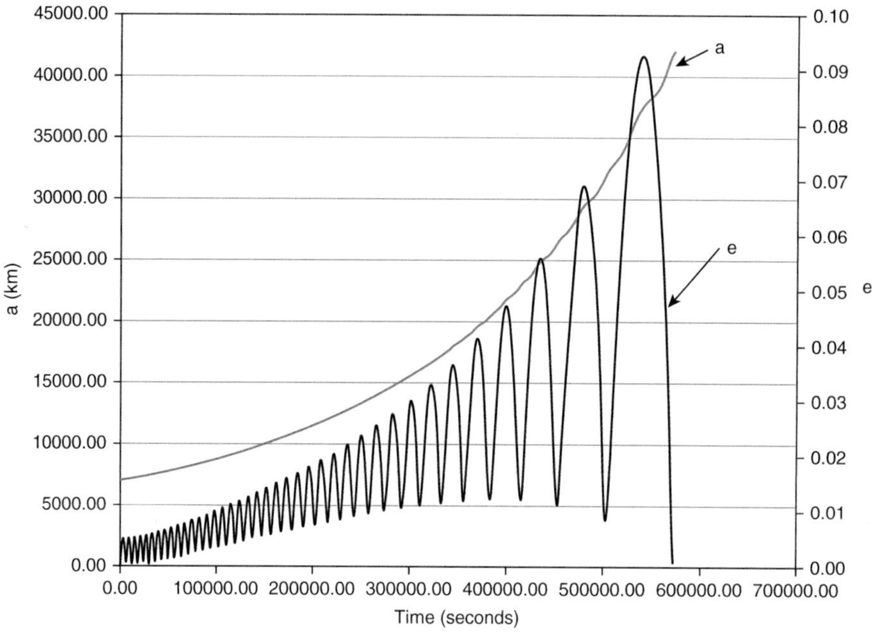

Figure 15.1 Evolution of the spacecraft orbit semimajor axis and eccentricity during optimal transfer to GEO.

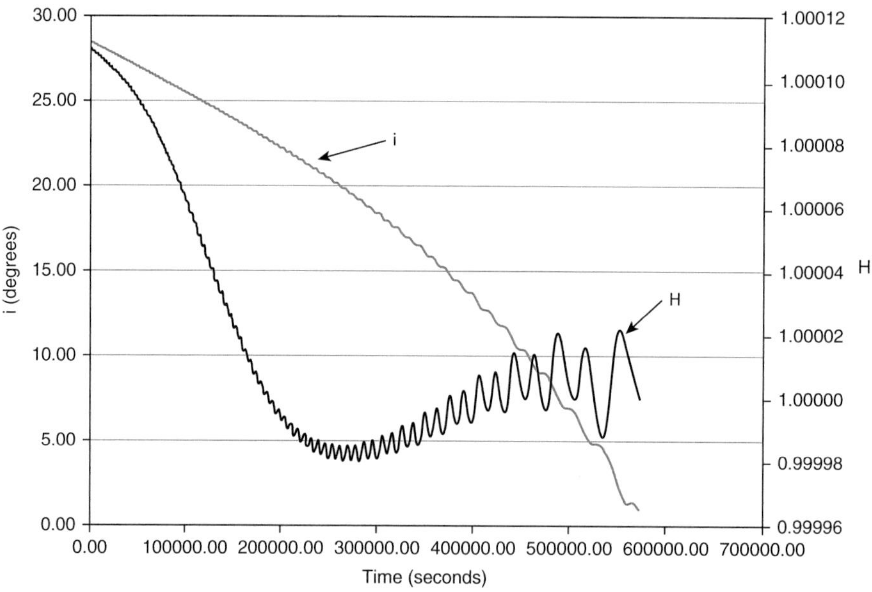

Figure 15.2 Evolution of the spacecraft orbit equatorial inclination and Hamiltonian variations during optimal transfer to GEO.

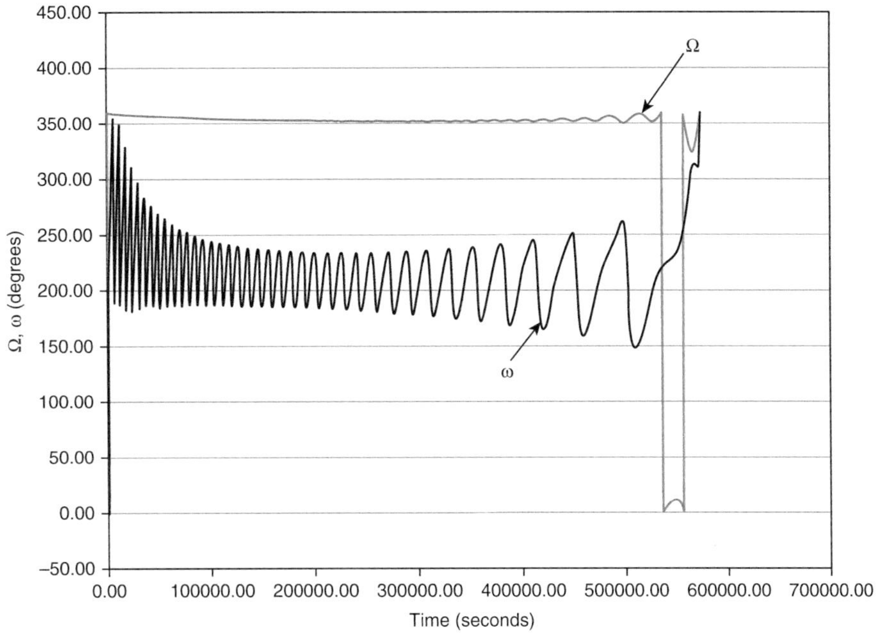

Figure 15.3 Evolution of the spacecraft orbit equatorial node and argument of perigee during optimal transfer to GEO.

Figures 15.5–15.7 show the variations of the six Lagrange multipliers with λ_a steadily decreasing as the target orbit is reached due to the decreased sensitivity of the final time to small variations in the current semimajor axis value. The λ_h and λ_k variations in Figure 15.6 are more complex, being associated with the varying eccentricity during the ascent to GEO.

A long-duration example spanning close to three months is examined next, from $a_0 =$ 32,000 km, $e_0 = 0.4$, $i_0 = 10$ deg, $\Omega_0 = 304$ deg, $\omega_0 = 180$ deg, $M_0 = 180$ deg, to $a_f = 42{,}164$ km, $e_f = 0$, $i_f = 4.84999999$ deg, $\Omega_f = 316.69999977$ deg, $\omega_f = 0$ deg, $L_f = 704.4999994$ rad, with $f_t = 1.5 \times 10^{-7}$ km/s^2 thrust acceleration, and with J_2, J_3 and J_4 turned on, as for the previous example. The epoch is 1 January 2008.

This GTO to GEO transfer requires a transfer time of 7,612,450.456 s using the method described in this chapter. The example is borrowed from Hallman, who generated it initially using Feistel's method with a transfer time of 7,612,576.73 s, or a difference of about 2 min with the solution generated here. In the absence of the Sun and Moon perturbations, Hallman finds a transfer time $t_f = 7{,}641{,}371.55$ s with closer agreement with this chapter's method $t_f = 7{,}641{,}374.97$ s, a difference of some 3 s. This particular transfer example requires some 480 min of additional thrusting when the Sun and Moon perturbations are accounted for. We provide the following solution for this example in the form of the initial values of the multipliers, namely $(\lambda_a)_0 = 0.115692295 \times 10^3$ s/km, $(\lambda_h)_0 = -0.698127582 \times 10^7$ s, $(\lambda_k)_0 = 0.479058028 \times 10^7$ s, $(\lambda_p)_0 = 0.115687414 \times 10^8$ s, $(\lambda_q)_0 = -0.418796315 \times 10^7$ s, $(\lambda_L)_0 = 0.752215425 \times 10^3$ s/rad and $t_f = 7{,}612{,}450.456$ s, or some 88 days. The final constraints a_f, h_f, k_f, p_f, q_f and

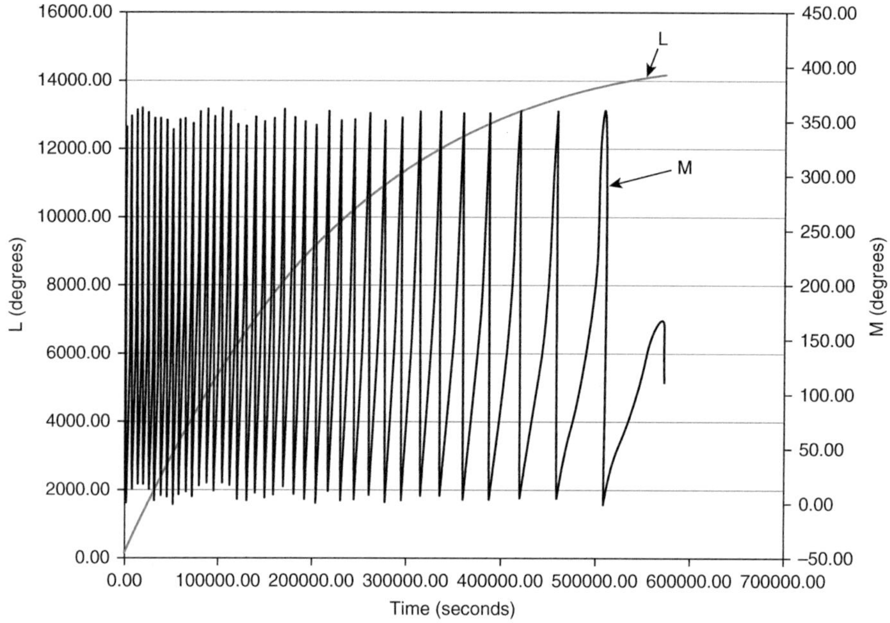

Figure 15.4 Evolution of the spacecraft orbit mean anomaly and accumulated true longitude during optimal transfer to GEO.

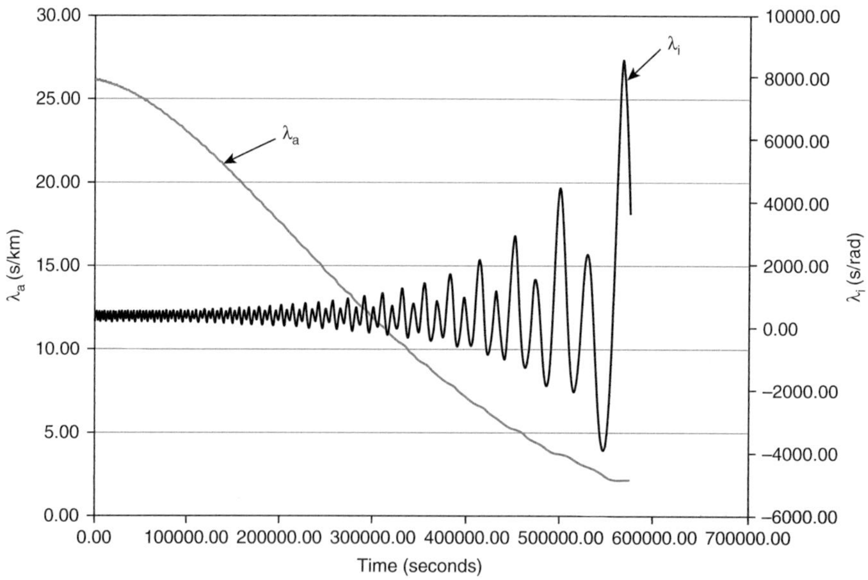

Figure 15.5 Evolution of the adjoints for the semimajor axis and true longitude during optimal transfer to GEO.

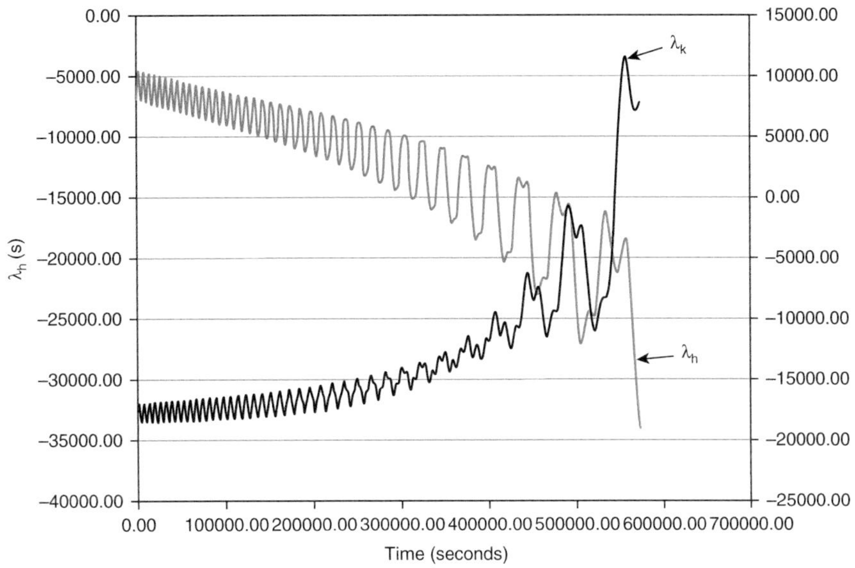

Figure 15.6 Evolution of the adjoints for the equinoctial elements h and k during optimal transfer to GEO.

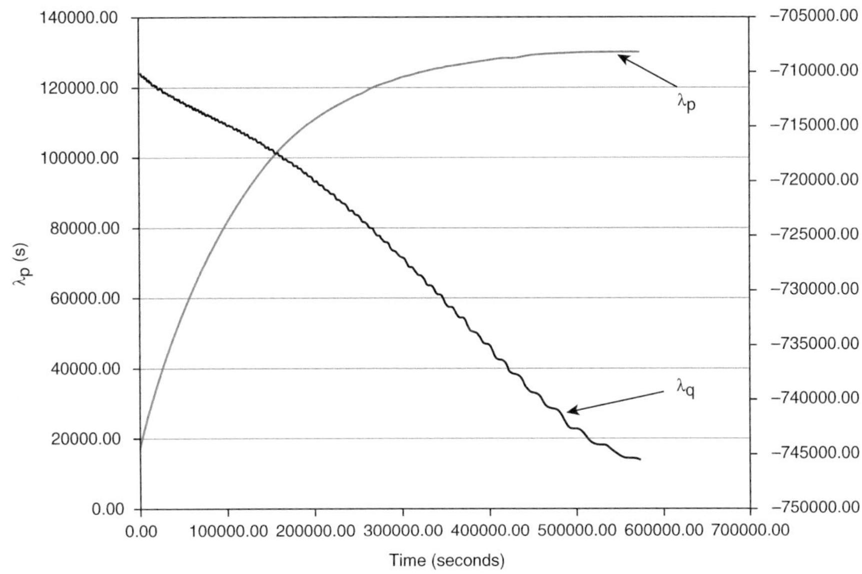

Figure 15.7 Evolution of the adjoints for the equinoctial elements p and q during optimal transfer to GEO.

L_f are met to within $-0.106563675 \times 10^{-6}$, $-0.592185139 \times 10^{-12}$, $0.139159869 \times 10^{-11}$, $-0.813758782 \times 10^{-13}$, $0.133365540 \times 10^{-12}$, $0.582986103 \times 10^{-9}$ and $-0.126230786 \times 10^{-7}$.

No attempt was made to compare the computational efficiency of the code based on the analysis developed in this chapter with the purely numerical method, as these codes are hosted on different platforms. However, it is expected that the present analytic approach may lead in general to somewhat faster runs simply because it avoids many of the frame and coordinate transformations inherent in the purely numerical codes, not to mention the required call to an outside ephemeris generator at each integration step.

15.6 Conclusion

A standalone and self-contained trajectory optimization software is developed in this chapter by describing the lunar and apparent solar orbits' motions according to de Pontécoulant's analytic theory. These two orbits are thus propagated in the mean sense by allowing the lunar ecliptic node to precess and its perigee to advance due to solar gravity perturbations, such that long-duration spacecraft transfers in Earth orbit spanning several months can be effectively simulated with high accuracy. The third-body perturbing accelerations are further resolved up to third order in the spacecraft radial distance using the expansion method and Breakwell's transformation scheme in [6], also used in Chapter 14 of this book, such that the variation-of-parameters equations are written in terms of these accelerations resolved directly in the rotating Euler–Hill frame. The adjoint differential equations also cast in terms of the spacecraft equinoctial elements are fully derived here for implementation and code generation. Simulations obtained by this present analysis are compared to results derived from the use of a method that converts the accelerations to the rotating frame after evaluation in the inertial system and repeated calls to an outside third-body ephemeris propagator at each integration step. This present approach is also more efficient from a computational point of view because both the state and adjoint differential equations are derived in their final form prior to simultaneous integration.

Further improvements could encompass the addition of the fourth-order terms in the third-body accelerations for transfers reaching very high altitudes beyond the GEO level, developed in the next chapter, and the consideration of equinoctial elements for the lunar and solar orbits as well.

Appendix 15.1 The Partial Derivatives of f_r^s, f_θ^s and f_h^s

The partial derivatives of f_r^s with respect to the equinoctial elements h, k, p, q and L:

$$\frac{\partial f_{r_1}^s}{\partial h} = 6\left[c^2_{\lambda_\odot-\Omega'}c_{\theta'}\frac{\partial c_{\theta'}}{\partial h} + c^2_{i'}s_{\theta'}\frac{\partial s_{\theta'}}{\partial h}s^2_{\lambda_\odot-\Omega'} + c_{i'}\left(\frac{\partial s_{\theta'}}{\partial h}c_{\theta'} + s_{\theta'}\frac{\partial c_{\theta'}}{\partial h}\right)\right.$$
$$\left.\times\, s_{\lambda_\odot-\Omega'}c_{\lambda_\odot-\Omega'}\right] \tag{A15.1}$$

$$\frac{\partial f_{r_2}^s}{\partial h} = 15\left[c^3_{\lambda_\odot-\Omega'}c^2_{\theta'}\frac{\partial c_{\theta'}}{\partial h} + c^3_{i'}s^2_{\theta'}\frac{\partial s_{\theta'}}{\partial h}s^3_{\lambda_\odot-\Omega'} + c^2_{\lambda_\odot-\Omega'}\left(2c_{\theta'}\frac{\partial c_{\theta'}}{\partial h}s_{\theta'}\right.\right.$$
$$\left.\left. + c^2_{\theta'}\frac{\partial s_{\theta'}}{\partial h}\right)c_{i'}s_{\lambda_\odot-\Omega'} + c^2_{i'}\left(2s_{\theta'}\frac{\partial s_{\theta'}}{\partial h}c_{\theta'} + s^2_{\theta'}\frac{\partial c_{\theta'}}{\partial h}\right)s^2_{\lambda_\odot-\Omega'}c_{\lambda_\odot-\Omega'}\right]$$
$$-3\left(c_{\lambda_\odot-\Omega'}\frac{\partial c_{\theta'}}{\partial h} + c_{i'}\frac{\partial s_{\theta'}}{\partial h}s_{\lambda_\odot-\Omega'}\right) \tag{A15.2}$$

$$\frac{\partial f_{r_3}^s}{\partial h} = 140\left[c_{\theta'}^3 \frac{\partial c_{\theta'}}{\partial h} c_{\lambda_\odot-\Omega'}^4 + c_{i'}\left(3c_{\theta'}^2 \frac{\partial c_{\theta'}}{\partial h} s_{\theta'} + c_{\theta'}^3 \frac{\partial s_{\theta'}}{\partial h}\right) s_{\lambda_\odot-\Omega'} c_{\lambda_\odot-\Omega'}^3 \right.$$
$$+ c_{i'}^2\left(2s_{\theta'} \frac{\partial s_{\theta'}}{\partial h} c_{\theta'}^2 + 2s_{\theta'}^2 c_{\theta'} \frac{\partial c_{\theta'}}{\partial h}\right) s_{\lambda_\odot-\Omega'}^2 c_{\lambda_\odot-\Omega'}^2 + c_{i'}^4 s_{\theta'}^3 \frac{\partial s_{\theta'}}{\partial h} s_{\lambda_\odot-\Omega'}^4$$
$$\left. + c_{i'}^3\left(3s_{\theta'}^2 \frac{\partial s_{\theta'}}{\partial h} c_{\theta'} + s_{\theta'}^3 \frac{\partial c_{\theta'}}{\partial h}\right) s_{\lambda_\odot-\Omega'}^3 c_{\lambda_\odot-\Omega'}\right] - 60\left[c_{\lambda_\odot-\Omega'}^2 c_{\theta'} \frac{\partial c_{\theta'}}{\partial h}\right.$$
$$\left. + c_{i'}^2 s_{\theta'} \frac{\partial s_{\theta'}}{\partial h} s_{\lambda_\odot-\Omega'}^2 + c_{i'}\left(\frac{\partial s_{\theta'}}{\partial h} c_{\theta'} + s_{\theta'} \frac{\partial c_{\theta'}}{\partial h}\right) s_{\lambda_\odot-\Omega'} c_{\lambda_\odot-\Omega'}\right] \qquad \text{(A15.3)}$$

$$\frac{\partial f_{r_1}^s}{\partial k} = 6\left[c_{\lambda_\odot-\Omega'}^2 c_{\theta'} \frac{\partial c_{\theta'}}{\partial k} + c_{i'}^2 s_{\theta'} \frac{\partial s_{\theta'}}{\partial k} s_{\lambda_\odot-\Omega'}^2 + c_{i'}\left(\frac{\partial s_{\theta'}}{\partial k} c_{\theta'} + s_{\theta'} \frac{\partial c_{\theta'}}{\partial k}\right) s_{\lambda_\odot-\Omega'} c_{\lambda_\odot-\Omega'}\right] \qquad \text{(A15.4)}$$

$$\frac{\partial f_{r_2}^s}{\partial k} = 15\left[c_{\lambda_\odot-\Omega'}^3 c_{\theta'}^2 \frac{\partial c_{\theta'}}{\partial k} + c_{i'}^3 s_{\theta'}^2 \frac{\partial s_{\theta'}}{\partial k} s_{\lambda_\odot-\Omega'}^3 + c_{\lambda_\odot-\Omega'}^2\left(2c_{\theta'} \frac{\partial c_{\theta'}}{\partial k} s_{\theta'} + c_{\theta'}^2 \frac{\partial s_{\theta'}}{\partial k}\right)\right.$$
$$\left. \times c_{i'} s_{\lambda_\odot-\Omega'} + c_{i'}^2\left(2s_{\theta'} \frac{\partial s_{\theta'}}{\partial k} c_{\theta'} + s_{\theta'}^2 \frac{\partial c_{\theta'}}{\partial k}\right) s_{\lambda_\odot-\Omega'}^2 c_{\lambda_\odot-\Omega'}\right]$$
$$- 3\left(c_{\lambda_\odot-\Omega'} \frac{\partial c_{\theta'}}{\partial k} + c_{i'} \frac{\partial s_{\theta'}}{\partial k} s_{\lambda_\odot-\Omega'}\right) \qquad \text{(A15.5)}$$

$$\frac{\partial f_{r_3}^s}{\partial k} = 140\left[c_{\theta'}^3 \frac{\partial c_{\theta'}}{\partial k} c_{\lambda_\odot-\Omega'}^4 + c_{i'}\left(3c_{\theta'}^2 \frac{\partial c_{\theta'}}{\partial k} s_{\theta'} + c_{\theta'}^3 \frac{\partial s_{\theta'}}{\partial k}\right) s_{\lambda_\odot-\Omega'} c_{\lambda_\odot-\Omega'}^3 \right.$$
$$+ c_{i'}^2\left(2s_{\theta'} \frac{\partial s_{\theta'}}{\partial k} c_{\theta'}^2 + 2s_{\theta'}^2 c_{\theta'} \frac{\partial c_{\theta'}}{\partial k}\right) s_{\lambda_\odot-\Omega'}^2 c_{\lambda_\odot-\Omega'}^2 + c_{i'}^4 s_{\theta'}^3 \frac{\partial s_{\theta'}}{\partial k} s_{\lambda_\odot-\Omega'}^4$$
$$\left. + c_{i'}^3\left(3s_{\theta'}^2 \frac{\partial s_{\theta'}}{\partial k} c_{\theta'} + s_{\theta'}^3 \frac{\partial c_{\theta'}}{\partial k}\right) s_{\lambda_\odot-\Omega'}^3 c_{\lambda_\odot-\Omega'}\right] - 60\left[c_{\lambda_\odot-\Omega'}^2 c_{\theta'} \frac{\partial c_{\theta'}}{\partial k}\right.$$
$$\left. + c_{i'}^2 s_{\theta'} \frac{\partial s_{\theta'}}{\partial k} s_{\lambda_\odot-\Omega'}^2 + c_{i'}\left(\frac{\partial s_{\theta'}}{\partial k} c_{\theta'} + s_{\theta'} \frac{\partial c_{\theta'}}{\partial k}\right) s_{\lambda_\odot-\Omega'} c_{\lambda_\odot-\Omega'}\right] \qquad \text{(A15.6)}$$

$$\frac{\partial f_{r_1}^s}{\partial p} = 6\left[c_{\lambda_\odot-\Omega'} \frac{\partial c_{\lambda_\odot-\Omega'}}{\partial p} c_{\theta'}^2 + c_{\lambda_\odot-\Omega'}^2 c_{\theta'} \frac{\partial c_{\theta'}}{\partial p} + \left(c_{i'} \frac{\partial c_{i'}}{\partial p} s_{\theta'}^2 + c_{i'}^2 s_{\theta'} \frac{\partial s_{\theta'}}{\partial p}\right) s_{\lambda_\odot-\Omega'}^2\right.$$
$$+ c_{i'}^2 s_{\theta'}^2 s_{\lambda_\odot-\Omega'} \frac{\partial s_{\lambda_\odot-\Omega'}}{\partial p} + \left(\frac{\partial c_{i'}}{\partial p} s_{\theta'} c_{\theta'} + c_{i'}\left\{\frac{\partial s_{\theta'}}{\partial p} c_{\theta'} + s_{\theta'} \frac{\partial c_{\theta'}}{\partial p}\right\}\right) s_{\lambda_\odot-\Omega'} c_{\lambda_\odot-\Omega'}$$
$$\left. + c_{i'} s_{\theta'} c_{\theta'}\left(\frac{\partial s_{\lambda_\odot-\Omega'}}{\partial p} c_{\lambda_\odot-\Omega'} + s_{\lambda_\odot-\Omega'} \frac{\partial c_{\lambda_\odot-\Omega'}}{\partial p}\right)\right] \qquad \text{(A15.7)}$$

$$\frac{\partial f_{r_2}^s}{\partial p} = 15\left[c_{\lambda_\odot-\Omega'}^2 \frac{\partial c_{\lambda_\odot-\Omega'}}{\partial p} c_{\theta'}^3 + c_{\lambda_\odot-\Omega'}^3 c_{\theta'}^2 \frac{\partial c_{\theta'}}{\partial p}\right.$$
$$+ \left(c_{i'}^2 \frac{\partial c_{i'}}{\partial p} s_{\theta'}^3 + c_{i'}^3 s_{\theta'}^2 \frac{\partial s_{\theta'}}{\partial p}\right) s_{\lambda_\odot-\Omega'}^3 + c_{i'}^3 s_{\theta'}^3 s_{\lambda_\odot-\Omega'}^2 \frac{\partial s_{\lambda_\odot-\Omega'}}{\partial p}$$
$$+ \left(\frac{\partial c_{i'}}{\partial p} c_{\theta'}^2 s_{\theta'} + c_{i'}\left\{2c_{\theta'} \frac{\partial c_{\theta'}}{\partial p} s_{\theta'} + c_{\theta'}^2 \frac{\partial s_{\theta'}}{\partial p}\right\}\right) c_{\lambda_\odot-\Omega'}^2 s_{\lambda_\odot-\Omega'}$$

$$+ c_{\theta'}^2 s_{\theta'} c_{i'} \left(2c_{\lambda_\odot-\Omega'} \frac{\partial c_{\lambda_\odot-\Omega'}}{\partial p} s_{\lambda_\odot-\Omega'} + c_{\lambda_\odot-\Omega'}^2 \frac{\partial s_{\lambda_\odot-\Omega'}}{\partial p} \right)$$
$$+ \left(2c_{i'} \frac{\partial c_{i'}}{\partial p} s_{\theta'}^2 c_{\theta'} + c_{i'}^2 \left\{ 2s_{\theta'} \frac{\partial s_{\theta'}}{\partial p} c_{\theta'} + s_{\theta'}^2 \frac{\partial c_{\theta'}}{\partial p} \right\} \right) s_{\lambda_\odot-\Omega'}^2 c_{\lambda_\odot-\Omega'}$$
$$+ c_{i'}^2 s_{\theta'}^2 c_{\theta'} \left(2s_{\lambda_\odot-\Omega'} \frac{\partial s_{\lambda_\odot-\Omega'}}{\partial p} c_{\lambda_\odot-\Omega'} + s_{\lambda_\odot-\Omega'}^2 \frac{\partial c_{\lambda_\odot-\Omega'}}{\partial p} \right) \Bigg]$$
$$- 3 \left(\frac{\partial c_{\lambda_\odot-\Omega'}}{\partial p} c_{\theta'} + c_{\lambda_\odot-\Omega'} \frac{\partial c_{\theta'}}{\partial p} + \left\{ \frac{\partial c_{i'}}{\partial p} s_{\theta'} + c_{i'} \frac{\partial s_{\theta'}}{\partial p} \right\} s_{\lambda_\odot-\Omega'} \right.$$
$$\left. + c_{i'} s_{\theta'} \frac{\partial s_{\lambda_\odot-\Omega'}}{\partial p} \right) \tag{A15.8}$$

$$\frac{\partial f_{r_3}^s}{\partial p} = 140 \left[c_{\theta'}^3 \frac{\partial c_{\theta'}}{\partial p} c_{\lambda_\odot-\Omega'}^4 + c_{\theta'}^4 c_{\lambda_\odot-\Omega'}^3 \frac{\partial c_{\lambda_\odot-\Omega'}}{\partial p} \right.$$
$$+ \left(\frac{\partial c_{i'}}{\partial p} c_{\theta'}^3 s_{\theta'} + c_{i'} \left\{ 3c_{\theta'}^2 \frac{\partial c_{\theta'}}{\partial p} s_{\theta'} + c_{\theta'}^3 \frac{\partial s_{\theta'}}{\partial p} \right\} \right) s_{\lambda_\odot-\Omega'} c_{\lambda_\odot-\Omega'}^3$$
$$+ c_{i'} c_{\theta'}^3 s_{\theta'} \left(\frac{\partial s_{\lambda_\odot-\Omega'}}{\partial p} c_{\lambda_\odot-\Omega'}^3 + 3 s_{\lambda_\odot-\Omega'} c_{\lambda_\odot-\Omega'}^2 \frac{\partial c_{\lambda_\odot-\Omega'}}{\partial p} \right)$$
$$+ 2 \left(c_{i'} \frac{\partial c_{i'}}{\partial p} s_{\theta'}^2 c_{\theta'}^2 + c_{i'}^2 \left\{ s_{\theta'} \frac{\partial s_{\theta'}}{\partial p} c_{\theta'}^2 + s_{\theta'}^2 c_{\theta'} \frac{\partial c_{\theta'}}{\partial p} \right\} \right) s_{\lambda_\odot-\Omega'}^2 c_{\lambda_\odot-\Omega'}^2$$
$$+ 2c_{i'}^2 s_{\theta'}^2 c_{\theta'}^2 \left(s_{\lambda_\odot-\Omega'} \frac{\partial s_{\lambda_\odot-\Omega'}}{\partial p} c_{\lambda_\odot-\Omega'}^2 + s_{\lambda_\odot-\Omega'}^2 c_{\lambda_\odot-\Omega'} \frac{\partial c_{\lambda_\odot-\Omega'}}{\partial p} \right)$$
$$+ \left(c_{i'}^3 \frac{\partial c_{i'}}{\partial p} s_{\theta'}^4 + c_{i'}^4 s_{\theta'}^3 \frac{\partial s_{\theta'}}{\partial p} \right) s_{\lambda_\odot-\Omega'}^4 + c_{i'}^4 s_{\theta'}^4 s_{\lambda_\odot-\Omega'}^3 \frac{\partial s_{\lambda_\odot-\Omega'}}{\partial p}$$
$$+ \left(3c_{i'}^2 \frac{\partial c_{i'}}{\partial p} s_{\theta'}^3 c_{\theta'} + c_{i'}^3 \left\{ 3s_{\theta'}^2 \frac{\partial s_{\theta'}}{\partial p} c_{\theta'} + s_{\theta'}^3 \frac{\partial c_{\theta'}}{\partial p} \right\} \right) s_{\lambda_\odot-\Omega'}^3 c_{\lambda_\odot-\Omega'}$$
$$\left. + c_{i'}^3 s_{\theta'}^3 c_{\theta'} \left(3s_{\lambda_\odot-\Omega'}^2 \frac{\partial s_{\lambda_\odot-\Omega'}}{\partial p} c_{\lambda_\odot-\Omega'} + s_{\lambda_\odot-\Omega'}^3 \frac{\partial c_{\lambda_\odot-\Omega'}}{\partial p} \right) \right]$$
$$- 60 \left[c_{\theta'} \frac{\partial c_{\theta'}}{\partial p} c_{\lambda_\odot-\Omega'}^2 + c_{\theta'}^2 c_{\lambda_\odot-\Omega'} \frac{\partial c_{\lambda_\odot-\Omega'}}{\partial p} + \left\{ c_{i'} \frac{\partial c_{i'}}{\partial p} s_{\theta'}^2 + c_{i'}^2 s_{\theta'} \frac{\partial s_{\theta'}}{\partial p} \right\} s_{\lambda_\odot-\Omega'}^2 \right.$$
$$+ c_{i'}^2 s_{\theta'}^2 s_{\lambda_\odot-\Omega'} \frac{\partial s_{\lambda_\odot-\Omega'}}{\partial p} + \left(\frac{\partial c_{i'}}{\partial p} s_{\theta'} c_{\theta'} + c_{i'} \left\{ \frac{\partial s_{\theta'}}{\partial p} c_{\theta'} + s_{\theta'} \frac{\partial c_{\theta'}}{\partial p} \right\} \right) s_{\lambda_\odot-\Omega'} c_{\lambda_\odot-\Omega'}$$
$$\left. + c_{i'} s_{\theta'} c_{\theta'} \left(\frac{\partial s_{\lambda_\odot-\Omega'}}{\partial p} c_{\lambda_\odot-\Omega'} + s_{\lambda_\odot-\Omega'} \frac{\partial c_{\lambda_\odot-\Omega'}}{\partial p} \right) \right] \tag{A15.9}$$

$$\frac{\partial f_{r_1}^s}{\partial q} = 6 \left[c_{\lambda_\odot-\Omega'} \frac{\partial c_{\lambda_\odot-\Omega'}}{\partial q} c_{\theta'}^2 + c_{\lambda_\odot-\Omega'}^2 \, c_{\theta'} \frac{\partial c_{\theta'}}{\partial q} \right.$$
$$+ \left(c_{i'} \frac{\partial c_{i'}}{\partial q} s_{\theta'}^2 + c_{i'}^2 s_{\theta'} \frac{\partial s_{\theta'}}{\partial q} \right) s_{\lambda_\odot-\Omega'}^2 + c_{i'}^2 s_{\theta'}^2 s_{\lambda_\odot-\Omega'} \frac{\partial s_{\lambda_\odot-\Omega'}}{\partial q}$$
$$+ \left(\frac{\partial c_{i'}}{\partial q} s_{\theta'} c_{\theta'} + c_{i'} \left\{ \frac{\partial s_{\theta'}}{\partial q} c_{\theta'} + s_{\theta'} \frac{\partial c_{\theta'}}{\partial q} \right\} \right) s_{\lambda_\odot-\Omega'} c_{\lambda_\odot-\Omega'}$$
$$\left. + c_{i'} s_{\theta'} c_{\theta'} \left(\frac{\partial s_{\lambda_\odot-\Omega'}}{\partial q} c_{\lambda_\odot-\Omega'} + s_{\lambda_\odot-\Omega'} \frac{\partial c_{\lambda_\odot-\Omega'}}{\partial q} \right) \right] \tag{A15.10}$$

$$
\begin{aligned}
\frac{\partial f^s_{r_2}}{\partial q} = 15\Bigg[& c^2_{\lambda_\odot-\Omega'}\frac{\partial c_{\lambda_\odot-\Omega'}}{\partial q}c^3_{\theta'} + c^3_{\lambda_\odot-\Omega'}\, c^2_{\theta'}\frac{\partial c_{\theta'}}{\partial q} \\
& + \left(c^2_{i'}\frac{\partial c_{i'}}{\partial q}s^3_{\theta'} + c^3_{i'}s^2_{\theta'}\frac{\partial s_{\theta'}}{\partial q}\right)s^3_{\lambda_\odot-\Omega'} + c^3_{i'}s^3_{\theta'}s^2_{\lambda_\odot-\Omega'}\frac{\partial s_{\lambda_\odot-\Omega'}}{\partial q} \\
& + \left(\frac{\partial c_{i'}}{\partial q}c^2_{\theta'}s_{\theta'} + c_{i'}\left\{2c_{\theta'}\frac{\partial c_{\theta'}}{\partial q}s_{\theta'} + c^2_{\theta'}\frac{\partial s_{\theta'}}{\partial q}\right\}\right)c^2_{\lambda_\odot-\Omega'}s_{\lambda_\odot-\Omega'} \\
& + c^2_{\theta'}s_{\theta'}c_{i'}\left(2c_{\lambda_\odot-\Omega'}\frac{\partial c_{\lambda_\odot-\Omega'}}{\partial q}s_{\lambda_\odot-\Omega'} + c^2_{\lambda_\odot-\Omega'}\frac{\partial s_{\lambda_\odot-\Omega'}}{\partial q}\right) \\
& + \left(2c_{i'}\frac{\partial c_{i'}}{\partial q}s^2_{\theta'}c_{\theta'} + c^2_{i'}\left\{2s_{\theta'}\frac{\partial s_{\theta'}}{\partial q}c_{\theta'} + s^2_{\theta'}\frac{\partial c_{\theta'}}{\partial q}\right\}\right)s^2_{\lambda_\odot-\Omega'}c_{\lambda_\odot-\Omega'} \\
& + c^2_{i'}s^2_{\theta'}c_{\theta'}\left(2s_{\lambda_\odot-\Omega'}\frac{\partial s_{\lambda_\odot-\Omega'}}{\partial q}c_{\lambda_\odot-\Omega'} + s^2_{\lambda_\odot-\Omega'}\frac{\partial c_{\lambda_\odot-\Omega'}}{\partial q}\right)\Bigg] \\
& - 3\left(\frac{\partial c_{\lambda_\odot-\Omega'}}{\partial q}c_{\theta'} + c_{\lambda_\odot-\Omega'}\,\frac{\partial c_{\theta'}}{\partial q} + \left\{\frac{\partial c_{i'}}{\partial q}s_{\theta'} + c_{i'}\frac{\partial s_{\theta'}}{\partial q}\right\}s_{\lambda_\odot-\Omega'}\right. \\
& \left. + c_{i'}s_{\theta'}\frac{\partial s_{\lambda_\odot-\Omega'}}{\partial q}\right) \qquad \text{(A15.11)}
\end{aligned}
$$

$$
\begin{aligned}
\frac{\partial f^s_{r_3}}{\partial q} = 140\Bigg[& c^3_{\theta'}\frac{\partial c_{\theta'}}{\partial q}c^4_{\lambda_\odot-\Omega'} + c^4_{\theta'}\, c^3_{\lambda_\odot-\Omega'}\frac{\partial c_{\lambda_\odot-\Omega'}}{\partial q} \\
& + \left(\frac{\partial c_{i'}}{\partial q}c^3_{\theta'}s_{\theta'} + c_{i'}\left\{3c^2_{\theta'}\frac{\partial c_{\theta'}}{\partial q}s_{\theta'} + c^3_{\theta'}\frac{\partial s_{\theta'}}{\partial q}\right\}\right)s_{\lambda_\odot-\Omega'}c^3_{\lambda_\odot-\Omega'} \\
& + c_{i'}c^3_{\theta'}s_{\theta'}\left(\frac{\partial s_{\lambda_\odot-\Omega'}}{\partial q}c^3_{\lambda_\odot-\Omega'} + 3s_{\lambda_\odot-\Omega'}c^2_{\lambda_\odot-\Omega'}\frac{\partial c_{\lambda_\odot-\Omega'}}{\partial q}\right) \\
& + 2\left(c_{i'}\frac{\partial c_{i'}}{\partial q}s^2_{\theta'}c^2_{\theta'} + c^2_{i'}\left\{s_{\theta'}\frac{\partial s_{\theta'}}{\partial q}c^2_{\theta'} + s^2_{\theta'}c_{\theta'}\frac{\partial c_{\theta'}}{\partial q}\right\}\right)s^2_{\lambda_\odot-\Omega'}c^2_{\lambda_\odot-\Omega'} \\
& + 2c^2_{i'}s^2_{\theta'}c^2_{\theta'}\left(s_{\lambda_\odot-\Omega'}\frac{\partial s_{\lambda_\odot-\Omega'}}{\partial q}c^2_{\lambda_\odot-\Omega'} + s^2_{\lambda_\odot-\Omega'}c_{\lambda_\odot-\Omega'}\frac{\partial c_{\lambda_\odot-\Omega'}}{\partial q}\right) \\
& + \left(c^3_{i'}\frac{\partial c_{i'}}{\partial q}s^4_{\theta'} + c^4_{i'}s^3_{\theta'}\frac{\partial s_{\theta'}}{\partial q}\right)s^4_{\lambda_\odot-\Omega'} + c^4_{i'}s^4_{\theta'}s^3_{\lambda_\odot-\Omega'}\frac{\partial s_{\lambda_\odot-\Omega'}}{\partial q} \\
& + \left(3c^2_{i'}\frac{\partial c_{i'}}{\partial q}s^3_{\theta'}c_{\theta'} + c^3_{i'}\left\{3s^2_{\theta'}\frac{\partial s_{\theta'}}{\partial q}c_{\theta'} + s^3_{\theta'}\frac{\partial c_{\theta'}}{\partial q}\right\}\right)s^3_{\lambda_\odot-\Omega'}c_{\lambda_\odot-\Omega'} \\
& + c^3_{i'}s^3_{\theta'}c_{\theta'}\left(3s^2_{\lambda_\odot-\Omega'}\frac{\partial s_{\lambda_\odot-\Omega'}}{\partial q}c_{\lambda_\odot-\Omega'} + s^3_{\lambda_\odot-\Omega'}\frac{\partial c_{\lambda_\odot-\Omega'}}{\partial q}\right)\Bigg] \\
& - 60\Bigg[c_{\theta'}\frac{\partial c_{\theta'}}{\partial q}c^2_{\lambda_\odot-\Omega'} + c^2_{\theta'}c_{\lambda_\odot-\Omega'}\frac{\partial c_{\lambda_\odot-\Omega'}}{\partial q} \\
& + \left\{c_{i'}\frac{\partial c_{i'}}{\partial q}s^2_{\theta'} + c^2_{i'}s_{\theta'}\frac{\partial s_{\theta'}}{\partial q}\right\}s^2_{\lambda_\odot-\Omega'} + c^2_{i'}s^2_{\theta'}s_{\lambda_\odot-\Omega'}\frac{\partial s_{\lambda_\odot-\Omega'}}{\partial q} \\
& + \left(\frac{\partial c_{i'}}{\partial q}s_{\theta'}c_{\theta'} + c_{i'}\left\{\frac{\partial s_{\theta'}}{\partial q}c_{\theta'} + s_{\theta'}\frac{\partial c_{\theta'}}{\partial q}\right\}\right)s_{\lambda_\odot-\Omega'}c_{\lambda_\odot-\Omega'} \\
& + c_{i'}s_{\theta'}c_{\theta'}\left(\frac{\partial s_{\lambda_\odot-\Omega'}}{\partial q}c_{\lambda_\odot-\Omega'} + s_{\lambda_\odot-\Omega'}\frac{\partial c_{\lambda_\odot-\Omega'}}{\partial q}\right)\Bigg] \qquad \text{(A15.12)}
\end{aligned}
$$

$$\frac{\partial f_{r_1}^s}{\partial L} = 6\left(c_{\lambda_\odot-\Omega'}^2 c_{\theta'}\frac{\partial c_{\theta'}}{\partial L} + c_{i'}^2 s_{\theta'}\frac{\partial s_{\theta'}}{\partial L}s_{\lambda_\odot-\Omega'}^2 + c_{i'}\left\{\frac{\partial s_{\theta'}}{\partial L}c_{\theta'} + s_{\theta'}\frac{\partial c_{\theta'}}{\partial L}\right\}s_{\lambda_\odot-\Omega'}c_{\lambda_\odot-\Omega'}\right) \tag{A15.13}$$

$$\begin{aligned}\frac{\partial f_{r_2}^s}{\partial L} = 15 &\left(c_{\lambda_\odot-\Omega'}^3 c_{\theta'}^2\frac{\partial c_{\theta'}}{\partial L} + c_{i'}^3 s_{\theta'}^2\frac{\partial s_{\theta'}}{\partial L}s_{\lambda_\odot-\Omega'}^3 + c_{\lambda_\odot-\Omega'}^2 s_{\lambda_\odot-\Omega'}c_{i'}\right.\\ &\left.\times\left\{2c_{\theta'}\frac{\partial c_{\theta'}}{\partial L}s_{\theta'} + c_{\theta'}^2\frac{\partial s_{\theta'}}{\partial L}\right\} + c_{i'}^2 s_{\lambda_\odot-\Omega'}^2 c_{\lambda_\odot-\Omega'}\left\{2s_{\theta'}\frac{\partial s_{\theta'}}{\partial L}c_{\theta'} + s_{\theta'}^2\frac{\partial c_{\theta'}}{\partial L}\right\}\right)\\ &- 3\left(c_{\lambda_\odot-\Omega'}\frac{\partial c_{\theta'}}{\partial L} + c_{i'}s_{\lambda_\odot-\Omega'}\frac{\partial s_{\theta'}}{\partial L}\right)\end{aligned} \tag{A15.14}$$

$$\begin{aligned}\frac{\partial f_{r_3}^s}{\partial L} = 140 &\left(c_{\theta'}^3\frac{\partial c_{\theta'}}{\partial L}c_{\lambda_\odot-\Omega'}^4 + c_{i'}\left\{3c_{\theta'}^2\frac{\partial c_{\theta'}}{\partial L}s_{\theta'} + c_{\theta'}^3\frac{\partial s_{\theta'}}{\partial L}\right\}s_{\lambda_\odot-\Omega'}c_{\lambda_\odot-\Omega'}^3\right.\\ &+ 2c_{i'}^2 s_{\lambda_\odot-\Omega'}^2 c_{\lambda_\odot-\Omega'}^2\left\{s_{\theta'}\frac{\partial s_{\theta'}}{\partial L}c_{\theta'}^2 + s_{\theta'}^2 c_{\theta'}\frac{\partial c_{\theta'}}{\partial L}\right\} + c_{i'}^4 s_{\lambda_\odot-\Omega'}^4 s_{\theta'}^3\frac{\partial s_{\theta'}}{\partial L}\\ &\left.+ c_{i'}^3\left\{3s_{\theta'}^2\frac{\partial s_{\theta'}}{\partial L}c_{\theta'} + s_{\theta'}^3\frac{\partial c_{\theta'}}{\partial L}\right\}s_{\lambda_\odot-\Omega'}^3 c_{\lambda_\odot-\Omega'}\right) - 60\left(c_{\lambda_\odot-\Omega'}^2 c_{\theta'}\frac{\partial c_{\theta'}}{\partial L}\right.\\ &\left.+ c_{i'}^2 s_{\theta'}\frac{\partial s_{\theta'}}{\partial L}s_{\lambda_\odot-\Omega'}^2 + c_{i'}\left\{\frac{\partial s_{\theta'}}{\partial L}c_{\theta'} + s_{\theta'}\frac{\partial c_{\theta'}}{\partial L}\right\}s_{\lambda_\odot-\Omega'}c_{\lambda_\odot-\Omega'}\right)\end{aligned} \tag{A15.15}$$

The partial derivatives of f_θ^s with respect to h, h, p, q and L:

$$\begin{aligned}\frac{\partial f_{\theta_1}^s}{\partial h} = &-\left(\frac{\partial s_{\theta'}}{\partial h}c_{\theta'} + s_{\theta'}\frac{\partial c_{\theta'}}{\partial h}\right)c_{\lambda_\odot-\Omega'}^2 - 2c_{i'}s_{\theta'}\frac{\partial s_{\theta'}}{\partial h}s_{\lambda_\odot-\Omega'}c_{\lambda_\odot-\Omega'}\\ &+ 2c_{i'}c_{\theta'}\frac{\partial c_{\theta'}}{\partial h}s_{\lambda_\odot-\Omega'}c_{\lambda_\odot-\Omega'} + c_{i'}^2\left(\frac{\partial s_{\theta'}}{\partial h}c_{\theta'} + s_{\theta'}\frac{\partial c_{\theta'}}{\partial h}\right)s_{\lambda_\odot-\Omega'}^2\end{aligned} \tag{A15.16}$$

$$\begin{aligned}\frac{\partial f_{\theta_2}^s}{\partial h} = 5&\left[\left(-\frac{\partial s_{\theta'}}{\partial h}c_{\theta'}^2 - 2s_{\theta'}c_{\theta'}\frac{\partial c_{\theta'}}{\partial h}\right)c_{\lambda_\odot-\Omega'}^3 + 3c_{i'}c_{\theta'}^2\frac{\partial c_{\theta'}}{\partial h}s_{\lambda_\odot-\Omega'}c_{\lambda_\odot-\Omega'}^2\right.\\ &- 3c_{i'}^2 s_{\theta'}^2\frac{\partial s_{\theta'}}{\partial h}c_{\lambda_\odot-\Omega'}s_{\lambda_\odot-\Omega'}^2 + c_{i'}^3\left(\frac{\partial c_{\theta'}}{\partial h}s_{\theta'}^2 + 2s_{\theta'}c_{\theta'}\frac{\partial s_{\theta'}}{\partial h}\right)s_{\lambda_\odot-\Omega'}^3\\ &- 2c_{i'}\left(2s_{\theta'}\frac{\partial s_{\theta'}}{\partial h}c_{\theta'} + s_{\theta'}^2\frac{\partial c_{\theta'}}{\partial h}\right)s_{\lambda_\odot-\Omega'}c_{\lambda_\odot-\Omega'}^2 + 2c_{i'}^2\left(\frac{\partial s_{\theta'}}{\partial h}c_{\theta'}^2\right.\\ &\left.\left.+ 2s_{\theta'}c_{\theta'}\frac{\partial c_{\theta'}}{\partial h}\right)s_{\lambda_\odot-\Omega'}^2 c_{\lambda_\odot-\Omega'}\right] - \left(-\frac{\partial s_{\theta'}}{\partial h}c_{\lambda_\odot-\Omega'} + c_{i'}\frac{\partial c_{\theta'}}{\partial h}s_{\lambda_\odot-\Omega'}\right)\end{aligned} \tag{A15.17}$$

$$\begin{aligned}\frac{\partial f_{\theta_3}^s}{\partial h} = 35&\left[\left(-\frac{\partial s_{\theta'}}{\partial h}c_{\theta'}^3 - 3s_{\theta'}c_{\theta'}^2\frac{\partial c_{\theta'}}{\partial h}\right)c_{\lambda_\odot-\Omega'}^4 - 4c_{i'}^3 s_{\theta'}^3\frac{\partial s_{\theta'}}{\partial h}c_{\lambda_\odot-\Omega'}s_{\lambda_\odot-\Omega'}^3\right.\\ &- 6c_{i'}\left(s_{\theta'}\frac{\partial s_{\theta'}}{\partial h}c_{\theta'}^2 + s_{\theta'}^2 c_{\theta'}\frac{\partial c_{\theta'}}{\partial h}\right)c_{\lambda_\odot-\Omega'}^3 s_{\lambda_\odot-\Omega'}\\ &- 3c_{i'}^2\left(3s_{\theta'}^2\frac{\partial s_{\theta'}}{\partial h}c_{\theta'} + s_{\theta'}^3\frac{\partial c_{\theta'}}{\partial h}\right)s_{\lambda_\odot-\Omega'}^2 c_{\lambda_\odot-\Omega'}^2 + 4c_{i'}c_{\theta'}^3\frac{\partial c_{\theta'}}{\partial h}s_{\lambda_\odot-\Omega'}c_{\lambda_\odot-\Omega'}^3\end{aligned}$$

$$+ c_{i'}^4 \left(\frac{\partial c_{\theta'}}{\partial h} s_{\theta'}^3 + 3 s_{\theta'}^2 c_{\theta'} \frac{\partial s_{\theta'}}{\partial h} \right) s_{\lambda_\odot - \Omega'}^4 + 3 c_{i'}^2 \left(3 c_{\theta'}^2 \frac{\partial c_{\theta'}}{\partial h} s_{\theta'} + c_{\theta'}^3 \frac{\partial s_{\theta'}}{\partial h} \right) c_{\lambda_\odot - \Omega'}^2 s_{\lambda_\odot - \Omega'}^2$$
$$+ 6 c_{i'}^3 \left(s_{\theta'} \frac{\partial s_{\theta'}}{\partial h} c_{\theta'}^2 + s_{\theta'}^2 c_{\theta'} \frac{\partial c_{\theta'}}{\partial h} \right) s_{\lambda_\odot - \Omega'}^3 c_{\lambda_\odot - \Omega'} \Bigg] - 15 \Bigg[\left(- \frac{\partial s_{\theta'}}{\partial h} c_{\theta'} - s_{\theta'} \frac{\partial c_{\theta'}}{\partial h} \right)$$
$$\times c_{\lambda_\odot - \Omega'}^2 - 2 c_{i'} s_{\theta'} \frac{\partial s_{\theta'}}{\partial h} s_{\lambda_\odot - \Omega'} c_{\lambda_\odot - \Omega'} + 2 c_{i'} c_{\theta'} \frac{\partial c_{\theta'}}{\partial h} s_{\lambda_\odot - \Omega'} c_{\lambda_\odot - \Omega'}$$
$$+ c_{i'}^2 \left(\frac{\partial s_{\theta'}}{\partial h} c_{\theta'} + s_{\theta'} \frac{\partial c_{\theta'}}{\partial h} \right) s_{\lambda_\odot - \Omega'}^2 \Bigg] \tag{A15.18}$$

$$\frac{\partial f_{\theta_1}^s}{\partial k} = - \left(\frac{\partial s_{\theta'}}{\partial k} c_{\theta'} + s_{\theta'} \frac{\partial c_{\theta'}}{\partial k} \right) c_{\lambda_\odot - \Omega'}^2 - 2 c_{i'} s_{\theta'} \frac{\partial s_{\theta'}}{\partial k} s_{\lambda_\odot - \Omega'} c_{\lambda_\odot - \Omega'}$$
$$+ 2 c_{i'} c_{\theta'} \frac{\partial c_{\theta'}}{\partial k} s_{\lambda_\odot - \Omega'} c_{\lambda_\odot - \Omega'} + c_{i'}^2 \left(\frac{\partial s_{\theta'}}{\partial k} c_{\theta'} + s_{\theta'} \frac{\partial c_{\theta'}}{\partial k} \right) s_{\lambda_\odot - \Omega'}^2 \tag{A15.19}$$

$$\frac{\partial f_{\theta_2}^s}{\partial k} = 5 \Bigg[\left(- \frac{\partial s_{\theta'}}{\partial k} c_{\theta'}^2 - 2 s_{\theta'} c_{\theta'} \frac{\partial c_{\theta'}}{\partial k} \right) c_{\lambda_\odot - \Omega'}^3$$
$$+ 3 c_{i'} c_{\theta'}^2 \frac{\partial c_{\theta'}}{\partial k} s_{\lambda_\odot - \Omega'} c_{\lambda_\odot - \Omega'}^2 - 3 c_{i'}^2 s_{\theta'}^2 \frac{\partial s_{\theta'}}{\partial k} c_{\lambda_\odot - \Omega'} s_{\lambda_\odot - \Omega'}^2$$
$$+ c_{i'}^3 \left(\frac{\partial c_{\theta'}}{\partial k} s_{\theta'}^2 + 2 s_{\theta'} c_{\theta'} \frac{\partial s_{\theta'}}{\partial k} \right) s_{\lambda_\odot - \Omega'}^3$$
$$- 2 c_{i'} \left(2 s_{\theta'} \frac{\partial s_{\theta'}}{\partial k} c_{\theta'} + s_{\theta'}^2 \frac{\partial c_{\theta'}}{\partial k} \right) s_{\lambda_\odot - \Omega'} c_{\lambda_\odot - \Omega'}^2 + 2 c_{i'}^2 \Bigg(\frac{\partial s_{\theta'}}{\partial k} c_{\theta'}^2$$
$$+ 2 s_{\theta'} c_{\theta'} \frac{\partial c_{\theta'}}{\partial k} \Bigg) s_{\lambda_\odot - \Omega'}^2 c_{\lambda_\odot - \Omega'} \Bigg] - \left(- \frac{\partial s_{\theta'}}{\partial k} c_{\lambda_\odot - \Omega'} + c_{i'} \frac{\partial c_{\theta'}}{\partial k} s_{\lambda_\odot - \Omega'} \right) \tag{A15.20}$$

$$\frac{\partial f_{\theta_3}^s}{\partial k} = 35 \Bigg[\left(- \frac{\partial s_{\theta'}}{\partial k} c_{\theta'}^3 - 3 s_{\theta'} c_{\theta'}^2 \frac{\partial c_{\theta'}}{\partial k} \right) c_{\lambda_\odot - \Omega'}^4$$
$$- 4 c_{i'}^3 s_{\theta'}^3 \frac{\partial s_{\theta'}}{\partial k} c_{\lambda_\odot - \Omega'} s_{\lambda_\odot - \Omega'}^3 - 6 c_{i'} \left(s_{\theta'} \frac{\partial s_{\theta'}}{\partial k} c_{\theta'}^2 + s_{\theta'}^2 c_{\theta'} \frac{\partial c_{\theta'}}{\partial k} \right) c_{\lambda_\odot - \Omega'}^3 s_{\lambda_\odot - \Omega'}$$
$$- 3 c_{i'}^2 \left(3 s_{\theta'}^2 \frac{\partial s_{\theta'}}{\partial k} c_{\theta'} + s_{\theta'}^3 \frac{\partial c_{\theta'}}{\partial k} \right) s_{\lambda_\odot - \Omega'}^2 c_{\lambda_\odot - \Omega'}^2 + 4 c_{i'} c_{\theta'}^3 \frac{\partial c_{\theta'}}{\partial k} s_{\lambda_\odot - \Omega'} c_{\lambda_\odot - \Omega'}^3$$
$$+ c_{i'}^4 \left(\frac{\partial c_{\theta'}}{\partial k} s_{\theta'}^3 + 3 s_{\theta'}^2 c_{\theta'} \frac{\partial s_{\theta'}}{\partial k} \right) s_{\lambda_\odot - \Omega'}^4 + 3 c_{i'}^2 \Bigg(3 c_{\theta'}^2 \frac{\partial c_{\theta'}}{\partial k} s_{\theta'}$$
$$+ c_{\theta'}^3 \frac{\partial s_{\theta'}}{\partial k} \Bigg) c_{\lambda_\odot - \Omega'}^2 s_{\lambda_\odot - \Omega'}^2 + 6 c_{i'}^3 \left(s_{\theta'} \frac{\partial s_{\theta'}}{\partial k} c_{\theta'}^2 + s_{\theta'}^2 c_{\theta'} \frac{\partial c_{\theta'}}{\partial k} \right) s_{\lambda_\odot - \Omega'}^3 c_{\lambda_\odot - \Omega'} \Bigg]$$
$$- 15 \Bigg[\left(- \frac{\partial s_{\theta'}}{\partial k} c_{\theta'} - s_{\theta'} \frac{\partial c_{\theta'}}{\partial k} \right) c_{\lambda_\odot - \Omega'}^2 - 2 c_{i'} s_{\theta'} \frac{\partial s_{\theta'}}{\partial k} s_{\lambda_\odot - \Omega'} c_{\lambda_\odot - \Omega'}$$
$$+ 2 c_{i'} c_{\theta'} \frac{\partial c_{\theta'}}{\partial k} s_{\lambda_\odot - \Omega'} c_{\lambda_\odot - \Omega'} + c_{i'}^2 \left(\frac{\partial s_{\theta'}}{\partial k} c_{\theta'} + s_{\theta'} \frac{\partial c_{\theta'}}{\partial k} \right) s_{\lambda_\odot - \Omega'}^2 \Bigg] \tag{A15.21}$$

$$\frac{\partial f_{\theta_1}^s}{\partial p} = \left(- \frac{\partial s_{\theta'}}{\partial p} c_{\theta'} - s_{\theta'} \frac{\partial c_{\theta'}}{\partial p} \right) c_{\lambda_\odot - \Omega'}^2 - 2 s_{\theta'} c_{\theta'} c_{\lambda_\odot - \Omega'} \frac{\partial c_{\lambda_\odot - \Omega'}}{\partial p}$$
$$- c_{i'} s_{\theta'}^2 \left(\frac{\partial s_{\lambda_\odot - \Omega'}}{\partial p} c_{\lambda_\odot - \Omega'} + s_{\lambda_\odot - \Omega'} \frac{\partial c_{\lambda_\odot - \Omega'}}{\partial p} \right) - \left(2 c_{i'} s_{\theta'} \frac{\partial s_{\theta'}}{\partial p} + \frac{\partial c_{i'}}{\partial p} s_{\theta'}^2 \right)$$

$$\times s_{\lambda_\odot-\Omega'}c_{\lambda_\odot-\Omega'} + \left(\frac{\partial c_{i'}}{\partial p}c_{\theta'}^2 + 2c_{i'}c_{\theta'}\frac{\partial c_{\theta'}}{\partial p}\right)s_{\lambda_\odot-\Omega'}c_{\lambda_\odot-\Omega'}$$

$$+ c_{i'}c_{\theta'}^2\left(\frac{\partial s_{\lambda_\odot-\Omega'}}{\partial p}c_{\lambda_\odot-\Omega'} + s_{\lambda_\odot-\Omega'}\frac{\partial c_{\lambda_\odot-\Omega'}}{\partial p}\right) + \left(2c_{i'}\frac{\partial c_{i'}}{\partial p}s_{\theta'}c_{\theta'}\right.$$

$$\left. + c_{i'}^2\left\{\frac{\partial s_{\theta'}}{\partial p}c_{\theta'} + s_{\theta'}\frac{\partial c_{\theta'}}{\partial p}\right\}\right)s_{\lambda_\odot-\Omega'}^2 + 2c_{i'}^2 s_{\theta'}c_{\theta'}s_{\lambda_\odot-\Omega'}\frac{\partial s_{\lambda_\odot-\Omega'}}{\partial p} \tag{A15.22}$$

$$\frac{\partial f_{\theta_2}^s}{\partial p} = 5\left[\left(-\frac{\partial s_{\theta'}}{\partial p}c_{\theta'}^2 - 2s_{\theta'}c_{\theta'}\frac{\partial c_{\theta'}}{\partial p}\right)c_{\lambda_\odot-\Omega'}^3 - 3s_{\theta'}c_{\theta'}^2 c_{\lambda_\odot-\Omega'}^2\frac{\partial c_{\lambda_\odot-\Omega'}}{\partial p}\right.$$

$$+ \left(\frac{\partial c_{i'}}{\partial p}c_{\theta'}^3 + 3c_{i'}c_{\theta'}^2\frac{\partial c_{\theta'}}{\partial p}\right)s_{\lambda_\odot-\Omega'}c_{\lambda_\odot-\Omega'}^2 + c_{i'}c_{\theta'}^3\left(\frac{\partial s_{\lambda_\odot-\Omega'}}{\partial p}c_{\lambda_\odot-\Omega'}^2\right.$$

$$\left. + 2s_{\lambda_\odot-\Omega'}c_{\lambda_\odot-\Omega'}\frac{\partial c_{\lambda_\odot-\Omega'}}{\partial p}\right) - \left(2c_{i'}\frac{\partial c_{i'}}{\partial p}s_{\theta'}^3 + 3c_{i'}^2 s_{\theta'}^2\frac{\partial s_{\theta'}}{\partial p}\right)c_{\lambda_\odot-\Omega'}s_{\lambda_\odot-\Omega'}^2$$

$$- c_{i'}^2 s_{\theta'}^3\left(\frac{\partial c_{\lambda_\odot-\Omega'}}{\partial p}s_{\lambda_\odot-\Omega'}^2 + 2c_{\lambda_\odot-\Omega'}s_{\lambda_\odot-\Omega'}\frac{\partial s_{\lambda_\odot-\Omega'}}{\partial p}\right)$$

$$+ \left(3c_{i'}^2\frac{\partial c_{i'}}{\partial p}s_{\theta'}^2 c_{\theta'} + c_{i'}^3\left\{\frac{\partial c_{\theta'}}{\partial p}s_{\theta'}^2 + 2s_{\theta'}c_{\theta'}\frac{\partial s_{\theta'}}{\partial p}\right\}\right)s_{\lambda_\odot-\Omega'}^3$$

$$+ 3c_{i'}^3 s_{\theta'}^2 c_{\theta'}s_{\lambda_\odot-\Omega'}^2\frac{\partial s_{\lambda_\odot-\Omega'}}{\partial p} - 2\left(\frac{\partial c_{i'}}{\partial p}s_{\theta'}^2 c_{\theta'} + c_{i'}\left\{2s_{\theta'}\frac{\partial s_{\theta'}}{\partial p}c_{\theta'} + s_{\theta'}^2\frac{\partial c_{\theta'}}{\partial p}\right\}\right)$$

$$\times s_{\lambda_\odot-\Omega'}c_{\lambda_\odot-\Omega'}^2 - 2c_{i'}s_{\theta'}^2 c_{\theta'}\left(\frac{\partial s_{\lambda_\odot-\Omega'}}{\partial p}c_{\lambda_\odot-\Omega'}^2 + 2s_{\lambda_\odot-\Omega'}c_{\lambda_\odot-\Omega'}\frac{\partial c_{\lambda_\odot-\Omega'}}{\partial p}\right)$$

$$+ 2\left(2c_{i'}\frac{\partial c_{i'}}{\partial p}s_{\theta'}c_{\theta'}^2 + c_{i'}^2\left\{\frac{\partial s_{\theta'}}{\partial p}c_{\theta'}^2 + 2s_{\theta'}c_{\theta'}\frac{\partial c_{\theta'}}{\partial p}\right\}\right)s_{\lambda_\odot-\Omega'}^2 c_{\lambda_\odot-\Omega'}$$

$$\left. + 2c_{i'}^2 s_{\theta'}c_{\theta'}^2\left(2s_{\lambda_\odot-\Omega'}\frac{\partial s_{\lambda_\odot-\Omega'}}{\partial p}c_{\lambda_\odot-\Omega'} + s_{\lambda_\odot-\Omega'}^2\frac{\partial c_{\lambda_\odot-\Omega'}}{\partial p}\right)\right]$$

$$- \left(-\frac{\partial s_{\theta'}}{\partial p}c_{\lambda_\odot-\Omega'} - s_{\theta'}\frac{\partial c_{\lambda_\odot-\Omega'}}{\partial p}\right.$$

$$\left. + \left\{\frac{\partial c_{i'}}{\partial p}c_{\theta'} + c_{i'}\frac{\partial c_{\theta'}}{\partial p}\right\}s_{\lambda_\odot-\Omega'} + c_{i'}c_{\theta'}\frac{\partial s_{\lambda_\odot-\Omega'}}{\partial p}\right) \tag{A15.23}$$

$$\frac{\partial f_{\theta_3}^s}{\partial p} = 35\left[\left(-\frac{\partial s_{\theta'}}{\partial p}c_{\theta'}^3 - 3s_{\theta'}c_{\theta'}^2\frac{\partial c_{\theta'}}{\partial p}\right)c_{\lambda_\odot-\Omega'}^4\right.$$

$$- 4s_{\theta'}c_{\theta'}^3 c_{\lambda_\odot-\Omega'}^3\frac{\partial c_{\lambda_\odot-\Omega'}}{\partial p} - \left(4s_{\theta'}^3\frac{\partial s_{\theta'}}{\partial p}c_{i'}^3 + 3s_{\theta'}^4 c_{i'}^2\frac{\partial c_{i'}}{\partial p}\right)c_{\lambda_\odot-\Omega'}s_{\lambda_\odot-\Omega'}^3$$

$$- c_{i'}^3 s_{\theta'}^4\left(\frac{\partial c_{\lambda_\odot-\Omega'}}{\partial p}s_{\lambda_\odot-\Omega'}^3 + 3c_{\lambda_\odot-\Omega'}s_{\lambda_\odot-\Omega'}^2\frac{\partial s_{\lambda_\odot-\Omega'}}{\partial p}\right)$$

$$- 3\left(\frac{\partial c_{i'}}{\partial p}s_{\theta'}^2 c_{\theta'}^2 + 2c_{i'}\left\{s_{\theta'}\frac{\partial s_{\theta'}}{\partial p}c_{\theta'}^2 + s_{\theta'}^2 c_{\theta'}\frac{\partial c_{\theta'}}{\partial p}\right\}\right)c_{\lambda_\odot-\Omega'}^3 s_{\lambda_\odot-\Omega'}$$

$$- 3c_{i'}s_{\theta'}^2 c_{\theta'}^2\left(3c_{\lambda_\odot-\Omega'}^2\frac{\partial c_{\lambda_\odot-\Omega'}}{\partial p}s_{\lambda_\odot-\Omega'} + c_{\lambda_\odot-\Omega'}^3\frac{\partial s_{\lambda_\odot-\Omega'}}{\partial p}\right)$$

$$- 3\left(2c_{i'}\frac{\partial c_{i'}}{\partial p}s_{\theta'}^3 c_{\theta'} + c_{i'}^2\left\{3s_{\theta'}^2\frac{\partial s_{\theta'}}{\partial p}c_{\theta'} + s_{\theta'}^3\frac{\partial c_{\theta'}}{\partial p}\right\}\right)s_{\lambda_\odot-\Omega'}^2 c_{\lambda_\odot-\Omega'}^2$$

$$
\begin{aligned}
&- 3c_{i'}^2 s_{\theta'}^3 c_{\theta'} \left(2 s_{\lambda_\odot-\Omega'} c_{\lambda_\odot-\Omega'}^2 \frac{\partial s_{\lambda_\odot-\Omega'}}{\partial p} + 2 s_{\lambda_\odot-\Omega'}^2 c_{\lambda_\odot-\Omega'} \frac{\partial c_{\lambda_\odot-\Omega'}}{\partial p} \right) \\
&+ \left(\frac{\partial c_{i'}}{\partial p} c_{\theta'}^4 + 4 c_{i'} c_{\theta'}^3 \frac{\partial c_{\theta'}}{\partial p} \right) s_{\lambda_\odot-\Omega'} c_{\lambda_\odot-\Omega'}^3 \\
&+ c_{i'} c_{\theta'}^4 \left(\frac{\partial s_{\lambda_\odot-\Omega'}}{\partial p} c_{\lambda_\odot-\Omega'}^3 + 3 s_{\lambda_\odot-\Omega'} c_{\lambda_\odot-\Omega'}^2 \frac{\partial c_{\lambda_\odot-\Omega'}}{\partial p} \right) \\
&+ \left(4 c_{i'}^3 \frac{\partial c_{i'}}{\partial p} s_{\theta'}^3 c_{\theta'} + c_{i'}^4 \left\{ \frac{\partial c_{\theta'}}{\partial p} s_{\theta'}^3 + 3 s_{\theta'}^2 c_{\theta'} \frac{\partial s_{\theta'}}{\partial p} \right\} \right) s_{\lambda_\odot-\Omega'}^4 + 4 c_{i'}^4 s_{\theta'}^3 c_{\theta'} s_{\lambda_\odot-\Omega'}^3 \frac{\partial s_{\lambda_\odot-\Omega'}}{\partial p} \\
&+ 3 \left(2 c_{i'} \frac{\partial c_{i'}}{\partial p} s_{\theta'} c_{\theta'}^3 + c_{i'}^2 \left\{ 3 c_{\theta'}^2 \frac{\partial c_{\theta'}}{\partial p} s_{\theta'} + c_{\theta'}^3 \frac{\partial s_{\theta'}}{\partial p} \right\} \right) c_{\lambda_\odot-\Omega'}^2 s_{\lambda_\odot-\Omega'}^2 \\
&+ 3 c_{i'}^2 s_{\theta'} c_{\theta'}^3 \left(2 c_{\lambda_\odot-\Omega'} \frac{\partial c_{\lambda_\odot-\Omega'}}{\partial p} s_{\lambda_\odot-\Omega'}^2 + 2 c_{\lambda_\odot-\Omega'}^2 s_{\lambda_\odot-\Omega'} \frac{\partial s_{\lambda_\odot-\Omega'}}{\partial p} \right) \\
&+ 3 \left(3 c_{i'}^2 \frac{\partial c_{i'}}{\partial p} s_{\theta'}^2 c_{\theta'}^2 + 2 c_{i'}^3 \left\{ s_{\theta'} \frac{\partial s_{\theta'}}{\partial p} c_{\theta'}^2 + s_{\theta'}^2 c_{\theta'} \frac{\partial c_{\theta'}}{\partial p} \right\} \right) s_{\lambda_\odot-\Omega'}^3 c_{\lambda_\odot-\Omega'} \\
&+ 3 c_{i'}^3 s_{\theta'}^2 c_{\theta'}^2 \left(3 s_{\lambda_\odot-\Omega'}^2 \frac{\partial s_{\lambda_\odot-\Omega'}}{\partial p} c_{\lambda_\odot-\Omega'} + s_{\lambda_\odot-\Omega'}^3 \frac{\partial c_{\lambda_\odot-\Omega'}}{\partial p} \right) \Bigg] \\
&- 15 \Bigg[- \left(\frac{\partial s_{\theta'}}{\partial p} c_{\theta'} + s_{\theta'} \frac{\partial c_{\theta'}}{\partial p} \right) c_{\lambda_\odot-\Omega'}^2 - 2 s_{\theta'} c_{\theta'} c_{\lambda_\odot-\Omega'} \frac{\partial c_{\lambda_\odot-\Omega'}}{\partial p} \\
&- \left(\frac{\partial c_{i'}}{\partial p} s_{\theta'}^2 + 2 c_{i'} s_{\theta'} \frac{\partial s_{\theta'}}{\partial p} \right) s_{\lambda_\odot-\Omega'} c_{\lambda_\odot-\Omega'} - c_{i'} s_{\theta'}^2 \left(\frac{\partial s_{\lambda_\odot-\Omega'}}{\partial p} c_{\lambda_\odot-\Omega'} + s_{\lambda_\odot-\Omega'} \frac{\partial c_{\lambda_\odot-\Omega'}}{\partial p} \right) \\
&+ \left(\frac{\partial c_{i'}}{\partial p} c_{\theta'}^2 + 2 c_{i'} c_{\theta'} \frac{\partial c_{\theta'}}{\partial p} \right) s_{\lambda_\odot-\Omega'} c_{\lambda_\odot-\Omega'} + c_{i'} c_{\theta'}^2 \left(\frac{\partial s_{\lambda_\odot-\Omega'}}{\partial p} c_{\lambda_\odot-\Omega'} + s_{\lambda_\odot-\Omega'} \frac{\partial c_{\lambda_\odot-\Omega'}}{\partial p} \right) \\
&+ \left(2 c_{i'} \frac{\partial c_{i'}}{\partial p} s_{\theta'} c_{\theta'} + c_{i'}^2 \left\{ \frac{\partial s_{\theta'}}{\partial p} c_{\theta'} + s_{\theta'} \frac{\partial c_{\theta'}}{\partial p} \right\} \right) s_{\lambda_\odot-\Omega'}^2 + 2 c_{i'}^2 s_{\theta'} c_{\theta'} s_{\lambda_\odot-\Omega'} \frac{\partial s_{\lambda_\odot-\Omega'}}{\partial p} \Bigg]
\end{aligned}
\tag{A15.24}
$$

$$
\begin{aligned}
\frac{\partial f_{\theta_1}^s}{\partial q} &= \left(- \frac{\partial s_{\theta'}}{\partial q} c_{\theta'} - s_{\theta'} \frac{\partial c_{\theta'}}{\partial q} \right) c_{\lambda_\odot-\Omega'}^2 - 2 s_{\theta'} c_{\theta'} c_{\lambda_\odot-\Omega'} \frac{\partial c_{\lambda_\odot-\Omega'}}{\partial q} \\
&- c_{i'} s_{\theta'}^2 \left(\frac{\partial s_{\lambda_\odot-\Omega'}}{\partial q} c_{\lambda_\odot-\Omega'} + s_{\lambda_\odot-\Omega'} \frac{\partial c_{\lambda_\odot-\Omega'}}{\partial q} \right) - \left(2 c_{i'} s_{\theta'} \frac{\partial s_{\theta'}}{\partial q} + \frac{\partial c_{i'}}{\partial q} s_{\theta'}^2 \right) s_{\lambda_\odot-\Omega'} c_{\lambda_\odot-\Omega'} \\
&+ \left(\frac{\partial c_{i'}}{\partial q} c_{\theta'}^2 + 2 c_{i'} c_{\theta'} \frac{\partial c_{\theta'}}{\partial q} \right) s_{\lambda_\odot-\Omega'} c_{\lambda_\odot-\Omega'} + c_{i'} c_{\theta'}^2 \left(\frac{\partial s_{\lambda_\odot-\Omega'}}{\partial q} c_{\lambda_\odot-\Omega'} + s_{\lambda_\odot-\Omega'} \frac{\partial c_{\lambda_\odot-\Omega'}}{\partial q} \right) \\
&+ \left(2 c_{i'} \frac{\partial c_{i'}}{\partial q} s_{\theta'} c_{\theta'} + c_{i'}^2 \left\{ \frac{\partial s_{\theta'}}{\partial q} c_{\theta'} + s_{\theta'} \frac{\partial c_{\theta'}}{\partial q} \right\} \right) s_{\lambda_\odot-\Omega'}^2 + 2 c_{i'}^2 s_{\theta'} c_{\theta'} s_{\lambda_\odot-\Omega'} \frac{\partial s_{\lambda_\odot-\Omega'}}{\partial q}
\end{aligned}
\tag{A15.25}
$$

$$
\begin{aligned}
\frac{\partial f_{\theta_2}^s}{\partial q} &= 5 \Bigg[\left(- \frac{\partial s_{\theta'}}{\partial q} c_{\theta'}^2 - 2 s_{\theta'} c_{\theta'} \frac{\partial c_{\theta'}}{\partial q} \right) c_{\lambda_\odot-\Omega'}^3 \\
&- 3 s_{\theta'} c_{\theta'}^2 c_{\lambda_\odot-\Omega'}^2 \frac{\partial c_{\lambda_\odot-\Omega'}}{\partial q} + \left(\frac{\partial c_{i'}}{\partial q} c_{\theta'}^3 + 3 c_{i'} c_{\theta'}^2 \frac{\partial c_{\theta'}}{\partial q} \right) s_{\lambda_\odot-\Omega'} c_{\lambda_\odot-\Omega'}^2 \\
&+ c_{i'} c_{\theta'}^3 \left(\frac{\partial s_{\lambda_\odot-\Omega'}}{\partial q} c_{\lambda_\odot-\Omega'}^2 + 2 s_{\lambda_\odot-\Omega'} c_{\lambda_\odot-\Omega'} \frac{\partial c_{\lambda_\odot-\Omega'}}{\partial q} \right) \\
&- \left(2 c_{i'} \frac{\partial c_{i'}}{\partial q} s_{\theta'}^3 + 3 c_{i'}^2 s_{\theta'}^2 \frac{\partial s_{\theta'}}{\partial q} \right) c_{\lambda_\odot-\Omega'} s_{\lambda_\odot-\Omega'}^2 - c_{i'}^2 s_{\theta'}^3 \left(\frac{\partial c_{\lambda_\odot-\Omega'}}{\partial q} s_{\lambda_\odot-\Omega'}^2 \right.
\end{aligned}
$$

$$
\begin{aligned}
&+2c_{\lambda_\odot-\Omega'}s_{\lambda_\odot-\Omega'}\frac{\partial s_{\lambda_\odot-\Omega'}}{\partial q}\Bigg)+\Bigg(3c_{i'}^2\frac{\partial c_{i'}}{\partial q}s_{\theta'}^2c_{\theta'}+c_{i'}^3\left\{\frac{\partial c_{\theta'}}{\partial q}s_{\theta'}^2\right.\\
&\left.+2s_{\theta'}c_{\theta'}\frac{\partial s_{\theta'}}{\partial q}\right\}\Bigg)s_{\lambda_\odot-\Omega'}^3+3c_{i'}^3s_{\theta'}^2c_{\theta'}s_{\lambda_\odot-\Omega'}^2\frac{\partial s_{\lambda_\odot-\Omega'}}{\partial q}\\
&-2\left(\frac{\partial c_{i'}}{\partial q}s_{\theta'}^2c_{\theta'}+c_{i'}\left\{2s_{\theta'}\frac{\partial s_{\theta'}}{\partial q}c_{\theta'}+s_{\theta'}^2\frac{\partial c_{\theta'}}{\partial q}\right\}\right)s_{\lambda_\odot-\Omega'}c_{\lambda_\odot-\Omega'}^2\\
&-2c_{i'}s_{\theta'}^2c_{\theta'}\left(\frac{\partial s_{\lambda_\odot-\Omega'}}{\partial q}c_{\lambda_\odot-\Omega'}^2+2s_{\lambda_\odot-\Omega'}c_{\lambda_\odot-\Omega'}\frac{\partial c_{\lambda_\odot-\Omega'}}{\partial q}\right)\\
&+2\left(2c_{i'}\frac{\partial c_{i'}}{\partial q}s_{\theta'}c_{\theta'}^2+c_{i'}^2\left\{\frac{\partial s_{\theta'}}{\partial q}c_{\theta'}^2+2s_{\theta'}c_{\theta'}\frac{\partial c_{\theta'}}{\partial q}\right\}\right)s_{\lambda_\odot-\Omega'}^2c_{\lambda_\odot-\Omega'}\\
&+2c_{i'}^2s_{\theta'}c_{\theta'}^2\left(2s_{\lambda_\odot-\Omega'}\frac{\partial s_{\lambda_\odot-\Omega'}}{\partial q}c_{\lambda_\odot-\Omega'}+s_{\lambda_\odot-\Omega'}^2\frac{\partial c_{\lambda_\odot-\Omega'}}{\partial q}\right)\Bigg]\\
&-\left(-\frac{\partial s_{\theta'}}{\partial q}c_{\lambda_\odot-\Omega'}-s_{\theta'}\frac{\partial c_{\lambda_\odot-\Omega'}}{\partial q}+\left\{\frac{\partial c_{i'}}{\partial q}c_{\theta'}+c_{i'}\frac{\partial c_{\theta'}}{\partial q}\right\}s_{\lambda_\odot-\Omega'}+c_{i'}c_{\theta'}\frac{\partial s_{\lambda_\odot-\Omega'}}{\partial q}\right)
\end{aligned}
\tag{A15.26}
$$

$$
\begin{aligned}
\frac{\partial f_{\theta_3}^s}{\partial q}=35\Bigg[&\left(-\frac{\partial s_{\theta'}}{\partial q}c_{\theta'}^3-3s_{\theta'}c_{\theta'}^2\frac{\partial c_{\theta'}}{\partial q}\right)c_{\lambda_\odot-\Omega'}^4\\
&-4s_{\theta'}c_{\theta'}^3c_{\lambda_\odot-\Omega'}^3\frac{\partial c_{\lambda_\odot-\Omega'}}{\partial q}-\left(4s_{\theta'}^3\frac{\partial s_{\theta'}}{\partial q}c_{i'}^3+3s_{\theta'}^4c_{i'}^2\frac{\partial c_{i'}}{\partial q}\right)c_{\lambda_\odot-\Omega'}s_{\lambda_\odot-\Omega'}^3\\
&-c_{i'}^3s_{\theta'}^4\left(\frac{\partial c_{\lambda_\odot-\Omega'}}{\partial q}s_{\lambda_\odot-\Omega'}^3+3c_{\lambda_\odot-\Omega'}s_{\lambda_\odot-\Omega'}^2\frac{\partial s_{\lambda_\odot-\Omega'}}{\partial q}\right)\\
&-3\left(\frac{\partial c_{i'}}{\partial q}s_{\theta'}^2c_{\theta'}^2+c_{i'}\left\{2s_{\theta'}\frac{\partial s_{\theta'}}{\partial q}c_{\theta'}^2+2s_{\theta'}^2c_{\theta'}\frac{\partial c_{\theta'}}{\partial q}\right\}\right)c_{\lambda_\odot-\Omega'}^3s_{\lambda_\odot-\Omega'}\\
&-3c_{i'}s_{\theta'}^2c_{\theta'}^2\left(3c_{\lambda_\odot-\Omega'}^2\frac{\partial c_{\lambda_\odot-\Omega'}}{\partial q}s_{\lambda_\odot-\Omega'}+c_{\lambda_\odot-\Omega'}^3\frac{\partial s_{\lambda_\odot-\Omega'}}{\partial q}\right)-3\Bigg(2c_{i'}\frac{\partial c_{i'}}{\partial q}s_{\theta'}^3c_{\theta'}\\
&+c_{i'}^2\left\{3s_{\theta'}^2\frac{\partial s_{\theta'}}{\partial q}c_{\theta'}+s_{\theta'}^3\frac{\partial c_{\theta'}}{\partial q}\right\}\Bigg)s_{\lambda_\odot-\Omega'}^2c_{\lambda_\odot-\Omega'}^2-3c_{i'}^2s_{\theta'}^3c_{\theta'}\Bigg(2s_{\lambda_\odot-\Omega'}c_{\lambda_\odot-\Omega'}^2\\
&\times\frac{\partial s_{\lambda_\odot-\Omega'}}{\partial q}+2s_{\lambda_\odot-\Omega'}^2c_{\lambda_\odot-\Omega'}\frac{\partial c_{\lambda_\odot-\Omega'}}{\partial q}\Bigg)+\left(\frac{\partial c_{i'}}{\partial q}c_{\theta'}^4+4c_{i'}c_{\theta'}^3\frac{\partial c_{\theta'}}{\partial q}\right)\\
&\times s_{\lambda_\odot-\Omega'}c_{\lambda_\odot-\Omega'}^3+c_{i'}c_{\theta'}^4\left(\frac{\partial s_{\lambda_\odot-\Omega'}}{\partial q}c_{\lambda_\odot-\Omega'}^3+3s_{\lambda_\odot-\Omega'}c_{\lambda_\odot-\Omega'}^2\frac{\partial c_{\lambda_\odot-\Omega'}}{\partial q}\right)\\
&+\left(4c_{i'}^3\frac{\partial c_{i'}}{\partial q}s_{\theta'}^3c_{\theta'}+c_{i'}^4\left\{\frac{\partial c_{\theta'}}{\partial q}s_{\theta'}^3+3s_{\theta'}^2c_{\theta'}\frac{\partial s_{\theta'}}{\partial q}\right\}\right)s_{\lambda_\odot-\Omega'}^4\\
&+4c_{i'}^4s_{\theta'}^3c_{\theta'}s_{\lambda_\odot-\Omega'}^3\frac{\partial s_{\lambda_\odot-\Omega'}}{\partial q}+3\Bigg(2c_{i'}\frac{\partial c_{i'}}{\partial q}s_{\theta'}c_{\theta'}^3+c_{i'}^2\Bigg\{3c_{\theta'}^2\frac{\partial c_{\theta'}}{\partial q}s_{\theta'}\\
&+c_{\theta'}^3\frac{\partial s_{\theta'}}{\partial q}\Bigg\}\Bigg)c_{\lambda_\odot-\Omega'}^2s_{\lambda_\odot-\Omega'}^2+3c_{i'}^2s_{\theta'}c_{\theta'}^3\Bigg(2c_{\lambda_\odot-\Omega'}\frac{\partial c_{\lambda_\odot-\Omega'}}{\partial q}s_{\lambda_\odot-\Omega'}^2\\
&+2c_{\lambda_\odot-\Omega'}^2s_{\lambda_\odot-\Omega'}\frac{\partial s_{\lambda_\odot-\Omega'}}{\partial q}\Bigg)+3\Bigg(3c_{i'}^2\frac{\partial c_{i'}}{\partial q}s_{\theta'}^2c_{\theta'}^2\\
&+2c_{i'}^3\left\{s_{\theta'}\frac{\partial s_{\theta'}}{\partial q}c_{\theta'}^2+s_{\theta'}^2c_{\theta'}\frac{\partial c_{\theta'}}{\partial q}\right\}\Bigg)s_{\lambda_\odot-\Omega'}^3c_{\lambda_\odot-\Omega'}
\end{aligned}
$$

$$
\begin{aligned}
&+3c_{i'}^3 s_{\theta'}^2 c_{\theta'}^2 \left(3s_{\lambda_\odot-\Omega'}^2 \frac{\partial s_{\lambda_\odot-\Omega'}}{\partial q} c_{\lambda_\odot-\Omega'} + s_{\lambda_\odot-\Omega'}^3 \frac{\partial c_{\lambda_\odot-\Omega'}}{\partial q}\right)\Bigg]\\
&-15\Bigg[-\left(\frac{\partial s_{\theta'}}{\partial q} c_{\theta'} + s_{\theta'} \frac{\partial c_{\theta'}}{\partial q}\right) c_{\lambda_\odot-\Omega'}^2 - 2 s_{\theta'} c_{\theta'} c_{\lambda_\odot-\Omega'} \frac{\partial c_{\lambda_\odot-\Omega'}}{\partial q}\\
&-\left(\frac{\partial c_{i'}}{\partial q} s_{\theta'}^2 + 2c_{i'} s_{\theta'} \frac{\partial s_{\theta'}}{\partial q}\right) s_{\lambda_\odot-\Omega'} c_{\lambda_\odot-\Omega'} - c_{i'} s_{\theta'}^2 \left(\frac{\partial s_{\lambda_\odot-\Omega'}}{\partial q} c_{\lambda_\odot-\Omega'}\right.\\
&\left.+ s_{\lambda_\odot-\Omega'} \frac{\partial c_{\lambda_\odot-\Omega'}}{\partial q}\right) + \left(\frac{\partial c_{i'}}{\partial q} c_{\theta'}^2 + 2c_{i'} c_{\theta'} \frac{\partial c_{\theta'}}{\partial q}\right) s_{\lambda_\odot-\Omega'} c_{\lambda_\odot-\Omega'}\\
&+ c_{i'} c_{\theta'}^2 \left(\frac{\partial s_{\lambda_\odot-\Omega'}}{\partial q} c_{\lambda_\odot-\Omega'} + s_{\lambda_\odot-\Omega'} \frac{\partial c_{\lambda_\odot-\Omega'}}{\partial q}\right) + \left(2c_{i'} \frac{\partial c_{i'}}{\partial q} s_{\theta'} c_{\theta'}\right.\\
&\left.+ c_{i'}^2 \left\{\frac{\partial s_{\theta'}}{\partial q} c_{\theta'} + s_{\theta'} \frac{\partial c_{\theta'}}{\partial q}\right\}\right) s_{\lambda_\odot-\Omega'}^2 + 2c_{i'}^2 s_{\theta'} c_{\theta'} s_{\lambda_\odot-\Omega'} \frac{\partial s_{\lambda_\odot-\Omega'}}{\partial q}\Bigg]
\end{aligned}
\tag{A15.27}
$$

$$
\begin{aligned}
\frac{\partial f_{\theta_1}^s}{\partial L} &= -\left(\frac{\partial s_{\theta'}}{\partial L} c_{\theta'} + s_{\theta'} \frac{\partial c_{\theta'}}{\partial L}\right) c_{\lambda_\odot-\Omega'}^2 - 2c_{i'} s_{\theta'} \frac{\partial s_{\theta'}}{\partial L} s_{\lambda_\odot-\Omega'} c_{\lambda_\odot-\Omega'}\\
&\quad + 2c_{i'} c_{\theta'} \frac{\partial c_{\theta'}}{\partial L} s_{\lambda_\odot-\Omega'} c_{\lambda_\odot-\Omega'} + c_{i'}^2 \left(\frac{\partial s_{\theta'}}{\partial L} c_{\theta'} + s_{\theta'} \frac{\partial c_{\theta'}}{\partial L}\right) s_{\lambda_\odot-\Omega'}^2
\end{aligned}
\tag{A15.28}
$$

$$
\begin{aligned}
\frac{\partial f_{\theta_2}^s}{\partial L} &= 5\Bigg[-\left(\frac{\partial s_{\theta'}}{\partial L} c_{\theta'}^2 + 2 s_{\theta'} c_{\theta'} \frac{\partial c_{\theta'}}{\partial L}\right) c_{\lambda_\odot-\Omega'}^3 + 3c_{i'} c_{\theta'}^2 \frac{\partial c_{\theta'}}{\partial L} s_{\lambda_\odot-\Omega'} c_{\lambda_\odot-\Omega'}^2\\
&\quad - 3c_{i'}^2 s_{\theta'}^2 \frac{\partial s_{\theta'}}{\partial L} c_{\lambda_\odot-\Omega'} s_{\lambda_\odot-\Omega'}^2 + c_{i'}^3 \left(\frac{\partial c_{\theta'}}{\partial L} s_{\theta'}^2 + 2 s_{\theta'} c_{\theta'} \frac{\partial s_{\theta'}}{\partial L}\right) s_{\lambda_\odot-\Omega'}^3\\
&\quad - 2c_{i'} \left(2s_{\theta'} \frac{\partial s_{\theta'}}{\partial L} c_{\theta'} + s_{\theta'}^2 \frac{\partial c_{\theta'}}{\partial L}\right) s_{\lambda_\odot-\Omega'} c_{\lambda_\odot-\Omega'}^2\\
&\quad + 2c_{i'}^2 \left(\frac{\partial s_{\theta'}}{\partial L} c_{\theta'}^2 + 2 s_{\theta'} c_{\theta'} \frac{\partial c_{\theta'}}{\partial L}\right) s_{\lambda_\odot-\Omega'}^2 c_{\lambda_\odot-\Omega'}\Bigg]\\
&\quad - \left(-\frac{\partial s_{\theta'}}{\partial L} c_{\lambda_\odot-\Omega'} + c_{i'} \frac{\partial c_{\theta'}}{\partial L} s_{\lambda_\odot-\Omega'}\right)
\end{aligned}
\tag{A15.29}
$$

$$
\begin{aligned}
\frac{\partial f_{\theta_3}^s}{\partial L} &= 35\Bigg[-\left(\frac{\partial s_{\theta'}}{\partial L} c_{\theta'}^3 + 3 s_{\theta'} c_{\theta'}^2 \frac{\partial c_{\theta'}}{\partial L}\right) c_{\lambda_\odot-\Omega'}^4 - 4c_{i'}^3 s_{\theta'}^3 \frac{\partial s_{\theta'}}{\partial L} c_{\lambda_\odot-\Omega'} s_{\lambda_\odot-\Omega'}^3\\
&\quad - 6c_{i'} \left(s_{\theta'} \frac{\partial s_{\theta'}}{\partial L} c_{\theta'}^2 + s_{\theta'}^2 c_{\theta'} \frac{\partial c_{\theta'}}{\partial L}\right) c_{\lambda_\odot-\Omega'}^3 s_{\lambda_\odot-\Omega'}\\
&\quad - 3c_{i'}^2 \left(3s_{\theta'}^2 \frac{\partial s_{\theta'}}{\partial L} c_{\theta'} + s_{\theta'}^3 \frac{\partial c_{\theta'}}{\partial L}\right) s_{\lambda_\odot-\Omega'}^2 c_{\lambda_\odot-\Omega'}^2 + 4c_{i'} c_{\theta'}^3 \frac{\partial c_{\theta'}}{\partial L} s_{\lambda_\odot-\Omega'} c_{\lambda_\odot-\Omega'}^3\\
&\quad + c_{i'}^4 \left(\frac{\partial c_{\theta'}}{\partial L} s_{\theta'}^3 + 3 s_{\theta'}^2 c_{\theta'} \frac{\partial s_{\theta'}}{\partial L}\right) s_{\lambda_\odot-\Omega'}^4 + 3c_{i'}^2 \left(3c_{\theta'}^2 \frac{\partial c_{\theta'}}{\partial L} s_{\theta'}\right.\\
&\quad \left.+ c_{\theta'}^3 \frac{\partial s_{\theta'}}{\partial L}\right) c_{\lambda_\odot-\Omega'}^2 s_{\lambda_\odot-\Omega'}^2 + 6c_{i'}^3 \left(s_{\theta'} \frac{\partial s_{\theta'}}{\partial L} c_{\theta'}^2 + s_{\theta'}^2 c_{\theta'} \frac{\partial c_{\theta'}}{\partial L}\right) s_{\lambda_\odot-\Omega'}^3 c_{\lambda_\odot-\Omega'}\Bigg]\\
&\quad - 15\Bigg[-\left(\frac{\partial s_{\theta'}}{\partial L} c_{\theta'} + s_{\theta'} \frac{\partial c_{\theta'}}{\partial L}\right) c_{\lambda_\odot-\Omega'}^2 - 2c_{i'} s_{\theta'} \frac{\partial s_{\theta'}}{\partial L} s_{\lambda_\odot-\Omega'} c_{\lambda_\odot-\Omega'}\\
&\quad + 2c_{i'} c_{\theta'} \frac{\partial c_{\theta'}}{\partial L} s_{\lambda_\odot-\Omega'} c_{\lambda_\odot-\Omega'} + c_{i'}^2 \left(\frac{\partial s_{\theta'}}{\partial L} c_{\theta'} + s_{\theta'} \frac{\partial c_{\theta'}}{\partial L}\right) s_{\lambda_\odot-\Omega'}^2\Bigg]
\end{aligned}
\tag{A15.30}
$$

The partial derivatives of f_h^s with respect to h, h, p, q and L:

$$\frac{\partial f_{h_1}^s}{\partial h} = s_{i'}\frac{\partial c_{\theta'}}{\partial h}s_{\lambda_\odot-\Omega'}c_{\lambda_\odot-\Omega'} + s_{i'}c_{i'}\frac{\partial s_{\theta'}}{\partial h}s^2_{\lambda_\odot-\Omega'} \tag{A15.31}$$

$$\begin{aligned}\frac{\partial f_{h_2}^s}{\partial h} &= 10s_{i'}\left(c_{\theta'}\frac{\partial c_{\theta'}}{\partial h}\right)s_{\lambda_\odot-\Omega'}c^2_{\lambda_\odot-\Omega'} + 10s_{i'}c^2_{i'}\left(s_{\theta'}\frac{\partial s_{\theta'}}{\partial h}\right)s^3_{\lambda_\odot-\Omega'}\\ &+ 10s_{i'}c_{i'}\left(\frac{\partial s_{\theta'}}{\partial h}c_{\theta'} + s_{\theta'}\frac{\partial c_{\theta'}}{\partial h}\right)s^2_{\lambda_\odot-\Omega'}c_{\lambda_\odot-\Omega'}\end{aligned} \tag{A15.32}$$

$$\begin{aligned}\frac{\partial f_{h_3}^s}{\partial h} &= -21s_{i'}\left(c^2_{\theta'}\frac{\partial c_{\theta'}}{\partial h}\right)s_{\lambda_\odot-\Omega'}c^3_{\lambda_\odot-\Omega'} - 21s_{i'}c^3_{i'}\left(s^2_{\theta'}\frac{\partial s_{\theta'}}{\partial h}\right)s^4_{\lambda_\odot-\Omega'}\\ &- 21s_{i'}c_{i'}\left(2c_{\theta'}\frac{\partial c_{\theta'}}{\partial h}s_{\theta'} + c^2_{\theta'}\frac{\partial s_{\theta'}}{\partial h}\right)s^2_{\lambda_\odot-\Omega'}c^2_{\lambda_\odot-\Omega'} - 21s_{i'}c^2_{i'}\left(2s_{\theta'}\frac{\partial s_{\theta'}}{\partial h}c_{\theta'}\right.\\ &\left.+ s^2_{\theta'}\frac{\partial c_{\theta'}}{\partial h}\right)s^3_{\lambda_\odot-\Omega'}c_{\lambda_\odot-\Omega'} + 3s_{i'}\frac{\partial c_{\theta'}}{\partial h}s_{\lambda_\odot-\Omega'}c_{\lambda_\odot-\Omega'} + 3s_{i'}c_{i'}\frac{\partial s_{\theta'}}{\partial h}s^2_{\lambda_\odot-\Omega'}\end{aligned} \tag{A15.33}$$

$$\frac{\partial f_{h_1}^s}{\partial k} = s_{i'}\frac{\partial c_{\theta'}}{\partial k}s_{\lambda_\odot-\Omega'}c_{\lambda_\odot-\Omega'} + s_{i'}c_{i'}\frac{\partial s_{\theta'}}{\partial k}s^2_{\lambda_\odot-\Omega'} \tag{A15.34}$$

$$\begin{aligned}\frac{\partial f_{h_2}^s}{\partial k} &= 10s_{i'}\left(c_{\theta'}\frac{\partial c_{\theta'}}{\partial k}\right)s_{\lambda_\odot-\Omega'}c^2_{\lambda_\odot-\Omega'} + 10s_{i'}c^2_{i'}\left(s_{\theta'}\frac{\partial s_{\theta'}}{\partial k}\right)s^3_{\lambda_\odot-\Omega'}\\ &+ 10s_{i'}c_{i'}\left(\frac{\partial s_{\theta'}}{\partial k}c_{\theta'} + s_{\theta'}\frac{\partial c_{\theta'}}{\partial k}\right)s^2_{\lambda_\odot-\Omega'}c_{\lambda_\odot-\Omega'}\end{aligned} \tag{A15.35}$$

$$\begin{aligned}\frac{\partial f_{h_3}^s}{\partial k} &= -21s_{i'}\left(c^2_{\theta'}\frac{\partial c_{\theta'}}{\partial k}\right)s_{\lambda_\odot-\Omega'}c^3_{\lambda_\odot-\Omega'} - 21s_{i'}c^3_{i'}\left(s^2_{\theta'}\frac{\partial s_{\theta'}}{\partial k}\right)s^4_{\lambda_\odot-\Omega'}\\ &- 21s_{i'}c_{i'}\left(2c_{\theta'}\frac{\partial c_{\theta'}}{\partial k}s_{\theta'} + c^2_{\theta'}\frac{\partial s_{\theta'}}{\partial k}\right)s^2_{\lambda_\odot-\Omega'}c^2_{\lambda_\odot-\Omega'}\\ &- 21s_{i'}c^2_{i'}\left(2s_{\theta'}\frac{\partial s_{\theta'}}{\partial k}c_{\theta'} + s^2_{\theta'}\frac{\partial c_{\theta'}}{\partial k}\right)s^3_{\lambda_\odot-\Omega'}c_{\lambda_\odot-\Omega'}\\ &+ 3s_{i'}\frac{\partial c_{\theta'}}{\partial k}s_{\lambda_\odot-\Omega'}c_{\lambda_\odot-\Omega'} + 3s_{i'}c_{i'}\frac{\partial s_{\theta'}}{\partial k}s^2_{\lambda_\odot-\Omega'}\end{aligned} \tag{A15.36}$$

$$\begin{aligned}\frac{\partial f_{h_1}^s}{\partial p} &= \left(\frac{\partial s_{i'}}{\partial p}c_{\theta'} + s_{i'}\frac{\partial c_{\theta'}}{\partial p}\right)s_{\lambda_\odot-\Omega'}c_{\lambda_\odot-\Omega'} + s_{i'}c_{\theta'}\left(\frac{\partial s_{\lambda_\odot-\Omega'}}{\partial p}c_{\lambda_\odot-\Omega'} + s_{\lambda_\odot-\Omega'}\frac{\partial c_{\lambda_\odot-\Omega'}}{\partial p}\right)\\ &+ \left(\frac{\partial s_{i'}}{\partial p}c_{i'} + s_{i'}\frac{\partial c_{i'}}{\partial p}\right)s_{\theta'}s^2_{\lambda_\odot-\Omega'} + s_{i'}c_{i'}\left(\frac{\partial s_{\theta'}}{\partial p}s^2_{\lambda_\odot-\Omega'} + 2s_{\theta'}s_{\lambda_\odot-\Omega'}\frac{\partial s_{\lambda_\odot-\Omega'}}{\partial p}\right)\end{aligned} \tag{A15.37}$$

$$\begin{aligned}\frac{\partial f_{h_2}^s}{\partial p} &= 5\left(\frac{\partial s_{i'}}{\partial p}c^2_{\theta'} + 2s_{i'}c_{\theta'}\frac{\partial c_{\theta'}}{\partial p}\right)s_{\lambda_\odot-\Omega'}c^2_{\lambda_\odot-\Omega'}\\ &+ 5s_{i'}c^2_{\theta'}\left(\frac{\partial s_{\lambda_\odot-\Omega'}}{\partial p}c^2_{\lambda_\odot-\Omega'} + 2s_{\lambda_\odot-\Omega'}c_{\lambda_\odot-\Omega'}\frac{\partial c_{\lambda_\odot-\Omega'}}{\partial p}\right)\\ &+ 5\left(\frac{\partial s_{i'}}{\partial p}c^2_{i'} + 2s_{i'}c_{i'}\frac{\partial c_{i'}}{\partial p}\right)s^2_{\theta'}s^3_{\lambda_\odot-\Omega'} + 10s_{i'}c^2_{i'}s_{\theta'}\frac{\partial s_{\theta'}}{\partial p}s^3_{\lambda_\odot-\Omega'}\end{aligned}$$

$$+ 15s_{i'}c_{i'}^2 s_{\theta'}^2 s_{\lambda_\odot-\Omega'}^2 \frac{\partial s_{\lambda_\odot-\Omega'}}{\partial p} + 10\left(\frac{\partial s_{i'}}{\partial p}c_{i'} + s_{i'}\frac{\partial c_{i'}}{\partial p}\right) s_{\theta'}c_{\theta'}s_{\lambda_\odot-\Omega'}^2 c_{\lambda_\odot-\Omega'}$$
$$+ 10s_{i'}c_{i'}\left(\frac{\partial s_{\theta'}}{\partial p}c_{\theta'} + s_{\theta'}\frac{\partial c_{\theta'}}{\partial p}\right) s_{\lambda_\odot-\Omega'}^2 c_{\lambda_\odot-\Omega'}$$
$$+ 10s_{i'}c_{i'}s_{\theta'}c_{\theta'}\left(2s_{\lambda_\odot-\Omega'}\frac{\partial s_{\lambda_\odot-\Omega'}}{\partial p}c_{\lambda_\odot-\Omega'} + s_{\lambda_\odot-\Omega'}^2\frac{\partial c_{\lambda_\odot-\Omega'}}{\partial p}\right)$$
$$- \frac{\partial s_{i'}}{\partial p}s_{\lambda_\odot-\Omega'} - s_{i'}\frac{\partial s_{\lambda_\odot-\Omega'}}{\partial p} \quad \text{(A15.38)}$$

$$\frac{\partial f_{h_3}^s}{\partial p} = -7\left(\frac{\partial s_{i'}}{\partial p}c_{\theta'}^3 + 3s_{i'}c_{\theta'}^2\frac{\partial c_{\theta'}}{\partial p}\right)s_{\lambda_\odot-\Omega'}c_{\lambda_\odot-\Omega'}^3 - 7s_{i'}c_{\theta'}^3\left(\frac{\partial s_{\lambda_\odot-\Omega'}}{\partial p}c_{\lambda_\odot-\Omega'}^3\right.$$
$$\left.+ 3s_{\lambda_\odot-\Omega'}c_{\lambda_\odot-\Omega'}^2\frac{\partial c_{\lambda_\odot-\Omega'}}{\partial p}\right) - 7\left(\frac{\partial s_{i'}}{\partial p}c_{i'}^3 + 3s_{i'}c_{i'}^2\frac{\partial c_{i'}}{\partial p}\right)s_{\theta'}^3 s_{\lambda_\odot-\Omega'}^4$$
$$- 21s_{i'}c_{i'}^3 s_{\theta'}^2\frac{\partial s_{\theta'}}{\partial p}s_{\lambda_\odot-\Omega'}^4 - 28s_{i'}c_{i'}^3 s_{\theta'}^3 s_{\lambda_\odot-\Omega'}^3\frac{\partial s_{\lambda_\odot-\Omega'}}{\partial p}$$
$$- 21\left(\frac{\partial s_{i'}}{\partial p}c_{i'} + s_{i'}\frac{\partial c_{i'}}{\partial p}\right)s_{\theta'}c_{\theta'}^2 s_{\lambda_\odot-\Omega'}^2 c_{\lambda_\odot-\Omega'}^2 - 21s_{i'}c_{i'}\left(2c_{\theta'}\frac{\partial c_{\theta'}}{\partial p}s_{\theta'}\right.$$
$$\left.+ c_{\theta'}^2\frac{\partial s_{\theta'}}{\partial p}\right)s_{\lambda_\odot-\Omega'}^2 c_{\lambda_\odot-\Omega'}^2 - 42s_{i'}c_{i'}s_{\theta'}c_{\theta'}^2\left(s_{\lambda_\odot-\Omega'}\frac{\partial s_{\lambda_\odot-\Omega'}}{\partial p}c_{\lambda_\odot-\Omega'}^2\right.$$
$$\left.+ s_{\lambda_\odot-\Omega'}^2 c_{\lambda_\odot-\Omega'}\frac{\partial c_{\lambda_\odot-\Omega'}}{\partial p}\right) - 21\left(\frac{\partial s_{i'}}{\partial p}c_{i'}^2 + 2s_{i'}c_{i'}\frac{\partial c_{i'}}{\partial p}\right)s_{\theta'}^2 c_{\theta'}s_{\lambda_\odot-\Omega'}^3 c_{\lambda_\odot-\Omega'}$$
$$- 21s_{i'}c_{i'}^2\left(2s_{\theta'}\frac{\partial s_{\theta'}}{\partial p}c_{\theta'} + s_{\theta'}^2\frac{\partial c_{\theta'}}{\partial p}\right)s_{\lambda_\odot-\Omega'}^3 c_{\lambda_\odot-\Omega'}$$
$$- 21s_{i'}c_{i'}^2 s_{\theta'}^2 c_{\theta'}\left(3s_{\lambda_\odot-\Omega'}^2\frac{\partial s_{\lambda_\odot-\Omega'}}{\partial p}c_{\lambda_\odot-\Omega'} + s_{\lambda_\odot-\Omega'}^3\frac{\partial c_{\lambda_\odot-\Omega'}}{\partial p}\right)$$
$$+ 3\left(\frac{\partial s_{i'}}{\partial p}c_{\theta'} + s_{i'}\frac{\partial c_{\theta'}}{\partial p}\right)s_{\lambda_\odot-\Omega'}c_{\lambda_\odot-\Omega'} + 3s_{i'}c_{\theta'}\left(\frac{\partial s_{\lambda_\odot-\Omega'}}{\partial p}c_{\lambda_\odot-\Omega'}\right.$$
$$\left.+ s_{\lambda_\odot-\Omega'}\frac{\partial c_{\lambda_\odot-\Omega'}}{\partial p}\right) + 3\left(\frac{\partial s_{i'}}{\partial p}c_{i'} + s_{i'}\frac{\partial c_{i'}}{\partial p}\right)s_{\theta'}s_{\lambda_\odot-\Omega'}^2 + 3s_{i'}c_{i'}\frac{\partial s_{\theta'}}{\partial p}s_{\lambda_\odot-\Omega'}^2$$
$$+ 6s_{i'}c_{i'}s_{\theta'}s_{\lambda_\odot-\Omega'}\frac{\partial s_{\lambda_\odot-\Omega'}}{\partial p} \quad \text{(A15.39)}$$

$$\frac{\partial f_{h_1}^s}{\partial q} = \left(\frac{\partial s_{i'}}{\partial q}c_{\theta'} + s_{i'}\frac{\partial c_{\theta'}}{\partial q}\right)s_{\lambda_\odot-\Omega'}c_{\lambda_\odot-\Omega'} + s_{i'}c_{\theta'}\left(\frac{\partial s_{\lambda_\odot-\Omega'}}{\partial q}c_{\lambda_\odot-\Omega'}\right.$$
$$\left.+ s_{\lambda_\odot-\Omega'}\frac{\partial c_{\lambda_\odot-\Omega'}}{\partial q}\right) + \left(\frac{\partial s_{i'}}{\partial q}c_{i'} + s_{i'}\frac{\partial c_{i'}}{\partial q}\right)s_{\theta'}s_{\lambda_\odot-\Omega'}^2$$
$$+ s_{i'}c_{i'}\left(\frac{\partial s_{\theta'}}{\partial q}s_{\lambda_\odot-\Omega'}^2 + 2s_{\theta'}s_{\lambda_\odot-\Omega'}\frac{\partial s_{\lambda_\odot-\Omega'}}{\partial q}\right) \quad \text{(A15.40)}$$

$$\frac{\partial f_{h_2}^s}{\partial q} = 5\left(\frac{\partial s_{i'}}{\partial q}c_{\theta'}^2 + 2s_{i'}c_{\theta'}\frac{\partial c_{\theta'}}{\partial q}\right)s_{\lambda_\odot-\Omega'}c_{\lambda_\odot-\Omega'}^2 + 5s_{i'}c_{\theta'}^2\left(\frac{\partial s_{\lambda_\odot-\Omega'}}{\partial q}c_{\lambda_\odot-\Omega'}^2\right.$$
$$\left.+ 2s_{\lambda_\odot-\Omega'}c_{\lambda_\odot-\Omega'}\frac{\partial c_{\lambda_\odot-\Omega'}}{\partial q}\right) + 5\left(\frac{\partial s_{i'}}{\partial q}c_{i'}^2 + 2s_{i'}c_{i'}\frac{\partial c_{i'}}{\partial q}\right)s_{\theta'}^2 s_{\lambda_\odot-\Omega'}^3$$

$$+ 10s_{i'}c_{i'}^2 s_{\theta'} \frac{\partial s_{\theta'}}{\partial q} s_{\lambda_\odot-\Omega'}^3 + 15s_{i'}c_{i'}^2 s_{\theta'}^2 s_{\lambda_\odot-\Omega'}^2 \frac{\partial s_{\lambda_\odot-\Omega'}}{\partial q}$$
$$+ 10\left(\frac{\partial s_{i'}}{\partial q}c_{i'} + s_{i'}\frac{\partial c_{i'}}{\partial q}\right) s_{\theta'}c_{\theta'}s_{\lambda_\odot-\Omega'}^2 c_{\lambda_\odot-\Omega'} + 10s_{i'}c_{i'}\left(\frac{\partial s_{\theta'}}{\partial q}c_{\theta'}\right.$$
$$\left.+ s_{\theta'}\frac{\partial c_{\theta'}}{\partial q}\right) s_{\lambda_\odot-\Omega'}^2 c_{\lambda_\odot-\Omega'} + 10s_{i'}c_{i'}s_{\theta'}c_{\theta'}\left(2s_{\lambda_\odot-\Omega'}\frac{\partial s_{\lambda_\odot-\Omega'}}{\partial q}c_{\lambda_\odot-\Omega'}\right.$$
$$\left.+ s_{\lambda_\odot-\Omega'}^2\frac{\partial c_{\lambda_\odot-\Omega'}}{\partial q}\right) - \frac{\partial s_{i'}}{\partial q}s_{\lambda_\odot-\Omega'} - s_{i'}\frac{\partial s_{\lambda_\odot-\Omega'}}{\partial q} \tag{A15.41}$$

$$\frac{\partial f_{h_3}^s}{\partial q} = -7\left(\frac{\partial s_{i'}}{\partial q}c_{\theta'}^3 + 3s_{i'}c_{\theta'}^2\frac{\partial c_{\theta'}}{\partial q}\right) s_{\lambda_\odot-\Omega'}c_{\lambda_\odot-\Omega'}^3$$
$$- 7s_{i'}c_{\theta'}^3\left(\frac{\partial s_{\lambda_\odot-\Omega'}}{\partial q}c_{\lambda_\odot-\Omega'}^3 + 3s_{\lambda_\odot-\Omega'}c_{\lambda_\odot-\Omega'}^2\frac{\partial c_{\lambda_\odot-\Omega'}}{\partial q}\right)$$
$$- 7\left(\frac{\partial s_{i'}}{\partial q}c_{i'}^3 + 3s_{i'}c_{i'}^2\frac{\partial c_{i'}}{\partial q}\right) s_{\theta'}^3 s_{\lambda_\odot-\Omega'}^4 - 21s_{i'}c_{i'}^3 s_{\theta'}^2\frac{\partial s_{\theta'}}{\partial q}s_{\lambda_\odot-\Omega'}^4$$
$$- 28s_{i'}c_{i'}^3 s_{\theta'}^3 s_{\lambda_\odot-\Omega'}^3\frac{\partial s_{\lambda_\odot-\Omega'}}{\partial q} - 21\left(\frac{\partial s_{i'}}{\partial q}c_{i'} + s_{i'}\frac{\partial c_{i'}}{\partial q}\right) s_{\theta'}c_{\theta'}^2 s_{\lambda_\odot-\Omega'}^2 c_{\lambda_\odot-\Omega'}^2$$
$$- 21s_{i'}c_{i'}\left(2c_{\theta'}\frac{\partial c_{\theta'}}{\partial q}s_{\theta'} + c_{\theta'}^2\frac{\partial s_{\theta'}}{\partial q}\right) s_{\lambda_\odot-\Omega'}^2 c_{\lambda_\odot-\Omega'}^2$$
$$- 42s_{i'}c_{i'}s_{\theta'}c_{\theta'}^2\left(s_{\lambda_\odot-\Omega'}\frac{\partial s_{\lambda_\odot-\Omega'}}{\partial q}c_{\lambda_\odot-\Omega'}^2 + s_{\lambda_\odot-\Omega'}^2 c_{\lambda_\odot-\Omega'}\frac{\partial c_{\lambda_\odot-\Omega'}}{\partial q}\right)$$
$$- 21\left(\frac{\partial s_{i'}}{\partial q}c_{i'}^2 + 2s_{i'}c_{i'}\frac{\partial c_{i'}}{\partial q}\right) s_{\theta'}^2 c_{\theta'} s_{\lambda_\odot-\Omega'}^3 c_{\lambda_\odot-\Omega'}$$
$$- 21s_{i'}c_{i'}^2\left(2s_{\theta'}\frac{\partial s_{\theta'}}{\partial q}c_{\theta'} + s_{\theta'}^2\frac{\partial c_{\theta'}}{\partial q}\right) s_{\lambda_\odot-\Omega'}^3 c_{\lambda_\odot-\Omega'}$$
$$- 21s_{i'}c_{i'}^2 s_{\theta'}^2 c_{\theta'}\left(3s_{\lambda_\odot-\Omega'}^2\frac{\partial s_{\lambda_\odot-\Omega'}}{\partial q}c_{\lambda_\odot-\Omega'} + s_{\lambda_\odot-\Omega'}^3\frac{\partial c_{\lambda_\odot-\Omega'}}{\partial q}\right)$$
$$+ 3\left(\frac{\partial s_{i'}}{\partial q}c_{\theta'} + s_{i'}\frac{\partial c_{\theta'}}{\partial q}\right) s_{\lambda_\odot-\Omega'}c_{\lambda_\odot-\Omega'}$$
$$+ 3s_{i'}c_{\theta'}\left(\frac{\partial s_{\lambda_\odot-\Omega'}}{\partial q}c_{\lambda_\odot-\Omega'} + s_{\lambda_\odot-\Omega'}\frac{\partial c_{\lambda_\odot-\Omega'}}{\partial q}\right)$$
$$+ 3\left(\frac{\partial s_{i'}}{\partial q}c_{i'} + s_{i'}\frac{\partial c_{i'}}{\partial q}\right) s_{\theta'}s_{\lambda_\odot-\Omega'}^2 + 3s_{i'}c_{i'}\frac{\partial s_{\theta'}}{\partial q}s_{\lambda_\odot-\Omega'}^2$$
$$+ 6s_{i'}c_{i'}s_{\theta'}s_{\lambda_\odot-\Omega'}\frac{\partial s_{\lambda_\odot-\Omega'}}{\partial q} \tag{A15.42}$$

$$\frac{\partial f_{h_1}^s}{\partial L} = s_{i'}\frac{\partial c_{\theta'}}{\partial L}s_{\lambda_\odot-\Omega'}c_{\lambda_\odot-\Omega'} + s_{i'}c_{i'}\frac{\partial s_{\theta'}}{\partial L}s_{\lambda_\odot-\Omega'}^2 \tag{A15.43}$$

$$\frac{\partial f_{h_2}^s}{\partial L} = 10s_{i'}c_{\theta'}\frac{\partial c_{\theta'}}{\partial L}s_{\lambda_\odot-\Omega'}c_{\lambda_\odot-\Omega'}^2 + 10s_{i'}c_{i'}^2 s_{\theta'}\frac{\partial s_{\theta'}}{\partial L}s_{\lambda_\odot-\Omega'}^3$$
$$+ 10s_{i'}c_{i'}\left(\frac{\partial s_{\theta'}}{\partial L}c_{\theta'} + s_{\theta'}\frac{\partial c_{\theta'}}{\partial L}\right) s_{\lambda_\odot-\Omega'}^2 c_{\lambda_\odot-\Omega'} \tag{A15.44}$$

$$
\begin{aligned}
\frac{\partial f^s_{h_3}}{\partial L} &= -21 s_{i'} c^2_{\theta'} \frac{\partial c_{\theta'}}{\partial L} s_{\lambda_\odot-\Omega'} c^3_{\lambda_\odot-\Omega'} - 21 s_{i'} c^3_{i'} s^2_{\theta'} \frac{\partial s_{\theta'}}{\partial L} s^4_{\lambda_\odot-\Omega'} \\
&- 21 s_{i'} c_{i'} \left(2 c_{\theta'} \frac{\partial c_{\theta'}}{\partial L} s_{\theta'} + c^2_{\theta'} \frac{\partial s_{\theta'}}{\partial L} \right) s^2_{\lambda_\odot-\Omega'} c^2_{\lambda_\odot-\Omega'} - 21 s_{i'} c^2_{i'} \left(2 s_{\theta'} \frac{\partial s_{\theta'}}{\partial L} c_{\theta'} \right. \\
&\left. + s^2_{\theta'} \frac{\partial c_{\theta'}}{\partial L} \right) s^3_{\lambda_\odot-\Omega'} c_{\lambda_\odot-\Omega'} + 3 s_{i'} \frac{\partial c_{\theta'}}{\partial L} s_{\lambda_\odot-\Omega'} c_{\lambda_\odot-\Omega'} + 3 s_{i'} c_{i'} \frac{\partial s_{\theta'}}{\partial L} s^2_{\lambda_\odot-\Omega'}
\end{aligned}
\tag{A15.45}
$$

References

[1] Kéchichian, J. A. (2010) *Analytic Expansions of Luni-Solar Gravity Perturbations Along Rotating Axes for Trajectory Optimization: Part 1: The Dynamic System*, Paper AAS 10–221, AAS/AIAA Space Flight Mechanics Meeting, San Diego, CA.

[2] Kéchichian, J. A. (1997) Trajectory Optimization using Nonsingular Orbital Elements and True Longitude, *Journal of Guidance, Control, and Dynamics*, Vol. 20, No. 5, 1003–1009.

[3] Kéchichian, J. A. (2008) Inclusion of Higher Order Harmonics in the Modeling of Optimal Low-Thrust Orbit Transfer, *The Journal of the Astronautical Sciences*, Vol. 56, No. 1, 41–70.

[4] Feistel, B. S. (2007) *Perturbing Accelerations in Low Thrust Trajectory Optimization*, M. S. Thesis, Department of Aerospace Engineering, University of Texas at Austin.

[5] Battin, R. H. (1987) *An Introduction to the Mathematics and Methods of Astrodynamics*, AIAA, New York, NY, 387–391.

[6] Breakwell, J. V. (1974) *Lecture Notes – Space Mechanics*, Stanford University, Department of Aeronautics and Astronautics, Palo Alto, CA.

16 Fourth-Order Expansions of the Luni-Solar Gravity Perturbations along Rotating Axes for Trajectory Optimization

16.1 Introduction

A fourth-order extension of the analytic form of the accelerations due to the luni-solar gravity perturbations along rotating axes is carried out in this chapter. These derivations are carried out in order to increase the accuracy of the dynamic modeling of perturbed optimal low-thrust transfers between general elliptic orbits, and enhance the fidelity of trajectory optimization software used in simulations and mission analyses. A set of rotating axes attached to the thrusting spacecraft is used such that both the thrust and perturbation accelerations due to Earth's geopotential and the luni-solar gravity are mathematically resolved along these axes prior to numerical integration of the actual trajectory. This Gaussian form of the state as well as the adjoint differential equations form a set of equations that are readily integrated and an iterative process is used to achieve convergence to a desired transfer. This analysis further reveals that further extensions to higher orders, say to the fifth order and beyond, are not needed to extract even more accuracy in the solutions because the minimum-time transfer solutions become fully stabilized in the sense that they do not exhibit any differences beyond a fraction of a second, or at most a few seconds even in the more extreme cases of very large orbits with apogee heights around 100,000 km with strong lunar influence.

The analysis presented in this chapter constitutes a further improvement in solution accuracy to earlier analyses carried out in the preceding chapters in the modeling of optimal low-thrust orbit transfers in general elliptic orbit. The initial contributions modeled the thrust acceleration along rotating axes in terms of nonsingular equinoctial elements in order to cast the system dynamic and adjoint differential equations in the Gaussian form for direct simultaneous precision integration [1] (see also Chapter 9). Zonal harmonics and even luni-solar gravity perturbations were later included in the modeling of the dynamics [2–4] (see also Chapters 13–15) to account for the precession of the spacecraft orbit and its in-plane and out-of-plane perturbations, thus arriving at even higher fidelity and accuracy in the simulation of long-duration low-thrust transfers. The consideration of the luni-solar gravity perturbations lead to the adoption of the expansion method shown in [5] in order to derive these perturbation accelerations along the rotating axes for final numerical integration of the full set of state and multiplier coupled differential equations. Thus the less efficient method in [6] of computing these accelerations first in the inertial system and later converting them to the rotating system and in terms of the equinoctial elements at each integration step while also calling an outside ephemeris generator to get the luni-solar state vectors is completely avoided. Instead, an analytic ephemeris for the sun's apparent and lunar orbits is used [7]. The derivations of the solar gravity acceleration components along the rotating axes are obtained in terms of the spacecraft orbit Eulerian angles with respect to the ecliptic system, while the corresponding lunar acceleration components are obtained in terms of the spacecraft orbit Eulerian angles with respect to the lunar orbit frame. Appropriate expressions of the direction cosines of the inertial axes relative to the ecliptic axes and the lunar frame axes, respectively [8], are used to finally arrive at the fully developed system of dynamic and adjoint equations. Simulations are carried out from high-energy elliptic orbits with apogees beyond the geosynchronous Earth orbit (GEO) altitude in order to examine the accuracy enhancements emanating from this fourth-order extension of the previous analysis shown in [3] and [4] and Chapters 14 and 15, and to further establish the need, or lack of it, to carry out even higher-order extensions, that is, to the fifth order and beyond in the spacecraft radial distance. The required derivations and corresponding simulation results are shown in the subsequent sections.

16.2 The Extension to Fourth Order of the Luni-Solar Gravity Perturbation Accelerations

As was shown in [3] and Chapter 14, the third-body gravity perturbation acceleration due to the Sun is given by

$$\begin{aligned}\mathbf{f}_{\odot} &= \mathbf{F} - \frac{\mu_{\odot}}{r_{\odot}^2}\hat{\mathbf{r}}_{\odot} = \frac{\mu_{\odot} r}{r_{\odot}^3}\left[3\left(\hat{\mathbf{r}}\,\hat{\mathbf{r}}_{\odot}\right)\hat{\mathbf{r}}_{\odot} - \hat{\mathbf{r}}\right] \\ &+ \frac{\mu_{\odot} r^2}{r_{\odot}^4}\left\{\frac{15}{2}\left(\hat{\mathbf{r}}\,\hat{\mathbf{r}}_{\odot}\right)^2\hat{\mathbf{r}}_{\odot} - \frac{3}{2}\left[\hat{\mathbf{r}}_{\odot} + 2\left(\hat{\mathbf{r}}\,\hat{\mathbf{r}}_{\odot}\right)\hat{\mathbf{r}}\right]\right\} \\ &+ \frac{\mu_{\odot}}{2r_{\odot}^5}r^3\left[35\left(\hat{\mathbf{r}}\,\hat{\mathbf{r}}_{\odot}\right)^3\hat{\mathbf{r}}_{\odot} - 15\left(\hat{\mathbf{r}}\,\hat{\mathbf{r}}_{\odot}\right)\hat{\mathbf{r}}_{\odot} - 15\left(\hat{\mathbf{r}}\,\hat{\mathbf{r}}_{\odot}\right)^2\hat{\mathbf{r}} + 3\hat{\mathbf{r}}\right] + \cdots \end{aligned} \tag{16.1}$$

where the vector function $\mathbf{F}$ itself is given by $\mathbf{F} = \nabla\left(\frac{\mu_{\odot}}{|\mathbf{r}_{\odot}-\mathbf{r}|}\right)$ with $\mu_{\odot}$. The gravity constant of the Sun. The denominator $|\mathbf{r}_{\odot} - \mathbf{r}|$ is expanded in terms of the Legendre polynomials in $c_{\phi} = \cos\phi$ for short, with $c_{\phi} = \hat{\mathbf{r}}\,\hat{\mathbf{r}}_{\odot} = \frac{\mathbf{r}}{r}\hat{\mathbf{r}}_{\odot}$ such that the expansion terms involve the small parameter $x' = \frac{r}{r_{\odot}}$ [3]. The vector function $\mathbf{F}$ is then reduced to the following form [3, and Chapter 14]:

$$\begin{aligned}\mathbf{F} = \nabla\Bigg\{&\frac{\mu_{\odot}}{r_{\odot}} + \frac{\mu_{\odot}}{r_{\odot}}\frac{\left(\mathbf{r}\,\hat{\mathbf{r}}_{\odot}\right)}{r_{\odot}} + \frac{\mu_{\odot}}{r_{\odot}}\left[\frac{3}{2}\frac{\left(\hat{\mathbf{r}}\,\hat{\mathbf{r}}_{\odot}\right)^2}{r_{\odot}^2} - \frac{1}{2}\frac{r^2}{r_{\odot}^2}\right] + \frac{\mu_{\odot}}{r_{\odot}}\left[\frac{5}{2}\frac{\left(\mathbf{r}\,\hat{\mathbf{r}}_{\odot}\right)^3}{r_{\odot}^3}\right. \\ &\left. - \frac{3}{2}\frac{\left(\mathbf{r}\,\hat{\mathbf{r}}_{\odot}\right)}{r_{\odot}^3}r^2\right] + \frac{\mu_{\odot}}{r_{\odot}}\left[\frac{35}{8}\frac{\left(\mathbf{r}\,\hat{\mathbf{r}}_{\odot}\right)^4}{r_{\odot}^4} - \frac{30}{8}\frac{\left(\mathbf{r}\,\hat{\mathbf{r}}_{\odot}\right)^2}{r_{\odot}^4}r^2 + \frac{3}{8}\frac{r^4}{r_{\odot}^4}\right] + \cdots\Bigg\}\end{aligned} \tag{16.2}$$

These equations were derived in [3] and Chapter 14, in this book as Equations (14.5) and (14.4) respectively. Figure 16.1, also borrowed from [3] and Chapter 14, shows the inertial and rotating coordinates as well as the geometry of the solar and lunar orbits, the spacecraft orbit and all the pertinent angles associated with the orbits. All the vectors shown in boldface as well as the unit vectors which are boldface and hatted are defined in [3] and Chapter 14. The various unit vectors and spacecraft, Sun and Moon orbit elements are also defined in [3] and Chapter 14, and therefore not repeated here for the sake of conciseness.

However, for the sake of completeness, the symbols appearing in Figure 16.1 are redefined here as follows: $(\hat{\mathbf{x}}, \hat{\mathbf{y}}, \hat{\mathbf{z}})$ represents the inertial equatorial frame centered at the Earth in O, with $\hat{\mathbf{x}}$ and $\hat{\mathbf{y}}$ in the equatorial plane and with $\hat{\mathbf{x}}$ pointing toward the vernal equinox γ, while $\hat{\mathbf{z}}$ is along the north pole. The rotating frame $(\hat{\mathbf{r}}, \hat{\boldsymbol{\theta}}, \hat{\mathbf{h}})$ is attached to the spacecraft at O' with $\hat{\mathbf{r}}$ along the radius $\mathbf{r}$ and $\hat{\boldsymbol{\theta}}$ in the instantaneous orbit plane and in the direction of motion. The angles, Ω, i, ω represent the spacecraft orbit Eulerian angles, θ^* its true anomaly and δ the spacecraft equatorial declination. The ecliptic frame is defined by $(\hat{\mathbf{x}}_e\hat{\mathbf{y}}_e\hat{\mathbf{z}}_e)$ with $\hat{\mathbf{x}}_e$ along $\hat{\mathbf{x}}$ and the angle $\varepsilon = i_s$ is the obliquity of the ecliptic. P stands for the perigee of the spacecraft orbit, while P_s represents the perigee of the sun's apparent orbit. $\mathbf{r}_\odot$ is the radius vector of the Sun respective to O, and $\phi_\odot$ the sun's latitude with respect to the equatorial plane. The sun's position is shown as $\theta_s = \omega_s + \theta_s^*$ with ω_s its argument of perigee and θ_s^* its true anomaly. The lunar orbit position is given by $\theta_m = \omega_m + \theta_m^*$ with $\lambda_\odot$ and λ_m standing, respectively, for the sun's equatorial

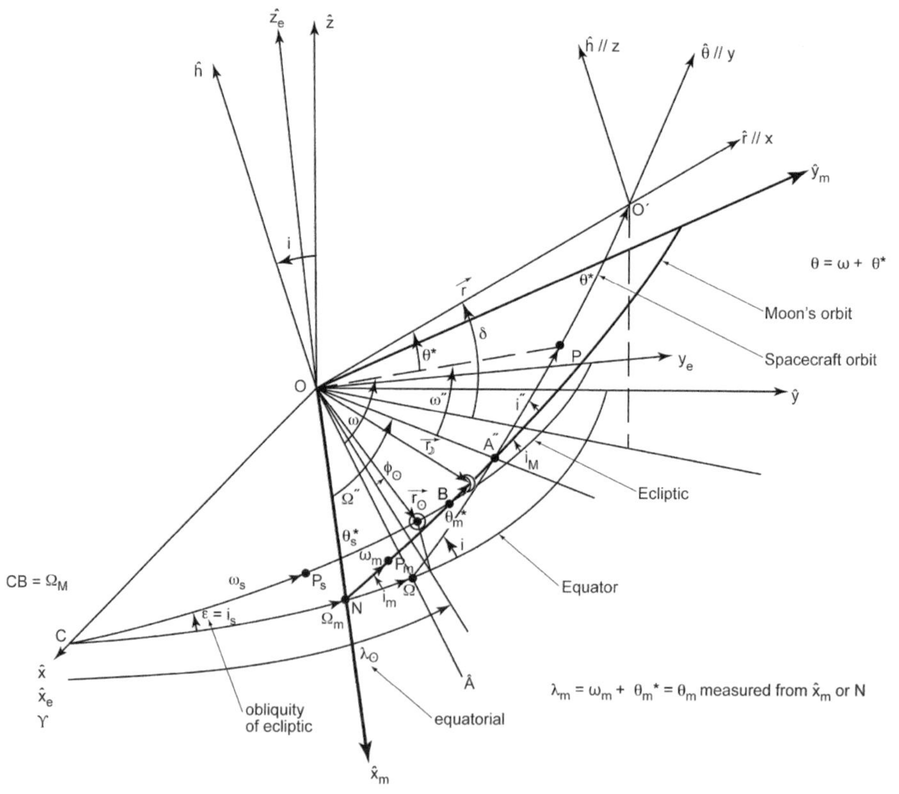

Figure 16.1 Rotating and inertial coordinate frames.

longitude, and the moon's longitude measured from its ascending node N. Finally, $\mathbf{r}_{\rightmoon}$ and i_M are the radial distance of the Moon respective to O, and its orbit ecliptic inclination.

The accelerations due to solar gravity and derived along the rotating axes $\hat{\mathbf{r}}, \hat{\boldsymbol{\theta}}, \hat{\mathbf{h}}$ to the third order in the spacecraft radial distance r have been derived in [3] and they are repeated below for clarity after removal of some typographical errors. They are also shown in Chapter 14 without the typographic errors:

$$\begin{aligned}
f_r = {} & r n_\odot^2 \left[3\left(c^2_{\lambda_\odot-\Omega'} c^2_{\theta'} + c^2_{i'} s^2_{\theta'} s^2_{\lambda_\odot-\Omega'} + 2 c_{i'} s_{\theta'} c_{\theta'} s_{\lambda_\odot-\Omega'} c_{\lambda_\odot-\Omega'}\right) - 1\right] \\
& + \frac{3}{2} r^2 \frac{n_\odot^2}{r_\odot}\left[5\left(c^3_{\lambda_\odot-\Omega'} c^3_{\theta'} + c^3_{i'} s^3_{\theta'} s^3_{\lambda_\odot-\Omega'} + 3 c^2_{\lambda_\odot-\Omega'} c^2_{\theta'} c_{i'} s_{\theta'} s_{\lambda_\odot-\Omega'}\right.\right. \\
& \left.\left. + 3 c^2_{i'} s^2_{\theta'} c_{\theta'} s^2_{\lambda_\odot-\Omega'} c_{\lambda_\odot-\Omega'}\right) - 3\left(c_{\lambda_\odot-\Omega'} c_{\theta'} + c_{i'} s_{\theta'} s_{\lambda_\odot-\Omega'}\right)\right] \\
& + \frac{r^3}{2} \frac{n_\odot^2}{r_\odot^2}\left[35\left(c^4_{\theta'} c^4_{\lambda_\odot-\Omega'} + 4 c_{i'} c^3_{\theta'} s_{\theta'} s_{\lambda_\odot-\Omega'} c^3_{\lambda_\odot-\Omega'}\right.\right. \\
& \left. + 4 c^2_{i'} s^2_{\theta'} c^2_{\theta'} s^2_{\lambda_\odot-\Omega'} c^2_{\lambda_\odot-\Omega'} + c^4_{i'} s^4_{\theta'} s^4_{\lambda_\odot-\Omega'} + 4 c^3_{i'} s^3_{\theta'} c_{\theta'} s^3_{\lambda_\odot-\Omega'} c_{\lambda_\odot-\Omega'}\right) \\
& \left. -30\left(c^2_{\lambda_\odot-\Omega'} c^2_{\theta'} + c^2_{i'} s^2_{\theta'} s^2_{\lambda_\odot-\Omega'} + 2 c_{i'} s_{\theta'} c_{\theta'} s_{\lambda_\odot-\Omega'} c_{\lambda_\odot-\Omega'}\right) + 3\right]
\end{aligned} \tag{16.3}$$

$$\begin{aligned}
f_\theta = {} & 3 r n_\odot^2 \left[-s_{\theta'} c_{\theta'} c^2_{\lambda_\odot-\Omega'} - c_{i'} s^2_{\theta'} s_{\lambda_\odot-\Omega'} c_{\lambda_\odot-\Omega'} + c_{i'} c^2_{\theta'} s_{\lambda_\odot-\Omega'} c_{\lambda_\odot-\Omega'}\right. \\
& \left. + c^2_{i'} s_{\theta'} c_{\theta'} s^2_{\lambda_\odot-\Omega'}\right] + \frac{3}{2} r^2 \frac{n_\odot^2}{r_\odot}\left[5\left(-s_{\theta'} c^2_{\theta'} c^3_{\lambda_\odot-\Omega'} + c_{i'} c^3_{\theta'} s_{\lambda_\odot-\Omega'} c^2_{\lambda_\odot-\Omega'}\right.\right. \\
& - c^2_{i'} s^3_{\theta'} c_{\lambda_\odot-\Omega'} s^2_{\lambda_\odot-\Omega'} + c^3_{i'} c_{\theta'} s^2_{\theta'} s^3_{\lambda_\odot-\Omega'} - 2 c_{i'} s^2_{\theta'} c_{\theta'} s_{\lambda_\odot-\Omega'} c^2_{\lambda_\odot-\Omega'} \\
& \left.\left. + 2 c^2_{i'} s_{\theta'} c^2_{\theta'} s^2_{\lambda_\odot-\Omega'} c_{\lambda_\odot-\Omega'}\right) - \left(-s_{\theta'} c_{\lambda_\odot-\Omega'} + c_{i'} c_{\theta'} s_{\lambda_\odot-\Omega'}\right)\right] \\
& + \frac{r^3}{2} \frac{n_\odot^2}{r_\odot^2}\left[35\left(-s_{\theta'} c^3_{\theta'} c^4_{\lambda_\odot-\Omega'} - s^4_{\theta'} c^3_{i'} c_{\lambda_\odot-\Omega'} s^3_{\lambda_\odot-\Omega'} - 3 c_{i'} s^2_{\theta'} c^2_{\theta'} c^3_{\lambda_\odot-\Omega'} s_{\lambda_\odot-\Omega'}\right.\right. \\
& - 3 c^2_{i'} s^3_{\theta'} c_{\theta'} s^2_{\lambda_\odot-\Omega'} c^2_{\lambda_\odot-\Omega'} + c_{i'} c^4_{\theta'} s_{\lambda_\odot-\Omega'} c^3_{\lambda_\odot-\Omega'} + c^4_{i'} c_{\theta'} s^3_{\theta'} s^4_{\lambda_\odot-\Omega'} \\
& \left. + 3 c^2_{i'} c^3_{\theta'} s_{\theta'} c^2_{\lambda_\odot-\Omega'} s^2_{\lambda_\odot-\Omega'} + 3 c^3_{i'} s^2_{\theta'} c^2_{\theta'} s^3_{\lambda_\odot-\Omega'} c_{\lambda_\odot-\Omega'}\right) - 15\left(-s_{\theta'} c_{\theta'} c^2_{\lambda_\odot-\Omega'}\right. \\
& \left.\left. - c_{i'} s^2_{\theta'} s_{\lambda_\odot-\Omega'} c_{\lambda_\odot-\Omega'} + c_{i'} c^2_{\theta'} s_{\lambda_\odot-\Omega'} c_{\lambda_\odot-\Omega'} + c^2_{i'} s_{\theta'} c_{\theta'} s^2_{\lambda_\odot-\Omega'}\right)\right]
\end{aligned} \tag{16.4}$$

$$\begin{aligned}
f_h = {} & -3 r n_\odot^2 \left[s_{i'} c_{\theta'} s_{\lambda_\odot-\Omega'} c_{\lambda_\odot-\Omega'} + s_{i'} c_{i'} s_{\theta'} s^2_{\lambda_\odot-\Omega'}\right] - \frac{3}{2} r^2 \\
& \times \frac{n_\odot^2}{r_\odot}\left[5 s_{i'} c^2_{\theta'} s_{\lambda_\odot-\Omega'} c^2_{\lambda_\odot-\Omega'} + 5 s_{i'} c^2_{i'} s^2_{\theta'} s^3_{\lambda_\odot-\Omega'} + 10 s_{i'} c_{i'} s_{\theta'} c_{\theta'} s^2_{\lambda_\odot-\Omega'} c_{\lambda_\odot-\Omega'}\right. \\
& \left. - s_{i'} s_{\lambda_\odot-\Omega'}\right] + \frac{5}{2} \frac{r^3 n_\odot^2}{r_\odot^2}\left[-7 s_{i'} c^3_{\theta'} s_{\lambda_\odot-\Omega'} c^3_{\lambda_\odot-\Omega'} - 7 s_{i'} c^3_{i'} s^3_{\theta'} s^4_{\lambda_\odot-\Omega'}\right. \\
& - 21 s_{i'} c_{i'} c^2_{\theta'} s_{\theta'} s^2_{\lambda_\odot-\Omega'} c^2_{\lambda_\odot-\Omega'} - 21 s_{i'} c^2_{i'} s^2_{\theta'} c_{\theta'} s^3_{\lambda_\odot-\Omega'} c_{\lambda_\odot-\Omega'} \\
& \left. + 3 s_{i'} c_{\theta'} s_{\lambda_\odot-\Omega'} c_{\lambda_\odot-\Omega'} + 3 s_{i'} c_{i'} s_{\theta'} s^2_{\lambda_\odot-\Omega'}\right]
\end{aligned} \tag{16.5}$$

These equations, shown in [3] and also in Chapter 14 as Equations (14.16)–(14.18) are written in terms of the spacecraft orbit Eulerian angles Ω', i', θ' with respect to the ecliptic system as depicted in Figure 2 of [3] and Figure 14.2 of Chapter 14, as well as the longitude of the Sun $\lambda_\odot$ along the ecliptic measured from the $\hat{\mathbf{x}}_e$ axis of the ecliptic triad $\hat{\mathbf{x}}_e, \hat{\mathbf{y}}_e, \hat{\mathbf{z}}_e$. Note that the sun's position unit vector $\hat{\mathbf{r}}_\odot = c_{\lambda_\odot}\hat{\mathbf{x}}_e + s_{\lambda_\odot}\hat{\mathbf{y}}_e$ is given as

$$\hat{\mathbf{r}}_\odot = \left(c_{\lambda_\odot-\Omega'}c_{\theta'} + c_{i'}s_{\theta'}s_{\lambda_\odot-\Omega'}\right)\hat{\mathbf{r}} + \left(-s_{\theta'}c_{\lambda_\odot-\Omega'} + c_{i'}c_{\theta'}s_{\lambda_\odot-\Omega'}\right)\hat{\boldsymbol{\theta}} - s_{i'}s_{\lambda_\odot-\Omega'}\hat{\mathbf{h}} \tag{16.6}$$

resolved along the rotating axes, such that the dot products $\hat{\mathbf{r}}_\odot\hat{\mathbf{r}}$, $\hat{\mathbf{r}}_\odot\hat{\boldsymbol{\theta}}$ and $\hat{\mathbf{r}}_\odot\hat{\mathbf{h}}$ used to reduce the accelerations expressions to their final forms in Equations (16.3)–(16.5) are repeated here for later use

$$\hat{\mathbf{r}}_\odot\hat{\mathbf{r}} = c_{\lambda_\odot-\Omega'}c_{\theta'} + c_{i'}s_{\theta'}s_{\lambda_\odot-\Omega'} \tag{16.7}$$

$$\hat{\mathbf{r}}_\odot\hat{\boldsymbol{\theta}} = -s_{\theta'}c_{\lambda_\odot-\Omega'} + c_{i'}c_{\theta'}s_{\lambda_\odot-\Omega'} \tag{16.8}$$

$$\hat{\mathbf{r}}_\odot\hat{\mathbf{h}} = -s_{i'}s_{\lambda_\odot-\Omega'} \tag{16.9}$$

We can now continue to expand $|\mathbf{r}_\odot - \mathbf{r}|^{-1}$ in terms of the Legendre polynomials to the fourth order following the procedure shown in [3] and Chapter 14 in order to add a new term to the vector function $\mathbf{F}$ such that, showing only the new term, $\mathbf{F}$ can be written as

$$\mathbf{F} = \nabla\left\{\cdots + \frac{\mu_\odot}{r_\odot}\left[\frac{63}{8}\frac{\left(\mathbf{r}\,\hat{\mathbf{r}}_\odot\right)^5}{r_\odot^5} - \frac{70}{8}\frac{\left(\mathbf{r}\,\hat{\mathbf{r}}_\odot\right)r^2}{r_\odot^5} + \frac{15}{8}\frac{\left(\mathbf{r}\,\hat{\mathbf{r}}_\odot\right)r^4}{r_\odot^5}\right]\right. \tag{16.10}$$

As before, $\mu_\odot$ is the Sun gravity constant, and $n_\odot^2 = \mu_\odot/r_\odot^3$. Taking the gradient and showing only the new fourth-order form, we have the perturbation acceleration

$$\mathbf{f}_\odot = \mathbf{F} - \frac{\mu_\odot}{r_\odot^2}\hat{\mathbf{r}}_\odot \tag{16.11}$$

$$\mathbf{f}_\odot = \cdots + \frac{\mu_\odot}{8r_\odot^6}r^4\left[315\left(\hat{\mathbf{r}}\,\hat{\mathbf{r}}_\odot\right)^4\hat{\mathbf{r}}_\odot - 210\left(\hat{\mathbf{r}}\,\hat{\mathbf{r}}_\odot\right)^2\hat{\mathbf{r}}_\odot\right.$$
$$\left. -140\left(\hat{\mathbf{r}}\,\hat{\mathbf{r}}_\odot\right)^3\hat{\mathbf{r}} + 15\hat{\mathbf{r}}_\odot + 60\left(\hat{\mathbf{r}}\,\hat{\mathbf{r}}_\odot\right)\hat{\mathbf{r}}\right] \tag{16.12}$$

Using the dot products in Equations (16.7)–(16.9), we can form the expressions for f_r, f_θ and f_h as in [3] and Chapter 14, such that with the s superscript for the Sun,

$$f_r^s = \mathbf{f}_\odot^s\hat{\mathbf{r}} \tag{16.13}$$

$$f_\theta^s = \mathbf{f}_\odot^s\hat{\boldsymbol{\theta}} \tag{16.14}$$

$$f_h^s = \mathbf{f}_\odot^s\hat{\mathbf{h}} \tag{16.15}$$

Thus, inclusive of terms up to order four, we have

$$f_r^s = n_\odot^2 r\left(f_r^s\right)_1 + \frac{3}{2}n_\odot^2\frac{r^2}{r_\odot}\left(f_r^s\right)_2 + \frac{1}{2}n_\odot^2\frac{r^3}{r_\odot^2}\left(f_r^s\right)_3 + \frac{5}{8}n_\odot^2\frac{r^4}{r_\odot^3}\left(f_r^s\right)_4 \tag{16.16}$$

$$f_\theta^s = 3n_\odot^2 r\left(f_\theta^s\right)_1 + \frac{3}{2}n_\odot^2\frac{r^2}{r_\odot}\left(f_\theta^s\right)_2 + \frac{1}{2}n_\odot^2\frac{r^3}{r_\odot^2}\left(f_\theta^s\right)_3 + \frac{15}{8}n_\odot^2\frac{r^4}{r_\odot^3}\left(f_\theta^s\right)_4 \tag{16.17}$$

$$f_h^s = -3n_{\odot}^2 r\left(f_h^s\right)_1 - \frac{3}{2}n_{\odot}^2\frac{r^2}{r_{\odot}}\left(f_h^s\right)_2 + \frac{5}{2}n_{\odot}^2\frac{r^3}{r_{\odot}^2}\left(f_h^s\right)_3 + \frac{15}{8}n_{\odot}^2\frac{r^4}{r_{\odot}^3}\left(f_h^s\right)_4 \tag{16.18}$$

These expressions have been derived in [3] and Chapter 14, except for the new fourth-order quantities $(f_r^s)_4$, $(f_\theta^s)_4$ and $(f_h^s)_4$ obtained here. After some algebra, and regrouping terms, these new additions can be cast as follows:

$$\begin{aligned}
\left(f_r^s\right)_4 = \Big[& 63\Big(c_{\theta'}^5 c_{\lambda_{\odot}-\Omega'}^5 + 5c_{i'}^4 s_{\theta'}^4 c_{\theta'} s_{\lambda_{\odot}-\Omega'}^4 c_{\lambda_{\odot}-\Omega'} + 10c_{i'}^2 s_{\theta'}^2 c_{\theta'}^3 s_{\lambda_{\odot}-\Omega'}^2 c_{\lambda_{\odot}-\Omega'}^3 \\
& + 5c_{i'} s_{\theta'} c_{\theta'}^4 s_{\lambda_{\odot}-\Omega'} c_{\lambda_{\odot}-\Omega'}^4 + c_{i'}^5 s_{\theta'}^5 s_{\lambda_{\odot}-\Omega'}^5 + 10c_{i'}^3 s_{\theta'}^3 c_{\theta'}^2 s_{\lambda_{\odot}-\Omega'}^3 c_{\lambda_{\odot}-\Omega'}^2\Big) \\
& - 70\Big(c_{\theta'}^3 c_{\lambda_{\odot}-\Omega'}^3 + 3c_{i'} s_{\theta'} c_{\theta'}^2 s_{\lambda_{\odot}-\Omega'} c_{\lambda_{\odot}-\Omega'}^2 + c_{i'}^3 s_{\theta'}^3 s_{\lambda_{\odot}-\Omega'}^3 \\
& + 3c_{i'}^2 s_{\theta'}^2 c_{\theta'} s_{\lambda_{\odot}-\Omega'}^2 c_{\lambda_{\odot}-\Omega'}\Big) + 15\left(c_{\lambda_{\odot}-\Omega'} c_{\theta'} + c_{i'} s_{\theta'} s_{\lambda_{\odot}-\Omega'}\right)\Big]
\end{aligned} \tag{16.19}$$

$$\begin{aligned}
\left(f_\theta^s\right)_4 = \Big[& 21\Big(-s_{\theta'} c_{\theta'}^4 c_{\lambda_{\odot}-\Omega'}^5 - c_{i'}^4 s_{\theta'}^5 s_{\lambda_{\odot}-\Omega'}^4 c_{\lambda_{\odot}-\Omega'} - 6c_{i'}^2 s_{\theta'}^3 c_{\theta'}^2 s_{\lambda_{\odot}-\Omega'}^2 c_{\lambda_{\odot}-\Omega'}^3 \\
& - 4c_{i'} s_{\theta'}^2 c_{\theta'}^3 s_{\lambda_{\odot}-\Omega'} c_{\lambda_{\odot}-\Omega'}^4 - 4c_{i'}^3 s_{\theta'}^4 c_{\theta'} s_{\lambda_{\odot}-\Omega'}^3 c_{\lambda_{\odot}-\Omega'}^2 + c_{i'} c_{\theta'}^5 s_{\lambda_{\odot}-\Omega'} c_{\lambda_{\odot}-\Omega'}^4 \\
& + c_{i'}^5 s_{\theta'}^4 c_{\theta'} s_{\lambda_{\odot}-\Omega'}^5 + 6c_{i'}^3 s_{\theta'}^2 c_{\theta'}^3 s_{\lambda_{\odot}-\Omega'}^3 c_{\lambda_{\odot}-\Omega'}^2 + 4c_{i'}^2 s_{\theta'} c_{\theta'}^4 s_{\lambda_{\odot}-\Omega'}^2 c_{\lambda_{\odot}-\Omega'}^3 \\
& + 4c_{i'}^4 s_{\theta'}^3 c_{\theta'}^2 s_{\lambda_{\odot}-\Omega'}^4 c_{\lambda_{\odot}-\Omega'}\Big) - 14\Big(-s_{\theta'} c_{\theta'}^2 c_{\lambda_{\odot}-\Omega'}^3 - c_{i'}^2 s_{\theta'}^3 s_{\lambda_{\odot}-\Omega'}^2 c_{\lambda_{\odot}-\Omega'} \\
& - 2c_{i'} s_{\theta'}^2 c_{\theta'} s_{\lambda_{\odot}-\Omega'} c_{\lambda_{\odot}-\Omega'}^2 + c_{i'} c_{\theta'}^3 s_{\lambda_{\odot}-\Omega'} c_{\lambda_{\odot}-\Omega'}^2 + c_{i'}^3 s_{\theta'}^2 c_{\theta'} s_{\lambda_{\odot}-\Omega'}^3 \\
& + 2c_{i'}^2 s_{\theta'} c_{\theta'}^2 s_{\lambda_{\odot}-\Omega'}^2 c_{\lambda_{\odot}-\Omega'}\Big) + \left(-s_{\theta'} c_{\lambda_{\odot}-\Omega'} + c_{i'} c_{\theta'} s_{\lambda_{\odot}-\Omega'}\right)\Big]
\end{aligned} \tag{16.20}$$

$$\begin{aligned}
\left(f_h^s\right)_4 = \Big[& -21 s_{i'} c_{\theta'}^4 s_{\lambda_{\odot}-\Omega'} c_{\lambda_{\odot}-\Omega'}^4 - 21 s_{i'} c_{i'}^4 s_{\theta'}^4 s_{\lambda_{\odot}-\Omega'}^5 \\
& - 126 s_{i'} c_{i'}^2 s_{\theta'}^2 c_{\theta'}^2 s_{\lambda_{\odot}-\Omega'}^3 c_{\lambda_{\odot}-\Omega'}^2 - 84 s_{i'} c_{i'} s_{\theta'} c_{\theta'}^3 s_{\lambda_{\odot}-\Omega'}^2 c_{\lambda_{\odot}-\Omega'}^3 \\
& - 84 s_{i'} c_{i'}^3 s_{\theta'}^3 c_{\theta'} s_{\lambda_{\odot}-\Omega'}^4 c_{\lambda_{\odot}-\Omega'} + 14 s_{i'} c_{\theta'}^2 s_{\lambda_{\odot}-\Omega'} c_{\lambda_{\odot}-\Omega'}^2 \\
& + 14 s_{i'} c_{i'}^2 s_{\theta'}^2 s_{\lambda_{\odot}-\Omega'}^3 + 28 s_{i'} c_{i'} s_{\theta'} c_{\theta'} s_{\lambda_{\odot}-\Omega'}^2 c_{\lambda_{\odot}-\Omega'} - s_{i'} s_{\lambda_{\odot}-\Omega'}\Big]
\end{aligned} \tag{16.21}$$

The system and adjoint differential equations for the equinoctial elements-based formulation are shown in Equations (1)–(3) of [3] and Equations (14.1)–(14.3) of Chapter 14, where the vectors $\mathbf{f}_{\odot}$ and $\mathbf{f}_{☽}$ for the solar and lunar accelerations have been expanded to third order in the spacecraft radial distance r along the rotating axes, resulting in the expressions for the f_r, f_θ and f_h components for both the Sun and the Moon, and the partials $\frac{\partial \mathbf{f}_{\odot}}{\partial \mathbf{z}}$ and $\frac{\partial \mathbf{f}_{☽}}{\partial \mathbf{z}}$ needed in the adjoint equations also fully derived up to the third order in r. The state vector $\mathbf{z} = (a, h, k, p, q, L)$, as stated earlier, consists of the equinoctial elements with L standing for the true longitude. We therefore need to add the fourth-order terms in the expressions for the $\partial f_r/\partial(a, h, k, p, q, L)$, $\partial f_\theta/\partial(a, h, k, p, q, L)$ and $\partial f_h/\partial(a, h, k, p, q, L)$, for both the solar and lunar accelerations components to complete this higher-order analysis. As in [3] and Chapter 14, the dependencies are such that for the solar case, $r = f(a, h, k, L)$, $i' = f(p, q)$, $\theta' = f(a, h, k, p, q, L)$, because $\theta' = \omega' + \theta^*$ with $\omega' = f(h, k, p, q)$ and $\theta^* = f(h, k, L)$, and finally $\lambda_{\odot} - \Omega' = f(p, q)$. For example, for the radial solar perturbation case,

$$\frac{\partial f_r^s}{\partial a} = \frac{\partial f_r^s}{\partial r}\frac{\partial r}{\partial a} = \cdots + \frac{5}{2}n_\odot^2 \frac{r^3}{r_\odot^3}\frac{\partial r}{\partial a}\left(f_r^s\right)_4 \tag{16.22}$$

$$\frac{\partial f_r^s}{\partial h} = \cdots + \frac{5}{2}n_\odot^2 \frac{r^3}{r_\odot^3}\frac{\partial r}{\partial h}\left(f_r^s\right)_4 + \frac{5}{8}n_\odot^2 \frac{r^4}{r_\odot^3}\frac{\partial \left(f_r^s\right)_4}{\partial h} \tag{16.23}$$

$$\frac{\partial f_r^s}{\partial k} = \cdots + \frac{5}{2}n_\odot^2 \frac{r^3}{r_\odot^3}\frac{\partial r}{\partial k}\left(f_r^s\right)_4 + \frac{5}{8}n_\odot^2 \frac{r^4}{r_\odot^3}\frac{\partial \left(f_r^s\right)_4}{\partial k} \tag{16.24}$$

$$\frac{\partial f_r^s}{\partial p} = \cdots + \frac{5}{8}n_\odot^2 \frac{r^4}{r_\odot^3}\frac{\partial \left(f_r^s\right)_4}{\partial p} \tag{16.25}$$

$$\frac{\partial f_r^s}{\partial q} = \cdots + \frac{5}{8}n_\odot^2 \frac{r^4}{r_\odot^3}\frac{\partial \left(f_r^s\right)_4}{\partial q} \tag{16.26}$$

$$\frac{\partial f_r^s}{\partial L} = \cdots + \frac{5}{2}n_\odot^2 \frac{r^3}{r_\odot^3}\frac{\partial r}{\partial L}\left(f_r^s\right)_4 + \frac{5}{8}n_\odot^2 \frac{r^4}{r_\odot^3}\frac{\partial \left(f_r^s\right)_4}{\partial L} \tag{16.27}$$

where only the new fourth-order terms are shown as the new additions to the expressions already derived in [3] and Chapter 14.

The relevant partials of the fourth-order radial component of the solar acceleration with respect to the equinoctials are shown in Appendix 16.1. These partial derivatives have been derived carefully by hand for a concise layout. A further check could consist of letting them be derived by Adifor, an automatic differentiation tool that computes the partial derivatives of expressions given under the form of a FORTRAN code

In a similar way, the normal component of the solar perturbation, namely f_θ^s, will generate the following six partial derivatives with respect to the six equinoctial elements of interest, showing once again the new additions emanating from the fourth-order expansion term:

$$\frac{\partial f_\theta^s}{\partial a} = \frac{\partial f_\theta^s}{\partial r}\frac{\partial r}{\partial a} = \cdots + \frac{15}{2}n_\odot^2 \frac{r^3}{r_\odot^3}\frac{\partial r}{\partial a}\left(f_\theta^s\right)_4 \tag{16.28}$$

$$\frac{\partial f_\theta^s}{\partial h} = \cdots + \frac{15}{2}n_\odot^2 \frac{r^3}{r_\odot^3}\frac{\partial r}{\partial h}\left(f_\theta^s\right)_4 + \frac{15}{8}n_\odot^2 \frac{r^4}{r_\odot^3}\frac{\partial \left(f_\theta^s\right)_4}{\partial h} \tag{16.29}$$

$$\frac{\partial f_\theta^s}{\partial k} = \cdots + \frac{15}{2}n_\odot^2 \frac{r^3}{r_\odot^3}\frac{\partial r}{\partial k}\left(f_\theta^s\right)_4 + \frac{15}{8}n_\odot^2 \frac{r^4}{r_\odot^3}\frac{\partial \left(f_\theta^s\right)_4}{\partial k} \tag{16.30}$$

$$\frac{\partial f_\theta^s}{\partial p} = \cdots + \frac{15}{8}n_\odot^2 \frac{r^4}{r_\odot^3}\frac{\partial \left(f_\theta^s\right)_4}{\partial p} \tag{16.31}$$

$$\frac{\partial f_\theta^s}{\partial q} = \cdots + \frac{15}{8}n_\odot^2 \frac{r^4}{r_\odot^3}\frac{\partial \left(f_\theta^s\right)_4}{\partial q} \tag{16.32}$$

$$\frac{\partial f_\theta^s}{\partial L} = \cdots + \frac{15}{2}n_\odot^2 \frac{r^3}{r_\odot^3}\frac{\partial r}{\partial L}\left(f_\theta^s\right)_4 + \frac{15}{8}n_\odot^2 \frac{r^4}{r_\odot^3}\frac{\partial \left(f_\theta^s\right)_4}{\partial L} \tag{16.33}$$

The new partials of $\left(f_\theta^s\right)_4$ with respect to h, k, p, q and L are shown in Appendix 16.1.

The out-of-plane component of the solar gravity perturbation f_h^s accepts the following partial derivatives with respect to the six equinoctial elements, where only the contributions from the fourth-order expansions are shown, as before:

$$\frac{\partial f_h^s}{\partial a} = \frac{\partial f_h^s}{\partial r}\frac{\partial r}{\partial a} = \cdots + \frac{15}{2}n_\odot^2\frac{r^3}{r_\odot^3}\frac{\partial r}{\partial a}\left(f_h^s\right)_4 \tag{16.34}$$

$$\frac{\partial f_h^s}{\partial h} = \cdots + \frac{15}{2}n_\odot^2\frac{r^3}{r_\odot^3}\frac{\partial r}{\partial h}\left(f_h^s\right)_4 + \frac{15}{8}n_\odot^2\frac{r^4}{r_\odot^3}\frac{\partial\left(f_h^s\right)_4}{\partial h} \tag{16.35}$$

$$\frac{\partial f_h^s}{\partial k} = \cdots + \frac{15}{2}n_\odot^2\frac{r^3}{r_\odot^3}\frac{\partial r}{\partial k}\left(f_h^s\right)_4 + \frac{15}{8}n_\odot^2\frac{r^4}{r_\odot^3}\frac{\partial\left(f_h^s\right)_4}{\partial k} \tag{16.36}$$

$$\frac{\partial f_h^s}{\partial p} = \cdots + \frac{15}{8}n_\odot^2\frac{r^4}{r_\odot^3}\frac{\partial\left(f_h^s\right)_4}{\partial p} \tag{16.37}$$

$$\frac{\partial f_h^s}{\partial q} = \cdots + \frac{15}{8}n_\odot^2\frac{r^4}{r_\odot^3}\frac{\partial\left(f_h^s\right)_4}{\partial q} \tag{16.38}$$

$$\frac{\partial f_h^s}{\partial L} = \cdots + \frac{15}{2}n_\odot^2\frac{r^3}{r_\odot^3}\frac{\partial r}{\partial L}\left(f_h^s\right)_4 + \frac{15}{8}n_\odot^2\frac{r^4}{r_\odot^3}\frac{\partial\left(f_h^s\right)_4}{\partial L} \tag{16.39}$$

The partials of $(f_h^s)_4$ with respect to the elements h, k, p, q and L are derived and shown in Appendix 16.1.

The trigometric functions and all the partial derivatives appearing in these equations have been developed in [3] and [4]and Chapters 14 and 15, with all the necessary details, and are not therefore repeated here. The lunar accelerations have the same form as the solar accelerations, except that the Eulerian angles Ω', i', ω', as well as θ' and ε, are replaced by Ω'', i'', ω'' and $\theta'' = \omega'' + \theta^*$, i_m, with also λ_m replacing $\lambda_\odot$ and n_m replacing $n_\odot$, while r_m is now used instead of $r_\odot$. It is also shown in [3] and Chapter 14, that Ω is now replaced by $\Omega - \Omega_m$ in the expressions of the partials of the three lunar acceleration components with respect to the six equinoctial elements as shown in [3] and Chapter 14. These three components of interest read as

$$f_r^m = n_m^2 r\left(f_r^m\right)_1 + \frac{3}{2}n_m^2\frac{r^2}{r_m}\left(f_r^m\right)_2 + \frac{1}{2}n_m^2\frac{r^3}{r_m^2}\left(f_r^m\right)_3 + \frac{5}{8}n_m^2\frac{r^4}{r_m^3}\left(f_r^m\right)_4 \tag{16.40}$$

$$f_\theta^m = 3n_m^2 r\left(f_\theta^m\right)_1 + \frac{3}{2}n_m^2\frac{r^2}{r_m}\left(f_\theta^m\right)_2 + \frac{1}{2}n_m^2\frac{r^3}{r_m^2}\left(f_\theta^m\right)_3 + \frac{15}{8}n_m^2\frac{r^4}{r_m^3}\left(f_\theta^m\right)_4 \tag{16.41}$$

$$f_h^m = -3n_m^2 r\left(f_h^m\right)_1 - \frac{3}{2}n_m^2\frac{r^2}{r_m}\left(f_h^m\right)_2 + \frac{5}{2}n_m^2\frac{r^3}{r_m^2}\left(f_h^m\right)_3 + \frac{15}{8}n_m^2\frac{r^4}{r_m^3}\left(f_h^m\right)_4 \tag{16.42}$$

The new additions are the fourth-order terms in each of the above equations, and they have the same form as for the expressions in Equations (16.19)–(16.21) except that the primed quantities are replaced by the double-primed ones, and $\lambda_\odot$ is replaced by λ_m as discussed earlier and also in [3] and [4] and in Chapters 14 and 15. The same observations are also true for the partials of f_r^m, f_θ^m and f_h^m with respect to the equinoctial elements as far as the substitutions are concerned. We do not therefore need to show these partials in detail.

16.3 Transfer Examples

The accuracy of this theory is tested by borrowing two transfer examples from W. Hallman of the Flight Mechanics Department at Aerospace, with initial orbits shown in

Table 16.1 Example 1 initial and final orbit parameters

a_o (km)	e_o	i_o (deg)	Ω_o (deg)
ω_o (deg)	M_o (deg)	h_o	k_o
p_o	q_o	L_o (deg)	
42,000	0.37	10	300
180	180	0.3204293994	−0.185
−0.0757674052	0.0437443318	660	
a_f (km)	e_f	i_f (deg)	Ω_f (deg)
ω_f (deg)	M_f (deg)	h_f	k_f
p_f	q_f	L_f (rad)	
42,164	0	4.858794229	315
0		0	0
−0.03	0.03	687	

Table 16.2 Example 2 initial and final orbit parameters

a_o (km)	e_o	i_o (deg)	Ω_o (deg)
ω_o (deg)	M_o (deg)	h_o	k_o
p_o	q_o	L_o (deg)	
73,264.0685	0.451982	10	300
180	180	0.3914278941	−0.2259910000
−0.0757674052	0.0437443318	660	
a_f (km)	e_f	i_f (deg)	Ω_f (deg)
ω_f (deg)	M_f (deg)	h_f	k_f
p_f	q_f	L_f (rad)	
42,164	0	4.858794229	315
0		0	0
−0.03	0.03	687	

Tables 16.1 and 16.2, respectively, with corresponding initial apogee heights of some 51,000 km and 100,000 km on the initial high elliptic orbit (HEO) orbits, and the common GEO target orbit, and the common final true longitude $L_f = 687$ rad. In Tables 16.1 and 16.2, L_o is shown in degrees but L_f in cumulative radians. Internally, the computer program integrates L in cumulative radians as the spacecraft executes several revolutions around the Earth. Because $L = \Omega + \omega + \theta^*$, it is clear that L_o is greater than 360 deg in these examples, namely 660 deg in both. L_o is shown in degrees in the tables for illustration purposes only. These long-duration examples start from apogee heights above the GEO orbit such that they experience increasing lunar perturbations which will allow us to validate the accuracy of this theory based on analytic expansion models. The epoch of January 1, 2008 at 0 h 0 min 0 s is selected corresponding to the following classical elements for the Sun and the Moon also borrowed from Hallman, referred to the Earth-centered-inertial (ECI) equatorial system, namely $a_s = 1.4960000 \times 10^8$ km, $e_s = 1.670575116 \times 10^{-2}$, $i_s = 23.43823951$ deg,

Table 16.3 Solutions for the initial values of the Lagrange multipliers and transfer times: example 1

	$(\lambda_a)_o$ (s/km) $(\lambda_q)_o$ (s)	$(\lambda_h)_o$ (s) $(\lambda_L)_o$ (s/rad)	$(\lambda_k)_o$ (s) t_f (s)	$(\lambda_p)_o$ (s)
To first order	-0.939968910×10^2 -0.276034568×10^6	-0.255082002×10^7 0.245753349×10^5	0.161842545×10^7 7,837,059.1090	0.316903190×10^7
To second order	-0.939927708×10^2 -0.275376699×10^6	-0.255041791×10^7 0.245724946×10^5	0.161799548×10^7 7,837,005.0097	0.316771475×10^7
To third order	-0.940299815×10^2 -0.275053255×10^6	-0.255005488×10^7 0.246419240×10^5	0.161770063×10^7 7,836,920.4768	0.316508069×10^7
To fourth order	-0.944785857×10^2 -0.275453398×10^6	-0.255487327×10^7 0.249042707×10^5	0.162084106×10^7 7,836,920.2372	0.317046352×10^7

$\Omega_s = 360$ deg, $\omega_s = 283.0780594$ deg, $M_s = 356.9565919$ deg, for the apparent solar orbit, and $a_m = 383,302.3623$ km, $e_m = 5.912456541 \times 10^{-2}$, $i_m = 27.95637345$ deg, $\Omega_m = 354.3328762$ deg, $\omega_m = 55.02609164$ deg, $M_m = 143.7023158$ deg for the lunar orbit. As in [3] and [4], and Chapters 14 and 15, we use $\mu_\odot = 0.13271244 \times 10^{12}$ km^3/s^2 and $\mu_{☽} = 0.490279 \times 10^4$ km^3/s^2 for the gravity constants of the Sun and the Moon. A continuous constant thrust acceleration $f_t = 1.5 \times 10^{-7}$ km/s^2 is used to transfer the spacecraft to the designated GEO orbit under the influence of the second zonal harmonic perturbation $J_2 = 1.08263 \times 10^{-3}$. Higher-order zonals such as J_3 and J_4 were not considered in the examples shown in this chapter because the emphasis is on the dynamical perturbations of the solar and lunar gravities. It is then sufficient to consider only J_2 here to account for the main effect of the Earth harmonics. The initial values of the six multipliers λ_a, λ_h, λ_k, λ_p, λ_q and λ_L and the transfer time t_f are searched on and slowly adjusted during the iterative process in order to satisfy the six end constraints a_f, h_f, k_f, p_f, q_f, L_f as well as $H_f = 1$ for a converged minimum-time solution between the fixed initial and final states.

The numerical integrations are carried out by using a Runge–Kutta 7/8 integration method as given by the Fehlberg coefficients, with relative and absolute error controls set to the 10^{-16} level for very high accuracy. The shooting method is used starting with guessed initial values for the multipliers and the NLP2 suite of software at Aerospace employed to slowly converge to the required initial values that effect the optimal transfers with rather tight tolerances on both the objective function and the end parameters.

The solutions for both examples 1 and 2 are shown in Tables 16.3 and 16.4 by first using the first-order expansion terms corresponding to the gravity gradient terms, and then adding the second-order terms, the third-order terms and finally the fourth-order terms. The examination of the transfer times t_f reveals that the solutions do indeed stabilize in the sense that t_f does not change by more than a quarter of a second between the third-order and fourth-order solutions for example 1. Example 2, which is much longer in transfer duration, or about 128 days versus some 90 days for the example 1 case, also starting from a much larger initial orbit apogee height of some 100,000 km, shows a difference of only 11 s between the third-order and fourth-order solutions. We

Table 16.4 Solutions for the initial values of the Lagrange multipliers and transfer times: example 2

	$(\lambda_a)_o$ (s/km) $(\lambda_q)_o$ (s)	$(\lambda_h)_o$ (s) $(\lambda_L)_o$ (s/rad)	$(\lambda_k)_o$ (s) t_f (s)	$(\lambda_p)_o$ (s)
To first order	-0.814957808×10^2 0.394218538×10^6	-0.340270299×10^7 0.707813467×10^5	0.224521249×10^7 11,027,935.6785	0.119531127×10^7
To second order	-0.814813577×10^2 0.395829784×10^6	-0.340127134×10^7 0.707860251×10^5	0.224341491×10^7 11,027,370.1101	0.119561006×10^7
To third order	-0.817800931×10^2 0.395160708×10^6	-0.340460680×10^7 0.719015585×10^5	0.224592448×10^7 11,027,173.7644	0.119366386×10^7
To fourth order	-0.827182270×10^2 0.396231489×10^6	-0.341784018×10^7 0.739407030×10^5	0.225556570×10^7 11,027,185.4446	0.119958355×10^7

can then assert that further expansions, say to the fifth order and beyond, are not necessary to continue extracting more accuracy from this analytic theory. The values of each of the Lagrange multipliers do not necessarily get closer and closer as the order of the expansions is increased because the order of these expansions is still very low, and the dynamic and multiplier system of equations experience substantial additions as the new terms from the next order are included. In the limit, if the expansions were allowed to grow to encompass higher-order terms, these further additions would get smaller and smaller and the values of each of the multipliers would get closer and closer.

A final comparison is carried out by letting Hallman solve these two transfers with his software based on Feistel's method [6], which uses a separate third-body ephemeris generator to extract the solar and lunar position vectors in inertial space. At each integration step, the corresponding inertial accelerations due to the Sun and the Moon are first converted in [6] to the rotating system and the equinoctial elements, to evaluate the differential equations for both the state and adjoint variables for further integration. Using Hallman's initial orbit elements for the apparent solar orbit and the moon's orbit, and assuming that they represent the mean elements, even though they are actually osculating elements, Tables 16.3 and 16.4 are thus generated. Hallman, on the other hand, finds the two solutions of examples 1 and 2, respectively, as $t_f = 7{,}836{,}959.4291$ and 11,026,977.8540 s, within some 40 and 207 s difference, respectively, with the solutions presented here. These differences of 40 and 207 s are almost entirely due to the fact that the initial values of the orbit elements of the Sun and the Moon borrowed from Hallman are assumed to be mean values which unfortunately they are not. The exact mean values are certainly sufficiently different from the ones assumed here to result in radial and angular differences in the position of the Moon in the order of several thousand kilometers over the long span of the transfers. A better comparison would have required the use of the true mean elements of the lunar orbit at the starting epoch, but these values were unfortunately not available. Nevertheless, it is clear from the stabilization of the transfer times as the expansion order to the fourth is considered, that the expansions are dynamically convergent rather rapidly and the true measure of their converging nature lies in the extremely smaller difference in flight time as compared to the third order, namely a fraction of a second and some 11 s for the examples at hand. Note that for example 1, in the absence of the luni-solar perturbations and with the J_2 perturbation on, the transfer time $t_f = 7,844,692.548$ s is about 2.158975 h shorter than the transfer accounting for both J_2 and the luni-solar gravity. For example 2, the J_2-only transfer is obtained as $t_f =$ 11,042,149.073 s or some 4.156554 h longer than for the solution accounting for J_2 and the luni-solar gravity, showing that in this case, the lunar gravity perturbation is helping adjust the transfer parameters, resulting in shorter transit time.

Hallman's luni-solar perturbations are derived from a polynomials-based ephemeris of the Sun and the Moon, and as such are representative of the exact solutions. The accuracy of the present analytic form of the third-body accelerations is evident in the fraction of a second and 11 s differences with respect to the third-order solutions in flight time as developed in the earlier papers in [3, 4] and also in Chapters 14 and 15 of this book, for the extreme example cases considered here.

The luni-solar perturbation accelerations derived from the fourth-order expansions, although adequate for solving low-thrust orbit transfers with high accuracy, are clearly limited to moderate time spans up to a few months. Third-body gravity perturbation effects spanning several years must use the exact ephemeris of the Sun and the Moon as well as the exact evaluations of the accelerations using inertial Cartesian coordinates for later transformation to the rotating system at each integration step, avoiding thereby the limitations of the expansion method used in this chapter.

The example 1 solutions in the form of the radial, normal and out-of-plane acceler-ations due to the Moon are shown in Figures 16.2–16.10 during the last phase of the transfer. For ease of reading and visualization, the acceleration components are depicted

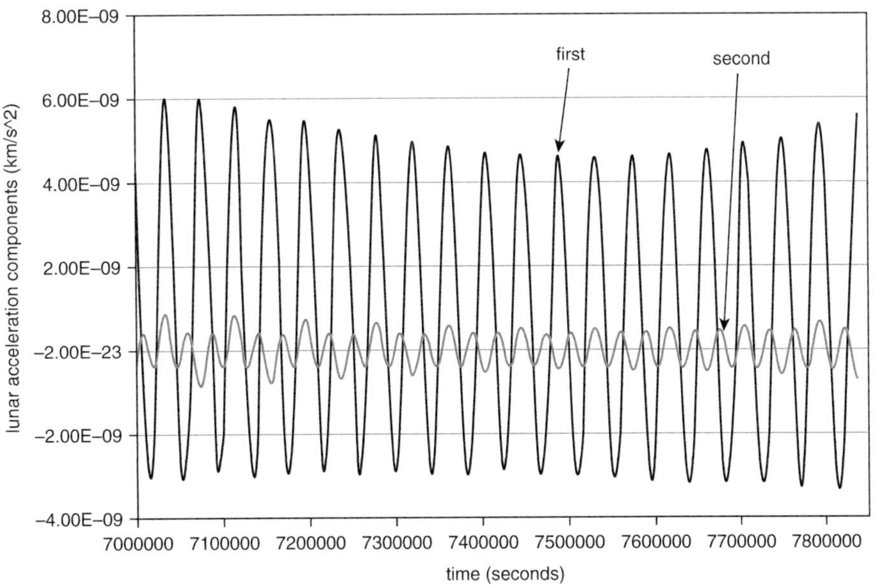

Figure 16.2 First- and second-order radial lunar acceleration components for 90-day transfer example.

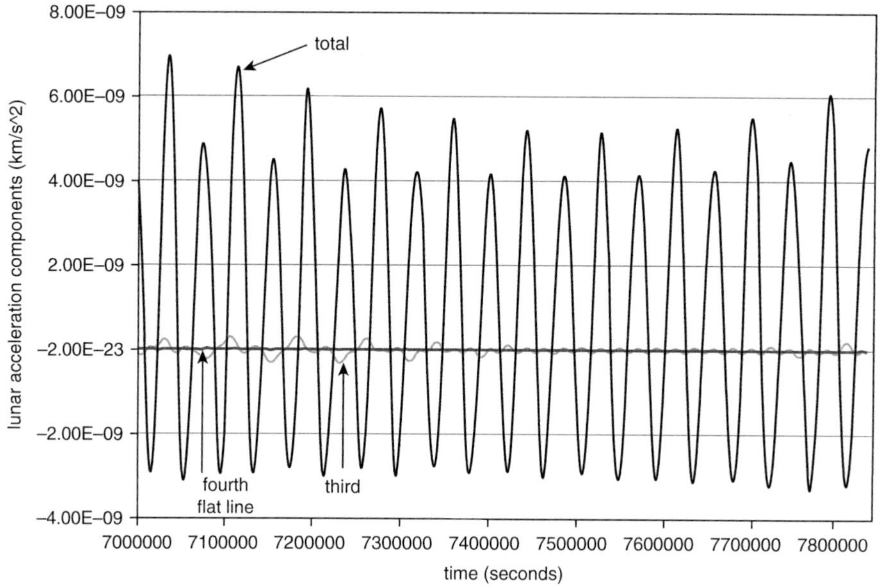

Figure 16.3 Third-, fourth- and total-order radial lunar acceleration components for 90-day transfer example.

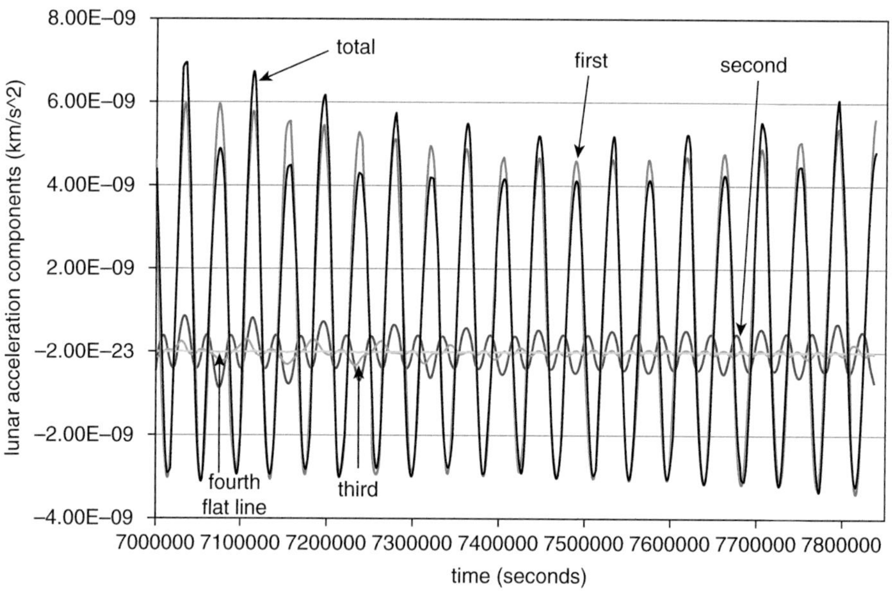

Figure 16.4 Radial components of lunar perturbation acceleration for 90-day transfer example.

with two as well as three curves at most in Figures 16.2 and 16.3 for the radial components, Figures 16.5 and 16.6 for the normal components, and Figures 16.8 and 16.9 for the out-of-plane components, while Figures 16.4, 16.7 and 16.10 show all curves simultaneously. These accelerations get smaller in magnitude as the order is increased

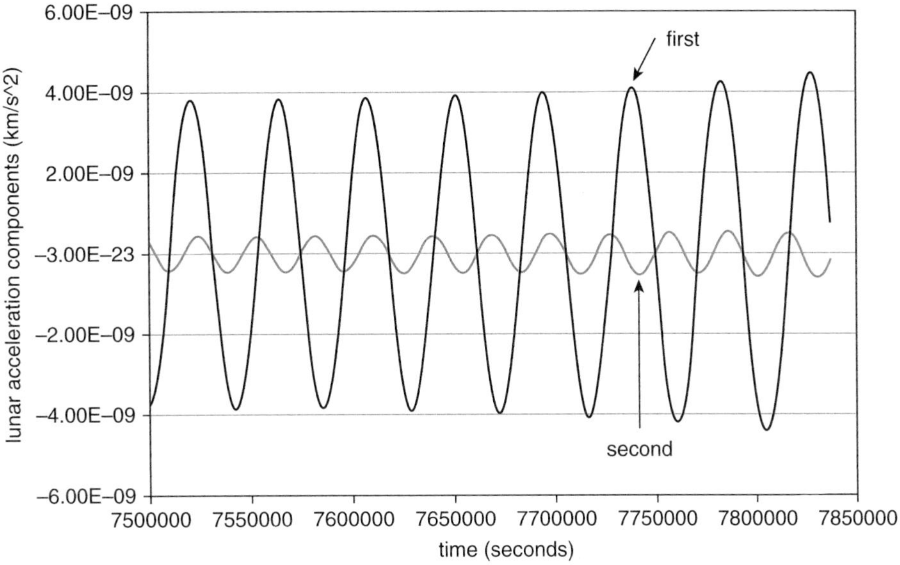

Figure 16.5 First- and second-order normal lunar acceleration components for 90-day transfer example.

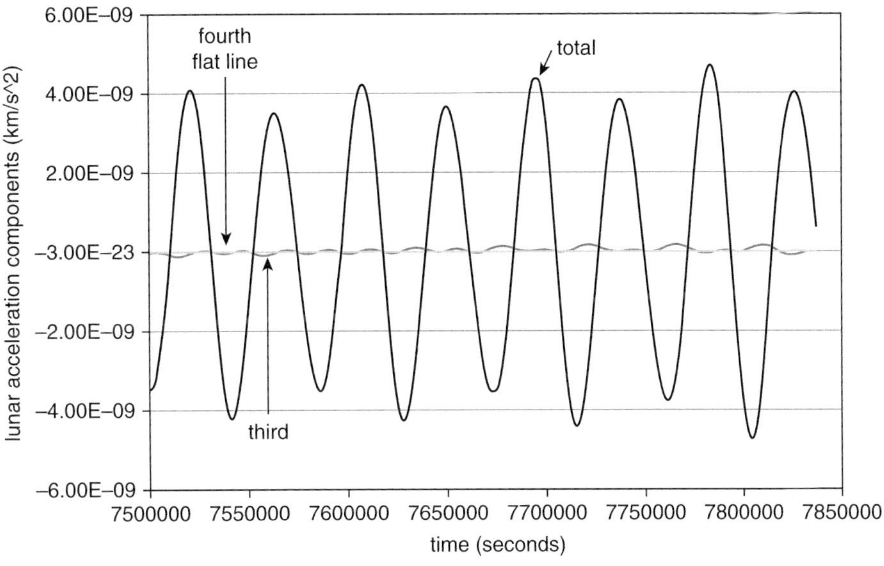

Figure 16.6 Third-, fourth- and total-order normal lunar acceleration components for 90-day transfer example.

from one to four in the sense that, say, the fourth-order contribution is only due to the fourth-order terms and not the cumulative effect of all the orders up to the fourth. The changes in the signs of the various perturbation accelerations are due to the rotation of the coordinate frame attached to the spacecraft as the latter completes each orbit on

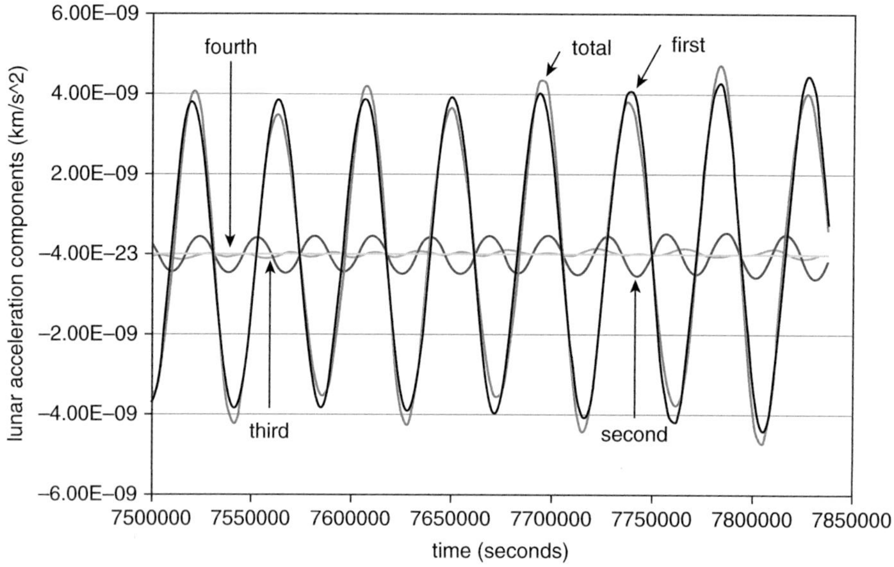

Figure 16.7 Normal components of lunar perturbation acceleration for 90-day transfer example.

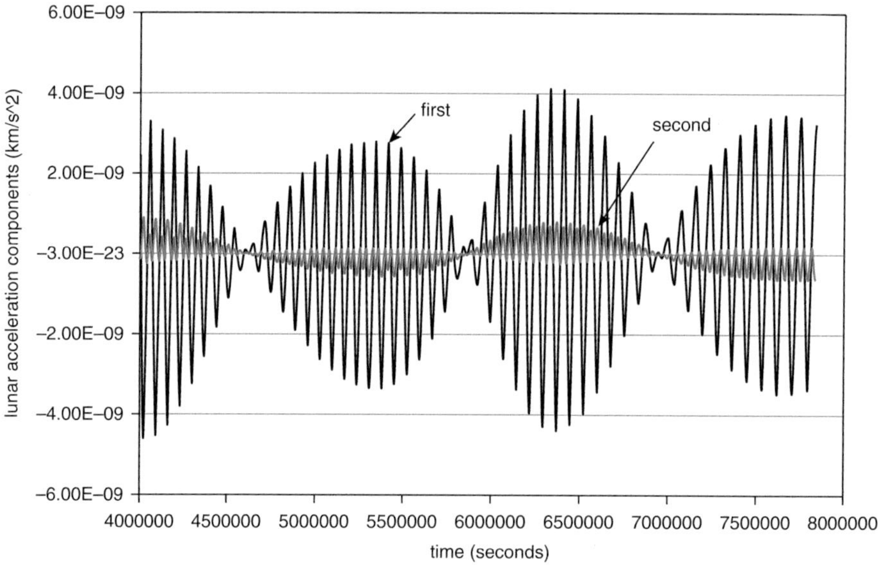

Figure 16.8 First- and second-order out-of-plane lunar acceleration components for 90-day transfer example.

its way to the final GEO target orbit. The cumulative total perturbation acceleration, meaning the total sum of all the individual contributions up to the fourth order, is also shown in these figures under the label "total." It is evident that the first-order contribution is responsible by itself to most of the total, with each higher-order term getting smaller as the order is increased. However, almost full accuracy is achieved by including

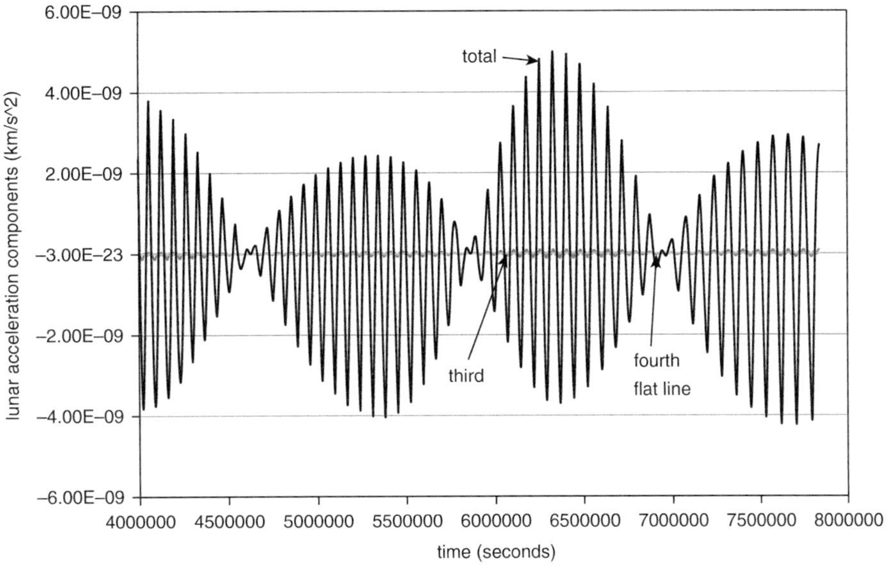

Figure 16.9 Third-, fourth- and total-order out-of-plane lunar acceleration components for 90-day transfer example.

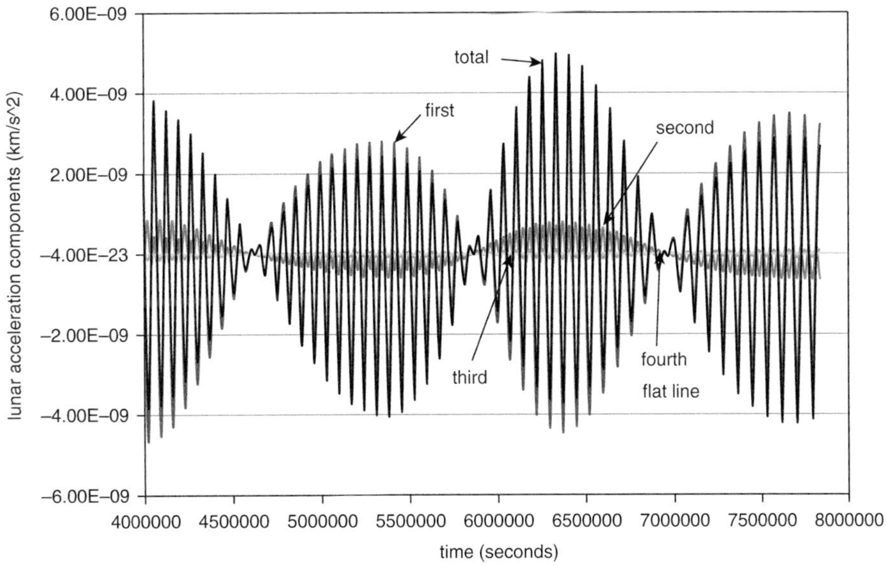

Figure 16.10 Out-of-plane components of lunar perturbation acceleration for 90-day transfer example.

the higher-order terms beyond the first order, in order to get the transfer time to within a fraction of a second, at least for example 1. Note that the range of the x or time axis in Figure 16.10 is much larger than in Figures 16.4 or 16.7. The long-period variations

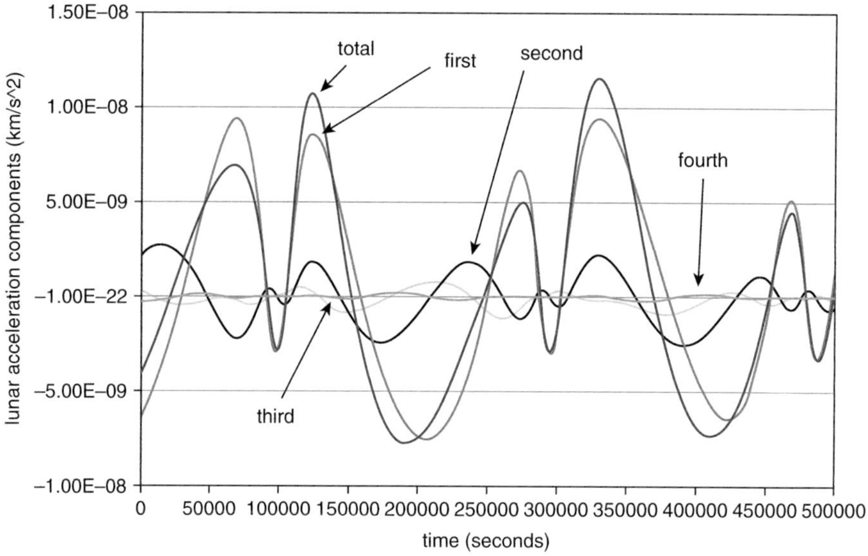

Figure 16.11 Radial components of lunar perturbation acceleration for 128-day transfer example during initial phase of transfer.

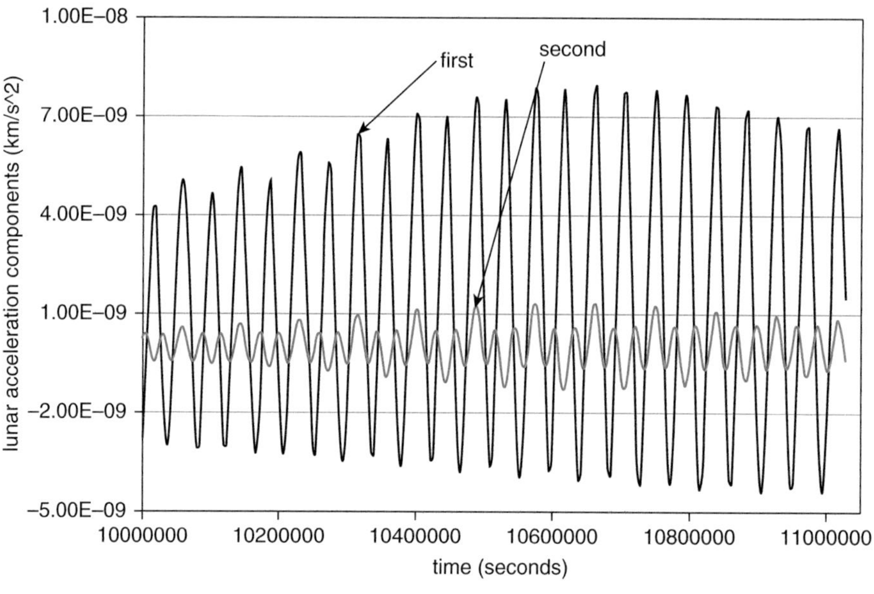

Figure 16.12 First- and second-order radial lunar acceleration components for 128-day transfer example during last phase of transfer.

of the out-of-plane accelerations in Figure 16.10 are most certainly due to the changing orbit plane orientation of the spacecraft relative to the plane of the lunar orbit. The remaining figures are for the case of example 2. Figures 16.11 and 16.14 show the variations of the radial components of the lunar accelerations as well as the total cumulative

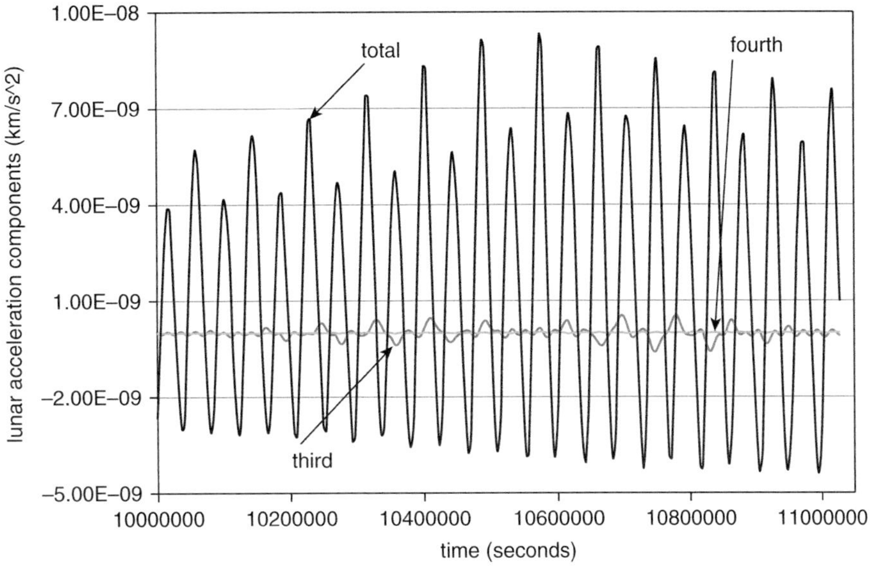

Figure 16.13 Third-, fourth- and total-order radial lunar acceleration components for 128-day transfer example during last phase of transfer.

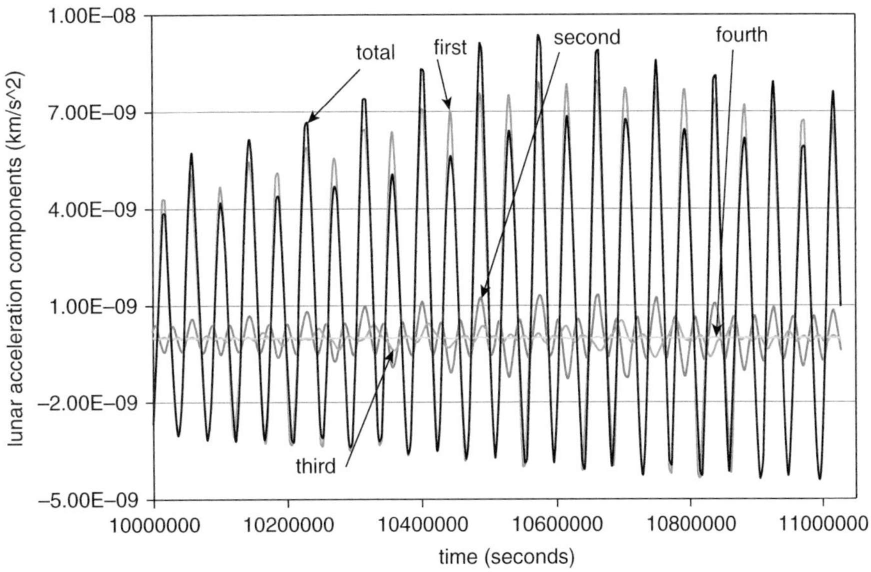

Figure 16.14 Radial components of lunar perturbation acceleration for 128-day transfer example during last phase of transfer.

acceleration during the first and last parts of the transfer. Figures 16.15 and 16.18 show the variations of the normal components, while Figures 16.19 and 16.22 are for the out-of-plane components during the first and last parts of the transfer as well. As for the case of example 1, only two or at most three curves are shown at a time in Figures 16.12

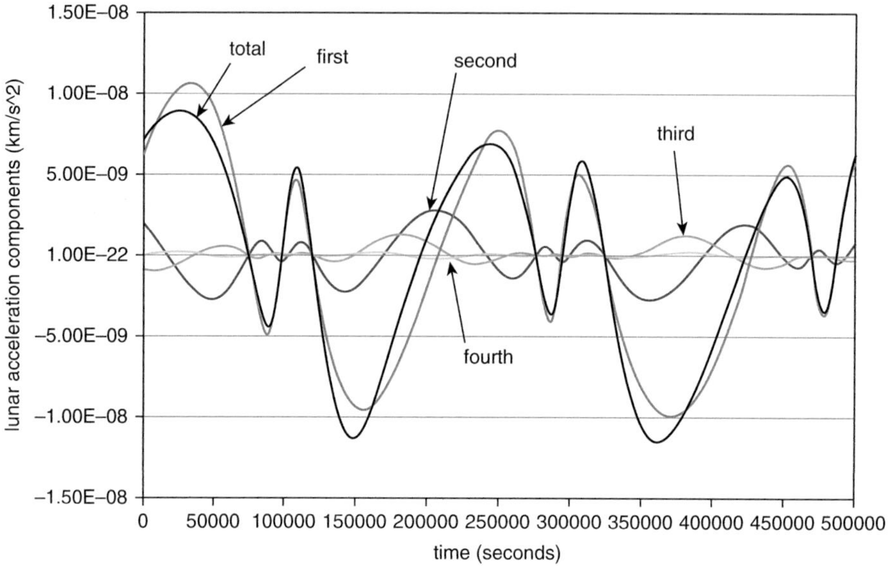

Figure 16.15 Normal components of lunar perturbation acceleration for 128-day transfer example during initial phase of transfer.

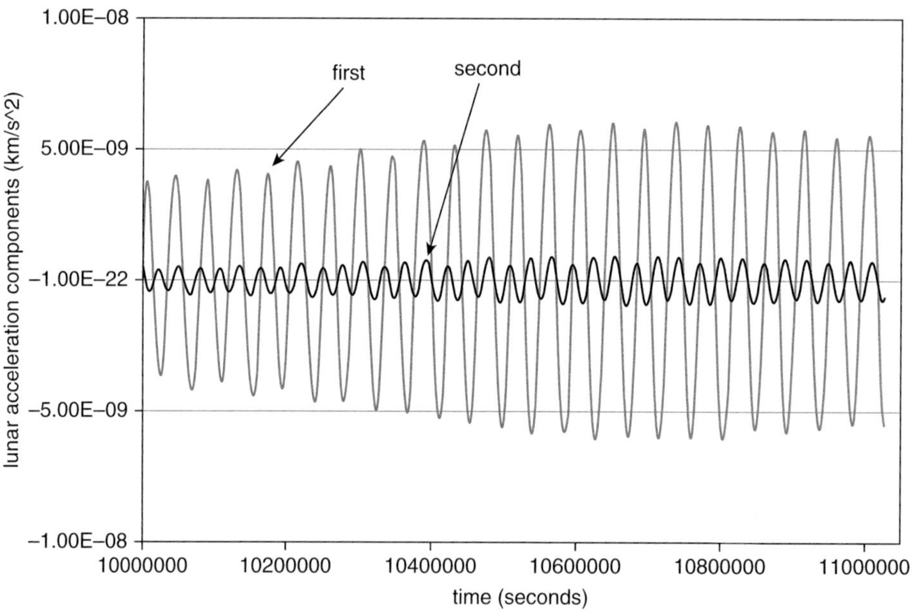

Figure 16.16 First- and second-order normal lunar acceleration components for 128-day transfer example during last phase of transfer.

and 16.13 for the radial acceleration components, Figures 16.16 and 16.17 for the normal components, and Figures 16.20 and 16.21 for the out-of-plane components. It is worthwhile noting that due to the nature of the expansion method used here, the various

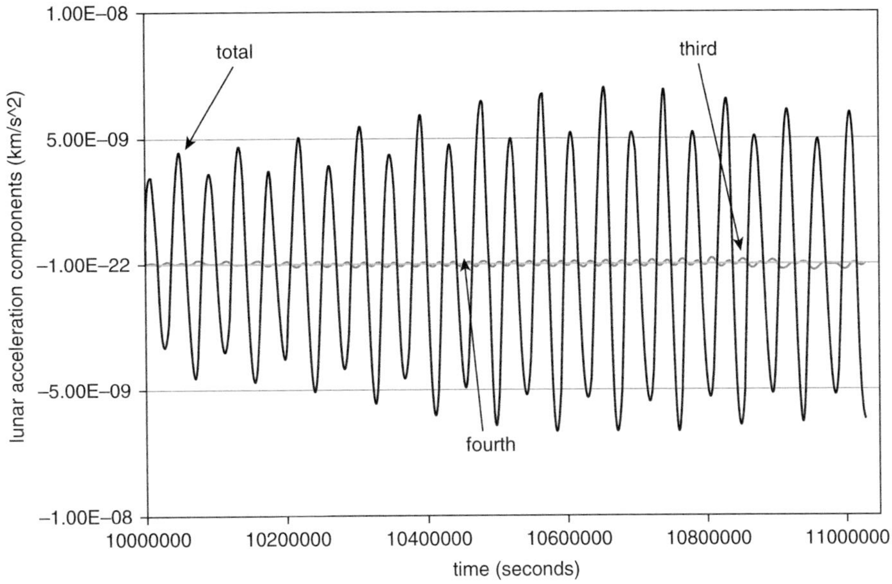

Figure 16.17 Third-, fourth- and total-order normal lunar acceleration components for 128-day transfer example during last phase of transfer.

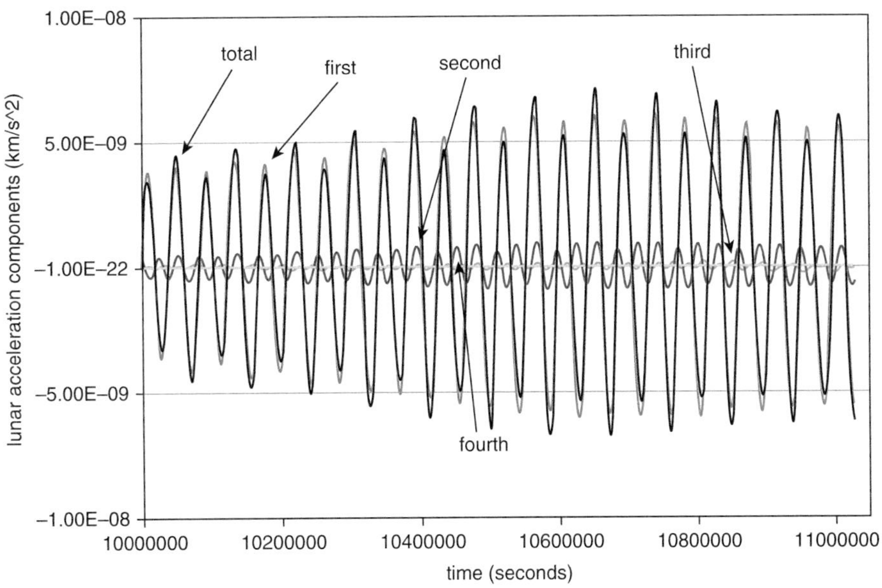

Figure 16.18 Normal components of lunar perturbation acceleration for 128-day transfer example during last phase of transfer.

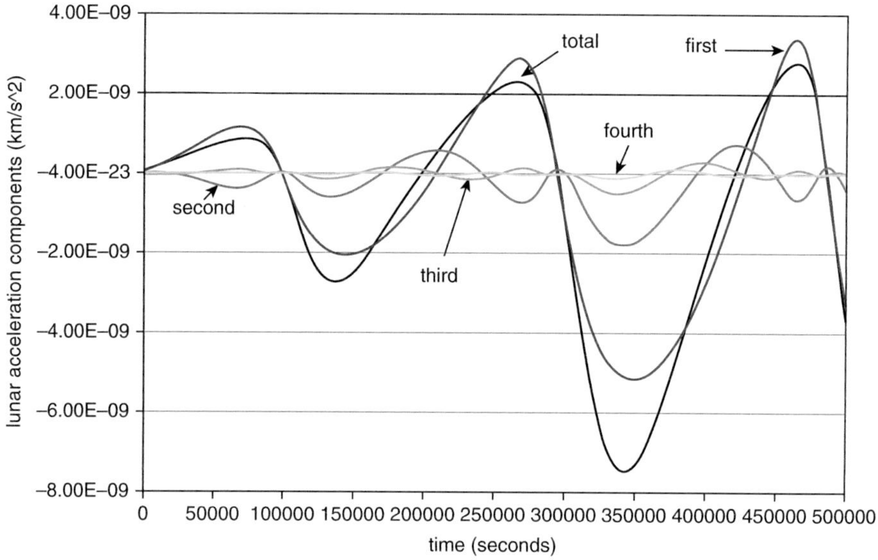

Figure 16.19 Out-of-plane components of lunar perturbation acceleration for 128-day transfer example during initial phase of transfer.

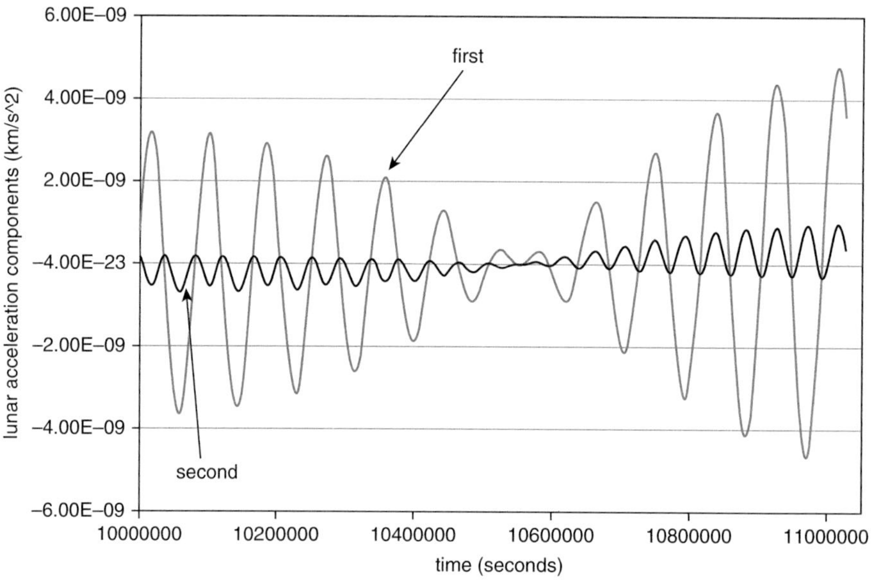

Figure 16.20 First- and second-order out-of-plane lunar acceleration components for 128-day transfer example during last phase of transfer.

order contributions are not necessarily in phase as they involve various sets of increasingly complicated trigonometric functions. The *x*-axis for flight time are left in units of seconds for compatibility reasons with the previous articles published by the author, as well as the preceding chapters, while the accelerations are left in units of km/s^2 for the

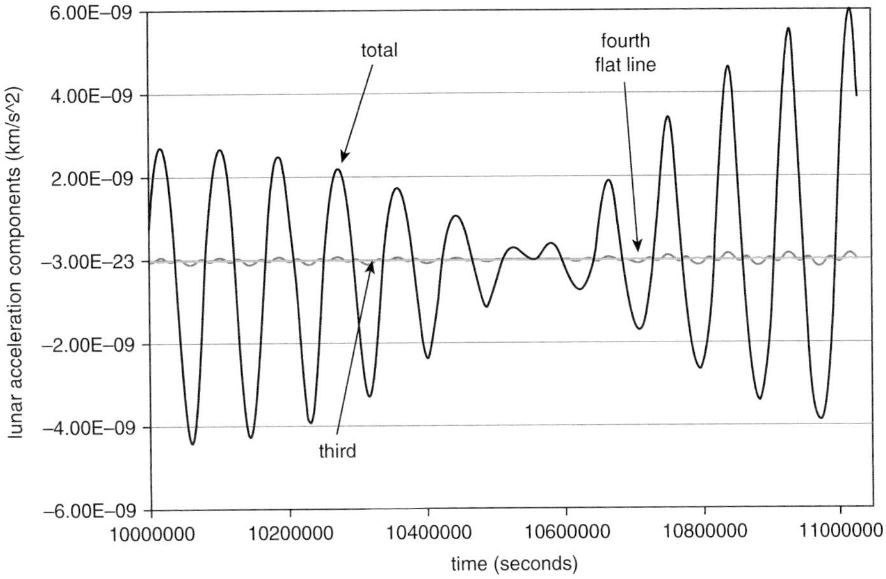

Figure 16.21 Third-, fourth- and total-order out-of-plane lunar acceleration components for 128-day transfer example during last phase of transfer.

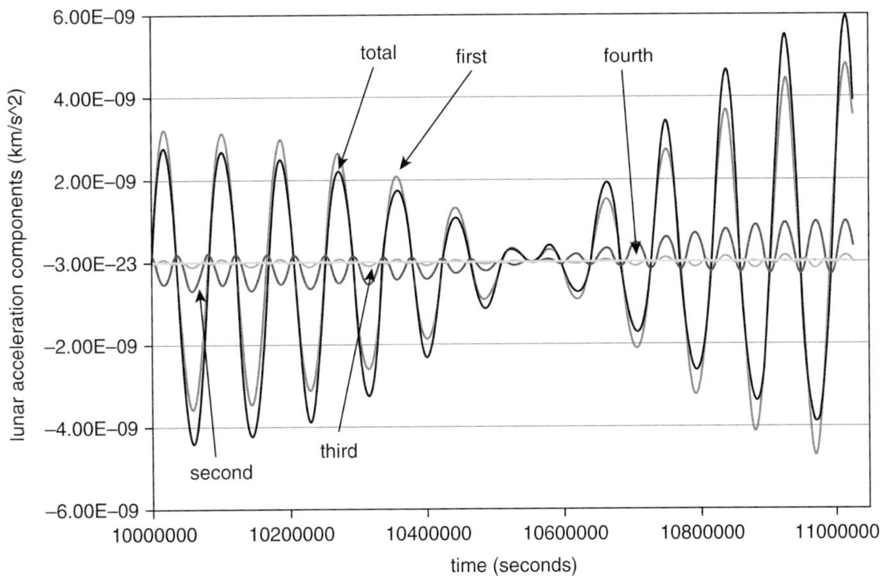

Figure 16.22 Out-of-plane components of lunar perturbation acceleration for 128-day transfer example during last phase of transfer.

same reason and also to show the very low level of the accelerations compared to the thrust acceleration and for uniformity with all the other units adopted in the chapter.

16.4 Conclusion

A dynamic modeling of the luni-solar gravity perturbation accelerations resolved along a set of rotating axes attached to a thrusting spacecraft in minimum-time transit between two general elliptic orbits, devised earlier and shown in the preceding chapters, has been extended to the fourth order in the spacecraft radial distance to the Earth center. The previous analysis based on the expansion method and including acceleration terms up to the third order has thus been improved upon from an accuracy point of view, and the solutions shown to be fully stabilized in the sense that the transfer times get to within a fraction of a second or at most a few seconds in very long-duration transits of the order of four months from very large initial elliptic orbits with 100,000 km apogee heights to the final GEO target. The simulations further reveal that any improvements toward the exact transfer time using higher-order expansions are insignificant and therefore unnecessary.

Appendix 16.1 The Partial Derivatives of $(f_r^s)_4$, $(f_\theta^s)_4$ and $(f_h^s)_4$

$$
\begin{aligned}
\frac{\partial\left(f_r^s\right)_4}{\partial h} = 63\Bigg[& 5c_{\theta'}^4\frac{\partial c_{\theta'}}{\partial h}c_{\lambda_\odot-\Omega'}^5 + 5c_{i'}^4\left(4s_{\theta'}^3\frac{\partial s_{\theta'}}{\partial h}c_{\theta'} + s_{\theta'}^4\frac{\partial c_{\theta'}}{\partial h}\right)s_{\lambda_\odot-\Omega'}^4 c_{\lambda_\odot-\Omega'} \\
& + 10c_{i'}^2\left(2s_{\theta'}\frac{\partial s_{\theta'}}{\partial h}c_{\theta'}^3 + 3c_{\theta'}^2\frac{\partial c_{\theta'}}{\partial h}s_{\theta'}^2\right)s_{\lambda_\odot-\Omega'}^2 c_{\lambda_\odot-\Omega'}^3 \\
& + 5c_{i'}\left(\frac{\partial s_{\theta'}}{\partial h}c_{\theta'}^4 + 4c_{\theta'}^3\frac{\partial c_{\theta'}}{\partial h}s_{\theta'}\right)s_{\lambda_\odot-\Omega'}c_{\lambda_\odot-\Omega'}^4 + 5c_{i'}^5 s_{\theta'}^4\frac{\partial s_{\theta'}}{\partial h}s_{\lambda_\odot-\Omega'}^5 \\
& + 10c_{i'}^3\left(3s_{\theta'}^2\frac{\partial s_{\theta'}}{\partial h}c_{\theta'}^2 + 2c_{\theta'}\frac{\partial c_{\theta'}}{\partial h}s_{\theta'}^3\right)s_{\lambda_\odot-\Omega'}^3 c_{\lambda_\odot-\Omega'}^2\Bigg] - 70\Bigg[3c_{\theta'}^2\frac{\partial c_{\theta'}}{\partial h}c_{\lambda_\odot-\Omega'}^3 \\
& + 3c_{i'}\left(\frac{\partial s_{\theta'}}{\partial h}c_{\theta'}^2 + 2c_{\theta'}\frac{\partial c_{\theta'}}{\partial h}s_{\theta'}\right)s_{\lambda_\odot-\Omega'}c_{\lambda_\odot-\Omega'}^2 + 3c_{i'}^3 s_{\theta'}^2\frac{\partial s_{\theta'}}{\partial h}s_{\lambda_\odot-\Omega'}^3 \\
& + 3c_{i'}^2\left(2s_{\theta'}\frac{\partial s_{\theta'}}{\partial h}c_{\theta'} + s_{\theta'}^2\frac{\partial c_{\theta'}}{\partial h}\right)s_{\lambda_\odot-\Omega'}^2 c_{\lambda_\odot-\Omega'}\Bigg] \\
& + 15\left(\frac{\partial c_{\theta'}}{\partial h}c_{\lambda_\odot-\Omega'} + c_{i'}\frac{\partial s_{\theta'}}{\partial h}s_{\lambda_\odot-\Omega'}\right)
\end{aligned} \tag{A16.1}
$$

$$
\begin{aligned}
\frac{\partial\left(f_r^s\right)_4}{\partial k} = 63\Bigg[& 5c_{\theta'}^4\frac{\partial c_{\theta'}}{\partial k}c_{\lambda_\odot-\Omega'}^5 + 5c_{i'}^4\left(4s_{\theta'}^3\frac{\partial s_{\theta'}}{\partial k}c_{\theta'} + s_{\theta'}^4\frac{\partial c_{\theta'}}{\partial k}\right)s_{\lambda_\odot-\Omega'}^4 c_{\lambda_\odot-\Omega'} \\
& + 10c_{i'}^2\left(2s_{\theta'}\frac{\partial s_{\theta'}}{\partial k}c_{\theta'}^3 + 3c_{\theta'}^2\frac{\partial c_{\theta'}}{\partial k}s_{\theta'}^2\right)s_{\lambda_\odot-\Omega'}^2 c_{\lambda_\odot-\Omega'}^3 \\
& + 5c_{i'}\left(\frac{\partial s_{\theta'}}{\partial k}c_{\theta'}^4 + 4c_{\theta'}^3\frac{\partial c_{\theta'}}{\partial k}s_{\theta'}\right)s_{\lambda_\odot-\Omega'}c_{\lambda_\odot-\Omega'}^4 + 5c_{i'}^5 s_{\theta'}^4\frac{\partial s_{\theta'}}{\partial k}s_{\lambda_\odot-\Omega'}^5 \\
& + 10c_{i'}^3\left(3s_{\theta'}^2\frac{\partial s_{\theta'}}{\partial k}c_{\theta'}^2 + 2c_{\theta'}\frac{\partial c_{\theta'}}{\partial k}s_{\theta'}^3\right)s_{\lambda_\odot-\Omega'}^3 c_{\lambda_\odot-\Omega'}^2\Bigg]
\end{aligned}
$$

$$-70\left[3c_{\theta'}^2\frac{\partial c_{\theta'}}{\partial k}c_{\lambda_\odot-\Omega'}^3+3c_{i'}\left(\frac{\partial s_{\theta'}}{\partial k}c_{\theta'}^2+2c_{\theta'}\frac{\partial c_{\theta'}}{\partial k}s_{\theta'}\right)s_{\lambda_\odot-\Omega'}c_{\lambda_\odot-\Omega'}^2\right.$$
$$\left.+3c_{i'}^3s_{\theta'}^2\frac{\partial s_{\theta'}}{\partial k}s_{\lambda_\odot-\Omega'}^3+3c_{i'}^2\left(2s_{\theta'}\frac{\partial s_{\theta'}}{\partial k}c_{\theta'}+s_{\theta'}^2\frac{\partial c_{\theta'}}{\partial k}\right)s_{\lambda_\odot-\Omega'}^2c_{\lambda_\odot-\Omega'}\right]$$
$$+15\left(\frac{\partial c_{\theta'}}{\partial k}c_{\lambda_\odot-\Omega'}+c_{i'}\frac{\partial s_{\theta'}}{\partial k}s_{\lambda_\odot-\Omega'}\right) \tag{A16.2}$$

$$\frac{\partial\left(f_r^s\right)_4}{\partial p}=63\left[5c_{\theta'}^4\frac{\partial c_{\theta'}}{\partial p}c_{\lambda_\odot-\Omega'}^5+5c_{\theta'}^5c_{\lambda_\odot-\Omega'}^4\frac{\partial c_{\lambda_\odot-\Omega'}}{\partial p}\right.$$
$$+20c_{i'}^3\frac{\partial c_{i'}}{\partial p}s_{\theta'}^4c_{\theta'}s_{\lambda_\odot-\Omega'}^4c_{\lambda_\odot-\Omega'}+5c_{i'}^4\left(4s_{\theta'}^3\frac{\partial s_{\theta'}}{\partial p}c_{\theta'}+s_{\theta'}^4\frac{\partial c_{\theta'}}{\partial p}\right)s_{\lambda_\odot-\Omega'}^4c_{\lambda_\odot-\Omega'}$$
$$+5c_{i'}^4s_{\theta'}^4c_{\theta'}\left(4s_{\lambda_\odot-\Omega'}^3\frac{\partial s_{\lambda_\odot-\Omega'}}{\partial p}c_{\lambda_\odot-\Omega'}+s_{\lambda_\odot-\Omega'}^4\frac{\partial c_{\lambda_\odot-\Omega'}}{\partial p}\right)$$
$$+20c_{i'}\frac{\partial c_{i'}}{\partial p}s_{\theta'}^2c_{\theta'}^3s_{\lambda_\odot-\Omega'}^2c_{\lambda_\odot-\Omega'}^3$$
$$+10c_{i'}^2\left(2s_{\theta'}\frac{\partial s_{\theta'}}{\partial p}c_{\theta'}^3+3c_{\theta'}^2\frac{\partial c_{\theta'}}{\partial p}s_{\theta'}^2\right)s_{\lambda_\odot-\Omega'}^2c_{\lambda_\odot-\Omega'}^3$$
$$+10c_{i'}^2s_{\theta'}^2c_{\theta'}^3\left(2s_{\lambda_\odot-\Omega'}\frac{\partial s_{\lambda_\odot-\Omega'}}{\partial p}c_{\lambda_\odot-\Omega'}^3+3c_{\lambda_\odot-\Omega'}^2\frac{\partial c_{\lambda_\odot-\Omega'}}{\partial p}s_{\lambda_\odot-\Omega'}^2\right)$$
$$+5\frac{\partial c_{i'}}{\partial p}s_{\theta'}c_{\theta'}^4s_{\lambda_\odot-\Omega'}c_{\lambda_\odot-\Omega'}^4+5c_{i'}\left(\frac{\partial s_{\theta'}}{\partial p}c_{\theta'}^4+4c_{\theta'}^3\frac{\partial c_{\theta'}}{\partial p}s_{\theta'}\right)s_{\lambda_\odot-\Omega'}c_{\lambda_\odot-\Omega'}^4$$
$$+5c_{i'}s_{\theta'}c_{\theta'}^4\left(\frac{\partial s_{\lambda_\odot-\Omega'}}{\partial p}c_{\lambda_\odot-\Omega'}^4+4c_{\lambda_\odot-\Omega'}^3\frac{\partial c_{\lambda_\odot-\Omega'}}{\partial p}s_{\lambda_\odot-\Omega'}\right)$$
$$+5c_{i'}^4\frac{\partial c_{i'}}{\partial p}s_{\theta'}^5s_{\lambda_\odot-\Omega'}^5+5c_{i'}^5s_{\theta'}^4\frac{\partial s_{\theta'}}{\partial p}s_{\lambda_\odot-\Omega'}^5+5c_{i'}^5s_{\theta'}^5s_{\lambda_\odot-\Omega'}^4\frac{\partial s_{\lambda_\odot-\Omega'}}{\partial p}$$
$$+30c_{i'}^2\frac{\partial c_{i'}}{\partial p}s_{\theta'}^3c_{\theta'}^2s_{\lambda_\odot-\Omega'}^3c_{\lambda_\odot-\Omega'}^2+10c_{i'}^3\left(3s_{\theta'}^2\frac{\partial s_{\theta'}}{\partial p}c_{\theta'}^2+2c_{\theta'}\frac{\partial c_{\theta'}}{\partial p}s_{\theta'}^3\right)$$
$$\times s_{\lambda_\odot-\Omega'}^3c_{\lambda_\odot-\Omega'}^2+10c_{i'}^3s_{\theta'}^3c_{\theta'}^2\left(3s_{\lambda_\odot-\Omega'}^2\frac{\partial s_{\lambda_\odot-\Omega'}}{\partial p}c_{\lambda_\odot-\Omega'}^2\right.$$
$$\left.\left.+2c_{\lambda_\odot-\Omega'}\frac{\partial c_{\lambda_\odot-\Omega'}}{\partial p}s_{\lambda_\odot-\Omega'}^3\right)\right]-70\left[3c_{\theta'}^2\frac{\partial c_{\theta'}}{\partial p}c_{\lambda_\odot-\Omega'}^3\right.$$
$$+3c_{\lambda_\odot-\Omega'}^2\frac{\partial c_{\lambda_\odot-\Omega'}}{\partial p}c_{\theta'}^3+3\frac{\partial c_{i'}}{\partial p}s_{\theta'}c_{\theta'}^2s_{\lambda_\odot-\Omega'}c_{\lambda_\odot-\Omega'}^2$$
$$+3c_{i'}\left(\frac{\partial s_{\theta'}}{\partial p}c_{\theta'}^2+2c_{\theta'}\frac{\partial c_{\theta'}}{\partial p}s_{\theta'}\right)s_{\lambda_\odot-\Omega'}c_{\lambda_\odot-\Omega'}^2$$
$$+3c_{i'}s_{\theta'}c_{\theta'}^2\left(\frac{\partial s_{\lambda_\odot-\Omega'}}{\partial p}c_{\lambda_\odot-\Omega'}^2+2c_{\lambda_\odot-\Omega'}\frac{\partial c_{\lambda_\odot-\Omega'}}{\partial p}s_{\lambda_\odot-\Omega'}\right)$$
$$+3c_{i'}^2\frac{\partial c_{i'}}{\partial p}s_{\theta'}^3s_{\lambda_\odot-\Omega'}^3+3c_{i'}^3s_{\theta'}^2\frac{\partial s_{\theta'}}{\partial p}s_{\lambda_\odot-\Omega'}^3+3c_{i'}^3s_{\theta'}^3s_{\lambda_\odot-\Omega'}^2\frac{\partial s_{\lambda_\odot-\Omega'}}{\partial p}$$
$$+6c_{i'}\frac{\partial c_{i'}}{\partial p}s_{\theta'}^2c_{\theta'}s_{\lambda_\odot-\Omega'}^2c_{\lambda_\odot-\Omega'}+3c_{i'}^2\left(2s_{\theta'}\frac{\partial s_{\theta'}}{\partial p}c_{\theta'}+s_{\theta'}^2\frac{\partial c_{\theta'}}{\partial p}\right)s_{\lambda_\odot-\Omega'}^2c_{\lambda_\odot-\Omega'}$$

$$+ 3c_{i'}^2 s_{\theta'}^2 c_{\theta'} \left(2s_{\lambda_\odot - \Omega'} \frac{\partial s_{\lambda_\odot - \Omega'}}{\partial p} c_{\lambda_\odot - \Omega'} + s_{\lambda_\odot - \Omega'}^2 \frac{\partial c_{\lambda_\odot - \Omega'}}{\partial p} \right) \Bigg]$$
$$+ 15 \left(\frac{\partial c_{\lambda_\odot - \Omega'}}{\partial p} c_{\theta'} + c_{\lambda_\odot - \Omega'} \frac{\partial c_{\theta'}}{\partial p} + \frac{\partial c_{i'}}{\partial p} s_{\theta'} s_{\lambda_\odot - \Omega'} \right.$$
$$\left. + c_{i'} \frac{\partial s_{\theta'}}{\partial p} s_{\lambda_\odot - \Omega'} + c_{i'} s_{\theta'} \frac{\partial s_{\lambda_\odot - \Omega'}}{\partial p} \right) \tag{A16.3}$$

$$\frac{\partial \left(f_r^s\right)_4}{\partial q} = 63 \Bigg[5c_{\theta'}^4 \frac{\partial c_{\theta'}}{\partial q} c_{\lambda_\odot - \Omega'}^5 + 5c_{\theta'}^5 c_{\lambda_\odot - \Omega'}^4 \frac{\partial c_{\lambda_\odot - \Omega'}}{\partial q} + 20c_{i'}^3 \frac{\partial c_{i'}}{\partial q} s_{\theta'}^4 c_{\theta'} s_{\lambda_\odot - \Omega'}^4 c_{\lambda_\odot - \Omega'}$$
$$+ 5c_{i'}^4 \left(4s_{\theta'}^3 \frac{\partial s_{\theta'}}{\partial q} c_{\theta'} + s_{\theta'}^4 \frac{\partial c_{\theta'}}{\partial q} \right) s_{\lambda_\odot - \Omega'}^4 c_{\lambda_\odot - \Omega'}$$
$$+ 5c_{i'}^4 s_{\theta'}^4 c_{\theta'} \left(4s_{\lambda_\odot - \Omega'}^3 \frac{\partial s_{\lambda_\odot - \Omega'}}{\partial q} c_{\lambda_\odot - \Omega'} + s_{\lambda_\odot - \Omega'}^4 \frac{\partial c_{\lambda_\odot - \Omega'}}{\partial q} \right)$$
$$+ 20c_{i'} \frac{\partial c_{i'}}{\partial q} s_{\theta'}^2 c_{\theta'}^3 s_{\lambda_\odot - \Omega'}^2 c_{\lambda_\odot - \Omega'}^3 + 10c_{i'}^2 \left(2s_{\theta'} \frac{\partial s_{\theta'}}{\partial q} c_{\theta'}^3 \right.$$
$$\left. + 3c_{\theta'}^2 \frac{\partial c_{\theta'}}{\partial q} s_{\theta'}^2 \right) s_{\lambda_\odot - \Omega'}^2 c_{\lambda_\odot - \Omega'}^3 + 10c_{i'}^2 s_{\theta'}^2 c_{\theta'}^3 \left(2s_{\lambda_\odot - \Omega'} \frac{\partial s_{\lambda_\odot - \Omega'}}{\partial q} c_{\lambda_\odot - \Omega'}^3 \right.$$
$$\left. + 3c_{\lambda_\odot - \Omega'}^2 \frac{\partial c_{\lambda_\odot - \Omega'}}{\partial q} s_{\lambda_\odot - \Omega'}^2 \right) + 5\frac{\partial c_{i'}}{\partial q} s_{\theta'} c_{\theta'}^4 s_{\lambda_\odot - \Omega'} c_{\lambda_\odot - \Omega'}^4$$
$$+ 5c_{i'} \left(\frac{\partial s_{\theta'}}{\partial q} c_{\theta'}^4 + 4c_{\theta'}^3 \frac{\partial c_{\theta'}}{\partial q} s_{\theta'} \right) s_{\lambda_\odot - \Omega'} c_{\lambda_\odot - \Omega'}^4 + 5c_{i'} s_{\theta'} c_{\theta'}^4$$
$$\times \left(\frac{\partial s_{\lambda_\odot - \Omega'}}{\partial q} c_{\lambda_\odot - \Omega'}^4 + 4c_{\lambda_\odot - \Omega'}^3 \frac{\partial c_{\lambda_\odot - \Omega'}}{\partial q} s_{\lambda_\odot - \Omega'} \right) + 5c_{i'}^4 \frac{\partial c_{i'}}{\partial q} s_{\theta'}^5 s_{\lambda_\odot - \Omega'}^5$$
$$+ 5c_{i'}^5 s_{\theta'}^4 \frac{\partial s_{\theta'}}{\partial q} s_{\lambda_\odot - \Omega'}^5 + 5c_{i'}^5 s_{\theta'}^5 s_{\lambda_\odot - \Omega'}^4 \frac{\partial s_{\lambda_\odot - \Omega'}}{\partial q} + 30c_{i'}^2 \frac{\partial c_{i'}}{\partial q} s_{\theta'}^3 c_{\theta'}^2 s_{\lambda_\odot - \Omega'}^3 c_{\lambda_\odot - \Omega'}^2$$
$$+ 10c_{i'}^3 \left(3s_{\theta'}^2 \frac{\partial s_{\theta'}}{\partial q} c_{\theta'}^2 + 2c_{\theta'} \frac{\partial c_{\theta'}}{\partial q} s_{\theta'}^3 \right) s_{\lambda_\odot - \Omega'}^3 c_{\lambda_\odot - \Omega'}^2$$
$$+ 10c_{i'}^3 s_{\theta'}^3 c_{\theta'}^2 \left(3s_{\lambda_\odot - \Omega'}^2 \frac{\partial s_{\lambda_\odot - \Omega'}}{\partial q} c_{\lambda_\odot - \Omega'}^2 + 2c_{\lambda_\odot - \Omega'} \frac{\partial c_{\lambda_\odot - \Omega'}}{\partial q} s_{\lambda_\odot - \Omega'}^3 \right) \Bigg]$$
$$- 70 \Bigg[3c_{\theta'}^2 \frac{\partial c_{\theta'}}{\partial q} c_{\lambda_\odot - \Omega'}^3 + 3c_{\lambda_\odot - \Omega'}^2 \frac{\partial c_{\lambda_\odot - \Omega'}}{\partial q} c_{\theta'}^3 + 3\frac{\partial c_{i'}}{\partial q} s_{\theta'} c_{\theta'}^2 s_{\lambda_\odot - \Omega'} c_{\lambda_\odot - \Omega'}^2$$
$$+ 3c_{i'} \left(\frac{\partial s_{\theta'}}{\partial q} c_{\theta'}^2 + 2c_{\theta'} \frac{\partial c_{\theta'}}{\partial q} s_{\theta'} \right) s_{\lambda_\odot - \Omega'} c_{\lambda_\odot - \Omega'}^2$$
$$+ 3c_{i'} s_{\theta'} c_{\theta'}^2 \left(\frac{\partial s_{\lambda_\odot - \Omega'}}{\partial q} c_{\lambda_\odot - \Omega'}^2 + 2c_{\lambda_\odot - \Omega'} \frac{\partial c_{\lambda_\odot - \Omega'}}{\partial q} s_{\lambda_\odot - \Omega'} \right)$$
$$+ 3c_{i'}^2 \frac{\partial c_{i'}}{\partial q} s_{\theta'}^3 s_{\lambda_\odot - \Omega'}^3 + 3c_{i'}^3 s_{\theta'}^2 \frac{\partial s_{\theta'}}{\partial q} s_{\lambda_\odot - \Omega'}^3 + 3c_{i'}^3 s_{\theta'}^3 s_{\lambda_\odot - \Omega'}^2 \frac{\partial s_{\lambda_\odot - \Omega'}}{\partial q}$$
$$+ 6c_{i'} \frac{\partial c_{i'}}{\partial q} s_{\theta'}^2 c_{\theta'} s_{\lambda_\odot - \Omega'}^2 c_{\lambda_\odot - \Omega'} + 3c_{i'}^2 \left(2s_{\theta'} \frac{\partial s_{\theta'}}{\partial q} c_{\theta'} + s_{\theta'}^2 \frac{\partial c_{\theta'}}{\partial q} \right) s_{\lambda_\odot - \Omega'}^2 c_{\lambda_\odot - \Omega'}$$
$$+ 3c_{i'}^2 s_{\theta'}^2 c_{\theta'} \left(2s_{\lambda_\odot - \Omega'} \frac{\partial s_{\lambda_\odot - \Omega'}}{\partial q} c_{\lambda_\odot - \Omega'} + s_{\lambda_\odot - \Omega'}^2 \frac{\partial c_{\lambda_\odot - \Omega'}}{\partial q} \right) \Bigg]$$

$$+\,15\left(\frac{\partial c_{\lambda_\odot-\Omega'}}{\partial q}c_{\theta'}+c_{\lambda_\odot-\Omega'}\frac{\partial c_{\theta'}}{\partial q}+\frac{\partial c_{i'}}{\partial q}s_{\theta'}s_{\lambda_\odot-\Omega'}\right.$$
$$\left.+\,c_{i'}\frac{\partial s_{\theta'}}{\partial q}s_{\lambda_\odot-\Omega'}+c_{i'}s_{\theta'}\frac{\partial s_{\lambda_\odot-\Omega'}}{\partial q}\right) \tag{A16.4}$$

$$\frac{\partial\left(f_r^s\right)_4}{\partial L}=63\left[5c_{\theta'}^4\frac{\partial c_{\theta'}}{\partial L}c_{\lambda_\odot-\Omega'}^5+5c_{i'}^4\left(4s_{\theta'}^3\frac{\partial s_{\theta'}}{\partial L}c_{\theta'}+s_{\theta'}^4\frac{\partial c_{\theta'}}{\partial L}\right)s_{\lambda_\odot-\Omega'}^4c_{\lambda_\odot-\Omega'}\right.$$
$$+\,10c_{i'}^2\left(2s_{\theta'}\frac{\partial s_{\theta'}}{\partial L}c_{\theta'}^3+3c_{\theta'}^2\frac{\partial c_{\theta'}}{\partial L}s_{\theta'}^2\right)s_{\lambda_\odot-\Omega'}^2c_{\lambda_\odot-\Omega'}^3$$
$$+\,5c_{i'}\left(\frac{\partial s_{\theta'}}{\partial L}c_{\theta'}^4+4c_{\theta'}^3\frac{\partial c_{\theta'}}{\partial L}s_{\theta'}\right)s_{\lambda_\odot-\Omega'}c_{\lambda_\odot-\Omega'}^4$$
$$\left.+\,5c_{i'}^5s_{\theta'}^4\frac{\partial s_{\theta'}}{\partial L}s_{\lambda_\odot-\Omega'}^5+10c_{i'}^3\left(3s_{\theta'}^2\frac{\partial s_{\theta'}}{\partial L}c_{\theta'}^2\;+2c_{\theta'}\frac{\partial c_{\theta'}}{\partial L}s_{\theta'}^3\right)s_{\lambda_\odot-\Omega'}^3c_{\lambda_\odot-\Omega'}^2\right]$$
$$-\,70\left[3c_{\theta'}^2\frac{\partial c_{\theta'}}{\partial L}\;c_{\lambda_\odot-\Omega'}^3+3c_{i'}\left(\frac{\partial s_{\theta'}}{\partial L}c_{\theta'}^2+2c_{\theta'}\frac{\partial c_{\theta'}}{\partial L}s_{\theta'}\right)s_{\lambda_\odot-\Omega'}c_{\lambda_\odot-\Omega'}^2\right.$$
$$\left.+\,3c_{i'}^3s_{\theta'}^2\frac{\partial s_{\theta'}}{\partial L}s_{\lambda_\odot-\Omega'}^3\;+3c_{i'}^2\left(2s_{\theta'}\frac{\partial s_{\theta'}}{\partial L}c_{\theta'}+s_{\theta'}^2\frac{\partial c_{\theta'}}{\partial L}\right)s_{\lambda_\odot-\Omega'}^2c_{\lambda_\odot-\Omega'}\right]$$
$$+\,15\left(\frac{\partial c_{\theta'}}{\partial L}c_{\lambda_\odot-\Omega'}+c_{i'}\frac{\partial s_{\theta'}}{\partial L}s_{\lambda_\odot-\Omega'}\right) \tag{A16.5}$$

$$\frac{\partial\left(f_\theta^s\right)_4}{\partial h}=21\left[\left(\frac{-\partial s_{\theta'}}{\partial h}c_{\theta'}^4-4c_{\theta'}^3\frac{\partial c_{\theta'}}{\partial h}s_{\theta'}\right)c_{\lambda_\odot-\Omega'}^5-5c_{i'}^4s_{\theta'}^4\frac{\partial s_{\theta'}}{\partial h}s_{\lambda_\odot-\Omega'}^4c_{\lambda_\odot-\Omega'}\right.$$
$$-\,6c_{i'}^2\left(3s_{\theta'}^2\frac{\partial s_{\theta'}}{\partial h}c_{\theta'}^2\;+2c_{\theta'}\frac{\partial c_{\theta'}}{\partial h}s_{\theta'}^3\right)s_{\lambda_\odot-\Omega'}^2c_{\lambda_\odot-\Omega'}^3$$
$$-\,4c_{i'}\left(2s_{\theta'}\frac{\partial s_{\theta'}}{\partial h}c_{\theta'}^3+3c_{\theta'}^2\frac{\partial c_{\theta'}}{\partial h}s_{\theta'}^2\right)s_{\lambda_\odot-\Omega'}c_{\lambda_\odot-\Omega'}^4$$
$$-\,4c_{i'}^3\left(4s_{\theta'}^3\frac{\partial s_{\theta'}}{\partial h}c_{\theta'}+s_{\theta'}^4\frac{\partial c_{\theta'}}{\partial h}\right)s_{\lambda_\odot-\Omega'}^3c_{\lambda_\odot-\Omega'}^2$$
$$+\,5c_{i'}c_{\theta'}^4\frac{\partial c_{\theta'}}{\partial h}s_{\lambda_\odot-\Omega'}c_{\lambda_\odot-\Omega'}^4+c_{i'}^5\left(4s_{\theta'}^3\frac{\partial s_{\theta'}}{\partial h}c_{\theta'}+s_{\theta'}^4\frac{\partial c_{\theta'}}{\partial h}\right)s_{\lambda_\odot-\Omega'}^5$$
$$+\,6c_{i'}^3\left(2s_{\theta'}\frac{\partial s_{\theta'}}{\partial h}c_{\theta'}^3+3c_{\theta'}^2\frac{\partial c_{\theta'}}{\partial h}s_{\theta'}^2\right)s_{\lambda_\odot-\Omega'}^3c_{\lambda_\odot-\Omega'}^2$$
$$+\,4c_{i'}^2\left(\frac{\partial s_{\theta'}}{\partial h}c_{\theta'}^4+4c_{\theta'}^3\frac{\partial c_{\theta'}}{\partial h}s_{\theta'}\right)s_{\lambda_\odot-\Omega'}^2c_{\lambda_\odot-\Omega'}^3$$
$$\left.+\,4c_{i'}^4\left(3s_{\theta'}^2\frac{\partial s_{\theta'}}{\partial h}c_{\theta'}^2+2c_{\theta'}\frac{\partial c_{\theta'}}{\partial h}s_{\theta'}^3\right)s_{\lambda_\odot-\Omega'}^4c_{\lambda_\odot-\Omega'}\right]$$
$$-\,14\left[\left(\frac{-\partial s_{\theta'}}{\partial h}c_{\theta'}^2-2c_{\theta'}\frac{\partial c_{\theta'}}{\partial h}s_{\theta'}\right)c_{\lambda_\odot-\Omega'}^3-3c_{i'}^2s_{\theta'}^2\frac{\partial s_{\theta'}}{\partial h}s_{\lambda_\odot-\Omega'}^2c_{\lambda_\odot-\Omega'}\right.$$
$$-\,2c_{i'}\left(2s_{\theta'}\frac{\partial s_{\theta'}}{\partial h}c_{\theta'}+s_{\theta'}^2\frac{\partial c_{\theta'}}{\partial h}\right)s_{\lambda_\odot-\Omega'}c_{\lambda_\odot-\Omega'}^2+3c_{i'}c_{\theta'}^2\frac{\partial c_{\theta'}}{\partial h}s_{\lambda_\odot-\Omega'}c_{\lambda_\odot-\Omega'}^2$$
$$+\,c_{i'}^3\left(2s_{\theta'}\frac{\partial s_{\theta'}}{\partial h}c_{\theta'}+s_{\theta'}^2\frac{\partial c_{\theta'}}{\partial h}\right)s_{\lambda_\odot-\Omega'}^3$$

$$
\begin{aligned}
&+2c_{i'}^2\left(\frac{\partial s_{\theta'}}{\partial h}c_{\theta'}^2+2c_{\theta'}\frac{\partial c_{\theta'}}{\partial h}s_{\theta'}\right)s_{\lambda_\odot-\Omega'}^2c_{\lambda_\odot-\Omega'}\Bigg]\\
&+\left(-\frac{\partial s_{\theta'}}{\partial h}c_{\lambda_\odot-\Omega'}+c_{i'}\frac{\partial c_{\theta'}}{\partial h}s_{\lambda_\odot-\Omega'}\right)
\end{aligned}
\tag{A16.6}
$$

$$
\begin{aligned}
\frac{\partial\left(f_\theta^s\right)_4}{\partial k}=21\Bigg[&\left(\frac{-\partial s_{\theta'}}{\partial k}c_{\theta'}^4-4c_{\theta'}^3\frac{\partial c_{\theta'}}{\partial k}s_{\theta'}\right)c_{\lambda_\odot-\Omega'}^5-5c_{i'}^4s_{\theta'}^4\frac{\partial s_{\theta'}}{\partial k}s_{\lambda_\odot-\Omega'}^4c_{\lambda_\odot-\Omega'}\\
&-6c_{i'}^2\left(3s_{\theta'}^2\frac{\partial s_{\theta'}}{\partial k}c_{\theta'}^2+2c_{\theta'}\frac{\partial c_{\theta'}}{\partial k}s_{\theta'}^3\right)s_{\lambda_\odot-\Omega'}^2c_{\lambda_\odot-\Omega'}^3\\
&-4c_{i'}\left(2s_{\theta'}\frac{\partial s_{\theta'}}{\partial k}c_{\theta'}^3+3c_{\theta'}^2\frac{\partial c_{\theta'}}{\partial k}s_{\theta'}^2\right)s_{\lambda_\odot-\Omega'}c_{\lambda_\odot-\Omega'}^4\\
&-4c_{i'}^3\left(4s_{\theta'}^3\frac{\partial s_{\theta'}}{\partial k}c_{\theta'}+s_{\theta'}^4\frac{\partial c_{\theta'}}{\partial k}\right)s_{\lambda_\odot-\Omega'}^3c_{\lambda_\odot-\Omega'}^2\\
&+5c_{i'}c_{\theta'}^4\frac{\partial c_{\theta'}}{\partial k}s_{\lambda_\odot-\Omega'}c_{\lambda_\odot-\Omega'}^4+c_{i'}^5\left(4s_{\theta'}^3\frac{\partial s_{\theta'}}{\partial k}c_{\theta'}+s_{\theta'}^4\frac{\partial c_{\theta'}}{\partial k}\right)s_{\lambda_\odot-\Omega'}^5\\
&+6c_{i'}^3\left(2s_{\theta'}\frac{\partial s_{\theta'}}{\partial k}c_{\theta'}^3+3c_{\theta'}^2\frac{\partial c_{\theta'}}{\partial k}s_{\theta'}^2\right)s_{\lambda_\odot-\Omega'}^3c_{\lambda_\odot-\Omega'}^2\\
&+4c_{i'}^2\left(\frac{\partial s_{\theta'}}{\partial k}c_{\theta'}^4+4c_{\theta'}^3\frac{\partial c_{\theta'}}{\partial k}s_{\theta'}\right)s_{\lambda_\odot-\Omega'}^2c_{\lambda_\odot-\Omega'}^3+4c_{i'}^4\left(3s_{\theta'}^2\frac{\partial s_{\theta'}}{\partial k}c_{\theta'}^2\right.\\
&\left.+2c_{\theta'}\frac{\partial c_{\theta'}}{\partial k}s_{\theta'}^3\right)s_{\lambda_\odot-\Omega'}^4c_{\lambda_\odot-\Omega'}\Bigg]-14\Bigg[\left(\frac{-\partial s_{\theta'}}{\partial k}c_{\theta'}^2-2c_{\theta'}\frac{\partial c_{\theta'}}{\partial k}s_{\theta'}\right)c_{\lambda_\odot-\Omega'}^3\\
&-3c_{i'}^2s_{\theta'}^2\frac{\partial s_{\theta'}}{\partial k}s_{\lambda_\odot-\Omega'}^2c_{\lambda_\odot-\Omega'}-2c_{i'}\left(2s_{\theta'}\frac{\partial s_{\theta'}}{\partial k}c_{\theta'}+s_{\theta'}^2\frac{\partial c_{\theta'}}{\partial k}\right)s_{\lambda_\odot-\Omega'}c_{\lambda_\odot-\Omega'}^2\\
&+3c_{i'}c_{\theta'}^2\frac{\partial c_{\theta'}}{\partial k}s_{\lambda_\odot-\Omega'}c_{\lambda_\odot-\Omega'}^2+c_{i'}^3\left(2s_{\theta'}\frac{\partial s_{\theta'}}{\partial k}c_{\theta'}+s_{\theta'}^2\frac{\partial c_{\theta'}}{\partial k}\right)s_{\lambda_\odot-\Omega'}^3\\
&+2c_{i'}^2\left(\frac{\partial s_{\theta'}}{\partial k}c_{\theta'}^2+2c_{\theta'}\frac{\partial c_{\theta'}}{\partial k}s_{\theta'}\right)s_{\lambda_\odot-\Omega'}^2c_{\lambda_\odot-\Omega'}\Bigg]\\
&+\left(-\frac{\partial s_{\theta'}}{\partial k}c_{\lambda_\odot-\Omega'}+c_{i'}\frac{\partial c_{\theta'}}{\partial k}s_{\lambda_\odot-\Omega'}\right)
\end{aligned}
\tag{A16.7}
$$

$$
\begin{aligned}
\frac{\partial\left(f_\theta^s\right)_4}{\partial p}=21\Bigg[&\frac{-\partial s_{\theta'}}{\partial p}c_{\theta'}^4c_{\lambda_\odot-\Omega'}^5-4c_{\theta'}^3\frac{\partial c_{\theta'}}{\partial p}s_{\theta'}c_{\lambda_\odot-\Omega'}^5-5s_{\theta'}c_{\theta'}^4c_{\lambda_\odot-\Omega'}^4\frac{\partial c_{\lambda_\odot-\Omega'}}{\partial p}\\
&-4c_{i'}^3\frac{\partial c_{i'}}{\partial p}s_{\theta'}^5s_{\lambda_\odot-\Omega'}^4c_{\lambda_\odot-\Omega'}-5c_{i'}^4s_{\theta'}^4\frac{\partial s_{\theta'}}{\partial p}s_{\lambda_\odot-\Omega'}^4c_{\lambda_\odot-\Omega'}\\
&-c_{i'}^4s_{\theta'}^5\left(4s_{\lambda_\odot-\Omega'}^3\frac{\partial s_{\lambda_\odot-\Omega'}}{\partial p}c_{\lambda_\odot-\Omega'}+s_{\lambda_\odot-\Omega'}^4\frac{\partial c_{\lambda_\odot-\Omega'}}{\partial p}\right)\\
&-12c_{i'}\frac{\partial c_{i'}}{\partial p}s_{\theta'}^3c_{\theta'}^2s_{\lambda_\odot-\Omega'}^2c_{\lambda_\odot-\Omega'}^3-6c_{i'}^2\left(3s_{\theta'}^2\frac{\partial s_{\theta'}}{\partial p}c_{\theta'}^2\right.\\
&\left.+2c_{\theta'}\frac{\partial c_{\theta'}}{\partial p}s_{\theta'}^3\right)s_{\lambda_\odot-\Omega'}^2c_{\lambda_\odot-\Omega'}^3\\
&-6c_{i'}^2s_{\theta'}^3c_{\theta'}^2\left(2s_{\lambda_\odot-\Omega'}\frac{\partial s_{\lambda_\odot-\Omega'}}{\partial p}c_{\lambda_\odot-\Omega'}^3+3c_{\lambda_\odot-\Omega'}^2\frac{\partial c_{\lambda_\odot-\Omega'}}{\partial p}s_{\lambda_\odot-\Omega'}^2\right)
\end{aligned}
$$

$$
\begin{aligned}
& -4\frac{\partial c_{i'}}{\partial p}s_{\theta'}^{2}c_{\theta'}^{3}s_{\lambda_{\odot}-\Omega'}c_{\lambda_{\odot}-\Omega'}^{4}-4c_{i'}\left(2s_{\theta'}\frac{\partial s_{\theta'}}{\partial p}c_{\theta'}^{3}+3c_{\theta'}^{2}\frac{\partial c_{\theta'}}{\partial p}s_{\theta'}^{2}\right)\\
& \times s_{\lambda_{\odot}-\Omega'}c_{\lambda_{\odot}-\Omega'}^{4}-4c_{i'}s_{\theta'}^{2}c_{\theta'}^{3}\left(\frac{\partial s_{\lambda_{\odot}-\Omega'}}{\partial p}c_{\lambda_{\odot}-\Omega'}^{4}+4c_{\lambda_{\odot}-\Omega'}^{3}\frac{\partial c_{\lambda_{\odot}-\Omega'}}{\partial p}s_{\lambda_{\odot}-\Omega'}\right)\\
& -12c_{i'}^{2}\frac{\partial c_{i'}}{\partial p}s_{\theta'}^{4}c_{\theta'}s_{\lambda_{\odot}-\Omega'}^{3}c_{\lambda_{\odot}-\Omega'}^{2}-4c_{i'}^{3}\left(4s_{\theta'}^{3}\frac{\partial s_{\theta'}}{\partial p}c_{\theta'}+s_{\theta'}^{4}\frac{\partial c_{\theta'}}{\partial p}\right)\\
& \times s_{\lambda_{\odot}-\Omega'}^{3}c_{\lambda_{\odot}-\Omega'}^{2}-4c_{i'}^{3}s_{\theta'}^{4}c_{\theta'}\left(3s_{\lambda_{\odot}-\Omega'}^{2}\frac{\partial s_{\lambda_{\odot}-\Omega'}}{\partial p}c_{\lambda_{\odot}-\Omega'}^{2}\right.\\
& \left.+2c_{\lambda_{\odot}-\Omega'}\frac{\partial c_{\lambda_{\odot}-\Omega'}}{\partial p}s_{\lambda_{\odot}-\Omega'}^{3}\right)+\frac{\partial c_{i'}}{\partial p}c_{\theta'}^{5}s_{\lambda_{\odot}-\Omega'}c_{\lambda_{\odot}-\Omega'}^{4}\\
& +5c_{i'}c_{\theta'}^{4}\frac{\partial c_{\theta'}}{\partial p}s_{\lambda_{\odot}-\Omega'}c_{\lambda_{\odot}-\Omega'}^{4}+c_{i'}c_{\theta'}^{5}\left(\frac{\partial s_{\lambda_{\odot}-\Omega'}}{\partial p}c_{\lambda_{\odot}-\Omega'}^{4}\right.\\
& \left.+4c_{\lambda_{\odot}-\Omega'}^{3}\frac{\partial c_{\lambda_{\odot}-\Omega'}}{\partial p}s_{\lambda_{\odot}-\Omega'}\right)+5c_{i'}^{4}\frac{\partial c_{i'}}{\partial p}s_{\theta'}^{4}c_{\theta'}s_{\lambda_{\odot}-\Omega'}^{5}\\
& +c_{i'}^{5}\left(4s_{\theta'}^{3}\frac{\partial s_{\theta'}}{\partial p}c_{\theta'}+s_{\theta'}^{4}\frac{\partial c_{\theta'}}{\partial p}\right)s_{\lambda_{\odot}-\Omega'}^{5}+5c_{i'}^{5}s_{\theta'}^{4}c_{\theta'}s_{\lambda_{\odot}-\Omega'}^{4}\frac{\partial s_{\lambda_{\odot}-\Omega'}}{\partial p}\\
& +18c_{i'}^{2}\frac{\partial c_{i'}}{\partial p}s_{\theta'}^{2}c_{\theta'}^{3}s_{\lambda_{\odot}-\Omega'}^{3}c_{\lambda_{\odot}-\Omega'}^{2}\\
& +6c_{i'}^{3}\left(2s_{\theta'}\frac{\partial s_{\theta'}}{\partial p}c_{\theta'}^{3}+3c_{\theta'}^{2}\frac{\partial c_{\theta'}}{\partial p}s_{\theta'}^{2}\right)s_{\lambda_{\odot}-\Omega'}^{3}c_{\lambda_{\odot}-\Omega'}^{2}\\
& +6c_{i'}^{3}s_{\theta'}^{2}c_{\theta'}^{3}\left(3s_{\lambda_{\odot}-\Omega'}^{2}\frac{\partial s_{\lambda_{\odot}-\Omega'}}{\partial p}c_{\lambda_{\odot}-\Omega'}^{2}+2c_{\lambda_{\odot}-\Omega'}\frac{\partial c_{\lambda_{\odot}-\Omega'}}{\partial p}s_{\lambda_{\odot}-\Omega'}^{3}\right)\\
& +8c_{i'}\frac{\partial c_{i'}}{\partial p}s_{\theta'}c_{\theta'}^{4}s_{\lambda_{\odot}-\Omega'}^{2}c_{\lambda_{\odot}-\Omega'}^{3}+4c_{i'}^{2}\left(\frac{\partial s_{\theta'}}{\partial p}c_{\theta'}^{4}+4c_{\theta'}^{3}\frac{\partial c_{\theta'}}{\partial p}s_{\theta'}\right)s_{\lambda_{\odot}-\Omega'}^{2}c_{\lambda_{\odot}-\Omega'}^{3}\\
& +4c_{i'}^{2}s_{\theta'}c_{\theta'}^{4}\left(2s_{\lambda_{\odot}-\Omega'}\frac{\partial s_{\lambda_{\odot}-\Omega'}}{\partial p}c_{\lambda_{\odot}-\Omega'}^{3}+3c_{\lambda_{\odot}-\Omega'}^{2}\frac{\partial c_{\lambda_{\odot}-\Omega'}}{\partial p}s_{\lambda_{\odot}-\Omega'}^{2}\right)\\
& +16c_{i'}^{3}\frac{\partial c_{i'}}{\partial p}s_{\theta'}^{3}c_{\theta'}^{2}s_{\lambda_{\odot}-\Omega'}^{4}c_{\lambda_{\odot}-\Omega'}+4c_{i'}^{4}\left(3s_{\theta'}^{2}\frac{\partial s_{\theta'}}{\partial p}c_{\theta'}^{2}+2c_{\theta'}\frac{\partial c_{\theta'}}{\partial p}s_{\theta'}^{3}\right)s_{\lambda_{\odot}-\Omega'}^{4}c_{\lambda_{\odot}-\Omega'}\\
& \left.+4c_{i'}^{4}s_{\theta'}^{3}c_{\theta'}^{2}\left(4s_{\lambda_{\odot}-\Omega'}^{3}\frac{\partial s_{\lambda_{\odot}-\Omega'}}{\partial p}c_{\lambda_{\odot}-\Omega'}+s_{\lambda_{\odot}-\Omega'}^{4}\frac{\partial c_{\lambda_{\odot}-\Omega'}}{\partial p}\right)\right]\\
& -14\left[-\left(\frac{\partial s_{\theta'}}{\partial p}c_{\theta'}^{2}+2c_{\theta'}\frac{\partial c_{\theta'}}{\partial p}s_{\theta'}\right)c_{\lambda_{\odot}-\Omega'}^{3}-3s_{\theta'}c_{\theta'}^{2}c_{\lambda_{\odot}-\Omega'}^{2}\frac{\partial c_{\lambda_{\odot}-\Omega'}}{\partial p}\right.\\
& -2c_{i'}\frac{\partial c_{i'}}{\partial p}s_{\theta'}^{3}s_{\lambda_{\odot}-\Omega'}^{2}c_{\lambda_{\odot}-\Omega'}-3c_{i'}^{2}s_{\theta'}^{2}\frac{\partial s_{\theta'}}{\partial p}s_{\lambda_{\odot}-\Omega'}^{2}c_{\lambda_{\odot}-\Omega'}\\
& -c_{i'}^{2}s_{\theta'}^{3}\left(2s_{\lambda_{\odot}-\Omega'}\frac{\partial s_{\lambda_{\odot}-\Omega'}}{\partial p}c_{\lambda_{\odot}-\Omega'}+s_{\lambda_{\odot}-\Omega'}^{2}\frac{\partial c_{\lambda_{\odot}-\Omega'}}{\partial p}\right)-2\frac{\partial c_{i'}}{\partial p}s_{\theta'}^{2}c_{\theta'}s_{\lambda_{\odot}-\Omega'}c_{\lambda_{\odot}-\Omega'}^{2}\\
& -2c_{i'}\left(2s_{\theta'}\frac{\partial s_{\theta'}}{\partial p}c_{\theta'}+s_{\theta'}^{2}\frac{\partial c_{\theta'}}{\partial p}\right)s_{\lambda_{\odot}-\Omega'}c_{\lambda_{\odot}-\Omega'}^{2}\\
& -2c_{i'}s_{\theta'}^{2}c_{\theta'}\left(\frac{\partial s_{\lambda_{\odot}-\Omega'}}{\partial p}c_{\lambda_{\odot}-\Omega'}^{2}+2c_{\lambda_{\odot}-\Omega'}\frac{\partial c_{\lambda_{\odot}-\Omega'}}{\partial p}s_{\lambda_{\odot}-\Omega'}\right)\\
& +\frac{\partial c_{i'}}{\partial p}c_{\theta'}^{3}s_{\lambda_{\odot}-\Omega'}c_{\lambda_{\odot}-\Omega'}^{2}+3c_{i'}c_{\theta'}^{2}\frac{\partial c_{\theta'}}{\partial p}s_{\lambda_{\odot}-\Omega'}c_{\lambda_{\odot}-\Omega'}^{2}
\end{aligned}
$$

$$
\begin{aligned}
& + c_{i'}c_{\theta'}^3\left(\frac{\partial s_{\lambda_\odot-\Omega'}}{\partial p}c_{\lambda_\odot-\Omega'}^2 + 2c_{\lambda_\odot-\Omega'}\frac{\partial c_{\lambda_\odot-\Omega'}}{\partial p}s_{\lambda_\odot-\Omega'}\right) \\
& + 3c_{i'}^2\frac{\partial c_{i'}}{\partial p}s_{\theta'}^2c_{\theta'}s_{\lambda_\odot-\Omega'}^3 + c_{i'}^3\left(2s_{\theta'}\frac{\partial s_{\theta'}}{\partial p}c_{\theta'} + s_{\theta'}^2\frac{\partial c_{\theta'}}{\partial p}\right)s_{\lambda_\odot-\Omega'}^3 \\
& + 3c_{i'}^3s_{\theta'}^2c_{\theta'}s_{\lambda_\odot-\Omega'}^2\frac{\partial s_{\lambda_\odot-\Omega'}}{\partial p} + 4c_{i'}\frac{\partial c_{i'}}{\partial p}s_{\theta'}c_{\theta'}^2s_{\lambda_\odot-\Omega'}^2c_{\lambda_\odot-\Omega'} \\
& + 2c_{i'}^2\left(\frac{\partial s_{\theta'}}{\partial p}c_{\theta'}^2 + 2c_{\theta'}\frac{\partial c_{\theta'}}{\partial p}s_{\theta'}\right)s_{\lambda_\odot-\Omega'}^2c_{\lambda_\odot-\Omega'} \\
& + 2c_{i'}^2s_{\theta'}c_{\theta'}^2\left(2s_{\lambda_\odot-\Omega'}\frac{\partial s_{\lambda_\odot-\Omega'}}{\partial p}c_{\lambda_\odot-\Omega'} + s_{\lambda_\odot-\Omega'}^2\frac{\partial c_{\lambda_\odot-\Omega'}}{\partial p}\right)\Bigg] \\
& + \left(-\frac{\partial s_{\theta'}}{\partial p}c_{\lambda_\odot-\Omega'} - s_{\theta'}\frac{\partial c_{\lambda_\odot-\Omega'}}{\partial p} + \frac{\partial c_{i'}}{\partial p}c_{\theta'}s_{\lambda_\odot-\Omega'}\right. \\
& \left. + c_{i'}\frac{\partial c_{\theta'}}{\partial p}s_{\lambda_\odot-\Omega'} + c_{i'}c_{\theta'}\frac{\partial s_{\lambda_\odot-\Omega'}}{\partial p}\right) \qquad \text{(A16.8)}
\end{aligned}
$$

$$
\begin{aligned}
\frac{\partial\left(f_\theta^s\right)_4}{\partial q} = {} & 21\Bigg[\frac{-\partial s_{\theta'}}{\partial q}c_{\theta'}^4c_{\lambda_\odot-\Omega'}^5 - 4c_{\theta'}^3\frac{\partial c_{\theta'}}{\partial q}s_{\theta'}c_{\lambda_\odot-\Omega'}^5 \\
& - 5s_{\theta'}c_{\theta'}^4c_{\lambda_\odot-\Omega'}^4\frac{\partial c_{\lambda_\odot-\Omega'}}{\partial q} - 4c_{i'}^3\frac{\partial c_{i'}}{\partial q}s_{\theta'}^5s_{\lambda_\odot-\Omega'}^4c_{\lambda_\odot-\Omega'} \\
& - 5c_{i'}^4s_{\theta'}^4\frac{\partial s_{\theta'}}{\partial q}s_{\lambda_\odot-\Omega'}^4c_{\lambda_\odot-\Omega'} - c_{i'}^4s_{\theta'}^5\left(4s_{\lambda_\odot-\Omega'}^3\frac{\partial s_{\lambda_\odot-\Omega'}}{\partial q}c_{\lambda_\odot-\Omega'} + s_{\lambda_\odot-\Omega'}^4\frac{\partial c_{\lambda_\odot-\Omega'}}{\partial q}\right) \\
& - 12c_{i'}\frac{\partial c_{i'}}{\partial q}s_{\theta'}^3c_{\theta'}^2s_{\lambda_\odot-\Omega'}^2c_{\lambda_\odot-\Omega'}^3 - 6c_{i'}^2\left(3s_{\theta'}^2\frac{\partial s_{\theta'}}{\partial q}c_{\theta'}^2 + 2c_{\theta'}\frac{\partial c_{\theta'}}{\partial q}s_{\theta'}^3\right)s_{\lambda_\odot-\Omega'}^2c_{\lambda_\odot-\Omega'}^3 \\
& - 6c_{i'}^2s_{\theta'}^3c_{\theta'}^2\left(2s_{\lambda_\odot-\Omega'}\frac{\partial s_{\lambda_\odot-\Omega'}}{\partial q}c_{\lambda_\odot-\Omega'}^3 + 3c_{\lambda_\odot-\Omega'}^2\frac{\partial c_{\lambda_\odot-\Omega'}}{\partial q}s_{\lambda_\odot-\Omega'}^2\right) \\
& - 4\frac{\partial c_{i'}}{\partial q}s_{\theta'}^2c_{\theta'}^3s_{\lambda_\odot-\Omega'}c_{\lambda_\odot-\Omega'}^4 - 4c_{i'}\left(2s_{\theta'}\frac{\partial s_{\theta'}}{\partial q}c_{\theta'}^3 + 3c_{\theta'}^2\frac{\partial c_{\theta'}}{\partial q}s_{\theta'}^2\right)s_{\lambda_\odot-\Omega'}c_{\lambda_\odot-\Omega'}^4 \\
& - 4c_{i'}s_{\theta'}^2c_{\theta'}^3\left(\frac{\partial s_{\lambda_\odot-\Omega'}}{\partial q}c_{\lambda_\odot-\Omega'}^4 + 4c_{\lambda_\odot-\Omega'}^3\frac{\partial c_{\lambda_\odot-\Omega'}}{\partial q}s_{\lambda_\odot-\Omega'}\right) \\
& - 12c_{i'}^2\frac{\partial c_{i'}}{\partial q}s_{\theta'}^4c_{\theta'}s_{\lambda_\odot-\Omega'}^3c_{\lambda_\odot-\Omega'}^2 - 4c_{i'}^3\left(4s_{\theta'}^3\frac{\partial s_{\theta'}}{\partial q}c_{\theta'} + s_{\theta'}^4\frac{\partial c_{\theta'}}{\partial q}\right) \\
& \times s_{\lambda_\odot-\Omega'}^3c_{\lambda_\odot-\Omega'}^2 - 4c_{i'}^3s_{\theta'}^4c_{\theta'}\left(3s_{\lambda_\odot-\Omega'}^2\frac{\partial s_{\lambda_\odot-\Omega'}}{\partial q}c_{\lambda_\odot-\Omega'}^2\right. \\
& \left. + 2c_{\lambda_\odot-\Omega'}\frac{\partial c_{\lambda_\odot-\Omega'}}{\partial q}s_{\lambda_\odot-\Omega'}^3\right) + \frac{\partial c_{i'}}{\partial q}c_{\theta'}^5s_{\lambda_\odot-\Omega'}c_{\lambda_\odot-\Omega'}^4 \\
& + 5c_{i'}c_{\theta'}^4\frac{\partial c_{\theta'}}{\partial q}s_{\lambda_\odot-\Omega'}c_{\lambda_\odot-\Omega'}^4 + c_{i'}c_{\theta'}^5\left(\frac{\partial s_{\lambda_\odot-\Omega'}}{\partial q}c_{\lambda_\odot-\Omega'}^4\right. \\
& \left. + 4c_{\lambda_\odot-\Omega'}^3\frac{\partial c_{\lambda_\odot-\Omega'}}{\partial q}s_{\lambda_\odot-\Omega'}\right) + 5c_{i'}^4\frac{\partial c_{i'}}{\partial q}s_{\theta'}^4c_{\theta'}s_{\lambda_\odot-\Omega'}^5 \\
& + c_{i'}^5\left(4s_{\theta'}^3\frac{\partial s_{\theta'}}{\partial q}c_{\theta'} + s_{\theta'}^4\frac{\partial c_{\theta'}}{\partial q}\right)s_{\lambda_\odot-\Omega'}^5 \\
& + 5c_{i'}^5s_{\theta'}^4c_{\theta'}s_{\lambda_\odot-\Omega'}^4\frac{\partial s_{\lambda_\odot-\Omega'}}{\partial q} + 18c_{i'}^2\frac{\partial c_{i'}}{\partial q}s_{\theta'}^2c_{\theta'}^3s_{\lambda_\odot-\Omega'}^3c_{\lambda_\odot-\Omega'}^2
\end{aligned}
$$

$$
\begin{aligned}
&+6c_{i'}^3\left(2s_{\theta'}\frac{\partial s_{\theta'}}{\partial q}c_{\theta'}^3+3c_{\theta'}^2\frac{\partial c_{\theta'}}{\partial q}s_{\theta'}^2\right)s_{\lambda_\odot-\Omega'}^3c_{\lambda_\odot-\Omega'}^2\\
&+6c_{i'}^3s_{\theta'}^2c_{\theta'}^3\left(3s_{\lambda_\odot-\Omega'}^2\frac{\partial s_{\lambda_\odot-\Omega'}}{\partial q}c_{\lambda_\odot-\Omega'}^2\right.\\
&\left.+2c_{\lambda_\odot-\Omega'}\frac{\partial c_{\lambda_\odot-\Omega'}}{\partial q}s_{\lambda_\odot-\Omega'}^3\right)+8c_{i'}\frac{\partial c_{i'}}{\partial q}s_{\theta'}c_{\theta'}^4s_{\lambda_\odot-\Omega'}^2c_{\lambda_\odot-\Omega'}^3\\
&+4c_{i'}^2\left(\frac{\partial s_{\theta'}}{\partial q}c_{\theta'}^4+4c_{\theta'}^3\frac{\partial c_{\theta'}}{\partial q}s_{\theta'}\right)s_{\lambda_\odot-\Omega'}^2c_{\lambda_\odot-\Omega'}^3\\
&+4c_{i'}^2s_{\theta'}c_{\theta'}^4\left(2s_{\lambda_\odot-\Omega'}\frac{\partial s_{\lambda_\odot-\Omega'}}{\partial q}c_{\lambda_\odot-\Omega'}^3+3c_{\lambda_\odot-\Omega'}^2\frac{\partial c_{\lambda_\odot-\Omega'}}{\partial q}s_{\lambda_\odot-\Omega'}^2\right)\\
&+16c_{i'}^3\frac{\partial c_{i'}}{\partial q}s_{\theta'}^3c_{\theta'}^2s_{\lambda_\odot-\Omega'}^4c_{\lambda_\odot-\Omega'}+4c_{i'}^4\left(3s_{\theta'}^2\frac{\partial s_{\theta'}}{\partial q}c_{\theta'}^2+2c_{\theta'}\frac{\partial c_{\theta'}}{\partial q}s_{\theta'}^3\right)\\
&\times s_{\lambda_\odot-\Omega'}^4c_{\lambda_\odot-\Omega'}+4c_{i'}^4s_{\theta'}^3c_{\theta'}^2\left(4s_{\lambda_\odot-\Omega'}^3\frac{\partial s_{\lambda_\odot-\Omega'}}{\partial q}c_{\lambda_\odot-\Omega'}+s_{\lambda_\odot-\Omega'}^4\frac{\partial c_{\lambda_\odot-\Omega'}}{\partial q}\right)\Bigg]\\
&-14\Bigg[-\left(\frac{\partial s_{\theta'}}{\partial q}c_{\theta'}^2+2c_{\theta'}\frac{\partial c_{\theta'}}{\partial q}s_{\theta'}\right)c_{\lambda_\odot-\Omega'}^3-3s_{\theta'}c_{\theta'}^2c_{\lambda_\odot-\Omega'}^2\frac{\partial c_{\lambda_\odot-\Omega'}}{\partial q}\\
&-2c_{i'}\frac{\partial c_{i'}}{\partial q}s_{\theta'}^3s_{\lambda_\odot-\Omega'}^2c_{\lambda_\odot-\Omega'}-3c_{i'}^2s_{\theta'}^2\frac{\partial s_{\theta'}}{\partial q}s_{\lambda_\odot-\Omega'}^2c_{\lambda_\odot-\Omega'}\\
&-c_{i'}^2s_{\theta'}^3\left(2s_{\lambda_\odot-\Omega'}\frac{\partial s_{\lambda_\odot-\Omega'}}{\partial q}c_{\lambda_\odot-\Omega'}+s_{\lambda_\odot-\Omega'}^2\frac{\partial c_{\lambda_\odot-\Omega'}}{\partial q}\right)\\
&-2\frac{\partial c_{i'}}{\partial q}s_{\theta'}^2c_{\theta'}s_{\lambda_\odot-\Omega'}c_{\lambda_\odot-\Omega'}^2-2c_{i'}\left(2s_{\theta'}\frac{\partial s_{\theta'}}{\partial q}c_{\theta'}+s_{\theta'}^2\frac{\partial c_{\theta'}}{\partial q}\right)s_{\lambda_\odot-\Omega'}c_{\lambda_\odot-\Omega'}^2\\
&-2c_{i'}s_{\theta'}^2c_{\theta'}\left(\frac{\partial s_{\lambda_\odot-\Omega'}}{\partial q}c_{\lambda_\odot-\Omega'}^2+2c_{\lambda_\odot-\Omega'}\frac{\partial c_{\lambda_\odot-\Omega'}}{\partial q}s_{\lambda_\odot-\Omega'}\right)\\
&+\frac{\partial c_{i'}}{\partial q}c_{\theta'}^3s_{\lambda_\odot-\Omega'}c_{\lambda_\odot-\Omega'}^2+3c_{i'}c_{\theta'}^2\frac{\partial c_{\theta'}}{\partial q}s_{\lambda_\odot-\Omega'}c_{\lambda_\odot-\Omega'}^2\\
&+c_{i'}c_{\theta'}^3\left(\frac{\partial s_{\lambda_\odot-\Omega'}}{\partial q}c_{\lambda_\odot-\Omega'}^2+2c_{\lambda_\odot-\Omega'}\frac{\partial c_{\lambda_\odot-\Omega'}}{\partial q}s_{\lambda_\odot-\Omega'}\right)\\
&+3c_{i'}^2\frac{\partial c_{i'}}{\partial q}s_{\theta'}^2c_{\theta'}s_{\lambda_\odot-\Omega'}^3+c_{i'}^3\left(2s_{\theta'}\frac{\partial s_{\theta'}}{\partial q}c_{\theta'}+s_{\theta'}^2\frac{\partial c_{\theta'}}{\partial q}\right)s_{\lambda_\odot-\Omega'}^3\\
&+3c_{i'}^3s_{\theta'}^2c_{\theta'}s_{\lambda_\odot-\Omega'}^2\frac{\partial s_{\lambda_\odot-\Omega'}}{\partial q}+4c_{i'}\frac{\partial c_{i'}}{\partial q}s_{\theta'}c_{\theta'}^2s_{\lambda_\odot-\Omega'}^2c_{\lambda_\odot-\Omega'}\\
&+2c_{i'}^2\left(\frac{\partial s_{\theta'}}{\partial q}c_{\theta'}^2+2c_{\theta'}\frac{\partial c_{\theta'}}{\partial q}s_{\theta'}\right)s_{\lambda_\odot-\Omega'}^2c_{\lambda_\odot-\Omega'}\\
&+2c_{i'}^2s_{\theta'}c_{\theta'}^2\left(2s_{\lambda_\odot-\Omega'}\frac{\partial s_{\lambda_\odot-\Omega'}}{\partial q}c_{\lambda_\odot-\Omega'}+s_{\lambda_\odot-\Omega'}^2\frac{\partial c_{\lambda_\odot-\Omega'}}{\partial q}\right)\Bigg]+\left(-\frac{\partial s_{\theta'}}{\partial q}c_{\lambda_\odot-\Omega'}\right.\\
&\left.-s_{\theta'}\frac{\partial c_{\lambda_\odot-\Omega'}}{\partial q}+\frac{\partial c_{i'}}{\partial q}c_{\theta'}s_{\lambda_\odot-\Omega'}+c_{i'}\frac{\partial c_{\theta'}}{\partial q}s_{\lambda_\odot-\Omega'}+c_{i'}c_{\theta'}\frac{\partial s_{\lambda_\odot-\Omega'}}{\partial q}\right)
\end{aligned}
\tag{A16.9}
$$

$$
\begin{aligned}
\frac{\partial\left(f_\theta^s\right)_4}{\partial L}=21\Bigg[&-\left(\frac{\partial s_{\theta'}}{\partial L}c_{\theta'}^4+4c_{\theta'}^3\frac{\partial c_{\theta'}}{\partial L}s_{\theta'}\right)c_{\lambda_\odot-\Omega'}^5-5c_{i'}^4s_{\theta'}^4\frac{\partial s_{\theta'}}{\partial L}s_{\lambda_\odot-\Omega'}^4c_{\lambda_\odot-\Omega'}\\
&-6c_{i'}^2\left(3s_{\theta'}^2\frac{\partial s_{\theta'}}{\partial L}c_{\theta'}^2+2c_{\theta'}\frac{\partial c_{\theta'}}{\partial L}s_{\theta'}^3\right)s_{\lambda_\odot-\Omega'}^2c_{\lambda_\odot-\Omega'}^3
\end{aligned}
$$

$$
\begin{aligned}
&- 4c_{i'}\left(2s_{\theta'}\frac{\partial s_{\theta'}}{\partial L}c_{\theta'}^3 + 3c_{\theta'}^2\frac{\partial c_{\theta'}}{\partial L}s_{\theta'}^2\right)s_{\lambda_\odot-\Omega'}c_{\lambda_\odot-\Omega'}^4 \\
&- 4c_{i'}^3\left(4s_{\theta'}^3\frac{\partial s_{\theta'}}{\partial L}c_{\theta'} + s_{\theta'}^4\frac{\partial c_{\theta'}}{\partial L}\right)s_{\lambda_\odot-\Omega'}^3c_{\lambda_\odot-\Omega'}^2 \\
&+ 5c_{i'}c_{\theta'}^4\frac{\partial c_{\theta'}}{\partial L}s_{\lambda_\odot-\Omega'}c_{\lambda_\odot-\Omega'}^4 + c_{i'}^5\left(4s_{\theta'}^3\frac{\partial s_{\theta'}}{\partial L}c_{\theta'} + s_{\theta'}^4\frac{\partial c_{\theta'}}{\partial L}\right)s_{\lambda_\odot-\Omega'}^5 \\
&+ 6c_{i'}^3\left(2s_{\theta'}\frac{\partial s_{\theta'}}{\partial L}c_{\theta'}^3 + 3c_{\theta'}^2\frac{\partial c_{\theta'}}{\partial L}s_{\theta'}^2\right)s_{\lambda_\odot-\Omega'}^3c_{\lambda_\odot-\Omega'}^2 \\
&+ 4c_{i'}^2\left(\frac{\partial s_{\theta'}}{\partial L}c_{\theta'}^4 + 4c_{\theta'}^3\frac{\partial c_{\theta'}}{\partial L}s_{\theta'}\right)s_{\lambda_\odot-\Omega'}^2c_{\lambda_\odot-\Omega'}^3 \\
&+ 4c_{i'}^4\left(3s_{\theta'}^2\frac{\partial s_{\theta'}}{\partial L}c_{\theta'}^2 + 2c_{\theta'}\frac{\partial c_{\theta'}}{\partial L}s_{\theta'}^3\right)s_{\lambda_\odot-\Omega'}^4c_{\lambda_\odot-\Omega'}\Bigg] \\
&- 14\Bigg[-\left(\frac{\partial s_{\theta'}}{\partial L}c_{\theta'}^2 + 2c_{\theta'}\frac{\partial c_{\theta'}}{\partial L}s_{\theta'}\right)c_{\lambda_\odot-\Omega'}^3 - 3c_{i'}^2s_{\theta'}^2\frac{\partial s_{\theta'}}{\partial L}s_{\lambda_\odot-\Omega'}^2c_{\lambda_\odot-\Omega'} \\
&- 2c_{i'}\left(2s_{\theta'}\frac{\partial s_{\theta'}}{\partial L}c_{\theta'} + s_{\theta'}^2\frac{\partial c_{\theta'}}{\partial L}\right)s_{\lambda_\odot-\Omega'}c_{\lambda_\odot-\Omega'}^2 \\
&+ 3c_{i'}c_{\theta'}^2\frac{\partial c_{\theta'}}{\partial L}s_{\lambda_\odot-\Omega'}c_{\lambda_\odot-\Omega'}^2 + c_{i'}^3\left(2s_{\theta'}\frac{\partial s_{\theta'}}{\partial L}c_{\theta'} + s_{\theta'}^2\frac{\partial c_{\theta'}}{\partial L}\right)s_{\lambda_\odot-\Omega'}^3 \\
&+ 2c_{i'}^2\left(\frac{\partial s_{\theta'}}{\partial L}c_{\theta'}^2 + 2c_{\theta'}\frac{\partial c_{\theta'}}{\partial L}s_{\theta'}\right)s_{\lambda_\odot-\Omega'}^2c_{\lambda_\odot-\Omega'}\Bigg] \\
&+ \left(-\frac{\partial s_{\theta'}}{\partial L}c_{\lambda_\odot-\Omega'} + c_{i'}\frac{\partial c_{\theta'}}{\partial L}s_{\lambda_\odot-\Omega'}\right)
\end{aligned}
\tag{A16.10}
$$

$$
\begin{aligned}
\frac{\partial\left(f_h^s\right)_4}{\partial h} &= -84s_{i'}c_{\theta'}^3\frac{\partial c_{\theta'}}{\partial h}s_{\lambda_\odot-\Omega'}c_{\lambda_\odot-\Omega'}^4 - 84s_{i'}c_{i'}^4s_{\theta'}^3\frac{\partial s_{\theta'}}{\partial h}s_{\lambda_\odot-\Omega'}^5 \\
&- 252s_{i'}c_{i'}^2\left(s_{\theta'}\frac{\partial s_{\theta'}}{\partial h}c_{\theta'}^2 + s_{\theta'}^2c_{\theta'}\frac{\partial c_{\theta'}}{\partial h}\right)s_{\lambda_\odot-\Omega'}^3c_{\lambda_\odot-\Omega'}^2 \\
&- 84s_{i'}c_{i'}\left(\frac{\partial s_{\theta'}}{\partial h}c_{\theta'}^3 + 3s_{\theta'}c_{\theta'}^2\frac{\partial c_{\theta'}}{\partial h}\right)s_{\lambda_\odot-\Omega'}^2c_{\lambda_\odot-\Omega'}^3 \\
&- 84s_{i'}c_{i'}^3\left(3s_{\theta'}^2\frac{\partial s_{\theta'}}{\partial h}c_{\theta'} + s_{\theta'}^3\frac{\partial c_{\theta'}}{\partial h}\right)s_{\lambda_\odot-\Omega'}^4c_{\lambda_\odot-\Omega'} \\
&+ 28s_{i'}c_{\theta'}\frac{\partial c_{\theta'}}{\partial h}s_{\lambda_\odot-\Omega'}c_{\lambda_\odot-\Omega'}^2 + 28s_{i'}c_{i'}^2s_{\theta'}\frac{\partial s_{\theta'}}{\partial h}s_{\lambda_\odot-\Omega'}^3 \\
&+ 28s_{i'}c_{i'}\left(\frac{\partial s_{\theta'}}{\partial h}c_{\theta'} + s_{\theta'}\frac{\partial c_{\theta'}}{\partial h}\right)s_{\lambda_\odot-\Omega'}^2c_{\lambda_\odot-\Omega'}
\end{aligned}
\tag{A16.11}
$$

$$
\begin{aligned}
\frac{\partial\left(f_h^s\right)_4}{\partial k} &= -84s_{i'}c_{\theta'}^3\frac{\partial c_{\theta'}}{\partial k}s_{\lambda_\odot-\Omega'}c_{\lambda_\odot-\Omega'}^4 - 84s_{i'}c_{i'}^4s_{\theta'}^3\frac{\partial s_{\theta'}}{\partial k}s_{\lambda_\odot-\Omega'}^5 \\
&- 252s_{i'}c_{i'}^2\left(s_{\theta'}\frac{\partial s_{\theta'}}{\partial k}c_{\theta'}^2 + s_{\theta'}^2c_{\theta'}\frac{\partial c_{\theta'}}{\partial k}\right)s_{\lambda_\odot-\Omega'}^3c_{\lambda_\odot-\Omega'}^2 \\
&- 84s_{i'}c_{i'}\left(\frac{\partial s_{\theta'}}{\partial k}c_{\theta'}^3 + 3s_{\theta'}c_{\theta'}^2\frac{\partial c_{\theta'}}{\partial k}\right)s_{\lambda_\odot-\Omega'}^2c_{\lambda_\odot-\Omega'}^3
\end{aligned}
$$

$$
\begin{aligned}
&- 84 s_{i'} c_{i'}^3 \left(3 s_{\theta'}^2 \frac{\partial s_{\theta'}}{\partial k} c_{\theta'} + s_{\theta'}^3 \frac{\partial c_{\theta'}}{\partial k} \right) s_{\lambda_\odot-\Omega'}^4 c_{\lambda_\odot-\Omega'} \\
&+ 28 s_{i'} c_{\theta'} \frac{\partial c_{\theta'}}{\partial k} s_{\lambda_\odot-\Omega'} c_{\lambda_\odot-\Omega'}^2 + 28 s_{i'} c_{i'}^2 s_{\theta'} \frac{\partial s_{\theta'}}{\partial k} s_{\lambda_\odot-\Omega'}^3 \\
&+ 28 s_{i'} c_{i'} \left(\frac{\partial s_{\theta'}}{\partial k} c_{\theta'} + s_{\theta'} \frac{\partial c_{\theta'}}{\partial k} \right) s_{\lambda_\odot-\Omega'}^2 c_{\lambda_\odot-\Omega'}
\end{aligned} \tag{A16.12}
$$

$$
\begin{aligned}
\frac{\partial \left(f_h^s\right)_4}{\partial p} = &-21 \frac{\partial s_{i'}}{\partial p} c_{\theta'}^4 s_{\lambda_\odot-\Omega'} c_{\lambda_\odot-\Omega'}^4 - 84 s_{i'} c_{\theta'}^3 \frac{\partial c_{\theta'}}{\partial p} s_{\lambda_\odot-\Omega'} c_{\lambda_\odot-\Omega'}^4 \\
&- 21 s_{i'} c_{\theta'}^4 \left(\frac{\partial s_{\lambda_\odot-\Omega'}}{\partial p} c_{\lambda_\odot-\Omega'}^4 + 4 s_{\lambda_\odot-\Omega'} c_{\lambda_\odot-\Omega'}^3 \frac{\partial c_{\lambda_\odot-\Omega'}}{\partial p} \right) \\
&- 21 \left(\frac{\partial s_{i'}}{\partial p} c_{i'}^4 + 4 s_{i'} c_{i'}^3 \frac{\partial c_{i'}}{\partial p} \right) s_{\theta'}^4 s_{\lambda_\odot-\Omega'}^5 - 84 s_{i'} c_{i'}^4 s_{\theta'}^3 \frac{\partial s_{\theta'}}{\partial p} s_{\lambda_\odot-\Omega'}^5 \\
&- 105 s_{i'} c_{i'}^4 s_{\theta'}^4 s_{\lambda_\odot-\Omega'}^4 \frac{\partial s_{\lambda_\odot-\Omega'}}{\partial p} - 126 \left(\frac{\partial s_{i'}}{\partial p} c_{i'}^2 + 2 s_{i'} c_{i'} \frac{\partial c_{i'}}{\partial p} \right) \\
&\times s_{\theta'}^2 c_{\theta'}^2 s_{\lambda_\odot-\Omega'}^3 c_{\lambda_\odot-\Omega'}^2 - 252 s_{i'} c_{i'}^2 \left(s_{\theta'} \frac{\partial s_{\theta'}}{\partial p} c_{\theta'}^2 + s_{\theta'}^2 c_{\theta'} \frac{\partial c_{\theta'}}{\partial p} \right) s_{\lambda_\odot-\Omega'}^3 c_{\lambda_\odot-\Omega'}^2 \\
&- 126 s_{i'} c_{i'}^2 s_{\theta'}^2 c_{\theta'}^2 \left(3 s_{\lambda_\odot-\Omega'}^2 \frac{\partial s_{\lambda_\odot-\Omega'}}{\partial p} c_{\lambda_\odot-\Omega'}^2 + 2 s_{\lambda_\odot-\Omega'}^3 c_{\lambda_\odot-\Omega'} \frac{\partial c_{\lambda_\odot-\Omega'}}{\partial p} \right) \\
&- 84 \left(\frac{\partial s_{i'}}{\partial p} c_{i'} + s_{i'} \frac{\partial c_{i'}}{\partial p} \right) s_{\theta'}^2 c_{\theta'}^3 s_{\lambda_\odot-\Omega'}^2 c_{\lambda_\odot-\Omega'}^3 \\
&- 84 s_{i'} c_{i'} \left(\frac{\partial s_{\theta'}}{\partial p} c_{\theta'}^3 + 3 s_{\theta'} c_{\theta'}^2 \frac{\partial c_{\theta'}}{\partial p} \right) s_{\lambda_\odot-\Omega'}^2 c_{\lambda_\odot-\Omega'}^3 \\
&- 84 s_{i'} c_{i'} s_{\theta'} c_{\theta'}^3 \left(2 s_{\lambda_\odot-\Omega'} \frac{\partial s_{\lambda_\odot-\Omega'}}{\partial p} c_{\lambda_\odot-\Omega'}^3 + 3 s_{\lambda_\odot-\Omega'}^2 c_{\lambda_\odot-\Omega'}^2 \frac{\partial c_{\lambda_\odot-\Omega'}}{\partial p} \right) \\
&- 84 \left(\frac{\partial s_{i'}}{\partial p} c_{i'}^3 + 3 s_{i'} c_{i'}^2 \frac{\partial c_{i'}}{\partial p} \right) s_{\theta'}^3 c_{\theta'} s_{\lambda_\odot-\Omega'}^4 c_{\lambda_\odot-\Omega'} \\
&- 84 s_{i'} c_{i'}^3 \left(3 s_{\theta'}^2 \frac{\partial s_{\theta'}}{\partial p} c_{\theta'} + s_{\theta'}^3 \frac{\partial c_{\theta'}}{\partial p} \right) s_{\lambda_\odot-\Omega'}^4 c_{\lambda_\odot-\Omega'} \\
&- 84 s_{i'} c_{i'}^3 s_{\theta'}^3 c_{\theta'} \left(4 s_{\lambda_\odot-\Omega'}^3 \frac{\partial s_{\lambda_\odot-\Omega'}}{\partial p} c_{\lambda_\odot-\Omega'} + s_{\lambda_\odot-\Omega'}^4 \frac{\partial c_{\lambda_\odot-\Omega'}}{\partial p} \right) \\
&+ 14 \frac{\partial s_{i'}}{\partial p} c_{\theta'}^2 s_{\lambda_\odot-\Omega'} c_{\lambda_\odot-\Omega'}^2 + 28 s_{i'} c_{\theta'} \frac{\partial c_{\theta'}}{\partial p} s_{\lambda_\odot-\Omega'} c_{\lambda_\odot-\Omega'}^2 \\
&+ 14 s_{i'} c_{\theta'}^2 \left(\frac{\partial s_{\lambda_\odot-\Omega'}}{\partial p} c_{\lambda_\odot-\Omega'}^2 + 2 s_{\lambda_\odot-\Omega'} c_{\lambda_\odot-\Omega'} \frac{\partial c_{\lambda_\odot-\Omega'}}{\partial p} \right) \\
&+ 14 \left(\frac{\partial s_{i'}}{\partial p} c_{i'}^2 + 2 s_{i'} c_{i'} \frac{\partial c_{i'}}{\partial p} \right) s_{\theta'}^2 s_{\lambda_\odot-\Omega'}^3 + 28 s_{i'} c_{i'}^2 s_{\theta'} \frac{\partial s_{\theta'}}{\partial p} s_{\lambda_\odot-\Omega'}^3 \\
&+ 42 s_{i'} c_{i'}^2 s_{\theta'}^2 s_{\lambda_\odot-\Omega'}^2 \frac{\partial s_{\lambda_\odot-\Omega'}}{\partial p} + 28 \left(\frac{\partial s_{i'}}{\partial p} c_{i'} + s_{i'} \frac{\partial c_{i'}}{\partial p} \right) s_{\theta'} c_{\theta'} s_{\lambda_\odot-\Omega'}^2 c_{\lambda_\odot-\Omega'} \\
&+ 28 s_{i'} c_{i'} \left(\frac{\partial s_{\theta'}}{\partial p} c_{\theta'} + s_{\theta'} \frac{\partial c_{\theta'}}{\partial p} \right) s_{\lambda_\odot-\Omega'}^2 c_{\lambda_\odot-\Omega'}
\end{aligned}
$$

$$+ 28 s_{i'} c_{i'} s_{\theta'} c_{\theta'} \left(2 s_{\lambda_\odot - \Omega'} \frac{\partial s_{\lambda_\odot - \Omega'}}{\partial p} c_{\lambda_\odot - \Omega'} + s^2_{\lambda_\odot - \Omega'} \frac{\partial c_{\lambda_\odot - \Omega'}}{\partial p} \right)$$
$$- \frac{\partial s_{i'}}{\partial p} s_{\lambda_\odot - \Omega'} - s_{i'} \frac{\partial s_{\lambda_\odot - \Omega'}}{\partial p} \tag{A16.13}$$

$$\frac{\partial \left(f_h^s\right)_4}{\partial q} = -21 \frac{\partial s_{i'}}{\partial q} c^4_{\theta'} s_{\lambda_\odot - \Omega'} c^4_{\lambda_\odot - \Omega'} - 84 s_{i'} c^3_{\theta'} \frac{\partial c_{\theta'}}{\partial q} s_{\lambda_\odot - \Omega'} c^4_{\lambda_\odot - \Omega'}$$
$$- 21 s_{i'} c^4_{\theta'} \left(\frac{\partial s_{\lambda_\odot - \Omega'}}{\partial q} c^4_{\lambda_\odot - \Omega'} + 4 s_{\lambda_\odot - \Omega'} c^3_{\lambda_\odot - \Omega'} \frac{\partial c_{\lambda_\odot - \Omega'}}{\partial q} \right)$$
$$- 21 \left(\frac{\partial s_{i'}}{\partial q} c^4_{i'} + 4 s_{i'} c^3_{i'} \frac{\partial c_{i'}}{\partial q} \right) s^4_{\theta'} s^5_{\lambda_\odot - \Omega'} - 84 s_{i'} c^4_{i'} s^3_{\theta'} \frac{\partial s_{\theta'}}{\partial q} s^5_{\lambda_\odot - \Omega'}$$
$$- 105 s_{i'} c^4_{i'} s^4_{\theta'} s^4_{\lambda_\odot - \Omega'} \frac{\partial s_{\lambda_\odot - \Omega'}}{\partial q} - 126 \left(\frac{\partial s_{i'}}{\partial q} c^2_{i'} + 2 s_{i'} c_{i'} \frac{\partial c_{i'}}{\partial q} \right) s^2_{\theta'} c^2_{\theta'} s^3_{\lambda_\odot - \Omega'} c^2_{\lambda_\odot - \Omega'}$$
$$- 252 s_{i'} c^2_{i'} \left(s_{\theta'} \frac{\partial s_{\theta'}}{\partial q} c^2_{\theta'} + s^2_{\theta'} c_{\theta'} \frac{\partial c_{\theta'}}{\partial q} \right) s^3_{\lambda_\odot - \Omega'} c^2_{\lambda_\odot - \Omega'} - 126 s_{i'} c^2_{i'} s^2_{\theta'} c^2_{\theta'}$$
$$\times \left(3 s^2_{\lambda_\odot - \Omega'} \frac{\partial s_{\lambda_\odot - \Omega'}}{\partial q} c^2_{\lambda_\odot - \Omega'} + 2 s^3_{\lambda_\odot - \Omega'} c_{\lambda_\odot - \Omega'} \frac{\partial c_{\lambda_\odot - \Omega'}}{\partial q} \right)$$
$$- 84 \left(\frac{\partial s_{i'}}{\partial q} c_{i'} + s_{i'} \frac{\partial c_{i'}}{\partial q} \right) s^2_{\theta'} c^3_{\theta'} s^2_{\lambda_\odot - \Omega'} c^3_{\lambda_\odot - \Omega'}$$
$$- 84 s_{i'} c_{i'} \left(\frac{\partial s_{\theta'}}{\partial q} c^3_{\theta'} + 3 s_{\theta'} c^2_{\theta'} \frac{\partial c_{\theta'}}{\partial q} \right) s^2_{\lambda_\odot - \Omega'} c^3_{\lambda_\odot - \Omega'}$$
$$- 84 s_{i'} c_{i'} s_{\theta'} c^3_{\theta'} \left(2 s_{\lambda_\odot - \Omega'} \frac{\partial s_{\lambda_\odot - \Omega'}}{\partial q} c^3_{\lambda_\odot - \Omega'} + 3 s^2_{\lambda_\odot - \Omega'} c^2_{\lambda_\odot - \Omega'} \frac{\partial c_{\lambda_\odot - \Omega'}}{\partial q} \right)$$
$$- 84 \left(\frac{\partial s_{i'}}{\partial q} c^3_{i'} + 3 s_{i'} c^2_{i'} \frac{\partial c_{i'}}{\partial q} \right) s^3_{\theta'} c_{\theta'} s^4_{\lambda_\odot - \Omega'} c_{\lambda_\odot - \Omega'} - 84 s_{i'} c^3_{i'} \left(3 s^2_{\theta'} \frac{\partial s_{\theta'}}{\partial q} c_{\theta'} + s^3_{\theta'} \frac{\partial c_{\theta'}}{\partial q} \right)$$
$$\times s^4_{\lambda_\odot - \Omega'} c_{\lambda_\odot - \Omega'} - 84 s_{i'} c^3_{i'} s^3_{\theta'} c_{\theta'} \left(4 s^3_{\lambda_\odot - \Omega'} \frac{\partial s_{\lambda_\odot - \Omega'}}{\partial q} c_{\lambda_\odot - \Omega'} + s^4_{\lambda_\odot - \Omega'} \frac{\partial c_{\lambda_\odot - \Omega'}}{\partial q} \right)$$
$$+ 14 \frac{\partial s_{i'}}{\partial q} c^2_{\theta'} s_{\lambda_\odot - \Omega'} c^2_{\lambda_\odot - \Omega'} + 28 s_{i'} c_{\theta'} \frac{\partial c_{\theta'}}{\partial q} s_{\lambda_\odot - \Omega'} c^2_{\lambda_\odot - \Omega'}$$
$$+ 14 s_{i'} c^2_{\theta'} \left(\frac{\partial s_{\lambda_\odot - \Omega'}}{\partial q} c^2_{\lambda_\odot - \Omega'} + 2 s_{\lambda_\odot - \Omega'} c_{\lambda_\odot - \Omega'} \frac{\partial c_{\lambda_\odot - \Omega'}}{\partial q} \right)$$
$$+ 14 \left(\frac{\partial s_{i'}}{\partial q} c^2_{i'} + 2 s_{i'} c_{i'} \frac{\partial c_{i'}}{\partial q} \right) s^2_{\theta'} s^3_{\lambda_\odot - \Omega'} + 28 s_{i'} c^2_{i'} s_{\theta'} \frac{\partial s_{\theta'}}{\partial q} s^3_{\lambda_\odot - \Omega'}$$
$$+ 42 s_{i'} c^2_{i'} s^2_{\theta'} s^2_{\lambda_\odot - \Omega'} \frac{\partial s_{\lambda_\odot - \Omega'}}{\partial q} + 28 \left(\frac{\partial s_{i'}}{\partial q} c_{i'} + s_{i'} \frac{\partial c_{i'}}{\partial q} \right) s_{\theta'} c_{\theta'} s^2_{\lambda_\odot - \Omega'} c_{\lambda_\odot - \Omega'}$$
$$+ 28 s_{i'} c_{i'} \left(\frac{\partial s_{\theta'}}{\partial q} c_{\theta'} + s_{\theta'} \frac{\partial c_{\theta'}}{\partial q} \right) s^2_{\lambda_\odot - \Omega'} c_{\lambda_\odot - \Omega'}$$
$$+ 28 s_{i'} c_{i'} s_{\theta'} c_{\theta'} \left(2 s_{\lambda_\odot - \Omega'} \frac{\partial s_{\lambda_\odot - \Omega'}}{\partial q} c_{\lambda_\odot - \Omega'} + s^2_{\lambda_\odot - \Omega'} \frac{\partial c_{\lambda_\odot - \Omega'}}{\partial q} \right)$$
$$- \frac{\partial s_{i'}}{\partial q} s_{\lambda_\odot - \Omega'} - s_{i'} \frac{\partial s_{\lambda_\odot - \Omega'}}{\partial q} \tag{A16.14}$$

$$\begin{aligned}\frac{\partial\left(f_h^s\right)_4}{\partial L} &= -84 s_{i'} c_{\theta'}^3 \frac{\partial c_{\theta'}}{\partial L} s_{\lambda_\odot-\Omega'} c_{\lambda_\odot-\Omega'}^4 - 84 s_{i'} c_{i'}^4 s_{\theta'}^3 \frac{\partial s_{\theta'}}{\partial L} s_{\lambda_\odot-\Omega'}^5 \\ &- 252 s_{i'} c_{i'}^2 \left(s_{\theta'} \frac{\partial s_{\theta'}}{\partial L} c_{\theta'}^2 + s_{\theta'}^2 c_{\theta'} \frac{\partial c_{\theta'}}{\partial L} \right) s_{\lambda_\odot-\Omega'}^3 c_{\lambda_\odot-\Omega'}^2 \\ &- 84 s_{i'} c_{i'} \left(\frac{\partial s_{\theta'}}{\partial L} c_{\theta'}^3 + 3 s_{\theta'} c_{\theta'}^2 \frac{\partial c_{\theta'}}{\partial L} \right) s_{\lambda_\odot-\Omega'}^2 c_{\lambda_\odot-\Omega'}^3 \\ &- 84 s_{i'} c_{i'}^3 \left(3 s_{\theta'}^2 \frac{\partial s_{\theta'}}{\partial L} c_{\theta'} + s_{\theta'}^3 \frac{\partial c_{\theta'}}{\partial L} \right) s_{\lambda_\odot-\Omega'}^4 c_{\lambda_\odot-\Omega'} \\ &+ 28 s_{i'} c_{\theta'} \frac{\partial c_{\theta'}}{\partial L} s_{\lambda_\odot-\Omega'} c_{\lambda_\odot-\Omega'}^2 + 28 s_{i'} c_{i'}^2 s_{\theta'} \frac{\partial s_{\theta'}}{\partial L} s_{\lambda_\odot-\Omega'}^3 \\ &+ 28 s_{i'} c_{i'} \left(\frac{\partial s_{\theta'}}{\partial L} c_{\theta'} + s_{\theta'} \frac{\partial c_{\theta'}}{\partial L} \right) s_{\lambda_\odot-\Omega'}^2 c_{\lambda_\odot-\Omega'} \end{aligned} \qquad \text{(A16.15)}$$

References

[1] Kéchichian, J. A. (1997) Trajectory Optimization Using Nonsingular Orbital Elements and True Longitude, *Journal of Guidance, Control, and Dynamics*, Vol. 20, No. 5, 1003–1009.

[2] Kéchichian, J. A. (2008) Inclusion of Higher Order Harmonics in the Modeling of Optimal Low-thrust Orbit Transfer, *Journal of the Astronautical Sciences*, Vol. 56, No. 1, 41–70.

[3] Kéchichian, J. A. (2010) *Analytic Expansions of Luni-Solar Gravity Perturbations along Rotating Axes for Trajectory Optimization: Part 1: The Dynamic System*, AAS Paper 10-221, AAS/AIAA Space Flight Mechanics Meeting, San Diego, CA.

[4] Kéchichian, J. A. (2010) *Analytic Expansions of Luni-Solar Gravity Perturbations Along Rotating Axes for Trajectory Optimization: Part 2: The Multipliers System and Simulations*, IAF Paper 10-C1.1.1, 61st International Astronautical Congress, Prague, Czech Republic.

[5] Battin, R. H. (1987) *An Introduction to the Mathematics and Methods of Astrodynamics*, AIAA, New York, NY, 387–391.

[6] Feistel, B. S. (2007) *Perturbing Accelerations in Low Thrust Trajectory Optimization*, M.S. Thesis, University of Texas at Austin.

[7] Smart, W. M. (1961) *Celestial Mechanics*, John Wiley & Sons, Inc., New York, NY, 263–290.

[8] Brouwer, D. and Clemence, G. M. (1961) *Methods of Celestial Mechanics*, Academic Press, New York, NY, 31–35.

Index

Printed in the United Kingdom by TJ Clays Ltd.